Practical Lubrication for Industrial Facilities

3rd Edition

Practical Lubrication for Industrial Facilities

3rd Edition

Compiled, Edited and Written by
Heinz P. Bloch
Kenneth E. Bannister

Published 2020 by River Publishers
River Publishers
Alsbjergvej 10, 9260 Gistrup, Denmark
www.riverpublishers.com

Distributed exclusively by Routledge
4 Park Square, Milton Park, Abingdon, Oxon OX14
4RN 605 Third Avenue, New York, NY 10158

First published in paperback 2024

Library of Congress Cataloging-in-Publication Data

Names: Bloch, Heinz P., 1933- editor | Bannister, Kenneth E., 1955- editor.
Title: Practical lubrication for industrial facilities / compiled and edited by Heinz P. Bloch, Kenneth E. Bannister.
Description: 3rd edition. | Lilburn, GA : Fairmont Press, Inc., 2016. | Includes bibliographical references and index.
Identifiers: LCCN 2016043250 | ISBN 9781003151357 (alk. paper) | ISBN 9788770222648 (electronic) | ISBN 9781138626799 (taylor & francis distribution : alk. paper)
Subjects: LCSH: Lubrication and lubricants--Handbooks, manuals, etc.
Classification: LCC TJ1075 .B57 2016 | DDC 621.8/9--dc23 LC record available at https://lccn.loc.gov/2016043250

Practical lubrication for industrial facilities -- 3rd ed., Heinz P. Bloch/Kenneth E. Bannister.
First published by Fairmont Press in 2017

Routledge is an imprint of the Taylor & Francis Group, an informa business

Publisher's Note
The publisher has gone to great lengths to ensure the quality of this reprint but points out that some imperfections in the original copies may be apparent.

ISBN 978-0-88173-761-5 (The Fairmont Press, Inc.)
ISBN 978-8-770-22264-8 (online)

ISBN: 978-1-138-62679-9 (hbk)
ISBN: 978-87-7004-450-9 (pbk)
ISBN: 978-1-003-15135-7 (ebk)

DOI: 10.1201/9781003151357

Dedications

To the memory of Tom Russo, an Australian engineer and inventor; in appreciation of John Williams, founder of Royal Purple, Inc.; and as further encouragement to an exceptional practitioner of advanced root cause failure analysis, Kenneth Bloch.
—Heinz P. Bloch

Dedicated to all those who advocate excellence in the field of lubrication and to the memory of Ted Monkiewicz, my dear departed friend and mentor.
—Kenneth E. Bannister

Contents

Acknowledgments

This text was compiled with the help of many individuals and companies whose assistance and cooperation is gratefully acknowledged. In addition to much material on lubricants from my principal contributors Exxon USA, Klüber-Summit, and Royal Purple, we received input in the form of illustrations, marketing bulletins, commercial literature, entire narrative segments, and even a full chapter from the following:

Ms. Judith Allen and the late Tom Russo (lube oil purification)
AESSEAL plc, Rotherham, UK, and Knoxville, TN (bearing protector seals). Special thanks to one of AESSEAL's directors, Mr. Alan Roddis, and to Chris Rehmann.
ASEA-BBC, Baden, Switzerland (gas and steam turbines)
Bijur Lubricating Corporation, Bennington, Vermont (lubricating equipment)
Miguel González Bonnin (oil mist and efficiency gains, if used on electric motors)
Cooper industries, Mount Vernon, Ohio (gas engines)
Dresser-Rand Company, Wellsville, New York (centrifugal and reciprocating compressors)
Don Ehlert, Lubrication Systems Company, Houston, Texas (segment on oil mist cost justification; also, experience updates)
Elliott Company, Jeannette, Pennsylvania (compressors)
Richard Ellis, Pearland, Texas (appendix material on lubrication programs)
Farval Lubrication Systems, Kinston, North Carolina, (centralized lubrication)
Jim Fitch, NORIA Corporation, Tulsa, Oklahoma (entire chapter on oil analysis)
General Electric Company, and Nuovo Pignone, Florence, Italy (gas turbines)
Glacier Metal Company Ltd., Northwood Hills/Middlesex, UK and Mystic, Connecticut (turbomachinery bearings)
Kingsbury inc., Philadelphia, Pennsylvania (turbomachinery bearings)
GHH/Borsig, Oberhausen and Berlin, Germany (compressors and gas turbines)
Mannesmann-Demag, Duisburg, Germany (centrifugal compressors)
Murray Turbomachinery Corporation, Burlington, Iowa (lubrication systems for turbines)
Oil-Rite Corporation, Manitowoc, Wisconsin (lubricating equipment)
Jim Partridge, Lufkin industries, Lufkin, Texas, (segment on gear lubrication)
Al Pate Jr., Klüber-Summit, Tyler, Texas (mineral oils and synthetic lubricants)
Safematic Oy, Muurame, Finland, and Alpharetta, Georgia (centralized lubrication systems)
Siemens Power Systems, Erlangen, Germany (steam turbines)
SKF USA, Kulpsville, Pennsylvania (lubrication of rolling element bearings)
Sulzer Turbomachinery Ltd., Winterthur, Switzerland, (dynamic compressors)
Torrington-Fafnir, Torrington, Connecticut (rolling element bearings)
Trico Manufacturing Corporation, Pewaukee, Wisconsin (lubricating equipment)
Tom Ward, Lubrication Systems Company, Houston, Texas (segment on oil mist lubrication)
Waukesha Bearings, Waukesha, Wisconsin (turbomachinery bearings)
John Williams, and Lee Culbertson, Royal Purple Company, Humble, Texas (synthetic lubricants; also, advanced synthetic grease technology)
Howard Marten Company, Ontario Canada (Lubricating Equipment)
Jane Alexander, Maintenance Technology magazine, ATP network, Illinois, USA
Fluid Defense Systems, Montgomery, Illinois, USA (Lubricant storage and dispensing equipment)
And to all other contributors in the book not mentioned above

Again, many thanks!

Forewords
To the Third Edition

Heinz P. Bloch

A bit of background. My involvement with machinery lubrication dates back to the 1950s when, as a 17-year-old apprentice in my hometown in Germany, I was held accountable for properly oiling and greasing the lead screws and guideways of the machine tools I was learning to operate. I had similar part-time duties in a New Jersey shop from about 1953 to 1958.

My responsibilities shifted when, after graduation from engineering college, I designed machines for the production of cotton-tipped applicators for Johnson & Johnson (1962). Understanding lubrication issues was further emphasized after I started my machinery reliability improvement career with (then) Esso Research & Engineering in 1965. The Esso/Exxon assignments often dealt with lubrication problems and what I had learned facilitated the transition to consulting work (in 1986).

In 1988, I assisted a Texas-based power generating station in upgrading its lubrication management. Trying to understand component degradation issues on pulverizer gears required tapping into the decade-long experience of Frank D. Myrick of Summit Oil Company in Tyler, Texas. Frank has long since retired, but in the years he served as the president of Summit (now Klueber-Summit, and part of Klueber Lubrication, an international lubricant manufacturing company headquartered in Munich, Germany, with offices in Londonderry/ New Hampshire, and Halifax/ UK), he often agreed with me on the need for a readable, practical reference text on industrial lubrication. It led me to compile the First Edition of this text in the mid-to-late 1990s.

If anything, the need for a practical text has become even greater in the past decades. Space-age solutions and "quick fixes" are being pursued. The teaching and understanding of the non-glamorous basics is being neglected more than ever. Many industrial companies have replaced the job of the lubrication specialist with the multi-task function of the jack-of-all-trades, or the contract employee who was hired on the basis of hourly wages and perceived (but untrue) overall savings. In many places, the procurement of industrial lubricants is no longer scrutinized by a competent reliability professional. Moreover, the task is often tackled without rigorous technical specifications. We have even seen the buying process entrusted to purchasing agents whose only objective was lowest initial cost per gallon, and "as-soon-as-possible" delivery scheduling. industry suffers from this downward trend and lubrication-related failure and downtime incident have cost us dearly.

Nevertheless, a handful of "Best-in-Class" industry performers do share a fundamental understanding of lube-related problems and procedures. They are the ones who often perform life cycle cost analyses and, based on the results, will find ample justification to selectively apply superior mineral or synthetic lubricants. These are the profitable, reliability-driven facilities that cherish and promote an understanding of the many interwoven facets of lubricant specification, selection, substitution, application, analysis, replacement, in-place purification, consolidation, handling and storage. May they continue to prosper, in good times and in difficult times.

How and why this book was compiled. Back to the scope and purpose of this book. In this brief foreword I also wanted to illustrate how readers will get the most out of this text. You really don't have to read every page, but you need to know how I have made good use of this type of material for about fifty years.

Years ago I set out to assemble practical and important lubrication and lubricant topics into a format that satisfies such principal requirements as technical relevance, readability, and applicability to the widely varying needs of modern industrial plants. Modern industrial plants include appliance makers, foundries and mining equipment makers, petroleum refineries and facilities that use air compressors. There is simply no end to the listings of industries and users whose lubrication requirements—and needs for lubricants—can vary greatly. This understanding caused me to tap into many available resources; these included Klueber Lubrication in Munich/Germany, Royal Purple in Porter/Texas and, especially, the Lube Marketing Department of my old employer, Exxon, in Houston/ Texas. These folks know their business and are among the many deserving of my sincere gratitude for allowing me to use so much of their outstanding, commercially available material.

The world's best manufacturers and formulators of lubricants are constantly seeking to improve products to keep pace with the development of higher-speed machinery, or equipment that is run at over 100 percent of name plate capacity, or machines that are being subjected to temperature extremes, extended oil drain intervals, or just plain simple abuse. In short, even as we read these introductory pages, new lubricants are in the process of being developed; they will go beyond the capabilities of today's already exceptional products. Still, few if any of the products described in this text have changed since the First Edition was compiled in the late 1990s. We should consider them "legacy brands," tried and tested and found worthy of consideration by a reliability-focused user.

Heinz P. Bloch
West Des Moines, Iowa
January 2009

Kenneth E. Bannister

in my early career, working as a mechanical design engineer in the research and development department of a stock form printing press manufacturing company I was tasked to design a unique multi use, open-style, right-angle drive, zero backlash gearbox. The printing press requiring this unusual type of gearbox was a revolutionary, world first, modular style printing press designed to accommodate quick-change cassette style print roll assemblies.

The gearbox was special in that it had to be mounted independently on the outside of each individual press tower frame and accommodate multiple gear train configurations, which meant each gearbox and gear train had to be independently lubricated; it was to be my first introduction to centralized oil lubrication.

Part of any OEM (original equipment manufacture) design process is to assess the reliable life of a machine's major components for warranty issues and spare parts requirements. As the gearbox was a critical item—old school designed using a basic calculator and logarithmic tables—corporate management decided to have all gearbox bearing life calculations checked and analyzed using the bearing company's newly installed mainframe computer (PC computers were not publicly available at this time). In addition, the analysis was to also determine the lubrication requirements for each different gear train configuration so the reservoir and lube pump sizing could be calculated. for an alternative analysis. The same information was sent to the Bijur Lubricating Corporation, the world's oldest and highly respected centralized lubrication application company, asking them to design and quote on a suitable recirculating oil system. The corresponding reports I received could not have been more different!

The bearing company gave a "triple-A" passing grade on the gearbox design and calculated lubricant requirements to achieve the maximum gearbox/gear train bearing life in accordance to its L10 design life rating. Their analysis concluded that approximately one liter of oil per minute would be required to pass over the bearings and gear faces. In comparison, Bijur Lubricating Corporation had calculated the total oil requirement for optimum bearing and gear train operation at a mere 10cc per minute!

At 100:1 ratio for both responses, I had a design dilemma to contend with. The bearing company's calculation required a super large reservoir with external coolers and a large gear pump, whose real estate surpassed the available space surrounding the gearbox. In contrast, the Bijur calculation requirement could be easily accommodated with a small gear pump fitted under the gearbox open cavity driven directly off the main shaft and a baffled reservoir built directly into the gearbox side cover, with no space penalty. With two impeccable sources giving such a different answer the question of whose solution to choose shifted from one of trusting the calculation to one of maximization versus optimization of the overall design life expectancy.

Upon further discussions with both bearing company and the Bijur Corporation, it became evident that the bearing company computer was utilizing a calculation designed to achieve maximum bearing life based on its L10 design life. Attaining this required the bearing to be rapidly cooled at a heat transfer rate requiring one liter flow per minute, whereas the Bijur solution was based on a calculation derived from over 60 successful years of lubricating gearboxes and gear trains in practical ambient operating conditions. Both calculations were correct but the practical answer depended on what practical life expectancy was required by the OEM for the print press bearings in service. As both solutions surpassed the five to seven year expected gearbox and gearing service design life, the Bijur solution was chosen and proved to be the optimal decision when the press was placed in service.

This particular gearbox design exercise opened up my eyes for the first time to the importance of understanding friction, lubrication, and wear and their affect on both equipment design and equipment maintenance. With this newfound knowledge I went on to enjoy working directly for a number of lubrication application companies in sales and application engineering, one of them being the Bijur Lubricating Corporation. For the past 30 years I have dedicated my life to helping companies implement best practice asset management programs that capitalize on practical lubrication programs to deliver uptime and effective life cycle management of their industrial equipment assets.

I am honored to collaborate with my legendary co-author Heinz P. Bloch on this third edition of the *Practical Lubrication for Industrial Facilities* handbook and would like to introduce you to some of the changes and updates that can be found in this third edition. Readers of previous editions will recognize the book has been increased from 19 chapters to 22 chapters in length, a number of chapters have been updated, some renamed and their order rearranged to give the reader a more logical flow to the updated and existing information.

To lubricate effectively, one must understand the physics in play when two surfaces interact and how oils and greases function in that environment. Chapter 1—Principles of Lubrication, has been updated with new graphics throughout and has been expanded in all areas to introduce the concepts of tribology, functions of lubricants, base oil types, lubricant additives, grease lubricant characteristics, and how lubricants can fail in service.

Chapter 2—Lubricant Categories and Properties has been updated to include a section on the categorization and grouping of base oils while Chapters 3—Lubricant Testing, and Chapter 5—Hydraulic fluids have received some minor updates with a new lubricant test chart and an introduction to environmentally hydraulic fluids, further expanded upon in the renamed Chapter 6—Food Grade and "Environment Friendly" Lubricants.

Because this is a text for lubrication practitioners, Chapter 10—Lubricant Delivery Systems expands considerably on the old Constant Level, Centralized, and Oil Mist Systems chapter to include and extended section of grease delivery systems and a step-by-step approach to implementing an effective manual lubrication program. The explosion in use of single point lubricators throughout industry in recent years warrants an expanded section on these very practical devices. In addition, this chapter has been updated to introduce the reader to all of the available centralized lubrication system types, their pros and cons, and a specific section on choosing a pump and controller type and how to maintain these systems.

Chapter 14—Lubricating Pumps, is a brand new chapter devoted to the practical lubrication of industrial pumps. This chapter takes a detailed look at the multiple types of bearings used in pumps and how grease and oil mist can be successfully employed to ensure pumps are kept running in the many difficult environments they must operate in.

in Chapter 15- Lubricating Paper Machines and forestry Equipment, a section on high performance (PFPE) Oil and grease lubricants has been added. Chapter 18—Lubricant Purchase, Handling & Storage, has been expanded to include a section on implementing a cradle-to-cradle lubricant management program that details how to implement a lubricant consolidation program, design and prepare a best practice lubricant storage area, and how to receive, handle and transfer lubricants throughout the plant with minimum contamination issues.

Chapter 19—Lube Oil Contamination, Filtration, and On-stream Purification Control has been expanded to look into contamination sources and the practical devices used to control and filter contamination. The also expanded Chapter 21—Lubricant Condition Testing—Oil Analysis further examines oil-sampling methods and introduces the reader to a practical seven-step process for implementing an industrial oil analysis program.

In conclusion, Chapter 22—Implementing a Quality Lubrication Management program is a new chapter that looks at developing a lubrication management program to the new ISO 55001 asset management standard now being reviewed and introduced throughout industry. This chapter also takes an in-depth look at the lubrication technician certification programs now available to industry and the body of knowledge required for certification, much of which is included in this newly expanded third edition!

Both Heinz and I thank you for purchasing this book and wish you the greatest success in applying its principles and information in the practical world!

Ken Bannister, July 2016

About the Author

Ken Bannister is a UK technical apprenticed and accredited mechanical design engineer with several engineering design patents to his credit. Ken is also a Certified Maintenance Reliability Professional and for the past 30 years has worked worldwide in a consulting practice helping clients implement practical and meaningful asset management and reliability programs; he is one of only a few asset management consultants worldwide holding expertise and accreditation in the field of tribology and lubrication failure management.

To date, Ken is the author of three maintenance books on lubrication, energy management and predictive maintenance and the lubrication section of the *Machinery's Handbook*, has close to 500 articles published, with over half dedicated to the field of lubrication practice, and is currently a contributing editor for a number of international industrial maintenance magazines.

Chapter 1
Principles of Lubrication

FRICTION

In the mid 1960s, the world took note of a remarkable ground breaking study performed for the British government by Mr. H. Peter Jost who for the first time quantified the effects of ineffective lubrication practices on Britain's GNP (Gross National Product). In that initial report, now referred to as the *Jost Report*, Jost coined the word Tribology to describe the science of lubrication, friction and wear. For the first time lubrication was internationally recognized as a bone fide science and practice in the corporate world of asset reliability, and for its significant positive fiscal impact on industry, and the GNP when practiced effectively.

The word Tribology is derived from the Greek word tribos meaning, "to rub" and is used to describe what happens when two mating surfaces move over one another. The resistive force causing this "rubbing" action is known as friction and was first recognized by Sir Isaac Newton in his laws of motion as an external force to motion that must be overcome for motion to occur.

The Webster's dictionary describes friction as "the force, which opposes the movement of one surface sliding or rolling over another with which it is in contact." Simply put, friction is the resistive force that retards motion.

Friction isn't necessarily a bad force, we employ frictional forces when we want to intentionally retard a body in motion when slowing down a rotating machine or auto mobile by applying a rough and soft consumable braking material with a high coefficient of friction against a smooth hard (less-consumable) surface. Friction becomes an undesirable force when it robs energy from an applied force used to intentionally move an object; estimates blame frictional forces for consuming over one third of the world's energy! When ignored in such cases, friction causes heat, wear, and sometimes, catastrophic failure of the bearing surfaces.

To understand friction we must recognize that there are two unchanging fundamental laws that govern friction:

1) Friction varies directly with load, and
2) Friction is independent of surface area

Figure 1-1 depicts the forces at play on two bodies at rest; to begin to move body A across body B we must first overcome its resistive frictional force in which the resistive force is a result of load N (the weight of body A) multiplied by the coefficient of friction. When bodies are rigid as shown, the frictional force is known as *solid* friction. Solid friction may be *static* or *kinetic*—the former encountered when initiating movement of body A from a rest position, the latter being a reduced force when body A is in motion. (Distinct from solid friction is *fluid friction*, a normally less resistive force that occurs between the molecules of gas or liquid in motion. As will be seen in later discussions, lubrication generally involves the substitution of high solid-to-solid friction for the much lower fluid friction.

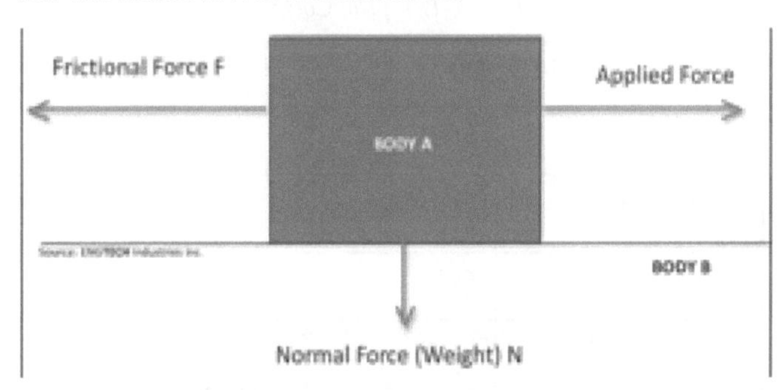

Figure 1-1. Forces on Bodies at Rest.

Example: if body A depicted in Fig 1-1 represented a wooden steamer trunk of books resting on a concrete floor (body B), using the formula F = μN we can calculate the initial (static) resistive force (F) we need to overcome to start the trunk moving across the floor. Therefore, if we assume the normal force (N) or weight of the loaded trunk is 100 lbs and the coefficient of friction (μ) of wood on dry concrete is 0.62 the applied force required to start the trunk moving would be 0.62 x 100 = 62lbs. As depicted in Fig 1-2, once the trunk has begun to

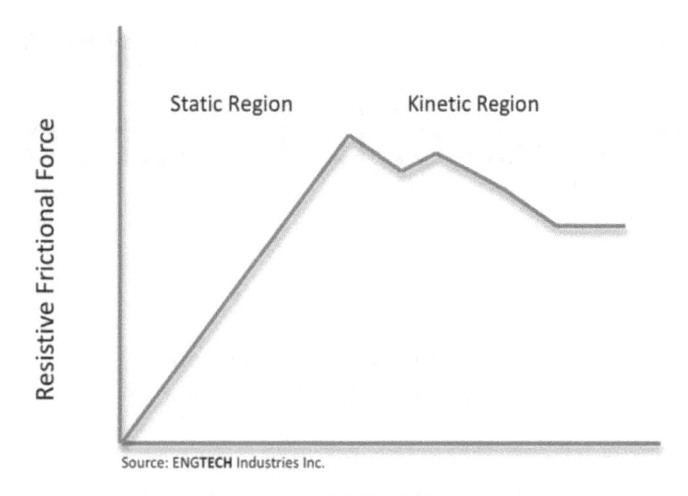

Figure 1-2. Solid Friction—Static Versus Kinetic

move the static friction barrier has been broken and the kinetic force required to keep the trunk moving reduces somewhat as long as the body remains moving.

The Coefficient Of Friction (COF), denoted by the Greek letter μ, (mu), is different for every material and fluid and values range from almost 0 to well over 1; the lower the value, the lower the resistance and motion retardation effect—see Table 1-1 for typical COF values. Therefore, whenever we want to produce work from moving parts, lower COF values are preferred as they require less energy expenditure to achieve movement,

Figure 1-3. Burnt Out Engine Bearing due to Frictional Heat caused by Lubricant Starvation.

or work. i.e the motor requires less amperage draw, or the engine requires less fuel to achieve the desired work performance. Obviously, we would expand enormous amounts of energy to move things around if we were only to allow surface-to-surface contact on all moving parts, to reduce these forces and overcome the large static and kinetic forces we must introduce a fluid film to separate the two moving parts.

Table 1-1. Typical Coefficient of Friction Values

Materials		Dry μ	Lubricated μ
Brass	Steel	0.35	0.19
Sintered Bronze	Steel	0.16	Self lubricated
Cast Iron	Oak	0.49	0.075
Graphite	Steel	0.1	Self lubricated
Iron	Iron	1.0	0.2
Teflon	Steel	0.05 – 0.2	No lube required
Teflon	Teflon	0.04	No lube required
Wood	Concrete	0.62	
Copper Lead	Steel	0.22	
Tire	Dry Road	1.0	0.6 (wet)
Tire	Snow	0.3	

Source: ENGTECH Industries Inc.

Causes of Solid Friction

Solid friction, or *sliding friction,* as it is sometimes known originates from two widely differing sources. The more obvious source is surface roughness; no machined surface, however polished, is ideally smooth. Though modern machinery is capable of producing finishes that approach perfection, microscopic irregularities inevitably exist. Minute protuberances on a surface are called *asperities,* and, when two solids rub together, interference between opposing asperities accounts for a considerable portion of the friction, especially if the surfaces are rough.

The other cause of sliding friction is the tendency of the flatter areas of the opposing surfaces to weld together at the high temperatures that occur under heavy loads. Rupture of the tiny bonds created in this manner is responsible for much of the friction that can occur between machine parts. On finely ground surfaces, in fact, these minute welds constitute a major source of potential frictional resistance.

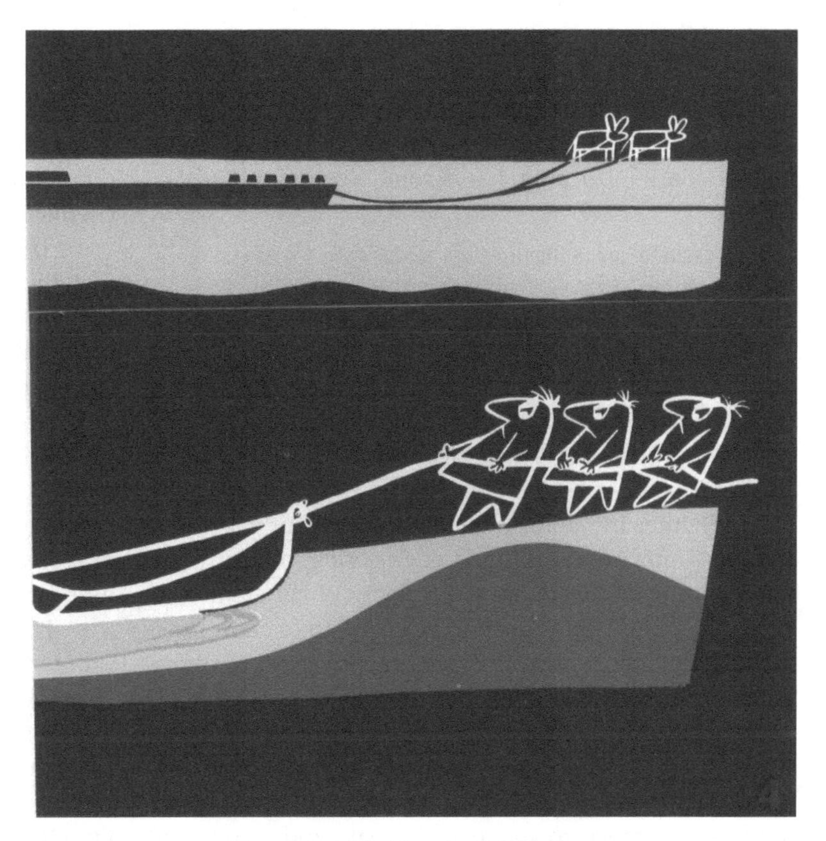
Figure 1-4. Fluid and solid friction.

Effect of Friction

Whenever friction is overcome, moreover, dislocation of the surface particles generates heat, and excessive temperatures developed in this way can be destructive. The same frictional heat that ignites a match is what "burns out" the bearings of an engine, Figure 1-4.

Additionally, where there is solid friction, there is wear: a loss of material due to the cutting action of opposing asperities and to the shearing apart of infinitesimal welded surfaces. In extreme cases, welding may actually cause seizure of the moving parts. Whether a piston ring, gear tooth, or journal is involved, the harmful effects of friction can hardly be overemphasized.

One of the tasks of the engineer is to control friction—to increase friction where friction is needed and to reduce it where it is objectionable. This discussion is concerned with the reduction of friction.

It has long been recognized that if a pair of sliding bodies are separated by a fluid or fluid-like film, the friction between them is greatly diminished. A barge can be towed through a canal much more easily than it can be dragged over, say, a sandy beach. Figure 1-3 should remind us of this fact.

Lubrication

The principle of supporting a sliding load on a friction-reducing film is known as lubrication. The substance of which the film is composed is a lubricant, and to apply it is to lubricate. When mechanical moving parts are present, the amount of lubrication required is dependent on the state of friction, which can manifest itself in three distinctive ways identified as:

1. Sliding Friction
2. Rolling Friction
3. Combination Friction—Sliding and Rolling

Sliding friction is common where any plain surface moves over one another evident in plain bearings where a journal moves within a sleeve. Sliding friction arrangements require the most lubricant, as the friction is evident over a larger surface contact area.

Rolling friction is found in all rolling element bearings that at one time were described as "friction-less" bearings. The contact surface is considerably smaller than in sliding friction bearings and thereby requires much smaller amounts of lubrication to provide a protective full fluid film.

Combination friction is unique to meshing gears due to the changing gear tooth profile when teeth mesh together and slide on one another until the opposing pitch surfaces meet and rolling friction takes over as they disengage. Certain types of gears such as hypoid gears and worm gears produce much higher degrees of sliding friction.

Whenever moving parts are employed, friction is ALWAYS present. Understanding friction helps us develop effective lubrication practices that in turn allow us to tame the harmful effects of friction and increase the component life cycle. These are not new concepts.

The principle of supporting a sliding load on a friction-reducing film is known as lubrication. The substance of which the film is composed is a lubricant, and to apply it is to lubricate. These are not new concepts, nor, in their essence, particularly involved ones. Farmers lubricated the axles of their ox carts with animal fat centuries ago.

But modern machinery has become many times more complicated since the days of the ox cart, and the demands placed upon the lubricant have become proportionally more exacting. Though the basic principle still prevails—the prevention of metal-to-metal contact by means of an intervening layer of fluid or fluid-like material—modern lubrication has become a complex study.

LUBRICANTS

The old adage "oil is oil, so any old oil will do!" may have had merit a hundred years ago, but in today's world of sophisticated machinery and demand for asset reliability, choosing the correct lubricant is now an important and informed decision.

Whether in the form of a liquid, solid or gas, modern lubricants are pure liquid engineering, and through the blending of additives to a variety of different base stocks they can be designed to perform up to eight functions simultaneously whilst operating in a host of differing environments.

Webster's dictionary describes a lubricant as "a substance (e.g. oil, grease, or soap) that when introduced between solid surfaces which move over one another reduces resistance to movement, heat production, and wear (i.e. friction and its effects) by forming a fluid film between the surfaces."

Essentially, a lubricant's function is to control and minimize the sacrificial harmful effects of moving surfaces passing over one another under load, and at speed. It does this in the following eight definitive ways:

Function 1—Control and minimize friction:

The primary function of every lubricant is to control and minimize the effects of friction.

When two solid surfaces passing over one another are allowed to come into contact under load, they "rub" together and produce dry friction requiring considerable energy to keep the surfaces moving. With no lubricant present to separate the moving surfaces from one another, surfaces quickly degrade and can weld or lock together resulting in a "seize." The indiscriminate sacrifices of wear surfaces produces rapid wear and loss of energy to heat resulting in poor performance, reduced reliability, and increased energy use.

The introduction of a lubricating film between the two wear surfaces creates a fluid barrier that prevents surface contact. Although a small amount of fluid friction is still present in the lubricant film, the energy required to move the surfaces over one another is but a small fraction of that required to overcome "surface to surface" *dry* friction.

Function 2—Control and minimize wear:

Knowing that a full lubricant film may not always be possible and that some metal to metal contact may occur under slow moving, heavy load, lubricant loss conditions, additives can be added to the lubricant that act as chemical "softening" agents on the metal surfaces.

The lubricant coats the surface with soft layers of metallic salts (sulfides and phosphate additives, and as the two surfaces slide over one another alternating load cycles can cause the softened high points (asperities) on each surface to collide with one another when film thickness is reduced. When the unit loading exceeds the sulfur phosphide film, a rupture can occur resulting a small metal-to-metal contact area. Localized heat builds up causing the two surfaces to "weld and break" that results in a small metal particulate, or asperity release into the lubricant film.

Many lubricants are designed to control wear by promoting minute surface degradation to allow asperity "tips" to be sacrificed easily with out "tearing" the parent metal, thereby minimizing surface wear under varying lubricant film conditions.

Function 3—Control and minimize heat:

Whenever friction and wear levels are controlled and minimized, the amount of heat is also reduced. Excessive heat can "cook" the lubricant and cause it to oxidize rendering it less effective; to combat this an anti-oxidant additive is added to the lubricant base stock.

Recirculating oil system and air/oil system designs take advantage of a lubricant's ability to transfer localized heat build up at a bearing load point and prevent any thermal runaway at the bearing surfaces. To facilitate the heat transfer/cooling process the oil may be pumped through a heat exchange unit (oil cooler), and/or reservoir baffle system.

Function 4—Control and minimize contamination:

As described above, a lubricant can become contaminated when wear asperities are introduced into the lubricant. Other forms of contamination, such as silica (dirt) can be introduced from the reservoir filling process when proper storage, transfer and cleanliness practices are not observed, or can enter the system when seals become compromised.

To combat contaminant solids, lubricant additives can be designed to coagulate particulate matter making them heavy enough to "drop out" into the sump bottom. Other additives can be added to attach to attach to asperities and stay colloidal, suspended in the lubricant ready to be extracted under pressure by an in-line system oil filter. Failure to refresh the oil filter on a regular basis will cause the contaminated lubricant to act as a "lapping" paste and accelerate the wear process in the bearing areas.

In the case of water or glycol contamination, additives are added to facilitate release of moisture in the sump or filter. These additives types are more prevalent in automotive oils.

Lubricants can also act to seal out contamination ingress around shafts, this is especially so in the case of a labyrinth seal that depends on grease to fill up a series of annular grooves cut into the non-moving shaft housing, designed to act as a live shaft seal.

Function 5—Control and minimize corrosion:

Oxygen may be a basic human life force, but can be fatal to a lubricant. When present it acts as catalyst to combine certain metals and organics that generate corrosive acids harmful to the bearing surfaces. If the wear surfaces are ferritic (iron based) the acids attack the metal and form rust on the bearing surface.

A lubricant is designed to "cling" to the metal surfaces and keep out moisture and oxygen from reacting with the surface. As not all lubricants are designed equal, if the bearing surfaces are iron based, a lubricant with anti corrosive additives must be employed.

Function 6—Control and minimize shock:

Most will be familiar to the quieting effect of adding lubricant to a gear train in which the lubricant acts as a hydraulic shock absorber between mating gears as they mesh. When poorly lubricated, meshing gears set up shock waves as they start to mesh resulting in a "chattering" sound that can fracture the gear teeth.

The very word "shock absorber" is synonymous with automobile suspension systems that employ hydraulic oil to dampen and absorb the effects of road shock on the vehicle.

Function 7—Control and transmit power:

In a typical hydraulic system, oil is used to transmit force and motion from a single source (usually a pump) into multiple sources, pistons, accumulators, etc.

Hydraulic oil is also used to transmit power in soft start devices such as fluid couplings, automatic transmissions and torque converters.

Function 8—Control and minimize energy consumption:

Effective lubrication practice dictates use of the Right lubricant, in the Right place, at the Right time, in the Right amount, using the Right method. Doing so, will ensure the asset is using the least amount of energy where moving parts are concerned.

In studies[1] conducted on behalf of various electric power companies, effective use of lubricants, delivery systems and lubrication methods resulted in an energy reduction of 7.3% when a synthetic lubricant was used to replaced a standard compressor oil, and a 17.92% energy reduction was achieved on a stamping press when the automated oil delivery system was "tuned" and a more appropriate oil was chosen!

All liquids will provide lubrication of a sort, but some do it a great deal better than others. The difference between one lubricating material and another is often the difference between successful operation of a machine and failure.

Mercury, for example, lacks the adhesive and metal-wetting properties that are desirable to keep a lubricant in intimate contact with the metal surface that it must protect. Alcohol, on the other hand (Figure 1-5), wets the metal surface readily, but is too thin to maintain a lubricating film of adequate thickness in conventional applications. Gas, a fluid-like medium, offers lubricat-

Table 1-2 Functions of a Lubricant

The function of a lubricant is to *CONTROL...*	1. Friction 2. Wear 3. Heat 4. Contamination 5. Corrosion 6. Shock 7. Power transmission 8. Energy consumption

ing possibilities—in fact, compressed air is used as a lubricant for very special purposes. But none of these fluids could be considered practical lubricants for the multitude of requirements ordinarily encountered.

BASE OIL TYPES

All finished "ready to use" lubricants are manufactured to a proprietary formula, each designed to combat friction according to the differing design, function, and operating conditions under which bearing surfaces must be kept separated.

Lubricating oil, whether in its liquid form or thickened form known as grease, is a blended product consisting of base oil stock and additive package. The base oil percentage of the finished oil can range from 75% up to 99% depending on how much additive package is required to enhance, suppress, or contribute new properties to the oil. Base oil stock is derived from three primary sources that classify the oil as an animal/vegetable, petroleum (mineral), or synthetic lubricant product.

Animal/Vegetable base oil

The industrial revolution ran on olive oil and rendered animal fat for a long time until the animal/vegetable oil's inability to arrest rapid acid formation under ever-increased speeds and loads was solved with the discovery of crude petroleum (mineral) oil. This oil classification is now generally reserved for cooking purposes only.

Petroleum base oil

Often defined as mineral based oil, petroleum base oil stock is overwhelmingly the most popular base oil stock in use today. For almost every situation, petroleum products have excelled as lubricants. Petroleum lubricants are high in metal-wetting capability and possess the body, or viscosity characteristic that a substantial oil film requires.

Petroleum oil is pumped or mined from the earth in a crude form that must be refined to remove impurities that include aromatic hydrocarbons, sulfur compounds, acids and wax, so as to improve the base oil's desirable properties that include viscosity index, pour point, and stability. Refining also separates the crude oil molecules by size and weight to produce a variety of petroleum based products of which lubricating base stock oil constitutes between 1-2% of a barrel of crude oil's yield (gasoline is the highest at approximately 25% yield).

Where in the world a crude oil originates will es-

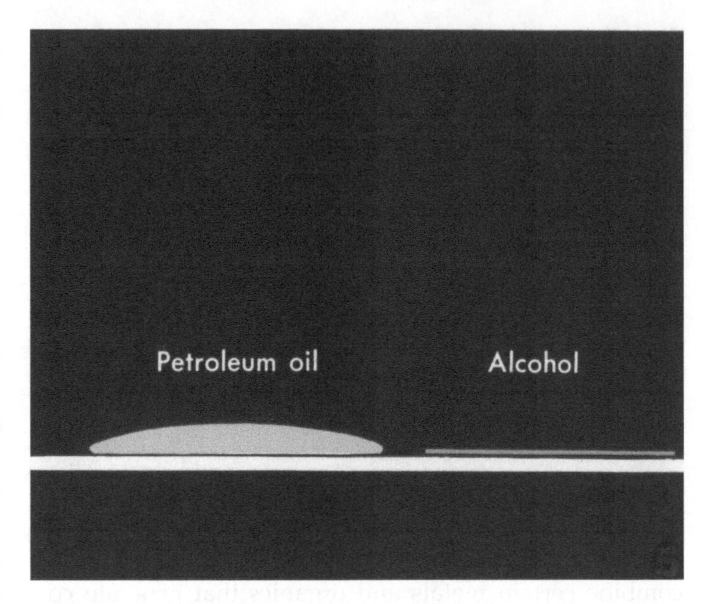

Figure 1-5. Petroleum oils make the best lubricating films.

tablish it's base oil properties and eventually what kind of service application the finished lubricating oil will be used in, determined by the levels of Paraffin and Naphthene present in the crude stock.

Paraffinic crude oils are generally found in the Mid Continental USA, the Middle East, and the British North Sea, and are favored for the manufacture of crankcase oils, Gear oils, bearing oils, turbine oils, and most hydraulic fluids. Paraffinic crudes constitute up to 60% paraffin and up to 10% wax content giving the oil an excellent Viscosity Index (VI) in the range of 95 to 105.

Naphthenic crude oils are generally found in the coastal US regions, and in the South Americas, and are favored for the manufacture of compressor oils, refrigerant oils, and locomotive oils. Naphthenic crudes constitute up to 75% naphthene and only show traces of wax giving it an improved (lower) pour point, but a lower flash point than its paraffinic sibling, and a less desirable Viscosity Index (VI) in the range of 30 to 70.

The basic petroleum lubricant is lubricating oil, often simply referred to as "oil." This complex mixture of hydrocarbon molecules represents one of the most important classifications of products derived from the refining of crude petroleum oils, and is readily available in a great variety of types and grades reviewed more closely in Chapters 4 to 9.

Synthetic base oil

Synthetic base oil stocks are man-made and designed to have an improved and more uniform molecular structure, allowing them to display more predictable fluid properties with the ability to work better under severe conditions less suited to petroleum oil stocks. Synthetic lubricants are reviewed more closely in Chapter 7.

BASE OIL PROPERTIES

The quality of any base stock oil is measured by its resulting properties that define how well the oil will perform in service, and what additives will be required to enhance its performance. There are five major properties identified in a base oil specification.

Viscosity

To understand how oil enters a bearing, picks up and carries the bearing load requires an understanding of viscosity. With lubricating oils, viscosity is arguably its most important property and much of the story of lubrication is built around it.

Defined by the oil's molecule size, viscosity is recognized as the oil's measure of resistance to flow. With larger molecule sizes, the oil's resistance to flow increases slowing down the flow rate. Higher viscosity, or "thicker" fluids like molasses will have a slow flow rate whereas lower viscosity, or "thinner" fluids like water will flow easily as shown in Figure 1-6.

Figure 1-6. High Viscosity Fluids like Molasses flow slower then low viscosity fluids like water

The viscosity of a particular fluid is not constant and can change based on the ambient temperature, Figure 1-7, and load. When the temperature increases the lubricant becomes thinner and the viscosity decreases. Inversely, as the temperature decreases, the lubricant thickens and the viscosity increases making it more difficult to flow. Therefore, a numerical number for viscosity is meaningless unless accompanied by the temperature to which it applies.

When a lubricant is subjected to extreme load, its viscosity will increase. This is a phenomenon experienced in the elastohydrodynamic lubrication (EHL) film state found in rolling element bearings when the ball or roller moves into the direct loading contact area known as the Hertzian contact area, causing it to elastically deform, trap and pressurize the lubricant momentarily raising its viscosity and causing it to change state from a fluid to a solid, and back again, as the roller or ball moves through the direct load area.

Base oils are rated with a viscosity number according to a recognized viscosity index numbering system. The most common used imperial system for industrial lubricating oils is the SUS (Saybolt Universal Seconds) rating index, which charts the viscosity rating at two different temperatures of 100F and 210F. Its equivalent metric rating is the ISO VG rating index that follows the Kinematic viscosity rating measured in centistokes @

Figure 1-7. Oil is thicker at lower temperatures, thinner at higher temperatures.

40C and 100C. For example, an ISO VG 220 gear oil is the equivalent viscosity to an SUS 1000 @100F gear. Different oil types have their own rating systems as shown in the Equivalence of Viscosity Grading System chart found in the Appendix section

Multi-grade Viscosity Oils

When it comes to for an automotive engine crankcase oils, choosing the right oil viscosity grade will depend on where you live and operate your vehicle. Northern climate use of vehicles in winter will subject oil to very cold temperatures on start up resulting in poor oil flow and lack of full film lubrication until the oil reaches operating temperature. The result is poor starting due to the oil thickness and drag on the engine, and excessive engine wear during the start up/warm up stage.

Up until 1952, all vehicle manufacturers recommended mono-grade (single viscosity grade) automotive crankcase lubricants for their vehicles and recommended the oil viscosity grade be changed to a lower viscosity number for winter driving. The viscosity grade choice would depend on the winter temperature the car would operate in.

Viscosity grades were set by the SAE (Society of Automotive Engineers) grade viscosity system that designates oil viscosity with a number followed by the letter W to designate winter use oil, eg.10W, 15W, 20W weight, or just a number on its own to designate summer grade oils, eg. 20, 30, 40, or 50 weight oil. All that changed in 1952 when Esso (now Exxon Mobil) introduced its Esso "Uniflo" product, the world's first multi-grade oil.

Multi-grades offered motorists the advantage of a single weight oil for all year round driving that delivered improved low temperature start-ups and high temperature performance.

The multi-grade designates its working range by identifying the base oil viscosity first that is designed for cold weather performance, followed by the viscosity the oil will perform as (or emulate) once it reaches operating temperature. For example, a 20W50 has relatively thin SAE20W winter base oil that will "thicken up" and act similar to a much thicker, viscous SAE50 weight oil at operating temperature to provide full film protection over a wider temperature range. This is achieved by blending polymeric viscosity improver additives to low-viscosity base oil; these additives are long string polymers that "curl up" like a ball at low temperatures and move freely amongst the oil molecules. Once the oil heats up, the polymer strings unfurl and expand to restrict the oil's flow and raise its apparent viscosity.

Viscosity Index

Viscosity Index, or VI, is a measure of oil's viscosity *change* due to temperature. Oils with higher VI rating are more desirable as they are more stable under changing temperature conditions, reflecting a narrower change in viscosity over a standard temperature range.

As previously noted, paraffinic oils have much higher VI ratings making them more stable and desirable where a wide operating temperatures range is experienced.

Oils can be grouped and classified by their VI property as depicted in Table 1-3.

Table 1-3. Viscosity Index rating chart

VI Rating	VI Group
<35	Low - LVI
35-80	Medium - MVI
80-110	High - HVI
>110	Very High - VHVI

Specific Gravity

Specific gravity rates the oil's density relative to water.

Flash Point

Flash point is used to determine a lubricant's volatility, measured by the lowest temperature a lubricant can be heated before its vapor, when mixed with air, will ignite but not sustain combustion. Paraffinic oils have higher flash points than naphthenic oils.

Pour Point

Pour point defines the lowest temperature at which oil will still pour, or flow. Because of their level of wax content, paraffinic oils are known to have a wax pour point, which is not as low as a naphthenic base oils that are described as having a viscosity pour point. Lower viscosity oils will have lower pour points.

An oil base stock is a canvas for the additive package that makes up the final lubricant blend designed for an intended application purpose.

LUBRICANT FILM REGIMES

To successfully combat friction and wear, a lubricant film must be present at all times between the mating bearing surfaces. The degree of protection, and subsequent bearing surface life, is directly relevant to the lubricant's working film thickness, load, speed, and lubricant viscosity The minimum working film thickness required to achieve full surface separation is also known as the Lambda—□ thickness ratio.

Because the degree of surface separation is dependent on the surface "roughness" (Ra), it must be determined by measuring the profile (peaks and valleys of the surface) of both mating surfaces and by defining a centerline through them so that the areas above and below the centerline are equal. The lambda ratio is then defined as the ratio of lubricating film thickness to surface roughness, which is a lubricant film thicker than the combined height of both surface asperities enough to completely separate both surfaces.

Figure 1-8 shows the lambda □ ratio thickness curve that depicts the relationship between the working film thickness and the resulting life expectancy of the lubricated component. Note that once the lambda ratio is greater that four, life expectancy remains constant. Figure 1-8 also makes reference to different film stages, or regimes, known as Hydrodynamic, Mixed Film and Boundary Layer discussed in the following sections.

In all, there are four distinct lubricant film regimes, each one describing a different relationship between two interacting surfaces as they slide over one another.

Regime 1—Hydrodynamic Lubrication—(HDL)

Basically, lubrication is governed by one of two principles: *hydrodynamic lubrication* and *boundary lubrication.* In the former, a continuous full-fluid film separates the sliding surfaces. In the latter, the oil film is not sufficient to prevent metal-to-metal contact.

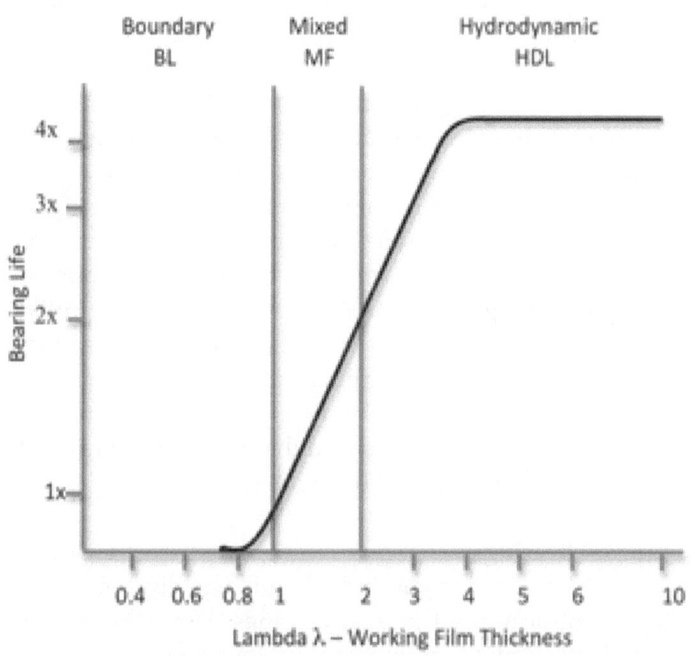

Figure 1-8. Oil Film Lambda Oil Thickness Ratio Curve

Hydrodynamic lubrication is the more common, and it is applicable to nearly all types of continuous sliding action where extreme pressures are not involved. Whether the sliding occurs on flat surfaces, as it does in most thrust bearings, or whether the surfaces are cylindrical, as in the case of journal (plain or sleeve) bearings, the principle is essentially the same, Figure 1-9.

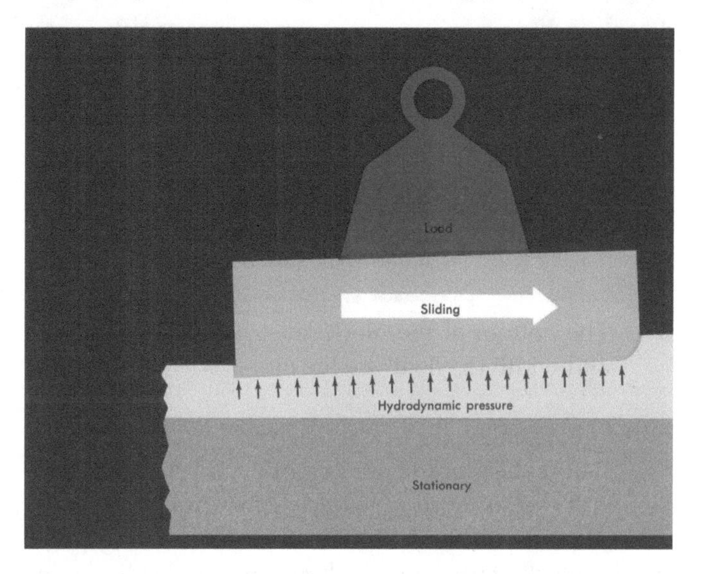

Figure 1-9. Sliding load supported by a wedge-shaped lubricating film.

Hydrodynamic Lubrication of Sliding Surfaces

It would be reasonable to suppose that, when one part slides on another, the protective oil film between them would be scraped away. Except under some conditions of reciprocating motion, this is not necessarily true at all. With the proper design, in fact, this very sliding motion constitutes the means of creating and maintaining that film.

Consider, for example, the case of a block that slides continuously on a flat surface. If hydrodynamic lubrication is to be effected, an oil of the correct viscosity must be applied at the leading edge of the block, and three design factors must be incorporated into the block:

1. The leading edge must not be sharp, but must be beveled or rounded to prevent scraping of the oil from the fixed surface.

2. The block must have a small degree of free motion to allow it to tilt and to lift slightly from the supporting surface, Figure 1-10.

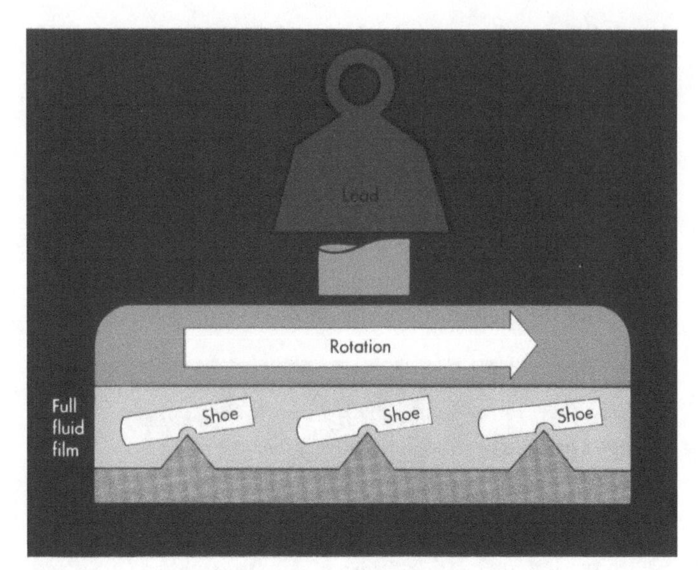

Figure 1-10. Shoe-type thrust bearing.

3. The bottom of the block must have sufficient area and width to "float" on the oil.

Full-fluid Film

Before the block is put in motion, it is in direct contact with the supporting surface. Initial friction is high, since there is no fluid film between the moving parts. As the block starts to slide, however, the leading edge encounters the supply of oil, and it is at this point that the significance of viscosity becomes apparent. Because the oil offers resistance to flow, it is not wholly displaced by the block. Instead, a thin layer of oil remains on the surface under the block, and the block, because of its rounded edge, rides up over it.

Effect of Viscosity

As the sliding block rises from the surface, more oil accumulates under it, until the oil film reaches equilibrium thickness. At this point, the oil is squeezed out from under the block as fast as it enters. Again, it is the viscosity of the oil that prevents excessive loss due to the squeezing action of the block's weight.

With the two surfaces completely separated, a full-fluid lubricating film has been established, and friction has dropped to a low value. Under these conditions, the block assumes of its own accord an inclined position, with the leading edge slightly higher than the trailing edge.

Fluid Wedge

This fortunate situation permits the formation of a wedge-shaped film, a condition essential to fluid-film lubrication. The convergent flow of oil under the block develops a pressure—hydrodynamic pressure—that supports the block. It can thus be said that fluid-film lubrication involves the "floating" of a sliding load on a body of oil created by the "pumping" action of the sliding motion, Figure 1-11.

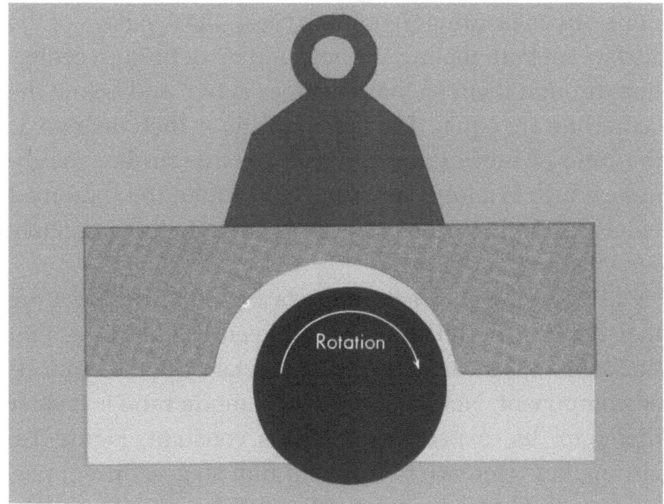

Figure 1-11. Rotation of journal "pumps" oil into the area of high pressure to carry the load.

BEARING LUBRICATION

Shoe-type Thrust Bearings

As was illustrated in Figure 1-10, many heavily loaded thrust bearings are designed in accordance with the principle illustrated by the sliding block. A disk, or thrust collar, rotates on a series of stationary blocks, or shoes, arranged in a circle beneath it. Each shoe is mounted on a pivot, rocker, or springs, so that it is free to tilt and to assume an angle favorable to efficient operation. The leading edge of each contact surface is slightly rounded, and oil is supplied to it from a reservoir.

Bearings of the type described serve to carry the tremendous axial loads imposed by vertically mounted hydro-electric generators. Rotation of the thrust collar produces a flow of oil between it and the shoes, so that the entire weight of the turbine and generator rotors and shaft is borne by the oil film. So closely does this design agree with theory, that it is said that the babbitt facing of the shoes may be crushed before the oil film fails.

Journal Bearings

The hydrodynamic principle is equally applicable to the lubrication of journal bearings. Here, the load is radial, and a slight clearance must be provided between the journal and its bearing to permit the formation of a wedge-shaped film.

Let it be assumed, for example, that a journal supports its bearing, as it does in the case of a plain-bearing railroad truck. The journal is an extension of the axle and, by means of the bearing, it carries its share of the load represented by the car.

All of the force exerted by the bearing against the journal is applied at the top of the journal—none against the bottom. When the car is at rest, the oil film between the bearing and the top of the journal has been squeezed out, leaving a thin residual coating that is probably not sufficient to prevent some metal-to-metal contact.

As in the case of the sliding block, lack of an adequate lubricating film gives rise to a high initial friction. As the journal begins to rotate, however, oil seeps into the bearing at the bottom, where the absence of load provides the greatest clearance. Some of the oil clings to the journal and is carried around to the upper side, dragging additional oil around with it.

In this manner, oil is "pumped" into the narrowing clearance at the top of the journal, where there is greatest need. The consequent flow of oil from an area of low pressure through a converging channel to an area of high pressure, as shown in Figure 1-11, produces a fluid wedge that lifts the bearing from the top of the journal, eliminating metal-to-metal contact.

When a state of equilibrium is reached, the magnitude of the entering flow displaces the bearing to one side, while the load on the bearing reduces the thickness of the film at the top. The situation is analogous to that of the inclined thrust-bearing shoe; in either case, the tapered channel essential to hydrodynamic lubrication is achieved automatically. The resulting distribution of hydrodynamic pressure is shown in Figure 1-12.

If the load were reversed, that is, if the bearing supported the journal, as is more generally the case, the relative position of the journal would be inverted. The low-pressure region would be at the top of the journal, and the protective film would be at the bottom.

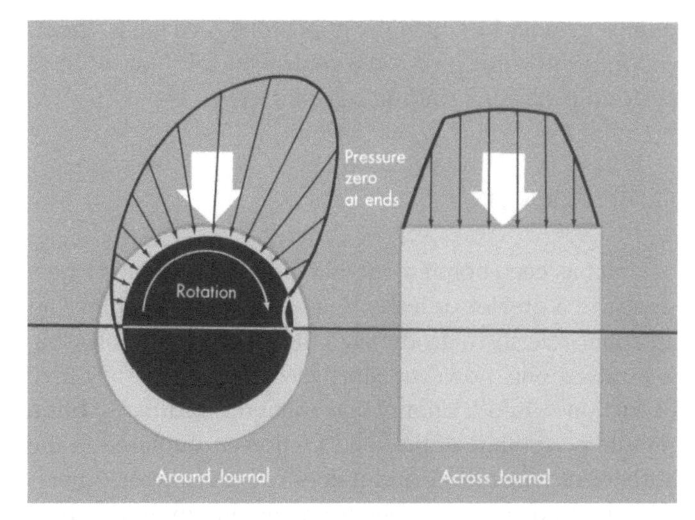

Figure 1-12. Oil pressure distribution diagrams.

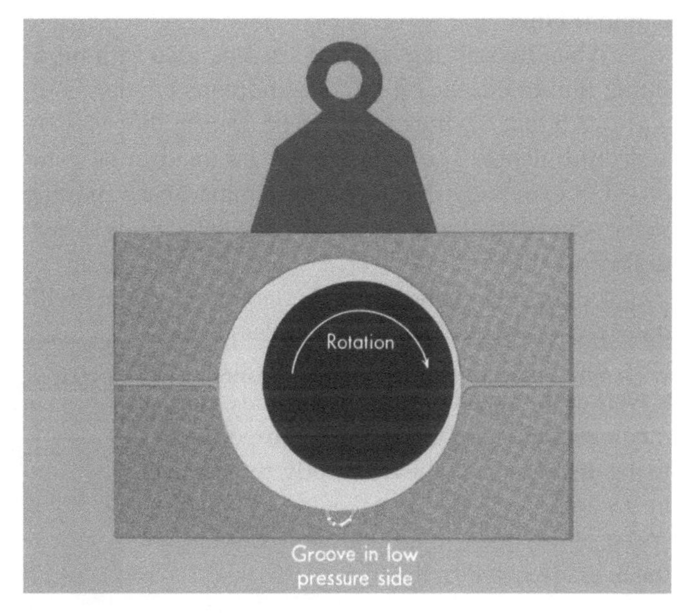

Figure 1-13. Edges of bearing face are rounded to prevent scraping of oil from journal.

Journal Bearing Design Requirements

The performance of a journal bearing is improved by certain elements of design. In addition to the allowance of sufficient clearance for a convergent flow of oil, the edges of the bearing face should be rounded somewhat, as shown in Figure 1-13, to prevent scraping of the oil from the journal. Like the leading edge of the thrust-bearing shoe, this edge should not be sharp.

Oil can enter the clearance space only from the low-pressure side of the bearing. Whatever the lubrication system, it must supply oil at this point. If the bearing is grooved to facilitate the distribution of oil across the face, the grooves must be cut in the low-pressure side. Grooves in the high-pressure side promote the discharge of oil from the critical area. They also reduce the effective bearing area, which increases the unit bearing load. No groove should extend clear to the end of the face.

Regime 2—Hydrostatic Lubrication (HSL)

Hydrostatic lubrication is a full film lubrication state set up when a lubricant is used to hydraulically separate a loaded surface and "float" one surface over another similar to that shown in Figure 1-10. This film regime is typically set up on precision machines and machine tools such as a plunge grinder in which the grindstone carriage is "floated" into the work piece on a controlled full film of oil to perform precision grinding on gears, etc.

FLUID FRICTION

It has been pointed out that viscosity, a property possessed in a greater or lesser degree by all fluids, plays an essential role in hydrodynamic lubrication. The blessing is a mixed one, however, since viscosity is itself a source of friction—fluid friction. Fluid friction is ordinarily but a minute percentage of the solid friction encountered in the absence of lubrication, and it does not cause wear. Nevertheless, fluid friction generates a certain amount of heat and drag, and it should be held to a minimum.

Laminar Flow

When two sliding surfaces are separated by a lubricating film of oil, the oil flows. Conditions are nearly always such that the flow is said to be *laminar,* that is, there is no turbulence. The film may be assumed to be composed of extremely thin layers, or laminae, each moving in the same direction but at a different velocity, as shown in Figure 1-14.

Under these conditions, the lamina in contact with the fixed body is likewise motionless. Similarly, the lamina adjacent to the moving body travels at the speed of the moving body. Intermediate laminae move at speeds proportional to the distance from the fixed body, the lamina in the middle of the film moving at half the speed of the body in motion. This is roughly the average speed of the film.

Shear Stress

Since the laminae travel at different speeds, each lamina must slide upon another, and a certain force is required to make it do so. Specific resistance to this force is known as *shear stress,* and the cumulative effect of shear stress is fluid friction. Viscosity is a function of shear stress, i.e., viscosity equals shear stress divided by shear rate. Therefore, fluid friction is directly related to viscosity.

Effect of Speed and Bearing Area

In a bearing, however, there are two additional factors that affect fluid friction, both elements of machine design. One is the relative *velocity* of the sliding surfaces, the other, their effective *area.* Unlike solid friction, which is independent of these factors, fluid friction is increased by greater speeds or areas of potential contact.

Again, unlike solid friction, fluid friction is not affected by load, Figure 1-15. Other considerations being the same, a heavier load, though it may reduce film thickness, has no effect on fluid friction.

Regime 3—Partial Film Lubrication

Mixed Film (MF) and Boundary Lubrication (BL)

This discussion of friction has so far been limited to full-fluid film lubrication. However, formation of a full-fluid film may be precluded by a number of factors, such as insufficient viscosity, a journal speed too slow to provided the necessary hydrodynamic pressure, a bearing area too restricted to support the load, or insufficient lubricant supply. Under these conditions, only partial lubrication in the form of mixed film lubrication or Boundary lubrication may be possible in which the resulting high bearing friction is a combination of fluid and solid friction, proportional to the severity of the operating conditions

Mixed film lubrication is often referred to as partial lubrication and is classified as an Intermediate lubrication regime when there is some lubricant present between two sliding surfaces, but not enough to fully separate the

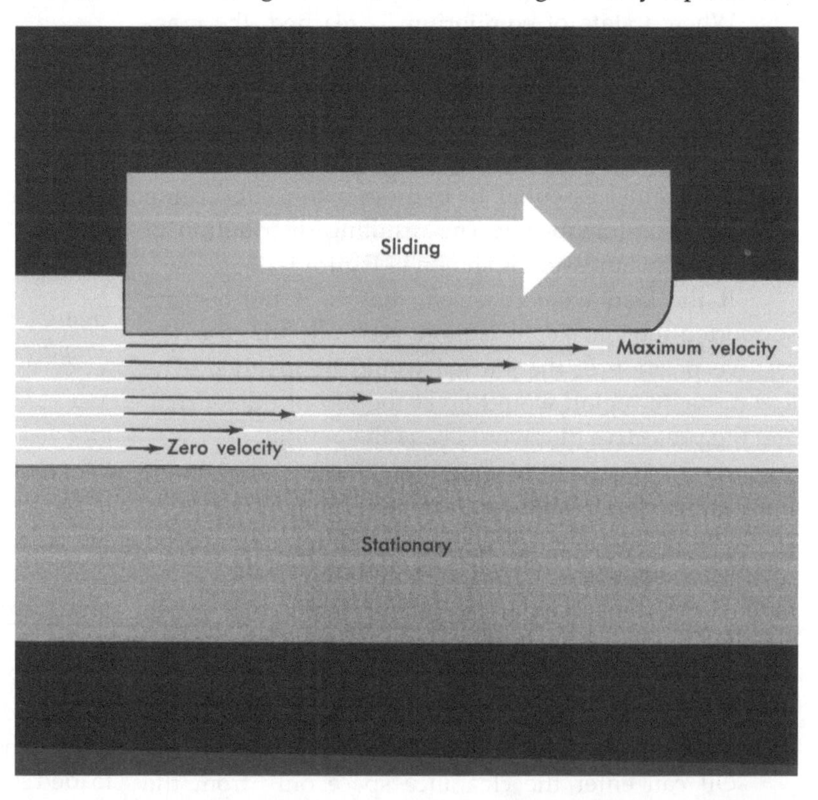

Figure 1-14. Fluid bearing friction is the drag imposed by one layer of oil sliding upon another.

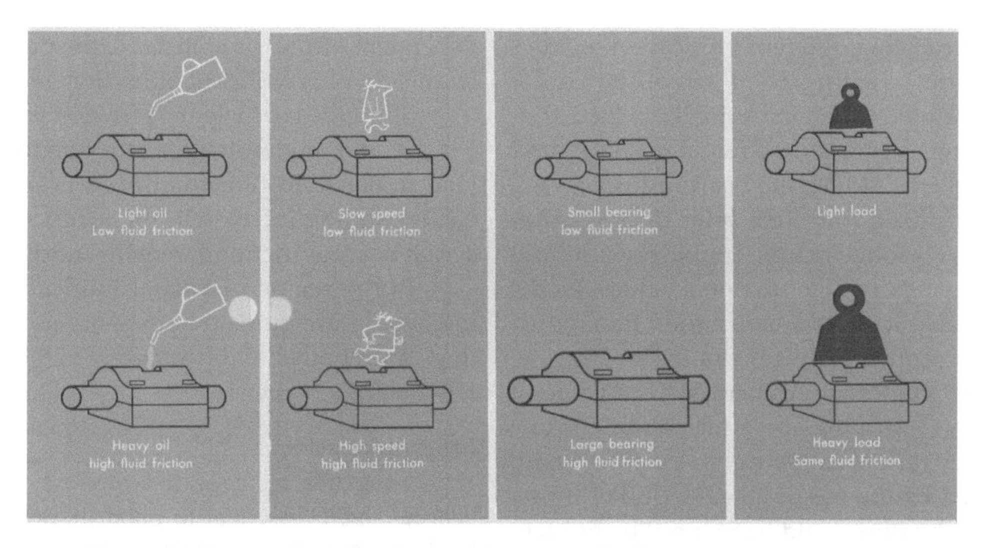

Figure 15. Factors that affect bearing friction under full-fluid-film lubrication.

surface allowing intermittent contact between the highest bearing surface points. This is also thought of as an "unstable" regime.

Boundary Lubrication is the least desirable lubrication regime with the highest coefficient of friction. Although a minimal amount of lubricant is present, the sliding surfaces are in full contact with one another at rest. With heavy load, slow moving machinery, boundary layer to mixed film regime may be the best condition achievable requiring a lubricant with EP (Extreme Pressure) and AW (Anti Wear) additives to offset the extreme bearing surface working condition. If insufficient lubricant is present, or an incorrect lubricant viscosity is used, a normally loaded bearing can stay in a boundary layer state when in full motion in which the surfaces will interfere with one another and cause rapid wear.

BEARING EFFICIENCY

Overall Bearing Friction

It is thus possible to relate all bearing friction, regardless of lubricating conditions, to oil viscosity, speed, and bearing area. Engineers express the situation mathematically with the formula:

$$F = (f)\, ZNA \qquad (1\text{-}1)$$

where

- F is the frictional drag imposed by the bearing;
- Z is oil viscosity;
- N is journal speed;
- A is the load-carrying area of the bearing.
- (f) is a symbol indicating that an unspecified mathematical relationship exists between the two sides of the equation.

Coefficient of Friction

It is customary to express frictional characteristics in terms of *coefficient of friction,* rather than friction itself. Coefficient of friction is more broadly applicable. It is *unit* friction, the actual friction divided by the force (or load) that presses the two sliding surfaces together. Accordingly, if both sides of equatio (1-1) are divided by the load L:

$$\frac{F}{L} = (f)\, ZN \frac{A}{L} \qquad (1\text{-}2)$$

Here, F/L is coefficient of friction and is represented by the symbol μ. Also, A/L is the reciprocal of pressure; or A/L - 1/P, where P is pressure, the force per unit area that the bearing exerts upon the oil. By substitution, Equation (1-2) can therefore be written:

$$\mu = (f) \frac{ZN}{P} \qquad (1\text{-}3)$$

This is the form that engineers customarily apply to bearing friction, the term ZN/P being known as a *parameter*—two or more variables combined in a single term.

ZN/P Curve

Equation (1-3) indicates only that a relationship exists, it does not define the relationship. In 1902, a Professor Richard Stribeck was the first to graphically define describe how the coefficient of friction changes for bearings experiencing different lubrication regimes as shown in Figure 1-16. The ZN/P curve shown depicts a typical example found in normally loaded sliding friction bearings such as those found in a shaft and sleeve bearing set up. At rest, the bearing surfaces are in a boundary or mixed film state prior to start up or shut down in which solid friction combines with fluid friction to yield generally high frictional values. As the shaft speed increases it will develop a fluid film reducing friction as it centrifugally centers in the bearing. Correspondingly, greater speed increases the value of the parameter ZN/P, driving operating conditions to a point on the curve farther to the right. A similar result could be achieved by the use of a heavier oil or by reducing pressure. Pressure could be reduced by lightening the load or by increasing the area of the bearing.

If these factors are further modified to increase the value of the parameter, the point of operation continues to the right, reaching the zone of perfect lubrication. This is an area in which a fluid film is fully established, and metal-to-metal contact is completely eliminated.

Beyond this region, additional increases in viscosity, speed, or bearing area reverse the previous trend. The greater fluid friction that they impose drives the operating position again to a region of high unit friction—now on the right portion of the curve.

Effect of Load on Fluid Friction

Within the range of full-fluid-film lubrication, it would appear, from Figure 1-16, that bearing friction could be reduced by increasing the bearing load or pressure. Actually, as pointed out earlier, fluid friction is independent of pressure. Instead, the property illustrated by this curve is *coefficient* of friction—not friction itself.

Since the coefficient of friction μ equals F/L, then F = μL, and any reduction of μ due to greater bearing load under fluid-film conditions is compensated by a corresponding increase in the load L. The value of the actual bearing friction F remains unchanged.

In the region of partial lubrication, however, an increase in pressure obviously brings about an *increase* in μ. Since both μ and L are greater, the bearing friction F is *markedly* higher.

Efficiency Factors

From this analysis, it is quite evident that proper bearing size is essential to good lubrication. For a given load and speed, the bearing should be large enough to permit the development of a full-fluid film, but not so large as to create excessive fluid friction (Figure 1-17). Clearance should be sufficient to prevent binding, but not so great as to allow excessive loss of oil from the area of high pressure. The relative position of the ZN/P curve for a loose-fitting bearing would be high and to the right, as shown in Figure 1-18, indicating the need for a relatively high-viscosity oil, with correspondingly high fluid friction.

Efficient operation also demands selection of an oil of the correct viscosity, an oil just heavy enough to provide bearing operation in the low-friction area of fluid-film lubrication. If speed is increased, a heavier oil is generally necessary. For a given application, moreover, a lighter oil would be indicated for lower ambient temperatures, while a heavier oil is more appropriate for high ambient temperatures. These relationships are indicated in Figure 1-19.

Temperature-Viscosity Relationships

To a certain extent, a lubricating oil has the ability to accommodate itself to variations in operating conditions.

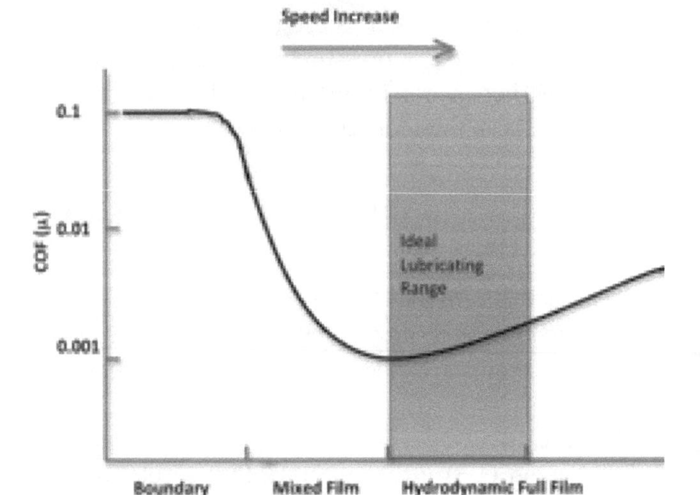

Figure 1-16. Stribeck bearing performance curve.

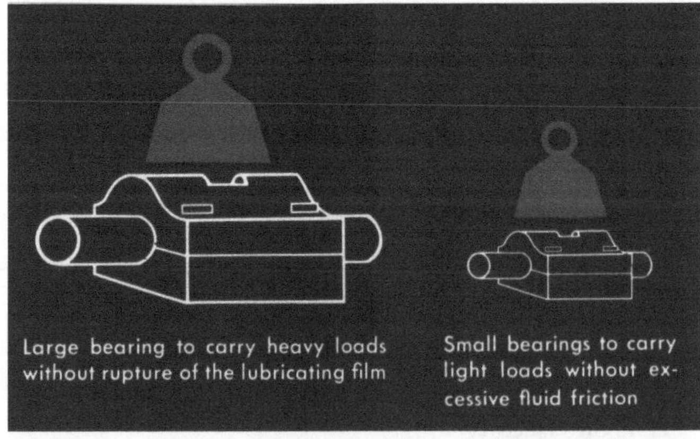

Figure 1-17. Bearing design should permit the development of a full-fluid film.

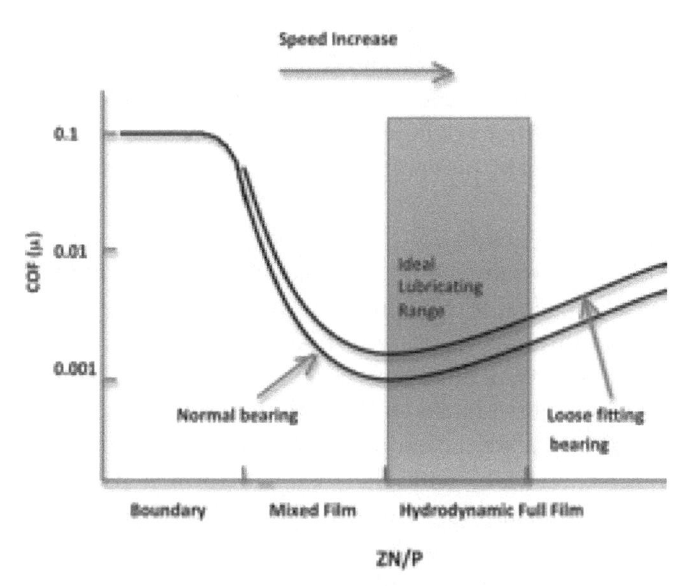

Figure 1-18. Loose-fitting bearings require high-viscosity oils.

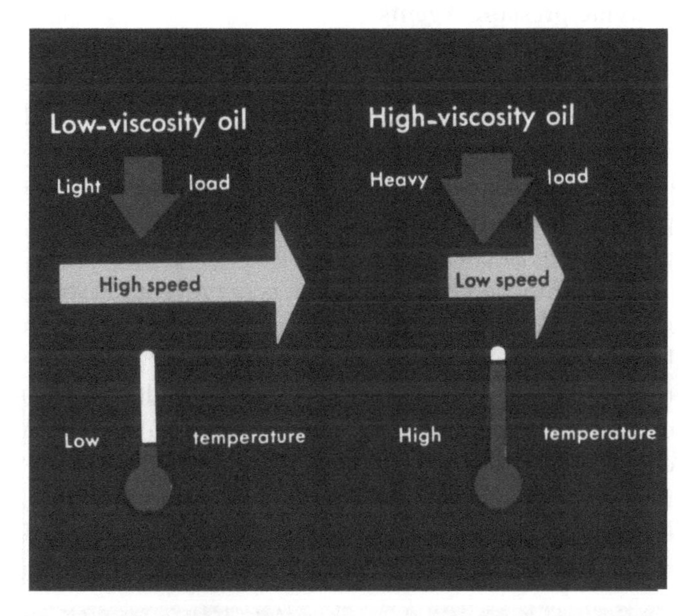

Figure 1-19. Relationships between oil viscosity, load, speed, and temperature.

If speed is increased, the greater frictional heat reduces the operating viscosity of the oil, making it better suited to the new conditions.

Similarly, an oil of excessive inherent viscosity induces higher operating temperatures and corresponding drops in operating viscosity. The equilibrium temperatures and viscosities reached in this way are higher, however, than if an oil of optimum viscosity had been applied. So the need for proper viscosity selection is by no means eliminated.

Oils vary, however, in the extent to which their viscosities change with temperature. An oil that thins out less at higher temperatures and that thickens less at lower temperatures is said to have a higher V.I. (viscosity index). For applications subject to wide variations in ambient temperature, a high-V.I. oil may be desirable, Figure 1-20.

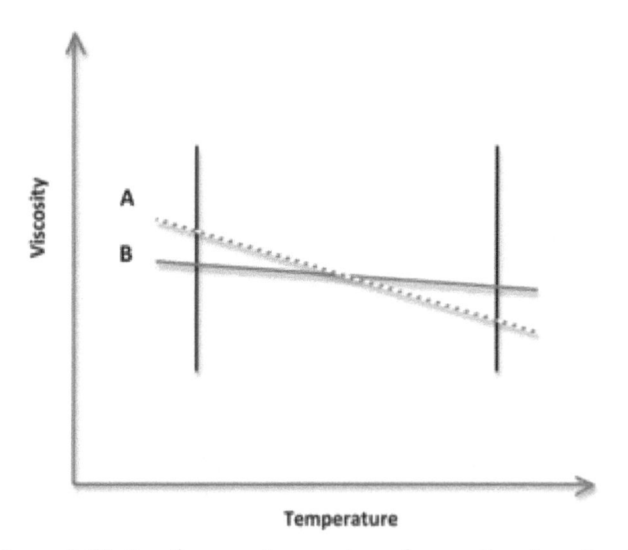

Figure 1-20. For the same temperature change, the viscosity of oil "B" changes much less than that of oil "A."

This is true, for example, of motor oils, which may operate over a 100°F temperature range. With an automobile engine, there is an obvious advantage in an oil that does not become sluggishly thick at low starting temperatures or dangerously thin at high operating temperatures. So good lubrication practices include consideration of the V.I. of the oil as well as its inherent viscosity.

As stated before, all of the factors that make hydrodynamic lubrication possible are not always present. Sometimes journal speeds are so slow or pressures so great that even a heavy oil will not prevent metal-to-metal contact. Or an oil heavy enough to resist certain shock loads might be unnecessarily heavy for normal loads. In other cases, stop-and-start operation or reversals of direction cause the collapse of any fluid film that may have been established. Also, the lubrication of certain heavily loaded gears—because of the small areas of tooth contact and the combined sliding and rolling action of the teeth—cannot be satisfied by ordinary viscosity provisions.

Since the various conditions described here are not conducive to hydrodynamic lubrication, they must be met with *boundary lubrication,* a method that is effective in the absence of a full-fluid film, Figure 1-21.

Additives for Heavier Loads

There are different degrees of severity under which boundary lubrication conditions prevail. Some are only

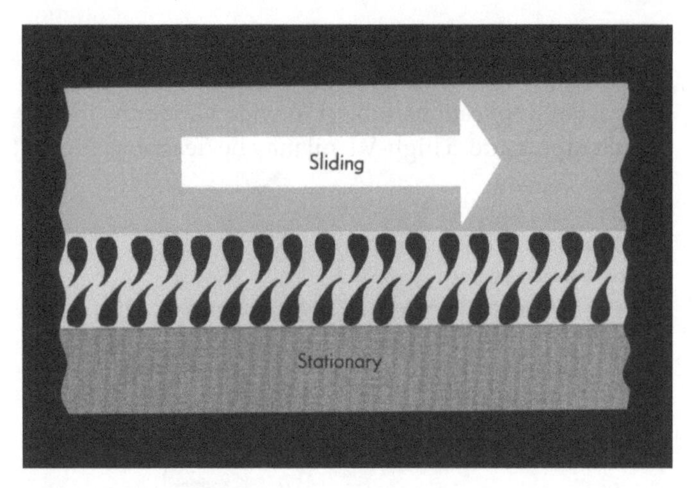

Figure 1-21. Sliding surfaces separated by a boundary lubricant of the polar type.

moderate, others extreme. Boundary conditions are met by a variety of special lubricants with properties corresponding to the severity of the particular application. These properties are derived from various additives contained in the oil, some singly, some in combination with other additives. Their effect is to increase the load-carrying ability of the oil.

Where loads are only mildly severe, an additive of the class known as *oiliness agents* or *film-strength additives* is applicable. Worm-gear and pneumatic-tool lubricants are often fortified with these types of agents. Where loads are moderately severe, *anti-wear agents* or *mild EP additives,* are used. These additives are particularly desirable in hydraulic oils and engine oils. For more heavily loaded parts, a more potent class of additives is required; these are called *extreme pressure* (EP) *agents,* Table 1-5.

Oiliness Agents

The reason for referring to oiliness agents as film-strength additives is that they increase the oil film's resistance to rupture. These additives are usually oils of animal or vegetable origin that have certain polar characteristics. A polar molecule of the oiliness type has a strong affinity both for the petroleum oil and for the metal surface with which it comes in contact. Such a molecule is not easily dislodged, even by heavy loads.

In action, these molecules appear to attach themselves securely, by their ends, to the sliding surfaces. Here they stand in erect alignment, like the nap of a rug, linking a minute layer of oil to the metal. Such an array serves as a buffer between the moving parts so that the surfaces, though close, do not actually touch one another. For mild boundary conditions, damage of the sliding parts can be effectively avoided in this way.

Lubricity is another term for oiliness, and both apply to a property of an oil that is wholly apart from viscosity. Oiliness and lubricity manifest themselves only under conditions of boundary lubrication, when they reduce friction by preventing breakdown of the film.

Anti-wear Agents

Anti-wear agents, also called *mild EP additives,* protect against friction and wear under moderate boundary conditions. These additives typically are organic phosphate materials such as zinc dithiophosphate and tricresyl phosphate. Unlike oiliness additives, which physically plate out on metal surfaces, anti-wear agents react chemically with the metal to form a protective coating that allows the moving parts to slide across each other with low friction and minimum loss of metal. These agents sometimes are called "anti-scuff" additives.

Extreme-pressure Agents

Under the extreme-pressure conditions created by very high loads, scoring and pitting of metal surfaces is a greater problem than frictional power losses, and seizure is the primary concern. These conditions require extreme-pressure (EP) agents, which are usually composed of active chemicals, such as derivatives of sulfur, phosphorus, or chlorine.

The function of the EP agent is to prevent the welding of mating surfaces that occurs at the exceedingly high local temperatures developed when opposing bodies are rubbed together under sufficient load. In EP lubrication, excessive temperatures initiate, on a minute scale, a chemical reaction between the additive and the metal surface. The new metallic compound is resistant to welding, thereby minimizing the friction that results from repeated formation and rupturing of tiny metallic bonds between

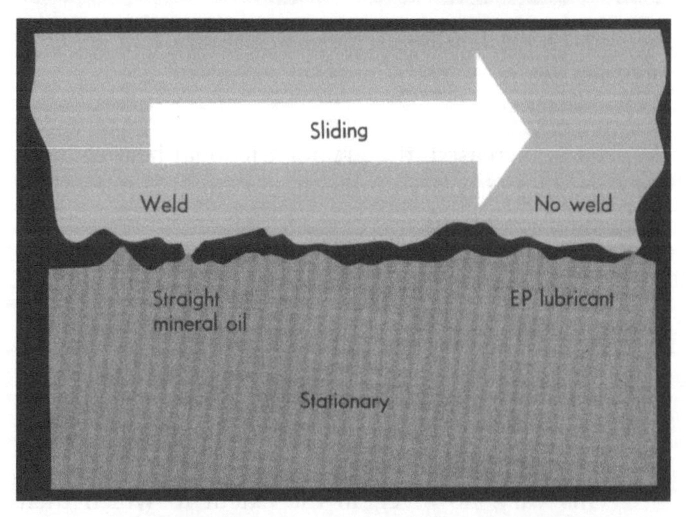

Figure 1-22. Extreme-pressure conditions.

the surfaces.

This form of protection is effective only under conditions of high local temperature. So an extreme-pressure agent is essentially an extreme-temperature additive.

Multiple Boundary Lubrication

Some operations cover not one but a range of boundary conditions. Of these conditions, the most severe may require an oil with a chemically active agent that is not operative in the milder boundary service. Local temperatures, though high, may not always be sufficient for chemical reaction. To cover certain multiple lubrication requirements, therefore, it is sometimes necessary to include more than a single additive: one for the more severe, another for the less severe service.

Incidental Effects of Boundary Lubricants

The question logically presents itself as to why all lubricating oils are not formulated with boundary-type additives. The basic reason is that this formulation is usually unnecessary; there is no justification for the additional expense of blending. Additionally, the polar characteristics of oiliness agents may increase the emulsibility of the oil, making it undesirable for applications requiring rapid oil-water separation. Some of the more potent EP additives, moreover, have a tendency to react with certain structural metals, a feature that might limit their applicability.

Stick-slip Lubrication

A special case of boundary lubrication occurs in connection with stick-slip motion. Remember, a slow or reciprocating action, such as that of a machine way, is destructive to a full-fluid film. Unless corrective measures are taken, the result is metal-to-metal contact, and the friction is solid, rather than fluid. Also remember that solid *static* friction is greater than solid *kinetic* friction, i.e., frictional drag drops after the part has been put in motion.

Machine carriages sometimes travel at very slow speeds. When the motive force is applied, the static friction must first be overcome, whereupon the carriage, encountering the lower kinetic friction, may jump ahead. Because of the slight resilience inherent in a machine, the carriage may then come to a stop, remaining at rest until the driving mechanism again brings sufficient force to bear. Continuation of this interrupted progress is known as *stick-slip* motion, and accurate machining may be difficult or impossible under these circumstances.

To prevent this chattering action, the characteristics of the lubricant must be such that kinetic friction is greater than static friction. This is the reverse of the situation ordinarily associated with solid friction. With a way lubricant compounded with special oiliness agents, the drag is *greater* when the part is in motion. The carriage is thus prevented from jumping ahead to relieve its driving force, and it proceeds smoothly throughout its stroke.

Regime 4—Elastohydrodynamic Lubrication (EHDL)

The foregoing discussion has covered what may be termed the classical cases of hydrodynamic and boundary lubrication. The former is characterized by very low friction and wear and dependence primarily on viscosity; the latter is characterized by contact of surface asperities, significantly greater friction and wear, and dependence on additives in the lubricant to supplement viscosity.

In addition to these two basic types of lubrication, there is an intermediate lubrication mode that is considered to be an extension of the classical hydrodynamic process. It is called *elasto-hydrodynamic* (EHD) *lubrication,* also known as EHL. It occurs primarily in rolling-contact bearings and in gears where non-conforming surfaces are subjected to very high loads that must be borne by small areas. An example of non-conforming surfaces is a ball within the relatively much larger race of a bearing (see Figure 1-23).

EHD lubrication is characterized by two phenomena:

1) the surfaces of the materials in contact momentarily deform elastically under pressure, thereby spreading the load over a greater area.

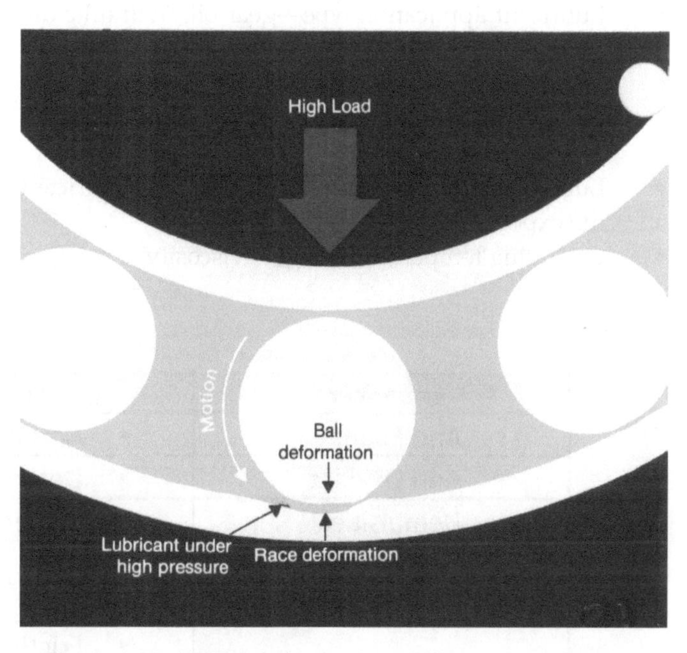

Figure 1-23. EHD lubrication in a rolling-contact bearing.

2) The viscosity of the lubricant momentarily increases dramatically at high pressure, thereby increasing load-carrying ability in the contact zone.

The combined effect of greatly increased viscosity and the expanded load-carrying area is to trap a thin but very dense film of oil between the surfaces. As the viscosity increases under high pressures, sufficient hydrodynamic force is generated to form a full-fluid film and separate the surfaces.

The repeated elastic deformation of bearing materials that occurs during EHD lubrication results in a far greater incidence of metal fatigue and eventual bearing failure than is seen in sliding, or plain, bearing operation. Even the best lubricant cannot prevent this type of failure.

ADDITIONAL LUBRICANT ADDITIVES

Earlier in this chapter we reviewed the different types and properties of Base oils that constitute 75% to 99% of the finished oil that must then be blended with an additive package designed to meet the needs of the operating conditions within which the finished lubricant must perform. Designing a lubricant requires the tribology chemist to begin with a performance specification sheet outlining parameters and conditions the lubricant must meet and exceed in its finished commercial form.

Typical design parameters will include:

- Lubricant application type—gear oil, hydraulic oil, air tool oil, engine oil, etc.
- Application requirements—load, speed, bearing surface motion (sliding, rolling, combination), delivery method
- Lubricant quality—Viscosity index (VI), lubricant life expectancy, selling point
- Operating temperature range—Viscosity
- Operating environment—moisture, chemicals,
- Biodegradability

These operational design parameters make up what is technically known as the "Tribological System" in which the lubricant must perform. Lubricant type, application, quality, and operating temperature range are primarily used to determine the appropriate base stock, which is then supplemented with a variety of lubricant additives to strengthen or modify the lubricant's characteristics to meet the finished design specification. Interestingly, lesser quality base stocks can be significantly bolstered and modified to meet specifications with additive package.

The Additive Role and Function

Additives play an important role in letting us know when the lubricant is no longer useful. Additives are sacrificial by nature, and once depleted the oil must be replaced or replenished again with the necessary additive(s). By monitoring and comparing the additive package "signature" of the virgin stock oil against a used oil sample through the use of oil analysis we can tell when the oil is degraded and ready to be changed/ replenished.

Oil additives serve three functions: to enhance, to promote new properties, and to suppress undesirable base oil properties. Different lubricant types require different formulation packages. Table 1-4 Oil Additives by Function, depicts which additive performs what function.

Additives can be organic or inorganic compounds and depending on their physical size, will dissolve in the oil (sub micron) or remain as suspended solids. These solids are often visible to the naked eye when decanting new oils from one container to another.

The Additive Package

In the mixed film section of this chapter we re-

Table 1-4. Oil Additives by Function

Enhancers	New Properties	Suppressants
• Anti Foam	• Anti Wear	• Pour Point
• Anti Oxidant	• Detergent	• Viscosity Improver
• Demulsifier	• Dispersant	
• Film Strength	• Dye	
• Rust Inhibitor	• EP	
	• Friction Modifier	

viewed in detail, Film strength (Oiliness), Anti-wear and Extreme pressure additives used to combat an alleviate boundary conditions. These additives, along with the ten additional additive types reviewed below make up the core additive package for most commercial lubricants.

Anti Foam Agent—when a fluid is moved quickly through a pumping action, it can entrain small air bubbles in the lubricant. These air bubbles are detrimental to a lubricant as air contains oxygen that will attack the base oil (see Anti Oxidant). Aerated fluids can also cause pump cavitation. Also known as de-foamants, or foam inhibitors, anti foam agents are primarily made up of that work to increase the lubricant's surface tension and enlarge bubble size allowing them to collapse more easily.

Anti Oxidant Agent—oxygen is base oil's primary enemy, especially at higher temperatures when in combination with contaminants such as water can lead to sludge and viscosity thickening, tar, varnish, and corrosive acid formation within the oil and on the bearing surfaces. Anti oxidant agents, also known as oxidation inhibitors, can successfully improve oxidation stability by more than 10 times by deactivating catalytic metallic contaminants and by decomposing any formed reactive hydroperoxides within the oil. The most common anti oxidant is Zinc dialkyldithiophophate, or ZDDP.

Demulsifier—also known as emulsion breakers, are used where water contamination is expected, and are designed to chemically prevent the formation of any water/oil emulsion by altering the surface tension of the oil allowing the water to separate easily and be drained off.

Detergents—used where combustion takes place as a chemical cleaner to keep combustion surfaces free from harmful deposits and to neutralize any combustion acids. Developed specifically for crankcase and compressor oils, detergent additives are made up from over base (alkaline) organic metallic soaps such as barium, calcium and magnesium.

Dispersants—again, used in crankcase and compressor oils, often in conjunction with detergents to chemically disperse and attach themselves to and suspend combustion and contaminant particles such as dirt, soot, glycol, depleted additives, to extract them by the oil filtration system.

Dye—used in transmission fluids and greases to better identify the product from other lubricants.

Friction Modifier—long chain polar additives that have an affinity for metal surfaces are added to crankcase and transmission oils to reduce the surface friction of lubricated parts in an effort to increase fuel economy

Pour Point Suppressant—used to prevent the formation of wax crystals in paraffinic mineral oils at low temperatures allowing the oil to pour at lower temperatures.

Rust Inhibitor—also known as a corrosion inhibitor, it forms a protective shield against water and corrosive acids to stop the formation of corrosion and rust on ferrous, copper, tin, and lead based metals.

Viscosity Improver—these employ long string polymers that expand as the oil temperature increases and serves the "thicken" the oil, or increase its viscosity. Added to increase an oil's serviceability over a wider temperature range in multi-grade form and used to bolster lower quality base oils that have lower viscosity index (VI) ratings.

Table 1-5. Oil Type Additive Package reference table, details which additives are used in the different oil types. This is for guideline purposes only; always consult with your lubricant supplier to determine what additives are actually used in the lubricant you have chosen to use.

Additive	Bearing Oil	Compressor Oil	Crankcase Oil	Gear Oil	Hydraulic Oil	Transmission Oil	Turbine Oil
Anti Foam	x	x	x	x	x	x	x
Anti Oxidant	x	x	x	x	x	x	x
Anti Wear	x	x	x	x	x	x	
Demulsifier	x	x		x	x		x
Detergent		x	x			x	
Dispersant		x	x			x	
Dye						x	
Extreme Pressure	x	x	x	x	x	x	
Friction Modifier			x			x	
Pour Point		x	x	x	x	x	
Rust Inhibitor	x	x	x	x	x	x	x
Viscosity Improver			x		x		

BREAK-IN

Though modern tools are capable of producing parts with close tolerances and highly polished surfaces, many machine elements are too rough, when new, to sustain the loads and speeds that they will ultimately carry. Frictional heat resulting from the initial roughness of mating parts may be sufficient to damage these parts even to the point of failure. This is why a new machine, or a machine with new parts, is sometimes operated below its rated capacity until the opposing asperities have been gradually worn to the required smoothness.

Under break-in conditions, it is sometimes necessary or advantageous to use a lubricant fortified with EP additives. The chemical interaction of these agents with the metal tends to remove asperities and leave a smoother, more polished surface. As the surface finish improves during initial run-in, the need for an EP lubricant may be reduced or eliminated, and it may then be appropriate to substitute a straight mineral oil or an EP oil with less chemical activity.

BEARING METALS

The break-in and operating characteristics of a journal bearing depend to a large extent upon the composition of the opposing surfaces. In the region of partial lubrication, friction is much less if the journal and bearing are of different metals. It is customary to mount a hard steel journal in a bearing lined with a softer material, such as bronze, silver, or babbitt.

There are several advantages in a combination of this sort. The softer metal, being more plastic, conforms readily to any irregularities of the journal surface, so that break-in is quicker and more nearly perfect. Because of the consequent closeness of fit, soft bearing metals have excellent wear properties. Moreover, in the event of lubrication failure, there is less danger of destructive temperatures. Friction is lower than it would be if steel, for example, bore directly against steel.

If temperature should rise excessively in spite of this protective feature, the bearing metal, with its lower melting point, would be the first to give way. Yielding of the bearing metal often prevents damage to the journal, and replacement of the bearing lining is a relatively simple matter. However, composition of bearing metals has no effect upon performance under full-film lubrication.

WEAR

Even with the most perfectly lubricated parts, some physical wear is to be expected. Sometimes wear is so slight as to be negligible, as in the case of many steam turbine bearings. Turbines used to generate power operate under relatively constant loads, speeds, and temperatures, a situation that leads to the most effective sort of lubrication.

Many other machines, however, operate under less ideal conditions. If they stop and start frequently, there will be interruptions of the lubricating film. Also, in any lubricating process, there is always the possibility of abrasive wear due to such contaminants as dirt and metallic wear particles. Wear is further promoted by overloading, idling of internal combustion engines, and other departures from optimal operating conditions.

Wear vs. Friction

Though wear and friction generally go hand-in-hand, there are extreme situations in which this is not so. Some slow-speed bearings are so heavily loaded, that an oil of the highest viscosity is required for complete lubrication. Because of the greater fluid friction, this lubricant imposes more bearing friction than a lighter lubricant would.

On the other hand, the lighter lubricant, since it would provide only partial lubrication, could not be considered suitable from the standpoint of protection of the metal surface. Some frictional advantage must be sacrificed in favor of an improvement in wear characteristics. Contrary to popular conception, therefore, less wear actually means more friction under extreme conditions such as this.

COMMON WEAR MECHANISMS

There are four common wear mechanisms that cause surface degradation and eventual loss of usefulness in bearing surfaces.

Abrasive Wear

Abrasive wear occurs when bearing surfaces run in a MF or BL regime. Abrasive wear can occur as a result of a 2-body or 3-body surface interaction. As seen in Figure 1-24, 2 and 3 body abrasive wear, the 2-body abrasion example depicts two surface points touching and cutting into the opposing sliding surface resulting in a scratched, grooved, or furrowed surface and a third body metal cutting being released into the lubricant. This third body is now free to also get caught between two surface points and add to the surface degradation. 3-body abrasion can also occur due to large particle (dirt) found or introduced into the lubricant form the lubrication process or machine operation.

Erosive wear is a form of abrasive wear caused by particles impacting a surface.

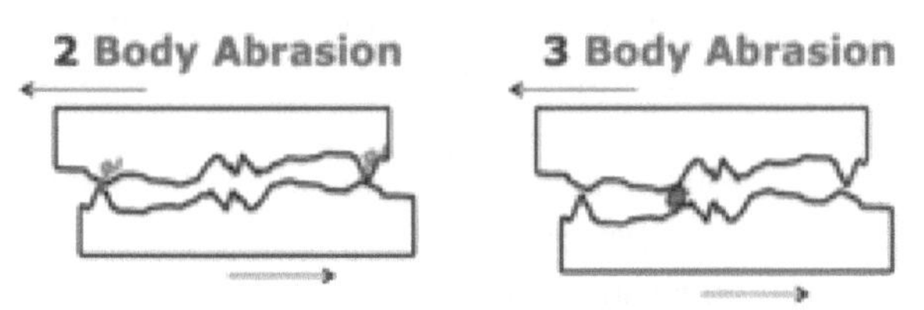

Figure 1-24. 2 and 3 Body Abrasive Wear
Courtesy ENGTECH Industries Inc.

Adhesive Wear

Adhesive wear usually occurs under highly loaded sliding friction when an incorrect viscosity lubricant is used, when EP and AW additives have depleted, or under heavy shock loading. As the surfaces come together they weld under the heat and load pressure, metal is transferred and torn apart under movement leaving a discreet, often jagged or smeared surfaces. Adhesive wear is also described as scuffing, shearing, or galling wear.

Fatigue Wear

Fatigue wear usually occurs in rolling friction surfaces that have experienced repeated long-term load cycles and stress that causes elastic deformation to the surfaces as in EHD lubrication. This long term action eventually results in small surface and sub surface cracking, which eventually travel up to and across the bearing surface resulting in surface delamination and surface pitting.

Corrosive Wear

Corrosive wear leaves an acid etched surface in the bearing contact surface area when surfaces corrode due to an oxidative chemical reaction that is accelerated in the presence of moisture contamination and heat. Corrosive wear is also known as acidic pitting and is caused by a lubricant that is moisture contaminated, one that is additive depleted, or use of a lubricant with no corrosion inhibitor additive.

GREASE LUBRICATION

Many situations exist in which lubrication can be accomplished more advantageously with grease than with oil. Most lubricating greases consist of petroleum oils thickened with special soaps that give them an unusual ability to stay in place. Grease is often used, therefore, in applications for which it is not practical to provide a continuous supply of oil.

Though the retentive properties of grease—also resistance to heat, water, extreme loads, and other adverse conditions—depend primarily on the proportion and type of soap, frictional characteristics themselves are related almost entirely to the oil content. Base-oil viscosity is a determining factor in the ability of the grease to provide a proper lubricating film.

The word grease is derived from the Latin word "Crassus" meaning "fat, dense, or thick," which adequately describes its consistency. Ancient Romans and Egyptians are both believed to have been amongst the first to create real grease by combining olive oil with lime (Calcium Carbonate) to make Calcium based grease.

Making grease is similar to making soap, both rely on a chemical reaction to take place between oil, fat or fatty acids (often present in the oil or added), and an alkali base material (referred to as the "thickener") to form soap like material, a reaction known as saponification.

Grease uses a variety of metal hydroxide alkaline to make and define the grease type, e.g. aluminum hydroxide makes aluminum grease, lithium hydroxide makes lithium grease, calcium hydroxide makes calcium grease, etc.

To give grease a wider temperature application range and enhanced properties a second thickener known as a complexing agent is added to the mix. This agent is a salt, usually of the same metal hydroxide used to originally thicken the grease. If lithium is used as the alkaline agent, lithium salts are then added to the mix to create lithium complex grease.

When applied, it is always the base oil, which constitutes 80% to 95% of the grease product that lubricates the bearing surfaces. The chemical soap fraction acts like a sponge to hold the oil and by default is designed to "wick" or release the oil into the bearing area when the bearing temperature rises. When the bearing ceases operation and/or the temperature cools, the grease will reverse its action and "wick" back the oil into suspension, acting as a semi-solid living reservoir for the oil.

Considering Grease as a Lubricant of Choice

Whenever rotating or interacting moving surfaces are encountered in a machine design, the designer must decide early on what lubricant is to be used.

Grease's unique capabilities give it both advantages and disadvantages over just oil as described in Table 1-6, Grease Versus Oil Comparison.

Note: Because oil is always the lubricating medium, all the decision factors involved in choosing the right viscosity and additives for the application will still apply. Before using specific grease, check with the grease vendor or manufacturer to ensure that the grease oil viscosity is suitable for your application.

Grease Characteristics

Greases are classified and characterized in many ways. These characteristics are identifiers that allow us to rate and compare the ability of various greases with one another and can be found on the lubricant specification sheet provide by the manufacturer. The most common characteristics are:

NLGI Consistency Rating

Greases are manufactured in a variety of consistencies. These consistencies are distinguished into 9 specific classifications using the NLGI - National Lubricating Grease Institute number rating system from a #000 grease representing a grease with a very fluid appearance at room temperature, all the way up to a #6 grease that appears block solid at room temperature. The most popular grease for use in grease guns is am NLGI #2 rated grease that appears soft looking at room temperature. The recommended grease consistency for use in an automated centralized grease lubrication system is NLGI #1 or lower.

NLGI consistency is determined in the laboratory using an ASTM (American Society of Testing Materials) D-217 cone penetration test in which grease is placed in a cup and a cone is dropped from a specified height at a room temperature of 77F and allowed to penetrate the grease for five seconds. The depth of penetration is then measured carefully in tenths of a millimeter and rated according to an NLGI classification chart that assigns a

Table 1-6. Grease Versus Oil Comparison

Grease	Oil
Advantages	
Stays where it is placed, easy to control	The most effective lubricant
Excellent Sealing capability - can be used as part of a labyrinth seal to keep out moisture and dirt	Excellent cleaning and flushing capability, especially in re-circulative systems
Can withstand heavy shock loads	Requires lower pump pressure and line size to move oil around in centralized systems
Requires less frequent applications than oil – better for remote locations and occasional use bearings	Can be used in re-circulative systems
	With correct application, no limit to machine speed
	Within centralizes systems, generally more stable than grease
Disadvantages	
Has a lower operating temperature than oil, can "cook" and harden in bearing in higher temperature applications	More difficult to control at bearing surface area
Introduced contamination from poor manual greasing process will not settle out and can harm bearing	Requires mechanical seals (e.g radial lip seal) to control leakage
Dependent on frictional heat to allow oil to wick into contact area – bearings run hotter than oil-only lubricated bearings	
Grease lubricated bearings consume more energy to overcome grease fluid friction	
Requires greater pump pressure and line size to move grease around in centralized systems	
Total loss only lubrication	

Courtesy ENGTECH Industries Inc.

rating number to penetration depth ranges e.g. if the cone penetration is between 265 (26.5mm) and 295 (29.5mm) the grease is classified as a NLGI #2

Note: If grease is classified with an EP letter rating prior to the NLGI number, e.g. EP2 grease, this will signify that the grease is formulated with an Extreme Pressure additive such as Molybdenum Disulfide for use in low speed high load applications.

Appearance

As we have seen in the NLGI rating system grease is classified by its appearance at room temperature and ranges from very fluid to soft or smooth to a block like appearance. Descriptions may vary slightly from manufacturer to manufacturer, e.g. soft versus smooth.

Color

Manufacturers often color grease with dyes for identification purposes. Grease can be black, green, blue, red, brown, or white—usually reserved for food grade products.

Thickener

The thickener tells us what alkali base (soap base) was used to manufacture the grease. Most popular thickeners are identified in Table 1-6. Grease Thickener Types.

Pumpability

Details the grease's ability to flow easily under pressure through a distribution system over a given temperature range

Slumpability

Sometimes referred to as "feedability," slumpability refers to grease's ability to relax in its reservoir under gravity and be fed into the pump.

Dropping Point

The temperature at which grease is liquid or soft enough to drip. Note: This is not the maximum operating temperature

Operating Temperature

The recommended optimum grease operating temperature range.

Water Resistance or Washout

Grease's ability to withstand the effects of water spray before it affects its ability to lubricate and prevent rust formation.

Shear Stability

As grease is "worked" or sheared in operating conditions its consistency may change. Greases better able to retain their consistency in working conditions have a better shear stability and are preferred. Greases that harden as they are worked are known as rheopectic grease, whereas greases that soften are known as thixotropic grease.

HOW LUBRICANTS FAIL IN SERVICE

In the lubrication world, temperature presents an unusual paradox. Lubricants require heat to help them flow efficiently over and around their bearing surfaces, yet if the temperature gets too hot, the lubricant is likely to undergo a chemical change that can drastically reduce its life expectancy. Inversely, if the temperature gets too cold, the lubricant will thicken and loose its ability to flow and efficiently lubricate its bearing surfaces. In addition

Table 1-7: Grease Thickener Types

Grease Thickener(s)	Operating Temperature Range
Lithium	-20 to 350 F
Lithium Complex	-50 to 375F
Calcium (Lime)	-50 to 230F
Calcium Complex	-20 to 350F
Sodium	0 to 250F
Barium	Up to 380F
Aluminum Complex	-50 to 350F
Bentonite	-20 to 500F

Note: temperature ranges shown are approximate; always check the manufacturers specification sheet for actual temperature range

to this paradox, irony is also at work at work, as lubricating oil is not only designed to separate and lubricate a bearing surface, it is also designed to absorb and carry away frictional heat from the bearing surface.

The old adage "oil is oil and grease is grease" may have been true in our long ago agrarian society, but in today's high speed complex industrial environments lubricants must be tailored and managed to their machine host's specific needs and operating environment if they are to guarantee asset availability and reliability.

When a lubrication film is insufficient to separate moving surfaces the surfaces will collide and transfer energy, resulting in a rapid heat build up and metal expansion further retarding motion until both surfaces eventually weld to one another and seize. To avoid this worst case scenario the ideal intent is to ensure the correct lubricant is introduced into the bearing surface area in sufficient quantity to provide full separation of moving surfaces in a consistent manner. This is to ensure bearing temperatures stay below the magic 210°F (100°C) operating temperature over which any bearing temperatures rise above will start to have adverse affects on the lubricant and the bearing life.

In cold temperature climates, hydrocarbon based oil can thicken to the point in which it will no longer pour, largely due to the wax content in the oil. More expensive hydro-treated and synthetic based oils will largely resolve this problem, or the user can choose to heat the reservoir containing the oil to a temperature that will allow it to flow again.

Heat Related Lubricant Failure

In the late 19th century, the Swedish Nobel laureate Svante Arrhenius, discovered a direct relationship between temperature change and chemical reaction rate in fluids, expressing this relationship in his Arrhenius equation. In the field of lubrication we know this equation as the Arrenhius rule used to express the temperature change dependent failure rate of lubricating oils in the following statement, "for every 18°F (10°C) increase in oil temperature, a lubricant's lifecycle is reduced by half." Inversely, lowering oil temperature by the same rate will double it's life cycle.

Two predominant failure mechanisms occur when oil is heated up, the most common categorized as oxidation failure and the lesser categorized as thermal failure.

Oxidation failure

Oxidation failure occurs when oxygen reacts with the lubricant base oil. Anti foaming and anti oxidant additive agents, if present in the oil, are designed to slow down the process but once depleted, the rate of oxidation will accelerate, especially in the presence of water and reactive bearing materials such as copper and iron, causing the oil to become oxidized. In an oxidized state, the hydrocarbon molecules in the oil will transform into a greasy sludge containing harmful corrosive acids. These will cause the oil to degrade and lose its lubricating properties and manifest themselves through an increase in lubricant viscosity, specific gravity, acidity (TAN), rapid additive depletion, a darkening of the oil, sour "rotten egg" odor and a varnishing of the bearing surfaces.

Thermal Failure

Thermal failure can occur due to transference of a localized or external heat source to the lubricant, or through the adiabatic compression (the thermodynamic process that occurs when entrained air compresses and heats up according to Boyle's law) of entrained air bubbles in pumps, bearings, and even in pressurized hydraulic and lubricating systems. The resulting heat causes the lubricant to decompose and its corresponding hydrogen loss creates carbon rich particles in the form of sludge and carbon deposits. These effects manifest themselves as a *decrease* in lubricant viscosity with a dark fluid with greasy suspensions that smell of burned food and evidence of coking and varnishing on the bearing surfaces.

Once a lubricant has failed, the molecular change state is usually irreversible requiring at the very least a lubricant change. If a lubricant failure is possible, this can be monitored through an oil analysis program. A lubricant's propensity to fail can be checked by subjecting a sample to a RPVOT (Rotary Pressure Vessel Oxidation Test) that simulates failure through a speeded up oxidation process and can give the maintainer an indication of a lubricant's suitability. This test can take a virgin oil sample or existing oil sample and predict its remaining life.

Practical Prevention of Lubricant Temperature—-related Failures

Whether your lubricant choice is grease or oil, once the correct one is chosen and employed it will require assistance from the maintainer to ensure the lubricant has a fighting chance to perform and deliver a reasonable life expectancy. This can be achieved by implementing some of the following best lubrication practices:

- Ensure the lubricant transfer delivery systems are immaculately clean so as not to allow the ingress of solid contamination that can create sludge, raise the lubricant viscosity and accelerate the oxidation process,

- Ensure the oil/grease delivery method/system is tuned to deliver the correct amount of lubrication in a timely manner. Over-lubrication will create fluid friction heat compounded by the bearing ball/rollers working overtime to mechanically push through the excess lubricant, both causing the lubricant to heat up rapidly. Under-lubrication can allow the bearing to go into boundary lubrication creating surface interaction frictional heat that can "cook" the lubricant,

- Ensure the oil reservoir level is between the Min and Max fluid level so as not to create cavitation in the oil pump or oil churn in the reservoir that can result in air bubbles accelerating the oxidation process,

- Ensure oil reservoirs are clean and free of debris. Dirt, dust and debris can create the effect of a thermal blanket and raise the temperature of the oil inside the reservoir,

- Ensure the integrity of shaft seals. Poor shaft seals lead to excessive lubricant leakage that can quickly lead to an under-lubrication state,

- Implement a lubricant test and control process to ensure incompatible lubricants are not mixed together in the same bearing space that can lead to a variety of detrimental conclusions, including overheating,

- Where hydrocarbon base lubricants are employed in cold weather climates, use timed block heaters or blanket wrap element heaters for reservoir and drum/pail heating. If a lubricant is to protect a bearing surface it must readily flow across the bearing.

Additive Depletion Failure

Lubricants are an integral part of any mechanism or machine and are an engineered choice. Their life cycle is dependent on machine/mechanism load, speed, ambient condition factors (temperature, humidity, water, dirt) and the level of maintenance and operator interaction.

Every finished lubricant product is a combination of base oil plus additive package, making every lubricant product not only unique but also consumable due to the additive package design. By nature, additives are designed to deplete as the oil is employed in service and their rate of depletion is wholly dependent on the additive type and working conditions. Additives can deplete, or fail, in three different ways:

Decomposition

Although it is called decomposition, additives actually experience numerous metamorphic changes caused by a variety of causes listed below:

- **Hydrolysis**—ZDDP additive + water and heat change to sulphuric acid and hydrogen sulphide

- **Neutralization**—Acid and Base (detergent additives) neutralize and change into salt and water

- **Oxidation**—ZDDP and hindered Phenol additives change into Polymers and precipitants

- **Thermal Degradation**—ZDDP and EP additives + heat change into Phosphates, Phosphides, Sulfates, Sulfides

Separation

Sometimes referred to as mass transfer, additives are lost, or transferred, due to the system design in the following manner:

- **Aggregate Adsorption**—Adsorptive filter media removes polar additives

- **Centrifugation**—heavy organic and metals separate under centrifugal forces

- **Condensation Settling**—additive becomes insoluble and settles

- **Evaporation**—vacuum dehydration

- **Filtration**—larger particles are captured in the oil filter. Better quality lubricants use sub micron particle additives that more easily pass through most inline filters

Absorption

Through action within the tribological system, or through contamination (dirt and water) ingression, additives can be attracted to system elements and be removed from the lubricant in the following manner:

- **Particle Scrubbing**—dirt particle attracts additive and traps additive in filter or drops out of suspension

- **Rubbing Contact**—Polarized EP and AW additives form soap like boundary films around themselves

- **Surface Adsorption**—Polarized additives attach to machine surfaces

- **Water Washing**—Water attracts the additive and settles on the reservoir bottom

As shown in the three depletion methods, additives deplete by performing their function as they were designed to do and due to the inefficiency of the system design and operating condition. Once the additives are depleted the oil is thought as consumed and must be changed or have its additives replenished. Failure to do so will cause a loss of protection to the machine surfaces and they will degrade rapidly.

References

1. Bannister, Kenneth E., "Energy Reduction Through Improved Maintenance Practices," Industrial Press, NY (1999)

Bibliography

Bannister, Kenneth E.,

Chapter 2
Lubricant Categories*

By way of introduction, a brief overview of the principal lubricant categories is offered and illustrated in this chapter. Lubricants are divided into the following groups

- gaseous,
- liquid,
- cohesive, and
- solid.

Among these, the gaseous lubricants are insignificant because construction costs for gas or air lubrication equipment are very high. Typical applications and industry sectors are given in Table 2-1. A somewhat more condensed summary of lubricant types is shown in Figure 2-1.

As would be logical to surmise, lubricants, in the global sense, should not only reduce friction and wear, but also

- dissipate heat,
- protect surfaces,
- conduct electricity,
- keep out foreign particles, and
- remove wear particles.

Different lubricants show different behavior regarding these requirements.

LIQUID LUBRICANTS

Liquid lubricants include

- fatty oils,
- mineral oils, and
- synthetic oils.

Their typical properties are summarized in Table 2-2.

The task of liquid lubricants is all-encompassing. It is to

- dissipate heat,
- protect surfaces,
- conduct electricity, and
- remove wear particles.

Fatty oils are not very efficient as lubricating oils. Even though their lubricity is usually quite good, their resistance to temperatures and oxidation is poor. Mineral oils are most frequently used as lubrication oils, but the importance of synthetic oils is constantly increasing. These oils offer the following advantages:

- higher oxidation stability,
- resistance to high and low temperatures,
- long-term and lifetime lubrication.

Anticorrosion and release agents are special products which also fulfill lubrication tasks.

COHESIVE LUBRICANTS

Cohesive lubricants include

- lubricating greases,
- lubricating pastes, and
- lubricating waxes.

Their task is to

- protect surfaces,
- conduct electricity, and
- keep out foreign particles.

Lubricating waxes are based on hydrocarbons of high molecular weight and are preferably used for boundary or partial lubrication at low speeds.

Lubricating greases are based on a base oil and a thickener imparting to them their cohesive structure. They can be used for elasto-hydrodynamic, boundary or partial lubrication.

Lubricating pastes contain a high percentage of solid lubricants. They are used in the case of boundary and partial lubrication, especially for clearance, transition and press fits. Cohesive lubricants are used when the lubricant should not flow off, because there is no adequate sealing and/or when resistance against liquids is required. These lubricant types play an increasingly important role, since it is possible to achieve long-term or lifetime lubrication with minimum quantities.

*Source: Klüber Lubrication North America, Londonderry, New Hampshire

Table 2-1. Typical applications and industry sectors for lubricants.

Application (● preferred application, ○ possible application) / Key words by lubricant type	For machine elements and components ...														When exposed to				In the following industrial sectors ...																		
	Rolling bearings	Plain bearings	Chains	Seals	Ropes	Nuts and bolts	Electrical contacts	Springs	Valves/taps	Gears	Shaft/hub connections	Hydraulic systems	Pneumatic systems	Electrical switches	Fluids	Oxygen	Tribo-corrosion	Vacuum	Electrical engineering	Automotive	Precision mechanics	Transport	Base Materials	Rubber	Wood	Plastics	Food	Metal	Furniture	Mechanical engineering	Communications	Paper	Cableways	Textile	Machine Tools	Energy	Environment
Cohesive lubricants																																					
Sealing grease				○					●						●			●									○			○		●		○			
"Bio" lubricating grease	●	●						○			○									○		○	○					○		○			○			●	●
Damping grease	○									●				●					●	●	●								○	○	○						
Running-in grease		○	○	○						●	○						●			○			●			○				○					○	●	●
Running-in paste										●	○						●			○			●					○		○					○	●	●
Fluid grease		○								●												○					○								○		
Gas valve grease				○					●																			●								●	
Oil for plastics		●												●					●		○					●			○		○						
Long-term oil		○					○			●		○							○	●			○				○			○	○					○	○
Food grade oil		●	●		○			○		●		●	●														●										
Anticorrosion lacquer		○			○	●		○		○	○												○					●		●			○		○	○	○
Anticorrosion oil	●	○	○		○	○		○		○	●								○	○	○	○	○					●		●	○		○		●	●	○
Anticorrosion wax emulsion		●	○			○		○			○								○	●		○						●		○	○		○		●	●	○
Lubricating wax emulsion		●	○	●	○	○		○			○								○	○	○	○		●		○		○	●	●	○		○	○	○	○	○
Low-temperature oil		●	●		○		●			●									○	●	●	○								○	○		●				○
Impregnating fluid		●																	●	●	●									●	●						
Impregnating oil		●																●	●	●	●									●	●						
Sliding agent	○	○		●					●							●							●					●									
Tap grease		○							●						●												●	●								○	
High-speed grease	●																																	○	●		
High-temperature grease	●	●	●		○		●	○	○	○	○		○	●	○		○		○	●	○	●	○	○	●			○		○		○		○		○	○
High-temperature long-term grease	●						●												●	○		●	○		○		●	○		○		○		○		○	○
High-temperature paste		○		○		●				○	○						○			○			●					○		○		○		○		●	○
Grease for plastics		●		○				●	○	○		○	○	●					●	●	○					●			○		○						
Copper paste		○				●														○			●					○		○					○	●	○
Long-term grease	●				●		●		●	○	○		○	●			○		●	●	○	○	○					○	○	○	○	○	○	○	○	○	○
	●	●		●		○		○	●	○	○		○		●												●										
Smooth running grease	●												●							○	○					○			○		○		●		●		
Assembly grease		●		●		○		○			●	○	○		○		●			●			○			●	○	○	●	○			○		○	○	○
Assembly paste		●	●	●		●				○	●						●			●			●					●	○	●			○		●	●	○
Lubricating paste		●		○		●			○	○	●				●		●		○	●	○	○	○					○	○	○		○	○	○	○	○	○
Lubricating wax		●	●			○					○								●	○	○					○		○	●	●							
Heavy-duty grease	●																						●													●	●
Low-temperature grease	●	○		○	●	○	●	○	●	○	○	○	●	●					○	●	○	○								○	○		●				
Release agent																								●		●		●									
Solid lubricants																																					
Tribo-system materials																																					
Sheets		●		●					●					●	●		●	●	○	●	●	●	○		○		●		○	○	○			●	●	○	○
Granules		●	○			○			○	●		○	○	○	●		○		○	●	●	●					○		●	○	●		○	○			
Semi-finished parts		●		○					○						●		●	●	○	●	●	●	●				●		○	○	○		○	●		○	○
Bearings		●													●		●	●	●	●	●	●					●	○	○	●	●		○	○		○	○
Two-component systems		●													●		●			●								○		●					●		
Tribo-system coatings																																					
Heat-hardening	●	●	●			●		●	○	○			●		●		●		●	●	●	●	○		○	○	○	●	●	●	●	○		●		○	○
Air-drying	●	●	●	●	○	●	○	●					●	○			●	○	●	●	●	●	○		○	○	○	●	●	●	●	○		●	●	○	○
Dry lubricants for tribo-systems																																					
Heat-hardening		●	○		○	●		○	○	○	○	○	●		●		●	○	●	●	●	●						●		●							
Air-drying		●	○	●	○	●		○	○	○	●	○	●			●	●	○	●	●	●	●	○	○	○	○	○	●	○	●	○	○		●	●	○	○

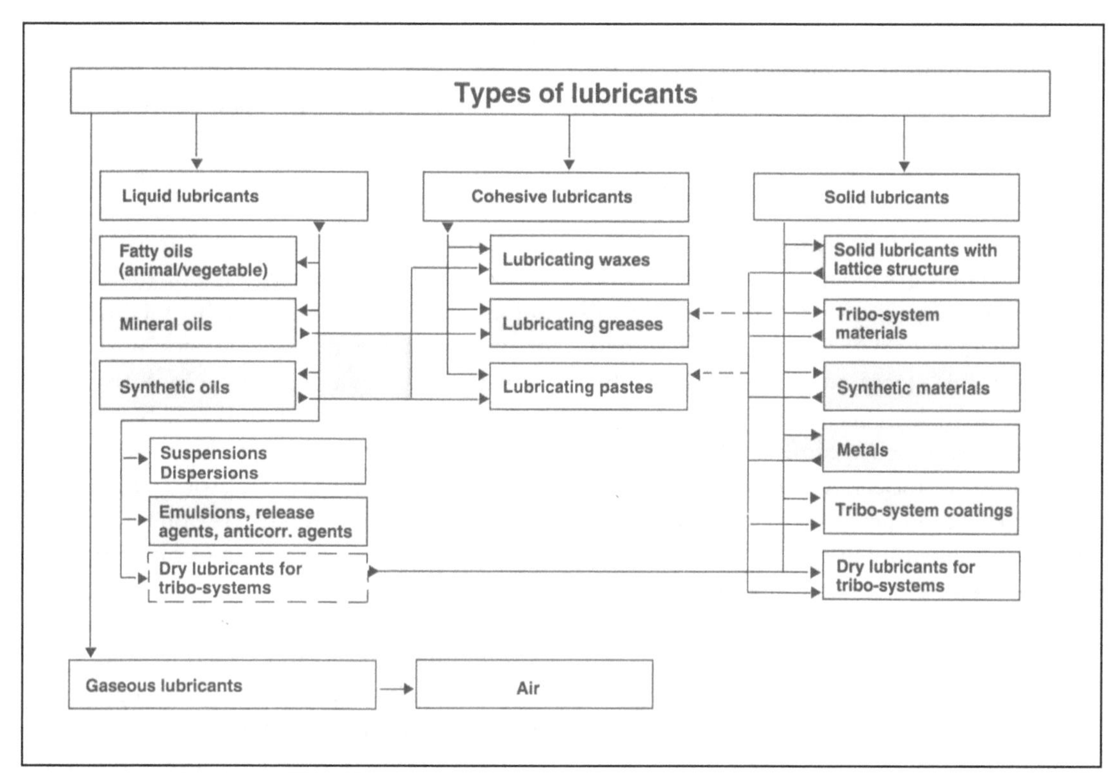

Figure 2-1. Types of lubricants summarized.

Table 2-2. Paraffinic versus Naphthenic base oils.

Paraffinic	Naphthenic
Found in Mid USA, North Sea, Middle East	Found In Southern USA, South America
< 30% Naphthene	< 25% Paraffin
Viscosity Index 95-105	Viscosity Index 65-75
Wax – up to 10%	Wax - Trace
Automotive engine oils Turbine Oils Gear Oils Bearing Oils Hydraulic Fluids	Locomotive oils Refrigerant Oils Compressor Oils Transformer oils

CATEGORIZING AND GROUPING BASE OILS

Base oil stocks are manufactured from crude oil extracted from many different oil fields throughout the world and refined using numerous methods based on the type of crude and the type of finished product required.

We can see from Table 2-3 that the majority of lubricating oils belong to the paraffinic family in which a higher Viscosity Index (VI) rating is more favorable. Their wax content levels do however; raise their cold temperature pour point, requiring additional processing at the refinery to extract them. Paraffinic stocks are refined using either a solvent extraction or hydroprocessing refining method and are said to have a wax pour point.

Because naphthenic stocks have only a trace of wax and have good dielectric properties they make excellent candidates for low temperature applications and electrical switching equipment. Naphthenic stocks are refined using only a solvent extraction method and are said to have a viscosity pour point.

Refining stocks

Solvent Extraction Process

Solvent extraction is performed using a multi-stage distillation/solvent extraction process. In stage one the crude is put through an atmospheric distillation process to remove the lighter and more volatile hydrocarbon fuel fractions that include benzene, gasoline, kerosene (jet fuel), and diesel. The stock is then charged into a vacuum distillation tower to take off (separate) the fluid by specific viscosity ranges. Next stage, a solvent is added to the stock to remove up to 85% of all unwanted large molecule aromatics, sulphur and nitrogen compounds. The primary solvent (usually N-Methyl pyrrolidone, or Furfural) is recovered and a second stage de-waxing solvent (usually Methyl-ethyl ketone or toluene) is added to facilitate the extraction of the wax crystals through cooling, solidifying, and filtering out unwanted wax. Once again, the solvent is recovered and the conventional base oil stock is now 70% to 85% pure and ready for additive formulation finishing, or further hydroprocessing.

These conventional base oil stocks are referred to as solvent refines, paraffinic **Group I base oils.**

Hydroprocessing

Similar to solvent extraction, hydroprocessing employs a multi-stage process that begins with an atmospheric/vacuum distillation process. Then, depending on the finished product requirements, the base stock, now referred to as *feedstock,* can be put through a two-stage, severe hydrocracking process or a severe hydrocracking/hydroisomerization process.

Table 2-3. Properties of typical base oils for industrial lubricants.

Properties / Oils	Mineral oils	Synth. hydro-carbon oils	Ester oile	Polyglycol oils	Polyphenyl ether oils	Silicone oils	Perfluoro alkylether
Density [g/ml] at 20°C	0.9	0.85	0.9	0.9 ... 1.1	1.2	0.9 ... 1.05	1.9
Viscosity Index (VI)	80 ... 100	130 ... 160	140 ... 175	150 ... 270	–20 ... –74	190 ... 500	50 ... 140
Pour point [°C]	–40 ... –10	–50 ... –30	–70 ... –37	–56 ... –23	–12 ... 21	–80 ... –30	–70 ... –30
Flash point [°C]	<250	<200	200... 230	150 ... 300	150 ... 340	150 ... 350	not flammable
Oxidation resistance	moderate	good	good	good	very good	very good	very good
Thermal stability	moderate	good	good	good	very good	very good	very good
Lubricity	good	good	good	very good	good to satisfactory	unsatistactory	good
Compatibility with elastomers, coatings, etc.	good	good	unsatis-factory	unsatis-factory to good	unsatis-factory	good	good

2-stage Severe Hydrocracking Process

To remove unwanted polar compounds containing sulphur, nitrogen and oxygen, and convert aromatics hydrocarbons to more desirable saturated hydrocarbons the distilled feedstock is charged into a hydrocracker. Here, the feedstock will react to a mixture of heat, pressure, and hydrogen gas in the presence of a catalyst.

The stock is once again distilled the lube oil is chill de-waxed and then passed through a high pressure hydro-treater to remove any last traces of araomatics and polar compounds.

This base stock is now better than 99% pure and ready for its additive package formulation and is referred to as a hydroprocess refined, paraffinic **Group II Base Oil**.

Severe Hydrocracking/Hydroisomerization Process

This process is identical to the severe hydrocracking process except that the chill de-waxing step is replaced with the more efficient hydroisomerization catalytic process. This efficiency yields very high VI base fluids up to 130 VI levels, with superior oxidation stability, excellent low temperature fluidity, High temperature stability, and low volatility / toxicity—similar characteristics to a Poly Alpha Olefin (PAO) synthetic lubricant!

These 99.9% pure base stocks are referred to as hydroprocess refined, paraffinic **Group III Base Oil**. Due to their synthetic like characteristics, in 1999, the US National Advertising Division of the Council of Better Business Bureaus ruled that because of their molecular conversion, a Group III-HydroIsomerized base oil can be referred to as a "synthetic" lubricant.

Group IV Base Oils are true synthetic PAO type base oils. (see Chapter 10—Synthetic Lubricants.)

Group V Base Oils are all other oils not mentioned above. These would include solvent refined naphthenic and non-PAO synthetic lubricants (See Chapter 7—Synthetic Lubricants).

SOLID LUBRICANTS

Solid lubricants include

- tribo-system materials,
- tribo-system coatings, and
- dry lubricants for tribosystems.

Their main task is to

- protect surfaces.

Solid lubricants also include synthetic, metallic or mineral powders, such as PTFE, copper, graphite and MoS_2. As powders are difficult to apply, they are mostly used as additives. Solid lubricants are normally used as dry lubricants operating under boundary lubrication conditions. If liquid or cohesive lubricants are incorporated in the tribo-system materials there can even be partial lubrication.

Solid lubricants are mainly used when the application of liquid or cohesive lubricants is not ideal for functional reasons or risk of contamination and when, at the same time, the lubrication properties of solid lubricants are sufficient.

LUBRICATING OILS

Lubricating oils consist of a base oil and additives which determine their performance characteristics. The base oil is responsible for the typical properties of an oil. The additives, however, determine its actual performance by influencing the base oil's

- oxidation stability,
- anticorrosion properties,
- wear protection,
- emergency lubrication properties,
- wetting behavior,
- emulsibility,
- stick-slip behavior,
- viscosity-temperature behavior.

The advantages of a lubricating oil as compared to a grease are improved heat dissipation from the friction point, and its excellent penetrating and wetting properties.

As mentioned in the preceding chapter, its main disadvantage is that a complex design is required to keep the oil at the friction point and prevent the danger of leakage.

Lubricating oils are used in a wide variety of elements and components, such as

- sliding bearings,
- chains,
- gears,
- hydraulic systems,
- pneumatic systems.

In addition to counteracting friction and wear, lubricating oils have other requirements to fulfill in various applications, e.g.,

- corrosion protection,
- neutrality to the applied materials,
- meet food regulations,
- resistance to temperatures,
- biodegradability.

Lubricating oils are applied in other primary or secondary applications as:

- running-in oils,
- slideway oils,
- hydraulic oils,
- instrument oils,
- compressor oils,
- heat carrier oils.

The main tasks, however, remain lubrication and protection against friction and wear.

TRIBOTECHNICAL DATA

Tribotechnical data are characteristics of mineral oils. These data are shown in Table 2-4. Within the framework of the intended application, they permit the selection of a lubricant suitable for the pertinent requirements (temperature, load and/or speed). In this regard, the viscosity grade selection (Table 2-5) is of primary importance.

Table 2-4. Tribotechnical data pertaining to lubricating oils.

Parameters	Test	Notes
Ash Content	DIN 51 575 DIN EN 7	The residual oxide or sulphate ash remaining after the combustion of an organic compound; the sulphate ash content is determined only for metallic-organic additives or for used oil.
Color / Color Code	DIN 51 411 ISO 2049	The specific color of the lubricant determined using a colorimeter.
Demulsability	DIN 51 589	Oil's ability to separate from water.
Density	DIN 51 757	The amount of lubricant required for a certain friction point is generally indicated by volume. The volume in mm density is the weight in g of the required lubrication.
Flash Point	DIN ISO 2592	Lowest temperature at which an oil gives off vapors that will ignite when a small flame is passed over the surface of the oil.
ISO VG	DIN 51 519	ISO VG is the abbreviation of ISO Viscosity Grade, characterizing the classification of fluid lubricants in terms of viscosity.
Pour Point	DIN ISO 3016	Lowest point at which an oil will pour or flow under certain conditions.
Viscosity	DIN 51 561	Indicates the resistance of a fluid substance to flow **Apparent Viscosity** = Viscosity taking density into account 1 Pa s = 1 N s/m2 **Kinematic Viscosity** = Viscosity – Density ratio mm2/s
Viscosity – Temperature relation (VT)	DIN 51 563	Flow properties of a lubricant in relation to temperature
Viscosity Index (VI)	DIN ISO 2909	Non-dimensional figure characterizing the change of viscosity as a function of the temperature

Table 2-5. ISO viscosity grades of fluid industrial lubricants.

ISO viscosity grade	Mid-point viscosity at 40 °C, mm²/s (cSt)	Kinematic viscosity limits at 40 °C, mm²/s (cSt) min.	max.
ISO VG 2	2.2	1.98	2.42
ISO VG 3	3.2	2.88	3.52
ISO VG 5	4.6	4.14	5.06
ISO VG 7	6.8	6.12	7.48
ISO VG 10	10	9.0	11.0
ISO VG 15	15	13.5	16.5
ISO VG 22	22	19.8	24.2
ISO VG 32	32	28.8	35.2
ISO VG 46	46	41.4	50.6
ISO VG 68	68	61.2	74.8
ISO VG 100	100	90.0	110
ISO VG 150	150	135	165
ISO VG 220	220	198	242
ISO VG 320	320	288	352
ISO VG 460	460	414	506
ISO VG 680	680	612	748
ISO VG 1000	1000	900	1100
ISO VG 1500	1500	1350	1650

Table 2-6. Tribo-technical data pertaining to lubricating greases.

Parameters	Test	Notes
Base Oil Viscosity	DIN 51 575 DIN EN 7	The residual oxide or sulphate ash remaining after the combustion of an organic compound; the sulphate ash content is determined only for metallic-organic additives or for used oil.
Base Oil Viscosity	DIN 51 561	Indicates the resistance of a fluid substance to flow **Apparent Viscosity** = Viscosity taking density into account 1 Pa s = 1 N s/m2 **Kinematic Viscosity** = Viscosity – Density ratio mm2/s
Density	DIN 51 757	The amount of lubricant required for a certain friction point is generally indicated by volume. The volume in mm density is the weight in g of the required lubrication.
Drop Point	DIN ISO 2176	The temperature at which grease drips off the testing unit in a non-decomposed condition.
Flow Pressure	DIN 51 805	Temperature-dependent pressure needed to force grease out of a nozzle. Used to indicate the lower service temperature.
Oil Separation	DIN 51 817	Determination of oil bleeding in percent by weight.
Oxidization Resistance	DIN 51 808	Resistance of the grease to absorb oxygen measured by the drop in pressure; used to indicate grease ageing.
Penetration	DIN ISO 2137	Used to determine the consistency of a lubricating grease by measuring the penetration depth of a metal cone into a grease-filled cup in tenths of mm. The ranges are classified as NLGI (National Lubricating Grease Institute) grades.
Water Resistance	DIN 51 807, Part 1	Statistic test to check the emulsification of the grease

Chapter 3

Lubricant Testing*

Virtually every lubricant product or other petroleum product is manufactured to specific performance standards and specifications. Each designed to exhibit certain properties or characteristics described by the manufacturer in their sales literature, data sheets, or related documents. These descriptions may range from somewhat superficial in common household products to the highly technical and sophisticated in specialty industrial products.

Whenever industrial users are involved in the selection, or faced with the optimized application of petroleum products, they will find themselves confronted by terms and descriptions that relate to these product properties and characteristics. Sometimes called "typical inspections," the properties of just one multi-purpose grease include such items as *worked penetration, dropping point, viscosity, oil separation, wheel bearing leakage, Timken OK load, four-ball wear test, water washout*, and *corrosion prevention rating*. The question is, what do these terms mean, and how important are they?

Since the scope and intent of our text is aimed at conveying practical knowledge to the reader, we must enable him or her to make comparisons among products. With this in mind, we have endeavored to describe the most important tests and their significance to the lubricant user. We deliberately opted to leave out many of the detailed descriptions of test apparatus and testing procedures of interest to the laboratory technician, but did include them in the few instances where clarity called for more detail.

Each test reviewed and described in this chapter is correctly referenced by its official test title (parameter) and test number as catalogued in Table 3-1. Lubricant Test Chart. Two testing bodies are referenced, these being the European test body represented by DIN – Deutsches Institut für Normung, or German national institute for standardization, and the ASTM, or American Society for Testing and Materials, whose membership now represents 95% of all countries worldwide.

Table 3-1. Lubricant Test Chart

Test Parameter	TEST	Notes
Air Entrainment	DIN 51 381	TUV Impinger test
Aniline Point	ASTM D 611 / D 1012	
Ash Content	ASTM D 874	For sulphated residue
Auto Ignition Temperature	ASTM D 2155	
Cloud Point	ASTM D 2500	
Copper Strip Corrosion	ASTM D 130	
Demulsibility	ASTM D 1401 / D 2711	
Grease Consistency	ASTM D 877 / D 1816	
Dilution of Crank Case Oils	ASTM D 322	
Dropping Point of Grease	ASTM D 566 / D 2265	
Flash / Fire Points – Open Cup	ASTM D 92	
Flash Point – Closed Cup	ASTM D 56 / D 93	
Foaming Characteristics - Oil	ASTM D 892	
Four Ball Wear Test	ASTM D 2266	
Four – Ball EP Test	ASTM D 2596	
Gravity	ASTM D 287	
Grease Consistency	ASTM D 217 / D 1403	
Interfacial Tension	ASTM D 971	
Neutralization Number	ASTM D 664 / D 974	
Octane Number	ASTM D 2699 / D 2700	
Oil Content of Petroleum Wax	ASTM D 721	
Oil Separation in Grease Storage	ASTM D 1742	
Oxidation Stability - Greases	ASTM D 942 / D 1402 / D 1261	
Pour Point and Cloud Point	ASTM D 97	
Power Factor	ASTM D 924	
Refractive Index	ASTM D 1218	
Rotary Bomb Oxidization Test	ASTM D 2272	RBOT
Rust Preventive Characteristics	ASTM D 665	
Timken Extreme Pressure Test	ASTM D 2509 / D 2782	
Vapor Pressure	ASTM D 323	
Viscosity	ASTM D 88 / D 445	Redwood, Engler
Viscosity Index	ASTM D 567 / D 2270	
Water Washout	ASTM D 1264	
Water and Sediment	ASTM D 95 / D 96 / D473	
Wax Melting Point	ASTM D 87 / D 127 / D 938	
Wheel Bearing Grease Leakage	ASTM D 1263	

*Source: Exxon Company U.S.A., Houston, Texas.

AIR ENTRAINMENT

DIN 51 381 TUV Impinger Test

Air entrained in a lubricating oil can disrupt the lubricating film and cause excessive wear of the surfaces involved. In hydraulic systems, because entrained air is compressible, it can cause erratic and inefficient operation of the system.

The term "air entrainment" refers to a dispersion of air bubbles in oil in which the bubbles are so small that they tend to rise very slowly to the air-oil interface. The presence of the bubbles gives the oil a hazy or cloudy appearance.

There is no standard method for testing the air entrainment characteristics of oil. The American Society for Testing Materials investigated various tests with the idea of standardizing on one. The German DIN 51 381 "TUV Impinger Test" is primarily used to test steam turbine oils and hydraulic fluids for their air release properties. It is the standard generally accepted in Europe and is being considered as the ASTM method.

Significance

Air entrainment consists of slow-rising bubbles dispersed throughout an oil and is to be distinguished from foaming, which consists of bubbles that rise quickly to the surface of the oil. Both of these conditions are undesirable in a lubricating system. However, it is often difficult to distinguish between them because of high flow rates and turbulence in the system. Relatively small amounts of air are involved in air entrainment while larger amounts are involved with foaming. These conditions are considered separate phenomena and are measured in separate laboratory tests. Entrained air is not a normal condition; it is primarily caused by mechanical problems. Some of these are:

Insufficient Reservoir Fluid Level: Air can be drawn into the pump suction along with the oil.

Systems Leaks: Air can be introduced into the oil at any point in the system where the pressure is below atmospheric pressure.

Improper Oil Addition Methods: If make-up oil is added in a manner that causes splashing, it is possible for air to become entrained in the oil.

Faulty System Design: Design faults involve such things as placement of oil return so that the returning oil splashes into the reservoir, or placement of the pump next to the return opening.

The general trend in hydraulic oil systems, turbine oil systems, and industrial circulating oil systems of every kind is to decrease reservoir size and increase flow rates and system pressures. This trend increases the tendency for air entrainment, thereby making the air release property of an oil more significant.

Some additives used to reduce foaming tend to increase the air entrainment tendency of an oil. The choice of an anti-foam additive requires striking a balance between these two undesirable phenomena.

ANILINE POINT

ASTM D 611 and ASTM D 1012

Many petroleum products, particularly the lighter ones, are effective solvents for a variety of other substances. The degree of solvent power of the petroleum product varies with the types of hydrocarbons included in it. Frequently it is desirable to know what this solvent power is, either as a favorable characteristic in process applications where good solvency is important, or as an unfavorable characteristic when the product may contact materials susceptible to its solvent action.

The aniline point determination is a simple test, easily performed in readily available equipment. In effect, it measures the solvent power of the petroleum product for aniline, an aromatic substance. The solvent powers for many other materials are related to the solvent power for aniline.

Aniline is at least partially soluble in almost all hydrocarbons, and its degree of solubility in any particular hydrocarbon increases as the temperature of the mixture is increased. When the temperature of complete solubility is reached, the mixture is a clear solution; at lower temperatures, the mixture is turbid. The test procedures make use of this characteristic by measuring the temperature at which the mixture clouds as it is cooled. The greater the solvent power of the hydrocarbon for aniline, the lower the temperature at which cloudiness first appears.

Usually, paraffinic hydrocarbons have the least solvency for aniline (and most other materials) and consequently have the highest aniline points. Aromatics have the greatest solvency and the lowest aniline points (usually well below room temperature), while naphthenic materials are intermediate between the paraffins and the aromatics.

Significance of Results

Aniline point is most significant for solvents, since it is one indication of solvent power. In general, the lower the aniline point of a product, the greater its solvent power. Other available laboratory tests measure the solvent power of the product for the specific type of substance with which it will be used. Two tests of this type, which have been standardized and accepted, are used to determine kauri-butanol value and nitro-cellulose diluting power. However, these latter tests are much more complicated than the aniline point determination, and small laboratories do not usually have the facilities for performing them. The choice of tests is usually dictated by which one correlates best in a particular application.

Aniline point is useful in predicting the ignition characteristics of diesel fuels. For this purpose, the aniline point is used in conjunction with the API gravity of the fuel to determine its "diesel index." This procedure is described later, "Cetane Number," since diesel index and cetane number are used for similar purposes. Diesel index, in turn, is useful in estimating the enrichment value of oils used for gas enrichment.

For a lubricating or hydraulic oil, aniline point is an indication of its tendency to cause softening and swelling of rubber parts contacted by the oil. The lower the aniline point, the greater the swelling tendency. Aniline point is also used as a factor in determining the relative compatibility of a rubber plasticizer with a rubber formulation.

ASH CONTENT

ASTM D 874 for Sulfated Residue

The ash content of a lubricating oil is related to the quantity of incombustible materials that may be present. Though a straight distillate mineral oil is, in itself, nearly ashless, certain lubricants are formulated with solutions of metallic additives that will not be completely burned. The ash left by these products may be appreciable. Used oils, moreover, may be contaminated with dirt and abraded metals that likewise appear as ash after the oil itself has been consumed.

The percentage of ash that remains after oil has been burned, therefore, gives an indication of the quantity of metallic additive, non-combustible solid contamination, or both, that the oil may contain. Significance of ash content depends on the type of oil, its condition, and the actual test by which it is evaluated.

The simplest method of determining the ash content of a lubricating oil is to burn a sample of known weight, applying sufficient heat to consume all of the combustibles. The weight of the residue that is left establishes a value for determining the percentage of ash. This procedure is described under the ASTM designation D 482.

In general, however, the preferred test is that specified under the ASTM method D 874 for "sulfated residue." Here, the oil is first strained to remove solid contaminants; then it is burned under controlled conditions. After the burning, the residue is treated with sulfuric acid to assure consistent degrees of oxidation of all components. Acid treatment improves the uniformity of the results, making them more reliable.

There is still another method, ASTM D 810, which also yields sulfated residue, but which serves primarily to determine the percentages of lead, iron, or copper. With lubricating oils, this sort of analysis is generally of lesser significance.

Significance of Results

Many oils for internal combustion engines are formulated with detergent additives based on metallic derivatives such as those of barium or calcium. These additives help to keep the engine clean. Being metallic, these materials appear in one chemical form or another in the ash.

For new oils of this type, therefore, sulfated ash may serve as a manufacturer's check on proper formulation. An abnormal ash may indicate a change in additive content and, hence, a departure from an established formulation.

For new oils of unknown formulation, sulfated ash is sometimes accepted as a rough indication of detergent level. The principle is based on the dubious assumption that a higher percent of ash implies a stronger concentration of detergent and, hence, an oil of greater cleanliness properties. As a means of evaluating detergency, however, the test for ash is far less reliable than the usual engine and field tests, the primary advantage of ash content lying in the expediency with which it is determined.

There are several reasons why the relationship between sulfated residue and detergency may be extremely distorted:

1. Detergency depends on the properties of the base oil as well as on the additive. Some combinations of base oil and additive are much more effective than others.

2. Detergents vary considerably in their potency, and some leave more ash than others. Detergents have been developed, in fact, that leave no ash at all.

3. Some of the ash may be contributed by additives other than detergents.

4. There appears to be a limit to the effective concentration of detergent. Nothing is gained by exceeding this limit, and a superabundance of detergent may actually reduce cleanliness.

Sulfated ash has also been used to determine additive depletion of used diesel oils. The assumption has been that the difference between the ash of the used oil and that of the new oil is related to the amount of detergent consumed in service. Here again, results may be misleading. Consumption of the additive does not ordinarily mean that it has been disposed of, but that its effectiveness has been exhausted in the performance of its function. The metallic elements may still be present and may appear in the residue in the same concentration as in the new oil.

Sulfated ash of used diesel oils has significance only of a very general nature. If it runs higher than that of a new oil, contamination with dirt or wear metals is suspected, and further analysis is required to identify the foreign material. If sulfated ash runs low, it may be attributed to faulty engine operation or a mechanical defect. With gasoline engines, a high sulfated ash may be caused by the presence of lead derived from the fuel.

AUTO-IGNITION TEMPERATURE

ASTM D 2155

All petroleum products will burn and, under certain conditions, their vapors will ignite with explosive force. For this to happen, however, the ratio of product vapor to air must be within certain limits.

When exposed to air, a certain amount of the liquid product evaporates, establishing a certain vapor-to-air ratio. As the temperature of the liquid increases, so does the evaporation, and thus the vapor/air ratio. Eventually a temperature is reached at which the vapor/air ratio will support combustion if an ignition source, such as a spark or flame, is present. This is the flash point of the product.

If no ignition source is present, as the temperature increases above the product's flash point, a temperature is reached at which the product will ignite spontaneously, without any external source of ignition. This temperature is the auto-ignition temperature of that fluid.

The auto-ignition temperature of a liquid petroleum product at atmospheric pressure is determined by the standard ASTM method D 2155 (which replaces the older ASTM D 286, discontinued in 1966).

Significance of Results

The auto-ignition temperature of a petroleum product is primarily significant as an indication of potential fire and explosion hazards associated with the product's use. The auto-ignition temperature may be used as a measure of the relative desirability of using one product over another in a high-temperature application. It is necessary to use a petroleum product with an auto-ignition temperature sufficiently above the temperature of the intended application to ensure that spontaneous ignition will not occur. Auto-ignition temperature thus places one—but by no means the only—limit on the performance of a product in a given application.

The auto-ignition temperature under a given set of conditions is the lowest temperature at which combustion of a petroleum product may occur spontaneously, without an external source of ignition. It is not to be confused with the flash point of a product, which is the lowest temperature at which a product will support momentary combustion, in the presence of an external ignition source.

The auto-ignition temperature of a product is a function of both the characteristics of the product and conditions of its environment. For example, the auto-ignition temperature of a substance is a function of such things as the pressure, fuel-to-air ratio, time allowed for the ignition to occur, and movement of the vapor-air mixture relative to the hot surface of the system container. Consequently, the auto-ignition temperature may vary considerably depending on the test conditions.

For a given product at atmospheric pressure, the auto-ignition temperature is always higher than the flash point. In fact, as a general rule for a family of similar compounds, the larger the component molecule, the higher the flash point, and the lower the auto-ignition temperature.

However, as the pressure of the system is increased, the auto-ignition temperature decreases, until a point is reached, usually at a pressure of several atmospheres, at which the auto-ignition temperature of a product under pressure may be less than the flash temperature of the product at atmospheric pressure. Thus, concern for the auto-ignition of a product increases as the pressure on the system increases.

As a general rule, the auto-ignition temperatures of many distillate products with similar boiling ranges can be related to hydrocarbon type. For example, aromatics usually have a higher auto-ignition temperature than do normal paraffins with similar boiling range.

The auto-ignition temperatures of isoparaffins and naphthalenes normally fall somewhere in between those of the aromatics and normal paraffins.

However, care should be used in attempting to extend this guideline. For example, increasing the aromatics content of a lube or hydraulic oil tends to reduce the auto-ignition temperature of the oil. Conversely, increasing the aromatic nature of a solvent tends to increase the auto-ignition temperature of the solvent.

BIODEGRADATION AND ECOTOXICITY

Biodegradation is the breakdown of a substance, e.g., hydraulic fluid, by living organisms into simpler substances, such as carbon dioxide (CO_2) and water. Most standard test methods for defining the degree of biodegradation of a substance use bacteria from a wastewater treatment system as the degrading organisms. This provides a relatively consistent source of bacteria, which is important, since the bacteria are the only variable in the test other than the test substance itself. A term that can be roughly defined as the opposite of biodegradability is *persistence*. A product is persistent if it does not degrade, or if it remains unchanged for long periods of time, i.e., years, decades.)

There are many tests for biodegradation. Depending on the test design, it can measure *primary* biodegradability or *ultimate* biodegradability. *Primary biodegradability* is a measure of the loss of a product, but it does not measure the degree of degradation, i.e., partial or complete (to CO_2 and water), or characterize the by-products of degradation. It merely determines the percentage of the product that disappears over the term of the test or, conversely, determines the time required to reach a certain percentage of loss. A popular primary biodegradation test in use today is the CEC-L-33-A-94, which measures disappearance of the test product and relates that to a biodegradation level. The assumption in this test is that all of the product that has disappeared is completely biodegraded. In actuality, this may not be the case, because the test does not measure *complete biodegradation,* but only the *loss* of the original product.

Ultimate biodegradability describes the percentage of the substance that undergoes complete degradation, i.e., degrades to CO_2 and water over the length of the test or, conversely, describes how long it takes to achieve a specified percentage of degradation. Two tests that are designed to measure ultimate biodegradability are the Modified Sturm Test (OECD 301B) and the EPA Shake Flask Test, both of which quantify CO_2 generated over 28 days (a standard test duration).

Thus, the terms *primary* and *ultimate* describe the *extent of biodegradation*. The *rate of biodegradation* is defined by the term *ready biodegradation*. A product is considered to be *readily biodegradable* if shown to degrade 60-70%, depending on the test used. Only a few tests measure ready biodegradability. The more commonly used include: the Modified Sturm Test (OECD 301 B); the Manometric Respirometry Test (OECD 301 F); and the Closed Bottle Test (OECD 301 D).

The final term to discuss here is *inherent biodegradability*. A product is considered *inherently biodegradable* if shown to degrade greater than 20%. However, unlike ready biodegradation tests, which run a specified 28 days, tests for inherent biodegradation have no defined test duration and are allowed to proceed as long as needed to achieve 20% degradation, or until it is clear that the product will never biodegrade to that extent. In the latter case, the product is then considered *persistent*. Evaluating a substance's *environmental toxicity* (ecotoxicity) can involve examining its effect on growth, reproduction, behavior, or lethality in test organisms. In general, ecotoxicity is measured using aquatic organisms like fish, aquatic insects, and algae. The most common endpoint for expressing *aquatic toxicity* in the laboratory is the LC_{50}, which is defined as the lethal concentration (LC) of a substance that produces death in 50% of the exposed organisms during a given period of time. Ecotoxicity data, properly developed, understood, and applied, are useful for evaluating the potential hazard of a material in the environment. Some of the most commonly used organisms for aquatic toxicity studies include rainbow trout, mysid shrimp, daphnids (water fleas), and green algae.

Significance of Test Results

Both the test method and the intended use of the data must be considered when evaluating the biodegradability and environmental toxicity of a product. Data from different test methods are generally not comparable, and data developed on different products by different labs should be evaluated with strict caution by experts in the field.

Comparable biodegradation data should be developed using a consistent inoculum (bacteria) source, and in the same time frame, due to the variation of bacterial populations over time. As with most laboratory test procedures, results cannot be directly extrapolated to natural settings.

Similarly, for environmental toxicity tests, comparative data should be developed using the same test procedures and the same organisms. Exposures experienced in the laboratory will not be replicated in nature. The natural environment is a large dynamic ecostructure, while the laboratory environment is static and limited in size. Further, if a contaminant enters a natural aquatic system, the event will most likely be random in concentration and frequency, unlike the laboratory environment, which depends on constant, measured contamination.

Biodegradation and environmental toxicity data help us to begin to better understand how to protect the world around us. There are discrete, clearly defined methods for testing products for environmental toxicity, and there are many different test methods for evaluating a product's potential persistence in the environment. However, at this time, in the United States, there is no standard set of universally accepted test procedures defined by government or industry to measure the environmental performance of a product.

Standard biodegradability and environmental toxicity tests are very simplistic in their approach, and the usefulness of the data is generally limited in scope. The test systems typically used will never be able to consider the myriad variables that occur in the environment. In order to truly evaluate the "environmental friendliness" of a product, other investigative approaches, such as Life Cycle Assessment, in which manufacturing, delivery, useful life, and disposal undergo equal scrutiny, should be considered. This is particularly critical when comparing different classes of products, e.g., mineral oil-based versus vegetable oil-based hydraulic fluids.

Too much emphasis should not be placed on the quantitative results from these tests. Environmental studies cannot merely be represented by the simple numerical values that are often used to support claims regarding the "friendliness" of a product. Rather, they need to be understood in the context within which they were developed, i.e., how and why the tests were done. If not considered in that limited context, the information could improperly represent the "friendliness" or "unfriendliness" of a product.

In summary, biodegradation and environmental toxicity test results are not directly comparable in the same way as tests for physical characteristics, such as viscosity. Finally, although they are very important in the overall evaluation of a product, they represent only part of the data important to a product's complete evaluation.

CLOUD POINT

ASTM D 2500

If chilled to sufficiently low temperature, distillate fuels can lose their fluid characteristics. This can result in loss of fuel supply. The time when cold weather causes fuel stoppages is precisely the time when fuel is needed most in residential and commercial heating units. Diesel powered equipment of all kinds is subject to failure due to poor low-temperature operability of the fuel. Consequently, it is generally necessary to know how cold a fuel can become before flow characteristics are adversely affected.

The most important indication of low-temperature flow characteristics of distillate fuels is *cloud point.* This is the temperature at which enough wax crystals are formed to give the fuel a cloudy appearance. (It should not be confused, however, with the turbid appearance that is sometimes caused by water dispersed in the fuel.)

Because of the effects of some additives on wax crystal formation, cloud point alone should not be used as an absolute minimum operating temperature. Some of these additives have been shown to lower the minimum operating temperature for specific fuels without affecting the base fuel's cloud point.

Significance

The cloud point of a distillate fuel is related to the fuel's ability to flow properly in cold operations. Some additives may permit successful operation with fuels at temperatures below their cloud points; however, for distillate fuels without additives, clogging of filters and small lines may occur due to wax crystal formation at temperatures near the fuel's cloud point.

COLOR SCALE COMPARISON

Several scales are used to measure color of petroleum products. Approximate conversion and comparison of the more common color scales can be accomplished through use of charts that are available from lubricant suppliers.

Color and Color Tests

Color is a term that is often misunderstood because it is a complex aggregate of human values and physical quantities. No two people have quite the same conception of color when it is allowed to assume its broader meanings. Hue, intensity, tone, purity, wavelength, opacity, and brightness are all directly or indirectly associated with color. It would be extremely difficult to depict mathematically all of these dimensions in a single index. Most attempts to define color do so in terms of only one or two factors, and any meaningful discussion of this index must be strictly confined to the dimensions it is able to represent.

Most of the color tests upon which these scales are based involve the same basic procedure. Light is transmitted simultaneously through standard colored glasses (or other standard reference material) and a given depth or thickness of the sample. The two light fields are compared visually and adjustments are made until a match is obtained. In some test, the volume of the sample is varied until the two fields match (as in the Saybolt test); in others, the light transmitted through a given depth or thickness of the sample is matched by using a series of glasses (as in the ASTM test). When the operator obtains a match, a color value is recorded. This color value corresponds to a point on the color scale associated with the particular color test.

These tests, by definition, involve only two qualities of the transmitted light—appearance, as compared with a standard, and intensity. These two dimensions are not sufficient to describe completely the color of the sample, and should be used only to indicate uniformity and freedom from contamination.

COMPOSITION ANALYSIS OF PETROLEUM HYDROCARBONS

The analysis of a petroleum hydrocarbon involves the identification or characterization of various components of the substance. This can be accomplished through a variety of techniques. If the amount of information required is great, the analysis can be an extremely complex undertaking. For example, the American Petroleum Institute Project No. 6, an analysis of a single petroleum sample, continued for about 25 years.

The kind of analyses used in quality control and routine laboratory inspections of petroleum products are much faster, of course. These short-cut methods can be carried out in a variety of ways, using different test procedures and different types of instruments. The choice of method depends upon the nature of the substance to be analyzed, and upon the type of information required.

Types of Analysis

The short-cut methods of analysis can generally be classified as either carbon-type or molecular-type. A carbon-type analysis is run when the distribution of the different sizes of molecules—as indicated by the number of carbon atoms in the nucleus—is required. For example, percentages of C_1, C_2, etc. molecules present in the substance can be determined by such an analysis. A molecular-type analysis is run when the object is to characterize the components according to the chemical *arrangement* of their molecules. Since there are several different ways of classifying the chemical structure of hydrocarbons, several different approaches are possible in molecular-type analyses. For example, a molecular-type analysis could be used to determine the relative percentages of the naphthenic, paraffinic, and aromatic components. Another analysis might simply determine the proportions of saturated and unsaturated compounds present.

General Methods and Instrumentation

The analysis of petroleum hydrocarbons is accomplished through use of a variety of instruments and techniques. The most common techniques go by such names as chromatography, mass spectrometry, ultraviolet and infrared absorption analysis, and precipitation analysis, according to the physical principle upon which each is based.

Chromatography is an analytical technique involving the flow of a gas or liquid, together with the material under analysis, over a special porous, insoluble, sorptive medium. As the flowing phase passes over the stationary phase, different hydrocarbon components are adsorbed preferentially by the medium. With some types of chromatography, these components are desorbed through a similar process, and they leave the chromatographic column in distinct individual patterns. These patterns can be detected and recorded, and with proper interpretation can provide an extremely accurate means of determining composition. Chromatography is used in both carbon-type and molecular-type analyses. There are a number of chromatographic methods, each named according to the technique of analysis. *Gas chromatography* refers to the general method that uses a gas as the flowing, or mobile, phase; *gas liquid chromatography,* a more specific term, describes the techniques of using gas as the flowing phase and a liquid as the stationary phase; etc.

Silica gel analysis is a liquid chromatographic method that also involves physical separation of the components of a substance. The technique is based on the fact that polar compounds are adsorbed more strongly by silica gel than are non-polar saturated compounds. A sample of material under test is passed through a column packed with silica gel. Alcohol, which is more strongly adsorbed than any hydrocarbon, follows the sample through the column, forcing the hydrocarbons out—saturates first, unsaturated compounds next, then aromatic compounds. Small samples of the emerging material are taken periodically, and the refractive index of each sample is measured. From this information, relative percentages can be determined. (Clay/silica gel analysis, a method designed for rubber process oils, uses both activated clay and silica gel to determine the proportion of asphaltene, aromatic, saturated, and polar compounds present.)

Fluorescent indicator analysis (FIA) is a refinement of silica gel analysis in which a mixture of fluorescent dyes is placed in a small layer in the silica gel column. The dyes separate selectively with the aromatics, olefins, and saturates in the sample. Under ultraviolet light, boundaries between these different fractions in the column are visible; the amount of each hydrocarbon-type present can be determined from the length of each dyed fraction.

Mass spectrometry identifies the components of a substance by taking advantage of the difference in behavior exhibited by molecules of different mass when subjected to electrical and magnetic fields. A particle stream of the test material is first ionized, then directed in a curved path by a combination of the electrical and magnetic fields. The heavier ions, having greater inertia, tend toward the outside of the curve The stream of particles is therefore split up into a "mass spectrum"—they are distributed across the path according to their masses. This differentiated stream is played across a detecting slot on the "target," and a record of the analysis is thus made. (When the target is a photographic plate, the instrument is referred to as a Mass "Spectroscope").

As might be expected, the Mass Spectrometer is most useful, at least for hydrocarbon analysis, in the determination of carbon-number distributions. However, because various types of material show distinct spectral patterns, the Mass Spectrometer is also used in molecular-type analysis.

Ultraviolet (UV) Absorption Analysis is a method in which the amount and pattern of ultraviolet light absorbed by the sample is taken as a "fingerprint" of the components. The analysis is carried out through use of a spectrophotometer, which measures the relative intensities of light in different parts of a spectrum. By comparing the UV-absorbance pattern of the test sample with patterns of known material, components of the sample may be characterized. Infrared (IR) Analysis is a similar method but utilizes a different radiation frequency range.

Precipitation Analysis is used primarily in the characterization of rubber process oils. Components are identified on the basis of their reaction with varying concentrations of sulfuric acid. Hydrocarbons are separated into asphaltenes, polar compounds, unsaturated compounds (which are further separated into two groups, First Acidiffins and Second Acidiffins), and saturated compounds.

CONSISTENCY OF GREASE (PENETRATION)

See "Grease Consistency."

COPPER STRIP CORROSION

ASTM D 130

Many types of industrial equipment have parts of copper or copper alloys. It is essential that any oil in contact with these parts be non-corrosive to them.

Though modern technology has made great progress in eliminating harmful materials from petroleum oils, corrosion is still a possibility to be considered. Certain sulfur derivatives in the oil are a likely source. In the earlier days of the petroleum industry, the presence of active sulfur might have been attributable to inadequate refining. Today, however, practical methods have been developed to overcome this problem, and straight mineral oils of high quality are essentially free of corrosive materials.

On the other hand, certain oil additives, such as some of the emulsifying and extreme pressure (EP) agents, contain sulfur compounds. In the higher-quality oils, including those for moderate EP conditions, these compounds are of a type that is harmless to copper. For the more severe EP applications, however, chemically active additives are required for the prevention of scoring and seizure. Though oils containing these additives may not be desirable in the presence of copper or copper alloys, they are indispensable to many applications involving steel parts. Automotive hypoid rear axles are an example of this type of application.

To evaluate the corrosive properties of oils to copper—also to check them for active sulfur-type EP additives—the copper strip corrosion test is a widely accepted procedure. This test—described under the ASTM test method D 130—is applicable to the determination of copper-corrosive properties of certain fuels and solvents as well it is not to be confused, however, with tests for the rust-inhibiting properties of petroleum oils. The copper strip test evaluates the copper-corrosive tendencies of the oil itself—not the ability of the oil to prevent corrosion from some other source.

Significance

In the lubrication of bronze bushings, bearings that contain copper, and bronze wheels for worm-gear reduction units, corrosive oils must be carefully avoided. Because of the use of bronze retainers, manufacturers of anti-friction bearings insist on non-corrosive oils for their products. Hydraulic fluids, insulating oils, and aviation instrument oils must also be non-corrosive. In the machining of non-ferrous metals, moreover, cutting fluids must be of a non-corrosive type. The copper strip corrosion test helps to determine the suitability of these oils for the type of service they may encounter. In addition, it may help to identify oils of the active chemical type formulated for severe EP application.

This test may serve also in the refinery to check finished products for conformity with specifications. It may be applied, too, to solvents or fuels for assurance that these products will not attack cuprous metals with which they come in contact. In addition, there are certain special tests for corrosiveness, including the silver strip corrosion test of diesel lubricants. This test is applicable to crankcase oils for engines with silver bearing metals.

In conducting the copper strip corrosion test, there are 3 variables that may affect test results:

1 time of exposure of the copper to the sample
2 temperature of the sample
3 interpretation of the appearance of the exposed sample.

It is reasonable to expect that these variables will be applied in such a way as to reflect the conditions to which the product is to be subjected.

There is nothing to be gained, for example, by testing the oil at 212°F if test results at 122°F give better correlation with actual service conditions. If service conditions are more severe, however, test results at the higher temperature may give a more reliable indication of the oil's performance characteristics. Similarly, selection of the critical ASTM classification must be based on experience gained in the type of service for which the product is formulated. A dark tarnish (Classification 3, Table 3-2) is wholly acceptable, for example, where it has been shown that this degree of copper discoloration is associated with safe performance of the tested product. The flexibility of the copper strip test makes it adaptable to a wide range of products and end uses.

DEMULSIBILITY

ASTM D 1401 and ASTM D 2711

In the petroleum industry, the term emulsion usually applies to an emulsion of oil and water. Though mutually soluble only to a slight degree, these substances can, under certain circumstances, be intimately dispersed in one another to form a homogeneous mixture. Such a mixture is an oil/water emulsion, and it is

Table 3-2. ASTM copper strip classification.

Classification	Designation		Description
1	Slight Tarnish	1a	Light orange, almost the same as a freshly polished strip
		1b	Dark orange
2	Moderate Tarnish	2a	Claret red
		2b	Lavender
		2c	Multi-colored with lavender blue and/or silver overlaid on claret red
		2d	Silvery
		2e	Brassy or gold
3	Dark Tarnish	3a	Magenta overcast on brassy strip
		3b	Multi-colored with red and green showing (peacock), but no gray
4	Corrosion	4a	Transparent black, dark gray or brown with peacock green barely showing
		4b	Graphite or lusterless black
		4c	Glassy or jet black

For each of the 4 ASTM designations, there are two or more standards considered to cover the same degree of tarnish or corrosion. Differences in the type of chemical reaction produce differences in discoloration. Even a completely non-corrosive oil will alter the appearance of the freshly polished strip.

usually milky or cloudy in appearance.

Commercial oils vary in emulsibility. A highly refined straight mineral oil resists emulsification. Even after it has been vigorously agitated with water, an oil of this type tends to separate rapidly from the water when the mixture is at rest. Emulsification can be promoted, however, by agitation and by the presence of certain contaminants or ingredients added to the oil. The more readily the emulsion can be formed and the greater its stability, the greater the emulsibility of the oil. Some products, such as soluble cutting fluids, require good emulsibility and are formulated with special emulsifying agents.

With many other products, however, such as turbine oils and crankcase oils, the opposite characteristic is desired. To facilitate the removal of entrained water, these products must resist emulsification. The more readily they break from an emulsion, the better their demulsibility.

Two tests for measuring demulsibility characteristics have been standardized by the ASTM. The older of the two is ASTM method D 1401, which was developed specifically for steam turbine oils having viscosities of 150-450 Saybolt seconds at 100°F. It can be used for oils of other viscosities if minor changes in the test procedure are made. This method is the one recommended for use with synthetic oils.

The second method, ASTM D 2711 is designed for use with R&O (rust and oxidation inhibited) oils. It can also be used for other types of oils, although minor modifications are required when testing EP (extreme pressure) oils.

Significance of Demulsibility Tests

In many applications, oil is exposed to contamination by water condensed from the atmosphere. With turbine oils, exposure is even more severe, since the oil tends to come in contact with condensed steam.

Water promotes the rusting of ferrous parts and accelerates oxidation of the oil. For effective removal of the water, the oil must have good demulsibility characteristics.

Steam cylinder oils that serve in closed systems require good demulsibility for the opposite reason: to facilitate removal of oil from the condensate, so that oil is kept out of the boiler. Hydraulic fluids, motor oils, gear oils, diesel engine oils, insulating oils, and many similar petroleum products must resist emulsification. Oil and water must separate rapidly and thoroughly.

Either of the ASTM methods is suitable for evaluating the demulsification properties both of inhibited and uninhibited oils. However, correlation with field performance is difficult. There are many cases where the circulating oil is operating satisfactorily in the field, but fails the demulsibility tests in the laboratory. Hence, it must be recognized that these laboratory test results should be used in conjunction with other facts in evaluating an oil's suitability for continued service.

DENSITY

Density is a numerical expression of the mass-to-volume relationship of a substance.

Density is important in volume-to-mass and mass-to-volume calculations, necessary in figuring freight rates, fuel loads, etc. Although it is not directly a criterion of quality, it is sometimes useful as an indicator of general hydrocarbon type in lubricants and fuels. For a given volatility, for example, aromatic hydrocarbons have a greater density than paraffins, naphthenic hydrocarbons usually being intermediate. Density data may also be used by manufacturers or their customers to monitor successive batches of a product as a check on uniformity of composition.

In the SI system (see the introduction) the official unit for density is kilograms/cubic meter (kg/m^3 at 15.C. Units formerly used were *density* in kilograms/cubic decimeter (kg/dm^3) at 15°C, *specific gravity* 60/60°F (mass per unit volume compared with that of water at the same temperature), and *API gravity* at 60°F (an arbitrary scale calibrated in degrees). An API gravity/specific gravity/density conversion chart appears in the appendix to this book.

Density may be determined by ASTM method D 1298, using a hydrometer graduated in units of density. Tables are available for conversion of observed density to that at 15°C.

See also Gravity.

DIELECTRIC STRENGTH

ASTM D 877 and D 1816

A dielectric is an electric insulating material, one that opposes a flow of current through it. There are two properties that contribute to this characteristic. One is resistivity, the specific resistance that a dielectric offers under moderate conditions of voltage. The other is dielectric strength, the ability to prevent arcing between two electrodes at high electric potentials. Though the two properties are not directly related, it so happens that commercial insulating materials of high dielectric strength also possess adequate resistivity. In the insulation of high-voltage electrical conductors, therefore, it is ordinarily dielectric strength that is of the greater concern.

Petroleum oil is an excellent dielectric and is used extensively in electrical equipment designed to be insulated with a liquid. Among the advantages that oil offers over solid insulation are the abilities to cool by circulation and to prevent corona. Corona is the result of ionization of air in the tiny voids that exist between a conductor and a solid insulating wrapper. Corona is destructive to certain types of solid insulation. By filling all of the space around a conductor, insulating oil eliminates the source of corona. Oil also has the high dielectric strength that good insulation requires.

At normal voltage gradients, conduction of electric current through a dielectric is negligible. The dielectric lacks the free charged particles that a conductor must have. If the voltage impressed on the dielectric is increased, however, the material becomes more highly ionized. Ions thus produced are free charged particles.

If a high enough voltage is applied, ions are produced in sufficient concentration to allow a discharge of current through the dielectric, and there is an arc. The minimum voltage required for arcing is the breakdown voltage of the dielectric incurred under the circumstances involved. When the dielectric break down, it undergoes a change in composition that permits it—temporarily, at least—to conduct electricity.

The magnitude of the breakdown voltage depends on many factors, such as the shape of the electrodes and the thickness and dielectric strength of the insulation between them. In accordance with the ASTM method D 877 or D 1816, the dielectric strength of an insulating oil is evaluated in terms of its breakdown voltage under a standard set of conditions. Because of the marked effect of contamination on test results, special care must be exercised in obtaining and handling the sample. The sample container and test cup must be absolutely clean and dry, and no foreign matter must come in contact with the oil.

In either case, the voltage noted at the specified end point is the breakdown voltage of the respective sample.

Significance of Test Results

Insulating oils find wide application in transformers, cables, terminal bushings, circuit breakers, and similar electrical equipment. Depending upon the installation the purpose of these oils may be to prevent electrical leakage and discharge, to cool, to eliminate corona effects, or to provide any combination of these functions. High dielectric strength is obviously an important insulating-oil property.

When new, a carefully refined petroleum oil can be expected to exhibit a high natural dielectric strength suitable for any of the conventional insulating purposes. Other properties of the oil, such as oxidation resistance, are therefore of greater significance.

In service, however, the oil eventually becomes contaminated with oxidation products, carbon particles, dirt, and water condensed from atmospheric moisture. Water is the principal offender. Though small quantities of water dissolved in the oil appear to have little influence on dielectric strength, *free* water has a pronounced effect. The dispersion of free water throughout the oil is promoted, moreover, by the presence of solid particles. These particles act as nuclei about which water droplets form. Dielectric strength is impaired also by dirt and oxidation sludges that may accumulate in the oil.

A relationship exists, therefore, between a drop in dielectric strength and the deterioration of an oil in service. Dielectric strength thus suggests itself as a method of evaluating the condition of a used insulating oil. In this application, a significant drop in dielectric strength may indicate serious water contamination, oxidation, or both.

If water is the only major contaminant, the oil can generally be reclaimed by drying. But, if the drop in dielectric strength is attributable to oxidation, the oil may already have deteriorated beyond a safe limit. By itself, therefore, dielectric strength is not ordinarily considered a sufficiently sensitive criterion of the suitability of a batch of oil for continued service. Power factor, neutralization number, and interfacial tension are test values that have found greater acceptance for this purpose.

DILUTION OF CRANK CASE OILS

ASTM D 322

Excessive crankcase dilution is associated with faulty operation of an internal combustion engine. It is caused by the seepage of raw and partially burned fuel from the combustion chamber past the piston into the crankcase, where it thins the crankcase oil. It is often desirable to know the extent to which a used oil has been diluted in this way. For motor oils from gasoline engines, dilution may be evaluated by the ASTM method D 322.

The procedure is to measure the percentage of fuel in the sample by removing the fuel from the oil. Since the fuel is considerably more volatile than the oil, the two can be separated by distillation.

To lower the distillation temperature and to make the test easier to run, a relatively large amount of water is added to the sample. Since the water and the sample are immiscible, the boiling point of the mixture is, at any instant, appreciably lower than that of the sample alone.

Because of its substantially higher volatility, the fuel is, to all intents and purposes, evaporated before the oil. A mixture of fuel vapor and water vapor passes into a condenser and is converted back to liquid. The fuel, which is lighter, floats on top of the water in a graduated trap. Here, the volume of condensed fuel can be observed before any significant distillation of the oil begins.

Significance of Results

This test for crankcase dilution is applicable to used motor oils from gasoline engines. Excessive dilution, as determined by test, is harmful in is own right, as well as being indicative of faulty engine performance.

In the first place, dilution is an obvious source of fuel waste. Another effect is to reduce the viscosity of the oil, which may seriously impair its lubricating value. A diluted oil may lack the body required to prevent wear, and it may not make a proper seal at the piston rings. Pistons and cylinders are especially vulnerable, since the oil on their wearing surfaces is subject to the direct washing action of the raw fuel.

Fuel may also reduce the oil's oxidation stability and may raise the oil level in the crankcase. An abnormally high level causes an increase in oil consumption and gives false readings as to the actual amount of lubricant present. Failure of a motor oil to lubricate as it should may be directly attributable to dilution.

As an indication of faulty performance, excessive crankcase dilution may be the symptom of an unsuitable fuel. If the fuel's volatility characteristics are too low, the fuel does not vaporize properly, and combustion is incomplete. The unburned portion of the fuel finds is way into the crankcase.

A similar effect may be produced, however, by incorrect operation or poor mechanical condition of the engine:

Too rich a fuel mixture—maladjustment of the carburetor or excessive choking may admit more fuel to the combustion chamber than can be burned with the amount of air present.

Too low an engine temperature—defective temperature control or short operating periods may keep the engine too cold for proper vaporization.

Inadequate breathing facilities—insufficient venting of the crankcase vapors may interfere with normal evaporation of the fuel from the crankcase. With older cars, the trouble may be caused by stoppage of the crankcase breather. On cars built after 1963, the positive crankcase ventilation system may be at fault.

Worn pistons, rings, or cylinders—excessive clearance between the pistons or rings and the cylinder walls facilitates the seepage of fuel into the crankcase.

Any of the deficiencies indicated by excessive crankcase dilution can be expected to jeopardize satisfactory engine performance.

With diesel engines, there is not the spread in volatility characteristics between fuel and lubricating oil that there is with gasoline engines. For this reason, there is no simple test for the crankcase dilution of diesel engine. The closest approximation is made by noting the reduced viscosity of the used oil as compared with that of the new oil and estimating what percentage of fuel dilution would cause such a viscosity reduction.

DISTILLATION

A chemically pure hydrocarbon, like any other pure liquid compound, boils at a certain temperature when atmospheric pressure is constant. However, almost all commercial fuels and solvents contain many different individual hydrocarbons, each of which boils at a different temperature. If the petroleum product is gradually heated, greater proportions of the lower-boiling constituents are in the first vapor formed, and the successively higher-boiling constituents are vaporized as the temperature is raised.

Thus, for any ordinary petroleum product, boiling takes place over a range of temperature rather than at a single temperature. This range is of great importance in fuel and solvent applications, and is the property measured in distillation tests.

Several ASTM tests are used for measuring the distillation range of petroleum products. These tests are basically similar, but differ in details of procedure. The following tests are widely used:

ASTM D 86-67: Distillation of Petroleum Products
ASTM D 216-54: Distillation of Natural Gasoline
ASTM D 850-70: Distillation of Industrial Aromatic Hydrocarbons
ASTM D 1078-70: Distillation Range of Lacquer Solvents and Dilutants

Significance of Results

For both fuels and solvents, distillation characteristics are important.

Automotive Gasoline: The entire distillation range is important in automotive fuels. The distillation characteristics of the "front end" (the most volatile portion, up to perhaps 30% evaporated), together with the vapor pressure of the gasoline (see discussion on Vapor Pressure), control its ability to give good cold-starting performance. However, these same characteristics also control its vapor-locking tendency. An improvement in cold-starting can entail a decrease in vapor lock protection.

The temperatures at which 50% and 90% of the fuel are evaporated are indications of warm-up characteristics. The lower these points, the better the warm-up. A low 50% point is also an indication of good acceleration. A low 90% point is desirable for completeness of combustion, uniformity of fuel distribution to the cylinders, and less formation of combustion chamber deposits.

Usually the volatility of a commercial gasoline is adjusted seasonally, and also in accordance with the climate in the region into which it is being shipped. In cold weather, a more volatile product is desired to provide better starting and warm-up. In warm weather, less volatility provides greater freedom from vapor lock.

Aviation Gasoline: In general, aviation gasolines have lower 90% points and final boiling points than automotive gasolines, but the significance of the various points on the distillation curve remain the same. A minimum limit on the sum of the 10% and 50% points is normally specified to control carburetor icing characteristics.

Diesel Fuel: Although diesel fuels have much lower volatility than gasoline, the effect of the various distillation points are similar. For example, the lower the initial boiling point for a given cetane number, the better its starting ability, but more chance of vapor lock or idling difficulties. Also, the higher the end point or final boiling point, the more chance there is of excessive smoking and deposits. The mid-boiling point (50% point) is related to fuel economy because, other things being equal, the higher the 50% point, the more Btu content and the better cetane number a diesel fuel possesses.

Burner Fuel: For burner fuels, ease of lighting depends on front end volatility. Smoking depends upon the final boiling point, with excessive smoke occurring if the final boiling point is too high.

Solvents: Many performance characteristics of solvents are related to distillation range. The initial boiling point is an indirect measure of flash point and, therefore, of safety and fire hazard. The spread between IBP and the 50% point is an index of "initial set" when used in a rubber or paint solvent.

The 50% point shows a rough correlation with evaporation rate; the lower the 50% point for certain classes of hydrocarbons, the faster the evaporation. If the dry point and the 95% point are close, there is little or no "tail" or slow-drying fractions. Also, useful indications of good fractionization of a solvent are the narrowness of the distillation range and the spread between IBP and 5% point and between 95% point and the dry point. The smaller the spread, the better.

DROPPING POINT OF GREASE

ASTM D 566 and ASTM D 2265

It is often desirable to know the temperature at which a particular lubricating grease becomes so hot as to lose its plastic consistency. Being a mixture of lubricating oil and thickener, grease has no distinct melting point in the way that homogeneous crystalline substances do. At some elevated temperature, however, the ordinary grease becomes sufficiently fluid to drip. This temperature is called the dropping point and can be determined by the ASTM Method D 566—"Dropping Point of Lubricating Grease" and ASTM Method D2265—"Dropping Point of Lubricating Grease of Wide Temperature Range."

Significance of Results

Since both these test are held under static conditions, the results have only limited significance with respect to service performance. Many other factors such as time exposed to high temperatures, changes from high to low temperatures, evaporation resistance and oxidation stability of the grease, frequency of relubrication, and the design of the lubricated mechanism are all influences that affect the maximum usable temperature for the grease.

Though both dropping point and consistency are related to temperature, the relationships follow no consistent pattern. The fact that a grease does not liquefy at a particular temperature gives no assurance that its consistency will be suitable at that temperature. However, the dropping point is useful in identifying the grease as to type and for establishing and maintaining bench marks for quality control.

One of the weaknesses of either procedure is that a drop of oil may separate and fall from the grease cup at a temperature below that at which the grease fluidizes. This would then give an erroneous indication of the actual temperature at which the grease becomes soft enough to flow from the cup.

ECOTOXICITY

(See "Biodegradation")

FLASH AND FIRE POINTS—OPEN CUP

ASTM D 92

The flash point and the fire point of a petroleum liquid are basically measurements of flammability. The flash point is the minimum temperature at which sufficient liquid is vaporized to create a mixture of fuel and air that will burn if ignited. As the name of the test implies, combustion at this temperature is only of an instant's duration. The fire point, however, runs somewhat higher. It is the minimum temperature at which vapor is generated at a rate sufficient to *sustain* combustion. In either case, combustion is possible only when the ratio of fuel vapor to air lies between certain limits. A mixture that is too lean or too rich will not burn.

The practice of testing for flash and fire points was originally applied to kerosene to indicate its potentiality as a fire hazard. Since then, the scope has been broadened to include lubricating oils and other petroleum products. Though it has become customary to report flash point (and sometimes fire point) in lubricating oil data, these properties are not as pertinent as they might appear. Only in special instances does a lubricating oil present any serious fire hazard. Being closely related to the vaporization characteristics of a petroleum product, however, flash and fire points give a rough indication of volatility and certain other properties.

The fire point of a conventional lubricating oil is so closely associated with its flash point, that it is generally omitted from inspection data. For the ordinary commercial product, the fire point runs about 50°F above the flash point. Fire and flash points are not to be confused, however, with auto-ignition temperature, which is an entirely different matter. Auto-ignition deals, not so much with volatility, as with the temperature necessary to precipitate a chemical reaction—combustion—without an external source of ignition. Though a more volatile petroleum product may be expected to have lower flash and fire points than one that is less volatile, its ASTM auto-ignition temperature is generally higher.

Significance of Test Results

To appreciate the significance of flash point and fire point test results, one must realize what the tests measure. It is necessary to understand how a combustible air-fuel mixture is created.

For all practical purposes, a petroleum liquid does not burn as such, but must first be vaporized. The vapor mixes with the oxygen in the air, and, when sufficient concentration of the vapor is reached, the mixture may be ignited, as by a spark or open flame. The mixture can be ignited only if the concentration of fuel vapor in the air is more than about 1% or less than about 6% by volume. A confined mixture containing more than 6% fuel vapor becomes a practical explosion hazard only if it is vented to admit a greater portion of air.

The significance of flash- and fire-point values lies in the dissimilarity that exists in the volatility characteristics of different petroleum liquids. Even among lubricating oils of comparable viscosity, there are appreciable variations in volatility, and hence in flash and fire point. In general, however, the storage and operating temperatures of lubricating oils are low enough to preclude any possibility of fire. Among the exceptions to this situation are such products as quenching and tempering oils, which come in direct contact with high-temperature metals. Heat-transfer oils, used for heating or cooling, may also reach temperatures in the flash- and fire-point ranges. Similarly, in the evaluation of roll oils, which are applied in steel mills to hot metal sheets from the annealing oven, fire hazard may likewise be a consideration. In many of these cases, however, auto-ignition temperature is of greater significance. At the auto-ignition temperature, as determined by test, fire is not mercy a possibility—it actually occurs spontaneously, i.e., without ignition from any outside source.

Since flash and fire point are also related to volatility, however, they offer a rough indication of the tendency of lubricating oils to evaporate in service. It should be apparent that lower flash and fire points imply a greater opportunity for evaporation loss. The relationship between test results and volatility is by no means conclusive, however. The comparison is distorted by several additional factors, the most important of which is probably the manner in which the oil is produced.

The relationship between flash and fire point, on the one hand, and volatility, on the other, is further distorted by differences in oil type. For a given viscosity, a paraffinic oil will exhibit higher flash and fire point than other types and may be recognized by these test results. Paraffinicity may also be indicated by a high viscosity index or by a high pour point.

Fire and flash points are perhaps of greater significance in the evaluation of used oils. If an oil undergoes a rise in flash or fire point in service, loss by evaporation is indicated. The more volatile components have been vaporized, leaving the less volatile ones behind; so an increase in viscosity is apparent. An excessive increase in viscosity may so alter lubricating properties that the oil is no longer suitable for its intended application.

If, on the other hand, the flash or fire points drop in service, contamination is to be suspected. This may happen to motor oils that become diluted with unburned fuel. Gasoline or heavier fuels in the crankcase reduce the viscosity of the oil, and bearings and other moving parts may be endangered by excessive thinning of the lubricant. These fuels, being more volatile than the oil, lower the flash and fire points of the mixture. So the flash- or fire-point test on used oils constitutes a relatively simple method for indicating the presence of dilution.

FLASH POINT-CLOSED CUP

ASTM D 56 and D93

All petroleum products will burn, and under certain conditions their vapors will ignite with explosive violence. However, in order for this to occur, the ratio of vapor to air must be within definite limits.

When a liquid petroleum product is exposed to air, some of it evaporates, causing a certain vapor/air concentration. As the temperature of the liquid product is raised, more and more evaporates, and the vapor/air ratio increases. Eventually, a temperature is reached at which the vapor/air ratio is high enough to support momentary combustion, if a source of ignition is present. This temperature is the flash point of the product.

For fuels and solvents, the flash point is usually determined by a "closed cup" method, one in which the product is heated in a covered container. This most closely approximates the conditions under which the products are handled in actual service. Products with flash points below room temperature must, of course, be cooled before the test is begun.

Two closed cup methods of determining flash point are widely used. They differ primarily in details of the equipment and in the specific fields of application. However, the tests are basically similar, and may be grouped together for descriptive purposes. The two tests are:

ASTM D 56 Flash Point by Means of the Tag Closed Tester

ASTM D 93 Flash Point by Means of the Pensky-Martens Closed Tester

The former test (Tag) is used for most fuels and solvents, including lacquer solvents and dilutants with low flash points. The latter test (Pensky-Martens) is ordinarily used for fuel oils but can also be used for cutback asphalts and other viscous materials and suspensions of solids.

Significance of Results

For a volatile petroleum solvent or fuel, flash point is primarily significant as an indication of the fire and explosion hazards associated with its use. If it is possible for any particular application to select a product whose flash point is above the highest expected ambient temperature, no special safety precautions are necessary. However, gasoline and some light solvents have flash points well below room temperature. When they are used, controlled ventilation and other measures are necessary to prevent the possibility of explosion or fire.

It should be remembered that flash point is the lowest temperature at which a product will support momentary combustion, *if a source of ignition is present*. As such, it should not be confused with auto-ignition temperature, which is the temperature at which combustion will take place spontaneously, *with no external source of ignition*. Products with low flash points often have high auto-ignition temperatures, and vice versa.

FOAMING CHARACTERISTICS OF LUBRICATING OILS

ASTM D 892

Foaming in an industrial oil system is a serious service condition that may interfere with satisfactory system performance and even lead to mechanical damage.

While straight mineral oils are not particularly prone to foaming, the presence of additives and the effects of compounding change the surface properties of the oils and increase their susceptibility to foaming when conditions are such as to mix air with the oils. Special additives impart foam resistance to the oils and enhance their ability to release trapped air quickly under conditions that would normally cause foaming.

The foaming characteristics of lubricating oils at specified temperatures are determined by the standard ASTM method D 892.

Significance

Foaming consist of bubbles that rise quickly to the surface of the oil, and is to be distinguished from air entrainment, consisting of slow-rising bubbles dispersed throughout the oil. Both these conditions are undesirable, and are often difficult to distinguish due to high flow rates and turbulence in the system. These two phenomena are affected by different factors and are considered in separate laboratory tests. The primary causes of foaming are mechanical—essentially an operating condition that tends to produce turbulence in the oil in the presence of air. The current trend in hydraulic oil systems, turbine oil systems, and industrial oil systems of every kind is to decrease reservoir sizes and increase flow rates. This trend increases the tendency for foaming in the oils.

Contamination of the oil with surface-active materials, such as rust preventatives, detergents, etc., can also cause foaming.

Foaming in an industrial oil is undesirable because the foam may overflow the reservoir and create a nuisance, and the foam will decrease the lubrication efficiency of the oil, which may lead to mechanical damage.

Antifoaming additives may be used in oils to decrease foaming tendencies of the oil. However, many such additives tend to increase the air entrainment characteristics of an oil, and their use requires striking a balance between these two undesirable phenomena.

FOUR-BALL WEAR TEST—ASTM D 2266
FOUR—BALL EP TEST—ASTM D 2596

(See "Load Carrying Ability")

GRAVITY
ASTM D 287

Practically all liquid petroleum products are handled and sold on a volume basis—by gallon, barrel, tank car, etc. Yet, in many cases, it is desirable to know the weight of the product. Gravity is an expression of the weight-to-volume relationship of a product.

Any petroleum product expands when it is heated, and its weight per unit volume therefore decreases. Because of this, gravity is usually reported at a standard temperature, although another temperature may actually have been used in the test. Tables are available for converting gravity figures from one temperature basis to another.

Gravity can be expressed on either of two scales. The "specific gravity" is defined as the ratio of the weight of a given volume of the product at 60°F to the weight of an equal volume of water at the same temperature.

In the petroleum industry, however, the API (American Petroleum Institue) gravity scale is more widely used. This is an arbitrary scale, calibrated in degrees, and related to specific gravity by the formula:

$$\text{API Gravity (degrees)} = \frac{141.5}{\text{specific gravity } 60/60°\text{F}} \quad 131.5$$

As a result of this relationship, the higher the specific gravity of a product, the lower is its API gravity. It is noteworthy that water, with a specific gravity of 1.000, has an API gravity of 10.0°.

Gravity (specific or API) is determined by floating a hydrometer in the liquid, and noting the point at which the liquid level intersects the hydrometer scale. Corrections must then be made in accordance with the temperature of the sample at the time of test.

Significance of Results

Gravity has little significance from a quality standpoint, although it is useful in the control of refinery operations. Its primary importance is in volume-to-weight and weight-to-volume calculations. These are necessary in figuring freight rates, aircraft and ship fuel loads, combustion efficiencies, etc.

To some extent, gravity serves in identifying the type of petroleum product. Paraffinic products have lower specific gravities (higher API gravities) than aromatic or naphthenic products of the same boiling range. Gravity data may also be used by manufacturers or by their customers to monitor successive batches of these products as a check on uniformity of product composition.

Gravity is important in process applications that depend on differences in gravity of the materials used. For example, petroleum products having higher specific gravities than 1.000 (that of water) are necessary in the field of wood preservation in order to permit separation of the materials involved. The specific gravity range of petroleum products is about 0.6 to 1.05.

Gravity is used in empirical estimates of thermal value, often in conjunction with the aniline point. With the exception of the above applications, gravity should not be used as an index of quality.

GREASE CONSISTENCY
ASTM D 217 and D 1403

The consistency of a lubricating grease is defined as its resistance to deformation under an applied force—in other words, is relative stiffness or hardness. The consistency of a grease is often important in determining its suitability for a given application.

Grease consistency is given a quantitative basis through measurement with the ASTM Cone Penetrometer. The method consists of allowing a weighted metal cone to sink into the surface of the grease, and measuring the depth to which the point falls below the surface. This depth, in tenths of millimeters, is recorded as the penetration, or penetration number, of the grease. The softer the grease, the higher its penetration.

The ASTM D 217 method recognizes five different categories of penetration, depending on the condition of the grease when the measurement is made. *Undisturbed* penetration is determined with the grease in its original container. *Unworked* penetration is the penetration of a sample which has received only minimum disturbance in being transferred from the sample can to the test cup. *Worked* penetration is the penetration of a grease sample that has been subjected to 60 double strokes in a standard grease worker (to be described). *Prolonged worked* penetration is measured on a sample that has been worked the specified number of strokes (more than 60), brought back to 77°F, then worked an additional 60 double strokes in the grease worker. *Block* penetration is the penetration of a block grease—a grease hard enough to hold its shape without a container.

All the above penetrations are determined on samples that have been brought to 77°F.

Significance

If a grease is too soft, it may not stay in place, resulting in poor lubrication. If a grease is too hard, it will not flow properly, and either fail to provide proper lubrication or cause difficulties in dispensing equipment. These statements sum up the reasons for classifying greases by consistency. Penetration numbers are useful for classifying greases according to the consistencies required for various types of service, and in controlling the consistency of a given grade of grease from batch to batch.

The National Lubricating Grease Institute has classified greases according to their worked penetrations. These NLGI grades, shown in Table 3-3, are used for selection of greases in various applications.

In comparing greases, worked and prolonged worked penetrations are generally the most useful values. The change in penetration between the 60-stroke value and prolonged worked value is a measure of grease stability. Prolonged worked penetrations should report the amount of working (10,000 and 100,000 strokes are most common) in order to be useful. Unworked penetrations often appear in specifications and in grease product data, but are of limited practical value. No significance can be attached to the difference between unworked and worked penetration. Undisturbed penetration is useful mainly in quality control.

INTERFACIAL TENSION

ASTM D 971

Molecules of a liquid have a certain attraction for one another. For some liquids, like mercury, this attraction is very great; for others, like alcohol, it is considerably less. Beneath the surface, the attractive forces are evenly distributed, since a molecule is drawn to the one above or below it as strongly as to the one at its side. But, at the surface, there are no similar molecules over it to attract the liquid upward; so the bonds between the molecules are concentrated in a lateral direction.

The strong mutual attraction between the surface molecules results in a phenomenon known as surface tension, and its effect is like that of a membrane stretched over the liquid face. Surface tension is an appreciable force, as anyone knows who has made the simple experiment of "floating" a needle on the top layer of still water.

Surface tension can be reduced, however, by the introduction of materials that weaken the links between the original molecules. Differences in surface tension can be measured, and these measurements sometimes serve as a guide to the condition of a used oil. Standard procedure for making these measurements is covered by the ASTM method D 971 for conducting the interfacial tension—IFT—test.

The IFT test is one for measuring the tension at the interface between two immiscible liquids: oil and

Table 3-3. NLGI grease grading system.

NLGI Grade	Worked Penetration Range mm/10
000	445-475
00	400-430
0	355-385
1	310-340
2	265-295
3	220-250
4	175-205
5	130-160
6	85-115

Grease Classification

1. Group I—General Purpose Greases
   Greases that are expected to give proper lubrication to bearings whose operating temperatures may vary from −40° C to 121° C (−40° F to 250° F).
2. Group II—High Temperature Greases
   Greases that are expected to give proper lubrication to bearings whose operating temperatures may vary from −18° C to 149° C (0° F to 300° F).
3. Group III—Medium Temperature Greases
   Greases that are expected to give proper lubrication to bearings whose operating temperatures may vary from 0° C to 93° C (32° F to 200° F).
4. Group IV—Low Temperature Greases
   Greases that are expected to give proper lubrication to a bearing whose operating temperature may go as low as −55° C (−67° F) or as high as 107° C (225° F).
5. Group V—Extreme High Temperature Greases
   Greases that are expected to give proper lubrication for comparatively short periods of time where bearing operating temperature may be as high as 232° C (450° F).

The requirements of the five groups of greases cover applications where the lubrication is not affected by extremely heavy loads, high speeds or excessive humidity. The tabulation of test requirements which follows is intended to establish for the five groups of greases those characteristics needed to meet the functional requirements of good rolling bearing greases.

Methods of Testing

Test Requirements	Method
ASTM Penetration Normal Worked	Method ASTM D-217
Oxidation	Method ASTM D-942
Low Temperature Torque	Method ASTM D-1478
Water Resistance	Method ASTM D-1264
Dropping Point	Method ASTM D-566
Evaporation	Method ASTM D-972
Dirt	Method Federal Test Method Standard No. 791-B Method 3005.3
Consistency, Stability	Method ASTM D-217
Abrasive Matter	Method ASTM D-1404
Free Acid and Free Alkali	Method ASTM D-128
Fillers	Method ASTM D-128
Moisture	Method ASTM D-128

Test Requirements

Tests	Group I	Group II	Group III	Group IV	Group V
ASTM Penetration Normal Worked[1]	250 to 350	200 to 300	220 to 300	260 to 320	250 to 310
Oxidation	10# max. drop in 500 hours	10# max. drop in 500 hours	10# max. drop in 500 hours	5# max. drop in 500 hours	5# max. drop in 100 hours at 121° C (250° F)
Low Temperature Torque	1 rev. in max. of 10 sec. at −40° C (−40° F)	—	—	1 rev. in max. of 5 sec. at −55° C (−67° F)	1 rev. in max. of 10 sec. at −40° C (−40° F)
Water Resistance	50% max. loss	—	50% max. loss	20% max. loss	50% max. loss
Dropping Point	149° C min. (300° F)	177° C min. (350° F)	149° C min. (300° F)	149° C min. (300° F)	232° C min. (450° F)
Evaporation	—	—	—	1.5% max. in 22 hours at 121° C (250° F)	4.0% max. in 22 hours at 204° C (400° F)
Dirt Max. Particles per cubic cm	5000/5-20 microns 2000/20-50 microns 50/50-75 microns None over 75 microns	Same as Group I	Same as Group I	Same as Group I	Same as Group I
Consistency Stability: max. ASTM Penetration after 100,000 strokes	Penetration shall not increase more than 100 points and in no case be more than 375	Penetration shall not increase more than 100 points and in no case be more than 350	Penetration shall not increase more than 100 points and in no case be more than 350	Penetration shall not increase more than 100 points and in no case be more than 375	Penetration shall not increase more than 100 points and in no case be more than 375

[1]National Lubricating Grease Institute Code — No. 0: 355-385; No. 1: 310-340; No. 2: 265-295; No. 3: 220-250; No. 4: 175-205

Figure 3-1. Grease Classification and Testing

distilled water. Ordinarily, oil and water do not mix, the oil floating on top of the water because it is less dense. At the interface, each liquid exhibits its own surface tension, the molecules of one having no great attraction for those of the other. To break through the interface, it is necessary to rupture the surface tensions both of the water and of the oil. However, if certain contaminants are added to the oil—such as soaps, dust particles, or the products of oil oxidation—the situation is altered. These contaminants are said to be hydrophilic, i.e., they have an affinity for the water molecules—as well as for the oil molecules. At the interface, the hydrophilic materials extend bonds across to the water, so that any vertical linkage between the liquids is strengthened, and the lateral linkage is weakened. The interface is less distinct, and the tension at the interface is reduced. The greater the concentration of hydrophilic materials, the less the tension. Since oxidation products tend to be hydrophilic, IFT test results are related to the degree of the oil's oxidation.

Significance of Results

For many purposes, large quantities of petroleum oil remain in service for very long periods. There is good reason, therefore, for checking the extent to which these oils have oxidized to determine their fitness for continued service. In some cases, neutralization number may provide a criterion by which used oils are evaluated. But, by the time the acid neutralization number has undergone a significant rise, oxidation has set in, and acids and sludges may be already formed.

It is felt by many people that the IFT test is more sensitive to incipient oxidation, that it anticipates oxidation before deterioration has reached serious proportions. Since the oxidation of an oil increases at an accelerated rate; an early warning of impending deterioration is advantageous.

The IFT test is frequently applied, therefore, to electric transformer oils, where oxidation is especially harmful. Acids formed by oxidation may attack the insulation, and oxidation sludges interfere with circulation and the cooling of the windings. Because of the importance of good quality, transformer oils may be checked periodically for IFT to determine the advisability of replacement. The critical IFT value is based on experience with the particular oil in service, and testing conditions must be uniform in all respects. Tests for power factor and dielectric strength are also related to the condition of the oil.

For new oils, IFT values have little significance, though they may be used for control purposes in oil manufacture. Additives added to the oil to improve its performance may grossly distort IFT test results, so that they bear no apparent relation to the oil's quality. For this reason, special consideration must be given if attempts are made to evaluate the condition of an inhibited steam turbine oil by the IFT method.

The IFT test itself is an extremely delicate one, and consistent results are not easily obtained. The test should be conducted only by an experienced person, and the apparatus must be scrupulously clean. A minute quantity of foreign matter can cause a tremendous increase in the oil's hydrophilic properties. The sample must be carefully filtered to remove all solid materials, which reduce the IFT value appreciably. Even under the most meticulous conditions, however, good reproducibility is difficult to achieve.

It has been found, moreover, that test results are affected by conditions outside of the laboratory. Prolonged storage of the oil sample may cause a drop in its IFT value; so may exposure to sunlight. Similarly, agitation of the sample may increase its hydrophilic properties, and, if the sample is not tightly sealed against air, there may be an increase in the oxidation products. While indifferent handling practices would not be expected to affect a new oil, they may cause the deterioration of a used oil to appear greater than it actually is.

LOAD-CARRYING ABILITY

For machine parts that encounter high unit loadings, the lubricant must be capable of maintaining a film that prevents metal-to-metal contact under the extreme pressures involved. Otherwise, scoring of the surfaces and possible failure of the parts will result. Special extreme pressure (EP) lubricants are required for such applications.

Several test machines have been constructed and test procedures established in attempts to approximate closely the conditions a lubricant will meet in field applications. Four widely used tests are the Timken machine, the FZG test, the 4-Ball EP test, and the 4-Ball friction and wear test.

Timken Machine—In the Timken test (ASTM D 2509), a rotating member is brought to bear against a stationary member with lubrication provided by the lubricant under test. The lubricant is evaluated on the basis of its ability to prevent scoring of the metal surfaces. The maximum load that can be applied without scoring is reported as the Timken OK load. The minimum load required to cause scoring is reported as the score load.

In addition to the OK and score loads, the actual pressure at the point of contact is sometimes reported. The area of the wear scar is determined using a Brinnell microscope; the unit loading in megapascals (or in psi) on the area of contact can then be calculated using the OK load.

Only very general conclusions can be drawn from the Timken EP test. Results should be related to additional information about the lubricant, such as antiwear properties, type of additive, and corrosion characteristics. Used in this way, Timken EP results can provide an experienced engineer with valuable information about the performance of one lubricant relative to another. In addition, the Timken EP test is often used in quality control of lubricants whose performance characteristics have already been established.

FZG Test—The FZG test is used in Europe to evaluate EP properties. Two sets of opposing gears are loaded in stages until failure of geartooth surfaces occurs. Results are reported in terms of the number of stages passed. Two standard sets of temperature and gear speed are used and should be stated along with the number of stages passed.

Four-Ball Wear Test and Four-Ball EP Test—Each of the four-ball tests is designed to evaluate a different load-carrying characteristic of lubricating oil or grease. Both use similar equipment and mechanical principles. Four 1/2-inch steel balls are arranged with one ball atop the three others. The three lower balls are clamped together to form a cradle, upon which the fourth ball rotates on a vertical axis.

The *four-ball wear test* (ASTM D 2266) is used to determine the relative wear-preventing properties of lubricants on sliding metal surfaces operating under boundary lubrication conditions. The test is carried out at specified speed, temperature, and load. At the end of a specified period the average diameter of the wear scars on the three lower balls is measured and reported.

Under standardized conditions, the four-ball wear test provides a means for comparing the relative antiwear properties of lubricants. Results of two tests run under different conditions cannot be compared, so operating conditions should always be reported. No correlation has yet been established with field service, so individual results should not be used to predict field performance.

The *four-ball EP test* (ASTM D 2596) is designed to evaluate performance under much higher unit loads than applied in the wear test, hence the designation EP (extreme pressure). The EP Tester is of slightly different design and construction than the Wear Tester. One steel ball is rotated against the other three at constant speed, but temperature is not controlled. The loading is increased at specified intervals until the rotating ball seizes and welds to the other balls. At the end of each interval, the scar diameters are measured and recorded.

Two values from the EP test are generally reported: load wear index (formerly called Mean Hertz Load) and weld point. *Load Wear Index (LWI)* is a measure of the ability of a lubricant to prevent wear at applied loads. *Weld point* is the lowest applied load at which either the rotating ball seizes and then welds to the three stationary balls, or at which extreme scoring of the three stationary balls occurs. It indicates the point at which the extreme pressure limit of the lubricant is exceeded.

The four-ball EP test is used in lubricant quality control, and to differentiate between lubricants having low, medium, or high extreme pressure qualities. Results do not necessarily correlate with actual service and should not be used to predict field performance unless other lubricant properties are also taken into consideration.

For comparison of the capabilities of various lubricants, the results of both four-ball tests should be considered, particularly if additives or grease thickeners are unknown or widely dissimilar. Lubricants with good extreme pressure properties may not be equally effective in reducing wear rates at less severe loads, and conversely.

NEUTRALIZATION NUMBER

ASTM D 664 and D 974

Depending on its source, additive content, refining procedure, or deterioration in service, a petroleum oil may exhibit certain acid or alkaline (base) characteristics. Data on the nature and extent of these characteristics may be derived from the product's neutralization number—or "neut number," as it is commonly known. The two principal methods for evaluating neut number are ASTM D 664 and ASTM D 974. Although respective test results are similar, they are not identical, and any reporting of results should include the method by which they are obtained.

Acidity and Alkalinity

Acidity and alkalinity are terms related to dissociation, a phenomenon of aqueous solutions. Dissociation is a form of ionization, the natural breaking up of some of the molecules into positive and negative ions. If the chemical composition of the aqueous solution is such that it yields more hydrogen ions (positive) than hydroxyl ions (negative), the solution is considered acid; an excess of hydroxyl ions on the other hand results in a solution that is considered to be basic or alkaline. The greater the excess, the more acid or alkaline the solution, as the case may be. If the hydrogen and hydroxyl ions are in equal concentration, the solution is—by definition—neutral.

Titration

Since acidity and alkalinity are opposing characteristics, an acid solution can be neutralized (or even made alkaline) by the addition of a base. The converse is also true. In either case, neutralization can be accomplished by titration, the gradual addition of a reagent until a specified end point is reached. The amount of acid or base materials in a solution can thus be measured in terms of the quantity of added reagent. Being non-aqueous, however, petroleum oils cannot truly be said to be acid or alkaline. Nevertheless, they can be modified to exhibit these properties by addition of water—plus alcohol to extract oil-soluble acid or alkaline compounds from the sample, and to dissolve them in the water. This principle is utilized in the determination of neutralization number.

pH

Actual acidity or alkalinity, on the other hand, can be expressed in accordance with the pH scale, where zero represents maximum acidity, 14 maximum alkalinity, and 7 neutrality. The pH value of a solution can be determined electrolytically. When two electrodes of different materials are immersed in the solution, a small electric potential (voltage) is generated between them, and the magnitude and polarity of this potential can be related directly to pH value.

Potentiometric Method

The potentiometric method for determining neut number (ASTM D 664) is based on the electrolytic principle, pH, as indicated potentiometrically, is recorded against added reagent. If the initial pH reading of the specially prepared sample lies between 4 and 11 (approximately), the sample may contain weak acids, weak bases, or an equilibrium combination of the two. It may be titrated to one end point with base to yield a total acid number, and then may be titrated to another end point with acid to yield a total base number.

If, on the other hand, the initial pH reading lies below 4 (approximately), the sample may be titrated with base up to this point to yield a strong acid number. It may also be titrated up to 11 (approximately) to yield a total acid number. Similarly, a sample whose initial pH reading lies above 11 (approximately) can be titrated with acid down to this value to yield a strong base number, and it can be titrated down to 4 (approximately) to yield a total base number.

End Points

Titration end points are not at fixed pH readings but at inflections that occur in the curve: reagent versus pH. Whether or not an end point represents a strictly neutral condition is of little significance. With test procedure carefully standardized, the results obtained in reaching an end point can be compared on an equal basis with other results obtained in the same way. A result reported simply as "neut number," moreover, may be assumed to be a total acid number. Although it is not provided for by ASTM procedure, the initial pH reading may also be reported.

Colorimetric Method

Under the colorimetric method for determining neut number (ASTM D 974), end point is identified by the change of a color indicator. This indicator exhibits one color above a specified pH value, another below. By this means, a total acid or strong base number can be determined with a *p*-naphtholbenzene indicator, while a strong acid number can be determined with a methyl orange indicator. Obviously, however, this method is not suitable for the investigation of dark-colored liquids.

Reporting the Results

Whatever the method, all acid numbers are expressed in milligrams of potassium hydroxide (KOH)—a base—required to "neutralize" a gram of sample. For reasons of uniformity, base numbers, which are obtained by titrating with hydrochloric acid (HCl), are expressed in the same units, the HCl being converted to the number of KOH units that it would neutralize.

Significance of Results

Because acidity is associated with corrosiveness, there has been a tendency to attribute undesirable properties to an oil that exhibits a high acid number or a low pH reading. This attitude is fostered by the fact that deterioration of an oil in service—oxidation—is ordinarily accompanied by an increase in acid test results. While this attitude is not in actual disagreement with fact, its oversimplification may be conducive to harmful misconceptions.

In the first place, petroleum oil is not an aqueous solution, and conventional interpretations of acidity and alkalinity do not apply. In the second place, the test results, while involving certain acid or alkaline implications, do not distinguish between those that are undesirable and those that are not. The ASTM Standards themselves include the statement that the test "method is not intended to measure an absolute acidic or basic property that can be used to predict performance of an oil under service conditions. No general relationship between bearing corrosion and acid or base number is known."

This is not to say, however, that neut number or pH reading are without significance. They are applied widely and effectively to turbine oils, insulating oils, and many other oils in critical service. With new oils, neutralization test results provide a useful check on consistency of product quality. With used oils, they may serve as a guide to mechanical condition, change in operating conditions, or product deterioration. A rise in acid number and/or a drop in base number or pH reading are generally indicative of increasing oxidation. They may also be related to depletion of an additive, many of which are alkaline.

It is impossible, however, to generalize about the limits to which the neutralization values of an oil in service may safely be allowed to go. Each combination of oil, machine, and type of service follows a pattern of its own. Only through experience with a particular set of conditions can it be determined at what neutralization value an oil is no longer suitable for service.

OCTANE NUMBER

ASTM D 2699 and D 2700

The octane number of a gasoline is a measure of its antiknock quality; that is, its ability to burn without causing the audible "knock" or "ping" in spark-ignition engines. While octane number is a common term, it is also widely misunderstood, primarily because there are several different methods of measuring this property. Motor Octane Number, Research Octane Number, and Road Octane Number are the three basic procedures. Each assesses antiknock quality of a given fuel under a particular set of conditions.

Octane Number in the Laboratory

In principle, the octane number of a fuel is a numerical expression of its tendency to prevent engine knock relative to a standard fuel. In the laboratory, this quantity is determined through use of the ASTM engine, a special, single-cylinder engine whose operating characteristics can be varied.

The fuel to be rated is first burned in the engine, and the air-fuel mixture is adjusted to produce *maximum knock,* which is measured by a sensing device known as a *knockmeter.* Next, the compression ratio of the engine is varied until a knock intensity of 55 is recorded on the knockmeter. The knock intensity of the test fuel under these conditions is then compared (by referring to charts) to the knock intensities of various reference fuels. The reference fuels are normally blends of two hydrocarbons—iso-octane, which resists knocking, and normal heptane, which knocks severely. Iso-octane is arbitrarily assigned an octane number of 100, while heptane is rated as zero.

The percentage, by volume, of iso-octane in the blend that matches the characteristics of the test fuel is designated as the Octane Number of the fuel. For example, if a blend of 90% iso-octane and 10% heptane will match the knock intensities of the "unknown" fuel, under the same conditions, the fuel would be assigned an octane number of 90. (For fuels having octane numbers above 100, the gasoline under test is compared with blends of iso-octane and tetraethyl lead, an effective antiknock agent. Such blends can have octane numbers considerably above 100.)

The general test procedure outlined above is the basis for two distinct laboratory methods of determining octane number, *Motor Octane Number* and *Research Octane Number.* Motor Octane Number, ASTM D 2700, is the name given to the octane rating as determined with the ASTM engine and a standard set of operating conditions that became widely known during the 1930's. Research Octane Number, ASTM D 2699, is a more recent method, and is determined under another set of conditions, the chief difference being the slower engine speed. The research method is therefore less severe than the motor method, and most gasolines have a higher octane number by the research method.

Road Octane Number

Laboratory octane ratings do not always provide an accurate prediction of how a fuel will behave in an automobile engine. A more reliable means of predicting antiknock quality is to test the gasolines in automobiles under varying condition of speed and load. There are several methods of determining this rating, which is known as *Road Octane Number;* each method compares the test fuel with various blends of iso-octane and heptane.

The Uniontown Procedure, one of the most common Road Octane methods, records the knock intensity at various speeds during acceleration. Knock ratings are recorded, either with instruments or by the human ear. The procedure is repeated, using various blends of iso-octane and heptane, until a reference fuel that produces the same knock characteristics is found. The test fuel is then assigned the same octane number as the reference blend.

The Modified Uniontown Procedure, another common method, depends on the human ear to establish where "trace knock" first occurs. A series of test runs, using various reference fuels of known octane numbers, is first made. For each blend, the spark advance setting that produces trace knock is determined, and the various settings are plotted into a curve. The Road Octane Number of the test fuel can then be assessed by referring to the curve to determine the octane number associated with the spark advance setting that produced trace knock with the test fuel.

Aviation Gasoline Knock Rating

The antiknock level of aviation gasoline is indicated by composite grade numbers, i.e., 80/87, 100/130, 115/145. In each case, the first number is the knock rating determined under conditions of lean air-fuel ratio by ASTM method D 614, while the second number is the rating under the supercharged-rich method, ASTM D 909. Values above 100 are expressed as Performance Numbers, which are related to the number of milliliters of tetraethyl lead in iso-octane.

Significance

Motor Octane Number is normally taken as an indication of a fuel's ability to prevent knocking at high engine speeds, while Research Octane Number measures low-speed knocking tendencies. It is the Road Octane Number, however, that an automobile engine will actually "see" in a given fuel with regard to knock characteristics.

The amount of technical literature devoted to octane numbers is immense, and many correlations exist among the three methods of determining octane numbers which, in the hands of experts, can be meaningful. For the motorist, however, the Road Octane Number of a gasoline offers the most practical prediction of whether the fuel is going to knock in his engine under the conditions to which the car is subjected.

OIL CONTENT OF PETROLEUM WAX
ASTM D 721

A major step in wax refining is the removal of oil; fully refined paraffin waxes usually contain less than 0.5% oil. Therefore, a measure of the oil content of a wax is also an indirect measure of the degree of refinement, and is a useful indicator of wax quality.

The ASTM D 721 test method is based on the low-temperature insolubility of wax in methyl ethyl ketone. A sample of the wax is dissolved in the solvent under heat, the solution is cooled to precipitate the wax and then filtered. The oil content of the filtrate is determined by evaporating the methyl ethyl ketone and weighing the residue. By definition, the oil content of a wax is that portion which is soluble in methyl ethyl ketone at –25°F.

Significance of Oil Content

The oil content of a petroleum wax is a criterion of purity and degree of refinement. Highly refined petroleum waxes have high purity and low oil content. This renders them suitable for many applications in the manufacture of drugs, other pharmaceuticals, and food packages. Crude scale waxes are not so highly refined and consequently contain more oil. Such waxes are suitable for applications where some odor or taste can be tolerated, and where higher oil content is permitted.

Oil content of microcrystalline wax can also be determined by ASTM D 721.

OIL SEPARATION IN GREASE STORAGE
ASTM D 1742

ASTM Method D 1742, "Oil Separation From Lubricating Grease During Storage," provides an indicator of the tendency of greases to separate oil while in containers in storage. The separation of a few ounces of oil at the top of a 35-lb container of grease may represent less than 1% of the total oil in the grease and is not detrimental to the performance of the product. However, this may produce housekeeping problems as well as cause loss of the user's confidence.

Significance

ASTM states that the test correlates directly with the oil separation that occurs in 35-pound grease pails in storage. The test is also indicative of the separation that may occur in other sizes of containers. This method is not suitable for greases softer than NLGI No. 1 consistency and is not intended to predict the bleeding tendencies of grease under dynamic service conditions.

Due to improved grease technology the problem of grease separation in containers rarely occurs today. Therefore, the relevance of this test to the service performance of modern greases is questionable. The test is primarily of value as a means of assuring batch-to-batch uniformity.

OXIDATION STABILITY—OILS
ASTM D 943

Oxidation is a form of deterioration to which all oils in service are exposed. It is a chemical reaction that occurs between portions of the oil and whatever oxygen may be present—usually the oxygen in the atmosphere. The oxidation of lubricating oils is accelerated by high temperatures, catalysts (such as copper), and the presence of water, acids, or solid contaminants. The rate of oxidation increases with time.

Oxidation tends to raise the viscosity of an oil. The products of oxidation are acid materials that lead to depositing of soft sludges or hard, varnish-like coatings. Paraffinic oils characteristically have greater oxidation resistance than naphthenic oils, although naphthenic oils are less likely to leave hard deposits. Whatever the net effect of oxidation, it is undesirable in any oil that lubricates on a long-term basis. Much has been done to improve oxidation resistance by the use of selected base stocks, special refining methods, and oxidation inhibitors. As might be expected, moreover, a great deal of study has been devoted to the means by which oxidation resistance of an oil may be evaluated.

A number of oxidation tests are in use. Some may be better related to a particular type of lubrication service than others. All are intended to simulate service conditions on an accelerated basis. At an elevated temperature, an oil sample is exposed to oxygen or air—and sometimes to water or catalysts—usually iron and/or copper. All of these factors make oxidation more rapid. Results are expressed in terms of the time required to produce a specified effect, the amount of sludge produced or oxygen consumed during a specified period.

One of the more common methods of examining steam turbine oils is the ASTM method D 943. This test is based on the time required for the development of a certain degree of oxidation under accelerated conditions; the greater the time, the higher the oil's rating. Here, oxidation is determined by an increase in the oil's acidity, a property measured by its acid neutralization number. (See discussion on *"Neutralization Number."*)

Significance of Results

Oxidation stability is an important factor in the prediction of an oil's performance. Without adequate oxidation stability, the service life of an oil may be extremely limited. Unless the oil is constantly replaced, there is a serious possibility of damage to lubricated parts. Acids formed by oxidation may be corrosive to metals with which the oil comes in contact. Sludges may become deposited on sliding surfaces, causing them to stick or wear; or they may plug oil screens or oil passages.

Oxidation stability is a prime requisite of oils serving in closed lubrication systems, where the oil is recirculated for extended periods. The higher the operating temperature, the greater the need for oxidation stability, especially if water, catalytic metals, or dirt are present. Resistance to oxidation is of special importance in a steam-turbine oil because of the serious consequences of turbine bearing failure. Gear oils, electric transformer oils, hydraulic fluids, heat-transfer oils, and many crankcase oils also require a high degree of oxidation stability.

Obviously, the ability to predict oxidation life by a test, and to do it with reasonable accuracy is highly desirable. There are certain factors, however, that make reliable test results difficult to obtain. In the first place, the tests themselves are very time-consuming; a method such as ASTM D 913 may require the better part of a year to complete. Prolonged though the test may be, moreover, its duration usually represents but a small fraction of the service life of the oil under investigation. A steam turbine oil, for example, may well last for a decade or more without serious deterioration. It is impossible to reproduce service conditions of this sort in the laboratory with a test even of several hundred hours' duration. And, in addition to the time factor, there are many other operational variables that cannot be duplicated under test conditions. Results can be distorted also by the presence of certain additives in the oil.

For these reasons, the correlation between oxidation test results and field experience leaves much to be desired. Test results are subject only to broad interpretations. It would be difficult to show, for example, that an oil with a 3000-hour ASTM test life gives better service than an oil with a 2500-hour test life. In evaluating the oxidation stability of an oil, primary consideration should be given to the record that it has established over the years in the type of service for which it is to be used.

OXIDATION STABILITY—GREASES

ASTM D 942 • 1P142, D 1402, and D 1261

The bomb oxidation test was developed in 1938 by the Norma-Hoffman Bearing Corporation. Its purpose was to evaluate the oxidation stability of a grease during the storage of machine parts to which it had been applied. It was not intended to predict either the stability of greases in service or their shelf life in commercial containers.

Method of Evaluation

Oxidation is a form of chemical deterioration to which no petroleum product is immune. Petroleum products vary appreciably in their resistance to oxidation, a property that can be evaluated in many ways for many purposes. In the case at hand, evaluation is related to the quantity of oxygen that reacts with a grease sample during a specified period under standard conditions. The oxidation rate is plotted as pressure drop vs. time, Figure 3-2.

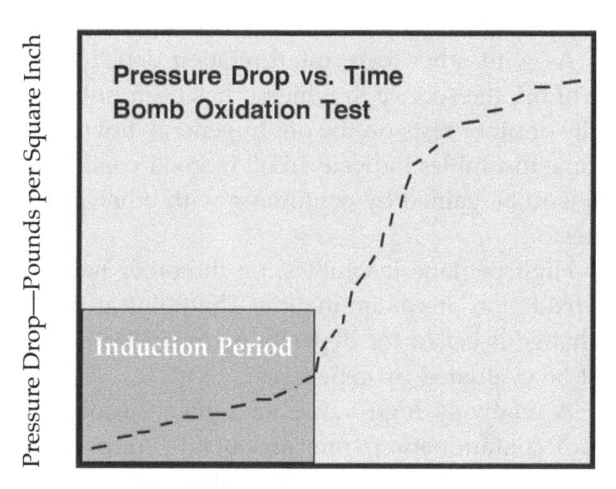

Figure 3-2. Bomb Oxidation Test—When pressure drop is plotted against time, the resulting curve will indicate a period of comparatively slow oxidation followed by a pronounced rise. The relatively flat portion at the beginning represents what is known as the "induction period," a phase during which oxidation is not ordinarily of serious magnitude. For practical reasons, it is not customary to continue the test beyond the induction period, its end being indicated by a sudden rise. Should the test be carried further, however, this rise would eventually taper off again as oxidation becomes complete. In some cases, test results have been expressed in terms of the duration of the induction period.

ASTM D 942 • 1P142

Procedure

Four grams of the grease to be tested are placed in each of five Pyrex dishes. These samples are then sealed and pressurized with oxygen (at 110 psi) in a heated bomb (210°F). The pressure is observed and recorded at stated intervals. The decrease in oxygen pressure determines the degree of oxidation after a given time period.

Significance of Results

A relationship exists between the pressure lost during this time and the amount of oxygen that has entered into chemical reaction with the grease. However, the drop in pressure is the *net* change resulting from absorption of oxygen and the release of gaseous products by the grease. Thus, this is a basic weakness of the test, since a grease that is being oxidized and at the same time is releasing gaseous products would appear to have greater oxidation resistance than is actually the case. This is a static test and is not intended to predict the stability of grease under dynamic conditions. Nor does it reflect oxidizing influence on bulk quantities in the original container. It more closely represents the conditions in a thin film of grease, as on pre-lubricated bearings or machine parts subjected to extended storage.

Certain machine parts are stored after an application of lubricating grease has been applied. This is particularly true of lubricated-for-life anti-friction bearings, which are greased by the manufacturer and then sealed. It is a common practice to make up these parts in advance and then stockpile them against future requirements.

There is hardly need to point out the damage that can be inflicted by a grease that deteriorates rapidly during this type of storage. The acidity associated with grease oxidation is corrosive to the highly sensitive bearing surfaces, and oxidation deposits may bind the bearing's action even before it has been put in operation. At best, a grease that has undergone significant deterioration in storage can hardly be in a condition to yield the long service life expected of it.

ASTM D 1402

Procedure

This test is run the same as ASTM D 942 is except that prepared copper strips are immersed on edge in each grease sample. Pressure readings are taken at 2-hour intervals over the duration of the test—until the pressure drops to 55 psi or for a specified time period if the pressure hasn't dropped to 55 psi during this time.

Significance of Results

The same limitations exist with this test as with ASTM D 942 since the determination of oxygen absorption rate as an indication of oxidation reaction is affected by the release of gaseous products from the grease.

Results from this test indicate the catalytic effect of copper and its alloys) in accelerating the oxidation of greases under static conditions. The results are not applicable to greases under dynamic conditions or when stored in commercial containers.

ASTM D 1261

Procedure

Each of two Pyrex dishes is filled with 4 grams of grease and a prepared copper strip is partially immersed in each grease sample. The procedure from this point on is the same as in ASTM D 942. At the end of the test time—24 hours—the copper strips are removed, washed, and examined for evidence of discoloration, etching, or corrosion.

Significance of Results

The effect of grease on copper parts of bearing assemblies with which the grease comes in contact is determined from the results of this test. In addition, some indication of the storage stability of greases which are in contact with copper may be found by visual inspection of the grease at the end of the test. The results do not apply to greases in contact with copper under dynamic service conditions.

In spite of the aforementioned limitations, these tests do have significant value. Many concerns find that the bomb test serves as an accurate check on uniformity of grease composition. Though test results may mean little by themselves, they are highly reproducible and highly repeatable. Results that are consistent from batch to batch give a good indication of product uniformity.

PENETRATION

(See "Grease Consistency.")

PENTANE AND TOLUENE INSOLUBLES

When a used oil is diluted sufficiently with pentane, certain oxidation resins that it normally holds in solution are precipitated out. In addition, the dilution helps to settle out materials suspended in the oil. Among the latter are insoluble oxidation resins and extraneous matter such as dirt, soot, and wear metals. All of the contaminants that can be separated from the oil by precipitation and settling are referred to as *pentane insolubles.*

The pentane insolubles may then be treated with a toluene solution which dissolves the oxidation resins. The extraneous matter left behind is called the *toluene insolubles.* The difference between the pentane insolubles and the toluene insolubles represents the quantity of oxidation resins in the oil. This is termed the *insoluble resins,* meaning insoluble in pentane.

Toluene has replaced benzene as the aromatic solvent in ASTM D 893 because of concern about the potential toxicity of benzene. Insoluble sludges are generally similar with the two solvents.

With detergent engine oils, a pentane-coagulant solution is customarily used instead of pentane. This precipitates material held in suspension by the detergent-dispersant which would not otherwise separate out.

As with other tests, interpretation depends on the type of oil, the service to which it has been put, and the results of other tests on the oil. In general, however, low pentane insolubles indicate an oil in good condition and little is to be gained by continuing with other phases of the test.

High pentane insolubles, on the other hand, indicate oxidation or contamination. The point at which an oil change is called for depends on many factors which must be evaluated by experience.

A relatively high value for toluene insolubles indicates contamination from an outside source such as soot from partially burned fuel; atmospheric dirt, the result of inadequate air filtration; tiny particles of metal produced by extreme wear. Emission spectrometry is often used to reveal the makeup of metal contamination: excessive lead, copper, or silver implies bearing wear; aluminum, piston wear; silicone, atmospheric dirt.

High insoluble resins mean a highly oxidized oil, which may result from excessive engine temperatures, contamination, an unsuitable oil, or excessive crankcase dilution. Loose-fitting piston rings, faulty injection, or low-temperature operation may allow raw fuel to enter the crankcase, where its oxidation adds to the amount of insoluble resins.

POUR POINT AND CLOUD POINT

ASTM D 97

It is often necessary to know how cold a particular petroleum oil can become before it loses its fluid characteristics. This information may be of considerable importance, for wide variations exist in this respect between different oils—even between oils of comparable viscosity.

If a lubricating oil is chilled sufficiently, it eventually reaches a temperature at which it will no longer flow under the influence of gravity. This condition may be brought about either by the thickening of the oil that always accompanies a reduction in temperature, or by the crystallization of waxy materials that are contained in the oil and that restrain the flow of the fluid portions. For many applications, an oil that does not flow of its own accord at low temperatures will not provide satisfactory lubrication. The extent to which an oil can be safely chilled is indicated by its *pour point,* the lowest temperature at which the undisturbed oil can be poured from a container.

The behavior of an oil at low temperature depends upon the type of crude from which it is refined, the method of refining, and the presence of additives. Paraffinic base stocks contain waxy components that remain completely in solution at ordinary temperatures. When the temperature drops, however, these waxy components start to crystallize, and they become fully crystallized at a temperature slightly below the pour point. At this temperature, the undisturbed oil will not generally flow under the influence of gravity.

Crystallization of the waxy component does not mean that the oil is actually solidified; flow is prevented by the crystalline structure. If this structure is ruptured by agitation, the oil will proceed to flow, even though its temperature remains somewhat below the pour point.

An oil that is predominantly naphthenic, on the other hand, reacts in a somewhat different manner. In addition to having a comparatively low wax content, a naphthenic oil thickens more than a paraffinic oil of comparable viscosity does when it is cooled. For these reasons, its pour point may be determined by the actual congealing of the entire body of oil instead of by the formation of waxy crystals. In such a case, agitation has little effect upon fluidity, unless it raises the temperature.

The pour point of a paraffinic oil may be lowered substantially by a refining process that removes the waxy component. For many lubricating oils, however, these components impart advantages in viscosity index and oxidation stability. Good performance generally establishes a limit beyond which the removal of these waxy component is inadvisable. It is possible, nevertheless, to lower the pour point of a paraffinic oil by the introduction of a pour depressant. Such an additive appears to stunt the growth of the individual crystals so that they offer less restriction to the fluid portions of the oil. It is hardly necessary to point out, however, that a pour depressant, as such, can have little, if any, effect upon a naphthenic oil.

Cloud point is the temperature, somewhat above the pour point, at which wax crystal formation gives the oil a cloudy appearance. Not all oils exhibit a cloud point and, although this property is related to pour point, it has little significance for lubricating oils. It is significant, however, for distillate fuels, and it is measured by ASTM D 2500.

Significance

The pour point of an oil is related to its ability to start lubricating when a cold machine is put in operation. Agitation by the pump will rupture any crystalline structure that may have formed, if the oil is not actually congealed, thereby restoring fluidity. But oil is usually supplied to the pump by gravity, and it can not be expected to reach the pump under these conditions, if the temperature is below the pour point. Passenger car engines and many machines that are stopped and started under low-temperature conditions require an oil that will flow readily when cold.

What is true of circulating lubrication systems, moreover, is equally true of gravity-feed oilers and hydraulic systems. A low pour point oil helps to provide full lubrication when the equipment is started and is easier to handle in cold weather. Low pour point is especially desirable in a transformer oil, which must circulate under all temperature conditions. The control of large aircraft is dependent upon hydraulic oils that must remain fluid after being exposed to extreme temperature drops. For these and similar applications, pour point is a very important consideration.

If the temperature of an oil does not drop below its pour point, the oil can be expected to flow without difficulty. It sometimes happens, however, that oil is stored for long periods at temperatures below the pour point. In some cases, the waxy crystalline structure that may be formed under these circumstances will not melt and redissolve when the temperature of the oil is raised back to the pour point. To pour the oil under these conditions, it is necessary to put the waxy crystals back in solution by heating the oil well above its pour point.

POWER FACTOR

ASTM D 924

Petroleum oils serve extensively as dielectrics for electrical power transmission equipment. Their primary functions are the cooling of coils (by circulation) and the prevention of arcing between conductors of high potential difference. In serving these purposes, any dielectric tends to introduce a degree of *dielectric loss,* a form of leakage equivalent to a flow of current through the dielectric from one conductor (wire or cable) to another. It is a leakage peculiar to a-c circuits. Though the loss associated with insulating oils is ordinarily a minor consideration, it could, under unusual conditions, assume a significant magnitude. In such a case, it would not only reduce the efficiency of the unit, but could cause a harmful rise in the unit's temperature.

Dielectric loss depends, among other things, on the nature and magnitude of the insulation's *impedance,* its opposition to the flow of alternating current through it. This is the current that is related to dielectric loss, and it increases as the impedance decreases. Only a portion of this current, however, is directly involved: a component equivalent to an *active* current. In a given a-c circuit, the loss is directly proportional to this active current. The ratio of active to *total* alternating current may vary—theoretically—from one to zero. This ratio is known as the *power factor* of the dielectric, and it can be considered to be an inherent dielectric property. Because of its effect on dielectric loss, the power factor of the dielectric should be as low as possible.

Though the power factor of an insulating oil is defined by the same mathematical expression as that of an a-c circuit, the two concepts should not be confused The overall power factor of a power-circuit affect *line* losses, rather than local dielectric losses, and the reduction of line losses requires a *high* power factor for the circuit.

Significance of Test Results

In a-c transmission cables, conductors of opposite polarity may extend for long distances in close proximity to each other. There is abundant opportunity for dielectric loss associated with the insulating material between the conductors. The higher the power factor of the insulation, the greater this loss will be.

For other applications, as in the insulation of transformers, dielectric loss is not appreciable, and small differences in power factor have little significance. A high-quality oil that is free of contamination can be expected to exhibit the low power factor that good performance requires.

In the evaluation of a used insulating oil's condition, however, power factor may be more meaningful. Here, the principal criterion is freedom from water and oxidation products—water that promotes the tendency to arc and oxidation sludges that interfere with cooling. Oxidation of the oil and contamination with water, dirt, or carbonized particles cause the power factor to rise.

Many engineers consider power factor to be a highly sensitive index of the oil's deterioration. If sufficient data on the performance of a particular oil in a particular service is available, it is possible to relate increases in power factor to degradation of the oil. In this way, power factor tests on a used oil may be helpful in estimating its remaining service life.

Tests for power factor frequently serve a useful purpose in the refinery as a check on uniformity of product quality. Consistent test values are indicative of consistent performance characteristics.

REFRACTIVE INDEX
ASIM D 1218

Uniformity of composition of highly refined petroleum products is of importance, especially in process applications such as those involving solvents or rubber process oils. Refractive index is one test often used either alone or in combination with other physical tests as an indication of uniformity.

Refractive index is the ratio of the velocity of light of a specified wavelength in air to its velocity in a substance under examination. When light is passed through different petroleum liquids, for example, the velocity of light will be different in each liquid. Several sources of light of constant wavelength are available, but the yellow D line of sodium (5893 Å) is the one most commonly used in this test. Since the numerical value of the refractive index of a liquid varies with wavelength and temperature, it must be reported along with the wavelength and temperature at which the test was run.

This test is intended for transparent and light-colored hydrocarbon liquids having refractive indices between 1.33 and 1.50. The method is capable of measuring refractive index with a reproducibility of ± 0.00006. It is generally not this accurate with liquids having ASTM colors darker than 4 (ASTM D 1500), or with liquids that are so volatile at the test temperature that a reading cannot be obtained before evaporation starts.

Significance of Results

The refractive index is easily measured and possesses good repeatability and reproducibility. It is sensitive to composition. This makes it an excellent spot test for uniformity of composition of solvents, rubber process oils, and other petroleum products. A general rule for petroleum products of equivalent molecular weight is that paraffins have relatively low refractive indices (approximately 1.37), aromatics have relatively high indices (approximately 1.50), and naphthalenes have intermediate indices (approximately 1.44).

Refractive index may be used in combination with other simple tests to estimate the distribution of carbon atom types in a process oil. Empirical refractive index charts relating viscosity, specific gravity, and refractive index have been prepared, and they make it possible to estimate the percent naphthenic, aromatic, and paraffinic carbon atoms present. This is an inexpensive and quick set of tests to run, in contrast to the more time-consuming clay/silica gel analysis, which is also used to determine hydrocarbon composition directly.

ROTARY BOMB OXIDATION TEST (RBOT)
ASTM D 2272

Oxidation is a form of chemical deterioration to which petroleum products are subject. Even though oxidation takes place at moderate temperatures, the reaction accelerates significantly at temperatures above 200°F.

In addition to the effect of high temperatures, oxidation may also be speeded by catalysts (such as copper) and the presence of water, acids, or, solid contaminants. Moreover, the peroxides that are the initial products of oxidation are themselves oxidizing agents, so that oxidation is a chain-reaction—the further it progresses the more rapid it becomes.

Even though subject to oxidation, many oils (such as turbine oils) give years of service without need for replacement. Petroleum products can be formulated to meet service and storage life requirements by: (1) proper selection of crude oil type; (2) thorough refining, which removes the more-oxidation-susceptible materials; and (3) addition of oxidation inhibitors.

A number of oxidation tests are currently being used. Some may be better related to a particular type of service than others. All are intended to simulate service conditions on an accelerated basis. The most familiar method is the "Oxidation Characteristics of Inhibited Steam-Turbine Oils," ASTM D 943. The long time (over 1000 hours) required to run this test makes it impractical for plant control work. ASTM Method D 2272, "Continuity of Steam-Turbine Oil Oxidation Stability by Rotating Bomb," on the other hand, allows rapid evaluation of the resistance of lubricants to oxidation and sludge formation, using accelerated test conditions that involve high temperatures, high-pressure oxygen, and the presence of water and catalytically active metals.

The "rotary bomb oxidation test" (RBOT) does not replace ASTM D 943, but is intended primarily as an aid in quality control during the manufacture of long-life circulating oils.

Significance of Results

The ASTM D 2272 procedure allows relative oxidation life of a turbine oil to be determined rapidly. Results are obtained by the RBOT test up to 1000 times faster than by the D 943 method. This speed makes the test practical for use as a product quality control measure, permitting decisions to be made within a matter of a few hours. The RBOT is also distinguished among oxidation tests by its good repeatability and reproducibility.

It should be remembered that the test is essentially a quality control device and no direct correlation has been established with other oxidation tests currently being used. For two oils of similar composition—both base stock and additive package—the RBOT test can be used as an indication of their relative oxidation stability.

RUST-PREVENTIVE CHARACTERISTICS
ASTM D 665

This test method was originally designed to indicate the ability of steam-turbine oils to prevent the rusting of ferrous parts, should water become mixed with the oil. While still used for this purpose, its application is now often extended to serve as an indication of rust preventive properties of other types of oils, particularly those used in circulating systems. It is a dynamic test, designed to simulate most of the conditions of actual operation.

In the method, a standard steel specimen is immersed in a mixture of the test oil and water under standard conditions and with constant stirring. At the end of a specified period, the steel specimen is examined for rust. Depending on the appearance of the specimen, the oil is rated as passing or failing.

Degrees of Rusting

An indication of the degree of rusting occurring in this test is sometimes desired. For such cases, the following classification is recommended:

LIGHT RUSTING—Rusting confined to not more than six spots, each of which is 1 mm or less in diameter.

MODERATE RUSTING—Rusting in excess of the preceding, but confined to less than 5% of the surface of the specimen.

SEVERE RUSTING—Rust covering more than 5 % of the surface of the specimen.

Reporting Results

Results obtained with a given oil are reported as "pass" or "fail." Since the test may be conducted with either distilled water or with synthetic sea water, and for varying periods of time, reports of results should always specify these conditions. For example: "Rust Test, ASTM D 665, Procedure B, 24 Hours—Pass."

Significance

When the lubricating oil of a turbine or other system is contaminated with water, rusting can result. Particles of rust in the oil can act as catalysts that tend to increase the rate of oil oxidation. Rust particles are abrasive, and cause wear and scoring of critical parts. In addition, rust particles can add to other contaminants in a circulating system, increasing the tendency toward the clogging of low-clearance members, such as servo valves, and increasing the probability of filter plugging.

In many cases, the rusting characteristics of the system in service are better than is indicated by testing a sample of the used oil, because the polar rust inhibitor "plates out" on the metal surfaces (which are therefore adequately protected). The sample of oil, being somewhat depleted of the inhibitor, will then allow greater rusting in the test than would occur in service.

The relative ability of an oil to prevent rusting can become a critical property in many applications. As noted, this test method was originally applied exclusively to steam turbine oils. However, the test is now frequently applied to other oils in different types of applications, whenever undesirable water contamination is a possibility.

SAPONIFICATION NUMBER

Many lubricating oils are "compounded" with fatty materials to increase their film strengths or water-displacing qualities. The degree of compounding is indicated by the saponification number of the oil, usually called its *sap number.* Sap number is commonly determined by ASTM D 94 or D 939, methods based on the fact that these fatty materials can be saponified—that is, converted to soap—by reaction with a base (alkali), usually at an elevated temperature.

A specified quantity of potassium hydroxide (KOH) is added to the prepared oil sample and the mixture heated to bring about reaction. The excess KOH is titrated to neutralization with hydrochloric acid, either colorimetrically (D 94) or potentiometrically (D 939). The sap number is reported as milligrams of KOH assimilated per gram of oil.

Other factors being the same, a higher degree of compounding will result in a higher sap number. For a given degree of compounding, however, some fatty materials show a higher sap number than others.

Even with a new oil, therefore, sap number cannot be translated directly into percentage of fatty materials unless their exact nature is known.

Considerable experience with a particular set of conditions and types of oil is needed to properly interpret the sap number of a used oil. While loss or decomposition of fatty materials is reflected in a drop of the sap number, oxidation of the mineral oil base may cause the sap number to rise. Test results may be further distorted by acid or metallic contaminants picked up in service.

It is advisable, therefore, to consider sap number in relation to neut number. Comparison of the two indicates what portion of the sap number is due to the presence of fatty materials and what portion to acids in the oil.

Sap number has little relevance for oils for internal combustion engines.

TIMKEN EXTREME PRESSURE TESTS

ASTM D 2509—Lubricating Greases

ASTM D 2782—Lubricating Fluids

(See "Load Carrying Ability.")

USP/NF TESTS FOR WHITE MINERAL OILS

The *US Pharmacopeia* and the *National Formulary*, publications by two independent associations of physicians and pharmacologists, contain specifications for white mineral oils. The US Pharmacopeia (USP) specification covers the more viscous "Mineral Oil," which is used primarily as a pharmaceutical aid or levigating agent. The National Formulary (NF) sets specifications for the less viscous "Light Mineral Oil," which is used as a vehicle in drug formulations. Both compendia have legal status, being recognized in federal statutes, especially the Federal Food, Drug and Cosmetic Act.

White mineral oils have certain physical properties that distinguish them from other petroleum products. Both the USP and NF describe them as colorless, transparent, oily liquids free or nearly free from fluorescence. When cold they are odorless and tasteless and develop only a slight petroleum odor when heated. They are insoluble in water and alcohol, soluble in volatile oils, and miscible with most fixed oils with the exception of castor oil.

Significance of Results

These tests are designed to establish standards that assure that the oils involved are pure, chemically inert, and free from potentially carcinogenic materials. Oils that meet these standards find use, not only in pharmaceuticals and cosmetics, but also in chemical, plastics, and packaging applications where they are considered as direct or indirect food additives as defined by the Federal Food and Drug Administration (FDA).

UV ABSORBANCE

FDA Method

Petroleum product applications often extend into areas other than the obvious ones. One such area is the direct or indirect application of a petroleum product to food. A direct food additive is one that is incorporated, in small quantities, into or onto food meant for human or animal consumption. An example would be the use of white mineral oil to coat raw fruits to protect them or to coat animal feeds to reduce dustiness. An indirect additive is one that has only incidental contact with food, as through contact with a packaging material.

The use of petroleum products as food additives falls under the jurisdiction of the Food and Drug Administration. A major concern in the regulation of food additives of petroleum origin is the potential contamination of food by polynuclear aromatic hydrocarbons—some of which are considered to be carcinogenic. In an attempt to assure the absence of carcinogens, the FDA has sanctioned the use of ultraviolet (UV) absorbance as a test for monitoring the polynuclear aromatics content.

Ultraviolet absorbance is a measure of the relative amount of ultraviolet light absorbed by a substance. Types of compounds can be characterized by the wavelength range of UV light that they absorb. As the wavelength of ultraviolet light is varied, broad peaks of absorbance occur at the wavelengths that are characteristic of the compounds present. Most polynuclear aromatics have their principal absorbances at wavelengths between 280 and 400 millimicrons. Most carcinogens absorb UV light in this range, but not all materials with UV absorbance between 280 and 400 millimicrons are carcinogenic.

Significance of Results

These results are compared with the corresponding UV absorbance limits set by the FDA for the specific regulation that applies to each UV. When UV absorbance of a petroleum product falls within these limits, the product is considered acceptable for the particular food application involved. The UV method described here represents the simplest case. The method becomes more complex as the aromatic concentration of the oil increases.

The UV test is not the only criterion the FDA has established for food additives of petroleum origin. There are often additional requirements for boiling range, color, odor, method of manufacture, US Pharmacopeia quality, and non-volatile residues.

VAPOR PRESSURE

ASTM D 323

All liquids are disposed to vaporize—that is, to become gases. This tendency is a manifestation of the material's liquid vapor pressure, the pressure exerted by molecules at the liquid surface in their attempt to escape and to penetrate their environment. For a given liquid, this pressure is a function purely of temperature. The liquid vapor pressure of water at is boiling temperature—212°F—for example, is 147 psi, the pressure of the atmosphere. The more volatile the liquid, the higher the liquid vapor pressure at a specified temperature, and the faster the vaporization. In the same dry atmosphere and at the same liquid temperature, gasoline evaporates much more readily than heating oil.

For a given temperature, therefore, the vapor pressure of a liquid is a measure of is volatility. This applies only to vapor pressure exerted by a *liquid.* Pressures exerted by vapor disassociated from the liquid are functions of *volume,* as well as temperature, and they cover a wide range of values less directly related to volatility. As used in engineering circles, the term vapor pressure means *liquid* vapor pressure.

Unlike water, a petroleum product usually comprises many different fractions, each with a composition and a vapor pressure of its own. The vapor pressure of the product is therefore a composite value that reflects the combined effect of the individual vapor pressures of the different fractions in accordance with their mole ratios. It is thus possible for two wholly different products to exhibit the same vapor pressure at the same temperature—provided the cumulative pressures exerted by the fractions are also the same. A narrow-cut distillate, for example, may exhibit the same vapor pressure as that of a dumbbell blend, where the effect of the heavy fractions is counterbalanced by that of the lighter ones.

When a petroleum product evaporates, the tendency is for the more volatile fractions to be released first, leaving a material of lower vapor pressure and lower volatility behind. This accounts for the progressive rise in distillation-curve temperature, boiling point being related to volatility. Distillation, which is another measure of volatility, was described earlier.

Vapor pressure is commonly measured in accordance with the ASTM method D 323 (Reid vapor pressure), which evaluates the vapor pressures of gasoline and other volatile petroleum products at 100°F.

Significance of Test Results

Reid vapor pressure has a special significance for gasoline, which contains a portion of high-volatility fractions such as butane, pentane, etc. These fractions exert a major influence on vapor-pressure test results. A high vapor pressure is accordingly an indication of the presence of these high-volatility fractions—components required for satisfactory starting in cold weather. Without them, it would be difficult or impossible to vaporize gasoline in sufficient concentration to produce a combustible air-fuel mixture at low temperatures.

On the other hand, vapor pressure may be too high. An excess of high-volatility actions in hot weather can lead to vapor *lock,* preventing delivery of fuel to the carburetor. This is the result of the partial vacuum that exists at the suction end of the fuel pump and that, along with high temperatures, increases the tendency of the fuel to vaporize. If the fuel vapor pressure is too high, vapors formed in the suction line will interrupt the flow of liquid fuel to the pump, causing the engine to stall.

While Reid vapor pressure is the principal factor in determining both the vapor-lock and the cold-starting characteristics of a gasoline, they are not the only criteria. Distillation data, which defines the overall volatility of the fuel, must also be considered.

The higher the vapor pressures of automotive and aviation gasolines, solvents, and other volatile petroleum products, the greater the possibility of evaporation loss and the greater the fire hazard. Sealed containers for high-vapor-pressure products require stronger construction to withstand the high internal pressure. In the refinery, moreover, vapor pressure tests serve as a means of establishing and maintaining gasoline quality.

VISCOSITY

ASTM D 88, D 445, Redwood, and Engler

Viscosity is probably the most significant physical property of a petroleum lubricating oil. It is the measure of the oil's flow characteristics. The thicker the oil, the higher its viscosity, and the greater its resistance to flow. The mechanics of establishing a proper lubricating film depend largely upon viscosity.

To evaluate the viscosity of an oil numerically, any of several standard tests may be applied. Though these tests differ to a greater or lesser extent in detail, they are essentially the same in principle. They all measure the time required for a specified quantity of oil at a specified temperature to flow by gravity through an orifice or constriction of specified dimensions. The thicker the oil, the longer the time required for its passage.

Close control of oil temperature is important. The viscosity of any petroleum oil increases when the oil is cooled and diminishes when it is heated. For this same reason, the viscosity value of an oil must always be accompanied by the temperature at which the viscosity was determined. The viscosity value by itself is meaningless.

The two most common methods of testing the viscosity of a lubricating oil are the Saybolt and the kinematic. Of these, the Saybolt (ASTM D 88) is the method more frequently encountered in conjunction with lubricating oils. However, the kinematic method (ASTM D 445) is generally considered to be more precise. There are also the Redwood and the Engler methods, which are widely used in Europe, but only to a limited extent in the United States. Each test method requires its own apparatus—viscosimeter (or viscometer).

Significance of Results

Viscosity is often the first consideration in the selection of a lubricating oil. For most effective lubrication, viscosity must conform to the speed, load, and temperature conditions of the bearing or other lubricated part. Higher speeds, lower pressures, or lower temperatures require an oil of a lower viscosity grade. An oil that is heavier than necessary introduces excessive fluid friction and creates unnecessary drag.

Lower speeds, higher pressures, or higher temperatures, on the other hand, require an oil of a higher viscosity grade. An oil that is too light lacks the film strength necessary to carry the load and to give adequate protection to the wearing surfaces. For these reasons, viscosity tests play a major role in determining the lubricating properties of an oil.

In addition to the direct and more obvious conclusions to be drawn from the viscosity rating of an oil, however, certain information of an indirect sort is also available. Since, to begin with, the viscosity of the lube oil cut is determined by its distillation temperature, it is apparent that viscosity and volatility are related properties. In a general way, the lighter the oil, the greater its volatility—the more susceptible it is to evaporation. Under high-temperature operating conditions, therefore, the volatility of an oil, as indicated by its viscosity, should be taken into consideration.

Though the significance of viscosity test results has been considered from the standpoint of new oils, these tests also have a place in the evaluation of used oils. Oils drained from crankcases, circulating systems, or gear boxes are often analyzed to determine their fitness for further service or to diagnose defects in machine performance.

An increase in viscosity during service may often indicate oxidation of the oil. Oxidation of the oil molecule increases its size, thereby thickening the oil. When oxidation has progressed to the point of causing a material rise in viscosity, appreciable deterioration has taken place.

VISCOSITY CLASSIFICATIONS COMPARISON

There are four common systems for classifying the viscosities of lubricating oils. It is frequently desirable to compare a grade in one system with a grade in another system, but this is often impossible because the standards in the different systems are not based on viscosities at the same temperature. The charts presented in Table 3-4 and Figure 3-3 are designed to overcome this problem by comparing the systems on the basis of viscosities at a single temperature—100°F, which is the base temperature for the ASTM viscosity grade system.

In order to convert all viscosities to 100°F, it is necessary to assume appropriate viscosity indices (VI's) for the oils involved. (Viscosity index of an oil is a measure of its resistance to change in viscosity as temperature changes.) The VI's assumed here are:

110 VI for crankcase oils (SAE)
90 VI for automotive gear oils (SAE)

Table 3-4. Numerical relationships among viscosity classification systems.

ASTM Industrial Oils	Viscosity, SSU at 100°F	SAE Crankcase Oils	Viscosity, SSU at 100°F (assumes 110 VI)	AGMA Industrial Gear Oils	Viscosity SSU at 100°F	SAE Gear Oils	Viscosity at 100°F (assumes 90 VI)
32	33-34						
36(A)	36-38						
40	40-43						
50(A)	46-50	5W	140 Max.			75	211 Max.
60	55-62						
75	72-83						
105	97-116						
150	136-165	10W	140-220				
215	193-235	20	168-366	1	193-234		
315	284-347	20W	202-500	2,2EP	284-347	80	211-575
465	417-510	30	366-560	3,3EP	417-510		
700	625-764	40	560-812	4,4EP	626-765		
1000	917-1121	50	812-1272	5,5EP	916-1122	90	768-1844
1500	1334-1631	60(B)	1272-1561	6,6EP	1335-1632		
2150	1918-2344	70(B)	1561-2085	7EP,7 Comp.	1919-2346		
						140	1844-4345
3150	2835-3465			8EP,8 Comp.	2837-3467		
4650	4169-5095			8A Comp.	4171-5098		
						250	4345 Min.
7000	6253-7643						

Notes: (A) Currently is not considered a standardized grade in the USA. (B) Not officially recognized by SAE.

*ASTM/BSI (American Society for Testing and Materials/British Standards Institution) "Viscosity Classification System for Industrial Fluid Lubricants."

SAE (Society of Automotive Engineers) "Crankcase Oil Viscosity Classification" and "Transmission and Axle Lubricant Classification."

AGMA (American Gear Manufacturers Association) "Specification— Lubrication of Industrial Enclosed Gear Drives."

These values are representative for the products involved in the respective classifications. Close comparison should not be attempted if the VI of the product differs appreciably from the values used.

Table 3-4 shows the numerical relationships; Figure 3-3 shows the graphical equivalents.

VISCOSITY INDEX

ASTM D 567 and D 2270

Liquids have a tendency to thin out when heated and to thicken when cooled. However, this response of viscosity to temperature changes is more pronounced in some liquids than in others.

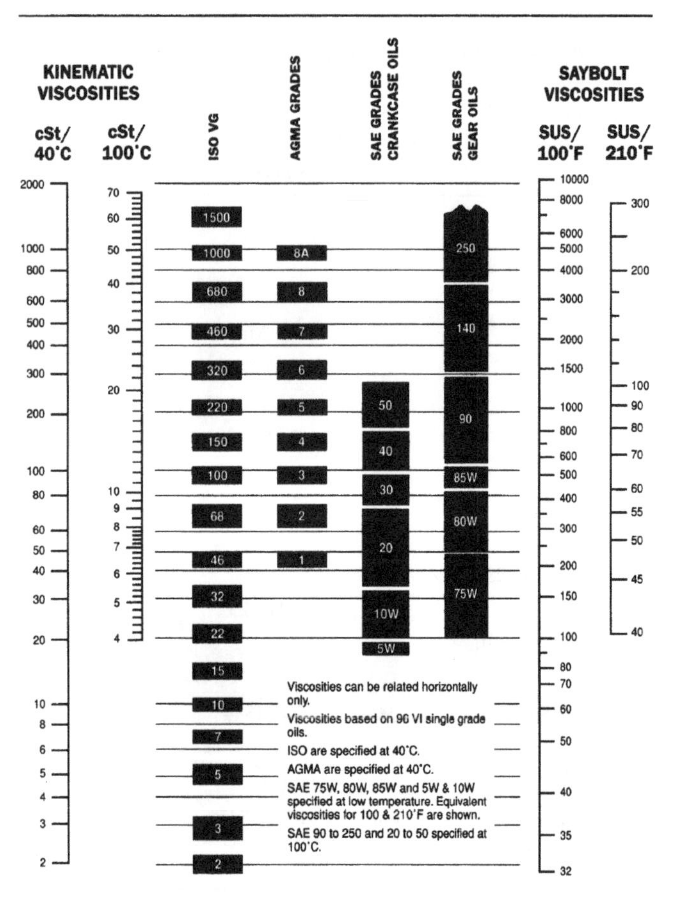

Figure 3-3. Viscosity classification equivalents. (See Appendix for enlarged illustration.)

Often, as with petroleum liquids, changes in viscosity can have marked effects upon the performance of a product, or upon its suitability for certain applications. The property of resisting changes in viscosity due to changes in temperature can be expressed as the viscosity index (V.I.). The viscosity index is an empirical, unitless number. The higher the V.I. of an oil, the less its viscosity changes with changes in temperature.

The Concept of Viscosity Index

One of the things that led to the development of a viscosity index was the early observation that, for oils of equal viscosities at a given temperature, a naphthenic oil thinned out more at higher temperatures than did a paraffinic oil. However, there existed no single parameter that could express this type of response to temperature changes.

The viscosity index system that was developed to do this was based upon comparison of the viscosity characteristics of an oil with those of so-called "standard" oils. A naphthenic oil in a series of grades with different viscosities at a given temperature, and whose viscosities changed a great deal with temperature, was arbitrarily assigned a V.I. of zero. A paraffinic series, whose viscosities changed less with temperature than most of the oils that were then available, was assigned a V.I. of 100. With accurate viscosity data on these two series of oils, the V.I. of any oil could be expressed as a percentage factor relating the viscosities at 100°F of the test oil, the zero-V.I. oil, and the 100-V.I. oil, all of which had the same viscosity at 210°F. This is illustrated by Figure 3-4 and is the basis for the formula,

$$\text{V.I.} = \frac{L - U}{L - H} \times 100$$

where L is the viscosity at 100°F of the zero-V.I. oil, H is the viscosity at 100°F of the 100-V.I. oil, and U is the viscosity at 100°F of the unknown (test) oil.

The ASTM Standards

This viscosity index system eventually became the ASTM standard D 567, which has been used for years in the petroleum industry.

ASTM D 567 is a satisfactory V.I. system for most petroleum products. However, for V.I.'s above about 125, mathematical inconsistencies arise which become more pronounced with higher V.I.'s. Because products with very high V.I.'s are becoming more common, a method (ASTM D 2270) that eliminates these inconsistencies has been developed.

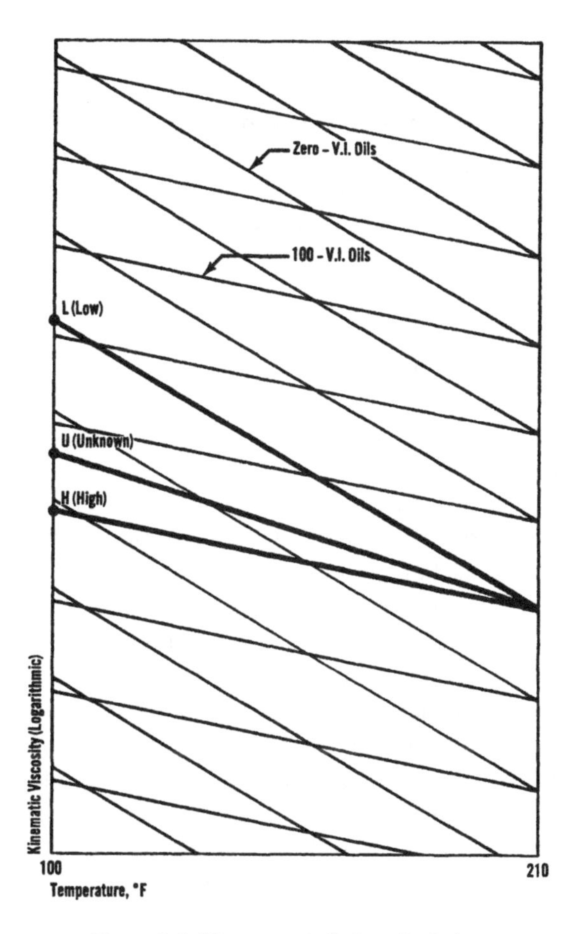

Figure 3-4. The concept of viscosity index.

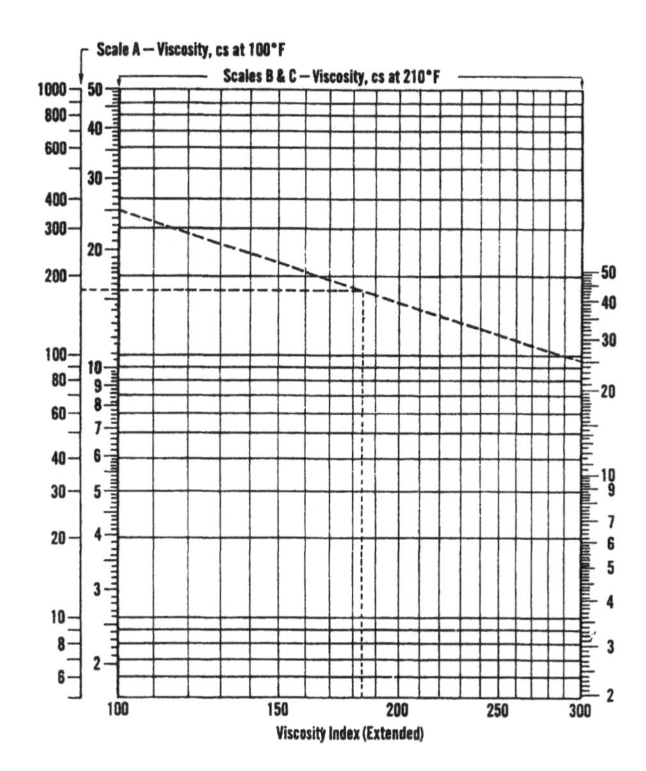

Figure 3-5. Chart for calculating V.I.'s above 100 from kinematic viscosity, based on ASTM D 2270. Dotted lines illustrate its use.

Calculating Viscosity Index

The viscosity index of an oil can be calculated from tables or charts included in the ASTM methods. For V.I.'s below 100, ASTM D 2270 and ASTM D 567 are identical, and either method may be used. For V.I.'s above 100, ASTM D 2270 should be used. Since ASTM D 2270 is suitable for all V.I.'s, it is the method now preferred by the leading petroleum companies.

The V.I. of an oil may also be determined with reasonable accuracy by means of special nomographs or charts developed from ASTM tables. A chart for V.I.'s above 100, as determined by ASTM D 2270, is shown in Figure 3-5.

Significance of Viscosity Index

Lubricating oils are subjected to wide ranges of temperatures in service. At high temperatures, the viscosity of an oil may drop to a point where the lubricating film is broken, resulting in metal-to-metal contact and severe wear. At the other extreme, the oil may become too viscous for proper circulation, or may set up such high viscous forces that proper operation of machinery is difficult. Consequently, many applications require an oil with a high-viscosity index.

In an automobile, for example, the crankcase oil must not be so thick at low starting temperatures as to impose excessive drag on the engine and to make cranking difficult. During the warm-up period, the oil must flow freely to provide full lubrication to all engine parts. After the oil has reached operating temperature, it must not thin out to the point where consumption is high or where the lubricating film can no longer carry its load.

Similarly, fluid in an aircraft hydraulic system may be exposed to temperatures of 100°F or more on the ground, and to temperatures well below zero at high altitudes. For proper operation under these varying conditions, the viscosity of the hydraulic fluid should remain relatively constant, which requires a high viscosity index.

As suggested by the relationship between naphthenic and paraffinic oils, the viscosity index of an oil can sometimes be taken as an indication of the type of base stock. A straight mineral oil with a high V.I.—80 or above—is probably paraffinic, while a V.I. below about 40 usually indicates a naphthenic base stock.

In general, however, this relationship between V.I. and type of base stock holds only for straight mineral oils. The refining techniques and the additives that are available today make it possible to produce naphthenic oils with many of the characteristics—including V.I.—of paraffinic oils. V.I., then, should be considered an indication of hydrocarbon composition only in the light of additional information.

WATER WASHOUT

ASIM D 1264

Lubricating greases are often used in applications that involve operation under wet conditions where water may enter the mechanism and mix with the grease. Therefore, the ability of a grease to resist washout becomes an important property in the maintenance of a satisfactory lubricating film, and tests for evaluating the effect of water on grease properties are of considerable interest.

Greases can be resistant to water in several ways. Some greases completely reject the admixture of water or may retain it only as occluded droplets with little change in structure. Unless these greases are adequately inhibited against rusting they may be unsuitable for lubrication under wet conditions since the "free" water could contact the metal surface and cause rusting.

Yet other greases that absorb water may be satisfactory under wet conditions. These types of grease absorb relatively large amounts of water by forming emulsions of water in oil. This absorption has little effect on the grease structure and leaves no "free" water to wet and rust the metal. Therefore, the grease continues to supply the proper lubrication while also acting as a rust preventive.

Other water-absorbing greases form thin fluid emulsions so that the grease structure is destroyed. These are useless for operation under wet conditions, and can be considered to have poor water resistance.

There are many effects that water has on grease, and no single test can cover them all. Many of the tests are useful tools; however, the results are subject to the personal judgment of the test operator and much skill is needed to interpret their meaning. ASTM Method D 1264, "Water Washout Characteristics of Lubricating Greases," is one method of evaluating this property.

Significance

Test results are useful for predicting the probable behavior of a grease in a shielded (not positively sealed) bearing exposed to the washing action of water. They are a measure of the solubility of a grease in water and give limited information on the effect of water on the grease structure. They say nothing about the rust preventive properties of the grease.

The test is a laboratory procedure and should not be considered equivalent to a service evaluation. Results on greases tested by this method may differ from service results because of differences in housing or seal design. Therefore a grease that proves unsatisfactory according to this test, may be satisfactory under service conditions if the housing or seal design is suitable.

WATER AND SEDIMENT

ASTM D 96, D 95, and D 473

Whether a petroleum fuel is burned in a boiler or in an engine, foreign matter in the fuel is undesirable. Excessive quantities of such impurities as water or solid contamination may interrupt the operation of the unit, and may also damage it.

The two most common impurities found in fuel oils are water and sediment, and several test procedures are available for measuring their concentrations. Water and sediment may be determined together by a centrifuge procedure. Water alone may be determined more accurately in most cases by distillation, and sediment alone may be determined with good accuracy by solvent extraction or by hot filtration.

The tests referred to are the following:

(1) ASTM D 96 Water and Sediment in Crude Oils
(2) ASTM D 95 Water in Petroleum Products and Other Bituminous Materials
(3) ASTM D 473 Sediment in Crude Petroleum Fuel Oil by Extraction

Significance of Results

Like many other tests, the determination of water and sediment gives results that must be interpreted in the light of a great deal of previous experience. It is obvious that large quantities of water and/or sediment can cause trouble in almost any application. However, different applications can tolerate different concentrations of impurities. In addition, the quantities of water and sediment determined by the various test procedures are not identical.

Therefore, for any particular application, it is necessary to determine the relation between the tolerance of that application and the results of one or more tests. When this has been done, these tests may be used as controls for that application.

It should be remembered that although a petroleum product may be clean when it leaves the refinery, it is possible for it to pick up contamination from the storage and handling equipment and practices, or as a result of condensation. Water and sediment are often picked up in tanks of ships and in other types of transportation or storage facilities.

WAX MELTING POINT

Melting Point (Plateau) of Petroleum Wax (ASTM D 87)
Drop Melting Point of Petroleum Wax (ASTM D 127)
Congealing Point of Petroleum Wax (ASTM D 938)

Each of the three test methods discussed here provide information about the transition between the solid and liquid states of petroleum waxes. The tests differ, however, by procedure and in the types of material to which they are applicable.

Both ASTM D 87 and ASTM D 127 are designed to determine the temperature at which most of the wax sample makes the transition between the liquid and the solid states. ASTM D 87 is applicable only to materials that show a "plateau" on their cooling curve. This plateau occurs when the temperature of a material passing into the solid phase remains constant for the time required to give up heat of fusion.

ASTM D 127 determines the temperature at which the material becomes sufficiently fluid to drip. The melting points of high-viscosity waxes that do not show a plateau can be determined by this method.

As determined by ASTM D 938, the congealing point is the temperature at which molten wax ceases to flow.

Significance

All three of the test methods are found in common specifications and buying guides among industries using large volumes of wax. The choice of a particular test method depends on the nature of the wax and the application.

Petroleum waxes are mixtures of hydrocarbon materials having different molecular weights. If these materials crystallize at about the same temperature, the cooling curve of the wax will show a plateau. ASTM D 87 is applicable to such waxes. Microcrystalline waxes, however, do not show a plateau in their cooling curve. The melting point of these waxes is usually reported by ASTM D 127. In general, ASTM D 127 is best suited for high-viscosity petroleum waxes.

The congealing point of a wax is usually slightly lower than either of its melting points. Congealing point (ASTM D 938) is often used when storage or application temperature is a critical factor, since it will give a more conservative estimate of the level at which temperatures should be maintained.

WHEEL BEARING GREASE LEAKAGE

ASTM D 1263

Under actual service conditions automotive front wheel bearings frequently operate at high temperatures. This is caused by a combination of heavy loads, high speeds, and the heat generated by braking. Because of this, greases used to lubricate these wheel bearings must be resistant to softening and leaking from the bearing. ASTM Method D 1263, "Leakage Tendencies of Automotive Wheel Bearing Greases," is an evaluation of leakage tendencies under prescribed laboratory test conditions.

Apparatus

The test apparatus consists of a special front wheel hub and spindle assembly encased in a thermostatically controlled air bath. Grease that leaks from the bearings is collected in the hub cap and leakage collector. Means for measuring both ambient and spindle temperatures are provided.

Significance

This test is an accelerated leakage test and is mainly a measure of the ability of a grease to be retained in the bearings at the test temperature. However, experienced operators can observe other changes in grease condition such as softening or slumping, but these are subjective judgments and not readily expressed in quantitative terms. There is no load or vibration applied to the bearings such as exists in normal wheel bearing service, and the test temperature of 220°F may be considerably lower than encountered in modern vehicles equipped with disc brakes. These factors are recognized and other test devices and procedures are under study by ASTM. The test is of primary value as a screening procedure to be used in conjunction with other stability tests in the development and evaluation of new grease formulations. Because of its limited sensitivity and precision, it permits differentiation only among products of distinctly different leakage characteristics.

Chapter 4

General Purpose R&O Oils

Please recall the foreword to this book and our note regarding brand names. The worldwide marketers of lube oils deal with occasional consolidation of producers. Exxon morphed into ExxonMobil and some legacy brands vanished altogether. Some lubricant manufacturers market certain lubricants in different regions of the world with brand designations that differ from each other. However, it is fair to say that their standard product formulations have remained relatively consistent and are often known under the older, original brand name. We found this brand consistency in place with most Exxon products and are comfortable with the decision to keep the various 1995-vintage brand names in this text.

If, however, the named lubricant is no longer available under its previous, widely used designation, the reader is encouraged to: (a) ask the supplier for the current name of a particular lubricant, and (b) compare the current "inspections"—the collective name for properties and ingredients—to those found widely accepted and described in this text. The next step would be to investigate the significance of these deviations by consulting Chapter 3, Lubricant Testing. Having done so will equip the reader to raise questions and request answers or explanations from the lubricant supplier.

That said, let's start by discussing R&O oils, the "workhorse lubricants."

The very first non-aqueous lubes were base oils—plain, non-additive base oils. But when the machinery was subjected to moisture, heat and oxygen, the oil oxidized. The introduction of moisture also led to rust, which began its corrosive creep. The result: breakdowns... blowouts... and, finally, replacement of expensive machinery.

But with the discovery of how to add certain ingredients to the base oil to help control rust and resist oxidation, lubricants developed broad, universal use.

R&O (rust and oxidation inhibition) has now become part of the language of lubrication for industrial machinery. R&O oils have become the workhorse lubricants in thousands of applications.

As of 1998, one such line of high quality lubricants, Exxon TERESSTIC®, has been in use for well over 50 years. Ongoing improvements to these products through the decades have given the TERESSTIC line an outstanding record of dependable service, whether as a hydraulic fluid, gear oil, heat transfer fluid or self-lubricating bearing oil.

ARE ALL "R&O" OILS THE SAME?

A common misconception is: "R&O oils are really all the same because they're mostly just oil." R&O products do contain "mostly oil," but the small concentration of carefully selected additives in proprietary basestocks that make up some superior lines of lubricants provide several key advantages:

- Base oil made from a dedicated crude source for most grades
- Refining by proprietary processes for optimized hydrocarbon composition
- Advanced systems for additive treatment, based or proprietary technology and understanding of the fundamentals of additive and lubricant behavior
- Reliable quality control procedures to ensure a highly consistent, superior product

These factors lift some oils above ordinary R&O lubes.

ADDITIVE FORMULATION

A key reason why TERESSTIC®, for instance, is not an average R&O oil: its state-of-the-art additive formulation. The additive formulation in the TERESSTIC® line is a sophisticated system of inhibitors designed for maximum potency. The combination of premium quality base oils plus advanced additive systems in the lubes provides qualities essential to premium R&O oils:

- Rust protection with film tenacity for persistent action
- Oxidation resistance for long service life of the oil without acid formation or sludge, even in the presence of catalytic metals
- Thermal stability to minimize deposit formation during prolonged exposure to high temperatures
- Demulsibility for rapid separation of water that becomes entrained in the lubrication system
- Foam and air entrainment control to ensure maximum lubrication efficiency
- High VI to avoid wide viscosity swings when variations in temperature occur
- A full range of viscosities to satisfy the wide range of machine conditions (Table 4-1)
- Low pour points to ensure oil flow at startup

DESIGNING A LINE OF R&O LUBRICANTS

The TERESSTIC® line of premium quality circulating oils was designed to lubricate and protect industrial machinery in a wide variety of applications. To meet the varied conditions of use, the TERESSTIC® line comprises a broad range of viscosity grades.

TERESSTIC® grades 32 through 100 are used primarily in turbine and other circulating oil systems (Figure 4-1)—hydraulic systems, compressors, pumps and general-purpose applications. TERESSTIC® grades 150 through 460 are used primarily in light-duty gear and higher temperature applications. Viscosity data on these products are shown in Table 4-2.

Before the creation of the ISO viscosity grading system, it was customary to use the SAE grading system when selecting lubes for industrial applications. Refer to the appendix for a review of the approximate relationships between the grading systems.

EXTREME PRESSURE (EP) R&O LUBRICANTS

Three related products of special interest are Exxon's TERESSTIC EP 32, 46, and 68. These have the same high-quality base oil selection, additive treatment and performance characteristics of the other TERESSTIC grades plus added anti-wear protection.

Certain geared turbines—steam and gas—are subject to shock loads and occasional overloading. This creates extreme pressure that can force the normal lubricating film out from between meshing gear teeth. The resulting grind of metal-to-metal contact can cause excessive wear. TERESSTIC EP is formulated with a non-zinc anti-wear additive to help reduce the possibility of metal-to-metal contact.

User experience proves these lubes to be effective in reducing wear rates in turbine gears and system components under extreme-pressure conditions.

And—like the other products in the line—TERESSTIC EP has an extremely effective oxidation inhibitor to help assure long, dependable operating life. All three TERESSTIC EP oils contain rust-inhibiting and anti-foam agents. They exhibit no rusting in either distilled water or synthetic sea water in the standard ASTM D 665 rust test procedure.

Table 4-1. Viscosity grade range for TERESSTIC® R&O lubricants.

Grade	32	32	46	68	77	100	150	220	320	460
ISO viscosity grade	32	32	46	68	—	100	150	220	320	460
Viscosity:										
cSt @ 40°C,										
ASTM D 445	32.3	30.0	44.2	65.1	78.7	105.0	157.5	214.0	306.0	432.0
cSt @ 100°C	5.3	5.2	6.5	8.4	9.5	12.1	15.7	18.7	23.8	30.3
SUS @ 100°F,										
ASTM D 2161	166	158	228	339	410	546	840	1123	1620	2284
SUS @ 210°F	44.2	43.8	48.1	58.5	66.5	80.0	93.0	115.0	149.0	

The TERESSTIC EP oils have extremely good demulsibility: any condensed moisture collecting in the lubricating system is readily shed by the oils. They also have a high viscosity index (VI), which allows more uniform operation of the system throughout a wide range of ambient and operating temperatures.

Dependable Turbine Lubrication

A few products have achieved a long record of reliable lubricant performance in the lubrication of steam turbines and gas turbines. For many years, the power industry has recognized the TERESSTIC line's ability to provide:

- Long life without need for changeout
- Prevention of acidity, sludge, deposits
- Excellent protection against rust and corrosion, even during shutdown
- Good demulsibility to shed water that enters the lubrication system
- Easy filterability without additive depletion
- Good foam control

An example of excellent performance value: TERESSTIC 32. Setting performance standards for turbine oil in the power industry, TERESSTIC 32 has been used in some cases for *over 30 years without changeout.* Results of controlled laboratory performance tests using TERESSTIC 32 are shown in Table 4-3.

Cleanliness Levels

Compressor lubrication can be one of the most demanding jobs for a lube oil because all compressors generate heat in the compressed gas. This heat directly

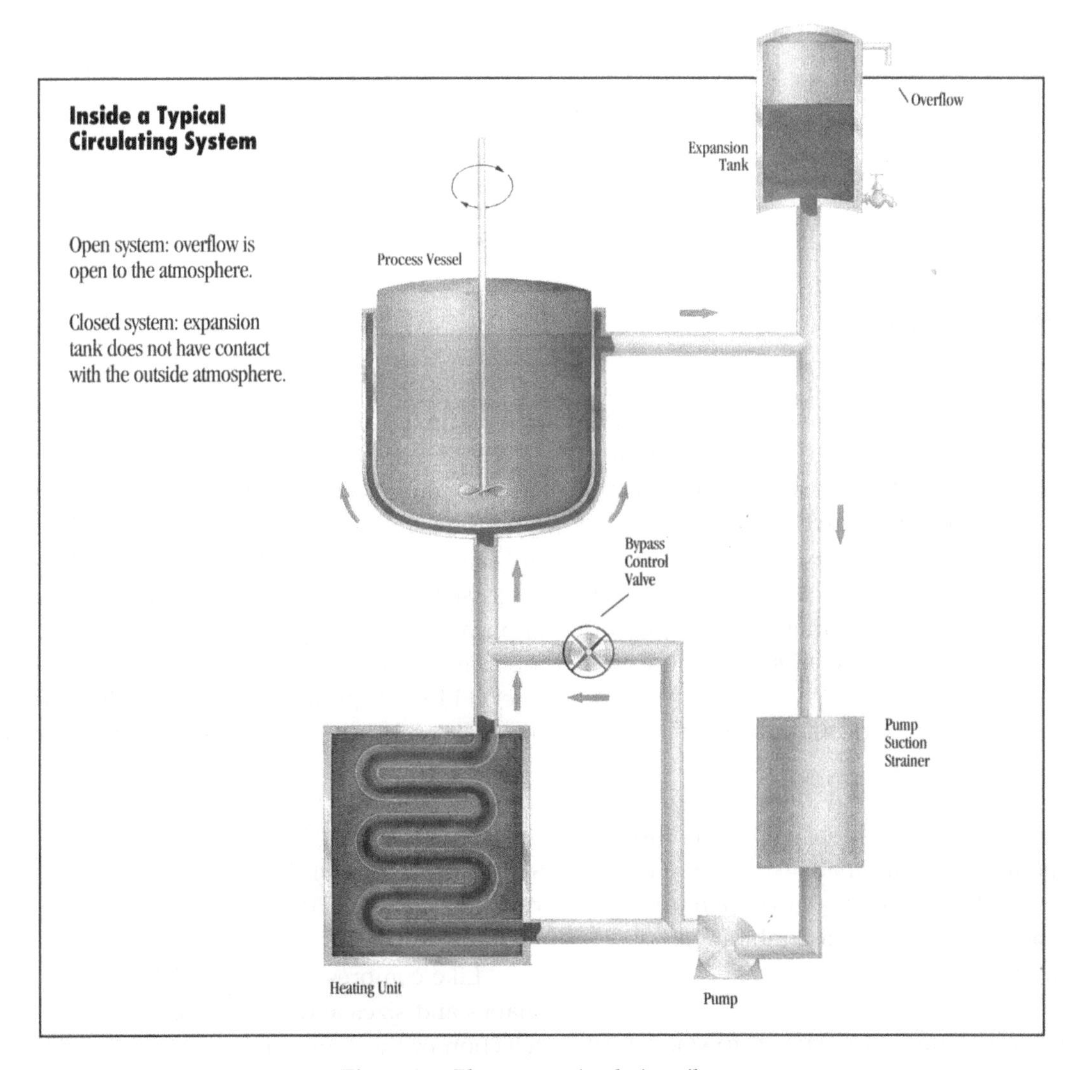

Figure 4-1. Elementary circulating oil system.

Table 4-2a. Typical inspections for TERESSTIC® R&O lubricants.

	TERESSTIC				
Grade	**32**	**33**	**46**	**68**	**77**
ISO viscosity grade	32	32	46	68	—
AGMA lubricant grade	—	—	1	2	—
Viscosity:					
cSt @ 40°C, ASTM D 445	32.2	30.0	44.2	65.1	78.7
cSt @ 100°C	5.3	5.2	6.5	8.4	9.5
SUS@100°F, ASTM D 2161	166	158	228	339	410
SUS @ 210°F	44.2	43.8	48.1	54.4	58.5
Viscosity index	97	97	97	97	97
Gravity, specific @ 15.6°C(60°F)	0.877	0.873	0.877	0.879	0.880
°API	29.9	30.6	29.9	29.5	29.3
Flash point, ASTM D 92					
COC, °C(°F)	206(403)	207(405)	212(414)	224(435)	227(441)
Pour point, ASTM D 97					
°C(°F)	-37(-35)	-21(-6)	-21(-6)	-21(-6)	-21(-6)
Rust test ASTM D 665 A/B					
Distilled water	pass	pass	pass	pass	pass
Synthetic sea water	pass	pass	pass	pass	pass
Copper strip corrosion, ASTM D 130					
3hrs@ 100°C(212°F)	1	1	1	1	1
FZG Gear test, A/8.3/90, DIN 51354,					
Failure load stage (FLS)	—	—	—	—	—
Foam, ASTM D 892	pass	pass	pass	pass	pass
Water Separability, ASTM D 1401	pass	pass	pass	pass	pass
Neutralization No.	0.06	0.06	0.06	0.06	0.06
Oxidation life, ASTM D 943, hrs	5000+	2500	3000	2500	2000
Four-ball wear test					
scar diameter, 10 kg, 1800 rpm,					
150°C(302°F), 30 min, mm	—	—	—	—	—

impacts lube oil life. The degree of impact depends upon the compressor type and the severity of operation. In some units—reciprocating or rotary type—the lubricant is *directly* exposed to the compressed gas.

This stress can cause rapid oxidative degradation and resultant formation of deposits and corrosive by-products leading to increased maintenance needs. But superior oils meet and beat the compressor lubrication challenge.

What makes a line of lubricants successful in compressor applications?

- Special resistance to oxidation under conditions of high temperature and intimate exposure to air
- Good demusibility during water condensation
- Long-lasting rust and corrosion inhibition
- Anti-foam properties

Proper lubricant selection is crucial to compressor life and service. TERESSTIC oils are a cost-effective option for many compressor applications.

The main types of compressors and the TERESSTIC grades generally used at some of the most profitable refineries, utilities, and petrochemical plants are shown in Figures 4-2 through 4-4. Specific grade selections are discussed in the text segment dealing with compressor applications and depend both on manufacturer recommendations and on the expected operational severity. The TERESSTIC grades are particularly suited for use in dynamic and rotary compressors and light-duty reciprocating units.

SUPERIOR R&O OILS COVER A WIDE RANGE OF PUMPS

Like compressors, industrial pumps come in many shapes and sizes and serve thousands of industries. The selection of the proper pump depends on: the nature of the liquid being pumped (its viscosity, lubricating value,

Table 4-2b. Typical inspections for TERESSTIC® R&O lubricants.

TERESSTIC					TERESSTIC EP		
100	**150**	**220**	**320**	**460**	**32**	**46**	**68**
100	150	220	320	460	32	46	68
3	4	5	6	—	—	1	2
105	157.5	214	306	432	30.3	44.2	64.8
12.1	15.7	18.7	23.8	30.3	5.2	6.5	8.4
546	840	1123	1620	2284	157	228	336
66.5	80	93	115	149	44	48.1	54.4
97	95	95	95	95	97	97	97
0.887	0.890	0.888	0.897	0.894	0.877	0.877	0.879
28.0	27.5	27.9	26.2	26.8	29 9	29 9	29 5
257(495)	266(511)	266(511)	277(531)	307(585)	210(410)	212(414)	224(435)
-18(0)	-18(0)	-18(0)	-9(16)	-9(16)	-21(-6)	-21(6)	-21(-6)
pass	pass	pass	pass	pass	pass	pass	pass
pass	—	—	—	—	pass	pass	pass
1	1	1	1	1	1	1	1
—	—	—	—	—	—	11	—
pass	pass	pass	pass	pass	pass	pass	pass
pass	pass	pass	pass	pass	pass	pass	pass
0.06	0.06	0.06	0.06	0.06	0.06	0.06	0.06
1500	1250	1000+	1000+	1000+	2500+	2500+	2500+
—	—	—	—	—	0.28	—	0.26

Table 4-3. TERESSTIC 32: Premium quality turbine oil

Test Methods	Typical Result
Oxidation life	
Turbine oil stability test, ASTM D 943	5000+ hours
Staeger oxidation test	3300+ hours
Universal oxidation test	800+ hours
CIGRE test, total oxidation product	Pass, 0.3g/100g oil
Rust protection, ASTM D 665 A/B	Pass
Distilled water	Pass
Synthetic sea water	Pass
Film tenacity test	Pass
CEICO test Pass	
Copper corrosion, ASTM D 130	Pass, 1A
Water separability	
Emulsion test @ 54°C(130°F), ASTM D 1401	Complete separation
Foam control	
Foam tendency/stability, ASTM D 892, ml	20/20
Viscosity index	97

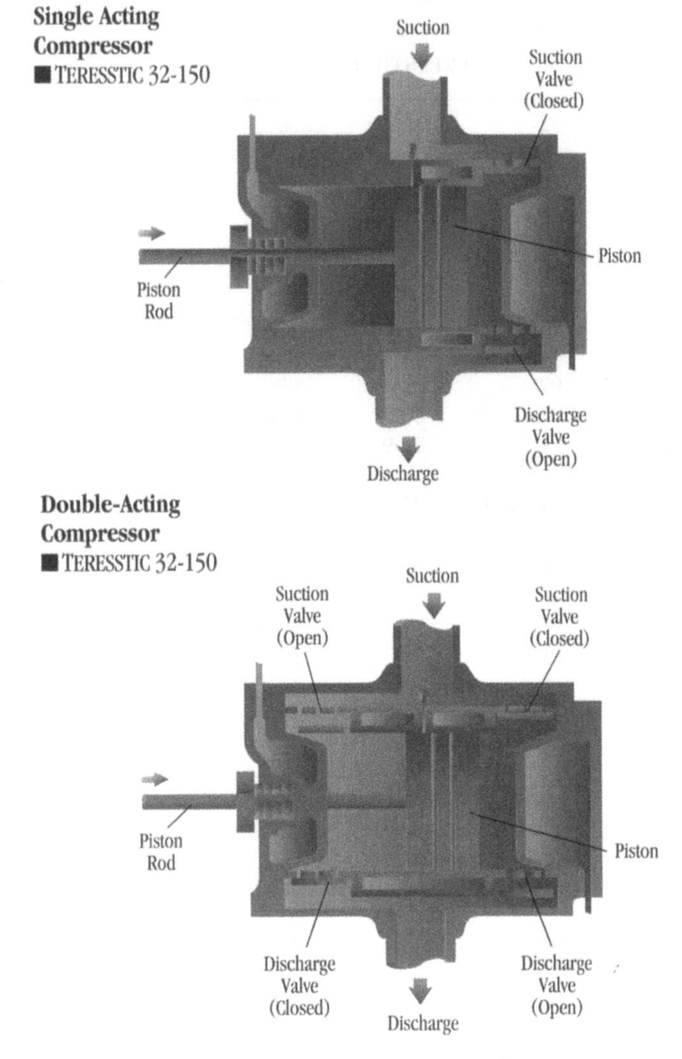

Figure 4-2. Reciprocating compressor applications.

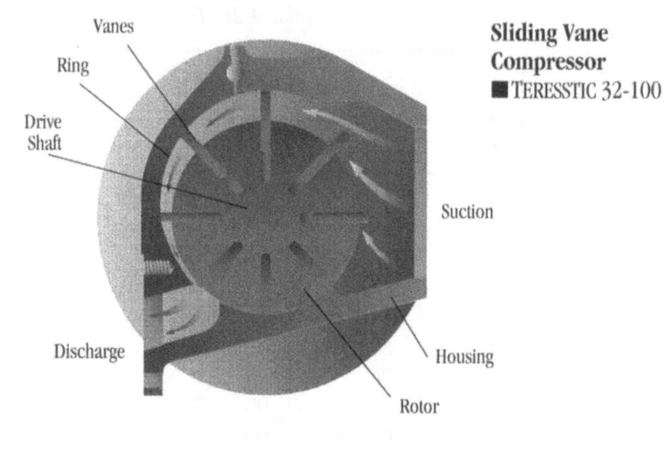

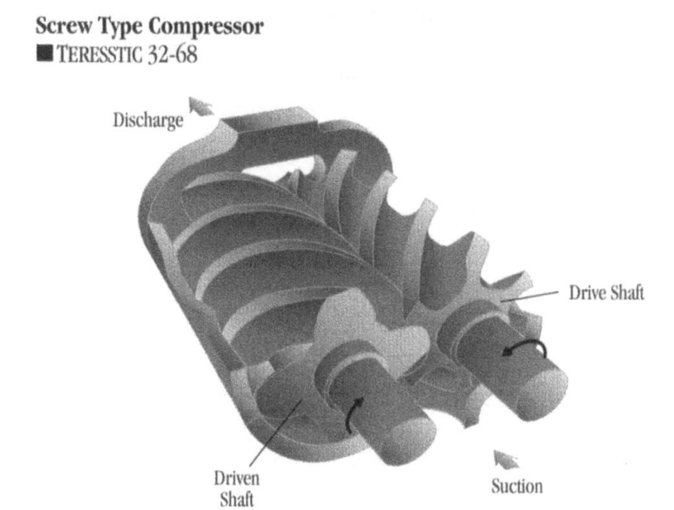

Figure 4-3. Rotary positive displacement applications.

density, volatility, corrosivity, toxicity, solids content), the pumping rate, desired pressure and the type of lubrication system to be used.

The centrifugal pump is the most widely used in the chemical and petroleum industries for transferring liquids of all types: raw materials, materials in process and finished products. Characterized by uniform (non-pulsating) flow and large capacity, the centrifugal pump is also used for water supply, boiler feed and condensate circulation and return.

Reciprocating and rotary pumps are particularly well adapted to low-capacity, high-pressure applications. They can deliver constant capacity against variable heads. Very close tolerances are required between internal rubbing surfaces in order to maintain volumetric efficiency, so the use of reciprocating or rotary pumps is generally restricted to liquids that have some lubricating qualities. Rotary and reciprocating pumps are used in fuel, lube circulating and hydraulic oil systems.

In some pump designs—hydraulic pumps, for example—the fluid being pumped also serves as the lubricant for the pump. In others, the lubricant is supplied by an external system sealed from the pumped liquid. The lubrication system delivers oil to the pump shaft bearings, packing, seals and gear reducers.

TERESSTIC products are well-suited to many pump operations. They can serve as the external lubricant for the pump itself, drive motor bearings or other lubricated parts in the pump system. *However, TERESSTIC and similar R&O oils should not be used in portable water systems.*

Recommendation of the right TERESSTIC product for a given pump can be determined by the pump manufacturer's specification or by contacting the appropriate sales office.

While some of the pumps being lubricated with TERESSTIC oils include all types of centrifugal, mixed, and axial flow units, many others fall into the positive

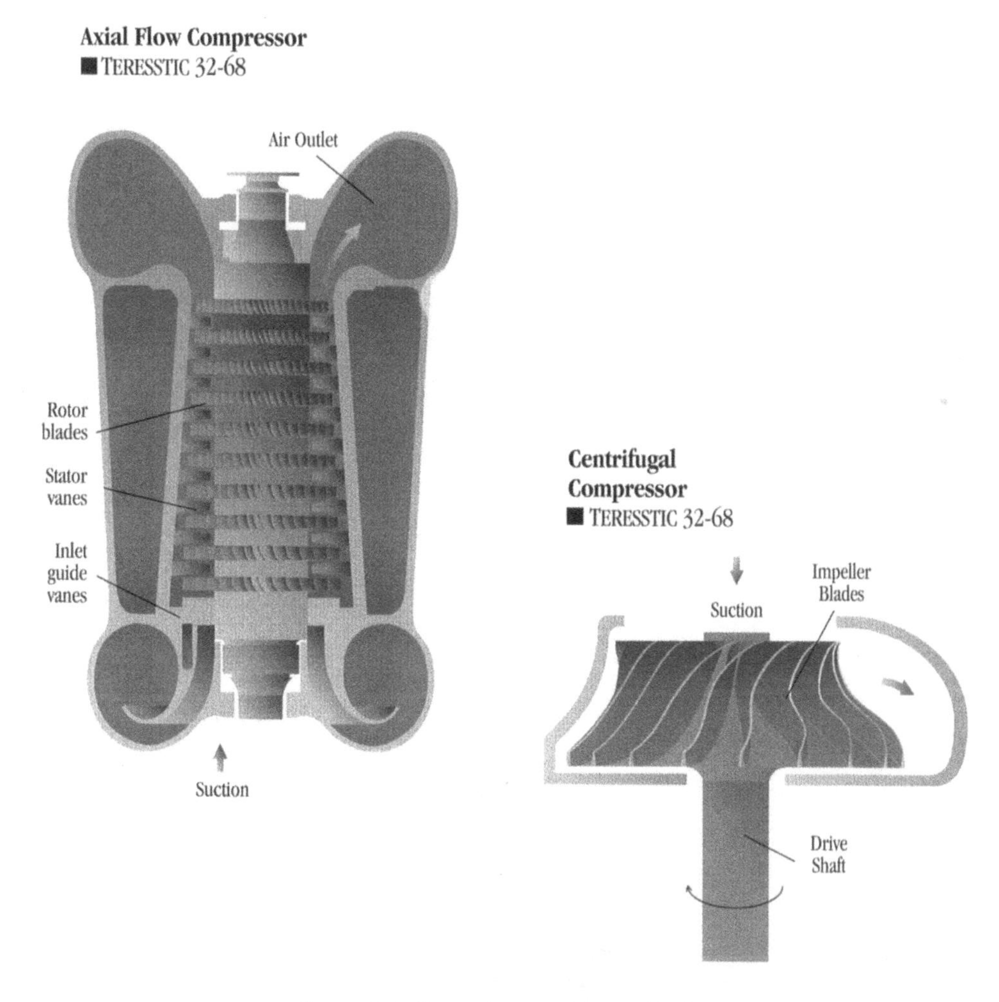

Figure 4-4. Dynamic compressor applications.

displacement category. These include piston, plunger, diaphragm, sliding vane, gear, lobe, and screw pumps.

HYDRAULIC APPLICATIONS FOR R&O OILS

In hydraulic power applications, the TERESSTIC product line has a long record of customer satisfaction. TERESSTIC oils, especially grades 32-150, are particularly well suited for general-purpose hydraulic systems that do not require anti-wear protection but do require a premium quality oil with long life, and for pump components requiring full hydro-dynamic film lubrication. Exxon's NUTO® H (see Chapter 5) hydraulic oils with special anti-wear features are normally recommended for equipment that has especially high loading at the pump component surfaces, such as vane pumps.

Selecting the proper viscosity grade is important for the most effective hydraulic system lubrication. Items to consider during selection: the expected ambient environment and the extremes of oil temperature expected in the system.

Figure 4-5 shows viscosity change of TERESSTIC hydraulic oils with temperature. It also shows the recommended viscosity range at operating temperatures, as well as recommended maximum viscosities at startup temperatures. The bar graph in Figure 4-6 indicates the operating ranges and minimum startup temperatures for the TERESSTIC grades (grades 32 through 150 are most commonly used as hydraulic fluids).

Here's how you would use the viscosity graph shown in Figure 4-5.

Example 1: ***What TERESSTIC grades are appropriate at a pump operating temperature of 90°C ?***
Locate 90°C on the bottom line and trace it upwards to the recommended viscosity range of 13-54 cSt. TERESSTIC grades 100 and above fall within that range.

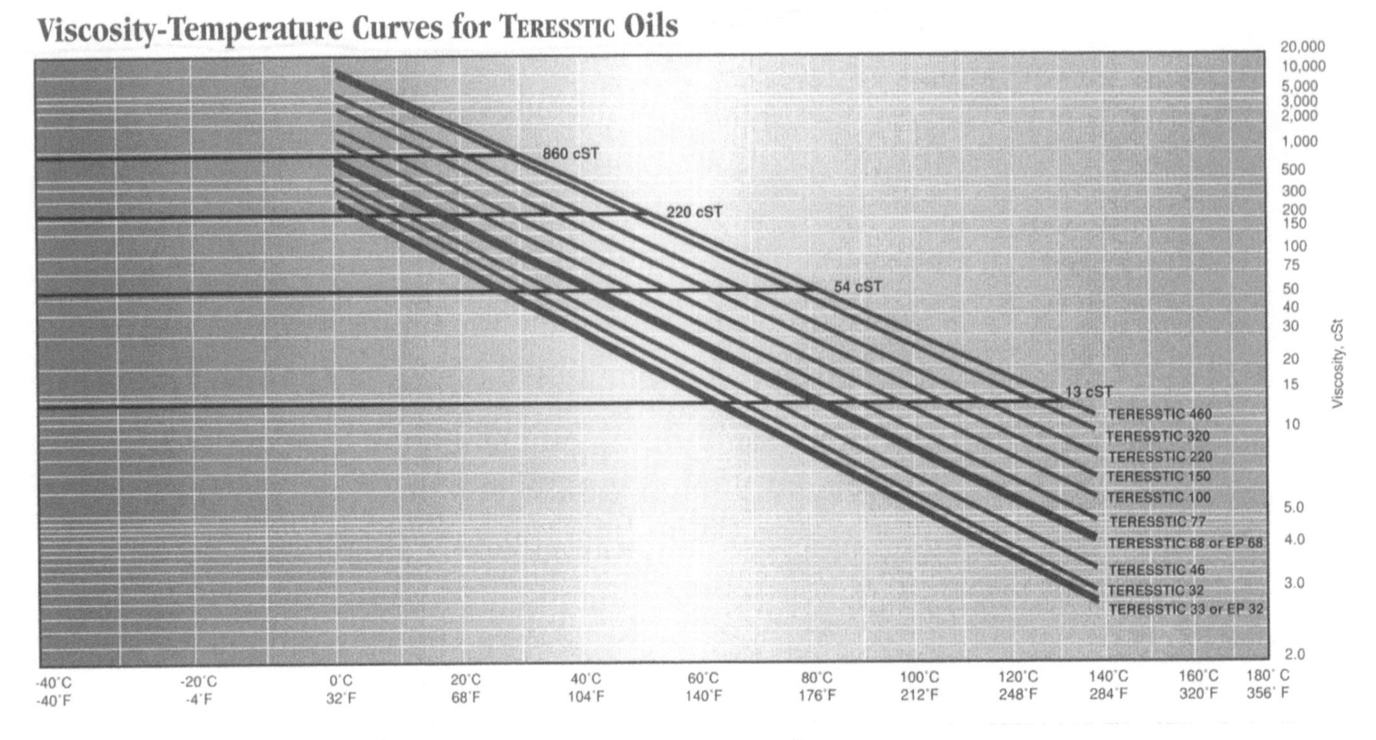

Figure 4-5. Viscosity-temperature curves for TERESSTIC® R&O oils widely used in industry.

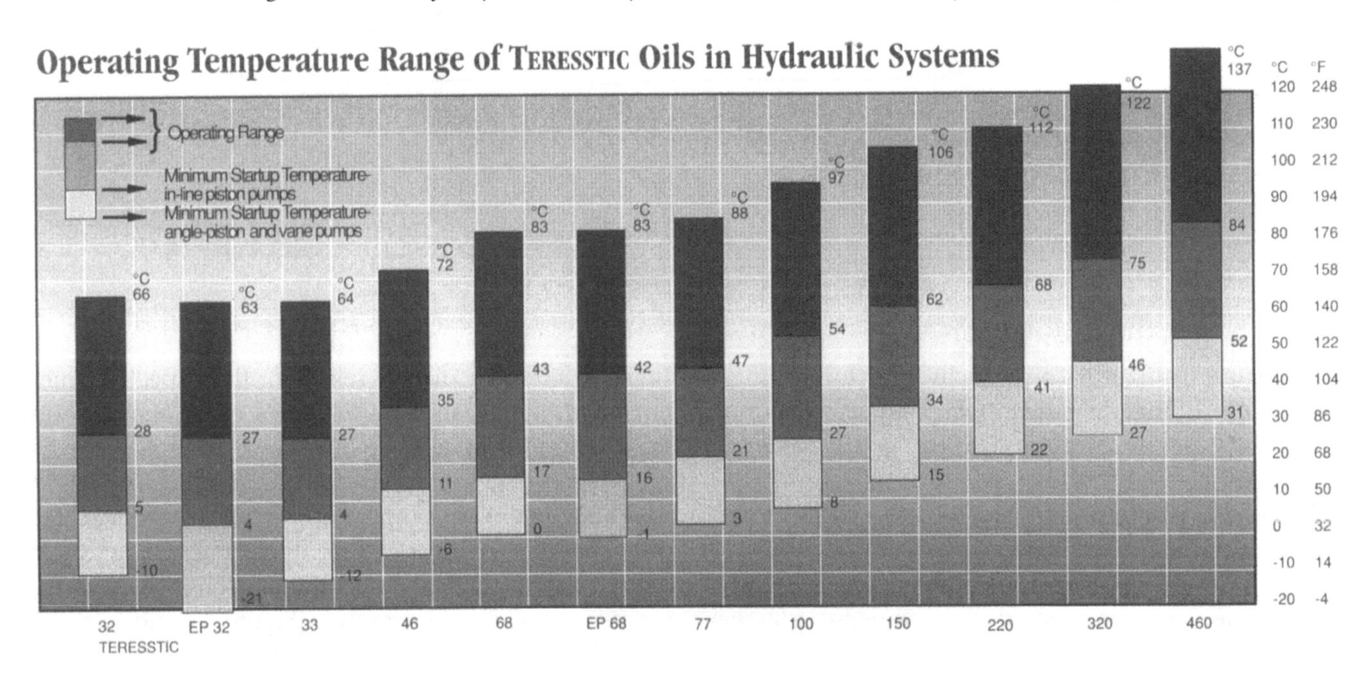

Figure 4-6. Operating temperature range of TERESSTIC® oils in hydraulic systems.

Example 2: ***What TERESSTIC grades have viscosities that fall below the allowable maximum at a vane pump startup temperature of 10°C?***

Locate 10°C on the bottom line and trace it upwards to the horizontal line representing 860 cSt, the maximum startup viscosity for a vane pump. TERESSTIC grades 100 and lower fall below 860 cSt.

Next, some guidelines. Avoid choosing too-low viscosity grades for your systems to:

- Maintain sufficient hydrodynamic film thickness
- Prevent excessive wear of moving parts
- Avoid excessive pump slippage or case drain and loss of pressure system response

Avoid choosing too-high viscosity grades for your system to:

- Eliminate startup problems at low temperatures and avoid high wear during startup
- Reduce pump cavitation tendency
- Achieve good response in hydraulic system devices
- Ensure good defoaming and good demulsibility
- Save energy

UNIVERSAL APPLICATION OF R&O OILS

The almost universal applicability of superior R&O lubricants is best demonstrated by the fact that TERESSTIC grades are often used for gear lubrication, or as heat transfer fluids, or in self-lubricated bearings.

For lubrication of enclosed gear drives in industrial equipment, the manufacturer's recommendation for the proper lubricant grade is usually indicated on the gear case as the American Gear Manufacturers Association (AGMA) number. As discussed later in our chapter dealing with gears, these AGMA specifications cover a wide range of viscosity and load-carrying grades, from light-duty to severe applications. The most widely used are the grades of circulating lubricants. They can easily be distributed to gears and bearings for lubrication and heat removal, and are readily filtered and cooled.

For light-duty applications where extreme-pressure properties are not required, AGMA recommends the use of an R&O lubricant. The TERESSTIC grades meet essentially all of the requirements of AGMA grades 1-6 (see Appendix for details).

Typical gear types suitable for TERESSTIC oil use include: spur gears, helical gears, double helical (herringbone) gears, bevel gears and spiral bevel gears.

TERESSTIC oils—with their long-lasting rust inhibitor system, technologically advanced oxidation prevention and good demulsibility and foam control—provide excellent gear protection and R&O performance. They enjoy a long record of trouble-free use.

A HEAT TRANSFER FLUID THAT KEEPS ITS COOL

TERESSTIC oils are made from petroleum components that are vacuum-fractionated to selectively remove lower volatility elements. The components are then fully refined and additive-treated to provide outstanding oxidation resistance. These characteristics make TERESSTIC products excellent heat transfer oils.

In many process applications—industrial heat-treating, chemical manufacturing and food processing—there are clear advantages to carrying heat by means of fluid transfer systems rather than using direct-fired heating systems. Using fluids for heat transfer allows closer control of the process temperatures, eliminates hot spots in vessel walls, permits heating several process vessels using one primary heat source and provides better economy in the overall heating operation.

The working fluid used in the system must have a high degree of oxidation stability, especially in open circuit systems where the expansion tank is open through a breather tube and air can contact the hot oil. The fluids must not develop sludge or allow carbonaceous deposits on the primary heat exchanger walls, pipes or the heated reaction kettle. For safety reasons, the fluid should have volatility, i.e., low vapor pressure at the heat transfer temperatures. It should also have high flash and fire point properties.

Several TERESSTIC grades, because of their excellent resistance to oxidation and thermal degradation, are ideally suited for use as high-quality heat transfer fluids. They are capable of providing long service in a heat transfer circulating system. The grades are listed in Table 4-4, which also shows general guidelines on:

- The maximum operating temperature—based upon safety considerations such as flash and fire points —for the fluid in the expansion tank, where some exposure to air may occur in open systems. (In less critical closed systems, considerably higher expansion tank temperatures may be used because no air contact occurs.)

- The maximum bulk oil temperature as it leaves the primary source heat exchange unit. (The actual film temperature or skin temperature next to the furnace tubes can be even higher, limited only by the ultimate thermal degradation property of the fluid.)

A brief tabulation of good practices in using heat transfer fluids might include a listing of general requirements, followed by a few memory joggers. Thus, in order to:

- Maximize the efficiency of the overall heat transfer operation

- Maximize the useful life of the fluid

- Minimize formation of deposits on walls and tubes

Table 4-4. Temperature guidelines for premium grade R&O oils in heat transfer applications.

Product	Flash Point	Maximum expansion tank temperature for open systems, COC,°C(°F)	Maximum bulk oil temperature for open or closed systems, °C(°F) °C(°F)
TERESSTIC 32	206(403)	185(365)	316(601)
TERESSTIC 46	212(414)	197(387)	316(601)
TERESSTIC 68	224(435)	209(408)	316(601)
TERESSTIC 77	227(441)	212(414)	316(601)
TERESSTIC 100	257(495)	242(468)	316(601)
TERESSTIC I50	266(511)	251(484)	316(601)

Remember to:

- Select a fluid of lowest feasible viscosity and the highest coefficient of heat transfer
- Use fluids that are effectively inhibited with non-volatile additive systems
- Avoid catalytic metals (i.e., copper alloys) in the system
- Do not use clay filters (i.e., avoid additive removal)
- Minimize air contact, avoid air leaks at pumps, etc.
- Do not mix different types of heat transfer fluids
- Do not use heating oil or solvent for flushing; use a mineral oil such as CORAY or FAXAM.
- Be watchful for oil leaks onto hot surfaces
- Use an oil with flash point at least 15°C (27°F) above expansion tank temperature
- Never operate either open or closed systems at temperatures near the auto-ignition point; keep the system below 360°C (680°F).

It is important to give close attention to flash points and maximum temperatures, as given in Table 4-4.

Figure 4-7. The extraordinary loads placed on the self-lubricating bearings of high-speed power tools can lead to premature tool failures without the protection of quality oils.

TERESSTIC USE IN SELF-LUBRICATING BEARINGS

Development of lubricated-for-life bearings has brought us electric shavers, kitchen appliances, power tools (Figure 4-7), automotive electric motors and a host of other conveniences we take for granted. The bearings, lightweight and complex in design, never have to be relubricated.

Self-lubricating bearings have evolved from the steadily emerging technology of powder metallurgy. Their success depends on having an oil impregnated into the metal of the bearing that is capable of lifelong, trouble-free operation.

A self-lubricating bearing is typically made in four steps:

(1) A selected metal or alloy is reduced to powder—usually by atomization of a molten stream—to a specified particle size and distribution.

(2) The powder is compacted in a mold to the dimensions of the intended finished part.

(3) The part is heated to sinter the powder, while leaving a controlled amount of internal space and capillary passageways.

(4) The part is impregnated with a lube-for-life such as TERESSTIC.

The oil absorbed into the pores of the sintered metal acts as a reservoir for lubricant when the part is in use. At startup a thin film on the surface provides initial lubrication. As the bearing warms up, the oil expands and is forced out of the pores into the space for journal/bearing lubrication. When rotation stops and the bearing and oil cool, the oil is drawn back into the pores of the bearing by capillary action. Conventional powder metallurgy (P/M) bearings can absorb about 10-30% by volume of oil. The thin film of oil and the small reservoir within the pores do the whole job of lubrication. There is no circulating system or oil reservoir in the conventional sense.

There are significant advantages to this technology:

- The finished part is less dense than a conventionally machined part; weight savings are of great importance in many applicators.
- Controlling of powder metallurgy adds strength and durability to the part.
- The P/M system eliminates the need for conventional machining, saving time and allowing fabrication of more complex shapes.
- Self-lubricating bearings simplify design considerations in the unit of machinery.
- High-quality self-lubricating bearings eliminate the need for maintenance or repair service and reduce the cost of warranty claims.

P/M parts are particularly effective where relatively light loading is present: home appliances, automotive accessory equipment, power tools, business machines and the like. Selection of high-quality oil is extremely important:

- The oil must be fully refined to provide maximum stability without forming gums, varnish or sludge that would block the porous structure.
- Additives in the oil should have high permanence but should not interact with the sintered metals or otherwise cause corrosion or deposits.
- The viscosity grade must provide the proper hydrodynamic lubrication under the conditions of use both at startup and at maximum operating temperature.

Based on field experience, TERESSTIC grades 68 and 77 have been used most frequently in making self-lubricating parts. In fact, TERESSTIC 77, a special grade not in the established ISO grade sequence, was specifically formulated for application in self-lubricating bearings.

Chapter 5

Hydraulic Fluids

A petroleum-base hydraulic fluid used in an industrial hydraulic system has many critical functions. It must serve not only as a medium for energy transmission, but as a lubricant, sealant, and heat transfer medium. The fluid must also maximize power and efficiency by minimizing wear and tear on the equipment.

But the specific needs of hydraulic systems and their components, Figures 5-1 and 5-2, may differ. Some require a fluid with greater oxidative or thermal stability, some need tougher anti-wear protection, some require extra lubricant stability in extreme-temperature environments, and some require the assurance of fire-resistant fluids. Yet others require a special assembly grease, compatible with the hydraulic oil.

A suitable lubricant will thus

- facilitate seal mounting
- improve the sealing effect
- reduce adhesive and starting friction
- reduce wear during operation
- be neutral to NBR, EPDM, FMP and PU materials
- have excellent load-carrying capacity
- show high affinity towards materials such as steel, plastics and elastomers

Assembly greases are shown in Table 5-1. For the food processing industry, the same company, Klüber, lists hydraulic oil selection criteria and characteristics in Table 5-2. (Note that Chapter 6 deals more extensively with food-grade lubricants.)

As mentioned above, hydraulic fluids not only act as the fluid-power medium, they lubricate system parts. Today's hydraulic pump units are subjected to high system pressures and pump speeds. This can create conditions of thin-film lubrication and cause eventual mechanical wear unless the fluid contains special protective additives.

Three main types of pumps (Figure 5-3) are found in hydraulic systems: gear pumps, piston pumps—both axial and radial—and vane pumps. Vane pumps are the most common and require the most anti-wear protection, due to the high contact pressures developed at the vane tip. Gear and piston pumps don't usually require anti-wear oils; however, the pump manufacturer should be consulted for specific requirements.

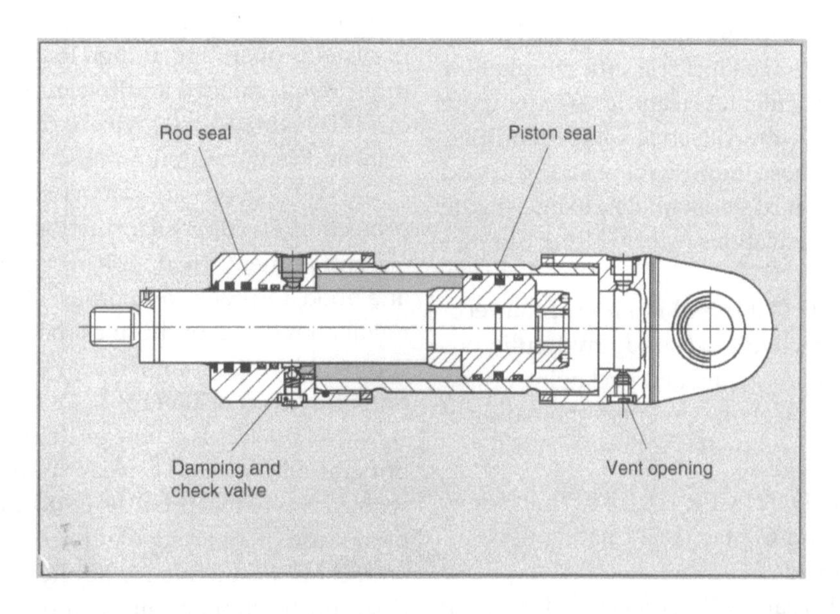

Figure 5-1. Hydraulic cylinder. (Source: Klüber Lubrication North America, Inc., Londonderry, New Hampshire)

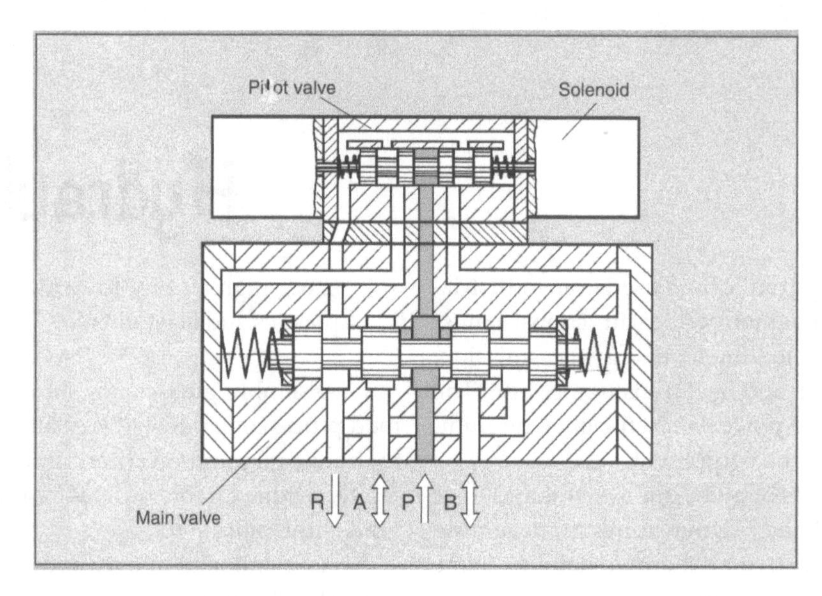

Figure 5-2. Electromagnetically operated 4/3-way valve with pilot valve. (Source: Klüber Lubrication North America, Inc., Londonderry, New Hampshire)

The anti-wear properties of a hydraulic fluid are typically tested by operation in an actual vane pump under overload conditions. Results are measured in terms of hours to failure or as the amount of wear (weight loss of the vanes and ring) after a specified number of hours of operation. Experience has shown that a good anti-wear fluid can reduce wear by 95% or more compared to conventional R&O oils.

Recall again that, occasionally, certain lubricants are marketed in different regions of the world with brand designations that differ from each other. If a lubricant is no longer available under its previous, widely used designation, the reader is encouraged to: (a) ask the supplier for the current name of a particular lubricant, and (b) compare the current "inspections"—the collective name for properties and ingredients--to those found widely accepted and described in this text. The next step would be to investigate the significance of these deviations by consulting Chapter 3, Lubricant Testing.

Exxon's NUTO® FG (Table 5-3) is a line of four economical, highly cost-effective food-grade hydraulic oils designed for use in food processing and packaging operations. It incorporates the following unique combination of features.

- Compliance with FDA 21 CFR 178.3570, "Lubricants With Incidental Food Contact (see Chapter 6)
- USDA H-1 approved
- Outstanding anti-wear (AW) properties, for pump protection
- Excellent extreme-pressure (EP) properties, for bearing protection
- Superior oxidation stability, for long, trouble-free life
- Suitable for hydraulic systems up to pressures of 3000 psi

NUTO H is the trademark for a line of premium-quality anti-wear hydraulic oils designed to meet the most stringent requirements of most major manufacturers and users of hydraulic equipment. The five grades meet the viscosity requirements of essentially all hydraulic systems. NUTO H is very effective in reducing vane and gear pump wear in systems operating at high loads, speeds, and temperatures. Its specialized additive makeup also allows the use of NUTO H in severe-service hydraulic systems employing axial and radial piston pumps.

NUTO H oils are characterized by outstanding rust protection, low deposit formation, good demulsibility, low air entrainment, oxidation resistance, low pour points, and good anti-foam properties. They are non-corrosive to metal alloys, except those containing silver, and are fully compatible with common seal materials. Typical inspections are given in Table 5-4.

NUTO HP is a line of high-performance, ashless, mineral-oil-based anti-wear hydraulic oils formulated with additives that provide reduced environmental impact in the case of an accidental release into the environment. NUTO HP is suitable for applications in woodland, marine, construction, mining, pulp and paper, and farming, as well as general industrial hydraulic applications where environmental concerns exist. The characteristics of Nuto

Table 15-1. Properties of hydraulic-compatible assembly greases. (Source: Klüber Lubrication North America, Inc., Londonderry, New Hampshire)

Selection criteria	Product name	Base oil/ thickener	Service temperature range (°C)	Density at 20°C (g/ml) DIN 51757 ≈	Color	Drop point DIN ISO 2176 (°C)	Speed factor ($n \bullet d_m$) ≈	Worked penetration DIN ISO 2137 $mm \bullet min^{-1}$	Consistency NLGI grade (0.1 mm) 51 818	Apparent viscosity, KL viscosity grade DIN	Notes
Assembly grease for hydraulic systems	Klüberplex BE 31 502	Mineral hydrocarbon oil/special Ca soap	-10 to 140	0.86	beige, brownish	>190	50 000	245 to 275	2/3	S	Suitable for the lubrication of NBR, FPM and PU elastomers and metal sliding components in hydraulic cylinders and valves; excellent adhesion and load-carrying capacity; USDA-H2 authorized, also suitable for hydraulic applications in the food processing industry
	SYNTHESO GLEP 1	Polyalkyglycol/ special Li soap	-50 to 150	0.97	beige, almost transparent	>220	—	280 to 310	—	M	Suitable for the lubrication of EPDM elastomers and metal sliding components in hydraulic elements; good load-carrying capacity and corrosion protection

Table 5-2. Hydraulic oil for the food processing industry—selection criteria and characteristics. (Source: Klüber Lubrication North America, Inc., Londonderry, New Hampshire)

Selection criteria	Product name	Base oil/ thickener	Service temperature range (°C) ≈	ISO VG DIN 51 519 ≈	Density at 20°C (g/ml) DIN 51757 ≈	Kinematic viscosity DIN 51561, (mm^2/s) at ≈ 20°C	Kinematic viscosity at 40°C	Viscosity index DIN ISO 2909 (VI)	Pour point DIN ISO 3016 (°C) ≈	Flash point DIN ISO 2592 (°C) ≈	Notes
Hydraulic oil for the food processing industry	Klüberoil 4 UH 1-68	Synthetic hydrocarbon oil, ester oil	-35 to 110	68	0.84	68	12	>160	-35	>200	For all hydraulic applications in the food processing and pharmaceutical industries; meets HLP requirements, is authorized in accordance with USDA H1 and complies with the German Regulations on Food Products and Associated Ancillaries (LMBG)

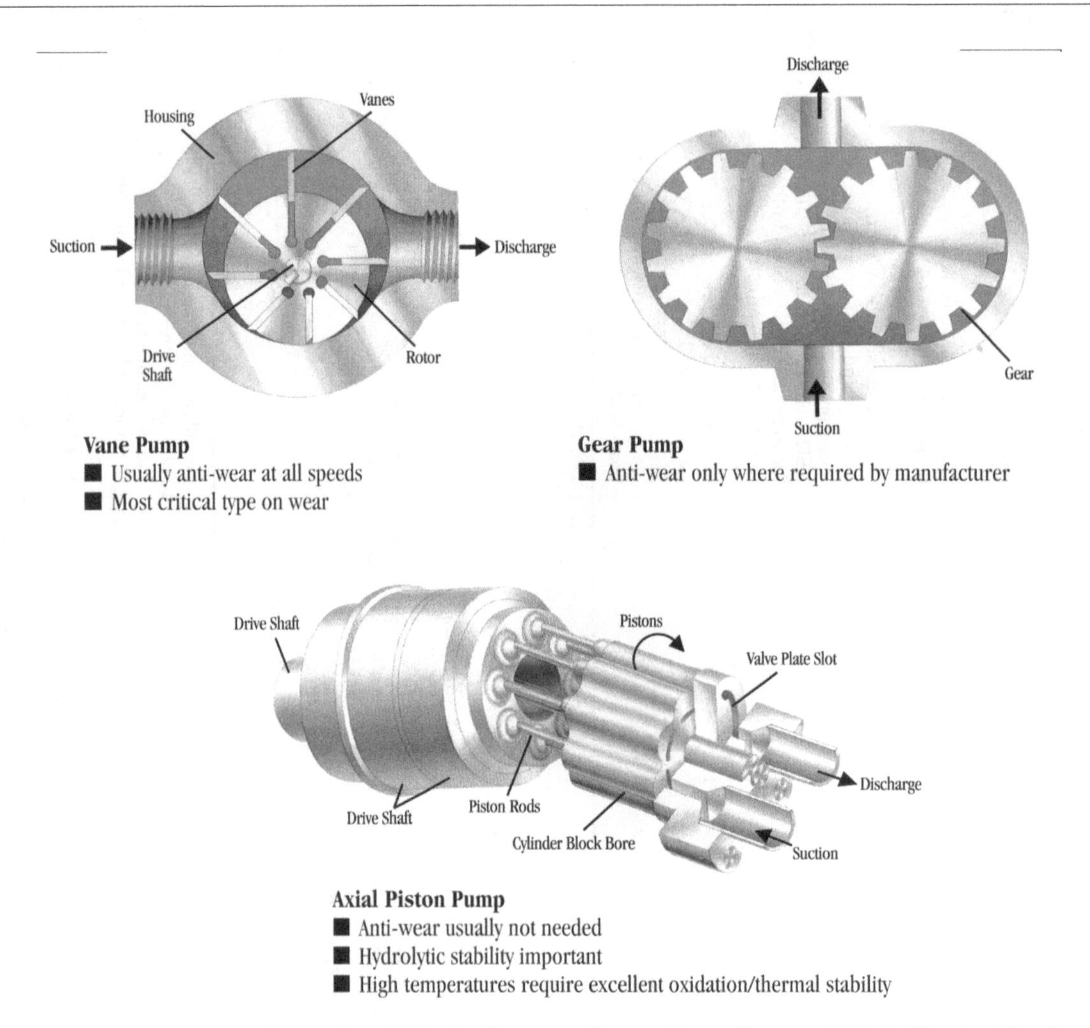

Figure 5-3. Main types of pumps found in hydraulic system. (Source: Exxon Company, U.S.A., Houston, Texas)

HP oils are give in Table 5-5. NUTO HP incorporates the following unique combination of features:

- Non-toxic as defined by the OECD 203 Fish Acute Toxicity Test
- Outstanding anti-wear (AW) properties, for pump protection
- Excellent extreme-pressure (EP) properties, for bearing protection
- Superior oxidation stability, for long, trouble-free life
- Available in ISO 32, 46, and 68 grades
- For use in all hydraulic systems
- Denison HF-0 approved
- Vickers M-2950-S, 1-286-S approved
- Cincinnati Milacron P68, P70, and P69 approved

Exxon uses HUMBLE Hydraulic M as the trademark for a line of mid-V.I. anti-wear hydraulic oils. These oils are designed for use in once-through applications or in equipment where oil consumption is high and temperatures are not excessive. In addition to anti-wear, HUMBLE Hydraulic M offers good rust and corrosion protection, good demulsibility and good anti-foam properties. Refer to Table 5-6 for typical inspections.

This company also produces a line of premium quality anti-wear hydraulic oils designed to meet the requirements of fluid power systems. HUMBLE Hydraulic H is formulated with a proven anti-wear additive which is effective in reducing wear in pumps. It has outstanding oxidation stability and excellent water separation, rust and corrosion prevention, and anti-foam properties. As noted in Table 5-7, HUMBLE Hydraulic H meets or exceeds the requirements of Denison HF-O, Vickers, Cincinnati Milacron, and USS 127.

HYDRAULIC OILS FOR EXTENDED TEMPERATURE RANGE

The marine, construction and public utility industries use hydraulic equipment—vane, piston and gear pumps

Table 5-4. Typical inspections for premium-quality anti-wear hydraulic oils (Exxon Nuto H)

Typical Inspections Grade	32	46	68	100	150
Gravity, °API					
specific @ 15.6°C (60°F)	29.9	29.6	29.2	28.8	28.6
Density, kg/m^3	0.877	0.878	0.881	0.883	0.884
lb/gal	7.303	7.311	7.336	7.352	7.360
Viscosity, cSt @ 40°C	32.2	46.0	66.1	99.0	154.1
cSt @ 100°C	5.3	6.5	8.5	11.1	14.9
SUS @ 100°F	166	230	343	516	810
SUS @ 210°F	44.2	48.1	54.8	64.3	79.3
Viscosity index	97	96	97	97	95
Pour point, °C	-37	-29	-27	-23	-21
°F	-35	-20	-17	-9	-6
Flash point, °C	206	216	222	241	266
°F	403	421	432	466	511
Color, ASTM D 1500	2.0	2.5	2.5	3.0	3.5
Neutralization number, ASTM D 974	0.40	0.40	0.40	0.40	0.40
Rust test, ASTM D 665					
A. Distilled water			no rust		
B. Synthetic sea water			no rust		
Hydrolytic stability, ASTM D 2619					
Cu mass loss, mg/cm^2	0.12	0.12	0.12	0.12	0.12
Denison HF-0 pump test			Approved		
Cincinnati Milacron Spec. No.	P68	P70	P69	—	—
Vane pump test, ASTM D 2882, total ring and vane wear, mg	25.0	25.0	25.0	25.0	25.0
35VQ25 vane pump test			Pass		
Oxidation life					
Turbine Oil Stability Test, ASTM D 943, hr	3000+	2500+	2500+	2500+	2500+
Rotary Bomb Oxidation Test, ASTM D 2772, min	350	350	350	350	350
FZG Spur Gear Test (FLS)	—	11	11	—	—

and high-pressure axial and radial piston pumps—in a wide range of environments...Boston Harbor during a harsh, icy winter... the Gulf of Mexico during a sweltering heat wave. Tough environments require a hydraulic oil that performs just as well in winter as in summer.

There's also a need for a lubricant that can operate across extended temperature ranges—higher and/or lower than average operating temperatures—without a dramatic change in viscosity. In other words, the lubricant should have a high viscosity index (VI). A high VI indicates a low tendency to thin or thicken with changes in temperature.

Table 5-5. Characteristics of NUTO HP hydraulic oils that provide reduced environmental impact.

Typical Inspections	32	46	68
Color, ASTM D 1500	0.5	0.5	0.5
Viscosity, cSt @ 40°C	33	45.5	68
Viscosity, cSt @ 100°C	5.4	6.6	8.5
Viscosity Index (VI)	95	95	95
Gravity, specific @ 15.6°C (60°F)	0.88	0.88	0.88
Flash point, °C (°F)	198 (388)	208 (406)	220 (428)
Pour point, °C (°F)	–45 (–49)	–33 (–27)	–27 (–16)
Rust test, ASTM D 665B	Pass	Pass	Pass
Copper Corrosion, ASTM D 130	1a	1a	1a
Water Separability, ASTM D 1401	40/40/0	40/40/0	40/40/0
Foaming Tendency/Stability, ASTM D 892, all sequences	20/0	20/0	20/0
Oxidation life, ASTM D 943, hrs	3000+	3000+	3000+
FZG test, failure load stage (FLS)	10	10	10
Cincinnati Milacron Thermal Stability Test	Pass	Pass	Pass

Table 5-6. Typical inspections for Exxon's "Humble Hydraulic M," a mid-V.I. anti-wear fluid.

Typical Inspections Grade	32	46	68	100
ISO viscosity grade	32	46	68	100
Viscosity, cSt@40°C	33	47	69	99
Viscosity index	60+	60+	60+	60+
Pour point, °C (°F)	-18 (0)	-18 (0)	-18 (0)	-18 (0)
Rust protection, ASTM D 665A&B	Pass	Pass	Pass	Pass
Vane pump test, ASTM D 2882	Pass	Pass	Pass	Pass
Foam, ASTM D 892, Seq. I, II, & III	Pass	Pass	Pass	Pass

That's why leading manufacturers develop premium products. UNIVIS N is a line of anti-wear hydraulic fluids designed for high performance in widely varying ambient temperature conditions. Superior quality components permit UNIVIS N to be used over a wider range of temperatures than conventional non-VI-improved hydraulic oils.

UNIVIS J is Exxon's trademark for two premium-quality hydraulic oils with unusually high viscosity indexes (V.I.). Because of its resistance to viscosity change with temperature, UNIVIS J is particularly recommended for equipment that is subject to wide temperature variations. Applications include hydrostatic transmissions and fluidic systems such as those on numerically controlled lathes, automatic screw machines, etc. UNIVIS J oils also can be used as a lubricant in fine instruments and other mechanisms where power input is limited and increases in torque due to lubricant thickening cannot be tolerated. In addition to high V.I., the UNIVIS J oils have long-lasting oxidation stability, low pour points, and excellent lubrication characteristics. UNIVIS J 13 contains a red dye to aid in leak detection. Refer to Table 5-8 for typical inspections.

UNIVIS N (Table 5-9) is the brand name for a line of premium-quality, high-viscosity-index (V.I.), anti-wear hydraulic oils designed to meet the all-season requirements of most major manufacturers and users of hydraulic

Table 5-7. Characteristic data for Exxon's "HUMBLE Hydraulic H."

Typical Inspections Grade	32	46	68	100
ISO viscosity grade	32	46	68	100
Viscosity, cSt@ 40°C	31.0	44.5	65.0	99.0
SUS @ 100°F	160	235	337	520
Viscosity index	90	90	90	90
Pour point, °C (°F)	-30 (-20)	-27 (-15)	-24 (-10)	-21 (-5)
Flash point, °C (°F)	198 (385)	207 (405)	219 (425)	231 (445)
Foam, ASTM D 892, Seq. I, II, III	Pass	Pass	Pass	Pass
Oxidation stability, ASTM D 943, hr	2000+	2000+	2000+	2000+
Hydrolytic stability, ASTM D 2619	Pass	Pass	Pass	Pass
Rust protection, ASTM D 665A&B	Pass	Pass	Pass	Pass
Vane pump test, ASTM D 2882	Pass	Pass	Pass	Pass
35VQ25 vane pump test	Pass	Pass	Pass	Pass
Meets or exceeds the requirements of:				
Denison HF-0	Yes	Yes	Yes	Yes
Cincinnati Milacron	P68	P70	P69	—

systems. The high V.I.s and low pour points of these oils help ensure pump startup at low temperatures, while maintaining oil viscosity at high ambient temperatures. The polymers used to thicken UNIVIS N are specially selected for their excellent shear stability. In addition, a very effective anti-wear additive provides pump protection even in severe-service hydraulic applications employing high-pressure axial and radial piston pumps. UNIVIS N 68 and 100 exhibit Stage 11 performance in the FZG Spur Gear Test. UNIVIS N also offers excellent rust protection, good demulsibility, oxidation resistance, and good anti-foam properties. It is non-corrosive to metal alloys, except those containing silver.

UNIVIS N grades 32, 46, 68, and 100 meet the requirements of Denison's HF-0 and HF-2 and Vickers' MS-2950-S specifications. Grades 32, 46, and 68, respectively, also are qualified against Cincinnati Milacron's P68, P70, and P69 specifications.

UNIVIS Special denotes a line of anti-stain hydraulic, bearing, and gear oils specially designed to minimize aluminum staining in cold-rolling aluminum operations while providing excellent equipment lubrication. The unique anti-stain properties of these oils derive primarily from their base oil, NORPAR®, the same aluminum rolling oil that has set the standard in the aluminum industry worldwide. The combination of NORPAR and specially selected additives minimizes the formation of stain-causing residue and oxidation products. Also, because UNIVIS Special oils are compatible with the rolling oil, they can help extend its life. Compared with conventional oils, UNIVIS Special has been proven to reduce downtime and increase productivity.

Owing to the high purity of the NORPAR base oil, and of its additives, the UNIVIS Special oils meet the requirements of FDA regulation 21 CFR 178.3570, "Lubricants for Incidental Contact with Food." All grades have high viscosity indexes, excellent anti-wear properties, rust-and-oxidation inhibition, and outstanding oxidation stability as demonstrated by their 2000+ hour lifetimes in the ASTM D 943 Oxidation Life test.

The four hydraulic oil grades, UNIVIS Special 22, 32, 46, and 68, are recommended for most piston and gear pump hydraulic applications and for most vane pump applications up to 3000 psig operating pressures. UNIVIS

Table 5-8. Characteristics of hydraulic oils with unusually high V.I. (Exxon Univis J).

Typical Inspections Grade	J13	J26
Gravity, °API 32.5	32.1	
specific at 15.6/15.6°C (60/60°F)	0.863	0.865
Viscosity,		
cSt @ 100°C	5.1	9.8
cSt @ 40°C	13.9	26.9
cSt@ -40°C	462	857
cSt @ -54°C	1800	—
SUS @ 210°F	43.2	59.0
SUS @ 100°F	75.8	134
Viscosity index	364	382
Color, ASTM D 1500	Red (dye)	0.5
Flash point, °C	104	103
°F	219	217
Pour point, °C	<-59	<-59
°F	<-75	<-75
Low temperature stability		
@-54°C	Pass	Pass

Table 5-9. Typical inspections for Exxon Univis N.

Typical Inspections **Grade**	**15**	**22**	**32**	**46**	**68**	**100**
Viscosity,						
cSt @ 40°C	16.4	22.0	34.9	46.0	73.8	104.7
cSt @ 100°C	4.1	5.0	6.9	8.5	12.1	15.7
SUS @ 100°F	87.5	113.9	177.0	233.0	376.0	535
SUS @ 210°F	39.8	43.0	49.3	54.7	67.9	82.3
Viscosity index	160	160	164	163	160	160
Pour point, °C	-45	-40	-40	-40	-35	-35
°F	-49	-40	-40	-40	-31	-31
Flash point, °C	150	180	194	198	204	218
°F	302	356	381	388	399	424
Rust test, ASTM D 665						
A. Distilled water	Pass	Pass	Pass	Pass	Pass	Pass
B. Synthetic sea water	Pass	Pass	Pass	Pass	Pass	Pass
Shear Stability, Bosch Injector, Modified ASTM D 3995, 250 cycles, viscosity loss, % @ 100°C	—	0.5	1.0	0.6	3.6	3.9
Oxidation Stability,						
Turbine oil, stability test, ASTM D 943, hr.	2000+	2000+	2000+	2000+	2000+	2000+

Special B 320 and B 320A are gear and bearing oil grades designed to provide long-term bearing and light-to-moderate-duty gear lubrication. UNIVIS Special EP 220 is designed for moderate-duty gear lubrication; with a 30-lb Timken OK load rating, it is intended for gear and bearing systems requiring EP protection. UNIVIS Special B 2200 is a high-viscosity bearing oil concentrate designed to restore the original viscosity of UNIVIS Special B 320 and B 320A when contaminated with low-viscosity aluminum rolling oil.

Where greater EP capability is desired in an anti-stain EP gear oil Exxon offers 3119, 3125 and 3126 EP Gear Oils, which have a 60-lb Timken OK load rating. However, these oils do not meet FDA requirements.

For characteristic data on UNIVIS S, refer to Table 5-10.

FIRE-RESISTANT HYDRAULIC FLUIDS

Mineral-oil hydraulic leaks in high-temperature operations can be costly—and potentially disastrous. If the fluid sprays or drips onto a hot surface, it can burst

Table 5-10. Characteristic data for UNIVIS Special hydraulic fluids.

Typical Inspections Grade	22	32	46	68	EP 220	B 320B	320A	B 2200
Viscosity,								
cSt @ 40°C	20.5	34.1	46.6	68.2	216.0	329.1	472	2122
cSt@ 100°C	6.0	8.8	11.1	14.7	27.3	45.3	60.0	169.7
SUS @ 100°F	105.3	170.8	233.3	341.5	1112	1682	2418	11,140
SUS @ 210°F	46.1	55.7	64.4	78.0	134	218	289	820
Viscosity index	270	262	241	228	162	197	199	195
Specific gravity @ 15.6 °C (60°F)	0.82	0.82	0.83	0.84	0.86	0.86	0.87	0.88
Flash point, COC, °C	118	118	118	118	137	118	118	127
°F	244	244	244	244	279	244	244	260
Pour point, °C	4	4	0	0	-18	4	4	-6
°F	40	40	32	32	0	40	40	21
Color, ASTM	0.5	0.5	0.5	0.5	0.5	0.5	0.5	0.5
Rust prevention, ASTM D 665B	Pass	Pass	Pass	Pass	Pass	Pass	Pass	Pass
Four-ball wear test, (40kg, 1200 rpm, 75°C. 1 hour), scar diameter, mm	0.42	0.42	0.43	0.43	0.40	0.49	0.48	0.47
Oxidation life, ASTM D 943, hr	2000+	2000+	2000+	2000+	2000+	2000+	2000+	—
Rotary Bomb Oxidation Test, life, minutes	200	175	200	217	117	82	85	—
Alcoa Aluminum Stain Test, Rating (10% dilution in rolling oil)	2	2	2	2	2	Light 3	Light 3	—
Timken OK load, ASTM D 2782, kg (lb)	—	—	—	—	14 (30)	—	—	—

Table 5-11. Comparison of Firexx and mineral oil properties.

	Nuto H 68	Firexx HF-A	Firexx HF-B	Firexx HF-C 46	Firexx HF-DU 68
Type of fluid	Premium mineral oil	HF-A: Oil-in-water emulsion	HF-B: Water-in-oil emulsion	HF-C: Water /glycol	HF-D: Synthetic (polyester)
Appearance	Amber fluid	Milky	Off-white	Clear red	Dark amber
Cost (mineral oil=1)	1	0.2-0.4	1.5	3	5
Fire resistance	Poor	Excellent	Excellent	Excellent	Good
Pump wear, ASTM D 2882 mg weight loss	25	Pump failure	164	<200	5.7
Relative wear protection	1	0.1	0.7	0.5	1.5
Pump guidelines	High pressure All pump types	Low pressure Piston, plunger Flooded suction Max. inlet size	Medium pressure Most pump types Flooded suction preferred	Medium pressure Most pump types Flooded suction preferred	High pressure All pump types
Operating Range °C (°F)	40-85 (103-183)	4-54 (40-130)	4-54 (40-130)	-18 to 60 (0-140)	-18 to 93 (0-200)
Freeze/thaw stability	Excellent	Poor	Moderate	Good	Excellent
Water content, %	0	95	39	42	0
Water hardness limit	N/A	250 ppm $CaCO_3$	N/A	N/A	N/A
Ease of recovery from water	Good	Good to Poor	Good	Poor	Good
Corrosion protection	Excellent	Fair	Good	Good	Excellent
ISO viscosity	68	N/A	N/A	46	68
Viscosity index	97	N/A	120-150	150	200
Specific gravity	0.88	1.005	0.928	1.09	0.90
Paint compatibility	Excellent	Very Good	Good	Moderate	Moderate
Seal compatibility	BN,N,V,T	NY,BN,V,T,P,N (B & E marginal)	BN,N,V,T,S	BN,N,V,T,S,B,E	NY,BN,N,V,T,S,P
Pour Point, °C (°F)	-27 (-17)	0 (32)	-37 (-35)	-45 (-50)	-26 (-15)
Flash Point, °C (°F)	222 (430)	—	—	—	306 (585)
Auto-ignition, °C (°F)	372 (700)	—	410 (770)	440 (825)	483 (900)

SEALS: BN—acrylonitrile-butadiene copolymer; N—polychloroprene; V—fluoroelastomer; T—PTFE; NY—nylon; P—polyurethane, B—butyl E—ethylene propylene copolymer; S—silicones

into flames and quickly propagate the fire.

Obviously, fire-resistant hydraulic fluids are designed to resist combustion. Specifically formulated to meet the stringent safety requirements of the mining and steel industries, these premium products help reduce fire hazards while providing excellent lubrication, foam resistance, and protection against rust and corrosion.

We will highlight four grades of fluids that offer a range of fire-resistant capabilities and lubricant properties:

Firexx HF-DU 68 (formerly Firexx HS 68)

At the top of the Exxon line, this synthetic polyol ester offers outstanding lubrication and pump protection. Firexx HF-DU 68 is recognized as a "less hazardous fluid" (HF-D) by the Factory Mutual group and has been approved under the FM2 regulation.

Firexx HF-C 46

This water/glycol fluid combines outstanding fire resistance with excellent performance at low temperatures and good resistance to corrosion. It is Factory Mutual approved (HF-C) as a reduced combustion hydraulic fluid, meets Denison HF-4 standards, and is recommended for applications with normal operating pressures up to 2,000 psi.

FIREXX HF-B

A pre-mixed invert (water-in-oil) emulsion, FIREXX HF-B provides superior anti-wear and anti-corrosion properties compared to oil-in-water emulsions. It meets Denison HF-3 standards as well as Factory Mutual Croup III (HF-B) 2N 3A3, Jeffrey Machine Co. #8, Lee-Norse Spec 100-5, MSHA 30-10-2, and USX 168.

FIREXX HF-A

For low-pressure applications where anti-wear properties are not critical and cost is a major concern, FIREXX HF-A oil-in-water emulsion provides outstanding fire resistance and excellent emulsion stability. FIREXX HF-A meets Factory Mutual Group IV (HF-A) 2N 3A3 standards, as well as MSHA 30-10-3 and Westfalia.

A comparison of FIREXX and mineral oil properties should be of interest. Refer to Table 5-11. Typical inspections are given in Table 5-12.

Thoroughness is the watchword when changing fluids in a hydraulic system. Sufficient time, thought and care can often mean the difference between successful operation and a system shutdown.

Table 5-13 presents general guidelines only. Consult the manufacturer for detailed instruction.

The values shown here are representative of current production. Some are controlled by manufacturing specifications, while others are not. All of them may vary within modest ranges.

Table 5-12. Typical inspections for FIREXX fire-resistant hydraulic fluids.

				FIREXX HF-A	
	FIREXX HF-DU 68	**FIREXX HF-C 46**	**FIREXX HF-B**	Concentrate	Finished Fluid
Appearance	Clear Amber Fluid	Clear Red Fluid	Milky White Fluid	B&C	Milky White Fluid
ISO Viscosity grade	68	46	—	—	—
Viscosity,					
SUS @ 100°F	316	—	438		—
SUS @ 150°F	—	—	145		—
SUS @ 175°F	—	—	100		
SUS @ 210°F	69	—	—		—
cSt @ 40°C	62.7	41	—	80	—
cSt @ 100°C	12.4	—	—		—
Viscosity index (VI)	200	150	135		—
Specific gravity @ 60°F	0.90	1.09	0.928		—
Pour point, °C(°F)	-37(-35)	-45(-49)	-37(-35)		0(32)
Flash point, °C(°F)	266(511)	—	N/A		N/A
Fire point, °C(°F)	306(585)	—	N/A		N/A
Auto-ignition temp., °C(°F)	483(900)	440(825)	410(770)		N/A
pH, neat	—	8.4	N/A		N/A
TAN, mg KOH/gm	2.5	—	N/A		N/A
Rust protection, ASTM D 665A	Pass	Pass	Pass		—

Table 5-13. Conversion procedures must be observed when changing to, or from, certain fire resistant hydraulic oils.

A SUMMARY OF CONVERSION PROCEDURES

OLD FLUID	NEW FLUID	Completely drain & clean the system.	Drain & blow out lines & valves with clean dry air.	Manually clean reservoir.	EXTREMELY important to remove residual fluid from the system!	Drain pumps, accumulators & cylinders to ensure removal of all fluid.	Dismantle, wash & clean strainers.	Drain & clean filter housings and replace filter elements
Mineral oil	HF-A	X (1)				X	X	X
	HF-B	X (1)	X	X		X	X	X
	HF-C	X (1)	X	X		X	X	X
	HF-DU	X (1)	X	X		X	X	X
HF-A High-water base (95/5) type fluid	Mineral oil	X (2)	X	X	X	X	X	X
	HF-B	X (3)						
	HF-C	X (3)						
	HF-DU	X (2)	X	X	X	X	X	X
HF-B Invert emulsion (water in oil) fluid	Mineral oil	X (2)	X	X	X	X	X	X
	HF-A	X						
	HF-C	X (3)	X	X		X	X	X
	HF-DU	X (2)	X	X	X	X	X	X
HF-C Water/glycol fluid	Mineral oil	X (2)	X	X	X	X	X	X
	HF-A	X (4)	X	X	X	X	X	X
	HF-B	X (2)(4)	X	X	X	X	X	X
	HF-DU	X (2)	X	X	X	X	X	X
HF-DU Synthetic (phosphate ester, polyol ester) fluid	Mineral oil	X	X	X		X	X	X
	HF-A	X (4)	X	X	X	X	X	X
	HF-B	X (4)	X	X	X	X	X	X
	HF-C	X (5)	X	X		X	X	X

(1) Small amounts of residual fluid should not interfere with performance but may reduce fire resistance.
(2) The use of low-pressure steam, high-pressure water or wiping are effective methods of cleaning.
(3) Small amounts of residual fluid usually do not interfere with fire-resistant properties but may adversely affect lubricating performance.
(4) Small amounts of water/glycol fluids or synthetics can cause severe phase separation.
(5) Small amounts of synthetic fluid may have an adverse effect on water/glycol fluids.

A SUMMARY OF CONVERSION PROCEDURES — continued

OLD FLUID (cont.)	NEW FLUID (cont.)	Establish compatibility of coatings inside of reservoir.	Obtain elastomer recommendations from mfg. Change seals where necessary.	Thoroughly drain & clean heat exchangers.	After cleaning, drain system & blow dry with clean air.	Close system & circulate minimum amt. of new fluid required to operate at low pressure.	Operate for 4-6 hours. Drain & refill with fresh fluid. Do not reuse flush fluid.	Inspect regularly for leakage, filter plugging & contamination.
Mineral oil	HF-A	X	X	X		X	X	X
	HF-B	X	X	X		X	X	X
	HF-C	X	X	X		X	X	X
	HF-DU	X	X	X		X	X	X
HF-A	Mineral oil	X	X	X	X	X	X	X
	HF-B					X	X	X
	HF-C					X	X	X
	HF-DU	X	X	X	X	X	X	X
HF-B	Mineral oil	X	X	X	X	X	X	X
	HF-A					X		X
	HF-C	X	X	X		X	X	X
	HF-DU	X	X	X	X	X	X	X
HF-C	Mineral oil	X	X	X	X	X	X	X
	HF-A			X		X		X
	HF-B		X	X		X	X	X
	HF-DU	X	X	X	X	X	X	X
HF-DU	Mineral oil	X	X	X	X	X	X	X
	HF-A			X		X		X
	HF-B		X	X		X	X	X
	HF-C		X	X		X	X	X

ENVIRONMENT-FRIENDLY HYDRAULIC FLUIDS

Hydraulic fluids, by tradition, are petroleum-based products. In recent years, industry has witnessed a move toward environment sustainability, especially where lubricants are concerned. Hydraulic systems are inherently susceptible to leakage due to their system design (high pressure versus seals) and require diligent maintenance to ensure they remain leak free. Unfortunately, hydraulic systems are often employed in environmentally sensitive situations, especially in mobile equipment and ship transportation. Government regulations have forced equipment and lubricant manufacturers to take a closer look at less toxic and more environmentally friendly hydraulic fluids. See Chapter 6—Food Grade and "Environment Friendly" Lubricants section on "Environment Friendly" hydraulic fluids.

Chapter 6

Food Grade and "Environment Friendly" Lubricants*

Today's food and food-associated processing plants are running faster and harder than ever before. Whether it is a can line, a dairy or a beverage bottler, plants can't afford to slow down. Rising costs, competitive pressures, and demanding production requirements are forcing food processing equipment of the type shown in Figures 6-1 and 6-2 to work harder, longer, and more efficiently. Lubricants used by the food industry had to meet not only these demanding performance requirements, but the stringent requirements of the Food and Drug Administration (FDA) and/or the United States Department of Agriculture (USDA). In response to this need, a few knowledgeable lubricant manufacturers have developed a complete line of food-grade lubricants that meet USDA and FDA requirements *and* the demanding performance requirements of the food processing industry.

Figure 6-1. Bottle filling line.

WHY USE FOOD-GRADE LUBRICANTS?

Prior to 1999, many lubricants that are used in the food processing industry were regulated by the FDA and USDA. USDA authorization is usually based on compliance with FDA regulations for direct and indirect food additives. Lubricants authorized by the USDA for use in certain food processing and other applications were typically defined by one of two USDA rating categories:

- USDA H-1—These lubricants could be used in equipment or applications in which the lubricant *may have incidental contact* with edible products.

Figure 6-2. Cheese packaging line.

*Source: Exxon Company, USA, Marketing Technical Services, Houston, Texas.

- USDA H-2—These lubricants were to be used only when there is *no possibility* of the lubricant coming in contact with edible products.

WHAT PERFORMANCE FEATURES ARE NEEDED?

The challenge in formulating lubricants for the food processing industry is to meet the necessary FDA and USDA food-grade requirements while also meeting the performance features needed to adequately protect food processing machinery. The required performance characteristics of a lubricant vary depending on the application, but key parameters often necessary for outstanding equipment protection are anti-wear, oxidation stability, extreme-pressure characteristics, and rust protection.

Please recall from earlier chapters our comments on brand names. Some lubricants are manufactured or marketed in different regions of the world with different brand designations. Yet, their standard product formulations have remained relatively consistent and are often known under the older, original names. This prompted our decision to keep the 1995-vintage brand names in this text.

If a specific lubricant is no longer available under its previous, widely used designation, you're encouraged to: (a) ask the supplier for the current name of a particular lubricant, and (b) compare the current "inspections"—the collective name for properties and ingredients—to those found widely accepted and described in this text. The significance of these deviations is highlighted in Chapter 3, Lubricant Testing. If the deviations are judged significant, consider requesting answers or explanations from the lubricant supplier.

Anti-Wear

Oils used in hydraulic systems are often subjected to high pressures and velocities. These forces can create conditions of thin-film lubrication and accelerated mechanical wear unless the fluid contains special protective additives. Each competent lubricant manufacturer has his own additives formulation. Take Exxon, for instance.

Exxon USDA H-1 rated oils that contain anti-wear additives include NUTO FG, TERESSTIC FG, UNIVIS SPECIAL MIST EP, and GLYCOLUBE FG which is a polyalkylene glycol synthetic oil and will be discussed separately. The anti-wear additives in these oils have been selected to provide peak performance in the equipment they are designed to lubricate. In addition, all of the anti-wear additives selected for these products meet the stringent requirements specified by a USDA H-1 approved rating.

Oxidation Stability

Oxidation stability is a measure of an oil's ability to resist oxidation, i.e., chemical deterioration, in the presence of air, heat, and other influences.

Oxidation resistance is an important quality in a lubricant. Insoluble oil and sludge resulting from oil oxidation can interfere with the performance of moving parts. Varnish and sludge can plug lines, screens, and filters and prevent equipment from operating efficiently. In addition, removing these contaminants can be very expensive and time-consuming.

Oxidation accelerates with time and increasing temperature. The deterioration process begins slowly, but speeds up as the oil nears the end of its useful life. Equipment metallurgy can also affect oxidation. Catalytic metals, such as copper and iron, which are commonly used in equipment, can also accelerate oxidation. The service life of an oil depends upon its ability to resist these influences.

Exxon's USDA H-1 rated lubricants have natural oxidation stability because they are formulated with extremely stable basestocks. In addition, the oxidation stability of many of these oils and greases is further enhanced with carefully selected additives.

Extreme-Pressure Protection

Extreme-pressure (EP) protection is a measure of an oil's ability to protect metal surfaces under heavy loads when the oil film has been pushed away or squeezed out by the mechanical action of gears or bearings. EP additives actually react with the metal surface to prevent welding, scuffing, and abrasion. Such additives, have to be carefully selected, however, because they can act as pro-oxidants, thus reducing the useful life of the oil.

Exxon USDA H-1 rated lubricants that demonstrate EP characteristics include NUTO FG (Table 6-1), UNIVIS SPECIAL MIST EP (Table 6-2), and two greases, FOODREX FG 1 and CARUM 330.

The EP additives in these oils have been selected to achieve the optimum balance between EP protection and oxidation life, while still meeting the requirements for a USDA H-1 lubricant.

Rust Protection

It is often difficult to keep lubrication systems free of water, particularly in the food industry where many machines are constantly washed down to keep the surface free of dirt and contaminants. Even under the most favorable conditions, rust is a possibility... and a potential problem.

Rust can score mating surfaces, form scale in piping, plug passages and damage valves and bearings. Ram shafts are sometimes exposed directly to the elements, and any

Table 6-1. Typical characteristics of food-grade hydraulic oils. (Source: Exxon Company, USA, Houston, Texas).

Typical Inspections

	NUTO FG 32	NUTO FG 46	NUTO FG 68	NUTO FG 100
Color, ASTM D 1500	0.5	0.5	0.5	0.5
Viscosity, cSt @ 40°C	31.0	46.5	68.2	102.1
Specific gravity @ 15.6°C(60°F)	0.867	0.870	0.873	0.876
Flash point, °C(°F)	200(392)	200(392)	200(392)	200(392)
Pour Point, °C(°F)	-23(-10)	-23(-10)	-20(-5)	-18(0)
Oxidation life,				
ASTM D 943, hrs	—	6000+	—	—
ASTM D 2272 (RBOT), min	232	254	240	265
Rust Test, ASTM D 665B	Pass	Pass	Pass	Pass
Copper corrosion, ASTM D 130	1a	1a	1a	1a
Water Separability, ASTM D 1401	Pass	Pass	Pass	Pass
FZG Test, Failure Load Stage (FLS)	—	10	—	—
Vickers V-104C Vane Pump Test,				
ASTM D 2882, total wt. loss, mg	—	2.5	—	—

pitting of their highly polished surfaces is likely to rupture the packing around them.

A competent supplier formulates all of its USDA H-1 food-grade lubricants with rust inhibitors to give extra protection against the destructive effects of water.

Next, we will examine a complete line of USDA H-1 rated food-grade lubricants, including hydraulic oils, gear oils, greases and can seaming lubricants. All of these products are formulated with basestocks and additives that meet the requirements specified by the USDA. In addition, all of these products are formulated to provide outstanding equipment protection. These USDA H-1 lubricants can be used in equipment or applications in which the lubricant *may have incidental contact* with edible products, Figure 6-3.

Figure 6-3. Margarine packaging plant.

Hydraulic Oils

NUTO FG (Table 6-1) is a line of super-premium hydraulic oils formulated with USP white oil basestocks. It is available in four viscosity grades (ISO 32, 46, 68, 100). Each grade provides outstanding wear protection for pumps, excellent extreme-pressure properties for bearings and lightly loaded gears, and superior oxidation stability for long, trouble-free life. NUTO FG is suitable for hydraulic systems up to pressures of 3000 psi.

Gear Oils

UNIVIS SPECIAL MIST EP (Table 6-2) is a line of premium synthetic gear oils formulated with polyisobutylene (PIB) basestocks. It is available in six viscosity grades (ISO 68, 100, 150, 220, 320, 460). These oils provide EP wear protection (30-lb Timken OK load) and have been successfully used for a number of years in the aluminum rolling industry. This

Table 6–2. Typical characteristics of food–grade gear oils. (Source: Exxon Company, USA, Houston, Texas.)

Typical Inspections	Univis Special Mist EP 68	Univis Special Mist EP 100	Univis Special Mist EP 150	Univis Special Mist EP 220	Univis Special Mist 320	Univis Special Mist 460
Viscosity,						
cSt @ 40°C	73.1	100.0	150 0	233.0	305.4	435.6
SUS@ 100°F	382	525	796	1245	1640	2350
Specific gravity @ 15.6°C (60°F)	0.852	0.858	0.861	0.864	0.866	0.870
Flash point, COC						
°C	146	148	148	150	158	170
°F	295	298	298	302	316	338
Pour Point						
°C	–39	–33	–33	–33	–30	–27
°F	–38	–27	–27	–27	–22	–17
Color, ASTM D 1500	0.5	0.5	0.5	0.5	0.5	0.5
Rust prevention, ASTM D 665A	Pass	Pass	Pass	Pass	Pass	Pass
Corrosion protection, ASTM D 130 1 hr @ 100°C	1A	1A	1A	1A	1A	1A
Timken OK load, ASTM D 2782, lb	30	30	30	30	30	30

particular lubricant is Anheuser-Busch Taste Test approved. Univis Special Mist EP also incorporates a mist suppressant and is suitable for use in mist lubrication systems.

Can Seamer Oil

Teresstic FG 150 (Table 6-3) is developed in close consultation with the can and beverage industries, is intended specifically for use in oil-lubricated can seamers. Formulated with USP white oil basestocks, Teresstic FG 150 incorporates a unique additive chemistry that provides outstanding anti-wear properties along with excellent rust protection, even in the presence of syrups and juices. It effectively emulsifies sugars and dry abrasives to prevent them from plating out on critical components. In addition to use in can seamers, Teresstic FG 150 can be used as a bearing and lightly loaded gear lubricant.

Table 6-3. Typical characteristics of food-grade can seamer oils. (Source: Exxon Company, U.S.A., Houston, Texas.)

Typical Inspections	
Teresstic FG 150	
Color, ASTM D 1500	0.5
Viscosity,	
cSt @ 40°C	146.8
cSt @ 100°C	14.3
Viscosity Index	97
Specific gravity @ 15.6°F (60°F)	0.879
Flash point, °C(°F)	200(392)
Rust test, ASTM D 665B	Pass

Refrigeration Oils

ZERO-POL S (Table 6-4) is a line of premium synthetic refrigeration lubricants formulated with polyalphaolefin (PAO) basestocks. It is available in two ISO viscosity grades (68 and 220). These oils have excellent thermal stability and extremely low pour points for use in refrigeration compressors in severe industrial service.

Greases

FOODREX FG 1 (Table 6-5) is a premium industrial grease formulated with an aluminum-complex thickener and USP white oil basestock. FOODREX FG 1 provides excellent water resistance and outstanding pumpability. It is white in color, has a smooth-tacky appearance and contains an extreme-pressure additive for carrying heavy loads. In addition, FOODREX FG 1 is KOSHER and PAREVE certified.

CARUM 330 grease (Table 6-6) is formulated with a calcium-complex thickener and USP white oil basestock. CARUM 330 provides excellent water resistance but is not recommended for use in central systems.

OVERVIEW OF USDA H-2 APPROVED LUBRICANTS

Exxon Company, U.S.A., offers a complete line of USDA H-2 food-grade lubricating oils and greases to meet nearly every USDA H-2 requirement found in the food processing industry. These products are to be used when there is no possibility of the lubricant coming in contact with edible products. Tables 6-7 through 6-9 summarize

Table 6-4. Typical characteristics of food-grade refrigeration oils. (Source: Exxon Company, U.S.A., Houston, Texas.)

Typical Inspections	**ZERO-POL S 68**	**ZERO-POL S 220**
Viscosity,		
cSt @ 40°C	68	240
cSt @ 100°C	10.4	27.2
Pour point, °C(°F)	–48(–54)	–45 (–49)
Cloud point, COC, °C(°F)	–51 (–60)	–51 (–60)
Flash point, COC, °C(°F)	254(489)	264(507)
Density @ 15°C, kg/m³	834	843

Table 6-5. Typical characteristics of food-grade greases formulated with aluminum-complex thickeners. (Source: Exxon Company, U.S.A., Houston, Texas.)

Typical Inspections	
FOODREX FG 1	
Penetration, ASTM D 217, worked 60X, mm/10	325
Thickener type	Aluminum–Complex
Color White	
Texture	Smooth, tacky
Dropping point, ASTM D 2265, °C(°F)	232(450)
Base oil viscosity	
SUS @ 100°F	800
SUS @ 210°F	75
cSt @ 40°C	60
cSt @ 100°C	14
Base oil viscosity index ASTM D 2270	90
Four–ball EP test, ASTM D 2596	
Weld point, kg	315
Load wear index, kg	39
Rust test, ASTM D 1743	Pass

Table 6-6. Typical characteristics of food-grade greases formulated with calcium-complex thickeners. (Source: Exxon Company, U.S.A., Houston, Texas.)

Typical Inspections	
CARUM 330	
Color Amber	
Texture	Buttery
Penetration, ASTM D 217, worked, 60 strokes, mm/10	325
Dropping Point, ASTM D 2265	
°C	260+
°F	500+
Base oil viscosity, cSt @ 40°C	70.4
Water washout, 79°C(175°F), ASTM D 1264, %	2
Corrosion protection, ASTM D 1743	1
Timken OK load, ASTM D 2509, kg(lb)	20(45)
Four–ball wear test, ASTM D 2266	
10kg, 1800 rpm, 75°C(167°F), 1 hr,	
scar diameter, mm	0.25
Oxidation test,	
ASTM D 942, kPa (psi) drop in 100 hrs	41.4 (6)

Table 6-7. Exxon USDA H-1 lubricants summary.

Lubricants	ISO/NLGI Grades	Typical Applications
NUTO FG	32, 46, 68, 100	hydraulic pumps, bearings, lightly loaded gears
UNIVIS SPECIAL MIST EP	68, 100, 150, 220, 320, 460	bearings, gear systems requiring EP protection of 30-lb Timken OK load or less
TERESSIC FG 150	150	can seamers, lightly loaded gears, bearings
ZERO-POL S	68, 220	refrigeration compressors
FOODREX FG 1 grease	1	bearings, gears, can seamers, most centrally lubricated grease systems
CARUM 330 grease	1	bearings, gears

Exxon USDA H-2 approved lubricants and their typical applications.

FOOD-GRADE POLYALKYLENE (PAG) SYNTHETIC LUBRICANTS

GLYCOLUBE FG is Exxon's line of extreme-pressure synthetic lubricants specially developed to provide superb lubricating performance in food processing and packaging machinery where incidental food contact may occur. Formulated with polyalkylene glycol (PAG) basestock and incorporating proven additive technology, GLYCOLUBE FG food-grade lubricants are designed for trouble-free performance and long service life. They are USDA H-1 compliant. In the manufacture of aluminum foil for the food industry, they offer the additional advantages of excellent low-stain and evaporative characteristics. Compared with food-grade white oils and non-food-grade mineral oils, GLYCOLUBE FG lubricants offer distinct advantages in oxidation and thermal stability, lubricity, wear protection and equipment cleanliness. Their extreme-pressure performance is comparable to that of commonly used sulfur- and phosphorus-containing EP gear lubricants.

The excellent oxidative and thermal stability of these lubricants assures long lubricant service life, even under heavy-load, high-temperature conditions. These performance features are highly cost-effective in their ability to

Table 6-8. Exxon USDA H-2 (food grade) oils summary.

Lubricants	Grades	Typical Applications
Aviation Oil EE	80, 100	aircraft piston engines
Coray	15, 22, 32, 46, 100, 150, 220	uninhibited naphthenic oils for once-through lubrication
Cylesstic TK	460, 680, 1000	worm gears and steam cylinders
Enmist	100, 220, 460	mist oils for gears and bearings
Esstic	32, 68, 150	naphthenic R&O oils for compressors and bearings
Faxam	22, 46	uninhibited paraffinic oils for once-through lubrication
Millcot K	220	adhesive oil for gears and bearings of textile and materials-handling equipment
Nuto H	32, 46, 68, 100, 150	premium-quality anti-wear hydraulic oils
Spartan EP	68, 100, 150, 220, 320 460, 680, 1000, 1500, 2200, 3200	premium-quality industrial gear oils
Spinesstic	10	high quality oil for lubrication of high-speed machine elements
Synesstic	68	synthetic (diester) lubrication for air compressors, hydraulic systems, and bearings
Teresstic	32	super-premium gas and steam turbine lubricant
Teresstic 32	33, 46, 68, 77, 100, 150, 220, 320, 460	premium circulating oils for bearings and turbines
Univis Special	22, 32, 46, 68	anti-stain hydraulic and bearings oils for the aluminum rolling industry
Univis Special	EP 220, B 320, B320A, B2200	anti-stain bearing and gear oils for the aluminum rolling industry
Zerice	22, 46, 68	uninhibited oils for refrigeration systems
Zerice N	22, 46, 68	premium quality inhibited oils designed specifically for ammonia refrigeration

*This list is inclusive of only those products for which Exxon has applied and received USDA H-2 approval. Other lubricants may meet the requirements of USDA H-2 and may have been formally approved.

Table 6-9. Exxon USDA H-2 greases* summary.

Lubricants	Soap Type	Typical Applications
Andok 260	sodium	synthetic-grease for extra-long life in ball and roller bearings
Beacon 325	lithium	synthetic-grease for bearing lubrication
Estan 1 and 2	calcium	extremely water resistant grease for bearings
Lidok EP 1 and EP 2	lithium	general-purpose EP grease for bearings and gears
Nebula EP 0 and EP 1	calcium-complex	extremely water resistant grease for bearings and gears
Ronex MP	lithium-complex	multi-purpose high-temperature grease
Unirex N 2	lithium-complex	electric motor bearings

significantly reduce lubricant consumption and maintenance shutdowns. Also, the unusually high viscosity indexes of Glycolube FG lubricants (187-220 vs. 90-100 for most petroleum gear lubricants) facilitate low-temperature startup and help maintain acceptable viscosity over a wide temperature range. This eliminates the need for seasonal lubricant changeovers and simplifies lubricant inventories.

Glycolube FG lubricants keep equipment cleaner than conventional lubricants. The highly stable polyalkylene glycol basestock has very low deposit-forming tendency, and its superior solvency keeps deposit-forming materials dispersed, thus preventing them from separating as sludge or contributing to the formation of varnish or lacquer. These lubricants are suitable for a wide range of applications and operating environments in hydraulic, bearing and gear drive systems.

They are suitable for use with most elastomeric materials used in seals and gaskets. Following is a partial listing of common elastomers compatible with GLYCOLUBE FG:

"Viton"	Butyl Rubber	Natural Red Rubber
"Kalrez"	Buna N	Natural Gum Rubber
Silicone	"Hycar"	Neoprene
Polysulfide	"Fluoraz"	"Hypalon"
EPR	Natural Black Rubber	"Aflas"
EPDM		"Teflon"

Testing has shown GLYCOLUBE FG lubricants to be compatible with silicone rubber 732 RTB and "Loctite" PST and 290 in direct lubricant contact and in exposure to the sealants between bonded copper surfaces.

Because of their high viscosity indexes, GLYCOLUBE FG lubricants are not classified by a single AGMA viscosity rating. A GLYCOLUBE FG lubricant will effectively span two or three AGMA petroleum lubricant ratings over the operating range of most gearboxes. Table 6-10 and Figure 6-4 can be used as guides to the proper selection of a GLYCOLUBE FG lubricant to replace an AGMA petroleum-base lubricant. If either the required viscosity at operating temperature or the AGMA rating of the current lubricant are known, one can readily determine the appropriate GLYCOLUBE FG grade.

Storage, Handling, and Changeover

GLYCOLUBE FG lubricants are stable, non-corrosive materials that can be stored in carbon steel tanks. Heated storage tanks can be employed for outside storage. Heated tanks and piping should be completely insulated. Preferably, GLYCOLUBE FG should not be in contact with industrial coatings during storage, since it may soften and lift such coatings. If coatings cannot be removed, clean all filters and strainers frequently, especially during initial use. Tanks previously used for petroleum products should be flushed clean before GLYCOLUBE FG is introduced, since it is slightly miscible with petroleum-base lubricants.

GLYCOLUBE FG is only slightly hygroscopic. If moisture

Table 6-10. Typical physical properties of a proven polyalkylene glycol (PAG) lubricant.

Typical Physical Properties

Property	**GLYCOLUBE FG 150**	**220**	**400**
Weight per Gallon, lb			
at 60°F (15.6°C)	8.35	8.38	8.39
at 68°F (20°C)	8.32	8.35	8.36
Specific Gravity, 20/20°C	0.9996	1.0036	1.0041
Viscosity, cSt			
at 100°F (37.8°C)	155	285	425
at 210°F (98.9°C)	22.9	40.7	59.5
Viscosity index	187	206	220
Pour Point, °F (°C)	–20 (–29)	–15 (–26)	–5 (–21)
Flash Point, ASTM D 93, °F(°C)	355(179)	405(207)	355(179)
Coefficient of Expansion at 55°C, per °C	0.00078	0.00081	0.00079
Additives			
Rust Inhibitor	Yes	Yes	Yes
Oxidation Inhibitor	Yes	Yes	Yes
Extreme Pressure	Yes	Yes	Yes
Turbine Oil Test, ASTM D 665A	Pass	Pass	Pass
Copper Corrosion, ASTM D 130	No Effect	No Effect	No Effect
Babbitt Corrosion (89 Sn/7.5 Sb/3. 5 Cu)	No Effect	No Effect	No Effect
Mist Test	Pass	Pass	Pass
FZG Spur Gear Test, Stages Passed (12 max)	12	12	12
Shell Four Ball Wear Test, ASTM D 2266			
Scar diameter, mm	0.38	0.38	0.38
Timken Test, lb			
OK Load	40	45	45
Score	45	50	50

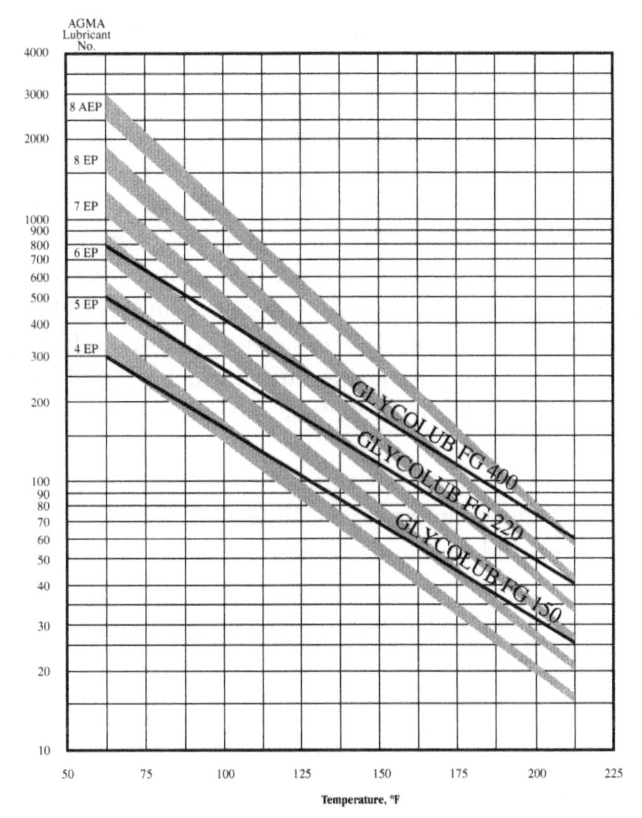

Selection Guide

GLYCOLUBE FG Lubricants	AGMA Grades Usually Replaced
GLYCOLUBE FG 150	3 – 6
GLYCOLUBE FG 220	5 – 7
GLYCOLUBE FG 400	6 – 8A

Figure 6-4. Viscosity ranges of GLYCOLUBE FG Lubricants vs. AGMA ratings of petroleum lubricants.

content is critical, take precautions to prevent atmospheric moisture from entering the storage tank. A desiccant unit can be installed on the vent line, or the tank can be blanketed with dry air or nitrogen. Where viscosities in excess of 500 cSt are to be handled, a rotary or gear pump is preferred. Transfer lines should be carbon steel and of adequate size to handle the desired flow and viscosity with a reasonable pressure drop in the line. A three-inch line should be provided for unloading bulk shipments.

When preparing to change over to GLYCOLUBE FG, it is important to determine its compatibility with the former lubricant. The supplier can assist in making this determination. If the two lubricants are shown to be incompatible, employ the following procedures ***before*** installing GLYCOLUBE FG. At a minimum, drain the old lubricant, clean the system to remove possible sludge and varnish, inspect seals and elastomers and replace the filters or clean the screens. If residual contamination is suspected, wipe or flush with a small amount of solvent or GLYCOLUBE FG; in new units, follow the same procedure to remove preservative or coating fluids.

After installing GLYCOLUBE FG, adjust the lubricators to deliver the manufacturer's recommended rate of lubricant. Check the filters or screens frequently during the early stages of operation, as GLYCOLUBE FG will likely loosen residual sludge, varnish and paint.

NOTE: In gearbox applications, after 24 hours of operation, the lubricant should be drained and the gearbox refilled.

It is also important to determine the compatibility of GLYCOLUBE FG with the elastomers and coatings in the system (see earlier discussion).

These food-grade lubricants meet the incidental-food-contact specifications of FDA Regulation 21 CFR 178.3570(a) and are USDA H-1 compliant. They are suitable for use in meat, poultry, and egg processing.

"ENVIRONMENT FRIENDLY" LUBRICANTS BENEFITS AND DRAWBACKS

There is a growing public interest in environmentally friendly, or "green," products, i.e., products that do not harm the environment during their manufacture, use, or disposal. Manufacturers and marketers have capitalized on this trend by introducing products claimed to be less harmful to the environment than competing products.

However, in the absence of standardized criteria, some companies have made untested and misleading claims regarding the environmental features of their products. For example, a manufacturer claimed that its plastic trash bags were biodegradable, but failed to note that such bags will not biodegrade under land-fill conditions.

A strong environmental commitment is a basic obligation that any business has to its customers and the community. This interest is not served by companies that make unproved and exaggerated environmental claims for their products or that fail to fully inform their customers of significant tradeoffs associated with environmentally oriented products. Until the establishment of meaningful environmental labeling standards, consumers should take a critical and questioning view of any product that is claimed to be "environmentally friendly."

This section specifically examines the environmental claims made by some lubricant manufacturers.

Ambiguity of Environmental Claims for Lubricants

The terms most often discussed with respect to environmentally friendly lubricants
are "biodegradable" and "non-toxic." Both of these terms

are ill-defined and severely situation dependent. There are many variables and little standardization in biodegradation testing. A given material may be found to be highly biodegradable under one set of test conditions and only moderately biodegradable under another. Thus, when a material is said to be "biodegradable," it is important to know the specific test circumstances. (For a detailed discussion, see the entry "Biodegradation," Chapter 3.)

The term "non-toxic" is similarly ambiguous. A material found to be non-toxic to one species may be toxic to another. Nevertheless, some lubricant manufacturers have claimed lubricants to be "non-toxic" on the basis of tests with only a single type of organism. The ambiguity of the term "non-toxic" is further exemplified by a recent case in which a lubricant had been reformulated to eliminate an EPA-identified y
component. However, the reformulated product, claimed to be less toxic, was found to contain an additive suspected of causing skin reactions in humans. This made the product OSHA hazardous.

Even assuming the validity of the environmental claims made for a product, there are important potential trade-offs that must realistically be considered. For example, the performance and useful life of a "green" lubricant may be significantly inferior to that of an alternative product, as discussed below.

Natural Base Oils

At first glance, the use of "natural" base oils, such as vegetable oils, as lubricants appears to be appropriate and prudent from an environmental perspective. Undoubtedly, the more rapid biodegradability of such oils versus petroleum oils is desirable in the event of accidental, routine, or excessive environmental exposure.

However, for most industrial lubricant applications, a number of additional factors must be considered. Foremost among these considerations is product performance. Vegetable oils have poor hydrolytic and oxidative stability; this may necessitate more frequent oil changes and result in significant disposal problems that may outweigh any environmental advantages. They also have relatively high pour points, which can impair low-temperature performance. Additives such as pour depressants and anti-oxidants may help compensate for these drawbacks, but they tend to reduce biodegradability and may increase the toxicity of the overall product to humans and the environment.

Additionally, vegetable oil-base lubricants are more susceptible to microbial action, which can both limit their storage life and rapidly degrade their performance in use.

As for the recyclability of "natural" oils, there is a practical problem here as well. Because these oils are not compatible with mineral oils, it may be difficult to find a recycler that will accept "natural" oils.

Conclusion

The development of environmentally friendly lubricants is an extremely worthwhile goal. However, in the absence of standardized test methods and guidelines, consumers would be well-advised to ask the following questions before purchasing any product claimed to be "environmentally friendly" or more environmentally responsible:

- Are the environmental claims made for the product valid and well-documented?
- What performance debits or other trade-offs are associated with the product?
- Do these trade-offs outweigh the environmental advantages of using the product?

"ENVIRONMENT FRIENDLY" HYDRAULIC FLUIDS: CONCEPTS AND CLAIMS

"Environmentally friendly" is a term used broadly today to identify products that are perceived to have little or no adverse effect on the environment, either through their manufacture, use, or disposal. However, while neutral impact on the environment is a commendable goal, almost all products brought to market affect the world around us.

The Federal Trade Commission (FTC), in fact, has explicitly discouraged use of the term "environmentally friendly" in product marketing. Because the term is not clearly defined, either legally or in practical terms, it is often misused and misunderstood. For example, conventional (petroleum-based) hydraulic fluids have frequently been considered to be "environmentally unfriendly" relative to vegetable oil-based fluids, due to their slower rate of biodegradation. However, the reduced useful life of vegetable oil-based products, the difficulties surrounding recycling them into lower uses, and the difficulty of disposal can potentially impact the environment as well. These considerations must be weighed in balancing the debits and credits of each lubricant type.

Generally, products represented as "environmentally friendly" have good biodegradability and low environmental toxicity, which, unfortunately, represent only a few of the properties necessary to fully describe the environmental compatibility of a product. These properties are often characterized by a single quantitative value, but there is no corresponding single reference value that is widely accepted against which to compare results. Further, the significance of the difference between any two values must be understood in the context of how and why the tests were done.

But let's consider a biodegradable hydraulic fluid, "UNIVIS BIO 40." This product (Table 6-11) was developed to meet the growing global demand for more "environmentally responsible" hydraulic fluids. It is a biodegradable*, vegetable-oil-based lubricant, with low toxicity, designed to meet the latest hydraulic equipment requirements. UNIVIS BIO 40 provides the high-performance characteristics of a premium quality conventional hydraulic oil with the added assurance of reduced environmental impact. UNIVIS BIO 40 can help achieve environmental objectives and, should accidental release occur, will lessen the damage and facilitate spill management.

Biodegradability

UNIVIS BIO 40 meets and exceeds the requirements of biological degradation as defined by the OECD guideline and the CEC L-33-A-94 primary biodegradation test method. The CEC procedure measures the natural biodegradability of a substance using non-acclimated, naturally occurring organisms. This test method tracks the disappearance of the hydraulic oil over a period of time using infrared techniques. UNIVIS BIO 40 is biodegradable at not less than 97% within 21 days, minimizing harm to soil or water by release of fluid.

Toxicity

UNIVIS BIO 40 is non-toxic as defined by the tests in the box below.

Oral Limit Test	OECD 401	Non-toxic (LD50>2000mg/kg)
Dermal Limit Test	OECD 402	Non-toxic (LD50>2000mg/kg)
Skin Irritation	OECD 404	Non-irritating to skin
Eye Irritation	OECD 405	Non-irritating to eyes
Ames Test		Not mutagenic

Table 6-11. Typical inspections for UNIVIS 40 biodegradable hydraulic fluid.

Typical Inspections

The values shown here are representative of current production. Some are controlled by manufacturing specifications, while others are not. All may vary within modest ranges.

UNIVIS BIO 40	
Kinematic Viscosity	
cSt @ 40°C	38
cSt @ 100°C	9
Viscosity Index	240
Specific Gravity @ 15.6°C	0.921
Flash Point, °C	230
Pour Point, °C	–42
Rust Protection, ASTM D 665B	Pass
Oxidation Stability, Rotary Bomb Oxidation Test (RBOT)	
ASTM D 2272, minutes	180
Demulsibility, ASTM D1401 (30 minutes) oil–water–emulsion	41–39–0
Slow Cool Data,	
Viscosity @ –30°C, cSt	
after one day	2300
after three days	2400
after seven days	2400
Load–Carrying Capacity, FZG pass stage	12

Anti-Wear

The excellent anti-wear characteristics of UNIVIS BIO 40 ensure extended pump life in hydraulic systems. In addition, UNIVIS BIO 40 exhibits excellent load-carrying ability, as demonstrated in the FZG gear test. These characteristics ensure exceptional protection against wear and scuffing.

High Viscosity Index

The high viscosity index of UNIVIS BIO 40 provides for minimal viscosity variation over a broad temperature range. Since high V.I. is a natural property of the basestock, the shear stability of the fluid is inherently superior to that of V.I.—improved oils.

Rust and Corrosion Protection

UNIVIS BIO 40 provides excellent rust and corrosion protection to help protect expensive system components.

Demulsibility

Demulsibility characteristics of UNIVIS BIO 40 ensure clean water separation.

Compatibility

UNIVIS BIO 40 is compatible with conventional mineral oils. It should be noted, however, that contamination of UNIVIS BIO 40 with other fluids may lead to a reduction in the biodegradability and other performance characteristics and could increase product toxicity. The degree of quality degradation will vary with the level and type of contamination.

UNIVIS BIO 40 is compatible with seals made of Nitrile, Viton and Acrylate. It is not suitable for use with Crude, Butyl or SBR elastomers.

UNIVIS BIO 40 can be used in industrial hydraulic applications in terrestrial and aquatic habitats where concerns exist about the release of these fluids to the environment.

Refer to Chapter 3 and the entry "Biodegradation" for further information on key terms and data.

COMPARING VEGETABLE BASED ENVIRONMENT FRIENDLY HYDRAULIC FLUIDS TO THE NEWER SYNTHETIC BASED FLUIDS

The original "environment friendly" hydraulic oil offerings were vegetable based products that utilized base oils derived from such products as soybean, canola and sunflower seed and are classified as Hydraulic Environmental Triglyceride (HETG) fluids.

There are currently four "environment friendly" hydraulic fluids available, one vegetable based product and three synthetic based products. As with all synthetics each type offers different characteristics suitable for different operating environments giving the equipment designer and end user some flexibility in their lubricant of choice. Table 6-12 defines the advantages and disadvantages of the four fundamental types of available "environment friendly" lubricants.

As with all lubricants, these environmental friendly fluids require careful consideration and testing before placing them in service. Their seal and hose compatibility, and operating temperature will play are big part in their choice. As with all lubricants, when replacing an already existing lubricant the correct choice of flushing oil is crucial to operational success requiring consultation with the oil company prior to use.

Use of a environmental friendly lubricant still required regular system maintenance to ensure a leak or spill does not occur in the first instance, and if a major leak or spill does occur, most authorities still require the spill to be reported. The difference with using environmental friendly lubricants is how the spill or leak is managed.

Table 6-12. "Environment Friendly" Hydraulic Fluid Comparison Table

Lubricant	Type	Base Oil	Advantages	Disadvantages
HETG Hydraulic Environmental Triglyceride	Vegetable	SoyaBean, Canola, Sunflower	• Excellent biodegradability • Non toxic • Excellent Lubricity • High VI (>200) • High Flash Point (>225°)	• Thickens at high and low temperatures • Rapid oxidization • Hydrolyze in water • Poor nitrile seal compatibility • Miscible with mineral oil (lowers biodegradability)
HEES Hydraulic Environmental Ester Synthetic	Synthetic	Synthetic Esters	• Excellent service life • Excellent thermal stability • Excellent oxidative stability • Good low temp flow	• Miscible with mineral oil (lowers biodegradability) • Hydrolyze in water • Poor nitrile seal compatibility • Expensive
HEPR Hydraulic Environmental Polyalphaolefin & Related	Synthetic	Polyalohaolefin (PAO)	• Excellent service life • Excellent oxidization stability • Good corrosion protection • Good lubricity • Good VI • Low pour point (-5 to -40°F)	• Only low PAOs are considered environment friendly • Poor nitrile seal life • Limited operating temperature range (-20 to 212°F) • Expensive
HEPG Hydraulic Environmental Poly Glycol	Synthetic	Polyalkylene Glycol (PAG)	• Excellent fire resistance • Good viscosity range available • Can be used as an anhydrous (water-free) lubricant	• Poor polyurethane seal compatibility • Poor nitrile seal compatibility • Limited operating temperature range (-5 to 180°F) • Biodegradability dependent on propylene/ethylene ratio

Chapter 7

Synthetic Lubricants*

Judicious application of properly formulated synthetic lubricants can benefit a wide spectrum of process machinery. This informed usage will drive down overall maintenance and downtime expenditures and can markedly improve plant profitability.

However, although synthetic lubricants have gained considerable acceptance in many forward-looking process plants world-wide, there are still misconceptions which impede the even wider acceptance many of these fluids so richly deserve. One of the erroneous understandings is that for a synthetic lubricant costing $45.00 per gallon to be justified, the drainage or replacement interval should be five times that of a mineral oil costing $9.00 per gallon. This reasoning does not take into account such savings as labor, energy, downtime avoidance, disposal of spent lubricants and equipment life extension.

A serious engineer or maintenance professional would be well advised to take a closer look at the profusion of authenticated case histories covering the widest possible spectrum of machinery. In 1990, one major chemical company documented yearly savings of $70,200 for 36 right-angle gear units driving cooling fans in a process plant. A refinery then saved $120,000 per year for Ljungstrom furnace air preheaters and greatly extended mean-time-between-repairs (MTBR) for Sundyne high speed gear units. On these Sundyne units, strangely enough, many users continue to use automatic transmission fluid (ATF). In this particular service, ATF is demonstrably inferior to properly formulated synthetic oils.

FORMULATIONS

The most knowledgeable formulators use a polyalphaolefin/diester blend. Additives are more readily soluble in diesters than in PAO. Therefore, PAO/diester blends are stable over a very wide temperature range. These superior synthetic base oils must be blended with additives to obtain the high level of performance required.

It should be emphasized that the additives represent by far the most important ingredients of properly formulated, high performance synthetic lubricants. Often, additives used in synthetic oil formulations are the same conventional additives used to formulate mineral oils, resulting in only marginal performance improvements. Truly significant performance improvements are obtained only when superior synthetic base oils are blended with superior synthetic additive technology.

The various proven PAO/diester blends contain synergistic additive systems identified with proprietary trade names (Synerlec, etc.). The synergism obtained in a competent additive blend combines all of the desirable performance properties plus the ability to ionically bond to bearing metals to reduce the coefficient of friction and greatly increase the oil film strength. The resulting tough, tenacious, slippery synthetic film makes equipment last longer, run cooler, quieter, smoother and more efficiently. Synergistic additive systems, in service, "micro-polish" bearing surfaces reducing bearing vibration, reducing friction and minimizing energy consumption. This gets us into the topic of "how and why."

How and Why

The most valuable synthetic lubricant types excel in high film strength and oxidation stability. However, while there are many high film strength oils on the market, these may not be appropriate for some process machine applications. High film strength oils based on extreme pressure (EP) technology and intended for gear lubrication may typically incorporate additives such as sulfur, phosphorus and chlorine which are corrosive at high temperatures and/or in moist environments. Sensitive to this fact, a reputable lubricant manufacturer thus would not offer an EP industrial oil with corrosive additives as a bearing lubricant for pumps, air compressors, steam turbines, high speed gear reducers and similar machinery.

At least one U.S. manufacturer of synthetic lubricants can lay claim to having pioneered the development of noncorrosive high film strength industrial

*Sources: Bloch, H.P., and Pate, A.R. (Jr.); "Consider Synthetic Lubricants for Process Machinery," *Hydrocarbon Processing*, January, 1995. Also: Bloch, H.P., and Williams, John B., "High Film Strength Synthetic Lubricants Find Application in Process Plant Machinery," *P/PM Technology*, April, 1994.

oils with outstanding water separation properties. Although such oils may not be critically important to the operating success of vast numbers of pumps, air compressors and turbines, which quite obviously have been running without high film strength oils for years, there are compelling reasons to look into the merits of superior lubricants. There is a considerable body of thoroughly evaluated evidence that *properly formulated* synthetic lubricants based on diesters, PAO, or a combination of these base stocks will result in significantly reduced bearing and gear operating temperatures.

Our advice to the serious maintenance professional is to ascertain the requirements needed for maximum performance in specific equipment. Look at the published specifications of various oils and determine their relative merits for the intended service. Make an informed decision based on the facts and then monitor the field performance. Chances are you will greatly increase equipment reliability by picking the *right* synthetic lube.

ORIGIN OF SYNTHETIC LUBES

Synthetic-based fluids, used in the production of synthetic lubricants, are manufactured from specific chemical compounds that are usually petroleum derived. The base fluids are made by chemically combining (synthesizing) various low molecular weight compounds to obtain a product with the desired properties. Thus, unlike petroleum oils which are complex mixtures of naturally occurring hydrocarbons, synthetic base fluids are man-made and have a controlled molecular structure with predictable properties. These are "generalized" in Table 7-1.

There is no typical synthetic lubricant. The major classes are as different from each other as they are from petroleum lubricants. Synthesized base fluids are classified as follows:

1. Synthesized hydrocarbons (polyalphaolefins)
2. Organic esters (diesters and polyol esters)
3. Polyglycols
4. Phosphate esters
5. Silicones
6. Blends

The first four base fluids account for more than 90% of the synthetic fluids used worldwide. The first three contain only atoms of carbon, hydrogen and oxygen. The first two are of greatest interest to machinery engineers in modern process plants.

EXAMINING SYNTHETIC LUBES

Understanding the principal features and attributes of the six base fluids will place the potential user in a position to prescreen applicable synthetics and to question suppliers whose offer or proposal seems at odds with these performance stipulations.

Synthetic Hydrocarbon Fluids

Synthetic hydrocarbon fluids (SHF), such as those with a polyalphaolefin (PAO) base, provide many of the best lubricating properties of petroleum oils but do not have their drawbacks. (Even the best petroleum oils contain waxes that gel at low temperatures and constituents that vaporize or readily oxidize at high temperatures.) The SHF base fluids are made by chemically combining various low molecular weight linear alpha olefins to obtain a product with the desired physical properties. They are similar to cross-branched paraffinic petroleum oils because they consist of fully saturated carbon and hydrogen.

These man-made fluids have a controlled molecular structure with predictable properties. They are available in several viscosity grades and range from products for low temperature applications to those recommended for high temperature uses. They are favored for their hydrolytic stability, chemical stability and low toxicity.

Organic Esters

Organic esters are either dibasic acid or polyol types. Dibasic acids have shear-stable viscosity over a wide temperature range (-90°F to 400°F), high film strength, good metal wetting properties and low vapor pressure at elevated temperatures. They easily accept additives, enhancing their use in many commercial applications and especially as compressor lubricants.

Polyol esters have many of the performance advantages of dibasic acid esters and can be used at even higher temperatures. They are used principally in high-temperature chain lubricants, for industrial turbines and in some aviation applications.

Polyglycols

Polyglycols were one of the first synthetic lubricants developed. The polyglycols can be manufactured from either ethylene oxide, propylene oxide or a mixture of both. The propylene oxide polymers tend to be hydrocarbon soluble and water insoluble, while the ethylene oxide tends to be water soluble and hydrocarbon insoluble. In many applications, the physical properties of the finished product can be engineered

*Table 7-1. Generalized properties of synthetic hydrocarbon lubes**

	Mineral Oils	Polyisobutenes	Polyalpha-olefins	Alkylated Aromatics	Polyalkylene Glycols	Perfluoroalkyl Ethers	Polyphenyl-ethers	Dicarboxylic Acid Esters	Neopentyl Polyesters	Trialkyl Phosphate Esters	Triakyl Phosphate Esters	Silicone Oils
Viscosity Temperature Behavior (VI)	Moderate	Poor	Very Good	Moderate	Very Good	Moderate	Poor	Very Good	Very Good	Poor	Excellent	Excellent
Low Temperature Behavior (Pourpoint)	Poor	Moderate	Excellent	Good	Good	Good	Poor	Excellent	Very Good	Moderate	Excellent	Excellent
Liquid Range	Moderate	Poor	Very Good	Good	Good	Excellent	Poor	Very Good	Very Good	Very Good	Good	Excellent
Oxidation Stability (Aging)	Moderate	Moderate	Very Good	Moderate	Good	Excellent	Very Good Good	Very Good/	Very Good	Very Good	Moderate	Very Good
Thermal Stability	Moderate	Moderate	Moderate	Moderate	Good	Excellent	Excellent	Good	Very Good	Very Good	Good	Very Good
Evaporation Loss, Volatility	Moderate	Moderate	Very Good	Good	Good	Excellent	Good	Excellent	Excellent	Very Good/	Very Good/	Very Good/
Fire Resistance, Flash Temperature	Poor	Poor	Poor	Poor	Moderate	Excellent	Moderate	Moderate	Moderate Very Good/	Excellent/ Very Good/	Excellent/	Good
Hydrolytic Stability	Excellent	Excellent	Excellent	Excellent	Good	Excellent	Excellent	Moderate	Moderate	Moderate	Good	Good
Corrosion Protection Properties	Excellent	Excellent	Excellent	Excellent	Good	Poor	Moderate	Moderate	Moderate	Moderate	Moderate	Good
Seal Material Compatibility	Good	Good	Very Good/	Good	Good	Excellent	Good	Moderate	Moderate	Poor	Poor	Good
Paint and Lacquer Compatibility	Excellent	Excellent	Excellent	Excellent	Moderate	Very Good/	Moderate	Moderate	Moderate	Poor	Poor	Good
Miscibility with Mineral Oil	—	Excellent	Excellent	Excellent	Poor	Poor	Good	Very Good/	Very Good/	Moderate	Moderate	Poor
Solubility of Additives	Excellent	Excellent	Very Good/	Excellent	Moderate	Poor	Very Good/	Very Good/	Very Good/	Excellent	Excellent	Poor
Lubricating Properties, Load Carrying Capacity	Good	Good	Good	Good	Very Good/	Excellent	Excellent	Very Good/	Very Good/	Excellent	Moderate	Poor
Toxicity	Good	Excellent	Excellent	Poor	Good	Excellent	Good	Good	Good Good	Very Good// Poor	Moderate/	Excellent
Biodegradability	Moderate	Poor	Good/ Moderate	Poor	Excellent Very Good/	Poor	Poor	Excellent Very Good/	Excellent Very Good/	Very Good/	Very Good/	Poor
Price Relation Against Mineral Oil	—	3-5	3-5	3-5	6-10	500	200-500	4-10	4-10	5-10	5-10	30-100

*Source: Technische Akademie Esslingen, Germany

by adjusting the ratio of ethylene oxide and propylene oxide in the final molecular structure.

Polyglycols have excellent viscosity and temperature properties and are used in applications from -40°F to 400°F and have low sludge-forming tendencies. A major application for polyglycol lubricants is in compressors that handle hydrocarbon gases. This is due to the nonhydrocarbon-diluting properties inherent in polyglycols. The polyglycols' affinity for water results in poor water separability.

Phosphate Esters

Phosphate esters are organic esters that, when used with carefully selected additives, provide a group of synthetic fluids that can be used where fire resistance is required. Even when ignited, the phosphate esters will continue to burn only if severe conditions required for ignition are maintained. Some phosphate esters are less stable in the presence of moisture and heat. The products of the resulting degradation are corrosive and will attack paints and rubbers. The poor viscosity index (VI) limits the operating temperature range for any given phosphate ester product.

Silicones have been in existence for many years and offer a number of advantages as lubricants. Silicones have good viscosity versus temperature performance, excellent heat resistance, oxidative stability and low volatility. Silicones are chemically inert and have good elastomer compatibility. Poor metal-to-metal lubricating properties and high cost limit their use to specialized applications where their unique properties and high performance can be justified.

Blends of the Synthetic Lubricants

Blends of the synthetic lubricants with each other or with petroleum lubricants have significant synergistic results. In fact, many of the synthetic lubricants being sold consist of a blend of two or more base materials to enhance the properties of the finished product.

Synthetic lubricants have been steadily gaining industrial acceptance since the late 1950s. In many applications today, they are the specified lubricant of the compressor manufacturer. This is especially true in rotary screw and rotary vane air compressors.

While the greatest industrial acceptance has been with air compression, many other industrial applications can be economically justified. Synthetic lubricants are currently being used in compressors processing such diverse materials as ammonia, hydrogen, hydrocarbon gases, natural gas, hydrogen chloride, nitrogen and numerous others.

Synthetic lubricants are not limited to compressors but are used in gear boxes, vacuum pumps, valves, diaphragm pumps and hydraulic systems. Synthetic lubricants are being used in applications that need more efficient, safe lubrication or where the environmental conditions preclude the use of traditional petroleum products.

PROPERTIES AND ADVANTAGES

Synthetic lubricant fluids provide many of the best lubricating properties of mineral oils but do not have their drawbacks. In fact, synthetics have these advantages over comparable petroleum-based lubricants:

- Improved thermal and oxidative stability
- More desirable viscosity-temperature characteristics
- Superior volatility characteristics
- Preferred frictional properties
- Better heat transfer properties
- Higher flash point and auto-ignition temperatures.

Experience clearly shows that these advantages result in the following economic benefits:

- Increased service life of the lubricant (typically four to eight times longer than petroleum lubricants)
- Less lubricant consumption due to its low volatility
- Reduced deposit formation as a result of good high-temperature oxidation stability
- Increased wear protection resulting in less frequent maintenance
- Reduced energy consumption because of increased lubricating efficiency
- Improved cold weather flow properties
- Reduced fire hazard resulting in lower insurance premiums
- Higher productivity, lower manufacturing costs and less downtime because machines run at higher speeds and loads with lower temperatures
- Longer machinery life because less wear results in more production during life of machine and tools.

Synthetic lubricant base stocks, while possessing many of the attributes needed for good lubrication, require fortification with additives relative to their intended use. An experienced formulator takes into consideration a range of requirements:

Dispersion of Contaminants

It is important to keep internally and externally generated oil-insoluble deposit-forming particles suspended in the oil. This mechanism reduces the tendency of deposits, which lower operating efficiency, to form in critical areas of machinery. Additives that impart dispersing characteristics are called "dispersants" and "detergents." A dispersant is distinguished from a detergent in that it is nonmetallic, does not leave an ash when the oil is burned and can keep larger quantities of contaminants in suspension.

Protecting the Metal Surface from Rust and Corrosion

Humidity (water) type rust and acid type corrosion must be inhibited for long surface life. An oil film itself is helpful but this film is easily replaced at the metal surface by water droplets and acidic constituents. Additives that have an affinity for a metal surface, more so than water or acids, are used in oils to prevent rust and corrosion and are generally referred to as simply "rust inhibitors."

Oxidative Stability

Oils tend to thicken in use, especially under conditions where they are exposed to the atmosphere or where oxygen is present. This phenomenon is chemically termed "oxidation."

Oxygen reacts with the oil molecule initiating a chain reaction that makes the molecule larger, thereby decreasing fluidity. Conditions that assist the oxidation process are heat, oxidation catalyzing chemicals, aeration and perhaps other mechanisms that allow the oxygen to easily attach itself. Additives that retard the oxidation process are termed "oxidation inhibitors."

Wear Prevention

Inevitably the metal surfaces being lubricated come in contact. Whenever the speed of relative motion is low enough, the oil film does not stay in place. This can also happen if the loading on either or both surfaces is such that the oil film tends to be squeezed out. When moving metal surfaces come in contact, certain wear particles are dislodged and wear begins. Additives that form a protective film on the surfaces are called "anti-wear agents."

Viscosity Index Improvers

Viscosity index improvers function to improve viscosity/temperature relationships, that is, to reduce the effect of temperature on viscosity change.

Foam Suppressants

Foam suppressants allow entrained air bubbles to collapse more readily when they reach the surface of the oil. They function by reducing surface tension of the oil film.

Oiliness Additives

Oiliness additives are materials that reduce the oil friction coefficient.

Surfactants

Surfactants improve the ability of the oil to "wet" the metal surface.

Alkalinity Agents

Alkalinity agents impart alkalinity or basicity to oils where this is a desirable feature.

Tackiness Agents

Tackiness agents impart stringiness or tackiness to an oil. This is sometimes desirable to improve adhesive qualities.

Obviously then, the lubricant supplier or formulator has to choose from a number of options. There are technical considerations to weigh and compromises to make. Close cooperation between supplier and user is helpful; formulator experience and integrity is essential.

CASE HISTORIES

The following are highlights from the many successful case histories of the late 1980's and 1990's.

Circulating Oil System for Furnace Air Preheaters

Several major refineries in the U.S. and Europe had experienced frequent bearing failures on these slow-rotating heat exchangers while operating on the manufacturer-recommended mineral oil. With bearing housings typically reaching temperatures around 270°F, the cooled and filtered mineral oil would still overheat to the point of coking. Bearing failures after six months of operation were the norm. After changing to a properly formulated synthetic, a lubricant with superior high-temperature capabilities and low volatility, bearing lives were extended to several years. As was mentioned in our introductory paragraph, one refinery alone has documented, in 1990, savings of at least $120,000 per year since changing lubricants.

Right-angle Gear Drives for Fin Fan Coolers

A European facility achieved a disappointing mean-time-between-failures (MTBF) of only 36 months on 36 hypoid gear sets in a difficult to reach, elevated area. In fact, using mineral oil (ISO VG 160), a drain interval of six months was necessary to obtain this MTBF. Each oil change required 12 man-hours and temporary scaffolding at a cost of $1,000. Change-over to an appropriate synthetic, i.e., a synthetic with optimized temperature stabilizers, wear reducers and oxidation inhibitors, has allowed drain intervals to be increased to two years while obtaining a simultaneous increase in equipment MTBF. Detailed calculations showed a net benefit of $1,950 per year per gear set. Combined yearly savings: $70,200 with no credit taken for power reduction or avoided production curtailments. That was in 1990—just project it to today!

Plant-wide Oil Mist Systems.

An oil mist lubrication system at a Southeast Texas chemical plant experienced an unscheduled shutdown as a result of cold weather. Twenty-seven mist reclassifiers in this system were affected. These reclassifiers provided lubrication to several fin fans, two electric motors and the rolling element bearings in 14 centrifugal pumps. Wax plugging of the mist reclassifiers brought on by the cold weather caused the unexpected shutdown. As a result, several bearings failed because of lubricant starvation. An ISO VG 68 grade conventional mineral oil was the source of the wax.

The oil mist system had to be isolated and blown out to avoid further bearing failures. In addition to the downtime costs, significant labor and hardware costs were required to restore the unit to normal operation.

For this reason, a synthetic wax-free lubricant replaced the mineral oil. Neither the oil feed rate nor the air-to-oil ratio required adjusting after switching to the synthetic.

Since converting, in 1980, to a diester-based oil mist system, the following has resulted:

- No cold weather plugging of the mist reclassifiers has been experienced.
- No lubricant incompatibility has been detected with other components of the oil mist system.
- The synthetic lubricant is providing proper bearing wear protection as evidenced by no increase in required maintenance for pumps, fans or motors served by the oil mist system.
- Downtime, labor and hardware replacement costs attributed to cold weather operational problems have been eliminated.
- Savings in contractor and plant manpower used to clean the reclassifiers have averaged close to $40,000 per year.
- Two failures of pumps and motors were assumed to be prevented via use of wax-free lubricant. The savings equaled $7,000 per year for each of the two pumps.

Total net credit has been estimated at over $100,000 per year. This does not include any process losses associated with equipment outages.

(For cost justsification calculations refer to Chapter 11.)

Pulverizing Mills in Coal-fired Generating Plant.

A large coal-fired power generating station in the southwestern U.S. was having lubrication problems with their coal pulverizing mill. The equipment, a bowl mill pulverizer, was experiencing the following problems lubricating the gears that drive the mill:

- The lubricant was losing viscosity and had to be changed every four to six months.
- Air entrainment in the lubricant was causing cavitation in the pumps that circulated the lubricant.
- The gears were experiencing an unacceptable level of wear as measured by a metals analysis on the lubricant.
- On very cold mornings, the lubricant was so viscous it had to be heated before the unit could be put in service.
- The petroleum-based lubricant's initial viscosity varied significantly.

After evaluating the options, it was decided that a synthetic-based lubricant offered the best solution. In cooperation with a major synthetic lubricant manufacturer, they decided a synthetic hydrocarbon base stock with the proper additive package would be the best choice. Additive package concentrations were evaluated in a number of bowl mills simultaneously to establish the optimum level and composition. Figure 7-1 shows the dramatic effect on metal gear wear accomplished over a 1,000-hour trial period.

The synthetic hydrocarbon base stock has proven to be extremely shear resistant. One particular bowl mill has been closely monitored during 54 months of operation (Figure 7-2) to establish viscosity stability. The data represent only operating hours, not total time elapsed, since the unit is not operated continuously. The performance has been excellent and lubricant life has exceeded 60 months.

The synthetic hydrocarbon lubricant was compared to two petroleum-based lubricants supplied by major oil companies. The tests were run on three bowl mills that had recently been reworked and tested. All three bowl mills were fed the same amount of coal during the test period. All three gear oils were the same ISO 320 viscosity grade.

The average current draws were:

Product	Amps
Petroleum #1	70
Petroleum #2	75
Synthetic hydrocarbon gear oil	68

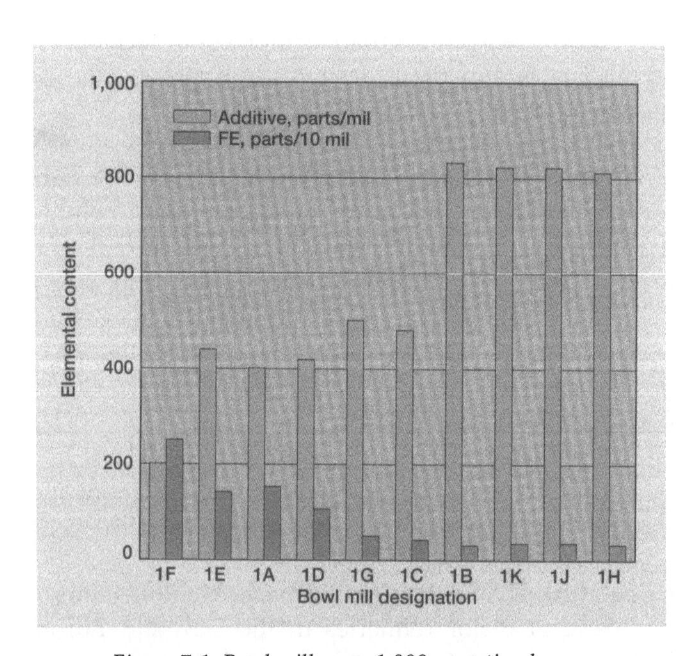

Figure 7-1. Bowl mill wear, 1,000 operating hours.

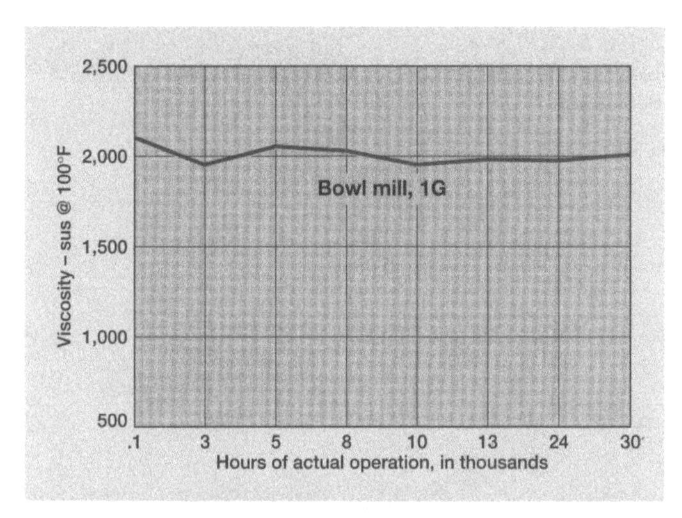

Figure 7-2. Viscosity stability of synthetic lubricant in bowl mill gear unit.

The lower amp difference shown by the synthetic hydrocarbon is the result of the lower coefficient of friction shown in Table 7-2.

Efficiency gains can be very sizeable and the resulting reduction in energy cost will often pay for the higher cost of synthetic lubricants within months. Table 7-3 shows a typical cost benefit analysis and additional data are given in Chapter 11.

As demonstrated, the synthetic hydrocarbon gear oil has solved the original problems and provided additional benefits not anticipated. The switch to synthetic lubricants has clearly improved performance and achieved significant savings in operating costs, as shown in the following tabulation.

The extended drain interval provides savings in three areas:

1.	**Lubricant consumption cost savings:**		
	Petroleum oil cost per gal	$12.00	
	Petroleum oil changes per yr	2	
	Volume of gear box, gal	300	
	Petroleum oil cost per yr		
	($4.00/gal)(2 changes/yr)(300 gal/unit) =	$7,200	
	Synthetic oil cost per gal	$48.00	
	Synthetic oil changes per yr	0.2	
	Volume of gear box, gal	300	
	Synthetic oil cost per yr		
	($16.00/gal)(0.2 changes/yr)(300 gal/yr) =	$2,880	
	Annual savings on lubricant cost—	**$4,320**	**per unit**
2.	**Reduced maintenance cost savings**		
	Petroleum oil changes per yr	2	
	Maintenance cost per change	$1,500	
	Petroleum oil maintenance cost per yr		
	(2 changes/yr)($500/change) =	$3,000	
	Synthetic oil changes per yr	0.2	
	Maintenance cost per change	$1,500	
	Synthetic oil maintenance cost per yr		
	(0.2 changes/yr)($500/change)	$300	
	Annual savings in scheduled maintenance costs—	**$2,700**	
3.	**Lubricant disposal costs**		
	Petroleum oil used per yr, gal	600	
	Disposal cost per gal	$1.00	
	Cost of disposal	$600	
	Synthetic oil used per yr, gal	60	
	Disposal cost per gal	$1.00	
	Cost of disposal	$60	
	Annual savings in disposal cost per year—	**$540**	

The reduction in energy consumption also provides significant savings:

Average annual power cost using petroleum oil lubricant	$49,920
Average annual power cost using synthetic lubricant	$46,815
Annual savings in power consumption—	**$3,105**

Table 7-2. Physical properties of ISO VG 320 gear oil.

	Petroleum	Synthetic hydrocarbon
Viscosity index	95	140
Thermal conductivity, Btu/hr/ft^2	0.071	0.085
Coefficient of friction	0.101	0.086
Pour point, °F	+5	–45

Table 7-3. Cost benefit analysis.

Synthetic Lubricant Price X Annual Volume	$......................................
Conventional Lubricant Price X Annual Volume	$......................................
NET COST DIFFERENCE (A)	$......................................
Credit For Enhanced Performance Of Synthetic Lube	
Extended Service Life	$......................................
Reduced Downtime	$......................................
Reduced Maintenance Cost	$......................................
Reduced Disposal Cost	$......................................
Reduced Energy Consumption	$......................................
Reduced Lubricant Inventory	$......................................
TOTAL OF CREDITS FOR SYNTHETICS LUBRICANT (B)	$......................................

Net Cost Difference (A)
– Total Credits (B)

Savings Through Use of Synthetic Lube (C)

The total annual savings for all of the above categories amount to $10,665. In addition, savings in reduced wear and thus fewer repairs are certain to be realized. A forward-looking process plant needs to explore the many opportunities for often substantial cost savings that can be achieved by judiciously applying properly formulated synthetic lubricants.

Returning to the questions raised at the beginning of this chapter, Exxon offers the following comments by way of summation.

- **Should a synthetic lubricant be used?**

Yes - if it is cost effective (increased productivity, extended lubricant life, etc.)

Yes - if a conventional lubricant has not worked (problem solver).

Yes - if it enhances safety or environmental aspects of an operation (higher flash & fire points, reduction of used lubricant requiring disposal).

Yes - if it reduces risk (failure to change out systems, reduced chance of misapplication through lubricant consolidation).

- **What type of synthetic lubricant?**

Key considerations here are temperature extremes in operation, material compatibility, equipment requirements and methods of its application.

- **What are the requirements for effective use of the selected synthetic lubricant?**

In selecting a lubricant for demanding lubricant applications, there are generally one or two key imperatives that must be satisfied for things to work. Temperature extremes, lubricant service life, extreme loads, safety and environmental aspects usually are the key drivers. One or more of the demands will drive the selection of a synthetic for a specific application.

Synthetic lubricants offer significant advantages over conventional lubricants under demanding conditions. Their judicious use has enabled users to capture the following benefits:

— Increased Productivity
— Enhanced Equipment Performance
— Cost Savings

— Enhanced Safety
— Enhanced Environmental Aspects

The decision to use a synthetic lubricant and selection of the best lubricant is a process with a multitude of interrelationships. Equipment manufacturers, lubricant suppliers, maintenance and engineering staff, along with your own experience, can aid in the worthwhile process of improving plant efficiency through proper application of synthetic lubricants.

NOTE: Certain synthetic lubricants have been formulated for specific machinery applications. Refer to "compressors" for information on synthetics using polyalkylene glycols, "pulp and paper" for data on a full synthetic used in saw mills, and "gas engines" for more information on ESTOR Elite, a low-ash formulation, full synthetic, allowing extended drain intervals.

SYNTHETIC LUBRICANTS FOR EXTREME PRESSURE AND TEMPERATURE

Using Exxon's SPARTAN® and SYNESSTIC® synthetic lubricants as an example, we will attempt to illustrate the merits of properly formulated industrial lubricants.

SPARTAN® Synthetic EP excels because it flows freely in arctic temperatures that "freeze" conventional mineral oils stiff, Figure 7-3. It keeps its viscosity at steel-mill temperatures, Figure 7-4, that turn mineral oils to watery liquids. At the same time, it resists oxidation and demonstrates excellent volatility control at high temperatures—for long-term, reliable service.

Exxon's SPARTAN Synthetic EP gear oil consists of seven ISO grades (100-1000) suitable for a wide range of industrial gears that are subject to severe operating conditions, such as high pressures, shock loading or extremes in temperatures. The long service life of SPARTAN Synthetic EP makes it the best choice whenever routine lubricant changes are difficult or costly. SPARTAN Synthetic EP also is an excellent lubricant for both plain and rolling-contact bearings. These oils use high-quality polyalphaolefin basestocks and a proprietary additive package that together meet or exceed the tough AGMA, U.S. Steel (224), Cincinnati-Milacron, and David Brown specifications for EP gear oils. Refer to Table 7-4 for typical inspections.

Polyalphaolefins Make the Difference

To understand polyalphaolefins, it helps to start with paraffins.

Long-chain paraffin molecules of 20 to 40 carbon atoms have many excellent properties such as oxidation stability and viscosity that does not change drastically as temperature goes up or down. Most important, they lubricate well because they cling to metal surfaces and slide past each other easily.

In fact, these molecules might be ideal lubricants except for one serious drawback: Somewhere around room temperature (depending on the length of the chain), they crystallize and pack together like sticks of dry spaghetti. The result is a solid matrix of wax.

In petroleum-base lubricants, paraffins work because they occur naturally attached to cyclic structures that crystallize at lower temperatures. For the wax crystals that do form, added pour-point depressants help

Figure 7-3. The arctic demands free-flowing synthetic lubricants.

Figure 7-4. Steel mill temperature environment calls for synthetic lubricants.

keep them from growing large enough to cause trouble.

But petroleum-base lubricants still have problems at temperature extremes: cyclic components get too thin at high temperatures and don't resist oxidation well—and even the best pour point depressants lose effectiveness at extremely low temperatures.

The ideal solution would be a paraffin that couldn't crystallize into wax. It should be exceptionally pure and uniform, with a narrow boiling range and virtually no variation in batch-to-batch properties.

That solution exists in the polyalphaolefin (PAO) basestocks used in superior synthetic EP industrial gear oils.

PAOs are specially synthesized branched paraffins with three to five 10-carbon chains united in a star-like structure. The shape virtually defies crystallization. The PAO molecules in SPARTAN Synthetic EP resist freezing down to -40°C(-40°F) or lower, and they come close to the ideal lubricant in other ways too: they excel at maintaining viscosity, resisting oxidation, and controlling high-temperature volatility.

With the addition of Exxon's proprietary additive package, SPARTAN Synthetic EP grades 150 and higher support a Timken OK load in excess of 100 lbs, compared to 60 lbs for a conventional petroleum-base EP gear oil.

Figures 7-5 through 7-8 convey the performance advantages that can be obtained from the many grades of this PAO-based synthetic EP industrial gear oil.

Oxidation of an oil—breakdown due to heat and oxygen—causes viscosity to increase, and it creates soft sludges and hard deposits that can lead to equipment failure. Among conventional petroleum-base products, a premium EP gear oil offers outstanding oxidation performance at operating conditions up to 93°C(200°F). As shown in Figure 7-5, SPARTAN Synthetic EP carries that performance to the extreme, staying clean up to 121°C(250°F).

Superior volatility control is illustrated in Figure 7-6, while pour point and viscosity characteristics can be compared in Figures 7-7 and 7-8. As can be seen, SPARTAN Synthetic EP is ideal for arctic and other cold-weather environments, because it keeps flowing even at -30°C (-22°F) and colder. Most conventional petroleum-base gear oils become too thick to pour below -10°C (14°F). Machines start easier and gear boxes run more efficiently with high quality synthetic EPs.

CASE HISTORIES INVOLVING PAO-BASED SYNTHETIC EP OILS

When a triple-race roller bearing on a crown roll fails, it can cost $25,000 or more to replace. One southeastern paper mill was replacing each bearing at least every two years. In one instance, a bearing lasted only nine months.

The problem arose from the crown roll's unusually low rotational speed. The machine (Figure 7-9) did not generate enough centrifugal force to keep the rollers pressed against the bearing's outer raceway, so rollers leaving the load zone stopped rotating. As each

Table 7-4. Typical inspections for SPARTAN *Synthetic EP oils.*

SPARTAN Synthetic EP Typical Inspections

The values shown here are representative of current production. Some are controlled by manufacturing specifications, while others are not. All of them may vary within modest ranges.

Grade	100	150	220	320	460	680	1000
AGMA number	3 EP	4 EP	5 EP	6 EP	7 EP	8 EP	8AEP
ISO viscosity grade	100	150	220	320	460	680	1000
Viscosity @ 40°C, cSt	106	151	220	327	427	680	989
@ 100°F, SUS	519	778	1149	1712	2241	3159	5154
@ 210°F, SUS	78	100	130	180	225	289	378
Viscosity index	151	150	151	162	167	160	154
API gravity	36.5	35.4	35.0	34.5	34.1	32.3	31.0
Demulsibility, ASTM D 2711							
Total free water, ml	86	85	85	84.5	84	85	8.4
% Water in oil	0.4	0.5	0.8	0.8	0.8	0.7	0.8
Emulsion, ml	0	0	0	0	0	0	0
Rust test, ASTM D 665A & B	Pass	Pass	Pass	Pass	Pass	Pass	Pass
Four-ball EP test, ASTM D 2783							
Weld point, kg	250	250	250	250	250	250	250
Load wear index	54	53	46	46	47	47	47
Copper corrosion, ASTM D 130, rating	1-A	1-A	1-A		1-A	1-A	1-A
Foam test, ASTM D 892, ml							
Sequence I	Tr/0						
Sequence II	Tr/0						
Sequence III	Tr/0						
Timken OK load, ASTM D 2782, kg (lb)	36(80)	45+(100+)	45+(100+)	45+(100+)	45+(100+)	45+(100+)	45+(100+)
Flash point, °C (°F)	231 (448)	236 (457)	249 (480)	250 (482)	254 (490)	270 (518)	280(536)
Pour point, °C (°F)	-55 (-67)	-52 (-62)	-46 (-50)	-46 (-50)	-40 (-40)	-30 (-22)	-20 (-4)
FZG test, pass	12 stages	12 stages	12 stages	12 stages	12 stages	12 stages	12 stages
Oxidation tests							
ASTM D 2893	Pass	Pass	Pass	Pass	Pass	Pass	Pass
USS S-200[1]	Pass	Pass	Pass	Pass	Pass	Pass	Pass
Mist test, ASTM D 3705							
% Stray mist	20	19	18	18	17	18	—
% Condensed mist	80	81	82	82	83	83	—

[1]Described in U.S. Steel 224 EP Gear Oil Specification.

roller reentered the load zone, it skidded like an airplane tire first touching ground. Damage to the bearing raceways quickly led to bearing failure.

The competitive petroleum-base EP gear oil used by the mill couldn't stop the destruction, and use of a higher viscosity oil was not possible: the oil is shared by the hydraulic system, which would have suffered startup problems with a higher viscosity.

The switch to SPARTAN Synthetic EP brought immediate benefits: its high viscosity index meant good fluidity at cold startup conditions. Also, its outstanding lubricity protected the expensive bearings. After two full years with the Exxon synthetic product, the bearings showed no evidence of unusual wear.

The second case history involved a 20-ton overhead crane at a major Midwestern steel company which had trouble every time the weather got cold. Located above an open railcar entryway, the gear boxes of the crane (Figure 7-10) were exposed to ambient temperatures, making the crane hard to start and difficult to keep running whenever the mercury dropped. When temperatures fell to -40°C (-40°F) one winter, the crane drew so much power trying to start that it blew the system's circuit breakers.

Based upon manufacturer's specifications, the crane's gear boxes were lubricated with a conventional ISO 320 EP gear oil. The pour point of this oil was -9°C (15°F), so the oil varied from stiff to solid in cold weather.

Working together, the technical services group of the oil manufacturer, steel-mill personnel, and the crane manufacturer determined that SPARTAN Synthetic EP ISO 220 would be an acceptable substitute for the petroleum-base gear oil.

SPARTAN Synthetic EP solved the problem. Its low

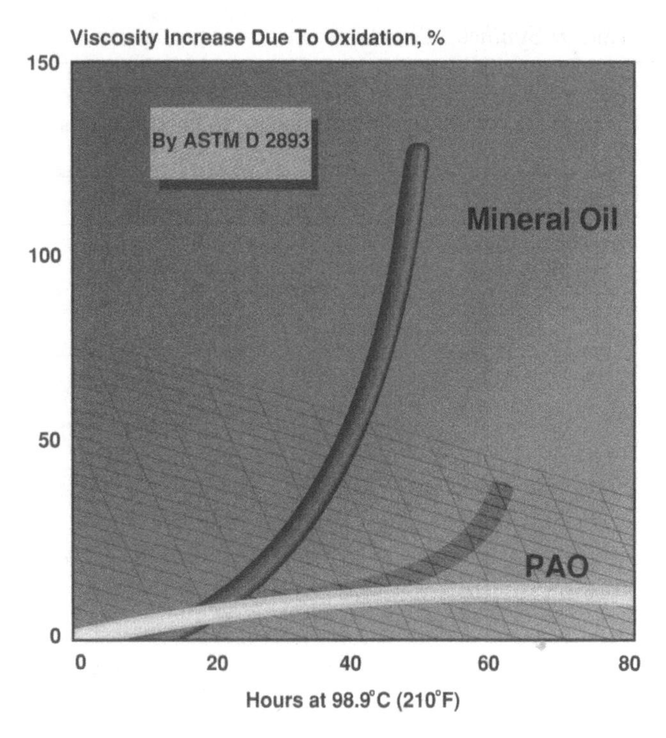

Figure 7-5. Viscosity increase due to oxidation, mineral oil vs. synthetic EP product. (Source: Exxon Company, USA, Houston, Texas.)

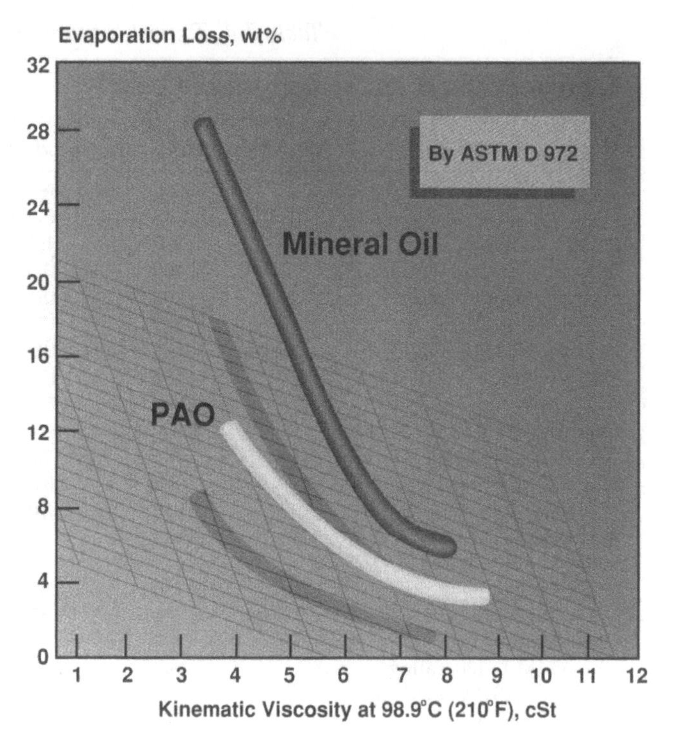

Figure 7-6. A superior synthetic EP oil will give superior volatility control to provide long-term lubrication effectiveness. Plus, the low volatility and high flash point compared to conventional gear oils give an added margin of safety at high operating temperatures. (Source: Exxon Company, USA, Houston, Texas.)

pour point made cold-weather startup easy, eliminating excess power drain. The high viscosity index provided good film thickness in summer temperatures, even at the lower viscosity grade. The high Timken OK Load rating ensured outstanding extreme-pressure protection. In addition, SPARTAN Synthetic EP appeared to provide energy savings, a subject touched on in the next segment of our text.

Finally, the papermaker's dilemma: To get more tonnage (Figure 7-11), run the machine faster, and turn up the heat. To get less downtime, slow the machine, and lower the heat.

SPARTAN Synthetic EP now helps a Southeastern mill run fast and hot *and* minimize downtime.

One big problem for this mill was the drive gears in the press section. Because of high ambient temperatures plus heavy loads, these gears operated continuously at 200°F. The conventional EP gear oil used by the mill oxidized to a viscous black sludgy material so fast that even oil changes every six months were not frequent enough to ensure that gears and bearings would stay adequately lubricated.

The mill switched to SPARTAN Synthetic EP. Two years later, when the original charge of the product was routinely tested, it still met the gear manufacturer's requirements.

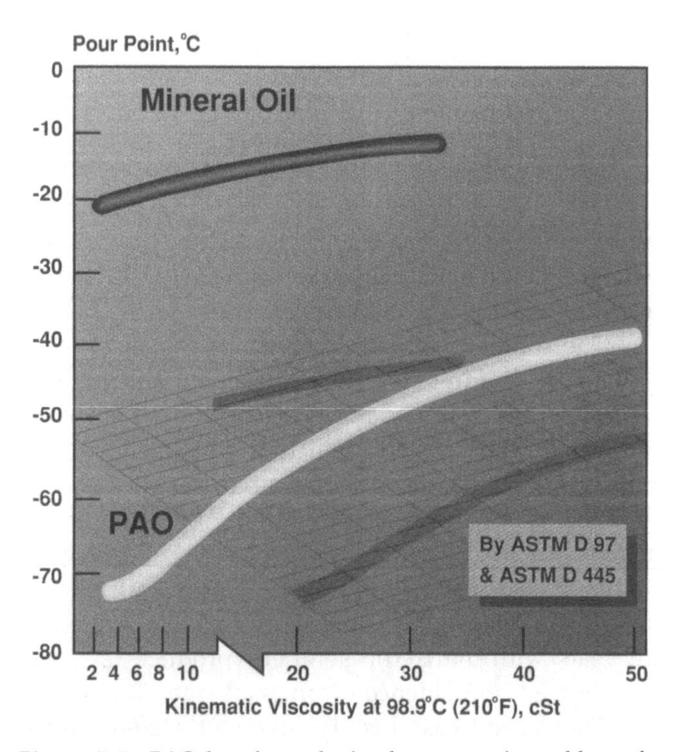

Figure 7-7. PAO-based synthetics have superior cold-weather performance. (Source: Exxon Company, USA, Houston, Texas.)

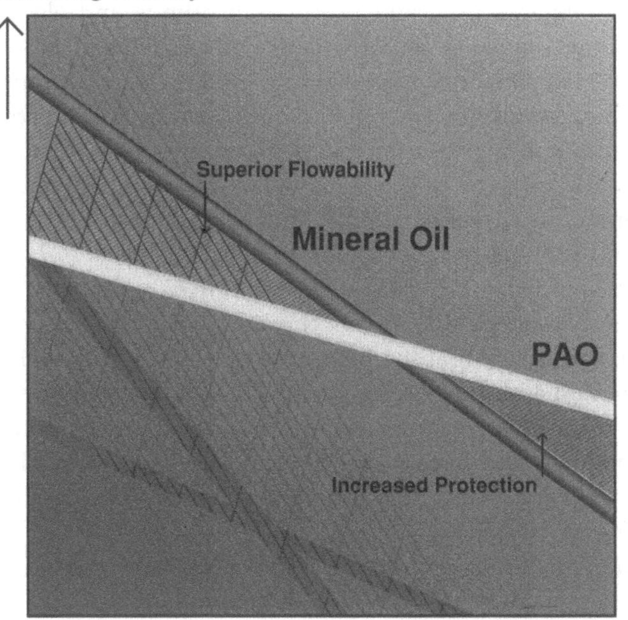

Figure 7-8. High-quality polyalphaolefin basestock helps keep viscosity stable over a wide range of temperatures. The viscosity index for SPARTAN Synthetic EP gear oils ranges from 150 to 167, compared to 90 to 100 for a conventional petroleum-base oil.

Figure 7-10. Overhead crane gear box exposed to severe ambient environment.

Figure 7-9. Paper machine rolls in severe duty service.

Figure 7-11. Tons of paper on a single roll.

The long life and excellent oxidation stability of SPARTAN Synthetic EP pay dividends: fewer oil changes mean lower costs, less downtime, and less material to dispose of, and continuous on-spec performance without viscosity increase means energy-efficient operation and maximum gear service life.

DIESTERS: ANOTHER SYNTHETICS OPTION

We had earlier considered SPARTAN Synthetic EPs, a line of six long-life, extreme-pressure industrial gear and bearing lubricants manufactured from synthesized hydrocarbons, predominantly polyalphaolefins. They are designed to provide outstanding performance under severe temperatures and loads. Applications include gear boxes, industrial differentials and highly-loaded rolling contact bearings. The thermal and oxidative stability of SPARTAN Synthetic EP, superior to that of conventional gear oils, provides excellent resistance to sludging and helps ensure long lubricant life, wen under the severe conditions encountered in small, high-temperature gear boxes. Because the synthesized hydrocarbons contain no wax, these lubricants can be used in mist lubricators and in low-temperature applications.

While there is some overlapping of applications for PAOs and diesters, Exxon states that their line of diester-based SYNESSTIC (Table 7-5) is formulated to give outstanding performance in air compressors, hydraulic systems, mist lubrication systems, air-cooled heat exchanger drives, and bearings in pumps and electric motors. SYNESSTIC is particularly well-suited for outside compressors that may be subjected to a wide range of temperatures. Changing from a conventional petroleum-base lubricant to SYNESSTIC can minimize the frustrating and costly problems of hot-running equipment, premature bearing failures, damaging deposit build-up, cold-weather wax plugging and the necessity for frequent oil changes. The superior lubricity of SYNESSTIC vs. comparable petroleum oils permits bearings to run cooler, thus extending the life of the bearings and the lubricant. In addition, its low volatility helps reduce lubricant consumption. SYNESSTIC lubricants are rust-and-oxidation inhibited and have excellent anti-wear properties. They have very low carbon-forming tendencies due to their diester base. The SYNESSTIC EP 220 grade provides extreme-pressure performance and meets or exceeds USS 224 and AGMA 250.04 gear oil requirements.

The application range of superior diester-base synthetics is best illustrated by briefly reviewing four case histories.

Table 7-5. SYNESSTIC typical inspections.

SYNESSTIC Typical Inspections

The values shown here are representative of current production. Some are controlled by manufacturing specifications, while others are not. All of them may vary within modest ranges.

Grade		32	68	100	150	EP 220
Viscosity						
cSt @ 40°C	ASTM D 445	30.0	65.0	96.0	143	215
cSt @ 100°C		5.5	7.5	10.1	13.6	27
SUS @ 100°F	ASTM D 2161	153	339	502	740	1108
SUS @ 210°F		44.7	51.4	60.6	73.0	132
Viscosity index	ASTM D 2270	120	77	88	89	160
Pour point, °C	ASTM D 97	-57	-37	-40	-34	-43
°F		-71	-35	-40	-30	-45
Flash Point, °C	ASTM D 93	266	271	252	249	174
°F		510	520	485	480	345
Autogenous ignition temperature, °C		385	393	413	418	393
°F		725	740	775	785	740
Specific gravity @ 15.6°C (600F)		0.926	0 957	0.963	0.959	0.884
Timken OK load, lb	ASTM D 2782	—	—	—	—	75
FZG gear test, stages		—	—	—	—	12
Demulsibility @ 54.4°C (130°F), ml oil/water/emulsion	ASTM D1401	40/40/0	40/40/0	40/40/0	40/40/0	40/40/0
Evaporative loss, 22 hr @ 98.9°C (210°F), mass %	ASTM D 972	1.7	0.20	0.30	0.30	—
Copper strip corrosion, 3 hrs @ 100°C (212°F)	ASTM D 130	1a	1a	1a	1a	1a
Rust prev., 24-hr dist. water	ASTM D 665	No rust	No rust	No rust	No rust	No rust
Panel coking, 6 hrs @ 316°C (600°F), mg deposit		9.8	9.6	6.0	9.9	—
Conradson carbon residue, mass %	ASTM D 189	0.00	0.00	0.00	0.00	—
Four-ball, 40 kg, 1800 rpm, 65.6°C (150°F), 30 min., scar diam., mm		0.30	0.40	0.40	0.40	—

- In one refinery, Synesstic synthetic eliminated bearing failures in a 4,000-HP electric motor and lowered oil temperatures by more than 50%.

- In a chemical plant, Synesstic 100 ended a cold-weather wax-plugging problem in an oil mist system and saved $9,375 the first year.

- In a British petrochemical operation, Synesstic 32 reduced valve overhauls in a reciprocating compressor from 24 per year to one; saved $4,300 per year in energy costs; and extended drain intervals by a factor of eight.

- In a major steel plant, Synesstic 100 virtually eliminated coke deposits that had caused four reciprocating compressors to be shut down every 1,500 hours for cleaning. The compressors now run 6,000 hours and longer with no problems.

Indeed, long after conventional mineral oils have blackened to coke and sludge, Synesstic synthetic lubricants run clean. Their superior film strength provides better wear performance than conventional fluids, so equipment can last longer. Their low pour points and outstanding high temperature stability can eliminate seasonal oil changes and reduce volatility losses and oil carryover. Synesstic fluids can even save money by saving energy.

Carbon-free Compressor Operation

A pair of 50-HP reciprocating air compressors in a chemical plant were in alternate service (one week continuously on, then off), using an ISO 150 mineral oil. Carbon deposits on discharge valves caused such operating problems that the machines required maintenance every three months.

In an operational test, one compressor was switched to Synesstic 100. After more than six months, discharge valves on this compressor were substantially cleaner than they were on the unit that used mineral oil for four months.

Synesstic 100 allowed compressor maintenance intervals to be doubled from three months to six, a significant saving in labor and material.

The comparison photos (Figure 7-12) tell the story.

High Film Strength for Better Wear Protection

Synesstic synthetic lubricants outperform both mineral oils and competitive synthetics in film strength and lubricity. That means less wear, reduced maintenance and longer operating life for your machinery.

One key to film strength is the polar molecules of the Synesstic diester base stock. These molecules line up on metallic surfaces like the nap of a carpet, creating a strong lubricant film that helps prevent metal-to-metal contact.

A proprietary additive package enhances these natural anti-wear properties, as can be shown in the four-ball wear test. Here, a steel test ball rotates on top of three other stationary steel test balls. At the end of the test, the average diameter of the wear scar on the three lower balls shows how much wear has occurred. The graph, Figure 7-13, shows that Synesstic surpasses competitive synthetic products in preventing wear.

As illustrated in Figure 7-14, many synthetic lubricants excel in wear protection tests over equivalent viscosity mineral oils. The bearing wear experience with an ISO grade 32 synthetic is typically similar to that of an ISO grade 68 mineral oil.

These facts helped a large Northeastern refinery. Here, the mineral oils traditionally used to lubricate the hot liquid pumps for the pipe stills could not prevent excess bearing wear. In fact, bearing failures caused more than 40% of total pump failures.

Introduction of Synesstic 100 essentially eliminated bear-bearing failures, reducing the occurrence to only 3% of all pump outages. The increase in the on-stream reliability of the distillation unit not only saved capital and labor costs to repair damaged equipment, it also improved the economic efficiency of the entire process.

Long-term Oxidation Resistance

Lubricants must be able to resist degradation when exposed to oxidizing conditions for long periods.

Figure 7-12. Photo on left - No. 1 discharge valve after 4 months' service with ISO 150 mineral oil. Photo on right - No. 1 discharge valve after 6 months' service with Synesstic 100.

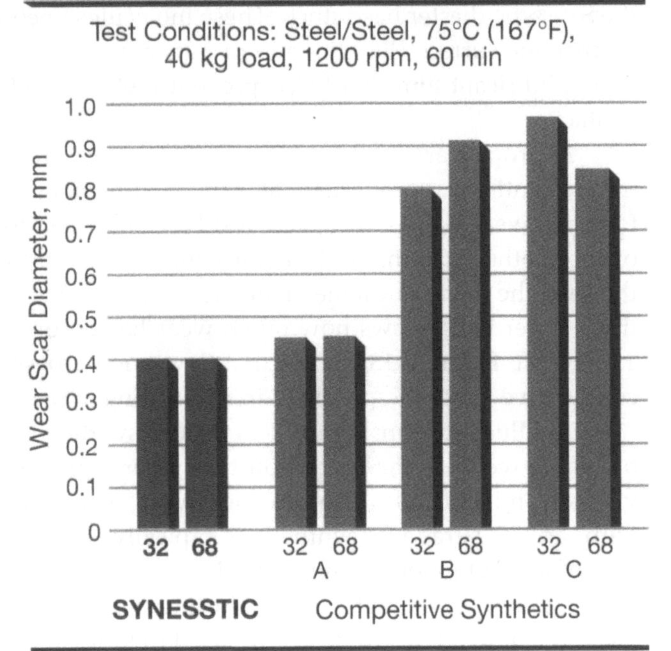

Figure 7-13. Wear test results, SYNESSTIC vs. competitive synthetics.

he test results below show the long-term stability of YNESSTIC fluids. In this severe laboratory (110°C copper atalyst, warm air current, as illustrated in Figure 7-15), ll of the oils showed initial control of oxidation. But nce the oxidation inhibitor in the conventional mineral il was consumed, that oil oxidized rapidly. Its degralation produced acidic by-products.

Unlike the mineral oil, all three SYNESSTIC grades esisted oxidation for 3,000 hours and more, evidence f the inherent chemical stability of this lubricant.

legligible Carbon Deposits

When mineral oils are heated enough, they break lown, leaving varnish, carbon and coke deposits that an be extremely damaging. SYNESSTIC synthetics offer lramatic improvements in thermal stability.

Results in the Panel Coker Test, Figure 7-16, illusrate the advantage. Hot oil was splashed onto these netal test panels at 260°C (500°F) and 274°C (525°F) in he presence of air for 6 hours, simulating a stressful pplication. The obvious carbon deposits on the test panels show the dramatic difference between SYNESSTIC and conventional mineral oils.

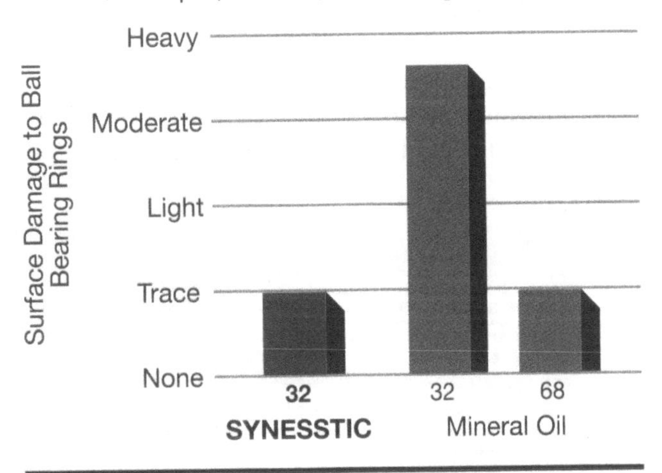

Figure 7-14. Laboratory studies with an instrumented bearing test rig demonstrate that a low viscosity SYNESSTIC can provide the same protection as a higher viscosity mineral oil. In the test illustrated in this graph, the additive packages of all three lubricants were identical—only the base changed. Because a lower viscosity grade can achieve the same degree of protection, you may be able to save energy and reduce costs.

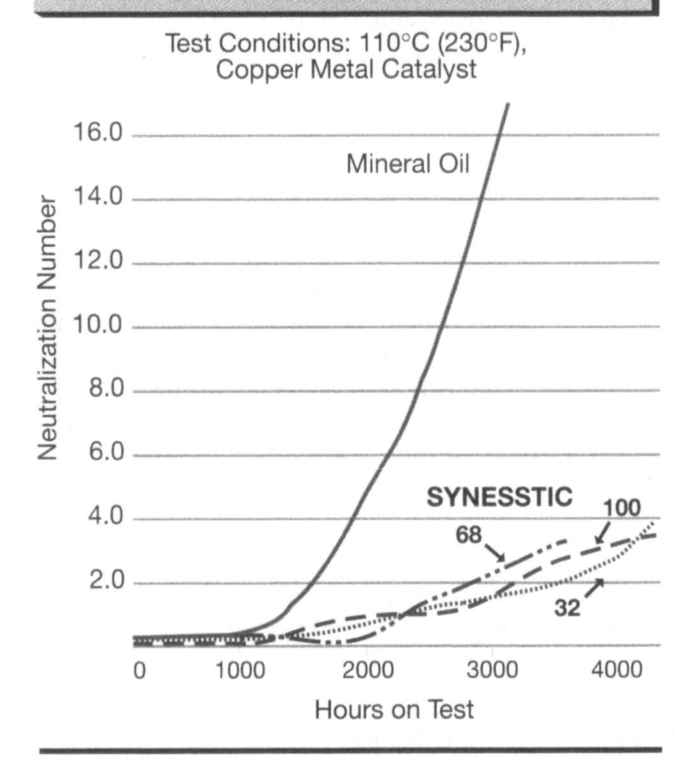

Figure 7-15. Oxidation tests show diester-base synthetics excel over mineral oils.

Low Pour Point Advantage

Synesstic fluids flow easily at temperatures where conventional lubricants almost refuse to budge. This means easy startup for intermittent and cold-weather operations. And it means you can eliminate wasteful and time-consuming seasonal oil changes.

Unlike petroleum oils, which typically contain some wax, Synesstic synthetics have no wax to hinder their flow.

Synesstic lubricants invariably show lower pour points than mineral oils of comparable grade, as shown in Figure 7-17 and 7-18. Although mineral oils generally undergo cold solvent treatment and filtration to remove most of the waxy hydrocarbon fractions, traces of wax that remain can freeze out at low temperatures. That translates to less-than-optimum lubrication for cold-weather and intermittent operations.

At low temperatures, wax crystals in conventional lubricants can clog mist reclassifier fittings. That means that expensive bearings and other machine elements may fail from lack of proper lubrication. Because Synesstic fluids contain no wax, low temperatures are no problem, and mist systems can operate dependably year-round.

Easy Cold Startup, Low Friction and Energy Savings

The microscope shows why Synesstic lubricants help machinery start so easily in the cold. Magnified 360 times (Figure 7-19), Synesstic 32 is free of wax at -18°C (0°F), while a typical mineral oil shows significant crystallization. Moreover, the diester-base lubricant reduces friction so effectively that the temperature of lubricated bearings remains much lower than it does in bearings lubricated with mineral oil. Lower temperature helps the lubricant last longer. And it means that bearings last longer too, because a cooler lubricant has a relatively higher effective viscosity and maintains a more reliable film thickness. Lower bearing temperatures also provide a better margin of safety against thermal fatigue effects.

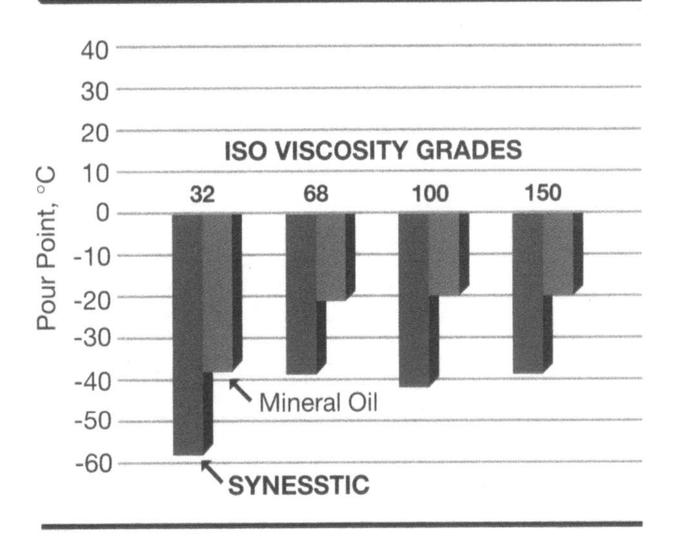

Figure 7-17. Pour point advantage of Synesstic diester lubricant over mineral oils.

Figure 7-20 compares the temperature rise in test bearings with three lubricants that differ only in

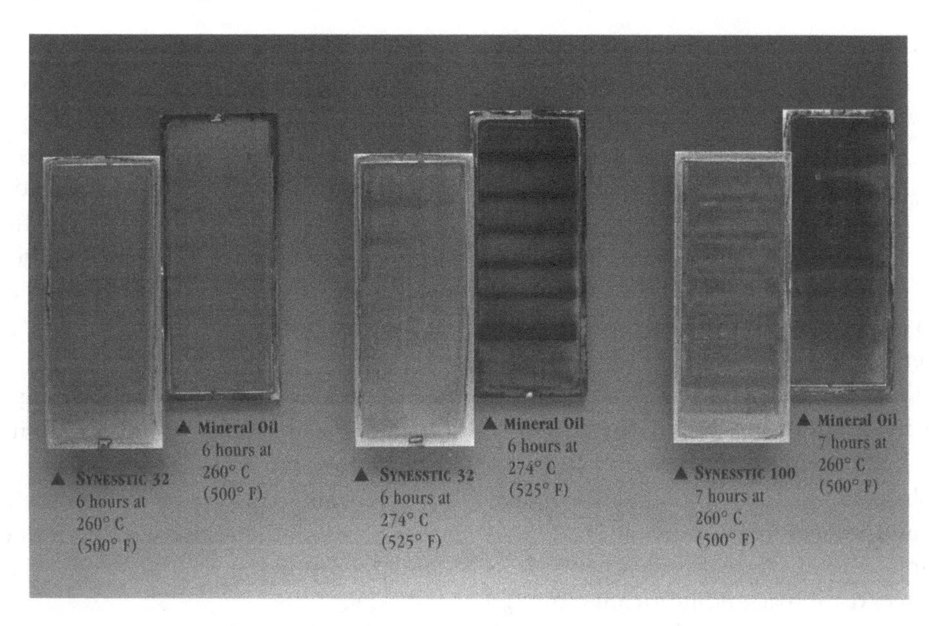

Figure 7-16. Negligible carbon deposits on the Synesstic test panels show how Synesstic lubricants help keep machine elements clean.

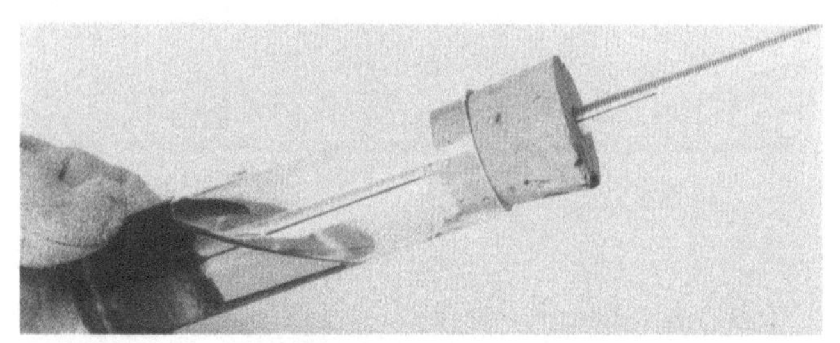

SYNESSTIC 32 @ -41°C (-42°F)

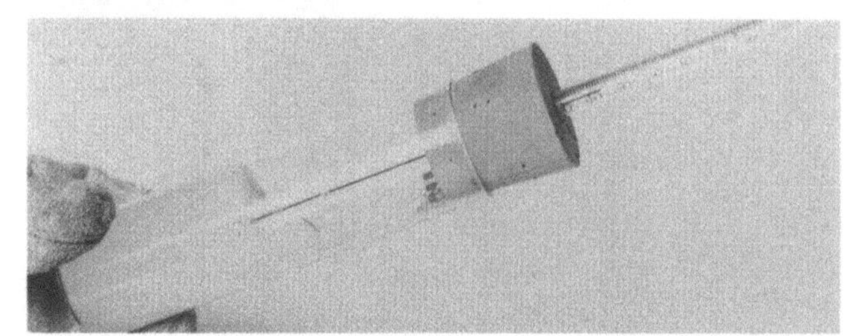

Mineral Oil ISO VG 32 @ -37°C (-34°F)

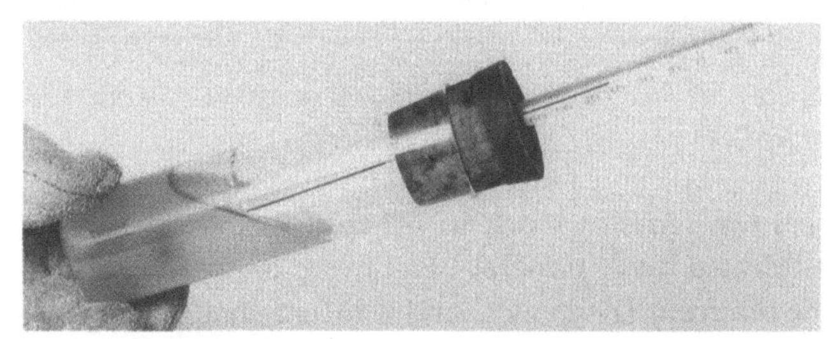

SYNESSTIC 100 @ -31°C (-24°F)

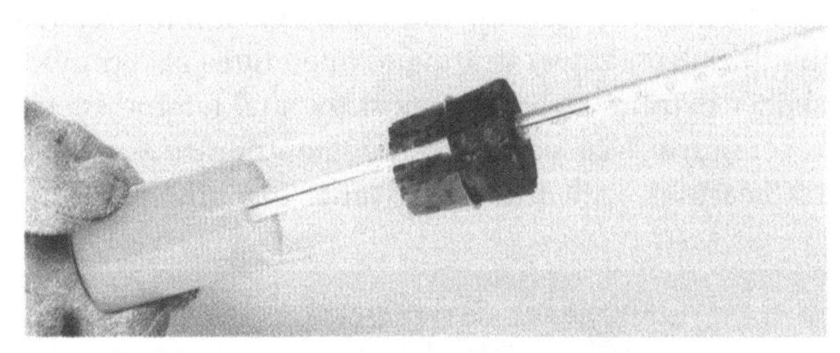

Mineral Oil ISO VG 100 @ -28°C (-18°F)

Figure 7-18. At comparative test temperatures, the mineral oils solidify while the SYNESSTIC 32 and 100 grades remain free flowing.

Figure 7-19. At –18°C, SYNESSTIC 32 (top photo) has no wax crystals, while a typical mineral oil (bottom photo) shows significant crystallization. The dark spots in the SYNESSTIC photo are air bubbles in the sample.

base stock (all have the same additive package). As expected, the lower the viscosity, the less the temperature increases as the load goes up. But a comparison of identical viscosity grades shows that the SYNESSTIC synthetic base stock stays significantly cooler than the petroleum-base lubricant.

Less friction and lower temperatures indicate that less energy is being wasted. As a result, energy bills for equipment switched to SYNESSTIC lubricants may go down as much as 7% or more, even if there is no switch to a lower viscosity grade.

A controlled test series showed that SYNESSTIC lubricants stay cooler and protect against wear, which means that SYNESSTIC 32 can be substituted for an ISO 68 mineral oil without loss of effective film thickness. Figure 7-21 shows the energy reduction in test applications using SYNESSTIC 32 instead of the 68-grade mineral oil. Energy savings ranged from 3% to 29% in these laboratory bearing tests.

Reduced Maintenance, Fuel Savings

At a Gulf Coast plant location, engineers ran a series of tests to measure the energy savings that could be achieved using SYNESSTIC synthetic lubricants. The tests were conducted in two different compressor types, with the following results:

- Ingersoll-Rand TVR-21 Reciprocating Compressor, a 1,200-HP compressor with six double-acting

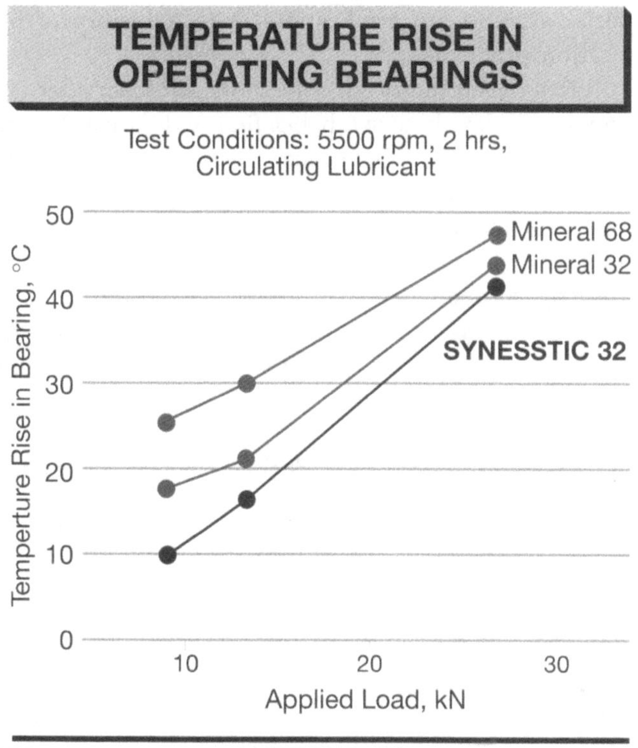

TEMPERATURE RISE IN OPERATING BEARINGS

Test Conditions: 5500 rpm, 2 hrs, Circulating Lubricant

Mineral 68
Mineral 32
SYNESSTIC 32
Temperture Rise in Bearing, °C
Applied Load, kN

Figure 7-20. Temperature rise in bearings lubricated with different oils.

compressor cylinders. In tests comparing SYNESSTIC 32 with a premium ISO 100 mineral oil, SYNESSTIC 32 achieved:

— 2.9% less power consumption
— $5,340 in natural gas fuel savings per year
— $6,160 in maintenance savings per year
— $6,596 net operating savings, including the higher cost of SYNESSTIC 32.

- Ingersoll-Rand Centac 4-stage Centrifugal Compressor, a 900-HP (670 kW) electric motor drive operating between 22,400 and 47,900 rpm. In this compressor, SYNESSTIC 32 was substituted for a premium ISO 32 mineral oil. SYNESSTIC achieved:

— 1.0% less power consumption
— $3,624 savings in electricity
— $1,936 net operating savings per year, including the cost of SYNESSTIC 32.

Anti-Foaming Properties Prevent Wear, Oil Carryover

Air entrainment and foaming can cause excessive wear, because air is a poor lubricant! In compressors, foaming can also lead to oil carryover, which removes oil from the system and causes lubricant starvation.

Tests show that SYNESSTIC fluids release air rapidly, whether it be entrained air (bubble size of 1 mm or less) or foam (bubble size larger than 1 mm). In the test shown in Figure 7-22, lubricants were heated to 75°C (165°F), stirred in a Waring Blender at high speed for one minute and placed in a glass cylinder, where the time to clear was measured. Regardless of viscosity,

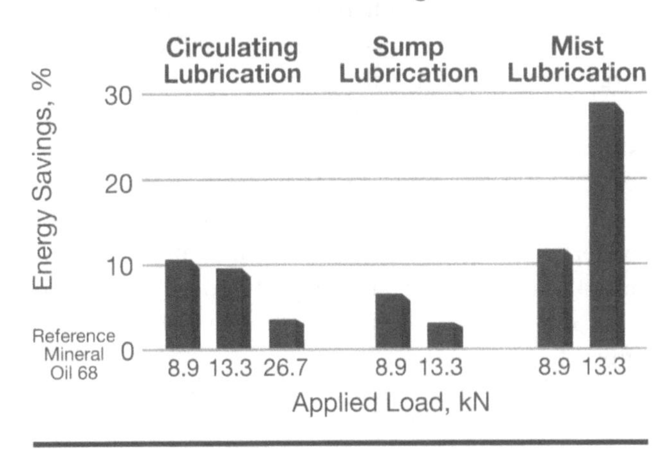

Figure 7-21. Frictional energy reduction achievable with SYNESSTIC 32 vs. mineral oil 68. Note that these two lubricants have the same wear protection quality.

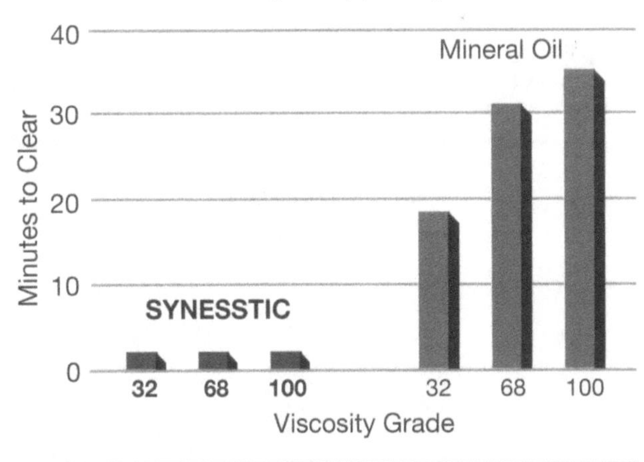

Figure 7-22. Diester-base lubricants will release entrained air much more rapidly than mineral oils.

SYNESSTIC released entrained air within two minutes; mineral-oil equivalents took nine to 18 times longer.

Composition Control Means Volatility Control, Safety and Long Life

Each viscosity grade of SYNESSTIC contains a selected molecular composition with well-defined volatility characteristics and no undesirable "light ends." The purity of this lubricant shows clearly in the narrow-cut distillation band of Figure 7-23, taken from a Gas Chromatograph Distillation (ASTM D 2887) evaluation.

Typical industrial-grade mineral oils show a much broader distillation band due to their natural origin. Mineral oils include significantly more light ends, which makes these lubricants more volatile at lower temperatures.

The relative absence of light-end volatility in SYNESSTIC translates to longer product life and improved safety: The flash point of Synesstic 32, for example, runs about 30°C (55°F) higher than a comparable mineral oil, and auto-ignition temperatures are 80°C (140°F) higher.

Field Experience with Diester Lubricants

After the initial laboratory work in the development of SYNESSTIC products, Exxon research personnel scaled up the testing with the aid of Exxon refinery and plant engineers. The project team evaluated SYNESSTIC formulations in a myriad of equipment types, paying special attention to applications with a history of lubrication problems.

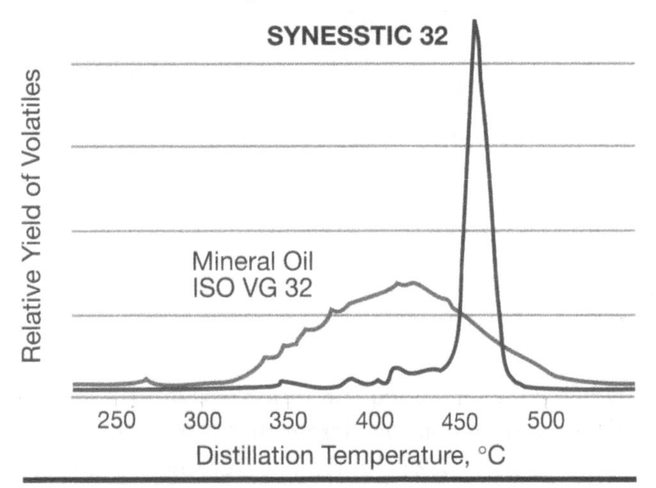

Figure 7-23. There is very little light-end volatility in this diester-base synthetic hydrocarbon lubricant.

This diester lubricant solved one problem after another, and today these synthetics find wide use at Exxon:

At one refinery complex, SYNESSTIC products are used in several hundred pieces of equipment, ranging from 1 to 800 HP, from 5 to 15,000 rpm, and from old reciprocating compressors to new high-speed pumps.

This company also uses these diester lubricants in oil mist systems in the refinery and tank field. In the late 1980s, some of the recorded maintenance savings at one location included:

- $46,000 per year by stopping failures of furnace air preheaters.
- $16,000 per year by eliminating valve repair on a gas engine reciprocating compressor.
- 350 work-hours per year by eliminating seasonal lube changes on fin-fan gear boxes.

It has been estimated, in the late 1980s, that the change to SYNESSTIC synthetic lubricants at this one refinery translates to net savings between $100,000 and $200,000 each year. Prorate that to today!

APPLICATION SUMMARY FOR DIESTER-BASE SYNTHETIC LUBRICANTS

Diester-base synthetics are cost justified whenever a plant requires a combination of exceptional oxidation resistance, outstanding high-temperature stability, low-temperature fluidity, deposit prevention and long-term cleanliness.

SYNESSTICS can extend drain intervals, as illustrated in Table 7-6. Moreover, these diester-base lubricants inevitably help prevent the high cost of machine servicing and replacement. They help avoid lubricant losses attributable to foaming and high-temperature evaporation; they can reduce energy costs and exhibit favorable viscosity-temperature relationships, Figure 7-24.

- SYNESSTIC lubricants perform exceptionally well in a wide variety of compressors. In reciprocating compressors, these products resist high temperatures, avoid fouling of discharge valves, extend drain intervals, reduce friction and wear, and offer potential energy savings. In tests of more than 50 makes and models of reciprocating compressors from all major vendors, SYNESSTIC synthetics invariably gave great improvement over petroleum-base lubricants (Figure 7-25).

Table 7-6. Diester-base synthetic lubricants extend oil drain intervals.

EQUIPMENT	METHOD	PREVIOUS	CURRENT
Rotary Screw Compressor	Circulation	2000 hours	8000 hours
Centrifuge Gear Case	Splash	3 months	24 months
Fin Fan Gear Unit	Splash	6 months	2 year plus
Parallel Shaft Gear Reducer	Splash	3 months	1 year plus
Sundyne® Gear Unit	Circulation	6 months	2 year plus
Reciprocating Pump	Splash	3 months	8 months plus

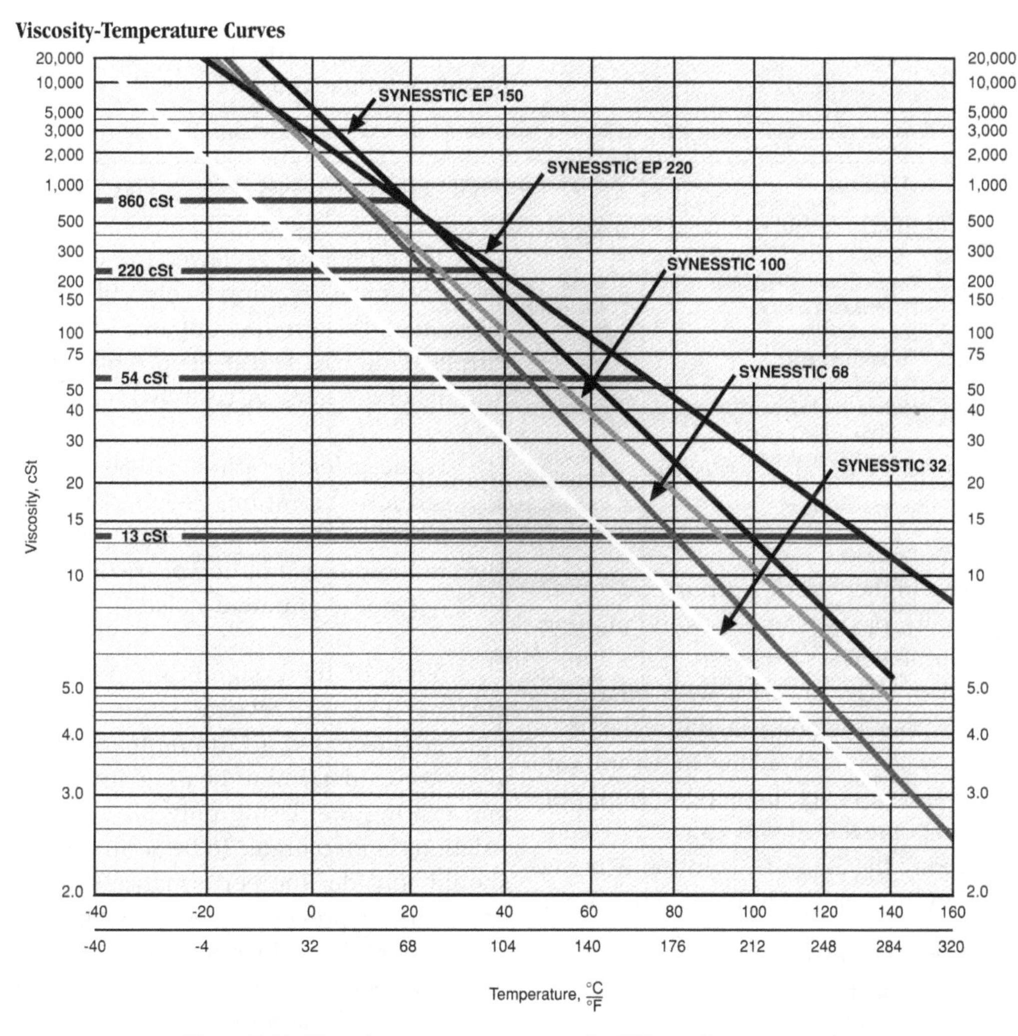

Figure 7-24. Viscosity-temperature curves for different SYNESSTIC grades.

In rotary compressors, where there is continuous intermingling of lubricant and gas, the excellent oxidation performance of SYNESSTIC synthetics helps prevent deposits and reduces downstream oil carryover. Compressed gas is cleaner, and the compressor consumes less lubricant.

- Oil mist lubrication is gaining popularity. Modern mist systems deliver uncontaminated lubricant, extending bearing life and reducing failures. At the same time, mist systems generally consume less total lubricant than traditional lubrication systems do. Machinery running with mist lubri-

Figure 7-25. Synesstic film strength offers improved bearing protection for all types of compressors.

cation produces higher work output and saves energy. And finally, mist systems have no moving parts to cause problems.

One problem that can arise with mist systems is cold-weather plugging in the reclassifier fittings, where wax present in the lubricant can precipitate out and obstruct the narrow passages. Synesstic products prevent this because they contain no wax. Today the Synesstic solution is being used in a variety of mist applications, including:

— Positive displacement pumps
— Furnace air preheaters
— Centrifugal pumps
— Fans and blowers
— Steam turbines
— Electric motors

- The good film strength and anti-wear protection of Synesstic synthetic lubricants deliver excellent field performance in high-speed integrally geared pumps and other equipment under moderate loads. Field experience shows that Synesstic 32 can be an effective and truly superior replacement for the automatic transmission fluids commonly found in high-speed gear boxes.

- While many hydraulic systems do not present extreme performance challenges in terms of oxidation, thermal stability, or wear resistance, Synesstic synthetics often solve operating problems that do exist. In particular, the cleanliness-promoting solvency of Synesstic can help overcome sluggish or erratic valve actuation and lead to faster, smoother system operation.

Oil change procedures and materials compatibilities are summarized in Tables 7-7 and 7-8, respectively.

HIGH FILM STRENGTH SYNTHETIC LUBRICANTS

The author has reviewed in excess of 150 user feedback documents relating to high film strength synthetic lubricants. The products of one such company, Royal Purple (Humble, Texas) have replaced competitive lubricants of the same viscosity with the following results:

1) reduced temperatures (10-60°F)
2) reduced vibrations (even critical bearings have had vibration reduced and stabilized to normal)
3) reduced consumption of frictional energy (10-30%)
4) reduced equipment noise

None of these significant performance improvements is viscosity-related, yet in new equipment design, engineers use a formula that depends only on oil viscosity to determine bearing load carrying capacity and bearing life. Using only oil viscosity in these calculations is inaccurate. To be accurate, design engineers should consider the performance contributions of other factors in preventing bearing failure. For example, design engineers do not take into consideration the different performance properties of base oils, such as paraffin oil, naphthenic oil, aromatic oil, synthetic oil, etc. More importantly, engineers do not recognize contributions from additive technology. These are more significant, and more important, than either viscosity or base oil in determining load-carrying capacity and bearing life. Royal Purple synthetic lubricants are modern formulations containing superior, unique additive technology making them super bearing lubricants, protecting bearings, and extending bearing life far beyond design engineers' calculations using viscosity

Table 7-7. Oil change procedures for SYNESSTIC diester lubricants.

Oil Change Procedures

SYNESSTIC synthetic lubricants have very good solvency characteristics, which helps keep machine components clean. However, this also means that you should observe certain precautions when changing from a mineral oil to SYNESSTIC. The following procedures are guidelines only; for help, contact your local supplier.

Once-through: Oil mist systems, reciprocating compressor cylinder lubrication, oil-injected screw compressors.

1. Drain mineral oil from reservoir and fill with the proper grade of SYNESSTIC.
2. If possible, flush SYNESSTIC through lube lines for five to ten minutes to remove old deposits.
3. If step 2 is not possible, operate lubrication system normally and observe lubrication points closely for any signs of plugging from displaced mineral-oil deposits.
4. When all deposits have been removed, return the machine to normal service.

Small Sumps: Pumps, steam turbines, small gear sets, other units with reservoirs of 10 gallons or less.

1. Drain mineral oil, inspect internal condition of the machine, and hand clean as much as practical.
2. Fill with proper grade of SYNESSTIC and run machine for one to two hours.
3. Drain SYNESSTIC synthetic into suitable container, refill and start up machine.
4. Take sample for oil analysis after two days.

Large Sumps: Gear units, reciprocating compressors, units with reservoirs of 10 gallons or more.

1. If possible, switch to SYNESSTIC after a machine overhaul. Ensure that most residual mineral oil has been removed, fill with proper grade of SYNESSTIC; and start up using normal procedures.
2. If step 1 is not possible, drain mineral oil; inspect internal condition; hand clean as much as practical.
3. Fill with proper grade of SYNESSTIC and start up the machine using normal procedures.
4. Take a sample for oil analysis after one and seven days.
5. If condition of lubricant deteriorates significantly between the one-day and seven-day sample, drain and refill with SYNESSTIC. Repeat step 4.

Circulating Oil: Centrifugal compressors, flooded rotary compressors, gear units, large pumps, etc.

1. Drain mineral oil from the system as completely as possible.
2. Clean the suction strainers of lube oil pumps and install new filter elements, if applicable.
3. Fill with the proper grade of SYNESSTIC and start up the system using normal procedures.
4. Monitor the filter condition carefully for any sign of plugging during the first week of operation.

only. These unique, proprietary bearing performance properties are important.

Oil Film "Toughness" Increased

Although Royal Purple uses a combination of PAO and diester base stocks, it may differ from the formulations of other suppliers. This company blends their products with Synerlec™, a proprietary, high load carrying, tough, tenacious, slippery synthetic film which is more important than oil viscosity/thickness in protecting bearings. Synerlec's™ tough film adheres "ionically" to bearings, giving superior protection, even under shock load conditions. Synerlec™ is so tough that a light ISO 32 grade oil with Synerlec™ protects bearings better than a heavy ISO 680 oil which depends only on viscosity for protection. Synerlec's™ tenacious adhesion to bearings prevents bearing wear during operation, and remains on bearing surfaces to prevent wear during start-up.

The most common explanations for premature bearing failures are misalignment, imbalance, extreme service duty, etc.—conditions that occur frequently in a world where more equipment is repaired with hammers than with lasers. Consequently, bearings frequently operate under stresses that exceed design standards, and frequently fail. These explanations are so easy to

Table 7-8. Compatibility with other lubricants, seals, paints, plastic and gases.

Other Lubricants

SYNESSTIC synthetic lubricants are completely compatible with all other mineral oil and synthetic-base industrial lubricants. From a performance standpoint, the oxidative stability of the resultant SYNESSTIC and mineral oil blend deteriorates as the mineral oil content increases.

Note that silicone-base oils are incompatible with both SYNESSTIC synthetics and mineral oils.

Seal Material	**Paints and Coatings**	**Gases**
Acceptable	*Acceptable*	*Recommended*
Fluorocarbon (Viton, PTFE)	Epoxy	*Inert*
Fluorosilicone rubber	Baked phenolic	Argon
Epichlorohydrin	Two-component urethane	Blast furnace gas
Polyacrylate rubber	Moisture-cured urethane	Carbon dioxide
Polyurethane		Helium
Medium- and high-nitrile rubber	*Marginally Acceptable*	Hydrogen
(Buna N, NBR) with more	Phenolic	Neon
than 30% acrylonitrile	Industrial latex	Nitrogen
	One-component urethane	Syngas
Unacceptable	Alkyls (baked finish preferred)	
Polysulfide		*Hydrocarbon*
Chloroprene rubber	*Not Recommended*	Acetylene
Butyl rubber	Acrylic	Butadiene
Natural rubber	Household latex	Butane
Ethylene/acrylic (polyacrylate)	Vinyl (PVC)	Butylene
Ethylene-propylene copolymer (EPM)	Varnish/lacquer	Coke oven gas
Ethylene-propylene terpolymer (EPDM)		Isobutane
Low nitrile rubber		Isobutylene
(Buna N, NBR) with less		Methane
than 30% acrylonitrile		Propane
Styrene-butadiene rubber		
(Buna S, SBR)		*Not Recommended*
Chlorosulfonated polyethylene		Ammonia
(Hypalon)		Chlorine
Isoprene synthetic		Freon
Silicone rubber		Hydrogen chloride
Butadiene rubber		Hydrogen sulfide
Urethane		Nitrous oxide
		Oxygen
		Sulfur dioxide

accept. They explain away the problems, especially if one accepts the untrue premise that one bearing lubricant is pretty much like another.

But bearings don't fail suddenly, they fail gradually. Predictive maintenance vibration technology can identify the first signs of bearing distress and accurately document the path to destruction. First, the bearing surfaces begin to smear or gall, creating rough surfaces that cause vibrations. Once bearing surfaces begin to deteriorate, a destructive pattern of wear, galling and vibration begins which feeds on itself until the bearings ultimately develop stress fractures and disintegrate.

Rolling element bearings will smear, gall or begin to fail, whenever the load-dependent stress exceeds the ability of the lubricant to protect the bearing surfaces. Where bearings have already begun to fail, Royal Purple's tough Synerlec™ film immediately arrests the galling process and allows damaged bearing surfaces to polish and heal over allowing bearings to remain in service. Synerlec™ greatly increases equipment reliability because bearings are protected from the kinds of stresses that regularly occur in the real world where operating conditions are less than ideal.

For example, Figure 7-26 shows the vibration trend data observed after converting an external washer filter in a pulp mill bleach plant from a mineral oil to a high film strength synthetic lubricant. Synthetic oils with greater film strength allow rolling element bearings to cross spall marks and other surface irregularities with reduced impact severity. As a result, some deteriorating bearings have been "nursed along" by a switch to high film strength synthetics.

Figure 7-27 shows vibration data from a multistage air blower at a fiber spinning plant. Conversion to a high film strength synthetic lubricant reduced vibration severity from 0.155 ips to 0.083 ips and reduced the bearing housing temperature by 20°C.

Another fiber spinning plant application was a 10 hp centrifugal pump. The overall vibration level was acceptable at 0.068 ips; however, bearing housing temperatures of 175°F were considered borderline when operating on a premium grade mineral oil (Figure 7-28a). After conversion to a high film strength synthetic lubricant, the overall vibration was reduced to 0.053 ips and the bearing housing temperature was reduced to 155°F (Figure 7-28b. In addition to the vibration and temperature reductions, motor amperage was reduced from 5.7 amps/phase on the premium mineral oil to 4.4 amps/phase with the synthetic lubricant.

Testing Corroborates Field Experience

In August 1992, Kingsbury, Inc., completed the testing of a high film strength ISO Grade 32 synthetic lubricant in their thrust bearing test machine. At low speeds and loads, there appeared to be little difference between this lubricant and identical premium grade mineral oils of the same viscosity. However, above 550 psi and 10,000 rpm, Kingsbury found the synthetic to be responsible for a 15°F reduction in bearing temperature and as much as a 10 percent decrease in frictional losses. As a result, Kingsbury approved this formulation for use with their bearings and recommended it for extreme service conditions.

In a parallel effort, a comparison was made of the lubricating properties of a specially formulated diester-base lubricant to those of a premium-grade

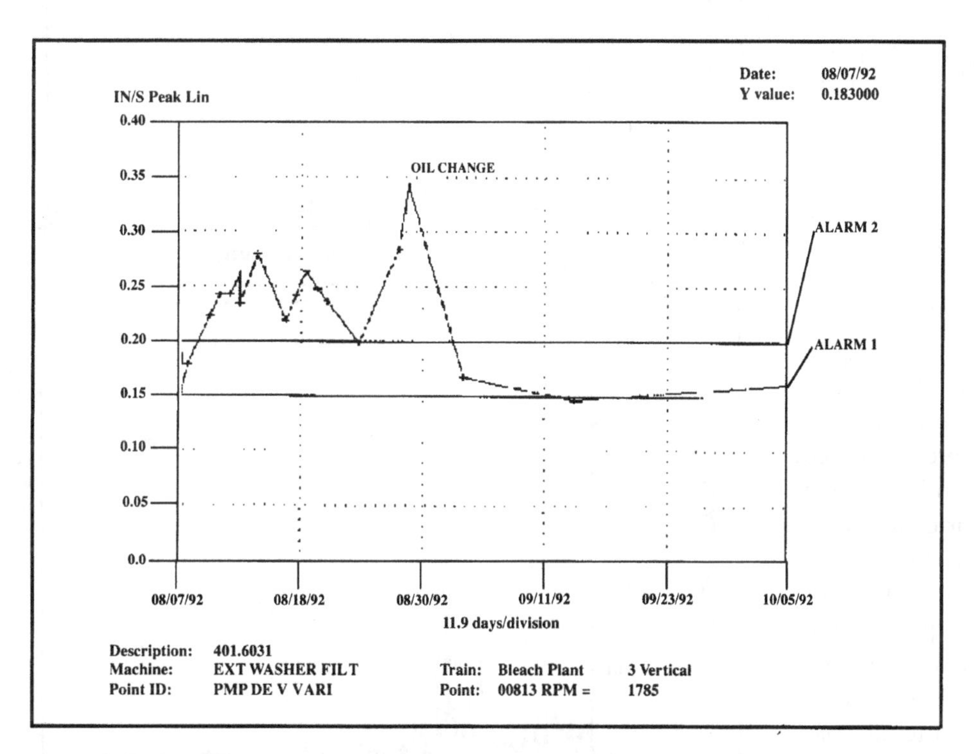

Figure 7-26. The vibration trend data observed after converting an external washer filter in a pulp mill bleach plant from a mineral oil to a high film strength synthetic lubrication.

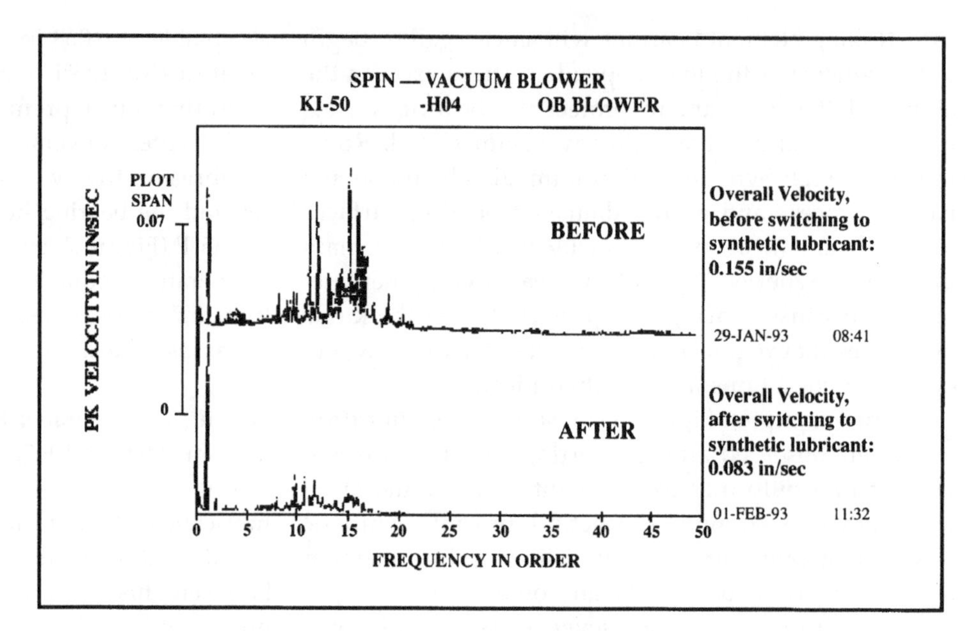

Figure 7-27. Vibration data from a multistage air blower at a fiber spinning plant. Conversion to a synthetic lubricant reduced vibration severity from 0.155 ips to 0.083 ips and reduced the bearing housing temperature by 20°C.

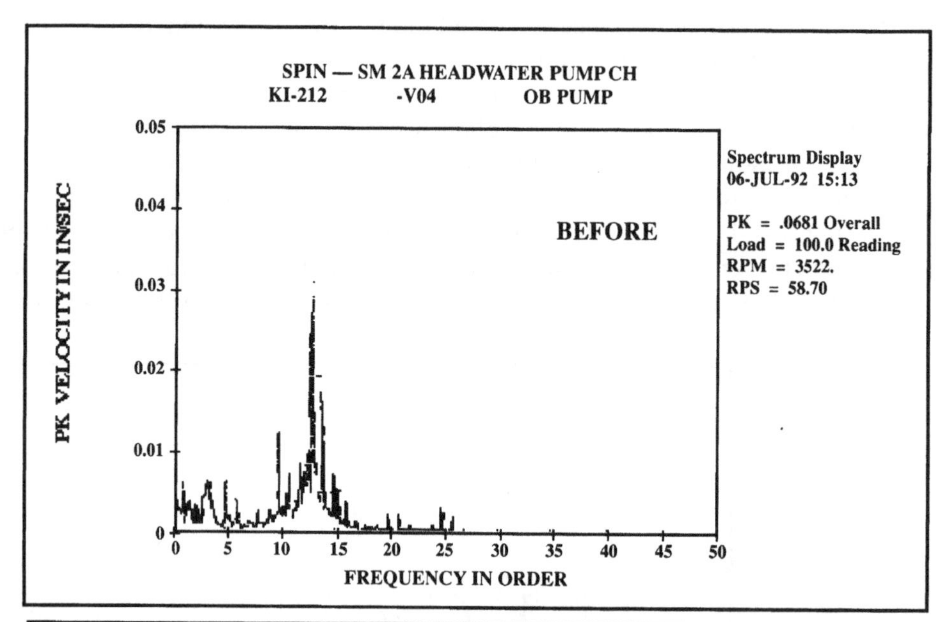

Figure 7-28(a). The overall vibration level of a 10 hp centrifugal pump was 0.068 ips; however, bearing housing temperatures of 175°F were borderline when operating on a premium grade mineral oil.

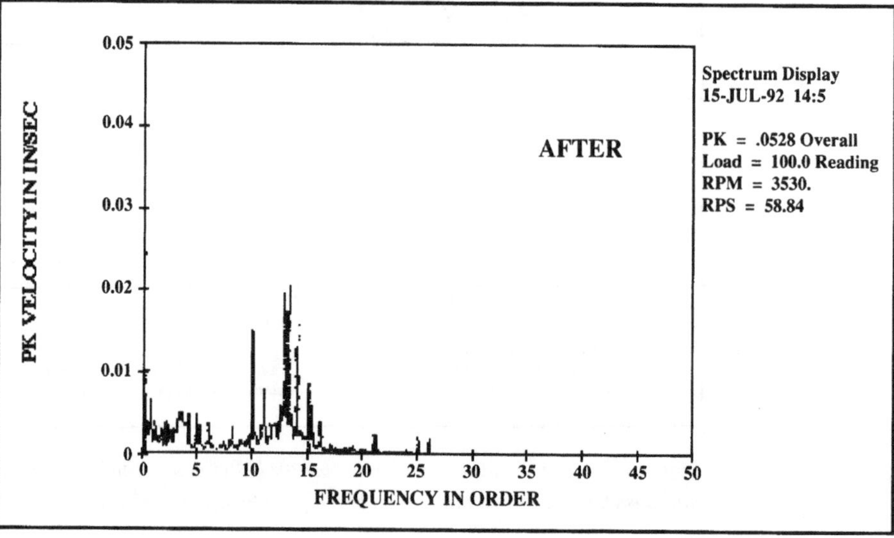

BELOW: Figure 7-28(b). After conversion to a high film strength synthetic lubricant, the overall vibration was reduced to 0.053 ips and the bearing housing temperature was reduced to 155°F. In addition to the vibration and temperature reductions, motor amperage was reduced from 5.7 amps/phase on the premium mineral oil to 4.4 amps/phase with the high film strength synthetic lubricant.

mineral oil currently in service in petrochemical plant equipment. Two synthetic lubricants and two mineral oils of different viscosities were compared. The test results indicated that the synthetic lubricant, having a viscosity of 32 cSt at 100°F, offered long-term contact surface protection equivalent to that of the base line mineral oil which had a viscosity of 68 cSt... without reducing bearing surface life below the theoretically predicted levels. The same wear protection was not achieved with a reduced viscosity mineral oil, prompting a major manufacturer of rolling element bearings to discontinue recommending the lower viscosity mineral oil for ball and roller bearings.

The lower viscosity synthetic lubricant provided projected energy savings of $75,000 per year when all applications were considered at a large petrochemical facility. Their power cost basis was $0.06/kWh.

Reduced vibration intensity will usually translate into an extension of equipment life. This is graphically illustrated in Figures 7-29 and 7-30.

Other Field Evaluations

It should again be noted that competent formulators of synthetic lubricants are able to provide large numbers of application reports, many of them with data supplied by the user's engineers. One such application report included rigorous cost justification data.

In this instance, engineers at the plastics production unit of a major U.S. Gulf Coast facility had reason to closely observe a 12,700 HP (9,475 kW) electric motor. This motor was driving an extruder through a large gear speed reducer and motor bearing temperatures were too high. As the density of the viscous hot melt product in the extrusion process increased, motor bearing temperatures would climb to 85C and higher. This required a throughput reduction ranging from 8 and 10 metric tons per hour so as to keep bearing temperatures below a vendor-prescribed maximum. To increase production it was decided to change the lubrication system from the existing splash cavity to a circulating system with greater oil flow.

But before the parts and equipment needed to revamp the lubrication system arrived on site, the electric motor bearing lubricant was changed from the

(Pictures taken with Tektronik 5113 Engine Analyzer)

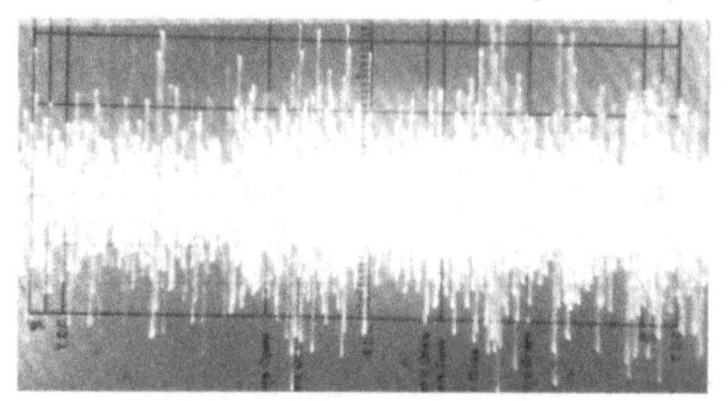

Gas engine turbocharger on premium grade oil.

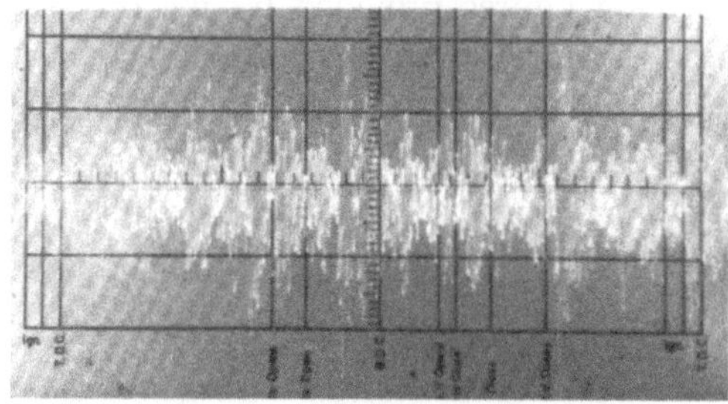

Same turbocharger after 2 days on Royal Purple.

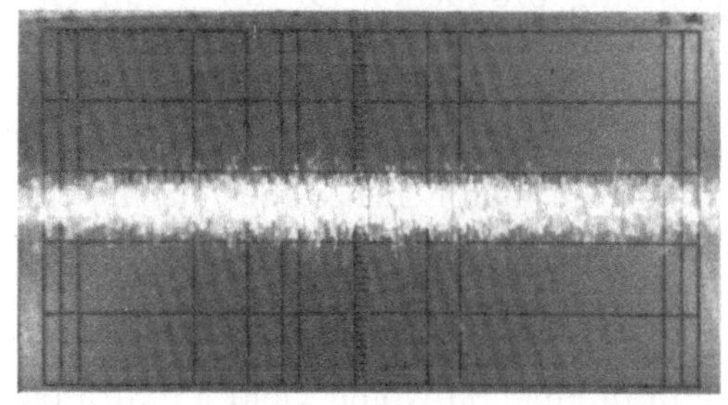

Turbocharger after 3 months on Royal Purple.

Figure 7-29. The vibration shock pulse activity of a compressor turbocharger before and after conversion to a high film strength synthetic lubricant. The turbocharger experienced a 3:1 reduction in vibration severity in the high-frequency spectrum.

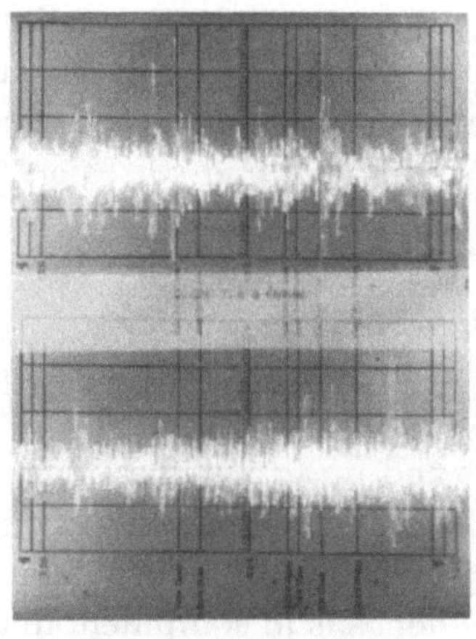

OIL PUMP BEFORE MARCH

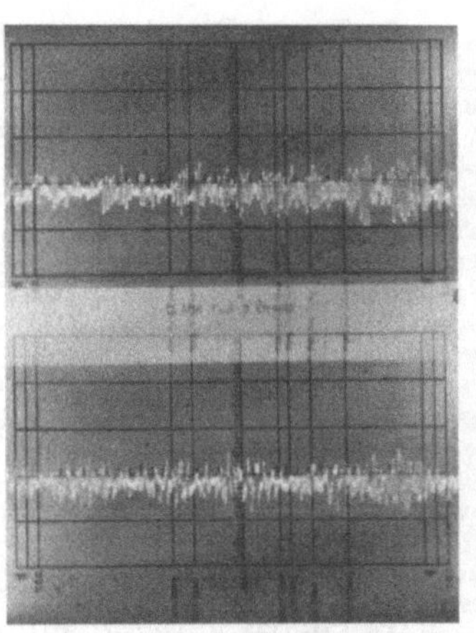

OIL PUMP AFTER MARCH

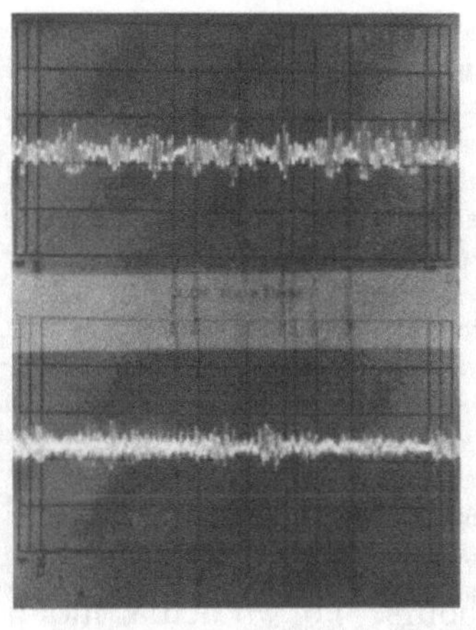

OIL PUMP ON 6/8/89

Figure 7-30. The vibration shock pulse activity of a compressor oil pump before and after conversion to a high film strength synthetic lubricant. The oil pump experienced a 5:1 reduction in both the outboard and inboard ends (top and bottom respectively). In this application a tenacious, yet slippery oil film has reduced vibration severity by "peening over" the asperities on the metal surfaces.

traditionally supplied synthetic oil to Royal Purple's Synfilm GT. Motor bearing temperatures and process density were then monitored for approximately 30 days. While processing the lighter product densities, motor bearing temperatures decreased between 6C and 8C. More importantly, when production changed to the highest density hot melt product, bearing temperatures increased only slightly and never exceeded 78C. Production was increased by 18,000 lbs per hour and bearing temperatures remained well within safe limits.

Based on figures provided by the process facility, the economic impact of this evaluation was calculated as follows; it serves as one more example of cost justification:

- User's assigned value per pound (453 grams) of product = 20 cents per lb.
- Increase obtained per hour = 18,000 lbs.
- Value of production increase = $3,600 per hour

Assuming production ran for a full eight hour day, this represents $28,800 of increased production value per day. In 2008, Synfilm GT cost $117 "per pail" and 4 pails (about 5 gallons, or 19 liters per pail) were required for the motor; hence the total investment in oil amounted to $508.

While the user's profit margin is not known, assuming even a minimal profit of 2% of the $28,800 increase in production would yield a return on investment of the total cost of the new oil of approximately 8 hours. The user company considered returning the equipment already purchased for the conversion to the circulating oil system; however, such a return was not possible. The new system was installed some 9 months subsequent to this evaluation. It is important to note that the decrease in motor bearing temperature and increase in production documented after the conversion to Synfilm GT was secured before the installation of the new oil system.

Soon after the demonstrated success of the new motor lubricant, the extruder gear speed reducer's conventional EP lubricant was replaced with Royal Purple Synergy gear oil. Extrusion process flow rates were carefully measured and a highly accurate "before vs. after" comparison conducted. It showed that a 3.1% energy reduction could be attributed to the oil change in the extruder gearbox.

Because the user's facility actually produced most of its electric power needs, they applied an imputed value of 7 cents per kWh and reported their savings as $174 per 100 hp per 1000 hours. For an actual motor power output of 12,700 hp, the savings were given as 127 times that figure, which is over $22,000 for each 1,000 hours of operation.

Parts Look Better

The same properties in Synerlec™ that allow galled bearing surfaces to heal-over also allow new bearings to properly mate with the race by "micro-polishing" both surfaces. This in-service micro-polishing smooths the bearing surfaces better than the manufacturing polishing process (shown in Figure 7-31 under 1500 magnification). Smoother bearing surfaces that mate properly increase the available load carrying area, effectively reducing the unit pressure load. Reduced loads greatly extend bearing life.

Molecular Composition Is Superior

Royal Purple oils are partially blended with large, high molecular weight synthetic oils. These big molecules keep parts from touching during operation, making them much more effective than viscosity in preventing bearing wear. Synthetic oils from this formulator contain molecules in the 1000/5100 molecular weight range and outperform petroleum oils which are in the low 300/600 molecular weight range. At the same viscosity, high molecular weight oils protect bearings better than low molecular weight oils, thereby extending bearing life.

Oil Dryness Is Enhanced

An extensive bearing fatigue life study by Grunberg & Scott shows that an oil contaminated with only 0.002% water (1 drop/quart) reduces bearing fatigue life 48%, regardless of viscosity. Royal Purple oils blended with "dry" hygrophobic synthetic oils (20 to 40 ppm water) protect bearings better than hygroscopic petroleum oils (400 to 6000 ppm water), regardless of viscosity. In service, Royal Purple oils separate rapidly and completely from water to remain dry.

Royal Purple's tough proprietary lubricating film displaces water from bearing surfaces and will not be washed off. Synerlec's™ superior anti-wear, anti-corrosion film protects bearings from wear and corrosion—even in very wet environments.

Oxidation Stability Is Vastly Superior

Royal Purple oils are many times more oxidation stable than competitive mineral and synthetic oils... 10 times longer in ASTM tests and up to 20 times longer in field service with no break-down. When oils begin to oxidize they lose lubricity and leave harmful lacquer, varnish, carbon, and sludge deposits in equipment that can interfere with efficient operation of bearings and

New machined bearing surface appears smooth until magnified 1500 times.

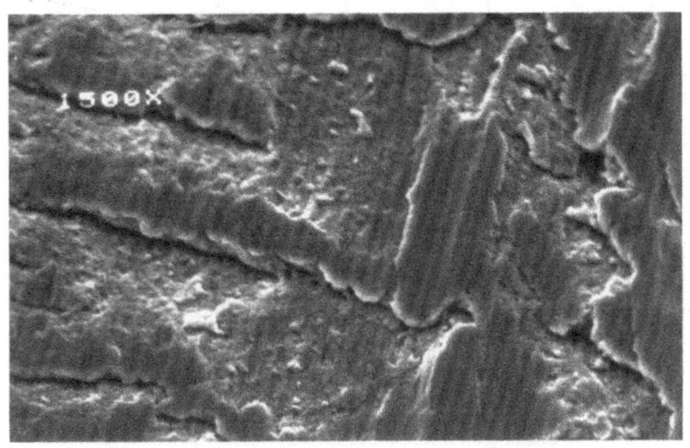

Same bearing surface "galled" using premium competitive synthetic oil (1500 power).

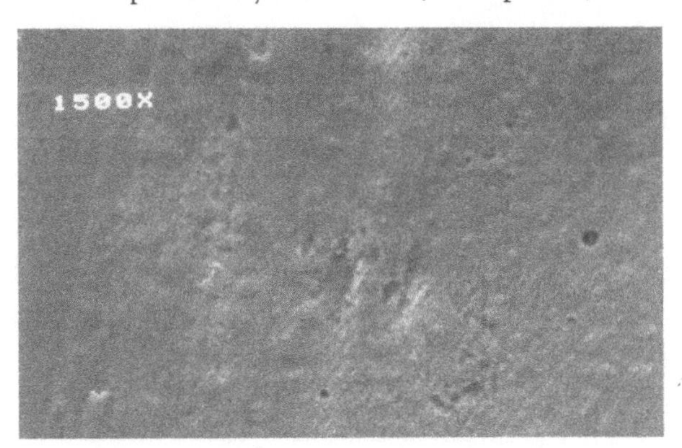

Same bearing surface in same severe service is "micro-polished" by Synerlec™. Surface is much smoother than new surface. (1500 power).

(Scanning Electron Microscope Photos by Det norske Veritas Industry, Inc., Houston Texas)

Figure 7-31. Micro-polishing effect obtained with a high film-strength synthetic lubricant. (Source: Royal Purple, Humble, Texas.)

equipment. Superior synthetic lubricants continue to perform like new, leaving no oxidation deposits, long after other oils have completely gelled. Royal Purple lubricants are formulated to perform in the real world where operating temperatures and oil drain intervals are neither timely nor consistent.

Summary

Even the best engineers can not design equipment to meet every conceivable operating condition users devise for their equipment. Bearings fail when subjected to stresses that exceed the ability of the lubricant to protect the bearing surfaces.

In 1980, one of the world's largest ethylene plants converted their 17 large oil-mist lubrication systems from a specially formulated mineral oil to a synthetic oil and found the yearly incremental cost of synthetic lube could be justified if only two pump repairs were avoided. This plant has successfully operated the oil mist systems ever since.

In a 1982 survey of experience with synthetic oils in refinery equipment, the researchers demonstrated that synthetic lubricants provided increased drain intervals, reduced maintenance, extended component life, and energy savings in a variety of production equipment. These evaluations also established the economic savings attainable through improved overall performance, even though the synthetic lubricant is more expensive than its mineral oil counterpart.

Finally, some synthetic lubricants, notably the ones with formulations based on PAO's and diesters, provide cost-effective lubrication in all service conditions where environmental concerns are of prime importance. Note, however, that not all formulations using the same base stock provide similar performance (Table 7-9).

Whether a specific oil, mineral or synthetic actually excels can only be determined by comparing specific performance properties in actual service. Our advice to the maintenance professional is to determine the performance requirements needed by specific equipment.

Look at the published specifications of various oils and determine their relative merits for the intended service, then monitor the field performance to confirm your expectations. Table 7-10 represents one set of specifications. Make it your goal to compare these data against data furnished by other experienced suppliers and request current updates on application experience.

Bibliography

1. Halliday, Kenneth R.; "Why, When and How to Use Synthetic Lubricants." Selco, Fort Worth, Texas, 1977.
2. Morrison, F.F., Zielinski, James R., "Effects of Synthetic Industrial Fluids on Ball Bearing Performance," ASME Paper 80-Pet-3, presented in New Orleans, Louisiana, Feb. 1980
3. Zielinski, James, and Perrault, Cary E.; "Survey of Commercial Experience With Diester-Based Synthetic Lubricants in Refinery Equipment," NPRA Paper AM-83-20, presented at the 1983 Annual Meeting of the National Petroleum Refiners Asso., San Francisco, March 20-22, 1983.
4. Douglas, Patrick J.; "An Environmental Case for Synthetic Lubricants," *Lubrication Engineering*, September, 1992, pp. 696-700.

Table 7-9. Although formulated from the same or similar base stocks, laboratory tests and field experience may show performance differences. Synthetic "Y" excels in this comparison.

Performance of Four Synthetic Air Compressor Lubricants

ASTM #	Test	W	X	Y	Z	Comments
D445	Viscosity @ 100°C (cs)	72.2	75.8	70.6	68.2	
D97	Pour point, °F	-40	-20	-40	-40	
D189	Residue (Conradson) %wt.	0.01	0.02	0.05	0.01	
D130	CU Strip Corrosion (2 hr @ 212°F)	1	1	1	1	All oils showed good corrosion protection.
D665	Rust test @60°C, fresh water	Pass	Pass	Pass	Pass	
-	Dry air oxidation, 312 hours @ 250°F, % increase in viscosity	5.1%	7.2%	4.0%	5.1%	If oil goes over 6% it fails. Y should have the longest life in compressor service.
D1401	Demulsibility @ 130°F, 1500 rpm, oil/water/emulsion in minutes	35-35-10 60 minutes	36-36-8 60 minutes	40-40-0 10 minutes	39-39-2 30 minutes	W and X failed. Y had best water separation, so air/oil separator and filters operate longer and more efficiently
D2783	4-Ball EP test (film strength)					Y has a significantly greater load bearing oil film to protect loaded bearing surfaces.
	Weld load, kg.	160	160	315	160	
	Load Wear Index, kg.	31.3	24.9	66.75	31.4	
D892	Foam test	Foam volume, ml	Foam volume, ml	Foam volume, ml	Foam volume, ml	All oils passed this test.
		5 min / 10 min	5 min / 10 min	5 min / 10 min	5 min / 10 min	
	Sequence I	5 / 0	30 / 0	30 / 0	0 / 0	
	Sequence II	50 / 0	5 / 0	20 / 0	20 / 0	
	Sequence III	0 / 0	20 / 0	20 / 0	0 / 0	
D2782	Timken O.K. Load test					Developed by the Timken Company to see how different oils protect loaded bearings. Y shows the greatest film strength. This should particularly improve thrust bearing life.
	Load lbs.	20	15	100+	15	
	Scar diameter, mm	weld	weld	2.8	weld	

Table 7-10. Typical lubricants selected for spur and bevel gears.

Selection criteria	Product name	Base oil / thickener	Service temperature range* (°C) ≈	ISO VG DIN 51 519	Density at 20°C (g/ml) DIN 51757 ≈	Kinematic viscosity DIN 51561, (mm²/s) at ≈ 20 °C	40 °C	100 °C	Viscosity index DIN ISO 2909 (VI)	Pour point DIN ISO 3016 (°C) ≈	Flash point DIN ISO 2592 (°C) ≈	Notes
Synthetic gear oils for high loads and temperatures between -40 and 140 °C	Klübersynth GEM 4-150	Synthetic hydrocarbon oil	-40 to 140	150	0.86	422	150	19	> 140	< -45	> 200	Miscible with mineral hydrocarbon oils; oil change intervals three times as long. Scuffing load stage > 12 in the special FZG test A/16.6/140
	Klübersynth GEM 4-220	Synthetic hydrocarbon oil	-40 to 140	220	0.86	627	220	26	> 150	< -40	> 200	
Synthetic gear oils for high loads and high temperatures up to 160 °C	Klübersynth GH 6-150	Polyglycol oil	-35 to 160	150	1.05	400	150	28	> 210	< -35	> 300	Not miscible with mineral hydro-carbon oils; check compatibility with sealing materials and paint coatings. Scuffing load stage > 12 in the special FZG test A/16.6/140
	Klübersynth GH 6-220	Polyglycol oil	-30 to 160	220	1.05	630	220	41	> 220	< -30	> 300	
Synthetic gear oils for extreme loads and hypoid gears	SYNTHESO D 150 EP	Polyglycol oil	-35 to 100	150	1.05	400	150	28	> 210	< -35	> 200	Not miscible with mineral hydrocarbon oils; check compatibility with sealing materials and paint coatings. Passed FZG test L-42
	SYNTHESO D 220 EP	Polyglycol oil	-30 to 100	220	1.05	550	220	38	> 210	< -35	> 200	
Gear oil for the food-processing industry	Klüberoil 4 UH 1-150	Synthetic hydrocarbon oil	-35 to 110	150	0.85	425	150	22	> 160	-35	> 200	Scuffing load stage 12 in the FZG test A/8.3/90, DIN 51345 Pt 2. USDA-H1 authorization
	Klüberoil 4 UH 1-220	Synthetic hydrocarbon oil	-30 to 110	220	0.85	670	220	33	> 170	-35	> 200	

Selection criteria	Product name	Base oil / thickener	Service temperature range* (°C) ≈	Density at 20°C (g/cm³) DIN 51757 ≈	Base oil viscosity DIN 51561 (mm²/s) ≈ 40°C	100°C	Colour	Drop point DIN ISO 2176 (°C)	Speed factor ($n \cdot d_m$) mm · min⁻¹ ≈	Worked penetration DIN ISO 2137 (0.1 mm)	Consis-tency NLGI grade DIN 51 818	Apparent viscosity, KL viscosity grade	Notes
Fluid gear grease for high loads	MICROLUBE GB 00	Min. hydro-carbon oil/silic.	0 to 100	0.9	600	38	reddish-brown	> 200	-	430 to 475	-	EL	Scuffing load stage >12 in the special FZG test A/2.76/50
Fluid synthetic gear grease for high loads, long-term and lifetime lubrication	Klübersynth GE 46-1200	Polyalkylene glycol / Li	-50 to 120	0.99	120	20	brown	> 160	-	400 to 430	00	EL	Scuffing load >12 in the FZG test A/8.3/90, DIN 51354 Pt 2
Fluid gear grease for the food-processing industry	Klübersynth UH 1 - 14 - 1600	Synthetic hydrocarbon oil Al compl. soap	-60 to 120	0.85	160	21	transpa-rent	> 220	500000	400 to 430	00	EL	Scuffing load stage 12 in the special FZG test A/2.76/50. USDA-H1 authorization

Chapter 8
Lubricating Greases

Today's new-generation greases are expected to do much more than lubricate. They must meet a wide range of demanding performance requirements, such as:

- Long, trouble-free service life, even at high temperatures
- Rust and corrosion prevention
- Dependable, low-temperature start-up
- Resistance to slingoff and water washoff
- Conformance to increasingly stringent industry and governmental standards

Lubricating greases consist of a lubricating oil, a thickener and one or more additives. The thickener is responsible for the characteristics of the grease (Table 8-1 and Figure 8-1). Complex greases generally have a higher drop point, are more resistant to oxidation, liquids and vapors. Synthetic thickeners are most resistant to temperature. Extensive testing is done to verify properties, Figure 8-2.

The *advantage* of a lubricating grease over an oil is that it remains at the friction point for a longer time and that less effort is required in terms of design.

Its *disadvantage* is that grease neither dissipates heat nor removes wear particles from the friction point.

Application

Lubricating greases are used to meet various requirements in machine elements and components, including

- valves
- seals
- springs
- gears
- threaded connections
- plain bearings
- chains
- contacts
- ropes
- switches
- screws
- rolling bearings
- shaft/hub connections

Lubricating greases can also fulfill various tasks apart from being effective against friction and wear. They may be required to

- be rapidly biodegradable
- conduct electric current
- be resistant to ambient media
- protect against corrosion
- be able to carry high loads
- be a low-noise grease
- comply with food regulations
- be resistant to temperatures
- be neutral to the materials involved

Other primary or secondary grease applications include:

- running-in grease
- fluid grease
- adhesive grease
- smooth running grease
- high-speed grease
- grease for underwater applications
- thermally conductive grease

Figure 8-1. Greases are graded according to their consistency, i.e., resistance to deformation under the application of force. Consistency is measured with a penetrometer—shown here—and is reported as the tenths of a millimeter that a standard cone will penetrate the test sample under conditions prescribed by ASTM D 217. The softer the grease, the higher the penetration number.

Table 8-1. Impact of thickener on the properties of greases. (Source: Klüber Lubrication North America, Londonderry, New Hampshire.

Thickener	Service temperature range (°C)		Drop point DIN ISO 2176 (°C)	Water resistance	Stability at high pressure	Preferred applications
	min. oil	synth. oil				
Aluminum	-20 to 70	—	120	good	satisf.	Gears, valves (coke gas)
Calcium	-30 to 50	—	100	very good	good	Labyrinth seals under impact of water
Lithium	-35 to 120	-60 to 160	170/200	good	satisf.	Rolling bearings, contacts
Sodium	-30 to 100	—	150/170	unsatisf.	satisf.	Gears
Al complex	-30 to 140	-60 to 160	>230	good	satisf.	Rolling and plain bearings (plastic bearings), small gears
Ba complex	-25 to 140	-60 to 160	>220	very good	very good	Rolling bearings, valves, plain bearings subject to mixed friction
Ca complex	-30 to 140	-60 to 160	>190	very good	very good	Rolling bearings, seals (high-speed grease), chain grease
Li complex	-40 to 140	-60 to 160	>220	good	satisf.	Rolling bearings, couplings
Na complex	-30 to 140	-60 to 160	>220	satisf.	satisf.	Rolling bearings (subject to vibration, tribocorrosion)
Bentonite	-40 to 140	-60 to 180	none	good	satisf.	Valves (silicone base for high vacuum), gears, contacts
Polyurea	-30 to 160	-40 to 180	none	good	satisf.	Rolling bearings (long-term and lifetime lubrication of bearings i.e., sealed bearings)
PTFE	—	-40 to 260	250	good	good	Rolling bearings, valves subject to aggressive media

Figure 8-2. A technical specialist at Exxon's Pittsburgh grease plant measures the acidity/alkalinity of grease components on a titrator.

Tribotechnical Data

A lubricating grease is characterized by its tribotechnical data, Table 8-2. It permits the selection of an adequate lubricant to suit the special requirements of an application (e.g., in terms of temperature, load, speed).

Greases are classified (graded by NLGI, the National Lubricating Grease Institute. For details, refer to the entry "Grease Consistency," in Chapter 3.

Incompatible Greases

Incompatibility occurs when a *mixture of greases* exhibits properties or performance significantly inferior to those of either grease before mixing. Some grease bases are intrinsically incompatible. Incompatibility may affect several performance properties such as lower heat resistance; change in consistency, usually softening; a decrease in shear stability or a change in chemical properties such as the formation of acids. (Ref. 1)

Electric motor bearings are often furnished with polyurea-type greases. Polyurea greases exhibit excellent corrosion protection and are thus well suited for equipment in storage. However, when mixed with lithium-base greases—typically used by operating facilities because of their attractive all-around properties—the grease mixture may give inferior protection than either grease by itself. Experience shows that six months is about the limit for the life of some grease mixtures.

Although Table 8-3 shows that certain greases are indeed compatible, mixing should be avoided. Refer to the chapter on electric motor lubrication for further information.

Table 8-3 also illustrates the need to periodically update our grease-related knowledge base. As an example, a Texas-based formulator of synthetic lubricants is now the single-source provider of grease for China's 125 mph (200 km/h) Star Bullet Trains. Prior to switching to this grease, the NSK-brand bearings in the Zhuzhou Electric

Table 8-2. Tribotechnical data of lubricating greases. (Source: Klüber Lubrication, North America, Londonderry, New Hampshire.

Characteristics	Test	Notes
Base oil/ thickener		Both raw materials provide information about the possible behavior of a lubricating grease. However, only the chemico-physical and the mechano-dynamic properties are characteristic of a grease's actual performance.
Density	DIN 51 757	The amount of lubricant required for a certain friction point is generally indicated by volume. The volume in mm^3 multiplied by the density equals the quantity of lubricant in g required for the friction point.
Base oil viscosity	DIN 51 561	Thin-bodied base oils ≤ 46 mm/s are preferred for increased speeds, low loads and low surface roughness. Thick-bodied base oils—220 mm/s are preferred for low speeds, high loads and/or increased surface roughness.
Color		The color is not a performance characteristic. It is usually determined by the raw materials and additives. Graphite and MoS_2 greases, for example, are black.
Drop point	DIN ISO 2176	Even though the drop point is not indicative of the upper service temperature, it should be at least 40°C above this limit.
Service temperature range		An adequate relubrication interval can be achieved within the specified range. Service temperatures are indicated on the basis of measurement and testing procedures as well as practical experience.
Speed factor (n • dm)		The speed factor is important for lubricating greases used in rolling bearings. It takes into consideration the varying internal friction of the base oil/ thickener combination. If a lubricating grease used in another field of applications has been attributed a certain speed factor, it is also suitable for rolling bearings.
Worked penetration	DIN ISO 2137	Measuring the worked penetration provides information about a grease's consistency, i.e. whether it is soft or hard.
Consistency	DIN 51 828	The classification of greases in NLGI grades is a simplified classification in terms of worked penetration.
Apparent viscosity		The apparent viscosity (measured in mPa • s) is the same as the lubricant's internal friction. Kluber has classified the apparent viscosity in 5 classes. Class EL (extra light) characterizes a smooth running grease suitable for applica-tions with an extremely low torque. If the measured value overlaps with class L (light), the value is indicated as EUL.

VISCOSITY GRADES

Viscosity class	Apparent viscosity (mPa•s)	Explanation
EL	≤ 2000	Extra-light lubricating grease for extremely low torques, e.g., smooth running grease
L	2000… 4000	Light lubricating grease for low torques or high speeds in rolling bearings, e.g., high-speed grease
M	4000… 8000	Medium lubricating grease for standard requirements at all grease lubrication points
S	8000… 20 000	Heavy lubricating grease for high loads or fluids, e.g., high-pressure or sealing greases
ES	≥ 20 000	Extra-heavy lubricating grease for high torques or increased securing effect, e.g., valve or optical greases

(*Continued*)

Table 8-2. (Continued).

CONSISTENCY CLASSES

Consistency class DIN 51 818 (NLGI)	Worked penetration DIN ISO 2137 (0.1 mm)	Texture	General application
000	445... 475	fluid	mainly for gear lubrication
00	400... 430	almost fluid	
0	355... 385	extremely soft	
1	310... 340	very soft	lubrication of rolling and plain bearings
2	265... 295	soft	
3	220... 250	medium	
4	175... 205	hard	sealing and barrier grease for labyrinths
5	130... 160	very hard	and valves
6	85... 115	extremely hard	

Table 8-3: General Grease Compatibility (Top) and Ultra-Performance Grease Compatibility Chart (Lower Portion). Source: Royal Purple, Porter, Texas

General Grease Compatibility Chart	Aluminum Complex	Barium	Calcium	Calcium 12-Hydroxy	Calcium Complex	Clay	Lithium	Lithium 12-Hydroxy	Lithium Complex	Polyurea	Silicone
Aluminum Complex	X	I	I	C	I	I	I	I	C	I	I
Barium	I	X	I	C	I	I	I	I	I	I	I
Calcium	X	I	X	C	I	I	I	I	C	I	I
Calcium 12-Hydroxy	C	C	C	X	B	C	C	C	C	I	I
Calcium Complex	I	I	I	B	X	I	I	I	C	C	I
Clay	I	I	C	C	I	X	I	I	I	I	I
Lithium	I	I	C	C	I	I	X	C	C	I	I
Lithium 12-Hydroxy	I	I	B	C	I	I	C	X	C	I	I
Lithium Complex	C	I	C	C	C	I	C	C	X	I	I
Polyurea	I	I	I	I	C	I	I	I	I	X	I
Silicone	I	I	I	I	I	I	I	I	I	I	X

C = Compatible

B = Borderline Compatible: Typically results in a light softening or hardening of the NLGI Grade and a lowering of the dropping point of the mixture of grease.

I = Incompatible: Typically results in a softening or hardening of greater than 1 1/2 the NLGI grade, a shift in the dropping point, and a possible reaction of additives or base oils.

Source: E.H. Meyers' paper entitled "Incompatibility of Greases" 49th Annual NLGI Meeting

Royal Purple Ultra-Performance® Grease Compatibility Chart	Aluminum Complex	Barium	Calcium	Calcium 12-Hydroxy	Calcium Complex	Clay	Lithium	Lithium 12-Hydroxy	Lithium Complex	Polyurea	Silicone
Operating Temps. <225°F	C	C	C	C	C	I	C	C	C	C	I
Operating Temps. 225-350°F	C	B	B	C	B	I	B	B	C	B	I
Operating Temps. >350°F	C	I	I	C	I	I	I	I	C	I	I

Note: Ultra-Performance® greases are more stable. This chart is generated from independent lab testing and field experience. Actual compatibility results may vary. It is recommended that bearings be purged of old grease per OEM instructions to ensure proper lubrication and performance.

Locomotives' 1,250 kW traction motors would overheat at 160 km/h. Now, they're reported to run much cooler.

The formulators of this ultra-performance grease (UPG #2) report they were the only supplier selected after extensive field tests on over 20 types of grease. During these tests, the Star Bullet Train set a new China speed record clocking 200 mph (321 km/h) and the bearings survived even this speed for extended periods of time. The formulator's ultra-performance grease uses an aluminum-complex base and exhibits a favorable combination of performance traits for both wet and hot environments.

Note that standard grease compatibility charts (Table 8-3, top portion) are intended to simply identify greases which, when mixed, can alter the consistency of the grease. The change in consistency is such that it can be measured with a cone penetration tester (Chapter 3), which means that traditional compatibility charts are based only on changes in this one physical property. Royal Purple prepared an alternative chart (Table 8-3, lower portion) that holds true for the practical application of UPG grease in single lube point applications.

Because this company has always recommended that multi-point automatic lubricators be purged before changing to its superior grease formulations, the compatibility issue could be academic. Also, this formulator has experience-based supplementary data in what the chart terms borderline cases.

In essence, then, the 1992 standard grease compatibility chart (top of Table 8-3) still applies, but more recent findings show the compatibilities of aluminum complex greases. Aluminum complex greases obviously are problem solvers; their performance is related to temperature and the lower portion of Table 8-3 pertains to these greases.

INDUSTRIAL GREASES AND TYPICAL PROPERTIES

Unirex EP

Unirex EP is typical of a premium, medium-base oil, lithium-complex-thickened lubricating grease for use in plain or anti-friction automotive and industrial bearings. This product is formulated in NLGI grades 1 and 2. Its formulation permits operation over a very wide temperature range. Both grades of Unirex EP are shear and oxidation stable, protect against rust and corrosion and resist softening at elevated temperatures. Unirex EP is specially formulated to resist the effects of water sprayoff and water washout. Both grades are also suitable for dispensing in long lines of centralized lubrication systems.

Table 8-4 highlights the properties of the two grades of grease. These are corroborated through rigorous testing, Figure 8-3.

Ronex Extra Duty

Ronex Extra Duty (Table 8-5) is a line of premium high-viscosity base-oil greases designed for heavy-duty

Table 8-4. Inspection and test data for Unirex EP greases

	Unirex EP 1	**Unirex EP 2**
NLGI grade	1	2
Thickener type	Lithium-complex	Lithium-complex
Color	Green	Green
Texture	Smooth & homogenous	Smooth & homogenous
Penetration, ASTM D 217		
Worked 60X mm/10	320	280
Change, 100,000, mm/10	±20	±20
Dropping Point, ASTM D 2265		
°C(°F)	260+(500+)	260+(500+)
Base oil viscosity:		
cSt @ 40°C	220	220
cSt @ 100°C	17	17
SUS @ 100°F	1100	1100
SUS @ 210°F	95	95
Timken OK load, kg(lb)	27(60)	27(60)
Four-ball wear test, ASTM D 2265		
Scar diam., mm	0.45	0.41
Corrosion prevention test, ASTM D 1743	Pass	Pass
Oxidation life, ASTM D 942,		
Pressure drop 100 hours, kPa(psi)	35(5)	35(5)

Figure 8-3. The Timken machine shown above simulates extreme pressure service conditions to measure the EP properties of lubricants.

applications such as those in paper and other rolling mills, construction and mining. Based on the proven lithium-complex thickener technology used in Exxon's RONEX MP multipurpose grease, RONEX Extra Duty greases have additives that provide higher load-carrying ability. This is partly due to the 2500 SUS viscosity of the base oil. They also offer excellent structural and oxidation stability and a high degree of water resistance.

Typical applications are paper machine wet end, press section and felt roll bearings, as well as multi-purpose lubrication—including couplings—in pulp and paper mills, construction and mining and other heavy industry.

RONEX Extra Duty is available in three grades: RONEX Extra Duty 1 and RONEX Extra Duty 2 (red NLGI Grade No. 1 and No. 2 greases) and RONEX Extra Duty Moly, a purple NLGI Grade No. 2 grease containing 3% molybdenum disulfide. All products are specially formulated with a tackiness additive to enhance retention on the lubricated part.

RONEX Extra Duty Moly is Exxon's primary recommendation for lubricating sliding and oscillating applications in off-road equipment such as fifth wheels and bucket pins.

UNIREX RS 460 (Table 8-6)

UNIREX RS 460 grease contains a specially blended high-viscosity synthetic base oil in a lithium-complex thickener. It is recommended for applications in plants requiring high-viscosity lubricants with good mobility at low temperatures. It is specifically recommended for

Table 8-5. Inspection and test data for RONEX greases frequently used in paper mills.

NLGI Grade	RONEX Extra Duty 1	RONEX Extra Duty 2	RONEX Extra Duty Moly
NLGI grade	1	2	2
Thickener type	Lithium-complex	Lithium-Complex	Lithium-complex
Color	Red*	Red	Purple
Tackiness	Yes	Yes	Yes
Dropping point, ASTM D 2265 °C(°F)	260+(500+)	260+(500+)	260+(500+)
Base oil viscosity:			
SUS @ 100°F, ASTM D 2161	2500	2500	2500
Timken OK load, ASTM D 2509 kg(lb)	27(60)	27(60)	27(60)
Corrosion prevention test, ASTM D 1743	Pass	Pass	Pass
Molybdenum disulfide %	—	—	3

*Also available in an "undyed" version in some package styles.

Table 8-6. Synthetic grease for use in paper mills.

Unires RS 460	
NLGI grade	1-1/2
Thickener type	Lithium-complex
Color	Olive
Penetration, ASTM D 217, mm/10	
Worked, 60X	300
Worked, 10,000X	298
Worked, 100,000X	283
Dropping Point, ASTM D 2265, °C(°F)	260+(500+)
Oil type	Polyalphaolefin
Base oil viscosity:	
cSt at 40°C ASTM D 445	443
SUS @ 100°F, ASTM D 2161	2054
Timken OK load, ASTM D 2509, kg(lb)	23(50)
Four-ball EP test, ASTM D 2296	
Load wear index, kg	63
Weld point, kg	400
Four-ball wear test, ASTM 2266	
20 kg, 1800 rpm, 54°C(130°F), 1 hr, scar diam, mm	0.6
Corrosion prevention test, ASTM D 1743, modified	
5% synthetic sea water	Pass
Copper strip corrosion, ASTM D 4048	1B
Water washout, ASTM D 1264, % loss @ 79°C(175°F)	7.0
Lubrication life, ASTM D 3336, hrs @ 163°C(325°F)	600

grease-lubricated dry-end felt rolls, wet-end process rolls and couplings, as well as for miscellaneous woodyard and mill-wide applications in paper mills. Unirex RS 460 is extremely shear-stable and resists the effects of water and corrosive atmospheres.

Polyrex

Polyrex is a super-premium, high-temperature, polyurea grease recommended for all ball bearings and low-loaded roller bearings. It is an excellent alternative to Unirex N at high temperatures. Long service life and superior structural stability at temperatures above 149°C(300°F) make Polyrex especially desirable in factory fill and sealed for-life applications. Polyrex grease's exceptional high-temperature performance was demonstrated in the ASTM D 3336 spindle test. It exhibited twice the life of any competitive polyurea grease (and of Unirex N) at 177°C(350°F). (See Table 8-7.)

Polyrex also performs well at -29°C(-20°F) and, where torque is not limiting, at temperatures as low as –40°C (–40°F). It exhibits exceptional structural stability, staying in grade even after extensive shearing. Other outstanding features include excellent anti-wear properties, water resistance and rust protection even in hot, corrosive marine environments. The reduced maintenance and extended lubrication intervals possible with this versatile, durable grease can be expected to reduce long-term lubrication costs.

Observe grease compatibility issues, mentioned earlier in this segment of our text.

Unirex N (also "EM"-electric motor grease)

Unirex N is the brand name for two premium-quality, multi-purpose greases suitable for long-life, high-temperature service in all types of bearings. These versatile greases have applications in a wide range of industries, including power plants. They are excellent for electric motors and most sealed-for-life bearings.

Unirex N greases have outstanding mechanical stability and excellent anti-wear properties. They have long lubrication life at recommended operating temperatures and are suitable for service down to -40°C(-40°F). They provide excellent rust protection, even in saline environments; they are excellent "EM" greases.

Unirex N greases (Table 8-8) are available in two consistencies. Unirex N 2 is an NLGI Grade No. 2. It is preferred for grease gun or hand-packing applications. Unirex N 3 may be used in applications, such as vertical

Table 8-7. High-temperature synthetic grease inspection and test data.

POLYREX	
NLGI grade	1-1/2
Thickener type	Polyurea
Color	Tan
Penetration, ASTM D 217, mm/10	
Unworked	300
Worked, 60X	305
Worked, 100,000X	340
Dropping Point, ASTM D 2265, °C(°F)	304+(580+)
Base oil viscosity:	
cSt at 40°C, ASTM D 445	115.0
cSt at 100°C	12.2
SUS @ 100°F, ASTM D 2161	600.0
SUS @ 210°F	68.5
Four-ball EP test, ASTM D 2596	
Load wear index kg	32
Weld Point, kg	160
Four-ball wear test, ASTM D 2266	
40kg, 1200 rpm, 75°C(167°F), 1 hr, scar diam, mm	0.45
Corrosion prevention test, ASTM D 1743	Pass
Modified, 5% synthetic sea water, 24 hrs	Pass
Water washout, ASTM D 1264	
% loss @ 79°C(175°F)	2.4
Lubrication life, ASTM D 3336	
hrs @ 177°C(350°F)	500+
Wheel bearing leakage, ASTM D 1263	
90g pack @ 105°C(220°F), g	0.4
60g pack @ 163°C(325°F), g	0.1

installations, which may require the higher consistency of the NLGI Grade No. 3.

UNIREX® SHP Synthetic-Base-Oil Greases

UNIREX SHP is a line of five super-premium synthetic base-oil greases intended for severe, heavy-duty service in a variety of industrial and automotive applications. Formulated with a high-viscosity-index synthetic base oil, an Exxon-developed lithium-complex thickener and a unique additive package, UNIREX SHP greases are designed to provide extra protection and longer service life over a wider temperature range than mineral-base greases of comparable base oil viscosity.

The versatility of UNIREX SHP across a broad range of applications and operating environments may enable equipment operators to consolidate lubricants and reduce inventory costs—and its extra reliability in severe-service operations can reduce downtime and extend equipment life.

There are paper mill and related applications for UNIREX SHP grades 100, 460, and 1500. UNIREX SHP 220 is a multi-purpose lubricant, suitable for automotive wheel bearings and chassis requiring an NLGI GC/LB grease and for industrial applications where long life and extended maintenance intervals are desired. UNIREX SHP 00 is recommended as a gear oil replacement in truck trailer wheel bearings to protect against seal leaks; it is also recommended for leaking industrial gear boxes.

Typical inspections for these five super-premium, synthetic base-oil greases are given in Table 8-9.

UNIREX S 2

UNIREX S 2 (Table 8-10) is formulated with a high-viscosity, low-volatility synthetic (ester) base oil to provide excellent high-temperature lubrication where frequent relubrication is impractical. It compares favorably with the more expensive silicone greases in many applications, with the added advantage of superior lubrication under high bearing loads. Designed for use at operating temperatures in the 177°-204°C(350°-400°F) range, UNIREX

Table 8-8. UNIREX N greases for electric motor bearings and similar applications.

	UNIREX N 2	UNIREX N 3
NLGI grade	2	3
Thickener type	Lithium-complex	Lithium-complex
Color	Green	Green
Penetration, ASTM D 217		
Worked, 60X	280	235
Worked, 100,000X, unit change	+5	+10
Dropping Point, ASTM D 2265, °C(°F)	260+ (500+)	260+ (500+)
Base oil viscosity:		
cSt at 40°C, ASTM D 445	115.0	115.0
cSt at 100°C	12.2	12.2
SUS @ 100°F, ASTM D 2161	600.0	600.0
SUS @ 210°F	68.5	68.5
Four-ball wear test, ASTM D 2266, scar diam, mm		
10 kg, 1800 rpm, 75°C(167°F), 1 hr	0.28	0.30
40 kg, 1200 rpm, 75°C(167°F), 1 hr	0.45	0.47
Corrosion prevention test		
ASTM D 1743	Pass	Pass
Modified 5% synthetic sea water 24 hrs	Pass	Pass
Lubrication life, ASTM D 3336		
Hrs @ 149°C(300°F)*	1000	1000
Hrs @ 163°C(325°F)*	500	500
Wheel bearing leakage, ASTM D 1263		
Modified, 60g pack @ 163°C(325°F), g	0.6	0.2
Oil separation, ASTM D 1742		
24 hrs @ 25°C(77°F),%	1.8	2.0

*Note: Lubrication life tests run on severe spindle rigs; data obtained on less severe rigs show up to 3, 000 hrs lubrication life at 149°C(300°F).

S 2 can be used at higher temperatures with suitable relubrication intervals. It also has excellent low-temperature properties, providing starting and running torque at -40°C(-40°F) and, in many applications, acceptable torque at -54°C (-65°F). UNIREX S 2 has good water resistance.

Proven applications include conveyor bearings in kilns and ovens, steel mill ladle bearings, jet aircraft starter clutch assemblies and bearings atop ovens in fiberglass manufacture.

While the equipment manufacturer or grease manufacturer's representative should have the final word regarding compatibility of UNIREX S 2 in specific applications, the compatibility chart for diesters, Table 7-8, can be consulted.

EXXON HI-SPEED COUPLING GREASE

EXXON HI-SPEED COUPLING GREASE, Table 8-11, is a high-quality grease formulated to lubricate all flexible couplings. In particular, it meets the lubrication needs of couplings operating at high speeds and with high centrifugal forces. It provides extreme-pressure protection and is suitable for operating temperatures between -40°C(-40°F) and 149°C (300°F). EXXON HI-SPEED Coupling Grease offers excellent resistance to oil separation, as indicated by the ASTM D 4425 test results (K36 typically = 0/24).

LIDOK EP

LIDOK EP, Table 8-12, is a line of three multi-purpose EP greases and one EP semi-fluid grease. LIDOK EP greases

Table 8-9. Typical inspections for UNIREX SHP synthetic-base-oil greases.

Typical Inspections

The values shown here are representative of current production. Some are controlled by manufacturing specifications, while others are not. All may vary within modest ranges.

	UNIREX SHP 100	UNIREX SHP 220	UNIREX SHP 460	UNIREX SHP 1500	UNIREX SHP 00
NLGI Grade	2	2	1.5	1	00
Soap type	Lithium Complex	Lithium Complex	Lithium Complex	Lithium Complex	Lithium
Complex					
Base oil	Synthetic	Synthetic	Synthetic	Synthetic	Synthetic
Base oil viscosity, ASTM D 445 @ 408°C, cSt	100	220	460	1500	460
Color	Red	Red	Off-White	Red	Red
Dropping point, ASTM D 2265, °F	520	550	550	550	550
Penetration ASTM D 217 60X mm/10	280	290	300	325	400
Viscosity index, ASTM D 2270	143	154	161	154	161
Timken OK load, ASTM D 2509, lb	Non-EP	65	60	60	60
4 ball wear test, ASTM D 2266, mm	0.38	0.44	0.42	—	—
4 ball EP weld ASTM D 2596, kg	250	315	315	—	—
Water washout, ASTM D 1264, %	11	3	2	—	—
Mobility, U.S. Steel, grams/minute @ -18°C(0°F)	—	11.6	7.5	—	—
Lubrication life, ASTM D 3336 hours @ 350°F	300	—	300	—	—
Corrosion prevention with 10% sea water, EMCOR rating	1,2	0,0	0,1	—	—

Table 8-10. UNIREX S @ grease formulated with synthetic ester-base oil for elevated load and temperature services.

UNIREX S 2	
NLGI grade	2
Thickener type	Lithium-complex
Color	Brown
Penetration, ASTM D 217	
Worked 60X mm/10	280
Worked 10 000X, unit change	+18
Dropping Point, ASTM D 2265, °C(°F)	260+ (500+)
Base oil viscosity, ASTM D 445	
cSt @ 40°C(SUS @ 100°F)	162(751)
cSt @ 100°C(SUS @ 210°F)	19.8(97.6)
Base oil viscosity index	160
Four-ball EP test, ASTM D 2596	
Load wear index, kg	31.4
Weld point, kg	160
Four-ball wear test, ASTM 2266, scar diam, mm	
10 kg, 1800 rpm, 75°C(167°F), 1 hr	0.53
40 kg, 1200 rpm, 75°C(167°F), 1 hr	0.68
Corrosion prevention test ASTM D 1743, distilled water	Pass
Water washout, ASTM D 1264, % loss @ 79°C(175°F)	0
Lubrication life, ASTM D 3336, 10,000 rpm	
Hrs @ 176°C(350°F)	1100

Table 8-11. Exxon high-speed coupling grease, typical inspections.

Exxon HI-SPEED Coupling Grease

Property	Test condition	Value
NLGI grade		1
Thickener type		Lithium-polymer
Color		Green
Penetration, ASTM D 217, mm/10	60X	336
	10,000X	370
Dropping Point, °C(°F), ASTM D 566		210(410)
Base oil viscosity, ASTM D 2161	SUS @ 100°F	2550
	SUS @ 100°F, after polymer addition	3300
	SUS @ 210°F	145
Base oil viscosity index, ASTM D 2270		85 min
Timken OK load, ASTM D 2509, kg (lb)		18(40)
Four-ball EP, ASTM D 2596	LWI, kg	51
	Weld point, kg	315
	Scar diam., mm	0.7
Corrosion prevention test, ASTM D 1743		Pass
Copper strip corrosion test, ASTM D 4048		lb
Centrifugal separation, ASTM D 4425	K36 = % oil separated/test hrs	K36=0/24

Table 8-12. LIDOK multi-purpose greases.

Grade	LIDOK EP 000	LIDOK EP 0	LIDOK EP 1	LIDOK EP 2
NLGI Grade	000	0	1	2
Thickener type	Lithium 12-OH	Lithium 12-OH	Lithium 12-OH	Lithium 12-OH
Color	Brown	Brown	Brown	Brown
Penetration, ASTM D 217				
Worked, 60X, mm/10	460	370	325	280
Worked 10 000X unit change	+12	+3	+4	+5
Dropping point, ASTM D 2265, °C(°F)	139(282)	171(340)	174(345)	177(350)
Base oil viscosity:				
cSt @ 40°C, ASTM D 445	310.0	150.0	150.0	150.0
cSt @ 100°C	19.0	12.8	12.8	12.8
SUS @ 100°F, ASTM D 2161	1600.0	800.0	800.0	800.0
SUS @ 210°F	97.0	70.8	70.8	70.8
Timken OK load, ASTM D 2509 kg(lb)	18(40)	18(40)	18(40)	18(40)
Four-ball EP test, ASTM D 2596				
Load wear index, kg	35	43	37	40
Weld Point, kg	250	250	250	250
Four-ball wear test, ASTM D 2266, scar diam, mm				
40 kg, 1200 rpm, 75°C(167°F), 1 hr	—	0.41	0.39	0.46
20 kg, 1800 rpm, 55°C(130°F), 1 hr	—	0.43	0.42	0.39
75 kg, 1800 rpm, 55°C(130°F), 1 hr	0.31	—	—	—
Corrosion prevention test, ASTM D 1743	Pass	Pass	Pass	Pass
Water washout, ASTM D 1264				
% loss @ 38°C(100°F)	—	10	7	3
% loss @ 79°C(175°F)	—	20	15	6
Wheel bearing leakage, ASTM D 1263, 104°C(220°F), g	—	—	5	0
Oil Separation, ASTM D 1742, %	*	*	4	3

*Grease passes through screen without separation

meet the lubrication requirements of plain and anti-friction bearings, gears and couplings in general industrial applications. Properties include good water resistance, high-temperature performance, resistance to mechanical breakdown, excellent oxidation resistance, good rust protection and fortified extreme-pressure properties. LIDOK EP contains no lead or other heavy metals.

Semi-fluid LIDOK EP 000 is designed primarily for use in the gear cases of underground mining machinery, where leakage can be a problem. It stays in place and is easily pumpable. It meets all requirements of Specification 100-4 of the Lee-Norse Company and U.S. Steel Requirements No. 373.

LIDOK EP 2 Moly

LIDOK EP 2 Moly is a multi-purpose lithium-base grease recommended for automotive and industrial applications. It contains an extreme-pressure (EP) additive to increase the load-carrying properties of the grease. Additionally, it contains molybdenum disulfide which enhances the antifriction properties under boundary lubricating conditions. LIDOK EP 2 Moly is recommended for heavily loaded, sliding or oscillating applications, including off-highway applications, Figure 8-4 and Table 8-13.

ANDOK B, C and 260 (Table 8-14)

ANDOK greases are specially formulated to provide exceptional service in rolling-contact bearings subjected to severe operating conditions. They have excellent oxidation resistance and protect against rust in damp locations. They also have excellent channeling characteristics, i.e., they are readily forced to the bearing sides during operation, leaving just the proper amount of grease to lubricate rolling elements. This significantly reduces torque and temperature rise in the bearing.

Applications of ANDOK B include sealed-for-life bearings, high-speed bearings and bearings operating at high temperatures. It is also used in both ball and roller bearings, particularly for factory/field replenishment by grease gun.

ANDOK C is used for factory fill of ball bearings and hand-packed replenishment in the field.

ANDOK 260 is an extra-long-life grease suitable for ball bearings *and* roller bearings.

BEACON P 290 AND 325 (Table 8-15)

BEACON is the brand name for two greases formulated for the lubrication of precision equipment at moderate and low temperatures. Both are made with base oils of extremely low viscosity. They are characterized by low starting and running torque at very cold temperatures. Both offer maximum lubrication protection to small gears, plain and anti-friction bearings and other parts of fine instruments, control mechanisms, small motors and generators.

BEACON P 290 is a lithium-soap, petroleum-base grease used in arctic environments to lubricate bearings in power tools, valve operator and similar instruments. It is formulated to be used at temperatures below -54°C(-65°F).

BEACON 325 is a lithium-soap, synthetic-oil-base grease. This makes it suitable over a wider temperature range—from as low as -54°C(-65°F) to as high as 120°C(250°F).

NEBULA EP (Table 8-16)

NEBULA EP is suitable for plain and anti-friction bearings—at high or low temperatures, high or low speeds, under heavy or light loading and wet or dry conditions. Its properties include anti-wear and EP protection, resistance to softening at high temperatures, water-resistance and good adhesion.

NEBULA EP 00 AND EP 0 have been used extensively

Figure 8-4. LIDOK EP 2 Moly is recommended for applications characterized by sliding or oscillating movements, such as the fifth wheel shown above.

Table 8-13. Typical inspections for lithium-base grease containing molybdenum.

LIDOK EP 2 Moly	
NLGI Grade	
Thickener type	Lithium 12-OH
Color	Dark gray
Texture	Smooth & buttery
Penetration, ASTM D 217 worked, 60X, mm/10280	
Dropping Point, ASTM D 2265, °C(°F)	175(347)
Base oil viscosity, ASTM D 445, cSt at 40°C	150
Timken OK load, kg(lb)	18 (40)
Water washout, ASTM D 1264, % loss @ 79°C(175°F)	6
Wheel bearing leakage, ASTM D 1263 104°C(220°F), g	0
Oil separation test, ASTM D 1742, %	3
Corrosion prevention test, ASTM D 1743	Pass

Table 8-14. Bearing grease formulated for moderate duty applications.

	ANDOK B	**ANDOK C**	**ANDOK 260**
NLGI grade	2	4	3
Thickener type	Sodium-complex	Sodium-complex	Sodium-complex
Color	Brown	Brown	Brown
Penetration, ASTM D 217, mm/10			
Unworked	270	180	260
Worked, 60X	285	205	260
Dropping Point, ASTM D 2265			
° C(°F)	245(475)	260(500)	190(375)
Base oil viscosity:			
cSt @ 40°C, ASTM D 445	59.2	93.0	105
cSt @ 100°C	6.2	8.8	11.1
SUS@ 100°F, ASTM D 2161	310.0	510.0	550.0
SUS @ 210°F	48.0	56.0	64.0
Lubrication life, ASTM 3336			
Hrs @ 120°F (250°F)	1500	1500	5000+
Hrs @ 150°F (300°F)	400	400	1000
Oxidation life, ASTM D 942			
ΔP after 500 hours, kPa(psi)	105(15)	105(15)	35(5)

in control valves in powerplants. NEBULA EP 0 is a soft grease for easier dispensing at low temperatures and may be used in all types of centralized systems. NEBULA EP 1 has the consistency characteristics that meet most requirements. It is recommended for all grease gun applications. NEBULA EP 2 exhibits excellent resistance to washout. It is recommended for wet applications, where fast-moving streams of water may dislodge a soft grease or where very high temperatures are a concern. NEBULA EP 1 and EP 2 should not be used in centralized lubricating systems.

RONEX MP

Although not an "industrial" grease in the true sense of the term, this versatile, premium NLGI Grade No. 2 multi-purpose grease has a wide range of automotive and industrial applications. It combines high-temperature performance with extreme-pressure properties, plus good water resistance, excellent oxidation stability, rust protection and resistance to chemical breakdown.

RONEX MP withstands the high temperatures of severe disc braking and provides extended trouble-free lubrication. It has passed the severe ASTM D 3428 Ball

Table 8-15. Typical grease used for instrument bearings and similar low-torque applications.

	Beacon P 290	Beacon 325
NLGI grade	1-1/2	2
Thickener type	Lithium	Lithium
Color	Tan	Light tan
Penetration, ASTM D 217		
worked, 60X mm/10	300	280
Dropping Point, ASTM D 2265, °C(°F)	160(320)	195(384)
Oil type	Mineral	Diester
Base oil viscosity:		
cSt @ 40°C, ASTM D 445	14.2	11.8
cSt @ 100°C	3.2	3.3
SUS @ 100°F ASTM D 2161	79.0	67.7
SUS @ 210°F	36.9	37.2
Water washout, ASTM D 1264		
% loss @ 79°C(175°F)	—	1.0
Oil separation, ASTM D 1742		
24 hrs @ 25°C(77°F), %	4.9	4.0

Table 8-16. Nebula greases are primarily applied in centralized systems and wet environments.

	Nebula EP 000	Nebula EP 0	Nebula EP 1	Nebula EP 2
NLGI	000	0	1	2
Thickener type	Calcium-complex	Calcium-complex	Calcium-complex	Calcium-complex
Color	Light tan	Dark tan	Dark tan	Dark tan
Penetration, ASTM D 217				
Worked, 60X, mm/10	415	370	325	280
Worked, 10,000X, unit change	NA	±5	±5	±5
Dropping point, ASTM D 2265, °C(°F)	260+ (500+)	260+(500+)	260+(500+)	260+(500+)
Oil type	PAO	Mineral oil	Mineral oil	Mineral Oil
Base oil viscosity:				
cSt @ 40°C, ASTM D 445	97 0	96.3	96.3	96.3
cSt @ 100°C	14.1	8.3	8.3	8.3
SUS@ 100°F, ASTM D 2161	590.0	510.0	510.0	510.0
SUS @ 210°F	84.8	56.0	56.0	56.0
Timken OK load, ASTM D 2509				
kg(lb)	18(40)	21(45)	21(45)	21(45)
Four-ball EP test, ASTM D 2596				
Load wear index, kg	44	50	54	63
Weld Point, kg	250	200	315	315
Four-ball wear test, ASTM D 2266				
40 kg, 1200 rpm, 75°C(167°F), 1 hr, scar diam, mm	0.4	0.67	0.67	0.67
Corrosion prevention test, ASTM D 1743	Pass	Pass	Pass	Pass
Water washout, ASTM D 1264				
% loss @ 79°C(175°F)	NA	2	2	1

Joint Test, which evaluates a grease's ability to provide minimum wear, minimum torque and protection against water contamination.

In industrial uses, Ronex MP is recommended for all types of bearings, gears and couplings where a multi-purpose, water-resistant EP grease is applicable.

Ronex MP meets or exceeds the requirements of the Mack MG-C extended lubrication internal specification, GM's specification 6031-M for chassis and wheel bearing lubrication, and NLGI GC-LB for chassis and wheel bearing lubrication as defined by ASTM D 4950. It also may be used in electric motors of NEMA (National Electric Manufacturers' Association) Insulation Class A & B types. Table 8-17 gives some of the more important characteristics of this grease.

LIDOK CG Moly (Table 8-18)

LIDOK CG Moly, *specially formulated as an automotive chassis grease,* has been tested and approved by Ford Motor Company for use under their Type M1C75B specification. It also meets the requirements of GM's specification 6031M. LIDOK CG Moly contains 4% polyethylene and 1% molybdenum disulfide.

ROLUBRICANT 1 and 2 (Table 8-19)

ROLUBRICANT 1 and 2 are lithium-base greases with extreme-pressure properties. They meet U.S. Steel Requirements 370 and 375. Specifically designed to meet the demanding needs of steel mills, ROLUBRICANT 1 and 2 resist water washout at medium-to-high operating temperatures. Both grades are suitable for centralized lubrication systems, although attention must be given to ROLUBRICANT 2 at low temperatures. As indicated in Figure 8-5, these greases excel by resisting water washout at high temperatures.

FOODREX FG 1 Grease

Covered earlier in this text (see Table 6-5), FOODREX FG 1 Grease is a premium grease specially formulated to meet the demands of the food and beverage industry. It is white in color and has a smooth-tacky appearance. FOODREX FG 1 Grease is an NLGI 1 grade consistency and contains an extreme-pressure additive for carrying heavy loads. All components of FOODREX FG 1 Grease are permitted under the U.S. Food and Drug Administration (FDA) Regulation 21 CFR 178.3570, "Lubricants With Incidental Food Contact." All components of Foodrex FG 1 Grease are acceptable to the U.S. Department of Agriculture for use as a lubricant with incidental food contact in establishments operating under the meat and poultry products inspection program. It is approved as a category "H-1" compound in the USDA list of Chemical Compounds. Additionally, FOODREX FG 1 Grease is Kosher and Pareve-certified.

CARUM 330

The typical inspections for CARUM 330 were given earlier in Table 6-6. This specially formulated grease is primarily used for lubricating food and beverage-processing machinery. It is highly resistant to water, steam, vegetable and fruit juices and carbonated beverages. CA-

Table 8-17. Multi-purpose, water-resistant EP grease, typical inspections.

RONEX MP		
NLGI grade		2
Thickener type		Lithium-complex
Color		Dark green
Penetration, ASTM D 217,	Worked, 60X, mm/10	285
	Worked, 100,000X, unit change	+ 10
Dropping point, ASTM D 2265, °C(°F)		260+(500+)
Base oil viscosity:	cSt at 40°C, ASTM D 445	146
	cSt at 100°C	14
	SUS @ 100°F, ASTM D 2161	767
	SUS @ 210°F	75
Timken OK load ASTM D 2509 kg(lb) 18(40)		
Four-ball EP test, ASTM D 2596,	Load wear index, kg	40
	Weld point kg	250
Four-ball wear test, ASTM D 2266		
40 kg, 1200 rpm, 75°C(167°F), 1 hr, scar diam, mm		0.40
Corrosion prevention test ASTM D 1743		Pass
Water washout ASTM D 1264, % loss @ 79°C(175°F)		4
Lubrication life, ASTM D 3336, hrs @ 163°C(325°F)		125
Wheel bearing leakage, ASTM D 1263 modified		
60g pack @ 163°C(325°F), g		1.5
Oil separation, ASTM D 1742, 24 hrs @ 25°C(77°F), %		3

Table 8-18. Specially formulated automotive chassis grease.

LIDOK CG Moly	
NLGI grade	2
Thickener type	Lithium 12-OH
Appearance / texture	Dark gray / butttery
Base oil type	Mineral oil
Dropping point, ASTM D 2265, °C	187
Penetration, ASTM D 217, 60X, mm / 10	285
Penetration change, ASTM D 217, 10,000X, %	12
Four-ball EP test, ASTM D 2596, kg	
LWI	42.6
Weld point	250
Four-ball wear test, ASTM D 2266, 10 kg, 1800 rpm	
75°C. 1 hr, scar diam., mm	0.37
Base oil viscosity, ASTM D 2161	
@ 100°F	800
@ 210°F	80.2
Base oil viscosity index, ASTM D 2270	90
Wheel bearing leakage, ASTM D 1263, modified, gm @ 325°F	2.1
Water washout, ASTM D 1264 % @ 175°F	4
Oil separation, ASTM D 1742 modified, ml	10
Oxidation stability, ASTM D 942, psi drop	
100 hrs	3
1000 hrs	27
Rust protection, ASTM D 1743	Pass (1,1,1)
Ball joint test, ASTM D 3428	Pass

RUM 330 contains a rust inhibitor and has good high-temperature and wear-preventive properties. Manufactured with a calcium-complex soap base, it offers the wear protection and load-carrying capacity that is characteristic of such greases.

All of the ingredients in CARUM 330 are permitted under FDA Regulations 21 CFR 178.3570, "Lubricants With Incidental Food Contact." CARUM 330 is acceptable to the USDA and is authorized for use as a lubricant with incidental food contact in establishments operating under the meat and poultry products inspection program. It is listed as a category "H1" compound in the USDA List of Chemical Compounds.

Figure 8-5. ROLUBRICANT 1 and 2 tenaciously resist water washout at high operating temperatures. This ideally suits them for use in the high-pressure, super-heated environments of steel mills. (Photo courtesy of Quamex Corporation.)

FIREXX Grease 1 and 2 (Table 8-20)

FIREXX Grease 1 AND 2 are flame-retardant, aluminum complex greases, Figure 8-6. They are excellent general-purpose greases that are highly resistant to combustion, providing additional time for plant personnel to respond to a potential fire. They are especially suitable for steel mills, underground railways, mines, welding areas or anywhere there is a need to minimize potential fire hazards. Both grades are adhesive and provide good high-temperature performance, rust and corrosion protection and excellent load-carrying capability and water resistance.

DYNAGEAR

DYNAGEAR (Table 8-21) represents an open gear lubricant that is formulated for cold temperature dispensability without the use of chlorinated solvents or

Table 8-19. Typical inspections for "ROLUBRICANT" grease used in steel plants.

	ROLUBRICANT 1	ROLUBRICANT 2
NLGI grade	1-1/2	2-1/2
Thickener type	Lithium 12-OH	Lithium 12-OH
Color	Black	Black
Penetration, ASTM D 217		
Worked, 60X, mm/10	300	265
Dropping Point, ASTM D 2265, °C(°F)	182(360)	186(366)
Base oil viscosity, ASTM D 2161		
SUS @ 100°F	1050	1200
SUS @ 210°F	74	83
Timken OK load, ASTM D 2509, kg(lb)	18(40)	18(40)
Four-ball EP test, ASTM D 2596		
Load wear index, kg	40	40
Weld point, kg	315	315
Four-ball wear test, ASTM D 2266		
20 kg, 1800 rpm, 54°C(130°F), 1 hr		
scar diam, mm	0.4	0.4
Corrosion prevention test, ASTM D 1743	Pass	Pass
Water washout, ASTM D 1264		
% loss @ 79°C(175°F)	4	3
Water sprayoff, ASTM D 4049		
276 kPa(psi), 38°C(100°F), % loss	10	3
Oxidation stability, ASTM D 942		
AP after 100 hours, kPa(psi)	42(6)	42(6)
Wheel bearing leakage, ASTM D 1263, %	3.2	0.7
Mobility, USS method		
Flow rate @ -18°C(0°F), g/min	5	1

Figure 8-6. This demonstration dramatically illustrates the flame-retardant capabilities of FIREXX Grease (center) compared with two conventional greases (lithium complex and aluminum complex, respectively). The FIREXX Grease sample resisted ignition and, once lit, extinguished itself within seconds. The photograph was taken approximately 20 seconds after igniting the conventional greases and 12 seconds after igniting the FIREXX Grease.

petroleum solvents. It offers an environmentally responsible lubricant for open gear applications (see Figure 8-7). DYNAGEAR can be dispensed down to -20°C(-4°F).

DYNAGEAR provides a tenacious lubricant film that firmly adheres to lubricant surfaces. The formulation is solvent-free; thus, run-off is less than for those formulations that contain solvents. This open gear lubricant is formulated with excellent anti-rust, anti-corrosion and oxidation protection properties. Powerful solid lubricants, graphite and molybdenum disulfide, are added to assist the load-carrying capability even at very low speeds. DYNAGEAR offers exceptional water resistance. Moderate amounts of water can be absorbed by these products with minimal effect on the NLGI consistency grade. Because the base oil used in DYNAGEAR does not contain asphalt,

Table 8-20. Fire-retardant, aluminum complex greases.

	FIREXX Grease 1	FIREXX Grease 2
NLGI grade	1	2
Color	Black	Black
Texture	Buttery	Buttery
Thickener type	Aluminum-complex	Aluminum-complex
Penetration, ASTM D 217		
Worked, 60X, mm/10	320	280
Dropping Point, ASTM D 2265,°C(°F)	212(415)	212(415)
Base oil viscosity:		
SUS @ 100°F, ASTM D 2161	1500	1500
Timken OK load, ASTM D 2509, kg(lb)	18(40)	18(40)
Four-ball EP test, ASTM D 2596		
Load wear index, kg	40	40
Weld point, kg	315	315
Four-ball wear test, ASTM D 2266		
40kg, 1200 rpm, 75°C(167°F)		
1 hr, scar diam, mm	0.45	0.45
Corrosion prevention test, ASTM D 1743	Pass	Pass
Water sprayoff, ASTM D 4049, % loss	15	6
Wheel bearing leakage, ASTM D 1263		
104°C(220°F), g	—	3.0
Mobility, USS method		
Flow rate @ -18°C(0°F) g/min	3	—

Table 8-21. Inspection and test data for DYNAGEAR open gear lubricant.

DYNAGEAR	
NLGI Grade	0
Thickener type	Lithium 12-OH
Base oil viscosity, ASTM D 445, cSt at 40°C	1000
Viscosity as applied, cp @ 100°C	40,000
Timken OK load, kg(lb)	28(60)
Four-ball EP test,	
Weld point, kg	800
Load wear index	140
Operating temperature range, °C(°F)	-40(-40) to +100(212)
Minimum dispensing temperature, °C(°F)	-20(-4)

Figure 8-7. Extensively field tested in equipment such as this, DYNAGEAR offers significant advantages over conventional asphalt-base products.

Figure 8-8. ARAPEN RB 320 is exceptionally resistant to shear, i.e., it retains consistency after prolonged working and is dependable over long hauls.

excessive lubricant build-up on gears and gear tooth roots is rare. This contributes to faster clean-up and reduces the likelihood of misalignment problems. Because the formulation does not contain solvents, flammability is equivalent to any other mineral-oil-base lubricant. Recommended operating and dispensing temperatures are shown in Table 8-21.

ARAPEN RB 320 (Table 8-22)

ARAPEN RB 320 is a long-life grease developed for the roller bearings of railroad car journals where no provision is made for in-service relubrication. It is fully approved against AAR Specifications M-942-88 for Journal Roller Bearing Grease for non-field-lubricated bearing applications.

ARAPEN RB 320 has high oxidation stability. It is resistant to heat and to deterioration in the presence of water and chemicals. It is also inhibited to give protection against rust. ARAPEN RB 320 will retain its consistency after prolonged working—as in the churning action of an antifriction bearing (see Fgure 8-8). It has little effect on elastomeric seal materials, and thus maintains good seal performance—a significant requirement for shop-to-shop wheel service.

ARAPEN RB 320 is used as the factory-fill lubricant by major manufacturers of railroad journal bearings.

ARAPEN RC 1 (Table 8-23)

ARAPEN RC 1 rail-flange grease is designed for use in onboard lubricators. ARAPEN RC 1 has also been successfully used in wayside lubricators. It has excellent water resistance and is inhibited to prevent corrosion. ARAPEN RC 1 is fortified with 3% moly, which provides residual lubrication. The product contains polymers that provide excellent adherence to lubricating surfaces and enhance track carry-down. The superior anti-wear characteristics of ARAPEN RC 1, compared with several competitive products, have been demonstrated in the Timken retention test, four-ball test and SRV. Figure 8-9 comments on the environmental impact of this railroad grease.

References

1. Kusnier, Walter J., "Mixing Incompatible Greases," *Plant Services*, June 1997, pp. 143-149.
2. Bloch, H.P., and Rizo, L.F., "Lubrication Strategies for Electric Motor Bearings in the Petrochemical and Refining Industries," presented at the NPRA Maintenance Conference, San Antonio, Texas, February, 1984.

Table 8-22. Long-life grease formulated for railroad car journals.

	ARAPEN RB 320	AAR M-942-88 Requirements
Thickener type	Calcium-lithium	—
Penetration, ASTM D 217		
Worked, 60X, mm/10	305	290-320
Worked, 100,000X, mm/10	+18	+25
Dropping Point, ASTM D 2265, °C(°F)	177(350)	163(325)
Base oil viscosity, ASTM D 2161		
SUS @ 100°F	840	750 min
Base oil viscosity index	100	80 min
Base oil flash point (open cup) °C(°F)	246(475)	171(340) min
Corrosion prevention test ASTM D 1743	Pass	Pass
Oxidation stability, ASTM D 942		
AP after 100 hours, kPa(psi)	1	10 max
AP after 500 hours, kPa(psi)	5	25 max
AAR elevated temperature roll stability,		
Penetration @ 82°C(180°F)		
Worked, mm/10	325	290-340
Dynamic mechanical stability		
Migration and distribution	1	1 or 2
Penetration after test, mm/10	307	280-320
Moisture, ASTM D 128, %	nil	0.10 max

Table 8-23. Rail-flange grease used to reduce wheel wear and noise.

ARAPEN RC 1	
NLGI grade	1
Thickener type	Lithium 12-OH
Tackiness	Yes
Penetration, ASTM D 217	
Worked, 60X, mm/10	325
Dropping Point, ASTM D 2265, °C(°F)	191 (375)
Base oil viscosity, ASTM D 2161	
SUS @ 100°F	500
Four-ball wear test, ASTM D 2266	
10 kg, 1800 rpm, 75°C(167°F), 1 hr, scar diam, mm	0.46
Timken retention test	
Scar diam	<1
Weight loss, mg	nil
Load, lb	25
Molybdenum disulfide, %	3

Chapter 9

Pastes, Waxes and Tribosystems

LUBRICATING PASTES

Lubricating pastes are cohesive lubricants made up of a base oil (mineral and/or synthetic oil), additives and solid lubricant particles. They are mainly applied under extreme conditions and prevent fretting corrosion, stick-slip and adhesive wear. Depending on their composition, lubricating pastes are resistant to water and water vapor and have good anti-corrosion characteristics. Metal-containing pastes may be suitable for service temperatures up to 1200°C.

Lubricating pastes can be classified in terms of:

- solid lubricant type (MoS_2, graphite, metals, PTFE, other plastics)
- base oil (synthetic oil, mineral oil and mixtures)
- application range (lubricating and assembly paste, high-temperature paste, conductive paste, etc.)
- special characteristics (color, EP properties, etc.)

The base oil and the solid lubricant particles have different tasks, depending on the type of paste:

Lubricating and Assembly Paste

The solid lubricant improves the base oil's lubricity.

High-temperature Pastes

The oil must distribute the solid lubricant particles over the friction surface. At temperatures of about 160 to 200°C all the base oil evaporates and leaves a coherent lubricant film on the friction surface.

Conductive Pastes

The solid lubricant particles contained in thermally and electrically conductive pastes compensate the insulating effect of the base oil. Conductive pastes must contain a certain percentage of solid lubricant powder.

Screw Pastes

These pastes are used to ensure precise assembly (tightening torque).

High-temperature Screw Compounds

The dry residue left after the base oil has evaporated must be "crumbly" to avoid sticking of the thread in the bore.

One manufacturer, Klüber, has developed a standard program based on many years of experience. Table 9-1 shows some of the available products and their fields of application, making it easier for the designer and the maintenance staff to select a suitable paste. For additional criteria, refer to Table 9-2.

Lubricating Waxes

These materials consist of a combination of synthetic hydrocarbons of high molecular weight plus additives. Wax emulsions also contain an emulsifier and water. In addition to lubricating greases and pastes, these coherent lubricants are gaining increasing importance.

Starting with a certain temperature, lubricating waxes typically change their structure from a coherent to a fluid state. The melting point depends on the waxes' ingredients, and their structure is reversible. If the tribological requirements are mainly about corrosion protection, a coherent structure is of advantage.

The *advantages* of lubricating waxes and wax emulsions over traditional lubricants are their excellent inherent lubricity and special anti-corrosion properties. In addition, they provide a non-tacky protective film when applied below their melting point. Their main *disadvantage* is the lack of heat dissipation until the incorporated water has evaporated. Lubricating waxes do not flow below their melting point, which is of special importance for relubrication.

The positive effects of waxes come obvious in their behavior. They

- are adhesive
- stick to metals
- exhibit polar properties
- protect against corrosion
- offer good lubricity
- protect against wear, and
- produce a dry film.

Their individual characteristics permit an application in the boundary and mixed friction regimes. In this context, the non-tacky wax film is an additional advantage for attracting less dust or dirt. This film ensures quasi-dry lubrication. When the friction point is heated up, the wax melts and is redistributed, whereas the perimeter areas remain below the melting point.

Application

Lubricating waxes and wax emulsions are especially interesting for the following machine elements and components:

- bolts
- seals
- dowels
- springs
- sliding points
- nails
- switches
- plug-in contacts
- screws
- ropes
- pins
- chains

Waxes are especially suitable for the following material pairings:

- Al alloys / ferrous metals, and
- Cu alloys / ferrous metals,

but also for all metallic materials, also when paired with elastomers, plastics or wood.

The main advantage is that they permit the fully automatic assembly of mass-produced parts. The wax film ensures clean and easy operations.

PARVAN is Exxon's brand name for a balanced line of high-quality, fully refined waxes. These waxes meet all applicable Food and Drug Administration requirements for food, health, and cosmetic-related uses. An FDA-approved oxidation inhibitor enhances the natural resistance of PARVAN to deterioration. For each grade, the three digits following the brand name indicate the typical

Table 9-1. Lubricating paste selection chart.

Application \ Product	Klüberpaste 46MR401	ALTEMP QNB50	UNIMOLY RAP	WOLFRAKOTE C	WOLFRAKOTE Top Paste
Machine screws				■	
Pressing-in of pins, bolts, bushings	■		■		
Plain bearings for medium to high speeds				■	
Slideways, cams, profile guides, sliding guides			■		
Threaded spindles, adjusting nuts, spline shafts	■				
Ball joints, bolt supports	■				
Clamping chucks		■			
Toothed gears	■				
Shaft seals, O-ring seals, V-ring seals, rubber cups	■				
Chains					■
Rolling bearings, very low sliding speeds	■				
Assembly aid	■				■

Table 9-2. Selection criteria for lubricating pastes.

Selection criteria Lubricating paste for	Product name	Base oil/ solid lubricant ≈	Service temperature range (°C) ≈	Density at 20.C (g/cm^3) DIN 51757 ≈	Color	Worked penetration DIN ISO 2137 (0.1 mm) ≈	Base oil viscosity DIN 51561 (mm^2/s)		Notes
							40°C	100°C	
High load, wide service temperature range, good resistance to water and humidity, not suitable for dry lubrication	Klüberpaste 46 MR 401	Synth. oil/ inorganic solid lubricant, soap thickener	–45 to 150	1.23	ivory	300 to 340	380	—	Threaded spindles, adjusting nuts, spline shafts, bolt and socket joints, bolt supports, toothed wheels, shaft seals, O-ring seals, V-ring seals, rubber cups, rolling bearings—very low sliding speed, assembly aid
High load, wide service temperature range, good resistance to water, not suitable for dry lubrication	ALTEMP Q NB 50	Min. oil/ Ba complex soap, inorganic solid lubricant	-15 to 150	1.40	beige	250 to 270	42 to 50	6 to 7	Clamping chucks
Low friction coefficient at low and high load high temperatures, suitable for dry lubrication	UNIMOLY RAP	Min. oil/MoS_2, graphite, inorganic solid lubricant	-10 to 450, dry lubrication from approx. 160	1.60	black	250 to 270	70	8	Pressing-in of pins, bolts, bushings, , slideways, cams profile guides, sliding guides
High humidity, high temperature suitable for dry lubrication	WOLFRA-COAT C	Synth. oil/ graphite inorganic and lubricant	-30 to 200, above that dry lubrication up to 1200	1.01	gray, copper-colored	270 to 310	120	12.5	Machine screws, plain bearings for low to medium sliding speeds
Low temperatures, high humidity, suitable for dry lubrication	WOLFRA-KOTE TOP Paste	Synth. hydrocarbon oil, graphite, inorganic solid lubricant	-25 to 200, above that dry lubrication up to 1000	1.30	light gray	300 to 330	800	38	Chains, assembly aid

melting or congealing point in degrees Fahrenheit (Table 9-3).

Sealite (Table 9-4) is the trademark for four additized waxes for cup and corrugated paperboard applications. All grades comply with Food and Drug Administration (FDA) regulations 21 CFR 176.170 and 21 CFR 176.180 for incidental food contact.

Sealite 128 and 133 are formulated for use as saturating or cascading waxes in corrugated board manufacture. They enhance the durability and moisture-resistance of corrugated containers. The polyethylene additive provides flexibility to the finished product. Sealite 133 is the premium, higher-melt-point product; Sealite 128 is the economy grade. Both compare favorably with competitive saturating waxes and in some cases exceed the performance of competitive products.

Sealite 142 and 145 are formulated as alternatives to conventional refined waxes for coating paper cups. They are blended with higher proportions of lower-melt-point waxes, along with a polyethylene additive. The two grades provide comparable performance to that of higher-melt unadditized waxes and can offer cost control advantages, as well as more assured availability compared with higher-melt waxes. The specification for three competitive waxes (Klüber) are given in Table 9-5.

Release Agents

Some branches of industry require petrochemical products for mold release purposes. These release agents consist of liquid hydrocarbons and/or solid lubricants, a solvent to carry the solid lubricants, and an emulsifier to ensure miscibility with water.

Water is either used for cooling or for obtaining a concentration suiting the individual application. The performance of a product depends on the adequate combination of its ingredients. In addition to the releasing effect, a product may also be required to ensure good lubricity or to protect mold or tool surfaces.

Table 9-3. Typical inspections for a balanced line of highly refined waxes (Exxon's PARVAN).

Typical Inspections Grade	127	129	131	137	138	142	145	147	152	154	158	161
Melting point, ASTM D 87, °F	126	130	131	136	137	142	144	147	—	—	—	—
Congealing point ASTM D 938, °F	—	—	—	—	—	—	—	—	151	153	157	159
Oil content, ASTM D 721 mass %	0.1	0.1	0.1	0.1	0.1	0.1	0.1	0.2	0.3	0.3	0.2	0.2
Saybolt color, ASTM D 156	+30	+30	+30	+30	+30	+30	+30	+30	+30	+30	+30	+30
Pounds per gallon @ 174°F	6.41	6.42	6.42	6.44	6.44	6.48	6.49	6.50	6.53	6.54	6.56	6.60
Blocking point, °F ASTM D 1465	—	—	—	—	—	118	121	—	—	—	—	—
Flash point, °F ASTM D 92	420	420	425	440	446	450	460	470	490	500	500	500
Kinematic viscosity, ASTM D 445, cSt@210°F	3.4	3.5	3.6	3.9	4.1	4.7	5.1	5.4	6.2	6.6	7.3	7.4
Needle penetration @ 100°F ASTM D 1321, mm/10	105	95	90	35	30	25	25	25	25	20	20	20
UV absorbance*	Pass											

*21CFR 172.886 and 21CFR 178.3710

Petrochemical release agents are applied in, for example:

- industrial baking tins
- pouring ladles
- tire molds (Figure 9-1)
- dies

The structure and ingredients of the various release agents depend on the requirements they have to meet, including:

- temperature resistance
- corrosion protection
- reduction of friction
- suitability for use in contact with food products
- neutrality towards rubber and plastics.

It may also be required to apply a product that ensures a separating effect and is at the same time neutral towards plastics.

Selection criteria and important physical characteristics are given in Table 9-6 for several Klüber products.

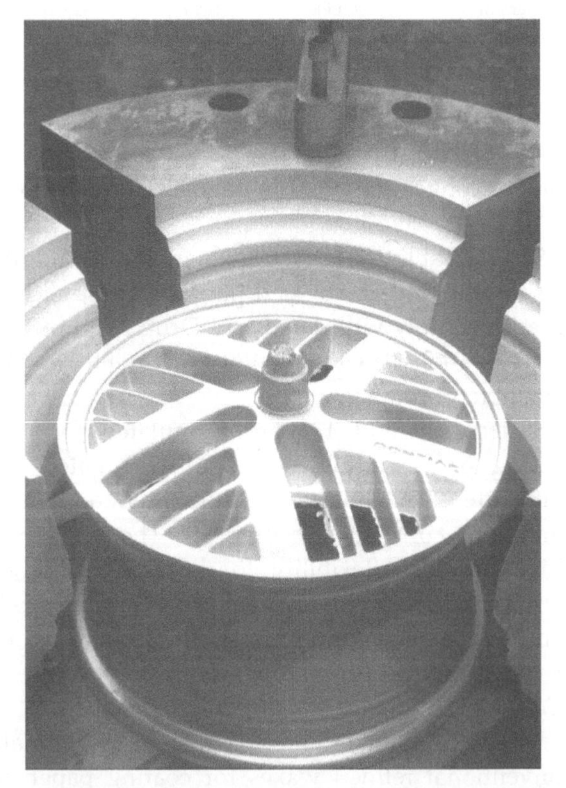

Figure 9-1. Releasing the mold of an aluminum rim.

Table 9-4. Waxes for cup and corrugated paperboard applications (Exxon SEALITE)

Typical Inspections Grade Grade	128	133	142	145
Melting point, ASTM D87, °C(°F)	53(127)	55(131)	—	—
Congealing point, ASTM D938, °C(°F)	53(128)	56(133)	61(141)	63(145)
Blocking point, ASTM D1465, °C(°F)	—	—	47(116)	49(120)
Oil content, ASTM D721, mass %	0.5	0.4	0.5	0.5
Viscosity @ 98.9°C, ASTM D445, cSt	4.7	5.4	5.1	5.2
Penetration, ASTM D1321, dmm				
@ 77°F	12	11	—	—
@ 100°F	—	—	40	45
Color, Saybolt, ASTM D156	+20	+20	+21	+21
Cloud point, °C (°F)	70(158)	71(160)	71(160)	71(160)

Exxon's TELURA line of release agents consists of 19 process oils. These oils are separated into the three categories, naphthenic, extracted naphthenic, and extracted paraffinic. The 19 oils differ in such parameters as viscosity, flash point, pour point, etc. The three categories differ in color and ultimate use.

Tribo-system Materials

Tribo-system materials are a combination of a lubricant and a base material to form a self-lubricating design element. By adding a lubricant, high-strength plastics are imparted better tribological characteristics (e.g. thermoplastics with incorporated lubrication). Plastics with good friction behavior can also be improved with strength-enhancing additives (e.g. PTFE compounds).

Providing the following advantages, tribo-system materials are a good alternative to traditional lubricants:

- simplified structural design because no extra lubrication is required
- lifetime lubrication without relubrication
- no contamination caused by the lubricant
- no corrosion
- excellent resistance to chemicals

PTFE based tribo-system lubricants are suitable for vacuum applications. They can be applied at low and high temperatures, prevent stick slip, protect against wear and ensure a low friction coefficient. When selecting a proper material and in the design phase it is important to take into account the peculiarities of plastic materials.

Tribo-system materials are used in many applications, for example:

- *machine tools*
  e.g. coating of slideways
- *packaging machines*
  e.g. plain bearings or sliding films in conveyors
- *compressors, pumps*
  e.g. piston rings, guide rings
- *medical equipment*
  e.g. plain bearings in sterilizers
- *textile industry*
  e.g. sliding guides in loom grippers
- *conveyor systems*
  e.g. guide roller bearings
- *valves*
  e.g. sliding rings and seals, also in drinking water valves

Tribo-system materials are illustrated in Figure 9-2; for selection criteria, refer to Table 9-7.

Table 9-5. Specifications for three waxes.

Lubricant type	Product name	Service temperature range (°C) ≈	Color	Drop point based on DIN ISO 2176 (°C) ≈	Needle penetration based on DIN ISO 2176	Water resistance based on DIN 51 807/1	Copper corrosion DIN 51 811	Steel corrosion DIN 50 017	Notes
Waxes	Klüberplus SK 02-295	-40 to 120	light yellow	75 to 80	20 ± 4 × 0.1 mm	0 at 50°C	0 (= no discoloration, and no corrosion	rating 0 after 30 cycles (720 h)	The product is rubbed onto the cold surface or spread over the surface heated to 50-70°C. If applied by immersion, let the excess wax flow back into the bath.
	Klüberplus SK 04-295	-40 to 120	light yellow	80	47 × 0.1 mm	0 at 50°C	—	rating 0 after 30 cycles	Especially suitable for lubrication of roller chains by immersion.

Lubricant type	Product name	Service temperature range (°C) ≈	Density at 20°C (g/ml) DIN 51757 ≈	Color	Ph Value ≈	Viscosity DIN 53 211 at 20°C 4 mm nozzle	Four-ball tester welding force DIN 51 350 Pt 4	Corrosion protection DIN 50 017 KFW towards St 14	Reichert wear indicator specific surface pressure N/cm^2 ≈	Notes
Wax emulsion	Klüberplus SK 01-205	-40 to 120*	0.98	ivory	8	20 to 30 s	≤ 1,200 N	no corrosion after ≤ 20 cycles	2,000	Ready-to-use. Mixing ratio of 1:1 to 1:5 with water possible if protection film is sufficient for application.

*As quasi-dry film free from water; otherwise 5 ... 95°C

Figure 9-2. Tribo-system materials (Klüberplast).

Tribo-system Coatings

Tribo-system coatings are procedures for the application of dry lubricants for tribosystems. The service life of tribo-system coatings depends on four main factors (see Figure 9-3): the dry lubricant used, the component's design, the load and stress factors and the manufacturing conditions. The coating costs are mainly determined by the coating technique. In most unfavorable cases, the coating may be five times as expensive as the material to be coated.

Dry lubricants for tribo-systems can be applied with most of the standard methods used for lacquers: by brush, spraying, immersion, centrifugation or tumble processing.

Klüberbond is the name of a process mainly used for the coating of metallic materials with dry lubricants.

For mass-produced parts the tumbling process

Table 9-6. Selection criteria for release agents.

Selection criteria	Fluid release agents	ISO VG DIN 51 519 mm^2/s	Density at 20°C (g/ml) DIN 51757	Kinematic viscosity DIN 51561, (mm^2/s) at 40°C	Viscosity index DIN ISO 2909 (VI)	Flash point DIN ISO 2592 (°C)	Pour point DIN ISO 3016 (°C)	Notes
Release agent for rubber, elastomers and plastics	FORMINOL TGK 680	680	1.06	600 to 660	—	—	-30	The product is miscible with water ≤ 40°C and forms a clear solution. The adequate dilution ratio (1:20, 1:1 or pure) depends on the requirements. FORMINOL TGK 680 is suitable for molds and tools and as an assembly aid.
Release oil for baking tins	PARALIQ 91	15	0.95	15	130	> 230	< -5	As a release agent PARALIQ 91 is applied at a very high dilution ratio. It meets the requirements of the German Law on Food Products and Associated Ancillaries (LMBG) and fully complies with the pertinent food regulations.
Mold release agent for Al pressure die	Klübertec HP 1-406	—	1.0	20 s**	—	—	—	Water-miscible mold release agent for Al pressure die casting. Recommended dilution ratio with water: 1:140 to 1:180
Mold release agent for tires (dry lubricant for tribo-systems)	Klübertop TP 10-961	—	0.93	—	—	30*	—	Mold release agent for the rubber industry, especially for tire molds. It is applied with a spray gun at a pressure of 2 bar. The recommended film thickness is 40 μm. The film is dry after 5 min. at 20°C and requires a burning time of 40 min. at 180°C or 80 min. at 160°C.

*DIN ISO 1516, closed cup, paints
**DIN 53 211, flow time, DIN cup (4mm)

Selection criteria	Coherent release agents	Base oil/ thickener	Service temperature range (°C) ≈	Density at 20°C (g/cm^3) DIN 51757 ≈	Color	Drop point DIN ISO 2176 (°C)	Worked penetration DIN ISO 2137 (0.1 mm)	Consistency NLGI grade DIN 51818	Apparent viscosity KL grade	Notes
Die grease for continuous Al casting	METALSTAR 820	Min. oil/ Li	—	0.9	light brown	> 160	265 to 295	2	L*	METALSTAR 820 is applied to the die in a thin and uniform layer. The generation of smoke or flame after initial casting is very low. The surface of the cast squares and bolts is of a good quality
Ladle dressing for the casting of non-ferrous metals	METALSTAR KS 210	Synth. hydrocarb. oil/solid lubricant	≤1,200 (dry lubr. above 200°C)	1.4	red	?200	280 to 300	—	M/S*	To ensure adequate distribution, METALSTAR KS 210 is applied to the heated ladle and dried for approx. 10...20 s above the oven. The drying time is 60 s if the product is applied to the cold ladle.
Die dressing for the casting of non-ferrous metals	METALSTAR KS 202	—	—	1.35	light blue	—	—	—	—	METALSTAR KS 202 has a fine structure and ensures good thermal insulation. It is applied as a priming, operating and coating dressing. It is mixed with water at a ratio between 1:5 and 1:10 and sprayed onto heated dies (200 ... 250°C)

*KL viscosity grades see under "lubricating greases"

Table 9-7. Selection criteria for Tribo-system materials.

Selection criteria	Product name	Service temperature range (°C) ≈	Color	Elongation at tear DIN 53 455	Indentation hardness DIN 53 456	p-v value (N/cm^2-m/min)	Friction coefficient (Tannert, 20°C, V_{max}= 0.243 mm/s		Notes
Hard opposing surface, hardness >35 HRC	Klüberplast LD	-240 to 260	red	170 - 190	44/45 14/15	2,150	0.05 to 0.06	0.04	Plain bearings, sheets for machine , tool slideways contact surfaces
Soft opposing surface, nonferrous metals, hardness <=35 HRC, for the processing ind. (USDA auth.)	Klüberplast J	-240 to 260	brown	160-180	33/34 11/22	1,500	0.04 to 0.06	0.03 to 0.04	Plain bearings, contact surfaces, surfaces, guideways, sheet bearings
Good lubricity under humidity, for drinking water and food applications (KTW USDA auth.)	Klüberplast W2	-240 to 260	black	140-160	43/44 12/13	2,150	0.06 to 0.07	0.05 to 0.06	Plain bearings, sealing and support washers, piston and guide rings, contact surfaces, sheet bearings
For mass-produced injection molded parts	Klüberdur KS 01-308	-50 to 90	white	21.4	118	—	0.04 at 0.486	—	Toothed wheels for small gears, bearing bushings, contact surfaces, sliding belts
For injection molded parts that are very resistant to wear and high temperatures	Klüberplast T 220	-50 to 300	black	—	245 —	—	—	—	Slideways subject to high thermal loads, toothed wheels, rolling bearing cages, electric safety components
Two-component materials with good adhesion and wear resistance	Klüberdur KM 01-854	-40 to 180	dark gray	—	100 —	—	0.05	0.18 at 120°C	Filling of lubrication holes in plain bearings, production and repair of slidways and guides. For plain bearings and contact surfaces.

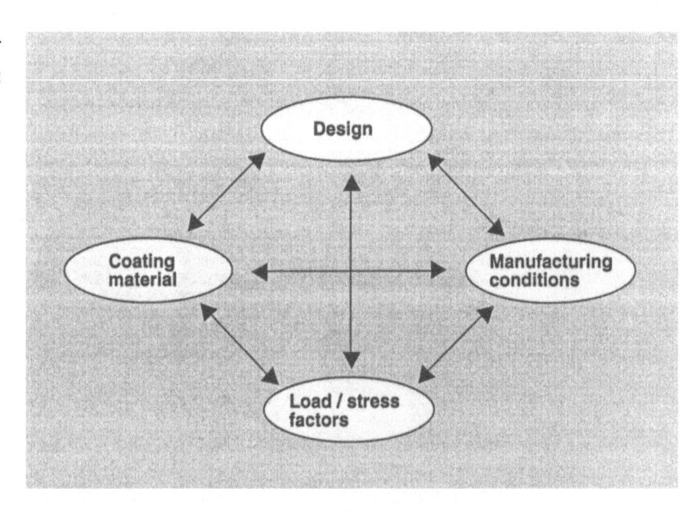

Figure 9-3. Factors influencing service life of tribo-system coatings.

is suitable provided that it is possible to entirely coat a component. This automated method provides a high degree of film thickness constancy. Tumble processing is a very cost-effective method for the coating of small to medium-size mass-produced parts.

Dry Lubricants for Tribo-systems (Table 9-8)

Dry lubricants consist of solid lubricants, a binding agent and a solvent. Their tribological behavior is determined by the type and quantity of solid lubricants. Wear resistance mainly depends on the binder. The solvent distributes the dry lubricant over the component and evaporates during the hardening process, not having any direct impact on the friction and wear behavior of the lubricating film. Dry lubricants for tribo-systems ensure a mostly coherent film between 3 and 15 μm thickness, depending on the applied product. They are characterized by an extremely wide service temperature range (between -180°C and 450°C) and excellent resistance to chemicals.

Their lubricating effect may be optimized by incorporating various solid lubricants. For example, products containing graphite have very good tribological behavior, and those containing MoS_2 are also suitable for vacuum applications. Dry lubricants with PTFE have a very low friction coefficient. In addition, adhesive friction is lower than sliding friction, which means that there is no stick-slip.

Dry lubricants for tribo-systems provide *advantages* wherever

- traditional lubricants cause contamination
- penetration may cause malfunctions
- service temperature limits of oils and greases are exceeded
- aggressive media, humidity and dust have an impact on the friction points
- lubricating oils and greases would impede the operational process

Table 9-8. Dry lubricants for tribo-systems.

Selection criteria	Product name	**Service temperature** range (°C) ≈	Color	Solid lubricant	**Burning-in temperature** (°C) **Hardening time** (minimum)	**Yield at a film thickness of** 10 μm, (m^2/l)	**Friction coefficient** (Tannert, 20°C, V_{max} = 0.243 mm/s F= 300 N)	Notes
Air-drying water-miscible	Klübertop G20-730	-40 to 80	black	PTFE	20/24 h	23	0.04	Springs, plastic parts, assembly aid
Air-drying, for elastomers	FLUOROPAN CO 60	-15 to 80	black	PTFE	20/24 h	30	0.04	Rotary shaft lip seals, O-ring seals, rubber seals, glass, non-ferrous metals, aluminum
Air-drying, as a release agent	Klübertop TP 08-852	-30 to 150	white	PTFE	20/24 h	11	0.04	Casting-in of tin hinge, assembly aid
Two components heat-hardening	FLUOROPAN 340 A/B	-40 to 230	black	PTFE	250/15 min	17	0.04	Shafts, armatures, control pistons and cylinders in pneumatic systems
Hygro-setting, excellent pressure resistance	UNIMOLY C 200*	-180 to 450	gray	MoS_2	20/30 min	8	0.10	Screws subject to high pressure toothed wheels, spindles, plain bearings, multiple spline shafts
Heat-hardening, excellent pressure resistance	UNIMOLY CP	-40 to 220	reddish-brown	MoS_2	180/60 min	16	0.12	Screws, nuts, washers, pressure springs, plain beariangs
Heat-hardening	Klübertop TG 05-371	-40 to 300	gray-black	C	250/15 min	40	0.07	Plain bearings, slide-ways, pistoncoatings

*Also available as a spray (CFC-free)

- uniform tribological conditions are required in a very wide temperature range
- the entire component requires protection against corrosion

The dry lubricant's effect is based on "transfer lubrication," a type of "erosion" of the top layers. If the lubricant layer is used up, the friction point will fail.

Application

As summarized in Table 9-9, dry lubricants for tribo-systems can be applied in various ways, such as immersion, spraying, tumbling or electrostatic coating. The surface to be coated must be treated as follows before applying the lubricant:

Cleaning of the components is mandatory to remove grease residues on the surface. Sand blasting or grinding ensure better adhesion. Phosphating improves the protection against corrosion. The selected binder system determines the hardening process.

Today, dry lubricants for tribosystems are used in many applications, for example (Table 9-10): Rolling bearings, bolts, screws, nuts, washers, springs, ropes, slideways, toothed gears and racks, O-ring seals, rotary shaft lip seals, threaded spindles, etc.

Some of these coated components are shown in Figures 9-4 through 9-6.

Table 9-9. Application methods for dry lubricants.

Process	Equipment requirements	Production costs	Reproducibility of film thickness
Application by brush	low	low	poor
Manual spraying	low	medium	poor-medium
Automatic spraying	high	medium-high	very good
Electrostatic spraying	high	medium	very good
Immersion by hand	low	low	medium
Immersion and centrifugation	high	low-medium	good
Immersion and tumbling	medium	low	good
Tumbling	medium	low-medium	good

Table 9-10. Coating processes.

Element	Process
Plastic O-ring seal	Klüberbond TP 06-008
Machine screws	Klüberbond TM 02-108
Nuts	Klüberbond TM 03-003
Washers	Klüberbond TM 03-003
Shafts	Klüberbond TP 04-107
Bolts	Klüberbond TP 02-113
Threaded spindles	Klüberbond TP 04-107
Springs	Klüberbond TM 03-105
Bearing components	SAWO-I

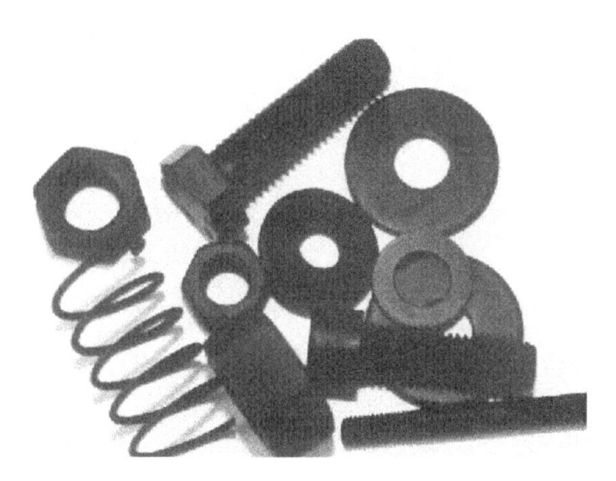

Figure 9-4. Nuts, bolts, washers and springs coated with Klübertop.

Figure 9-5. Nuts coated with Klübertop.

Figure 9-6. Coated components of an element rolling bearing.

Chapter 10

Lubricant Delivery Systems

MANUAL LUBRICATION SYSTEMS

In the beginning there was oil. The industrial revolution of the 19th century witnessed the harnessing of the power of water, steam and electricity with a collection of never before seen motive power devices. Beautifully crafted engines were built with close tolerance large bearing surface areas, requiring constant lubrication. To satisfy the bearing's continual appetite for oil, a device called an "oil cup" was employed. The "oil cup" is a metal, or glass and metal reservoir device, varying in size from a thimble to a teacup, which is screwed into the bearing's oil entrance port and used to gravity feed oil by constricting its flow through a valve type orifice. Variations of the oil cup were fitted with wicks to further slow the oil flow, or with a brush to "paint" the oil on to chains.

The year was 1916, and driven by the repetitive task of having to continually refill oil cups on his family business's die cast machines, Arthur Gulborg was motivated to develop a less taxing process that employed the new grease lubricant medium, resulting in his invention of the grease gun. Using a screw type grease gun design built to mate up to a bearing point fitting made from a braided metal hose complete with proprietary end fitting; he appropriately named the invention the "Alemite High-Pressure Lubricating System," after the family business name. Initially pitched to the military for lubricating their transport trucks, the Alemite system received a large boost in sales when they introduced the "button-head' style grease fitting, as shown in Figure 10-1, and were adopted as standard equipment on passenger cars in the early 1920s.

At this same time Oscar Zerk working for a competitive company in Cleveland, Ohio chose to go down a different path and invented a different style of lubrication fitting that only located the grease gun coupler, rather than positively lock them together as in the Alemite button style fitting, with the locating method attaining its seal by the operator "pushing" the coupler on to the fitting. This allowed the grease fitting or nipple , see Figure 10-2, to be designed significantly smaller, resulting in more options for the equipment designer.

1924 was arguably the most pivotal year in the history of lubrication; Joseph Bijur invented and marketed the World's first engineered centralized oil lubrication

Figure 10-1. Button head manual grease fitting and fill adaptor

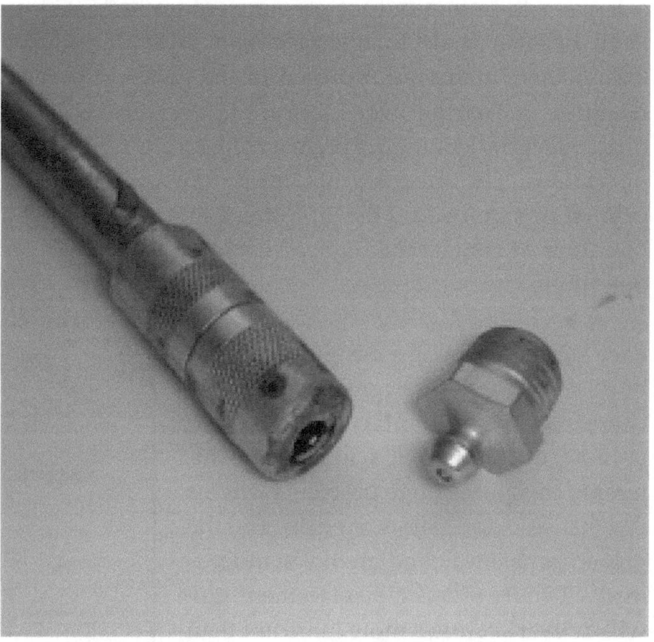

Figure 10-2. Standard Zerk style grease fitting and fill adaptor

system, and the Alemite Die Casting and Manufacturing Company of Chicago, Illinois purchased the Allyne Zerk Company of Cleveland, Ohio whose marriage was to have a huge impact on the industrial world with the marketing of its push-type single point manual grease system—still the most practiced method of lubrication today.

The Manual Grease Gun—A Lethal Weapon?

Moving back to the 21st Century, we will find little change to Gulborg and Zerk's original inventions. The standard grease gun of today is a slightly more sophisticated hydraulic device employing a lever or pistol grip to perform the pumping action. Newer developments utilize compressed air or electric motor to energize the pumping action of the handgun. The SAE J534 Zerk grease fitting is faithful to its original push-style design and remains the industry standard for grease fittings.

A grease gun's simple design inherently makes for an intuitive device - what could be simpler? Just locate the coupler on the grease nipple, and pump! Unfortunately, its simplicity and ease of use has lulled virtually every grease gun operator into the immediate belief they can effectively grease any bearing without the need for training or understanding. The following facts explain why the seemingly benign grease gun, is a "Lethal Weapon" in the hands of an untrained operator.

Fact #1: Because grease guns are hydraulic devices, depending on their internal design they are able to achieve astounding output pressures of up to 15,000psi! With bearing seals unable to retain pressures higher than 500psi, many are ruptured in the greasing process resulting in bearing over lubrication, over heating, and dirt collection leading to premature bearing failure and downtime. The grease gun output pressure is rarely printed or stamped on the gun, check the accompanying literature or contact the manufacturer to determine.

Fact #2: Differing grease guns are not built to the same design specifications, therefore their displacement output volume, or "shot" size, will likely be different. This poses huge problems when a PM task asks for 2 shots of grease and the grease guns employed are not standardized - 2 shots of a 3cc displacement gun will deliver six times more lubricant than 2 shots of a ½ cc displacement gun. Over lubricated bearings create internal fluid friction that leads to overheating and premature bearing failure. If the volumetric displacement cannot be found on the gun or in the accompanying literature, displace 10 shots into a test tube measure and divide the results by 10 to assess the correct shot size and clearly stamp or indelibly mark the shot size in CC's or Cu.Ins. on the side of the grease gun's barrel.

Fact #3: Many grease gun manufacturers confuse the issue by marketing the grease gun reservoir capacity and "shot" displacement by weight in grams and/or ounces. Grease manufacturers use totally different ingredients and formulations for each grease type resulting in differing specific gravity ratings. For example, a standard grease gun tube of Klüber Barrierta L55-2 grease weighs 400 grams, whereas most standard #2 greases similar to a Mobil/Esso Unirex grease weigh in at 300 grams. This means similar volumes can have totally differing weights resulting once again in inconsistent lubricant delivery. To combat this problem always view displacement in volumetric terms and review Fact #2 for ascertaining gun displacement size. Table 10-1 depicts the varying pressure and displacement of different sizes and styles of hand grease guns.

Fact #4: Manual-bearing lubrication requires application consistency and discipline if it is to be effective. The lubrication PM must be written in clear objective language stating the exact lubricant to be displaced along with the exact volumetric requirement. Undisciplined grease gun operators will virtually always over lubricate in the mistaken belief that "more is better." The Lubrication PM task must be completed on schedule, all the time if the bearing is not to be under or over lubricated.

Fact #5: Most untrained grease gun operators do not wipe the grease gun coupling and grease nipple

Table 10-1. Typical grease gun specification data

Grease Gun	ID #	Capacity	Delivery	Pressure
	F104	3oz bulk	1oz-25 stroke	**2500psi**
	525	14oz cartridge	1oz-33 stroke 1oz-60 stroke	**4500psi** **10,000psi**
	4015	14oz cartridge	1oz-7 stroke	**1700psi**
	6268	12oz bulk	1oz-24 stroke	**15,000psi**

Courtesy ENGTECH Industries Inc. and Alemite Corporation

clean with a lint free cloth, before and after greasing each lubrication point. Bearings are precision devices manufactured in laboratory clean "white rooms"; they will perform poorly and rapidly fail if injected with dirt from a dirty grease nipple or gun coupler. Figure 10-3 shows a dirty grease nipple in an over-lubricated bearing.

Fact #6: Hard to reach bearing points are often plumbed to a remote easy access area of the equipment, many of these "blind" lubrication points are ganged together on a manifold block. Unless clearly identified and managed, these points will likely be over lubricated.

Without proper training and program management, the use of manual lubrication has, and will continue to do so, wreak havoc on equipment reliability and availability.

Figure 10-3. Dirty grease fitting and over lubricated bearing

Implementing an Effective Manual Lubrication Program

In the hands of properly trained and managed lubrication technicians, manual grease lubrication systems provide a very inexpensive and effective method of sustaining and increasing bearing life. Achieving this is a matter of following a simple 6-step program:

Step 1: *Perform a Lubricant Consolidation Program* (See Chapter 22—Implementing a Quality Lubrication Management Program). Minimize the amount of grease lubricants in current use to facilitate the management of grease lubricant application.

Step 2: *Standardize all Grease Guns*. To facilitate the change management process, allow the maintenance staff to take home and keep the old grease guns. Replace old guns with new, identical specification grease guns, these can be cartridge or buk fill style so long as there is only one style chosen. Once chosen, perform a volume displacement check and mark each gun with the pressure and volume displacement rating. Grease guns are now available with transparent bodies as see in Figure 10-4 that allow the operator to view the grease tube and immediately identify the grease in use.

Step 3: *Determine Bearing Requirements*. To reflect the new grease gun volume displacement per shot, all PM tasks will need to be updated. An engineering study may be required to determine if the current bearing requirements and schedule application periods are correct.

Step 4: *Introduce Engineered Distribution Devices*. To facilitate the lubrication process and assure consistent lubrication metering every time, investigate replacing the Zerk fittings or non-engineered "ganged" fittings shown in Figure 10-5 with an engineered progressive divider lubrication distribution block. In a progressive divider style distribution block seen in Figure 10-6 and Figure 10-7, dozens of points can all be lubricated with an exact engineered amount of lubricant in a matter of seconds through one zerk fitting on the block. Grease is pumped into the block until the "tell tale" pin indicates a

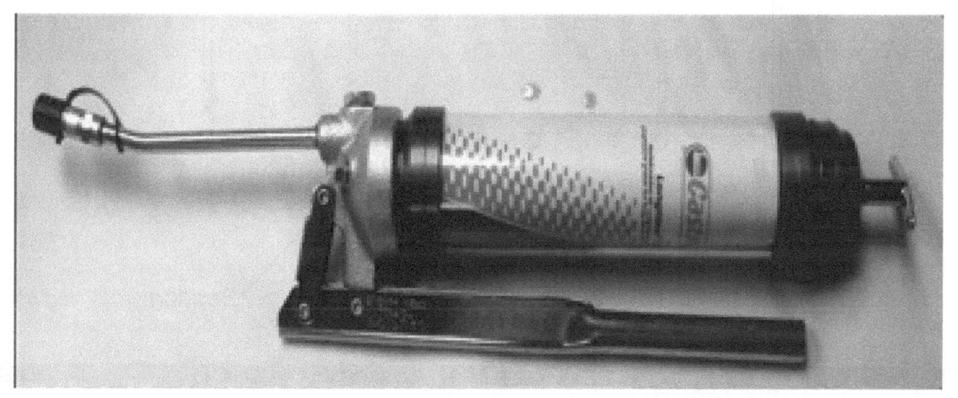

Figure 10-4. See-through reservoir grease gun

complete cycle of the block has taken place and all points have received an engineered lubricant amount. As part of the sales process, the engineered system vendors will usually perform the bearing requirement analysis as part of amount of purchase process.

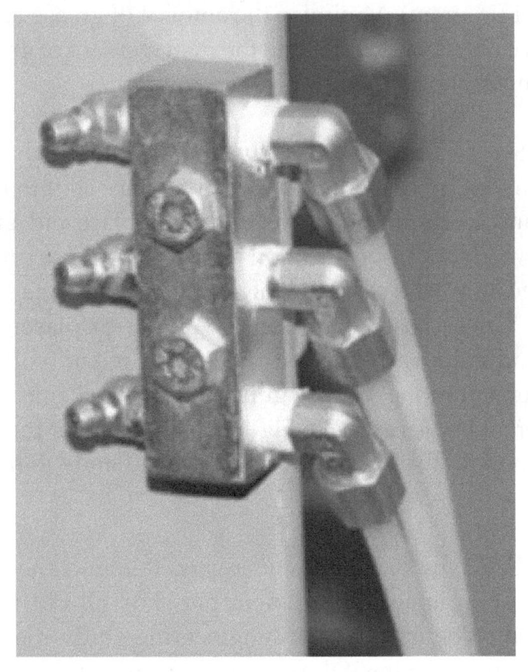

Figure 10-5. Non-engineered triple gang" grease block

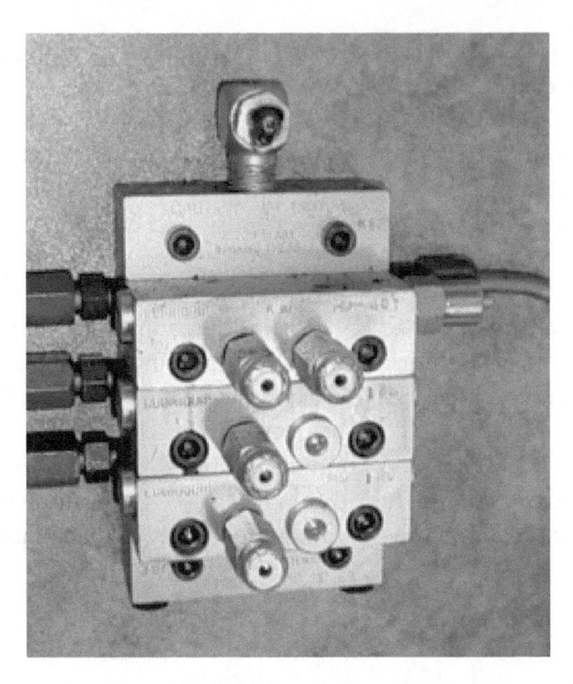

Figure 10-6. Lubriquip/Trabon manual progressive divider valve block (Source: ENGTECH Industries Inc.)

An added advantage to these blocks is the ability to hook up directly to an automated lubrication pump at a later date.

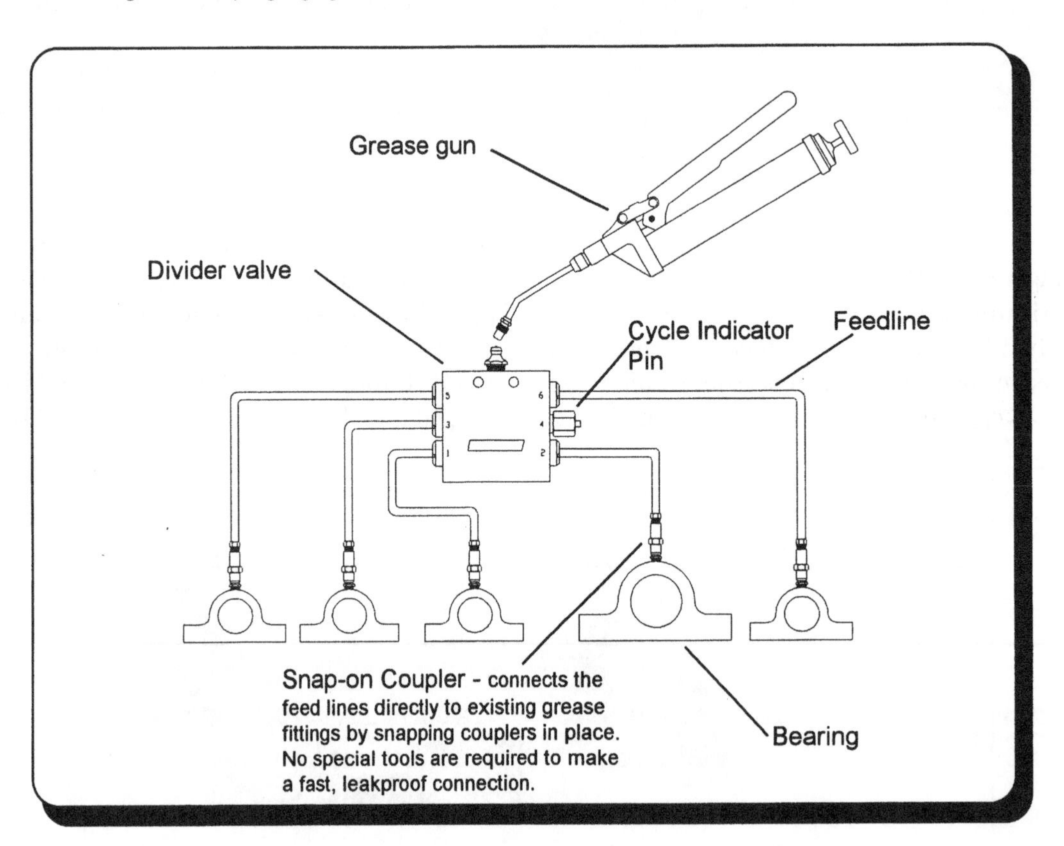

Figure 10-7. Simple centralized grease lubrication system. (Source: Lincoln Company, St. Louis, Missouri.)

In an engineered block system, the system operating sequence follows three steps:

1. The lubricant is delivered to the divider valve block through a machine mounted or hand operated grease gun
2. The positive displacement divider valve dispenses grease in measured amounts directly to each bearing through attached feed lines
3. The cycle pin indicator visually signals the completion of lubricant flow to the bearings

Step 5: *Color Code*. Managing with color is excellent for ensuring lubricants are not mixed. Different greases are assigned their own color and dedicated grease guns are marked with the assigned color, or purchased in that color. If grease is to be applied singularly at the bearing point the grease point can then be marked with its appropriate grease color using a simple grease point collar tag shown in Figure 10-8, further ensuring that lubricants are not mixed.

Figure 10-8. Grease point colored collar ID and dust protector tag

Step 6: *Perform Operator Training*. In house training can be provided from both vendors and Lubrication management experts who must provide training tailored to your plant environment and the specific needs of the individuals to be trained.

Manual lubrication systems can work effectively but will take considerable effort to set up correctly. Constant monitoring will be required to ensure the grease guns are only replaced with like specification guns and lubrication frequency and grease type remain consistent with existing staff and any new staff requiring training in to the lubrication program.

AUTOMATED SINGLE POINT LUBRICATION (SPL) DEVICES AND SYSTEMS

Anyone who has visited the engine room of a steam ship, steam train engine, or has witnessed an old steam engine at work will have seen and appreciated the beautiful brass and glass chambers sitting full of oil directly atop each major bearing point. These pioneering single point automatic lubricating devices were elegantly simple; the majority operated by cracking open a tapered valve to a determined point to allow oil to flow by gravity into the bearing cavity, or onto an intermediate transfer wick or brush in contact with the bearing surface.

In 1872, while working as an oiler for the Michigan Central railroad, Elijah McCoy changed the game considerably when he invented the world's first automatically pressurized (non-gravity activated) lubricator that used steam from the engine to activate and force feed lubricant from the device to the bearing surface. So successful was McCoy's patented lubricator that the railroad companies shunned all other designs in favor of McCoy's, coining the phrase by wanting only "the real McCoy!"

Using a similar design, grease can also be successfully dispensed using gravity and the aid of spring tensioned follower plate inserted in the lubricant reservoir chamber. Grease is filled from the bottom of the reservoir via a nipple that allows grease to hydraulically push against the follower plate and spring to load the chamber. Once loaded, the grease is expelled by the loaded spring pushing the grease against the hydraulic back pressure set up by the bearing clearance; the bigger the bearing, the bigger the clearance, the more grease is expelled into the bearing area—see Figure 13-10. At higher altitudes where the atmospheric pressure is lower, the grease can be expected to expel at a faster rate.

Although expressly designed as a one bearing, single point system, the gravity lubricator could, and was, often coupled to a single inlet, multiple outlet manifold to lubricate multiple points simultaneously, albeit in a non-metered manner.

As a testament to their original design these devices are still available for sale, with many original lubricators still in use today largely due to their simplicity, quality of manufacture, and ability to be refilled easily by the user as seen in Figure 10-9. Their modern predecessors are in a different league, and are sophisticated devices that use chemical, electro-chemical, and electro-mechanical pumps controlled by built in electronics to move oil and grease at pressures great enough to use progressive styled metered divider block delivery systems.

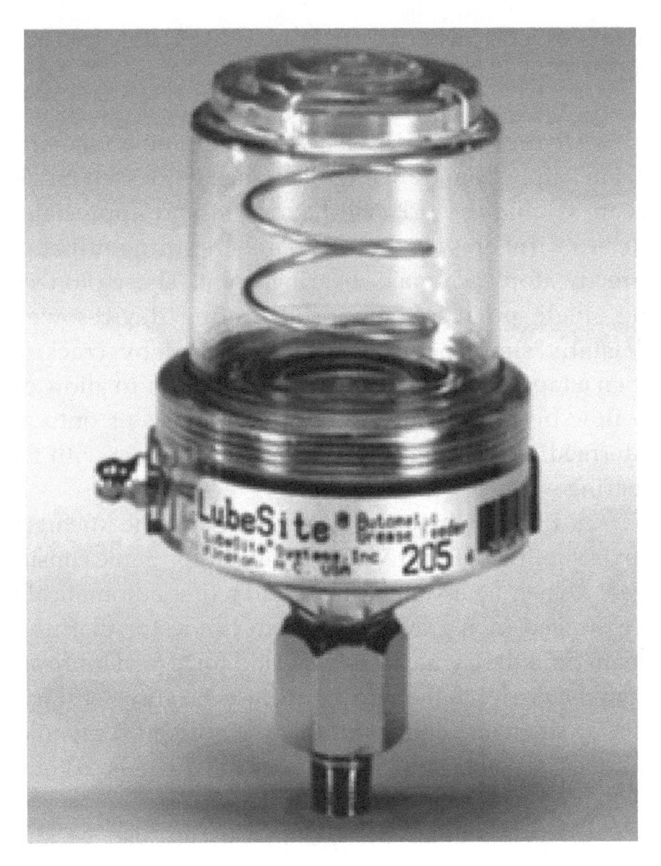

Figure 10-9. Single point spring loaded grease unit

Figure 10-10. Filling a late 1800s original Single Point Lubricator with oil—still in use in 2016

Constant Level Oilers (bottle oilers)

Many small machines, and especially centrifugal pumps employ constant level oilers, more commonly referred to as bottle oilers." Unfortunately, some of the most popular brands of constant level oilers will maintain their level setting only if the dynamic, or operating pressure inside the bearing housing remains the same as the static, or non-operating equilibrium pressure that existed when the bottle oiler was being set up on the non-running machine. Since these pressures may be different due to changing pressure drop conditions across bearing housing vents, breathers, or bearing housing seals, it would be prudent to use only pressure-balanced constant level oilers, Figure 10-11.

When the liquid in the bearing recedes because of liquid consumption, the liquid seal on the inside of the lubricator is temporarily broken. This allows air from the air intake port (smaller threaded connection in Figure 10-11) to enter the lubricator reservoir, releasing the liquid until a seal and proper level are again established. For reference, a liquid level line is scribed on the base. The unit is being refilled through a top filler cap. It should be noted that the reservoir need not be removed for refilling. A shut-off valve holds the liquid in the reservoir when the filler cap is removed. After the cap is tightly screwed down again, the lubricator resumes normal functioning. This particular oiler becomes a pressure balanced device when the air vent is piped back to bearing housing, thereby equalizing any existing pressure or vacuum.

Numerous variants of these highly reliable lubricators are available. Figure 10-12 shows the device fitted with a low level safety switch; Figure 10-13 depicts the same constant level oiler with a large sight glass for viewing the liquid level and condition of the liquid. Used in conjunction with properly selected face-type bearing housing seals, pressure-balanced constant-level lubricators with integral sight glasses allow for the hermetic sealing of virtually any bearing environment. Pressure balance is very important. Sadly, it is often overlooked by both machine manufacturers and owner-users. Lack of pressure balance often causes insufficient oil levels in bearing housings. Hermetic sealing refers to the exclusion of atmospheric contaminants, including of course dirt, dust and water vapor. It is these contaminants which are largely responsible for bearing degradation and damage. Hermetically sealed bearing housings no longer incorporate open vents, breathers, expansion chambers, desiccant cartridges, check

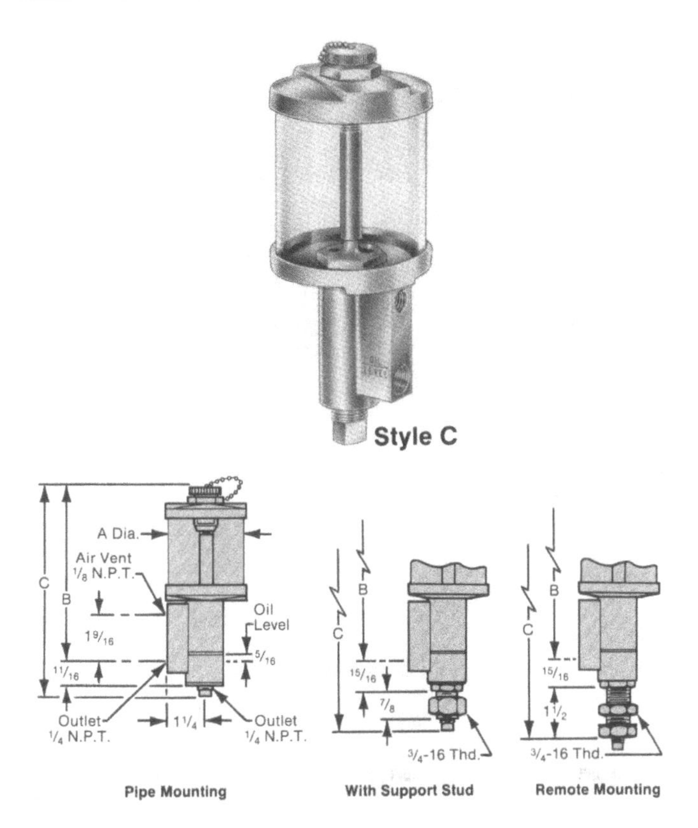

Figure 10-11. Piping "Air Vent" back into the equipment bearing housing produces a reliable pressure—balanced lubricator. (Source: Oil-Rite Corporation, Manitowoc, Wisconsin.)

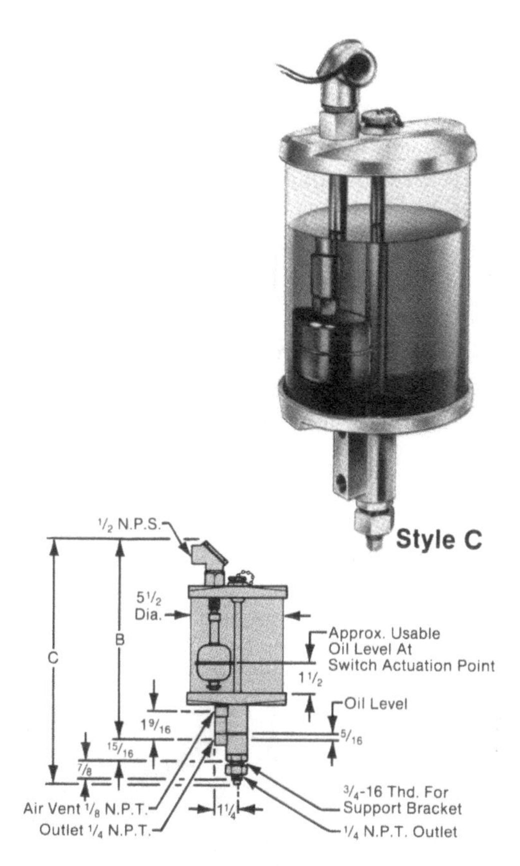

Figure 10-12. Pressure-balanced lubricator with low level safety (warning or shut-down) switch. (Source: Oil-Rite Corporation, Manitowoc, Wisconsin.)

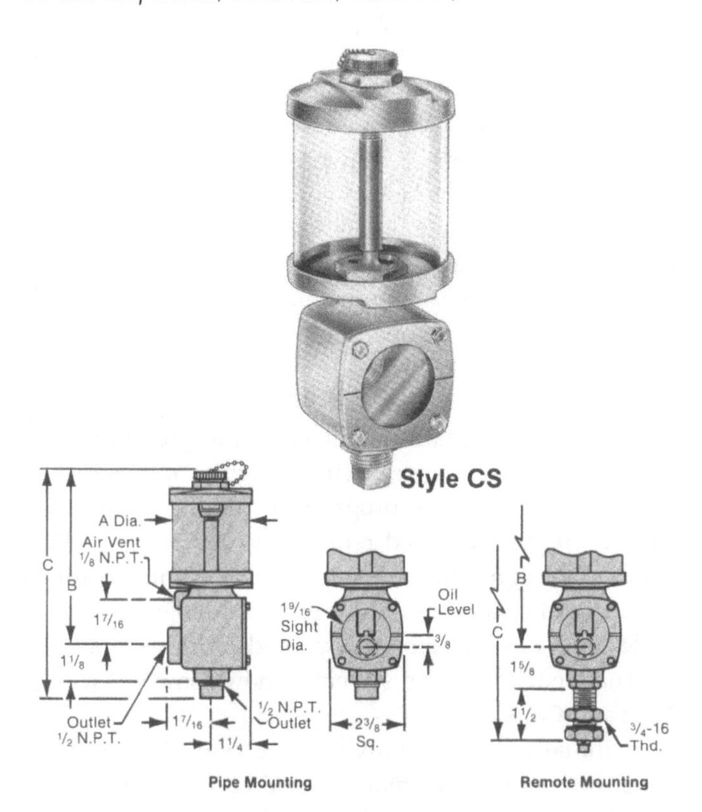

Figure 10-13. Pressure-balanced constant level oiler with integral sight glass. (Source: Oil-Rite Corporation, Manitowoc, Wisconsin.)

valve-type vents, filter inserts, or similar components offered to the average maintenance community.

How Modern SPLs Function

Altogether, there are five basic styles of single point lubricators. These include the gravity lubricator seen above, that can be refilled over and over again requiring a preventive strategy to ensure the unit never runs out of lubricant; the chemical activated lubricator—the first truly automated single point unit on. The electro chemical unit that introduced an on/off switch that gave the user more control over the lubricant delivery term and rate, and the electro-mechanical unit that allows for refills of lubricant flow and rate control and can be hooked up to a progressive divider style distribution block.

Chemical Activated SPLs

One of the earliest innovators of the modern styled SPL's is the German company Perma, who developed an inexpensive disposable chemical activated SPL in the early 1960s. As seen in the cutaway Figure 10-14, this simple unit uses a chemical reaction to develop a gas contained within a sealed expandable bellows unit.

The maintainer activates the unit by releasing a fixed chemical charge pellet into the bellows that is forced to react with an electrolyte to produce an expandable gas. As the gas slowly expands within the bellows it pushes the lubricant out of the unit into the bearing area. Different chemical charge amounts are used to vary the dispensing time from days to months, depending on the bearing's needs. Once the pellet is released, the unit cannot be deactivated.

With a 4-oz (120ml) reservoir, this style of unit, in its current updated version, continues to be one of the most popular SPL units offered for sale for the dispensing of grease.

Figure 10-14. Cutaway of Perma classic chemical activated SPL. (Source: Perma USA)

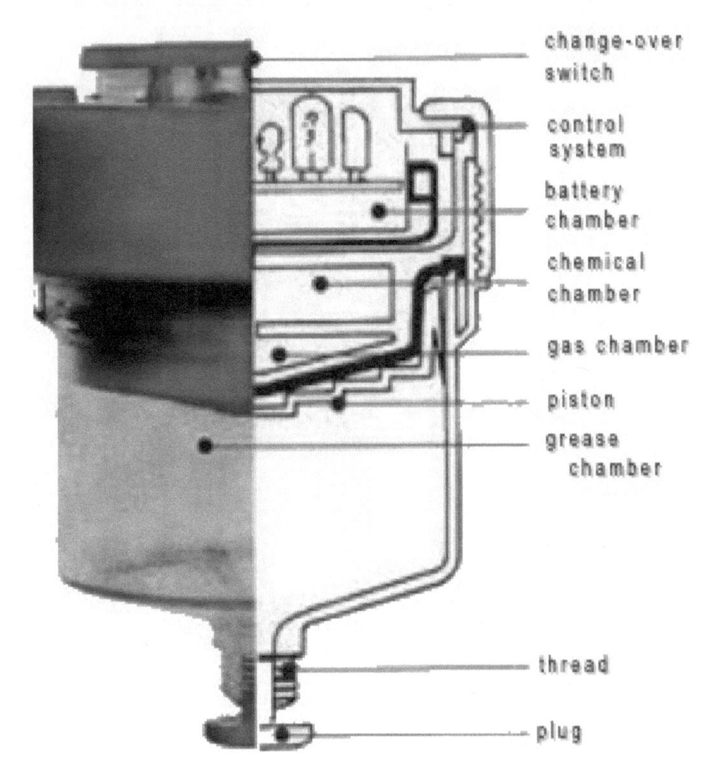

Figure 10-15. Programmable electro-chemical SPL (Source: ATS Electrolube)

Electro-Chemical Activated SPLs

Moving forward to the early 1980s, the first electronic controllable SPL is designed by ATS Electrolube. Using an electro-chemical reactor cell the user activates the unit via a series of time selector switches connected to a battery operated electronic circuit board (See Figure 10-15).

Once activated, a pulsed electrical current is sent through a contained electrolyte causing an electro-chemical reaction. This reaction creates an inert nitrogen gas to form inside a hermetically sealed bellow gas chamber that pushes against the oil or grease charge contained in the lube chamber section of the unit. Unlike the chemical style unit, the discharge can be controlled or turned off completely by the circuit board selector switches. Like it's chemical predecessor, this is also a single use disposable unit, as the bellows cannot be collapsed. Modern modified variations of this unit style now allow for a refillable reservoir with only the power unit and bellows requiring replacement when exhausted. Offered in many differing reservoir sizes, the most popular size unit offered continues to be the 4oz (120ml) unit.

Electro-Mechanical Activated SPLs

One of the major drawbacks of all the units addressed thus far is the lack capability to deliver controlled lubrication to multiple points. The original gravity lubricators only ever had a single point of control and residual line pressure, whereas the chemical and electro-chemical lubricators only developed between 50 and 60 psi. The latest generation of SPLs address this issue with its battery operated rotary mechanism driving a positive displacement pump that can deliver output pressures from 350 psi to 900 psi—more than enough to move a small multiple outlet series progressive divider valve built into the pump, or piped remotely to the pump. Unlike previous designs, the core unit is reusable with refreshed lubricant and batteries. Still very affordable, these new style units offer a viable centralized lubrication system alternative to the bigger systems it now competes with.

Pros and Cons

Initially, the early gravity units were designed to relieve the continuous attention required from the lubricator whose job was now to check on the units reservoir levels and fill them as necessary while the bearings reaped

Figure 10-16. Electro Mechanical SPL. (Source: Howard Marten Co.)

the benefit of continued lubrication in small amounts. The advent of the newer auto style units allowed out of sight bearings such as those found on overhead cranes or roof top HVAC fan units to receive continued lubrication for weeks or months, again relieving the burden on the lubricator whose role now changed to checking and marking the lubricator reservoir and performing a unit change out when required. For no capital outlay, these low cost units have extended the life of many bearings over the years.

Because individual units are inexpensive and convenient, many maintenance departments fall into the trap of using them on every bearing with a grease nipple. Performance notwithstanding, widespread indiscriminate use can get very expensive, very quickly. Their use should be monitored against the cost of putting in place the more expensive electro-mechanical SPL with hard piped divider delivery systems, or one of the more robust standard type of centralized lubrication delivery systems that can deliver lubricant to hundreds of points simultaneously.

Tips for Setting Up and Using SPLs

Tip #1: When using disposable units and disposable cores their disposal must follow municipal and state guidelines for hazardous waste (chemicals and batteries).

Tip #2: When using manual, chemical and electro-chemical units the user must understand that both styles of units rely on atmospheric backpressure to control the flow from the lubricator. Using the manufacturer's recommended settings is fine for units used below 1000 feet elevation. Above 1000 feet elevation the settings change approximately 5% for every 1000 ft. As the air thins at higher elevations, the backpressure is reduced and the unit flows at a faster rate, which can mean over-lubrication of the bearing and an empty reservoir sooner than anticipated. Know your elevation and follow the manufacturer's charts for settings at different elevations.

Tip #3: Working in Northern climates means hot summer weather and cold winter weather that will affect the lubricant viscosity in all styles of units. Using a #2 grade grease on a cold day in an outside location e.g. rooftop and fan units, will stall the unit and starve the bearing of its lubrication. Enter a seasonal PM in the computerized maintenance management (CMMS) or work order system to change out #2 grease units to a #1 or #0 grease unit in the late fall, and another PM to change back to a #2 grease in the spring.

Tip #4: Always pre-prime the lines with a grease gun fresh lubricant to the bearing on any new installation before screwing the lubricator in place.

Tip #5: all units require switching on and priming to the outlet point prior to installation to ensure continuity of lubricant to the bearing. This may mean switching on the unit 12-24 hours prior to intended use. Check the manufacturer's recommendations for the correct start up of units.

Tip #6: Always clearly mark the date of installation on the unit in large visible letters (See Figure 10-17). Startup dates allow the maintainer to check the actual delivery time versus the setting time and adjust accordingly.

Tip #7: At each PM check of the unit mark the reservoir level to ensure the unit has actually delivered lubricant since the last check.

SPLs aren't for every application, but they are an important part of every plant's lubrication management program and are here to stay.

Figure 10-17. Electro-Chemical Lubricator with Date in Service clearly indicated by the installer

AUTOMATED CENTRALIZED LUBRICATION SYSTEMS

Whenever there are multiple lubrication points required, the most accurate, efficient and cost effective way to lubricate is to employ an automated centralized lubrication system. Centralized systems can even be fully automated to serve entire plants with a choice that includes single line resistance, pump-to-point, series progressive, single and dual line parallel (Figure 10-18), and mist systems all available to the user.

Properly engineered, centralized greasing systems are ideal for a wide range of lubrication requirements. Modular in design and easily expandable, they are suitable for machinery with just a few lubrication points as well as installations covering complete production plants and involving thousands of points. Systems of the type shown in Figure 10-19 are designed for the periodic lubri-

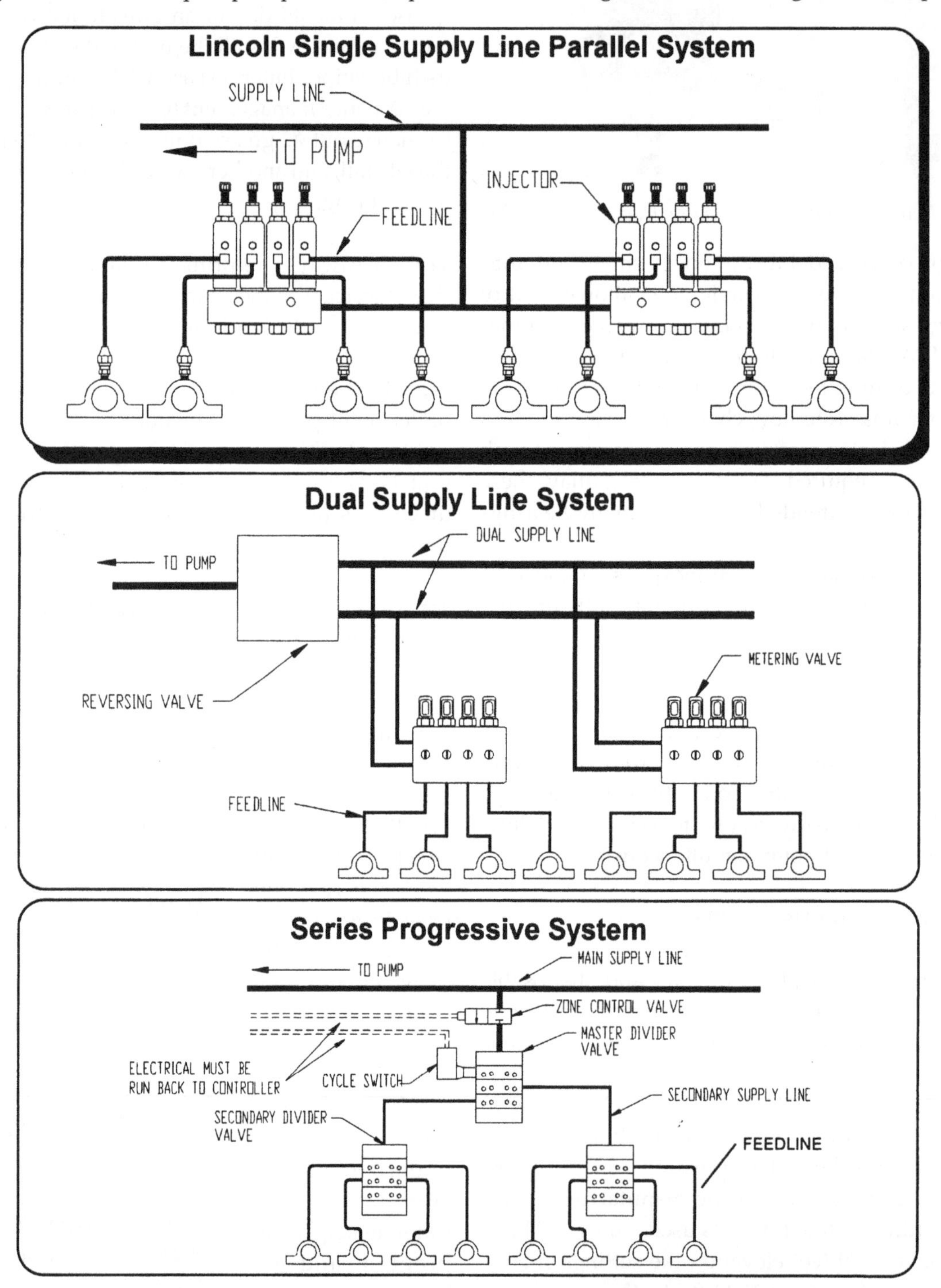

Figure 10-18. Different versions of automated centralized grease/oil systems. (Source: Lincoln Company, St. Louis, Missouri.)

cation of anti-friction rollers and sleeve bearings, guides, open gears and joints.

Depending on plant and equipment configuration, engineered automatic lubrication systems consist of a single- or multi-channel control center (Figure 10-19, Item 1), one or more pumping centers or pumping stations (Item 2), appropriate supply lines (3), dosing modules (4), supply tubing (5) and a remote controlled shut-off valve (6). Different size dosing modules are used to optimally serve bearings of varying configurations and dimensions. The dosing modules themselves are individually adjustable to provide an exact amount of lubricant and to thus avoid over-lubrication. A pressure sensing switch (7) completes the system.

The control center starts up a pump, which feeds lubricant from the barrel through the main supply line to the dosing modules. When pressure in the system rises to a preset level the pressure switch near the end of the line transmits an impulse to the control center, which then stops the pumps and depressurizes the pipeline. The control center now begins measuring the new pumping interval. If for some reason the pressure during pumping does not rise to the preset level at the pressure switch, an alarm is activated and the lubrication center will not operate until the problem has been rectified and the alarm subsequently reset.

Special multi-channel controllers are available with state-of-art automatic lubrication systems. These have the ability to provide lubrication to installations requiring a variety of lube types, or consistencies. Even different timing intervals can be controlled from single multi-channel controller locations. These systems have proven their functional and mechanical dependability in operating environments ranging from –35°C to 150°C. One Finnish manufacturer tests every type of grease supplied by user / client companies under these temperature extremes and leaves no reliability-related issues open for questioning.

Cost Studies Prove Favorable Economics of Automated Lubrication Systems

Direct contact with user companies in Finland in 1996 proved revealing and educational. In one mill alone, 3798 lubrication points were covered by two-line automatic grease systems. More recent installations have opted for equally reliable, flexible, but less expensive one-

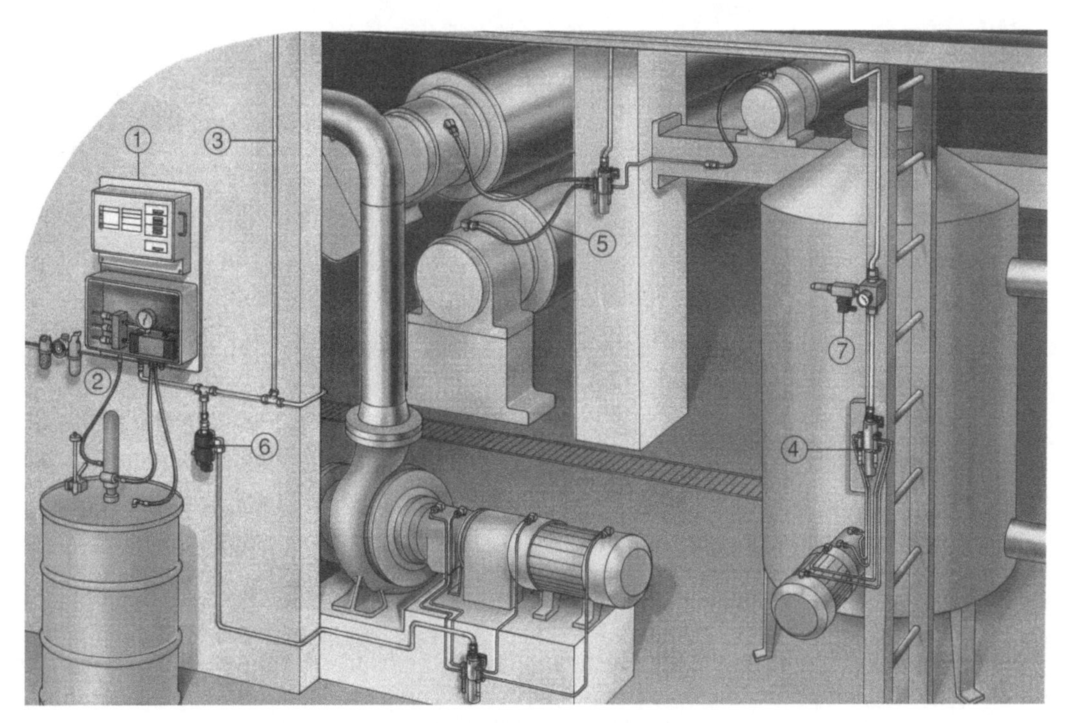

Figure 10-19. Safematic SG1-grease lubrication system, consisting of:

1. *Control center*
2. *Pumping center*
3. *Main supply line*
4. *Dosing assembly*
5. *Lubrication line*
6. *Remote-controlled shut-off valve*
7. *Pressure switch*

(Source: Safematic, Inc., Alpharetta, Georgia, and Muurame, Finland.)

Figure 10-20. Graco/Trabon 250 point Centralized grease system using a drum pump with control station and progressive divider distribution on a large walking beam furnace. Source: ENGTECH Industries Inc.

line systems. Washers, agitators, pumps, electric motors, soot blowers, barking drums, chippers, screens, presses, conveyors and other equipment are automatically lubricated at this facility. A machine which before lube automation had five to ten lubrication-related bearing failures per year now experiences none. We were advised that this translates into 30-60 hours of additional machine time and profit gains of $90 000-$180 000 annually. The plant reports a decrease in total maintenance downtime from 470 hours before lube automation to 148 hours per year after the implementation of automatic lubrication on just one major machine. Plant-wide maintenance costs have been reduced by 23% over a period of six years. Grease consumption is now only 85% of the amount used previously with manual "hit-and-miss" lubrication. Over-greasing has been eliminated and the ranks of lubrication/preventive-maintenance workers have been reduced from 20 to now only 12 technicians.

This explains why since the early 1990s this mill-wide reliability and availability upgrade approach has been employed in many retrofit as well as grass-roots installations.

We found it interesting to note that in Scandinavia alone, there are over 200 paper machines equipped with automated wet-end lubrication systems. Payback for these systems, both originally supplied as well as retrofitted, typically ranges in the half to three year time frame. This might be one of the explanation why Scandinavian paper producers, whose workers have higher incomes that most of their American counterparts, are profitable and able to compete in the world markets. Automated lubrication has consistently yielded increased plant uptimes ranging from 0.1% to 0.5%.

Elements of a Quality Two-header Lubrication System

A modern two-header system is characterized by its versatility. Modular in design, reliable in operation and capable of accepting a wide range of dosing modules it is suited to just about all industrial requirements—from lubrication of the smallest joint to the largest of roller bearings.

Key Features of Single-header Lubrication Systems

Modern single-header systems utilize an improved spool technology in combination with traditional technology now in use in dual-line systems. One such design leaves the spool ports open during the pressurization, thus eliminating any grease separation risk.

Comparing Manual and Automatic Grease Lubrication Provisions

Three principal disadvantages of manual lubrication are generally cited:

- Long relubrication intervals allow dirt and moisture to penetrate the bearing seals. Well over 50% of all bearings experience significantly reduced service life as a result of contamination.
- Over-lubrication occurring during grease replenishment, which causes excessive friction and short-term excessive temperatures. These temperature excursions cause oxidation of the oil portion of the grease.
- Under-lubrication occurring as the previously applied lube charge is being depleted, and prior to the next regreasing event.

In contrast, automated lubrication has significant technical advantages. Time and again, statistics compiled by major bearing manufacturers have shown lubrication-related distress responsible for at least 50%, and perhaps as much as 70% of all bearing failure events worldwide. Thoroughly well-engineered automatic lubrication systems, applying either oil or grease, are now available to forward-looking, bottom line-oriented user companies. In paper, pulp, refining and steel plants (Figure 10-21), these systems are ensuring that:

- The time elapse between relubrication events is optimized.
- Accurately predetermined, metered amounts of lubricant enter the bearing "on time" and displace contaminants.
- The integrity of bearing seals is safeguarded.
- Supervisory instrumentation and associated means of monitoring are available at the point of lubrication for critical bearings.

Figure 10-21. Modern steel mills use automatic grease lubrication systems.

CIRCULATING LUBRICATION SYSTEMS

Circulating (liquid oil) lubrication systems are typically required in equipment where the oil performs cooling or heating duties in addition to its original purpose, which, of course, is lubrication of parts. Some such applications include paper machine drying sections and steel industry rolling mills.

A typical circulating system used in the paper and steel industries is shown in Figures 10-22 and 10-23. It consists of a circulation lubrication center (1), comprised of a stainless steel reservoir, twin filters and pumps, and one or two cooler or heater sets. Each of these elements would normally be furnished with supervisory instrumentation. A well-planned system would further include oil flow meter groups (2), pressure piping (3) and return piping (4).

A modern, closed circulating system utilizes the lubrication center (see also Figure 10-19) so that each lubrication point receives the correct amount of high quality, clean oil at the required temperature. Such a system would be sized to accommodate the exact requirements of the equipment it serves. Moreover, each flow meter would again be equipped with appropriate alarms or similar annunciation devices. It should be noted that a good flow meter offers easy calibration and readability. Flow calibration in accordance with the viscosity characteristics of the oil should be possible over a fairly wide range without requiring meter replacement.

OPEN, CENTRALIZED OIL LUBRICATION SYSTEMS

Open, centralized oil lubrication systems are designed for the cyclic lubrication of industrial conveyors, guides, and other heavy duty machinery. Figure 10-24 illustrates the principal components of one such system, comprising a control center (1), pumping station (2), main supply line (3), dosing module (4), branch tubing (5), appropriately configured spray nozzles (6, 7 & 8), and a remote-controlled shut-off valve (9). Experience shows that open, centralized systems reduce energy consumption and unscheduled downtime. These systems can operate in widely varying temperature environments. Many different lubricants can be accommodated and delivered at the point of usage as a clean, metered quantity. Greatly reduced component wear and a three-fold overall increase in machine life are not

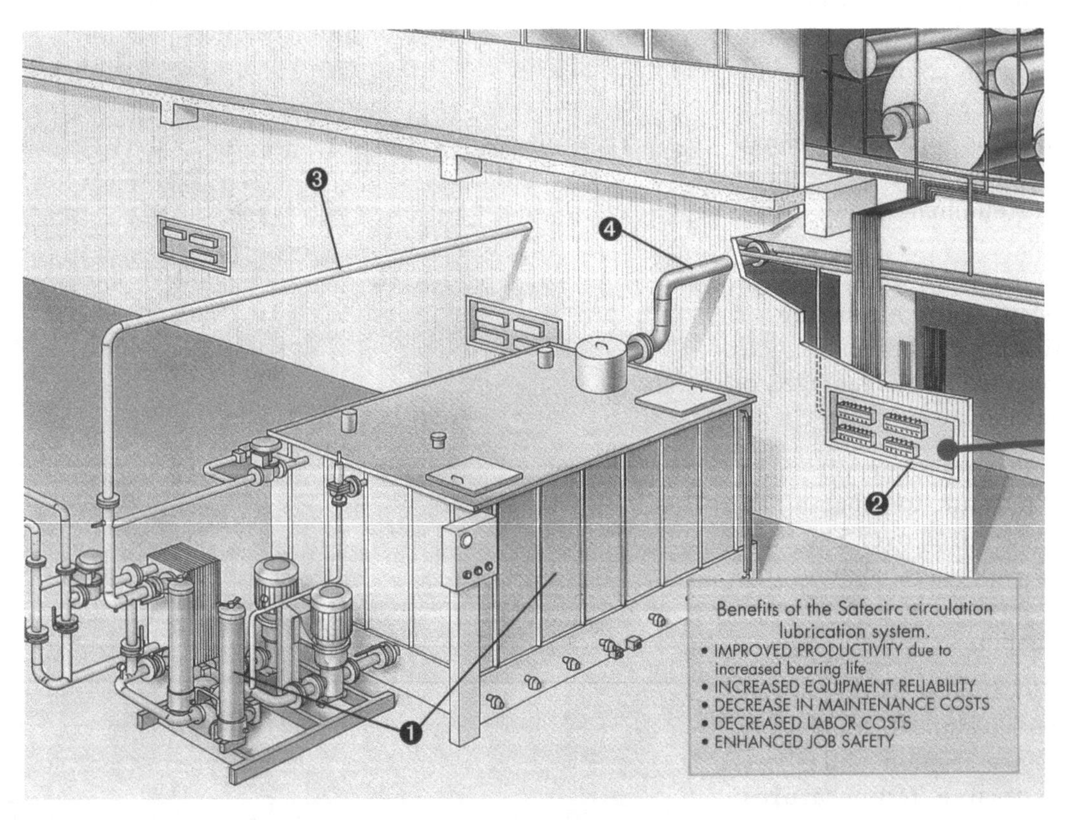

Figure 10-22. Circulating lubrication system comprised of filters/cooler/pumps and reservoir (1), oil flow metering modules (2), pressure piping (3), and return piping (4). (Source: Safematic, Inc., Alpharetta, Georgia, and Muurame, Finland.)

Figure 10-23. Large recirculating oil—Circoil™ pumping unit used of large paper mill applications. Source: Howard Marten Company

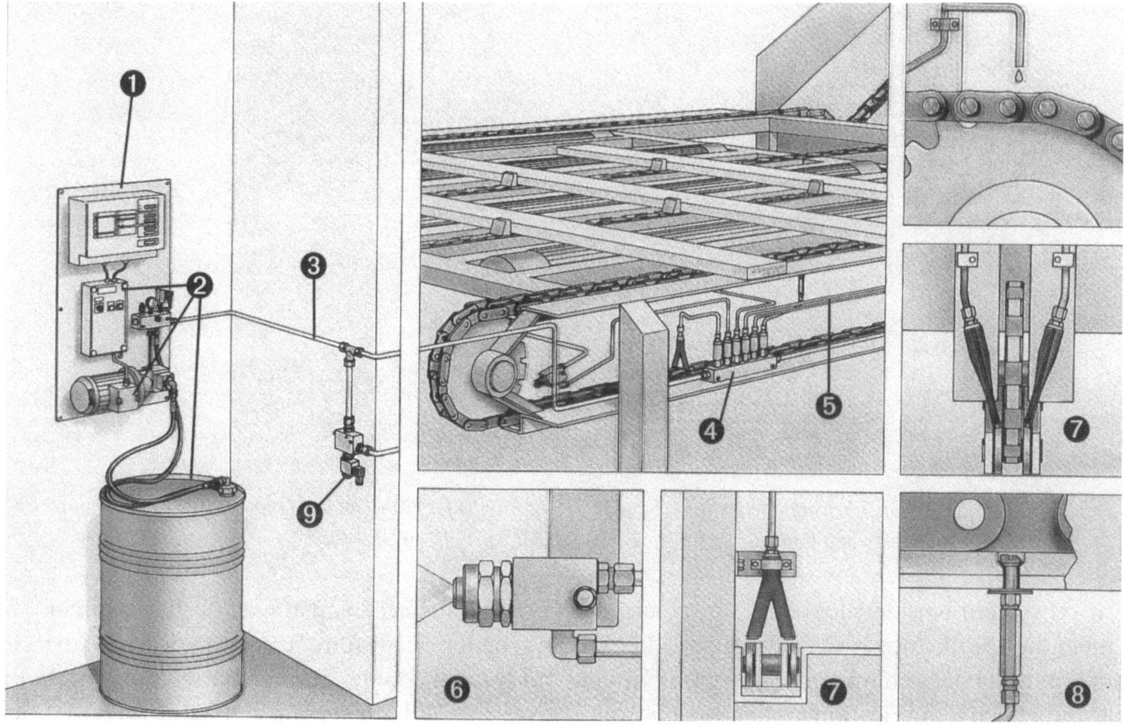

Figure 10-24. Open, centralized oil lubrication system. (Source: Safematic, Inc., Alpharetta, Georgia, and Muurame, Finland.)

unusual on forest product machinery (Figure 10-25) and other equipment.

AIR-OIL LUBRICATION SYSTEMS

In high speed rotating equipment such as machine spindles, turbines, blowers, etc., bearing and gearing speeds can approach DN surface speed values approaching 2.2 million DN (DN value is calculated by multiplying the bearing diameter in mm (D) by the rotational speed of the spindle in RPM (N)).

Traditional wet sump oil and grease lubrication methods cope poorly in these high-speed environments as they struggle to dissipate the additional heat load created by speed and fluid friction. This will often result in significant reductions in lubricant life, energy loss, and machine speed capability. Traditional mist lubrication systems allow for higher rotating speeds over traditional oil lubrication systems but are not able to provide the exact metering requirements needed for extended bearing service life, and as mist is in a micro droplet form it is susceptible to becoming airborne in the plant environment and is seen as a health and safety problem by some companies.

Figure 10-25. Forest product processing machinery using open, centralized lubrication. (Source: Safematic, Inc., Alpharetta, Georgia, and Muurame, Finland.)

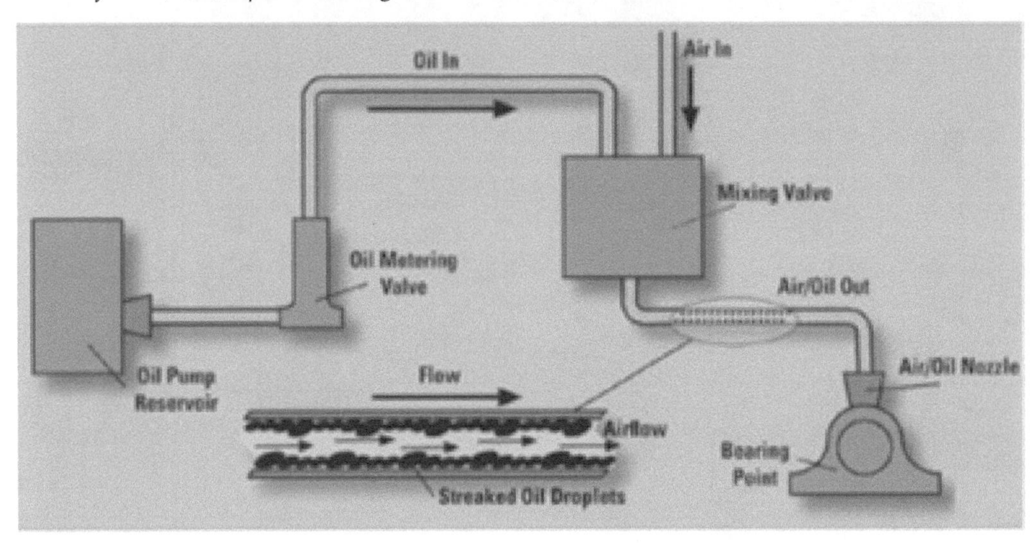

Figure 10-26. Air Oil lubrication schematic. (Source: Ken Bannister for Lubrication Management & Technology magazine's "Anatomy of a Centralized Lubrications System" article series

The air-oil system was developed as a total loss oil system to meet the specific needs of the high-speed bearing environment and has been refined to the point where it is a viable and effective lubrication system for any rotary or linear machine application.

A relatively new lubrication system design, the air-oil lubrication system can be regarded as a hybrid system that utilizes the metering capability of existing single line resistance, positive displacement injector, and progressive divider delivery systems. Oil is metered in the traditional manner in minute quantities to a mixing block connected to a clean and dry compressed air supply. The individual oil drops are dispensed into a small diameter (usually 4mm–3/16 inch diameter) delivery tube where the oil is "streaked" by the air into macro droplets and transported along the inner walls of the delivery tube to a dispensing nozzle located at each individual lubrication point. The small nozzle diameter creates a venturi effect allowing the air and the droplet to envelop around the bearing surface in an almost oil-free manner. The air forces the lubricant film across the bearing surfaces creating additional cooling and a positive pressure in the bearing area that aids in sealing out external contaminants such as coolant and dirt. (Figure 10-26)

Air-oil lubrication systems, Figures 10-27 and 10-28, are designed to provide continuous metered flow in minimum, exact quantities to points of application. There are many options and configurations; they typically include pressure switches, pressure gauges, check valves, and controllers ranging from elementary to the most advanced electronic models. Air flows can be fixed or adjustable to each point of application.

Depending on selection criteria, air-oil lubricator devices use mixing chambers remote or integral with the metering valve. The four-section air-oil valve assembly illustrated in Figure 10-30 features integral air-oil mixing valves. The associated pumping stations can be either electrically or pneumatically powered.

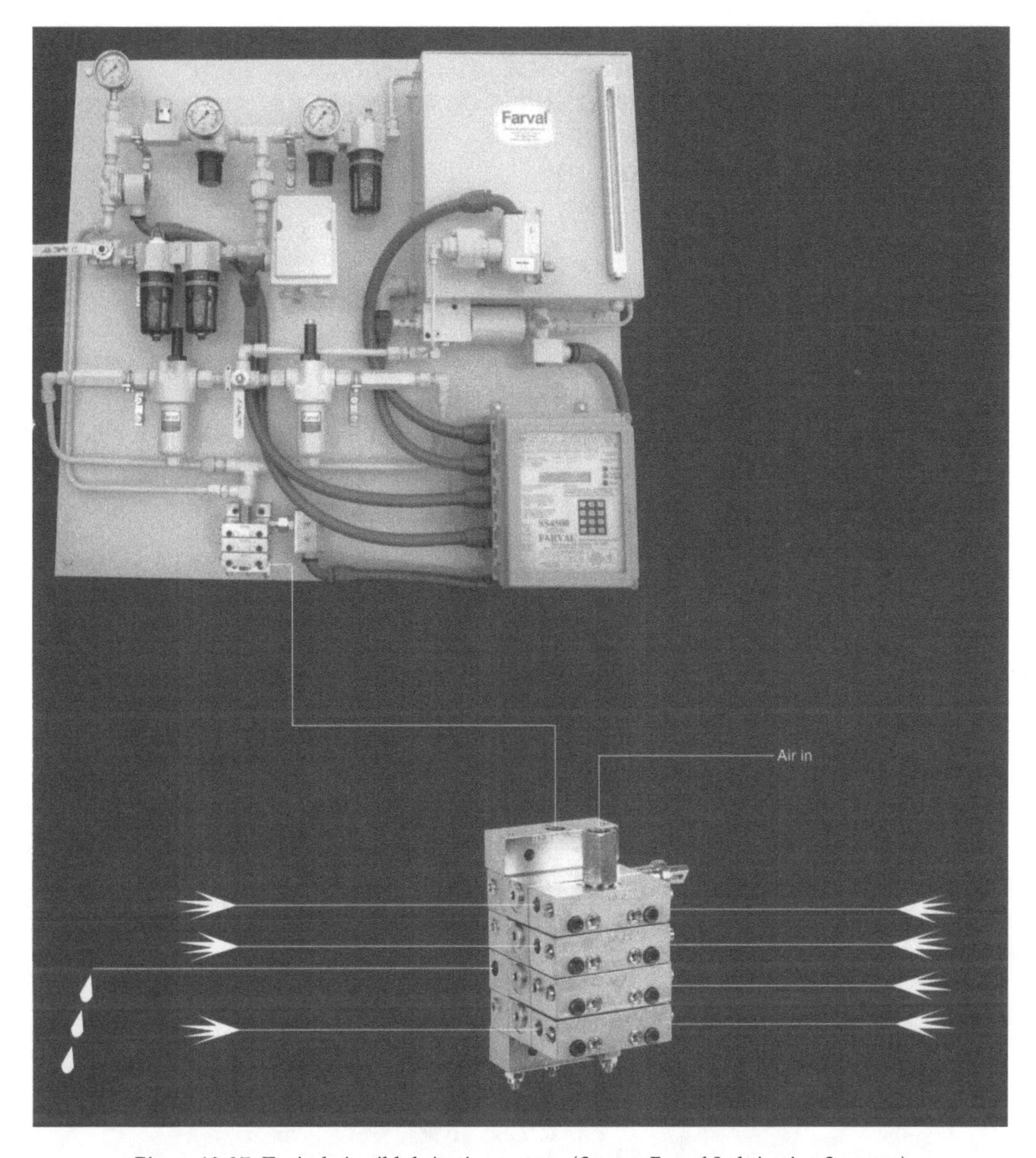

Figure 10-27. Typical air-oil lubrication system. (Source: Farval Lubrication Systems.)

The units depicted in Figures 10-31 and 10-32 operate on the principle of filtered air entering the main lubricator. Separate lines carry air and oil to the oil delivery control unit inside the manifold block assembly. Coaxial distribution delivery tubes separately carry oil and air near the points of application. Air and oil are then mixed at the nozzle assembly and controlled spray droplets (not mist) generated at the ends of the distribution line nozzles. The system is monitored for low oil level and low oil and air operating pressure.

Pros and Cons

When set up correctly the air-oil system boasts an impressive resume of benefits that include:

- Highest possible bearing surface speeds,
- Bearing temperature rise over ambient of less than half that of traditional lubrication methods,
- Energy consumption reduction,
- Health and safety approved system,
- Continuous fresh oil delivery,
- Up to 90% lubricant use reduction compared to mist lubrication systems,
- Up to 99% lubricant use reduction compared to grease systems,
- Extended bearing and seal life expectancy,
- Small environmental footprint in comparison to other traditional lubrication cleanup/disposals methods.

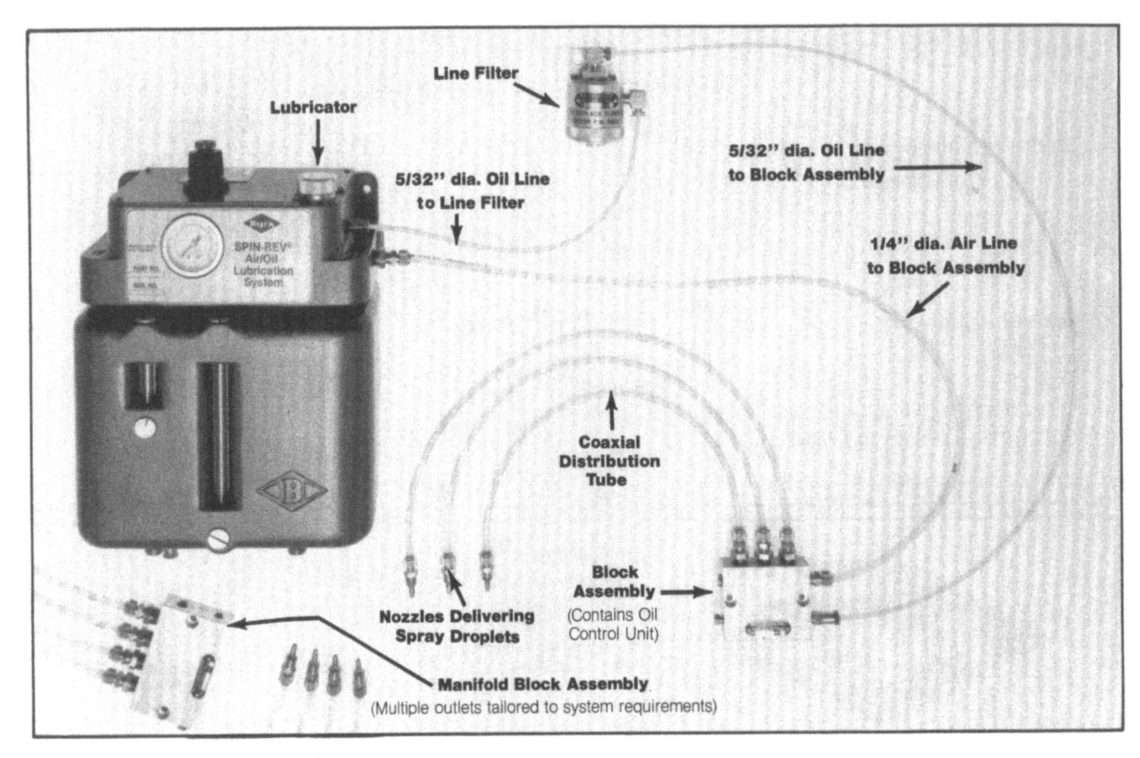

Figure 10-28. Small air-oil lubrication system (Source: Bijur Lubricating Corporation.)

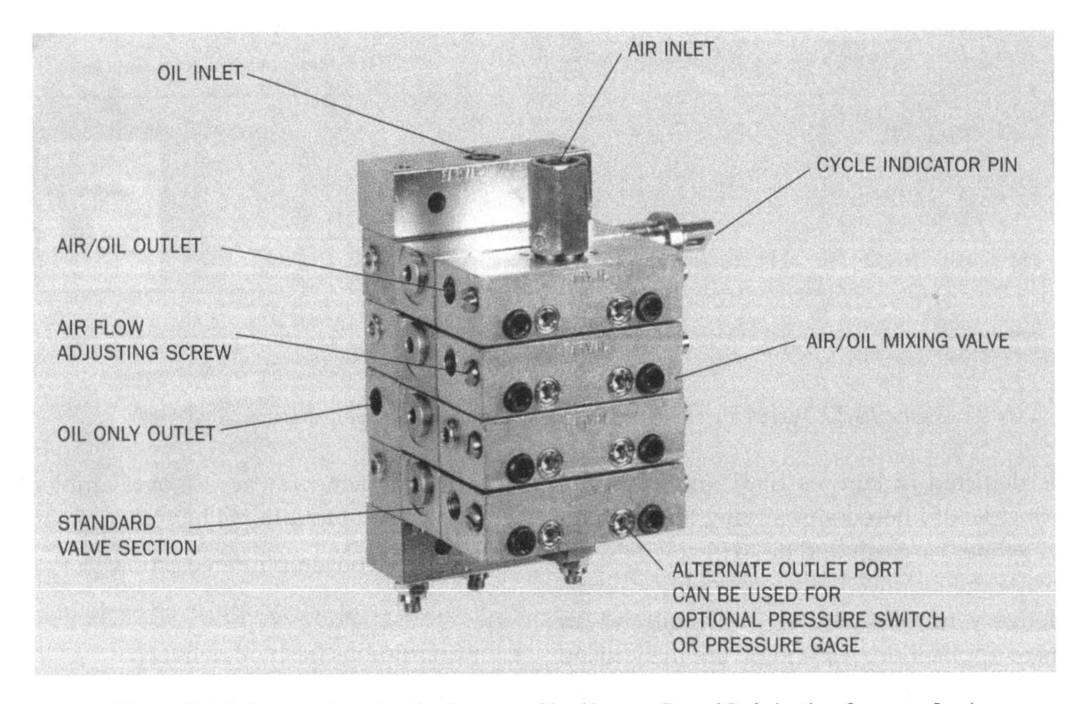

Figure 10-29. Four-section air-oil valve assembly. (Source: Farval Lubrication Systems, Inc.)

Because the air-oil system is a hybrid system, it requires the purchase of a traditional lubrication system and the additional costs of adding mixing control valves and a clean and dry air supply network to each mix valve assembly. Due to the small aperture of the delivery nozzle, solids additives such as moly and PTFE are discouraged from use as they can "bridge" across the nozzle, form deposits and cause an oil starvation situation. The air-oil system is easier to incorporate in the machine design stage and much more difficult to design as an add-on system later on.

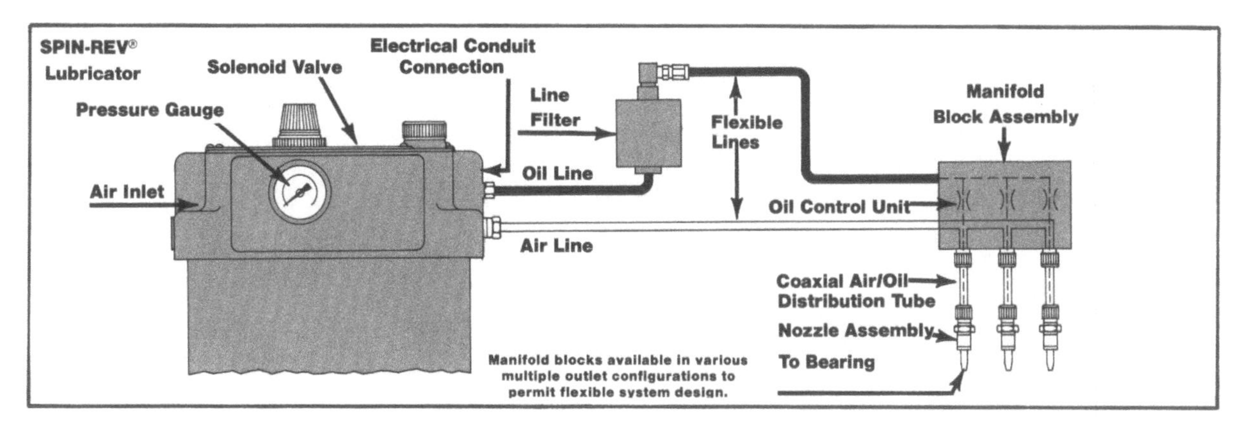

Figure 10-30. Air-oil lubrication systems with coaxial delivery tubing. (Source: Bijur Lubricating Corporation.)

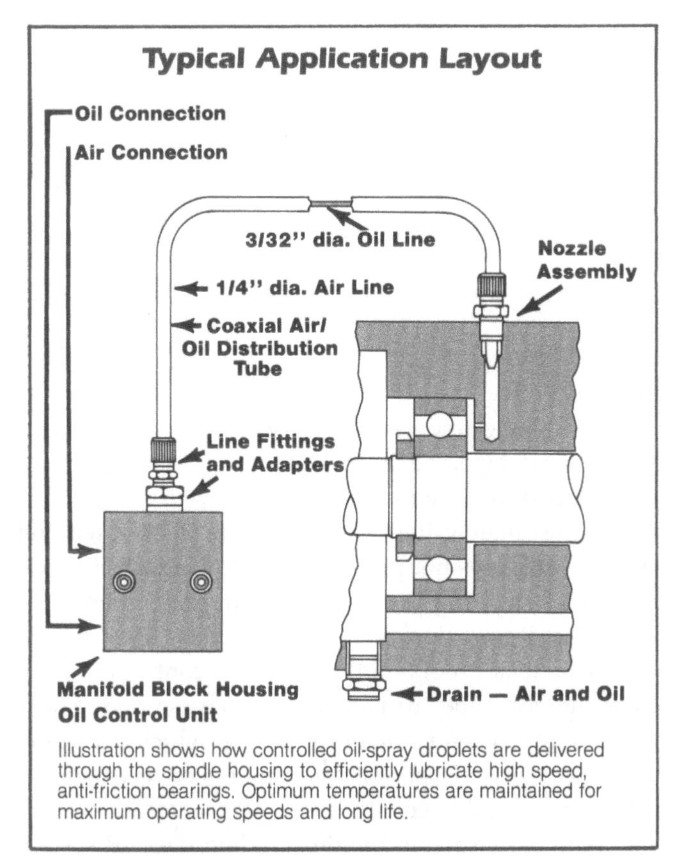

Fig 10-31. Typical application layout for air-oil lubrication. (Source: Bijur Lubricating Corporation.)

SINGLE LINE RESISTANCE (SLR) SYSTEMS

The single line resistance system is a fully engineered central system designed to pump oil manually in a single shot (total loss) method, in an automated cyclical (total loss) manner, or in a continuous (recirculative) manner. The system is a low pressure system engineered to deliver an apportioned amount of lubricant to every single bearing point when the pump is operated, and can be designed to accommodate up to 200 delivery points in a single pump system design.

Today, this system is arguably the most copied lubrication system on the market and was originally designed for automobile lubrication, and later on adapted for small to medium machine tools and manufacturing equipment. Introduced to market in 1923, the system was a US design by Joseph Bijur of the Bijur Lubricating Corporation—who today remains a leader in the manufacture of lubrication delivery systems.

How the System Works

In a typical total loss one shot or cyclic system shown in Figure 10-32, a piston pump delivers lubricant through a 5/32 or 3/16 diameter line at a pressure of approximately 60 psi. Each lubrication point is proportioned controlled by individual metering devices connected in series. Pressured lubricant is forced through the metering unit's check valve and flows into the bearing point. As the system pressure subsides, the check valve closes and the pump resets itself ready for the next operation, the dormant system retaining a residual system line pressure of between 2-5 psi.

In a continuous recirculating system, a gear pump delivers continuous oil flow through a flow proportioning device known as a control unit.

Both meter and control units pictured in Figure 10-33 appear identical from the outside but differ considerably in how they are constructed. A meter unit contains a metering pin of a controlled diameter that floats in an accurately reamed cylindrical passage producing an annular orifice of known flow rate. The clearance between the pin and the cylinder wall determines the meter unit flow rate designation. A control unit used in continuous systems has no check valve and uses a helical screw to meter the flow.

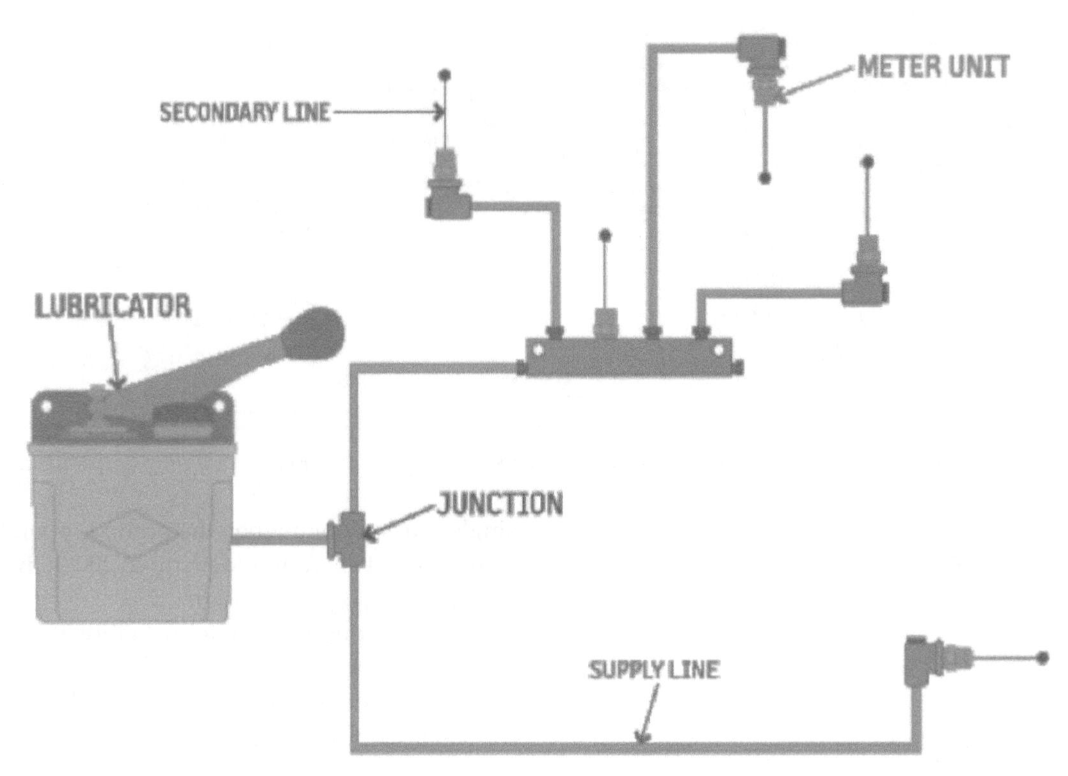

Figure 10-32. Single Line Resistance System. Source: Bijur Lubricating Corporation

Figure 10-33. SLR metering units. Source: ENGTECH Industries Inc.

Pros and Cons

The single line resistance system is a simple, inexpensive, engineered lubrication system designed for small to medium sized machinery.

Unfortunately, the system can only be used with oil with oil, and does not produce a control signal. Because the system metering units are piped in series (hence the single line designation), care must be taken to ensure all fittings are leak proof and that meter or control units are never allowed to be "drilled" out to increase rate of flow as both have the effect of a broken line, which allows oil to take the path of least resistance and effectively "starve" all bearings simultaneously.

With basic care and understanding the SLR system will provide excellent bearing life cycle management at a minimal cost.

SINGLE SUPPLY LINE PARALLEL SYSTEM (POSITIVE DISPLACEMENT INJECTOR)

Known also as a single line parallel system, the Positive Displacement Injector (PDI) system was first developed and introduced to industry by the Lincoln Industrial Corporation in 1937 to accurately displace metered quantities of oil or grease in a cyclical manner to small/

medium sized industrial equipment.

In contrast to Single Line Resistance and Progressive divider type systems, each metering valve, or point, can be set independently, adjusted, or easily changed, without affecting the system design. This enables additional injectors (lube points) to be added into the system at a later date, without the need to re engineer the entire system.

How the System Works

All PDI systems utilize either a pull handle manual or automated pump to force oil or grease into the main line and injectors (all connected together in a single line) to a pressure greater than 800 psi or 55 Bar.

In fully automated systems, a pressure switch located at the very end of the main line is set up to shut off the pump once line pressure is achieved. In manual pump systems, a pressure gauge is often employed enabling the operator to see the built up line pressure and discontinue pumping once suitable line pressure is achieved.

Each lubrication point requires its own injector and is connected directly to the lubrication point via a secondary delivery line as shown in Figure 10-34. As lubricant is pumped into the injector under pressure, a fixed displacement piston is hydraulically moved against spring pressure to discharge a fixed lubricant amount into the bearing point.

With line pressure achieved, all injectors have simultaneously discharged, and the pumping action is ceased. To reset the injectors, pressured lubricant is diverted through a reservoir relief valve and allowed to "backflow" into the pump's reservoir. As this occurs, the injectors are spring returned allowing lubricant to flow from the loading chamber into the firing chamber, ready for the next lubrication cycle. Once a predetermined time has passed the entire operation repeats itself.

Pros and Cons

Because it can be used with oil and grease, does not require much system engineering, and can easily have additional points added to the system, the PDI system has long enjoyed a reputation as both a versatile and universal system.

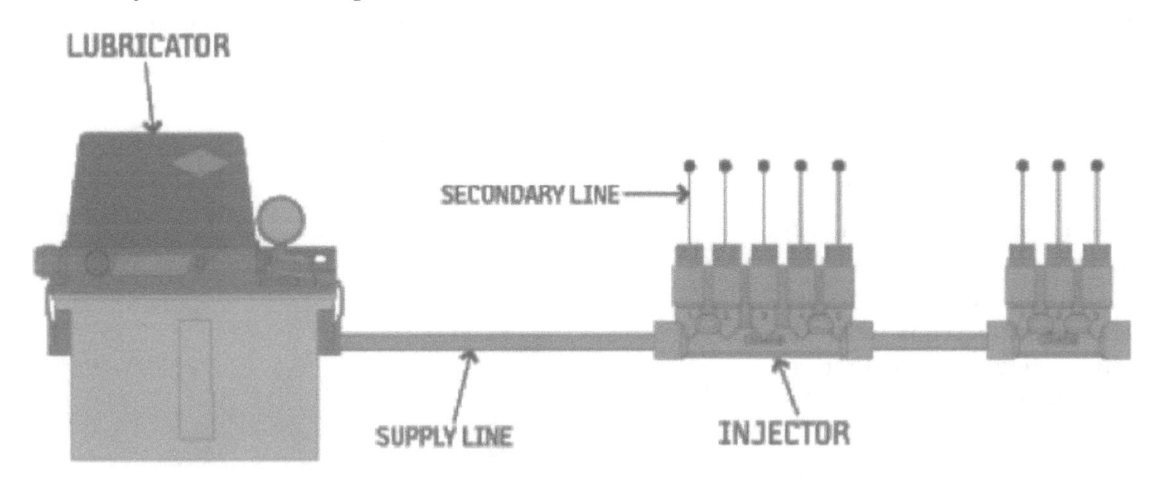

Figure 10-34. Single Line Parallel (PDI System). Source: Bijur Lubricating Corporation

Figure 10-35. Lincoln style Positive displacement injectors

PDI systems that use fixed injector displacement caps are preferred over injector types that allow the user/operator to readily adjust the piston output via an external adjustment wheel or lever on the side of the injector. User/operator adjustable injectors are easily tampered with and can easily lead to an over or under lubrication condition unless they access controlled.

Although a main open line failure can easily be detected through a time out switch located at the end, no secondary line failure device is available. Users must perform system line integrity checks as part of their PM program.

DUAL SUPPLY LINE PARALLEL SYSTEM

First introduced by the Farval Lubrication company, now part of the Bijur-Delimon corporation, the dual supply line lubrication system—also known also as the twin line parallel system—was designed to accurately displace and deliver oil or grease to as little as 20 lubrication points up to many hundreds of points, over great distances, from a single pumping station. The system's heavy duty construction, and its use of small bore piping and tubing made it an ideal choice for automated lubrication in medium/large sized industrial equipment typically employed by the steel, mining, pulp and paper, power generation and petro-chemical industries.

The dual line system bears many similarities to the PDI single line parallel or positive displacement injector system, in that each metering valve, or point, can be set independently, or easily adjusted while in operation. This unique feature also enables additional injectors (lube points) to be added into the system at a later date, without the need to re engineer the entire system.

How the System Works

As suggested by its name, the system employs two main lubrication lines that run in parallel from the pump to the last lubrication point through a series of lubrication delivery valves, shown in Figure 10-36. Once the pump is activated, line pressure is built up on the pressure or delivery supply line to fire the lubrication point injectors while simultaneously venting the second return line back through a reversing valve to the reservoir.

Dual line injectors differ to single line injectors in that they do not use a spring arrangement to fire and load the injector, but rather employs a dual acting hydraulic spool valve set up to feed two separate lubrication points (one per each pressure cycle). Once an end of line pressure switch signals that a preset line pressure has been reached and all injectors have fired, the system has completed one pressure cycle, or one half-lubrication cycle. The reversing valve is now actuated to its changeover position to allow the previous venting line to become the new primary pressure line and the process is repeated to complete one full-lubrication cycle.

The system can operate in both a manual mode with a pull handle pump or full automatic mode as shown in Figure 10-37.

Pros and Cons

Because it can be used with oil and grease, the system engineering is not demanding and will easily accommodate the addition or reduction of system points post installation.

As with single line systems, the adjustable injectors are easily tampered with and can lead to an over or under lubrication condition unless they are access controlled.

A pressure line failure is easily detected through a time out switch located at the end of line; no secondary line failure device is available. Users must perform system line integrity checks as part of their PM program.

SERIES PROGRESSIVE DIVIDER SYSTEMS

The highest engineered of all lubrication systems, the series progressive system is designed to pump oil or

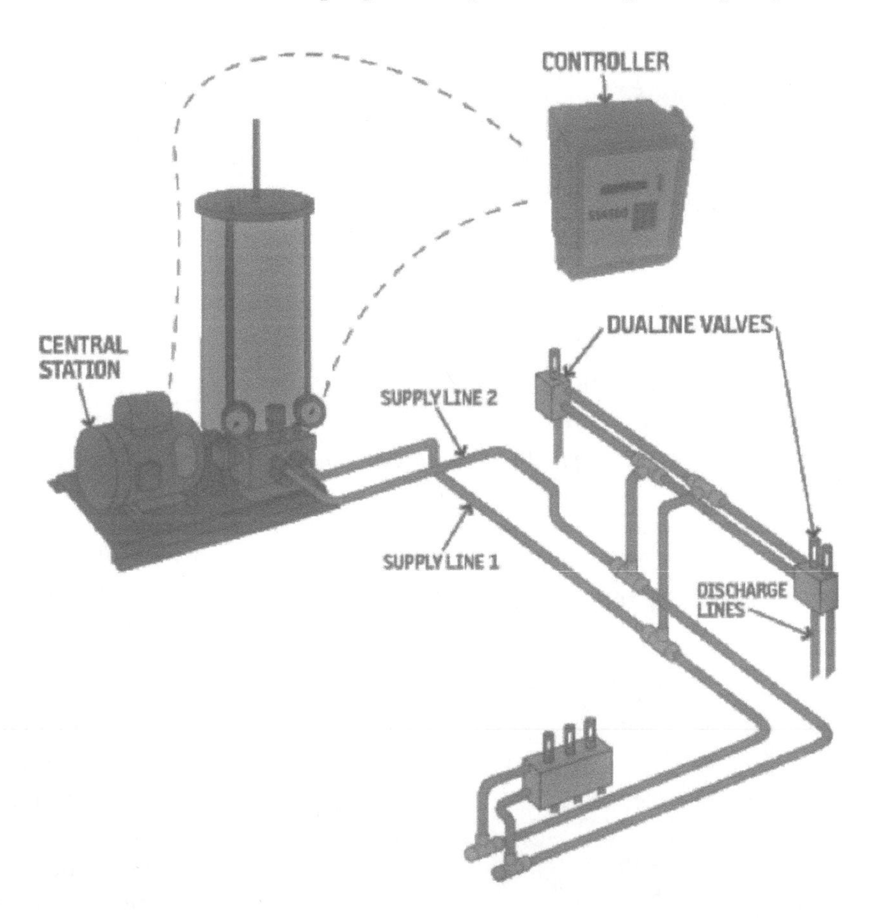

Figure 10-36. Dual Supply Line Parallel System. Source: Bijur Lubricating Corporation

Figure 10-37. Dual Line Farval pumping system. Source: ENGTECH Industries Inc.

grease in either a cyclical (total loss) or continuous (recirculative) manner. Its divider block design is engineered to positively deliver an exact displaced amount of lubricant to every single bearing point, to operate in severe environments and accommodate upwards of 200 delivery points in a single pump system design.

Today, most lubrication OEM's offer a version of this most popular system type, with the original system designed and developed in the US by the Lubriquip organization in the earlier part of the 20th century and marketed under the Trabon name continues to be sold today.

How the System Works

A lubricant pump is connected to an engineered network of series progressive divider blocks, and via a controller is allowed to pump lubricant in a continuous or controlled cyclical systematic manner to each divider block as depicted in Figure 10-38.

Divider blocks are built in a modular manner and contain a series of lapped hydraulically actuated spool valves sized for varying displacements—Figure 10-39. The ability to "cross-port" a valve results in "doubling" the delivery of lubricant on one side of the valve only. The valves are progressively linked together in series causing each valve to "shuttle" over to one side of the block in a progressive pattern, then to "shuttle" back

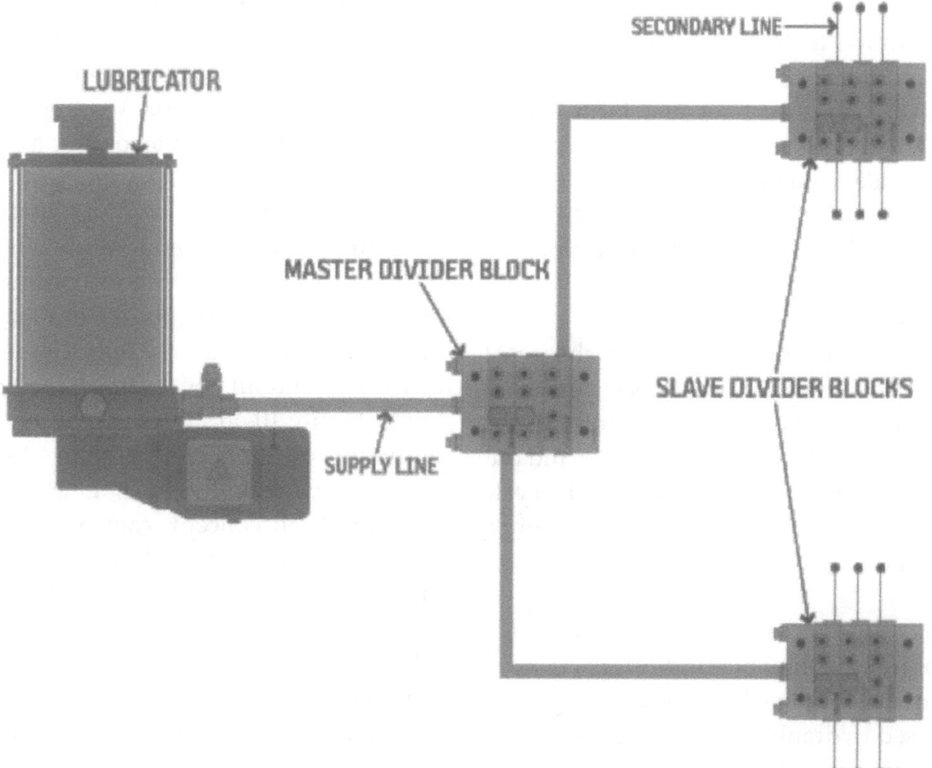

Figure 10-38. Series Progressive System. Source: Bijur Lubricating Corporation

Figure 10-39. Trabon Divider block with indicator pins. Source ENGTECH Industries Inc.

to their original position as the lubricant continues to be pumped through the block. Because of its hydraulic nature, as the valve is shuttled back and forth it both displaces a lubricant charge on one end of the valve to the bearing point while simultaneously filling the void on the other side of the valve in preparation for displacement once the valves "shuttle" back to the other side.

Monitoring the System

To ensure every point has delivered a charge of lubricant the first valve in the divider block can be attached to a cycle pin indicator that visually indicates the block has completed one full cycle by moving in and out one time. Attaching a counter/ timer control to the cycle pin will indicate if no delivery has taken place within a given time period and signify a broken main delivery line.

Blocked and crimped lines can be visually detected via simple mechanical overpressure indicators connected at the secondary delivery line block outlet. Whenever a "restriction" caused hydraulic backpressure is sensed, a visual indicator pin "pops" up to indicate the exact line/bearing point requiring maintenance. If no electronic alarm sensors are used, operations and/or maintenance must perform regular visual checks for visual alarm indications.

Pros and Cons

With basic TLC this system type will provide trouble free service for many years due to it's engineering and tamper proof design. Blocks are modular, and replacement parts are relatively inexpensive to purchase, stock, and replace. As depicted in the divider valve photo, this system can even be utilized with a manual grease gun to deliver an engineered amount to each bearing point and can be automated at a later date. (Photo clearly shows block cycle pin and four overpressure indicators in "run" position).

The down side of the system is the difficulty to add points once the initial system is installed (although achievable), and in comparison to other single line systems series progressive systems demand more system engineering. However, the plus side to this is every system is engineered by the vendor on your behalf, and provided with schematics and a Bill of Material to include in your CMMS and/or maintenance files.

PUMP TO POINT LUBRICATION SYSTEM

Pump to point systems, sometimes referred to as multipoint systems, owe their pedigree to the industrial revolution when a cam shaft set in an oil reservoir was connected for the first time via a pitman arm to a rotary or rocking motion shaft as part of a steam engine bearing lubrication system. A cam, attached to a series of in line rocker arms, in turn was attached to a series of individual pistons tasked to draw oil from the reservoir to lubricate the camshaft and rockers and pump oil to individual bearing points on the machine via copper lines; this now meant the lubrication individual need only watch and fill one reservoir feeding 6-8 lubrication points. Figure 10-40 shows two different styled box cam multipoint oil lubricators employed on a 19th century Victorian textile mill steam engine.

Simple in concept, cam box lubricators were expensive to manufacture, limited to oil only, and could only accommodate a small number of bearing points. The next evolution of the pump to point design provided an independent air driven pump with a reservoir mounted atop of the pump, with the ability to dispense oil or grease through a series of ported outlets mounted around the periphery of the pump chamber. As the pump piston is actuated--usually from an electronic

Figure 10-40. Steam engine box cam pump to point lubricator. Source: ENGTECH Industries Inc.

timer or counter--it passes by each "metered by restriction" outlet, positively displacing lubricant into each line piped to the individual bearing points. The air valve turns off and the spring loaded pump returns drawing lubricant into the firing chamber as it resets for the next lubrication cycle. (See Figure 10-41).

These systems are inexpensive to purchase and install, and are most popular for small to medium size machinery with less than 40 points. They are also very popular for use as chassis lubrication systems for trucks and tractors with on board compressors, used to lubricate the fifth wheel, shackles, and steering components while the vehicle is in motion.

Pros and Cons

Early cam box units were oil only devices constrained to their designed number of points that could easily be individually adjusted for flow to bearing points. No line breakage or blockage protection devices are available.

Later pump to point systems can be used with oil and grease and system engineering is not demanding. Within the confines of the pump size it will accommodate the addition or reduction of system points post installation. Flexibility of lubricant delivery can be adjusted by the frequency of the pump operation. A failure top pump incident can be detected via the controller, but secondary line protection is difficult to implement.

Figure 10-41. Modern Interlube pump to point/multipoint lubricator. Source: Interlube Corporation

BLACK OIL FORMATION IN BEARING HOUSINGS

Virtually any bearing housing of the type or design shown in Figure 10-42 is at some risk of experiencing "black oil." Although lubricant degradation is among the many major bearing failure topics, not many articles have dealt extensively with oil ring-related contamination and "black oil" formation over short periods of time. Needless to say, any manner of repeated short-term degradation of the lube oil impairs bearing reliability. Equipment outages result and time and money will inevitably be spent. That's why the subject of "black oil" should be important to us.

Bearing housings with or without oil rings? By the author's estimate, about 45% of oil-lubricated pump bearing housings are furnished with oil rings dipping into an oil sump, as shown on the right side of Figure 10-42. Perhaps another 45% of oil-lubricated pumps are designed for operation without oil rings. In the typical no-oil-ring design shown on the left side of Figure 10-42, the oil level must reach to the center of the lowermost ball, roller or other bearing element. This level location requirement would be difficult to achieve if the bearings supporting the shaft were to have different diameters. Also, at higher rolling element speeds, ploughing through an oil bath may generate too much heat, prompting the equipment manufacturer to lower the oil level and decide in favor of one or two oil rings.

In the estimated remaining 10% of oil-lubricated pump bearings the lubricant is probably applied by pump-around ("pressurized") means, or the housings incorporate flinger discs, or pure oil mist is being used. In some applications, oil rings are also needed for the purpose of keeping the oil volume more uniformly mixed, i.e., to prevent stratification. Stratification is a term that describes hot oil floating upward and staying near the top of the oil sump.

No oil rings are needed at slow-to-moderate shaft peripheral velocities. These velocities are proportional to and conveniently expressed as DN-values; DN is obtained by multiplying the shaft diameter (in inches) by shaft rpm. For example, a 70 mm (~2.75 inch) bearing operating with a shaft turning at 1,800 rpm would have a DN-value of 4,960; values below 6,000 are somewhat arbitrarily considered low-to-moderate DN numbers. In DN < 6,000 designs, both housing geometry and constant level lubricator settings are generally selected to allow lubricating oil to reach the center of the lowermost bearing ball or rolling element.

However, if one were to keep the oil level at the center of the lowermost bearing ball while increasing the DN value above 6,000, potentially excessive temperatures would likely result. These temperature concerns led to the decision to lower the oil levels in many of the larger API-compliant pumps operating at 3,000 and 3,600 rpm. In these pump bearing housings oil levels are customarily set well below the periphery of even the lowermost bearing ball. Because oil is then no longer flooding a portion of the lowermost part of the bearing, mechanical means must be employed to feed, lift, spray, or splash the lubri-

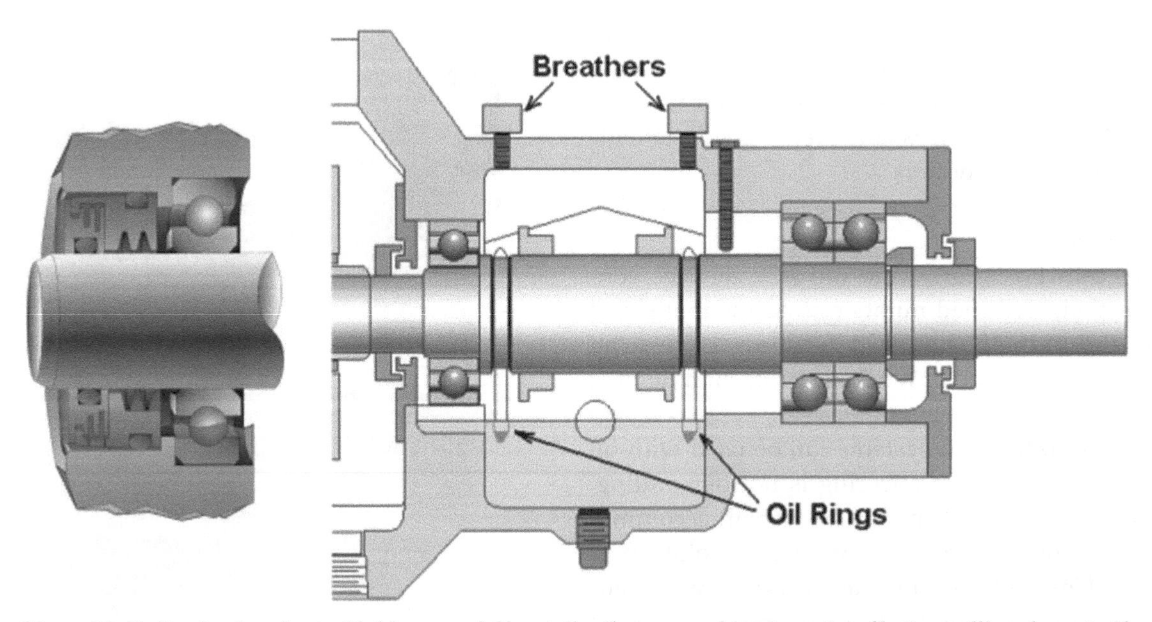

Figure 10-42. Bearing housings with (shown on left) a static oil sump reaching to center of bottom rolling element. Alternatively, a lowered oil level (shown on right) will require one or two oil rings dipping in the oil so as to "lift" oil into the bearings.

cant into the pump bearings. Oil rings are often the least expensive means for lifting oil into the bearings. But oil rings have limitations, and some of these are either little known or not well publicized.

DN-value limitations and attempts to improve on troublesome oil ring lubrication. Oil ring instability ("wobbling") is not a new phenomenon. To ensure proper operation, the Wilcock & Booser text ("W&B," see bibliography) cites surface velocity limits around 3,500 to 4,000 fpm (~18-20 m/s) with water cooling. Stipulating water cooling implies the need for maintaining constant viscosity by closely controlling oil temperature. Without water cooling of the lubricant, W&B advise staying well inside the stable limit for oil rings and not to exceed shaft peripheral velocities of 2,000 to 2,500 fpm (~10-13 m/s). Another source, a major multi-national corporation's "Lube Marketing Course" text, suggests using a DN value of 6,000 as the threshold of instability for oil rings. As a precautionary rule, both of these authoritative texts warn that oil rings in field situations tend to become unstable whenever DN enters the region from 6,000 to perhaps 8,000.

With its DN value of 7,200, a 2-inch shaft at 3,600 rpm would thus operate in the risky or instability-prone zone, whereas equipment with a 3-inch shaft operating at 1,800 rpm (DN = 5,400) might use oil rings without undue risk of instability. As just one more example, a 3-inch (76 mm) diameter shaft at 3,600 rpm would operate with a shaft peripheral velocity of $(\pi D/12)(3{,}600) = 2{,}827$ fpm (~14.4 m/s). The fact that a pump manufacturer can point to satisfactory test stand experience at higher peripheral velocities is readily acknowledged, but field situations represent the "real world" where shaft horizontality and oil viscosity, depth of oil ring immersion, bore finish and out-of-roundness are rarely perfect. We can thus opt for using either the DN < 6,000 or the Surface Velocity < 2,000 fpm "rule of thumb." Either way, the vendor's test stand experience is of academic value and the field experience that led to these rules of thumb trumps all else.

When industry received many reports of "black oil," one major pump manufacturer decided to look into the matter of oil ring instability and random events of black oil being experienced. The resulting "Investigations into the Contamination of Lubricating Oil in Rolling Element Pump Bearing Assemblies" were published by a Sulzer-Bingham team (see bibliography) at a pump industry user conference in 2000. The scope of both report and lecture was to give an analysis of factors affecting the short-term contamination of liquid lubricants in ring-oil lubricated rolling element bearings. "Short-term" was defined as intervals ranging from one hour to several weeks. The analysis dealt with the pump manufacturer's standard range of centrifugal pumps, and short-term oil degradation had been reported in a number of installations. However, although the pump manufacturer's report was compiled in 1999, several large petrochemical corporations were still struggling with the black oil issue in 2008.

At the 2000 conference, the presenter described the tests conducted and discussed the results. He then made a number of recommendations to which this author would like to add relevant comments in brackets.

- Unless slightly preloaded, back-to-back mounted angular contact thrust bearings allow the unloaded side to skid. The unloaded bearing then gets quite hot. [Comment: The pump manufacturer had supplied back-to-back angular contact bearings and had inserted a thin shim between adjacent bearing inner rings. Competent users have, since the late 1960s, insisted on using matched pairs of lightly preloaded or flush-ground back-to-back angular contact bearings. Whenever possible, shims are avoided.]

- Much of the contamination originated from metal particles being worn off the bronze oil rings (Figure 10-43, right side) which were erratically hitting the adjacent components.
  [Comment: Abrasive wear has been known to occur with unstable oil rings. Factors contributing to instability are discussed below.]

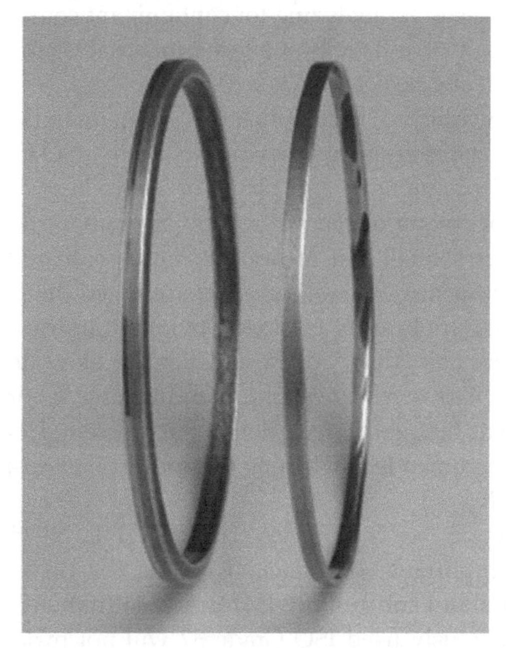

Figure 10-43. Worn oil ring on right (Source: TRICO Corporation, Pewaukee, Winconsin.)

- Instead of metallic oil rings, plastic oil rings should be used.
  [Comment: High-performance polymers are indeed less likely to suffer from abrasive wear than metallic oil rings.]

- Instead of using mineral oils that were either too light (ISO Grade 32) or too heavy(ISO Grades 68 and higher), ISO Grade 46 lubricants should be used.
  [Comment: With ISO Grades 32 and lower, the resulting oil film is often too thin to prevent metal-to-metal contact in rolling element bearings for centrifugal pumps. Bearing manufacturer SKF commented on the matter in several commercial publications]

The various recommendations and our comments deserve to be further examined in terms of root cause problem identification.

Examining the causes of darkened oil. We know from experience that closely observing both new and "used" oil rings will prove revealing. Abrasive wear of oil rings is easily recognized; a previously chamfered edge is now razor-sharp or burred (Figure 10-43), or an originally straw-colored lubricant has recently turned dark. Debris from many copper-containing alloys leaves a grayish color; pure overheating generally produces floating carbon particles that cause the oil to turn black. One could also measure the "as installed" oil ring width and later compare it with its "as-found" width. Measuring will avoid the cost of analyzing for contaminant composition. Needless to say, flaked-off oil ring material is suspended in the oil and bearing life is cut short.

It was noted, in 2000, that this pump manufacturer's prior practice of using low viscosity "thin" ISO Grade 32 oil, while acceptable for "plain" or sleeve bearings, was risky for rolling element bearings. In many of the manufacturer's small and mid-sized pumps with rolling element bearings, thin oil simply exacerbated the problem of premature bearing failures. Obviously, to use a much more viscous "thick" oil would tend to slow down an oil ring. However, the pump manufacturer reported on what was thought the solution to the problem. His report now recommended switching to another oil viscosity, ISO Grade 46.

Again, we look at experience. While oil viscosity is an important parameter, tweaking the oil viscosity selection and substituting ISO Grade 46 (mineral oil) for the previously used ISO Grade 32 will not make much difference. Relatively minor changes in ambient temperatures negate the effect of small viscosity changes; many temperature vs. viscosity charts included in this text confirm the validity of this concern. Moreover, SKF had determined, well before 2000, that film thickness and film strength limitations rendered ISO Grade 32 mineral oils unsuitable for rolling element bearings in many centrifugal pumps. In its centrifugal pump handbook, this bearing manufacturer asked users to restrict mineral oil ISO Grade 46 lubricants to bearings operating at temperatures not exceeding 70°C (158°F) and recommended ISO Grade 68 (again assuming mineral oil) for bearing operating temperatures not exceeding 80°C (176°F).

It should be noted how, in slower speed bearings that are not using oil rings, black oil formation usually happens after the oil level has dropped by a fraction of an inch, essentially low enough to just barely reach (in horizontally arranged shaft systems) the bearing's outer ring shoulder. The outer ring shoulder is, occasionally, described as the edge (at the 6 o'clock position) of the bearing outer ring bore.

However, there have also been reported instances of black oil formation at higher speeds and/or while using bearing protector seals. In one particular model with a dynamic O-ring contacting the sharp edges of a stationary groove (see Chapter 11), O-ring degradation has caused lube oil contamination. It is not known if this eventuality was assessed in the research studies of the pump manufacturer making the black oil presentation in 2000.

Oil level and oil application concerns must be addressed. There have been many instances where the reservoir bulbs of constant level lubricator assemblies showed adequate levels of oil which, understandably, led technician-operators to assume the oil level in the bearing housing was at the correct height. Hydraulic laws should convince us that such an assumption is not always correct. The lube oil level in a bearing housing cannot coincide with the level in the base of the constant level lubricator if the internal bearing housing pressure is different from atmospheric pressure. It should be noted that, in a typical non-balanced constant level lubricator base (Figure 10-44), the oil level is contacted by ambient air. Obeying the laws of physics, slightly elevated pressures in the bearing housing will drive the oil level lower. Moreover, ambient air in most industrial locations carries water vapor and airborne contaminants, either of which will decrease bearing life.

Allowing a slightly higher pressure on one side of a bearing vs. the other side of the same bearing will cause the oil level on one side to vary from the level on the other side. Attempts to equalize pressure are made by milling a small slot (perhaps 1/4 inch or ~6 mm wide and 1/8th inch or ~3 mm deep) into the lowermost location of the housing bore. This was done at the radial, but not the thrust bearing, depicted in Figure 10-16. The expectation of pressures being equalized may thus be thwarted on the thrust bearing side, especially if the bearing(s) are so-called angular contact types. In some bearings (and especially in angular contact types), the slanted bearing cage (ball separator) inclination creates windage or fan effects that promote unequal oil levels.

Using a properly pressure-balanced constant level oiler assembly (Figures 10-12 & 10-13) goes a long way toward curing the problem. A suitably sized balance line back to the bearing housing ensures this pressure equalization. Together with magnetically closed dual-face bearing housing protector seals, a pressure-balanced constant level lubricator makes a fully enclosed oil-lubricated bearing environment feasible. (An advanced rotating labyrinth bearing protector seal will often be almost as good as a dual-face magnetic seal; Chapter 12 highlights both).

Compared to non-balanced constant level models, the typical incremental cost of an average-size pressure-balanced constant level lubricator is around $40. Just think if one retrofitted 200 pumps with advanced bearing protector seals and pressure-balanced lubricators. Calculate the value of avoiding even a single unscheduled pump downtime event in each of the next five years and imagine the peace of mind those incremental dollars would buy. That would be a tangible and verifiable reliability improvement step for thousands of pumps.

From an operating and maintenance perspective, don't overlook that there must always be a partial vacuum in the upper part of the lubricator bulb. The constant level lubricator cannot possibly function if it has been filled to the top, as shown in Figure 10-44. Also, if caulking is used to cement a transparent bulb to a metal body, it must be realized that this caulking has a finite life. Once it develops tiny aging cracks, rainwater may enter via capillary action. Finally, traditional constant lubricator assemblies are direction-sensitive and should be mounted on the correct side of the bearing housing. A manufacturer's instruction manuals usually describe this requirement, as do publications found in the bibliography.

Figure 10-44. Conventional "non-balanced" constant level lubricator, shown here in overfilled condition.

Letting physics help you understand the limitations of oil rings. Assume the rolling element bearings are routinely furnished with oil rings and the manufacturer is either unable or unwilling to offer superior lube application methods in pumps or equipment with high shaft (or bearing bore) DN-values. In those instances, the polymer oil rings recommended by Bradshaw[24] will be a (minor) step in the right direction. Still, it should be kept in mind that all oil rings have limitations that can be explained by simple physics:

(a) a DN limitation, i.e. certain rpm-times-oil ring-bore values should not be exceeded;
(b) oil rings are viscosity-sensitive, and the 2000 conference presentation merely confirms what Archimedes knew in antiquity;
(c) oil rings are immersion-sensitive, and disputing it would be denying the effects of viscous drag. This drag is approximately proportional to the velocity;
(d) oil rings must not be out-of-round or "slightly oval." Common sense will tell us that, and a nice research grant would (hopefully) prove it to even the most challenged-by-physics. In the meantime, users should not allow more than 0.002" (0.05 mm) ring eccentricity (per Wilcock);
(e) oil rings will run down-hill; they tend to malfunction if the total shaft assembly is not truly horizontal—unless one disputes that gravity exists and believes Sir Isaac Newton's findings are wrong. For the rest of us, simple physics apply.

Industrial facilities today have access to and benefit from superb laser-type shaft alignment tools. However, to achieve good alignment the installer typically shims up one end of the pump, thereby jeopardizing shaft horizontality. Also, the user will experience occasions where, in spite of lubricant delivery via "constant level" lubricators, the oil level is actually lower than the set point indicated on the constant level lubricator assembly. Understanding how these lubricators function is very important and will save you much pain.

Needed: A better choice than oil rings. Most 3,000 and 3,600 rpm pumps obtain splash lubrication through the action of oil rings. Yet, with oil rings having inherent shortcomings that often make them a poor choice for risk-averse plants, flinger discs securely mounted on the shaft seem to offer a better choice than oil rings. Many reputable European pump manufacturers avoid oil rings and use metal flinger discs instead; some have done so for many decades. Since flinger discs are secured to the shaft, they are not subject to the influence of shaft horizontality, oil viscosity, depth of immersion, shaft surface finish and ring concentricity. These five factors inevitably vary from pump to pump; they result in an infinite combination of variables and some of these combinations tend to make oil rings prone to malfunction.

Visualize also that oil rings operate at slippage conditions relative to the shaft surface and that abrasive wear often results from this slippage. Slivers or particles of oil ring material have ruined thousands of bearings. Fortunately and unlike oil rings (sometimes called "slinger" rings), flinger discs (Figure 10-45) are securely fastened to the shaft and avoid slippage and abrasion problems. They are often made of a suitable elastomer and diameters can be trimmed to suit a particular bearing housing. The elastomer will deflect so as to allow insertion through narrow bearing housing bores. However, when metal flinger discs are made with diameters larger than the bearing housing bore, a bearing mounting cartridge may be required to allow assembly without interference. Mounting cartridges are precision-made and add to pump cost.

So, the vagaries of constant level lubricators and the very existence of "black oil" illustrate that the basic laws of physics apply to even the most insignificant part of machinery.

Figure 10-45. Flexible flinger disc mounted on shaft. (Source: TRICO Corporation, Pewaukee, WI)

Musings on future lube applications. Some day an enterprising inventor or pump manufacturer may turn his back on cost-cutting and will return to ingenuity. He may then develop a better alternative to oil rings or even flinger discs for rolling element bearings. It may be a smart device just short of the well-proven oil mist, and less costly than the widely known and often necessary pressurized oil pump-around systems. Perhaps reconfiguring the shaft to serve as the rotor of a progressive cavity pump, or utilizing the principle of magnetic coupling might lead to a housing-internal means of picking up some oil, pressurizing a small stream of this oil, and then spraying it into the bearing rolling elements.

INJECTOR PUMP SYSTEMS

Injector pump systems (Figures 10-46 and 10-47) are used for dispensing adjustable quantities of oil, grease (or, occasionally, other liquids) for such varied duties as

- point lubrication
- application of cutting and cooling oils
- air line lubrication
- oils for punching and stamping

Other liquids, although not within the general framework of this text, would include liquids for food additives, application of adhesives, inks, printing fluids, solvents, dyes, medical applications and chemicals in processing units. Both liquid and spray applications are feasible. Also, manual, air, and motor-operated configurations and arrangements are available. A motor-driven grease lubricator is shown in Figure 10-48, although purely manual systems (Figure 10-49) can be cost-effective as well.

Properly engineered, injector pump systems will have pulling capacities in the vicinity of 20 inches (500 mm) Hg. They will not exhibit "after-drip," will operate in any position, eliminate vapor lock and cavitation, and require no priming. Most of these systems are available in single-feed or multiple-feed outlets. In general, each injector will be adjustable from zero to around 0.01 cubic inches of liquid per cycle. Used in conjunction with a solid state timer, injector pump systems can be cycled (pulsed) from perhaps once per second to once per day.

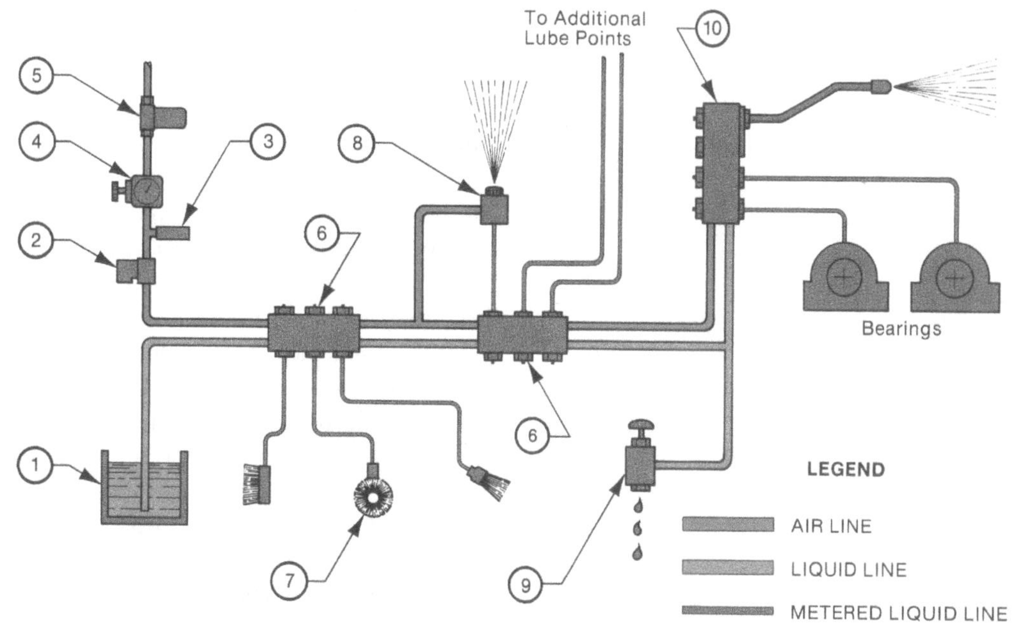

1. Reservoir
2. 3-Way Solenoid Valve
3. Pressure Switch
4. Air Regulator
5. Air Filter
6. 3-Feed Air Operated Injector Pump
7. Liquid Application Brushes
8. Remote Spray Valve
9. Manually Operated Injector Pump
10. 2-Feed Air Operated Injector Pump with Single Feed Spray Valve

Figure 10-46. Air-operated injector pump system. (Source: Oil-Rite Corporation, Manitowoc, Wisconsin.)

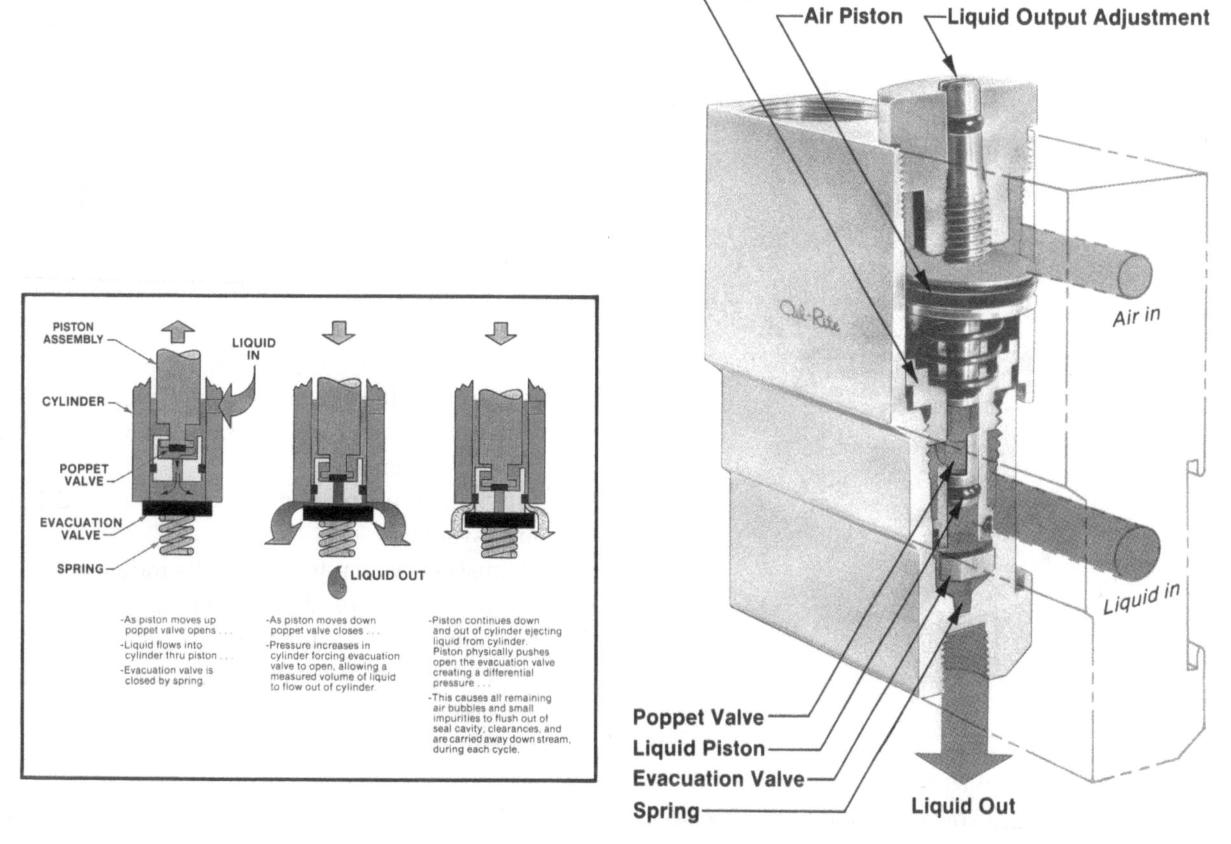

Figure 10-47. Cut-away view of "Oil Rite" air-operated injector pump.

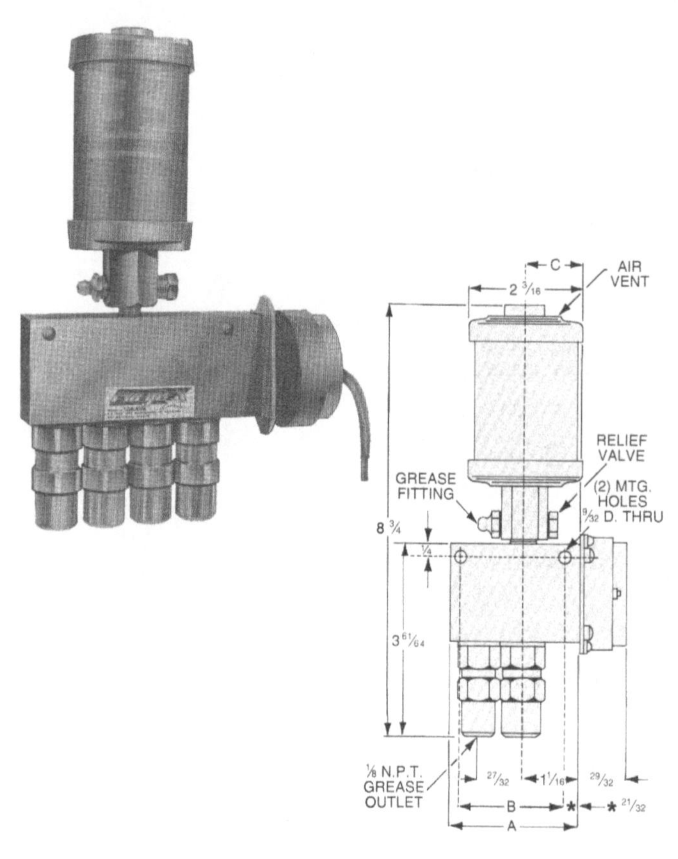

Figure 10-48. Motor-powered grease injector assembly. (Source: Oil-Rite Corporation, Manitowoc, Wisconsin.)

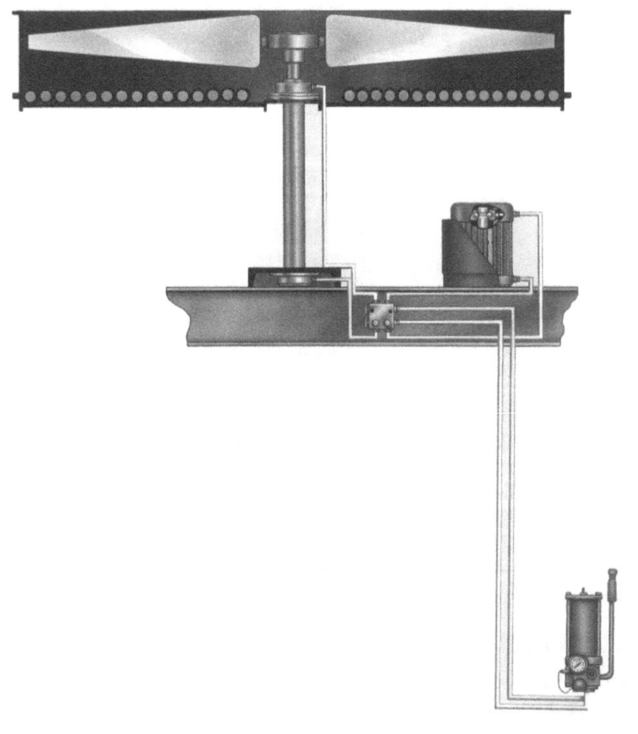

Figure 10-49. Hand-operated injector pump system serving a cooling fan assembly.

OIL MIST LUBRICATION TECHNOLOGY AND APPLICATIONS*

Oil mist systems have come a long way since their introduction to the European textile industry over 70 years ago. Initially designed to resolve high-speed spindle bearing failures that would not respond to conventional oil and grease lubrication tactics, oil mist first saw favor in the North American market in the 1960s/70s with the use of oil mist lubrication in the refining and petrochemical industries dates back to the 1960s when companies such as Exxon and Chevron began to apply oil mist to pump bearings.[1] By the early 1970s oil mist lubrication was being applied to rolling element bearings of electric motors in the refining industry.[2] In the mid-1990s, 77% of the major, multi-location US refining companies had at least one large-scale oil mist system in at least one of their refineries.[3] The use of oil mist in the hydrocarbon processing and other industries, such as pulp/paper, world-wide is growing because of oil mist lubrication's proven performance in delivering improved machinery reliability, reducing maintenance costs and providing a fast pay-back on the investment in the oil mist lubrication system.

Oil mist technology has kept pace with advances made by process industries and the mist systems being designed and installed today are far superior and more efficient than those installed in the 1980s. Some of the advances and new applications that are utilized in today's systems are:

- Microprocessor controlled central oil mist generators compatible with central distributive control systems
- More efficient and effective distribution system design practices
- Improved oil mist manifolds
- Environmentally clean mist collection containers
- Drain leg designs and components, which eliminate waste and venting
- Efficient, environmentally clean, closed-loop oil mist systems
- Demisting system for the textile industry
- Miniaturized, closed-loop lubricators
- Portable mist density monitors
- New applications for oil mist:
  —Rotary lobe blowers
  —Defibrator press

*Source: T.K. Ward, Lubrication Systems Company, Houston, Texas. Adapted by permission.

Each of these advances will be described in this segment of our text.

BENEFITS AND DESCRIPTION OF AN OIL MIST LUBRICATION SYSTEM

As oil mist became more mainstream and moved into the large manufacturing industries, its open vent system design—combined with the ability for operators and maintainers to easily adjust the air flow pressure at will—often created an oily environment for workers. Unfortunately, an ever-increasing awareness of workplace environment safety forced a ban on both well-designed and poor designed oil mist systems in many plants.

With the recent introduction of micro process control, vortex style mist generation and closed loop system design, coupled with airborne mist detection systems, one of the most effective lubrication system designs has now witnessed a dramatic resurrection.

How the System Works

In the original oil mist system design a pressure controlled, filtered compressed air supply is passed through a venturi. Oil is then siphoned from a reservoir by the quickened airflow directly following the venturi and is directed at a baffle plate causing the oil to atomize into very fine droplets known as "dry mist." Any thing less than a 1part oil to 200,000 part oil ratio (approximately 1.5 microns or 0.00006" diameter) falls back to the reservoir as heavy oil particles; mist then is then piped into a 2" diameter scheduled pipe header at between 5-40 inch water pressure creating a velocity of up to 24 feet per second. As it approaches the bearing point, the mist is then passed through a series of mist metering fittings that allow an engineered amount of oil to enter the bearings. Depending on the fitting type used, the "dry mist" or partially reclassified mist then "envelops" and "wets" the entire bearing surface area with a thin lubricant film while imparting a partial positive pressure within the bearing housing that works to prevent any contamination influx.

In this original design, the mist was allowed to vent to atmosphere. New system designs employ sloped lines to carry reclassified oil within the lines directly back to the reservoir, and more importantly, employ a closed loop design with drain legs and components designed to capture coalescent waste and eliminate open venting. An improved Vortex air chamber design to replace the old venturi design has resulted in a more efficient mist generation that can now carry over distances of 600 feet—three times that of the original system design—translating into the ability to lubricate substantially more points from a single mist generation unit. The coalesced oil return improvements have turned a once total loss system design into a much more effective partial recovery system design using less lubricant.

Oil mist is a centralized lubrication system that continuously produces, conveys and delivers mist lubrication to bearings and metal surfaces. Oil mist lubrication has been shown to significantly reduce the number of lubrication related bearing failures when compared to oil splash and grease lubrication. The successful application of oil mist has been documented in technical papers, trade journal articles, maintenance magazines and bearing maintenance catalogs.

A brief review of specific experience will prove interesting. For instance, in one large petrochemical plant, bearing failures in two similar units with a population of about 200 pumps each were compared. One had mist, the other had conventional [oil splash] lubrication. The unit on oil mist had about 85 percent fewer bearing failures.[4] Figure 10-50 is a graphic illustration of typical findings.

In a comprehensive research study, oil mist lubricated bearings were found to run cooler by about 10°C compared to oil sump lubricated bearings. The oil mist lubricated bearings also ran with about 25% less friction than oil sump lubricated bearings.[5] Also, since the air under pressure in the housing escapes through the housing enclosures or vents, the entrance of moisture and grit is retarded. In addition, oil mist lubrication continuously supplies only clean, fresh oil to the bearings. The two factors combine toward full life expectancy. Because the bearings require very little lubricant, the oil consumption is comparatively small.[6] Recall, again, that with "wet sump" (purge mist) the mist floats in the space above the liquid oil. In the "dry sump" (pure mist) application

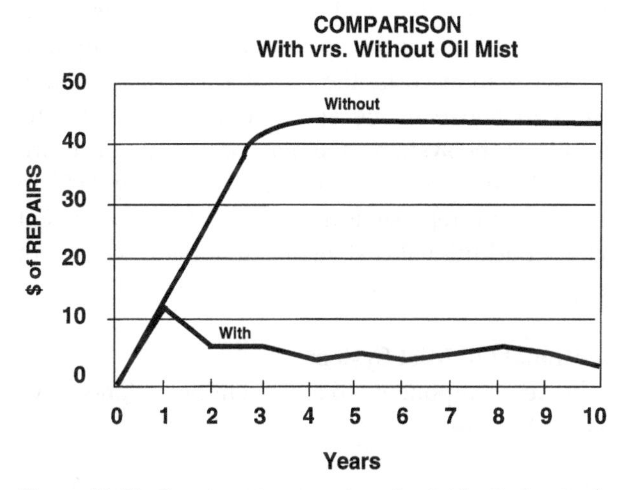

Figure 10-50. Repair cost comparison for 2 identical petrochemical facilities, with vs. without oil mist lubrication.

method, oil mist is routed as shown in Figure 10-51. Oil is applied in the form of tiny droplets carried by air. These tiny droplets coalesce and "plate out" on the bearing's rolling elements.

By the mid-1970s, sufficient experience had accrued to single out dry sump oil mist methods as best suited for plant-wide petrochemical complexes.

In addition to technical and failure statistics, plant owners also turn to cost reduction and return on investment to justify use of oil mist systems. One of the largest cost savings attributed to the use of oil mist lubrication is reduced equipment repair and lower maintenance cost.

Figure 10-26 shows what happened to pump repair costs attributable to bearing failures after oil mist was installed on a US refinery crude unit in early 1990. After the mist system was fully commissioned and brought online, these repair costs dropped by over 90%.

Similarly, Figure 10-27 shows annual pump bearing replacement costs for three process units in a US refinery. The data represent the average annual repair costs for the two-year periods immediately before and after oil mist was installed on these units. An overall 65% reduction in costs was measured.

There are other factors that add to the justification for use of oil mist. These include:

- Reduced lubricant consumption
- Greater manpower flexibility
- Reduced spare parts inventory
- Low mist system maintenance requirements
- Higher equipment availability

Oil mist systems are extremely reliable and have a fifteen to twenty year useful life without major overhaul. Oil mist systems can be installed with new projects or retrofitted to existing facilities. When savings are compared to total installation costs, the payback period normally calculates to between one and two years.[8] Given the 20-year life of the systems, the discounted rate of return (DCF) on the investment in oil mist is typically 50% to 100%, meaning it represents a very attractive investment project. (For additional cost justification data, see end of chapter.)

Conventional Oil Mist System

The key components of a conventional, "one-way" oil mist system are:

- *Central oil mist generator* and *oil supply tank*. (Figure 10-54)

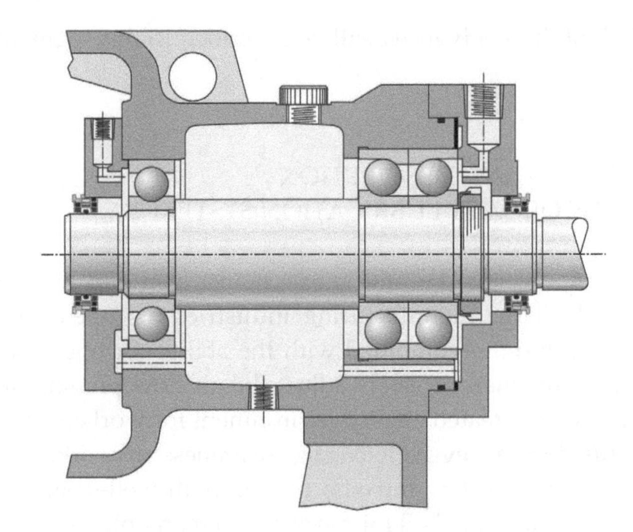

Figure 10-51. Dry sump oil mist applied between the bearing housing seal and bearing (per API-610). The mist is vented from the bottom (drain location) of the bearing housing.

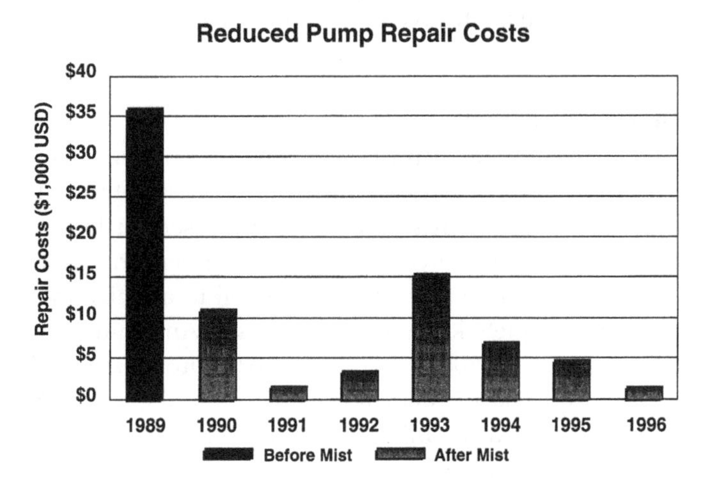

Figure 10-52. Pump repair cost statistics from a crude oil processing unit.

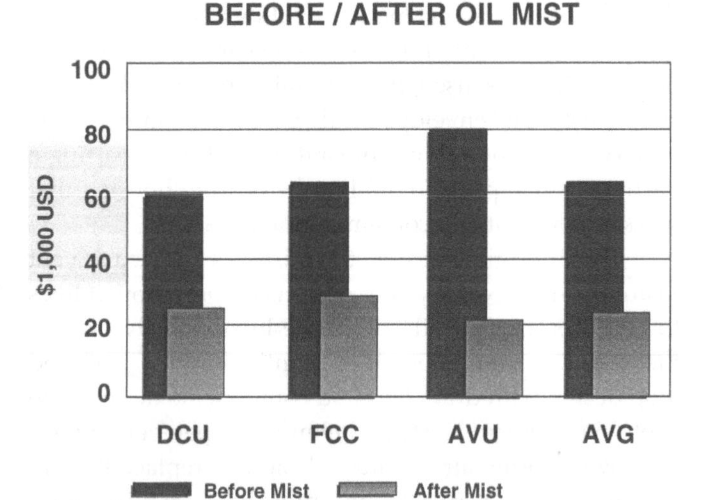

Figure 10-53. Annual pump bearing replacement costs for three process units with a US refinery.

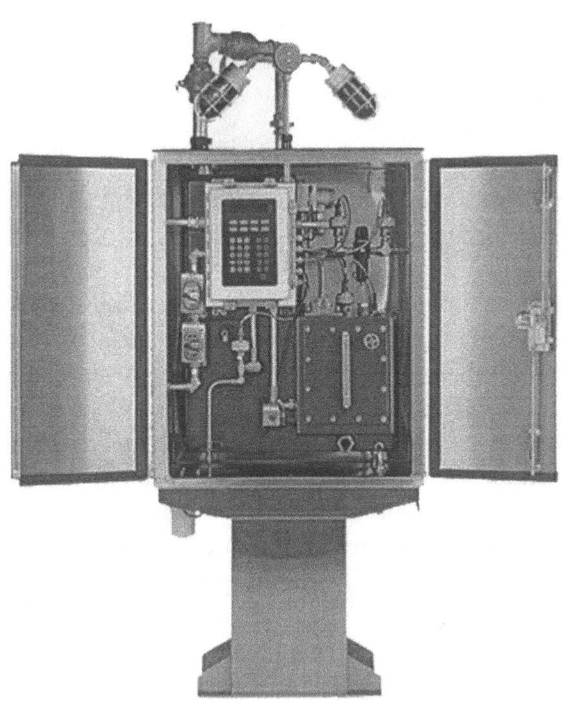

Figure 10-54. Central oil mist generator and supply tank. (Source: Lubrication Systems Company, Houston, Texas.)

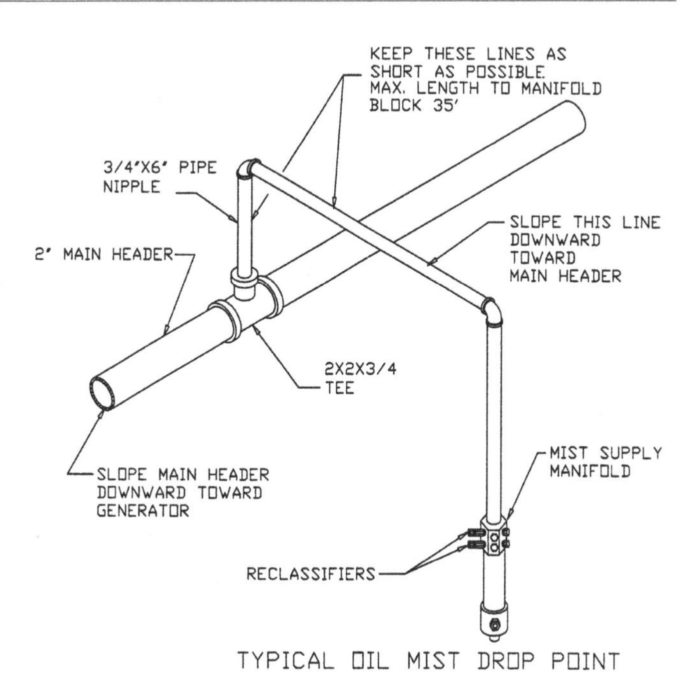

Figure 10-55. Oil mist distribution piping. (Source: Lubrication Systems Company, Houston, Texas.)

- *Distribution piping* to convey the mist. (Figure 10-55)
- *Collecting containers* at the equipment receiving the oil mist. (Figure 10-56)
- *Mist manifolds* to divide and direct the mist. (Figure 10-57)
- *Stainless steel tubing* to direct mist to each application point. (Figure 10-58, 10-60)
- *Reclassifiers* to measure and apply the mist. (Figure 10-55, 10-58)
- *Drain lines* to an oily water sewer or some type of collection container. (Figure 10-58, 10-59)

Each of these types of components existed in oil mist systems that proved successful in 1975, but significant design improvements to each have made the 1999 and 2008 systems even more effective, efficient, reliable and environmentally friendly.

New Central Mist Generator Design

The heart of the system is the generator (Figure 10-54) which utilizes the energy of compressed air, typically from the instrument air system, to atomize oil into micron-size particles. In modern, large-scale systems the mist generator is fully monitored and microprocessor controlled. Solid-state pressure and temperature transducers and level sensing devices have replaced the old-style electro-mechanical switches. Rather than using gauges, all monitored variables are displayed on demand by an alpha-numeric panel that not only shows typical gauge values but also provides messages describing the operating condition.

Figure 10-56. Automated drain leg assembly incorporating drain lines, collection container, and air-operated switch and pump assemblies. (Source: Lubrication Systems Company, Houston, Texas.)

The control panel is password-protected, meaning only those operators trained and authorized have the capability to set and adjust operating and alarm conditions.

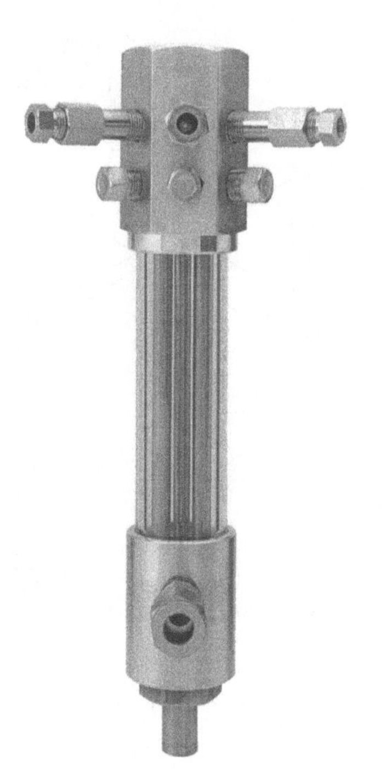

Figure 10-57. Oil mist manifold. (Source: Lubrication Systems Company, Houston, Texas.)

Therefore, the possibility for making well intended but incorrect adjustments is minimized.

This state-of-the-art central mist generator has the following properties:

- Meets Class 1, Division 2, Group B/C/D, standards.
- Continuously monitors at least eight (8) operating variables including mist density.
- Factory plus user set control ranges allow for establishing sequential and system specific alarm limits.
- Alarm save and recall function allows for efficient and effective troubleshooting.
- Independently set and monitored mist and regulated air pressure control ranges to safeguard against improper troubleshooting and alarm elimination.
- Independent 4 to 20 milli-amp current signal which allow for external monitoring of each operating variable.
- Large capacity, nine-gallon (35 liter), internal reservoir, constructed of stainless steel, aluminum, or painted carbon steel.

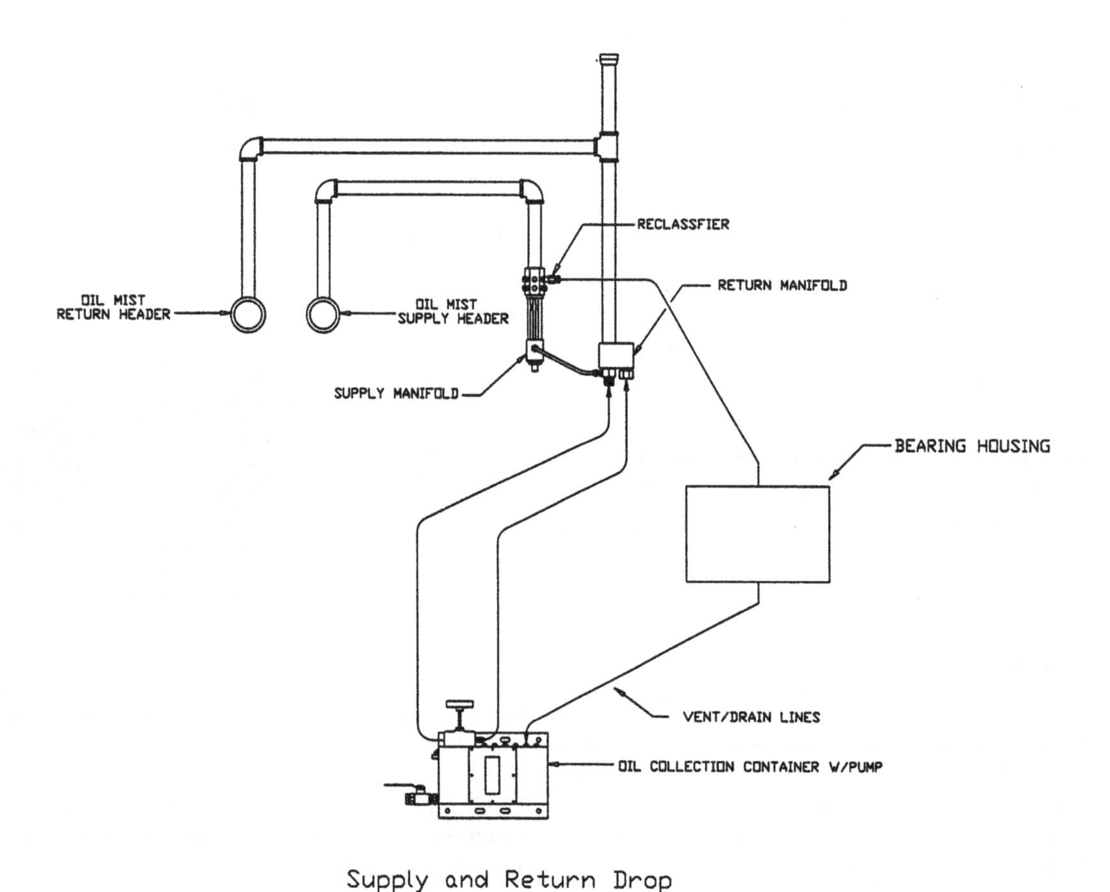

Figure 10-58. Closed-loop oil mist system schematic. (Source: Lubrication Systems Company, Houston, Texas.)

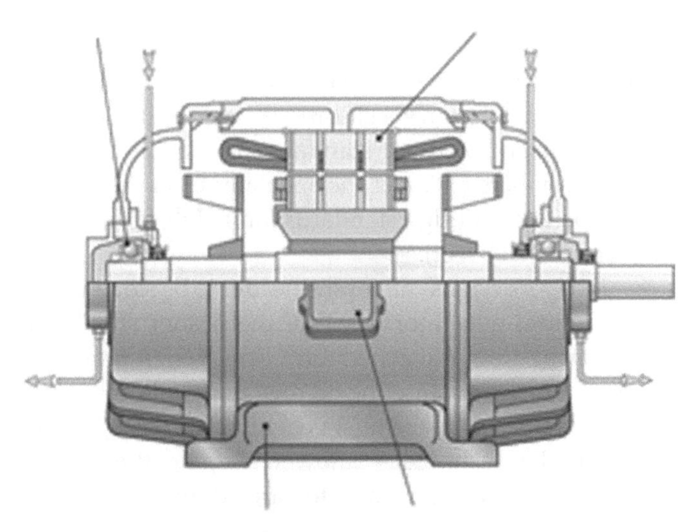

Figure 10-59. Oil mist lubrication applied to electric motor. Note that no liquid-oil sump exists in this dry-sump ("pure mist") application. (Source: AESSEAL plc, Rotherham, UK, and Knoxville, TN.)

Controls and Alarms

In addition to improved reliability, the microprocessor control of the new mist generator units provides for customizing operating set points and alarm limits to exact user requirements. High temperature cutout controls are factory set. The user sets operating conditions and alarm limits for all monitored variables, meaning the system can be optimized for that particular application. For example, users can tailor the unit to mist very heavy viscosity lubricants.

Fault conditions are annunciated locally in three ways:

1. External status lights switch from green to red.
2. Individual panel indicators located on the control pad change from green to red.
3. Alpha-numeric message appears on the display panel.

In addition to the local annunciation, remote alarm contacts are available for communication to control centers:

1. Common remote alarm contact (dry FORM-C).
2. Individual 4-20 mA current conditioning circuits for each of the eight operating functions.
3. RS-232 communication port.

Troubleshooting Assistance and Alarm Interlock

Another superior design feature of new central mist generators is the troubleshooting capability. These modules identify and distinguish the first alarm condition from all secondary alarms. The first alarm continues to be identified and annunciated on the alpha-numeric panel even when secondary alarms occur. This is extremely helpful when troubleshooting and searching for the root cause of a problem.

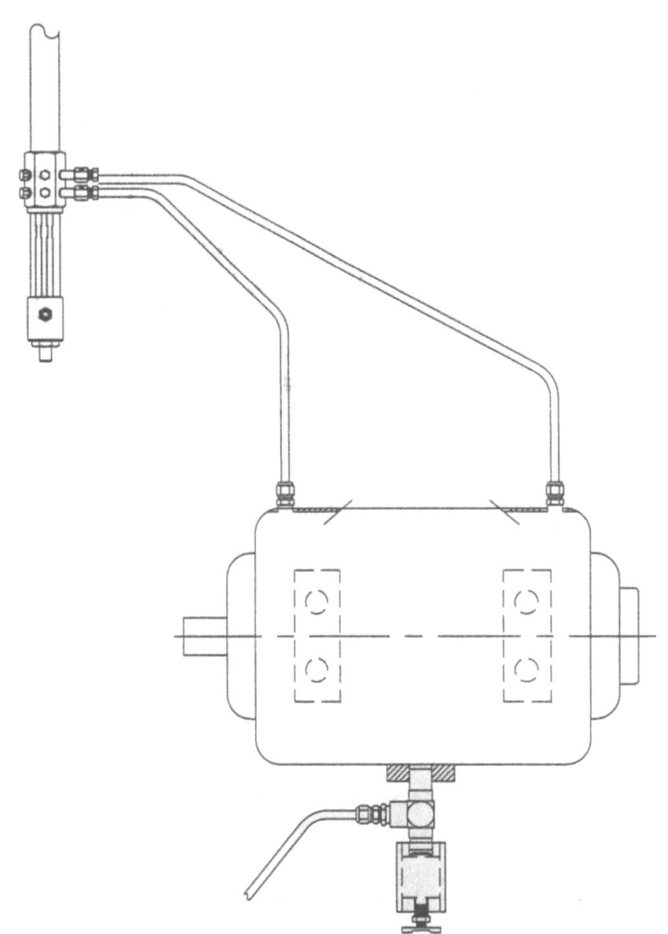

Figure 10-60. Mist manifold and stainless steel tubing directing oil mist to two application points in a bearing housing.

Alarms are also interlocked. For example, the low mist pressure alarm is linked, using control logic, with the air heater. If mist pressure falls to the low alarm setting, this being the first fault, the air heater control is then disabled and the air heater is de-energized. This results in a secondary fault, low air temperature, but because of the capture of the first fault the operator inspecting the unit will know how to best search for the root cause of the failure.

The interlocking control logic also ensures that an alarm condition cannot be avoided or ignored by improper adjustment of another related variable. For example, if header pressure increases because of plugging reclassifiers, a condition that can occur if paraffinic based lube oil is used in colder climates, on old style mist generators the high mist pressure alarm can be corrected by simply lowering the supply air pressure. The alarm is eliminated but the problem, plugged reclassifiers, remains. This false correction is not possible with the newer oil mist generators. Manually lowering the regulated air supply pressure will send this channel into fault condition and thus trigger another alarm. The operator or maintenance person must find the root cause of the problem and not simply be satisfied by changing the red status light to green.

Internal Reservoir Design

The internal reservoir of the new oil mist generators is compartmentalized and equipped with baffles. This design allows for efficient heating of the oil and elimination for the possibility of coking. The design also allows for settling of any contaminants that may be present in the lube oil. These reservoirs have a bottom that slopes to a low point where a bulls-eye sight glass allows for inspection of the oil. There is also a low point, valved drain port to draw off contaminants. The internal reservoirs of the new oil mist generators are much more than a simple rectangular shaped container as used on older units. A United States Patent[9] describes these features in greater detail.

Distribution Header System Design

Sloping and Distances

The oil mist produced in the central oil mist generator is transported throughout the process unit through header pipe. Typically this is 2-inch schedule 40 galvanized, threaded and coupled piping. In the hydrocarbon processing industry the header pipe is normally installed in overhead pipe racks. The header pipe must be installed without traps or sags as pressure in the header is only .050 bar (20 inches of water column).

The latest design specifications state that the oil mist can be transported using standard installation practices up to 180 meters (600 feet) horizontally from the central generator. Older design standards limited this run length to 60 meters (200 feet).

Today's designs call for the header to be sloped back to the central generator; none of the header should be sloped away from the generator.[10] This design promotes oil usage efficiency since the oil mist that coalesces in the header is returned to the mist generator for reuse (Figure 10-29). Older technology allowed sloping the header away from the mist generator towards drain legs.[11]

Automated Drain Legs

Where elevation changes do not allow for sloping back to the generator, drain legs (Figure 10-30) are installed. The drain leg prevents accumulation of oil in the main header because such accumulation of oil would block the flow of mist downstream of the trap. In drain legs utilizing the prior art, the collected coalesced lubricant has been manually drained, either to a sewer or a container which needed to be manually emptied, while the drain leg continuously vented oil mist to atmosphere. Today's distribution systems utilize automated drain leg assemblies that do not require manual operation and can be fully integrated into closed-loop systems.

These assemblies are equipped with an air-activated level switch and pump. They collect the coalesced oil and

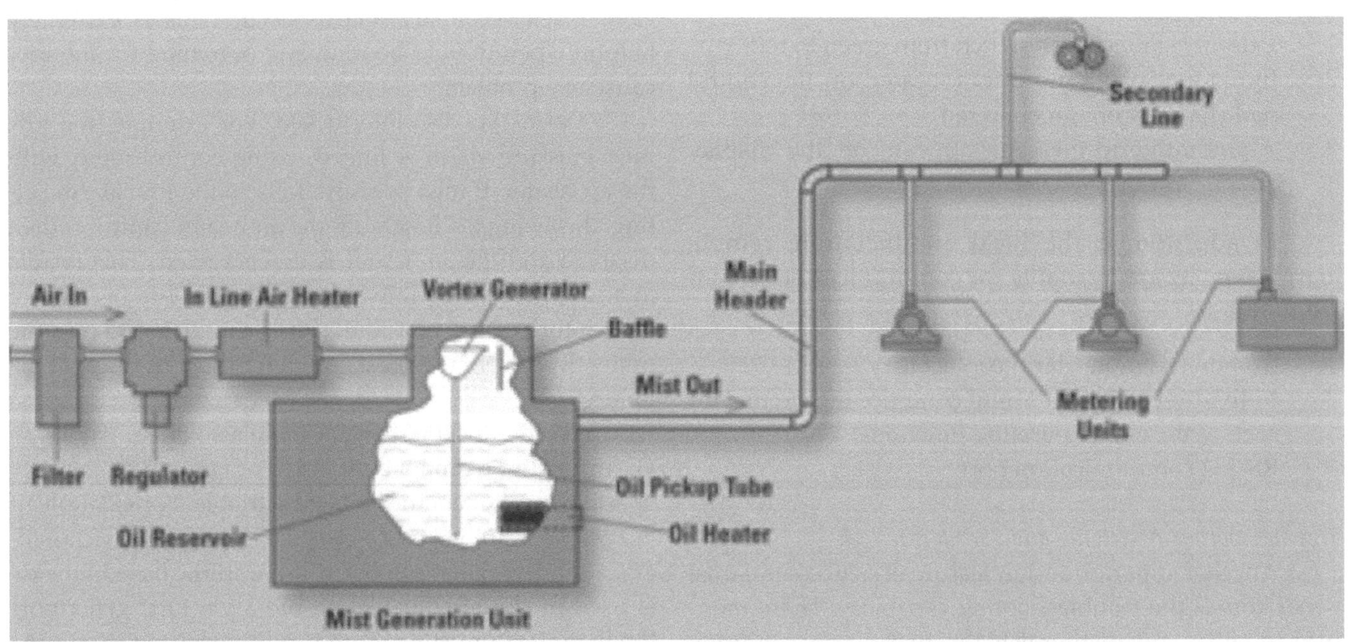

Figure 10-61. Typical Mist System Design. Source Ken Bannister for Lubrication Management & Technology magazine's "Anatomy of a Centralized Lubrication System" article series

automatically pump that oil overhead to a point in the distribution header that does slope back to the central generator. [12]

These automated drain leg assemblies can also be retrofitted to older, once-through mist systems, thus enhancing oil recovery and minimizing oil flow to sewers.

Mist Manifold

Above the equipment that is to receive oil mist, a "drop" is installed. Drop piping is normally 3/4-inch galvanized steel. The drop rises from the top of the 2-inch header so that liquid oil and any particulate contamination are not carried to the lubricated equipment. The drop terminates in a manifold assembly that divides the mist flow to the individual lube points that receive oil mist. Often the reclassifier or application fitting, an orifice metering device, is mounted in the manifold block. From the manifold block, stainless steel tubing is used to direct the mist to the application point.

In older mist systems the manifold was typically a rectangular metal block with ports drilled for mist flow. Most of these blocks were equipped with a snap acting valve for draining of the collected coalesced oil but the level of the collected oil was not visible. Modern mist manifolds (Figure 10-57) contain a high temperature glass viewing chamber that allows for visual monitoring of the level of collected, coalesced oil.[13]

Operators can see when oil needs to be drained. Draining is accomplished through an internally ported push valve that is channeled to a vent port on the side of the manifold. This port is tubed via a return manifold and into a collection container. When the manifold is drained, the flow of mist can be seen in the viewing chamber. Thus, mist flow is inspected without venting to atmosphere.

Closed-loop Oil Mist Systems

The performance and reliability of a properly designed, installed and maintained once-through oil mist system should not be subject to debate. These systems have proven their worth and have delivered the intended benefits. Their only shortcoming has been related to housekeeping and venting of excess mist in a world highly focused on and energized about environmental matters. Modern closed-loop oil mist system design and technology has been implemented since the early 1990s and is covered by patents.[14]

Return Header System

Parallel to the mist distribution header a return header pipe/system is installed (Figure 10-58). Sloping direction is the same as for the supply header but the return header drains into an oil supply tank/demisting vessel. The return header is constructed of 2 inch galvanized schedule 40 pipe threaded and coupled; the same as the main header.

The return header operates at atmospheric pressure. It is not under vacuum. The design of the demisting vessel ensures that only the flow of air from the central mist generator supplies the motive force, allowing excess oil mist to flow through the return header. Liquid oil travels, by gravity, back to the demisting vessel.

Pure Mist Application per API-610

Placement of reclassifiers can remain in the manifold when closed-loop technology is adapted to existing systems and equipment. Pumps purchased to API-610 Standards, Eighth Edition and later, will have bearing housings equipped with 1/4-inch NPS connections on the housing end covers. See Figures 10-59 (electric motor) and 10-60 (general) for details.

This design is compatible with the closed-loop design as mist is directed through the bearing and the tendency for mist to flow out through the seals is minimized. The flow of mist is continuous and vents through the bottom drain port of the bearing housing, as does coalesced oil.

Purge Mist Application

To eliminate mist system venting with purge mist/wet sump applications, a unique purge mist vent/fill assembly, Figures 10-62 and 10-63, has been developed. It is easily affixed to bearing housings. It eliminates uncontrolled venting, oil accumulation on and around the equipment receiving purge oil mist and problems associated with re-filling reservoirs with oil.[15]

The device, Figure 10-64, provides clog-free mist flow into and from the cavity receiving purge mist. It incorporates a screw on/off cap with internal porting into the mist penetration tube. This porting is designed so that when the cap is removed mist is not flowing into the housing and creating backpressure that causes problems when adding oil. The 38-mm (1-1/2-inch) wide funnel mouth makes for easy, spill-free addition of oil.

The internal porting allows for the controlled venting of escaping purge mist through tubing into the companion oil level sight and constant level oiler assemblies. The oil level sight assembly protects against overfilling from both coalesced oil mist and liquid oil addition and provides for visual inspection of the oil level in the bearing housing.

The constant level oiler assembly (Figure 10-62) also should reduce the risk of overfilling and will add oil if oil

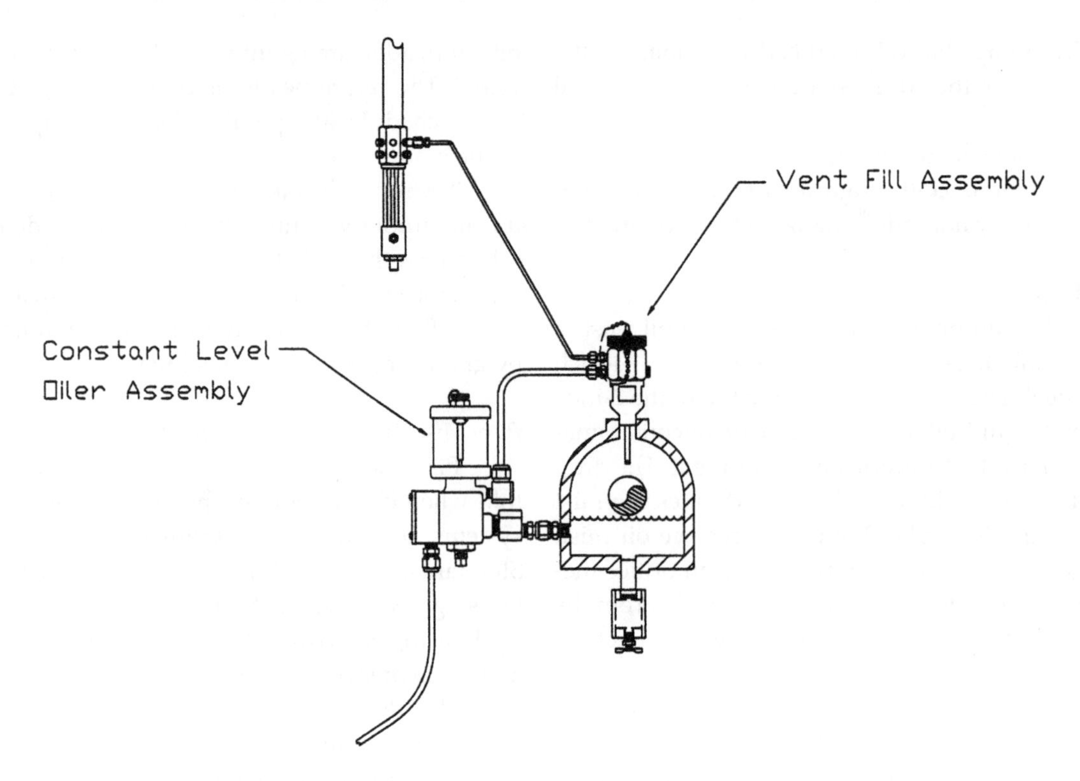

Figure 10-62. Purge mist application using constant level oiler assembly. (Source: Lubrication Systems Company, Houston, Texas.)

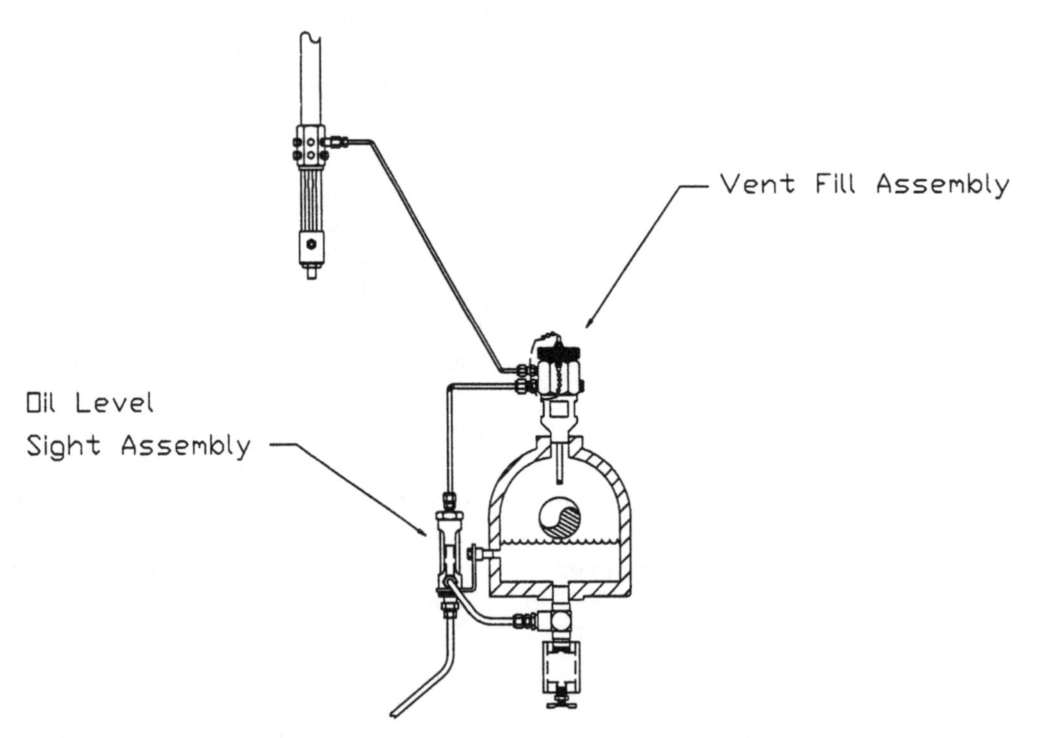

Figure 10-63. Purge mist (wet sump) application using oil level sight glass assembly. (Source: Lubrication Systems Company, Houston, Texas.)

is lost, for example, through seals.[16]

With these newer devices excess oil mist and overflow oil is directed to the collection container. Thus, even with purge mist, venting at the equipment and oil drainage to base plates and foundations is eliminated with today's technology.

Oil Collection Container

This recently commercialized container (Figure 10-59) is mounted to the equipment foundation. Oil mist plus coalesced oil flow into the container. An internal overflow tube with liquid seal prevents the container from filling with oil if it is not emptied. If the container overfilled, oil would rise into the bearing housing and block the flow of mist. The excess mist from the bearing housing that does not coalesce in the container travels through the container and tubing to the overhead return header. The container is also equipped with a manually operated pump that pushes collected liquid oil through piping into the overhead return header.[17]

The 3.8-liter (1-gallon) capacity of the container means it needs to be emptied only once per month. Because the container needs to be evacuated infrequently, incorporating automated, air operated level controlled pumps has not been considered cost effective. In fact, the monthly interval is significantly longer than that required for emptying the traditional sight bottle located under dry sump mist applications.

Return Drop

The oil collection container is connected via stainless steel tubing to the return manifold assembly. Internal porting with a check valve accommodates the continuous flow of mist to the return header and the intermittent pumping of liquid oil to the same return header. The vertical return line is not simply a length of pipe but rather a tube within a pipe. This arrangement allows for efficient field installation and provides for a more compact, less cumbersome piping arrangement. This design also allows for the simultaneous pumping of oil and the continuous flow of mist without creating blockage of mist flow or unwanted system backpressure.[18]

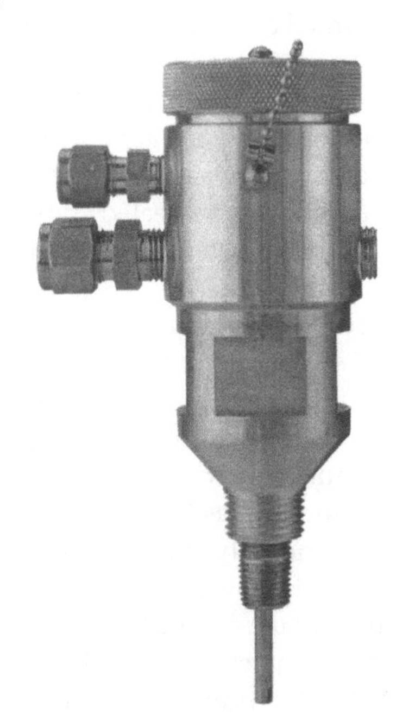

Figure 10-64. Purge mist vent-fill assembly. (Source: Lubrication Systems Company, Houston, Texas.)

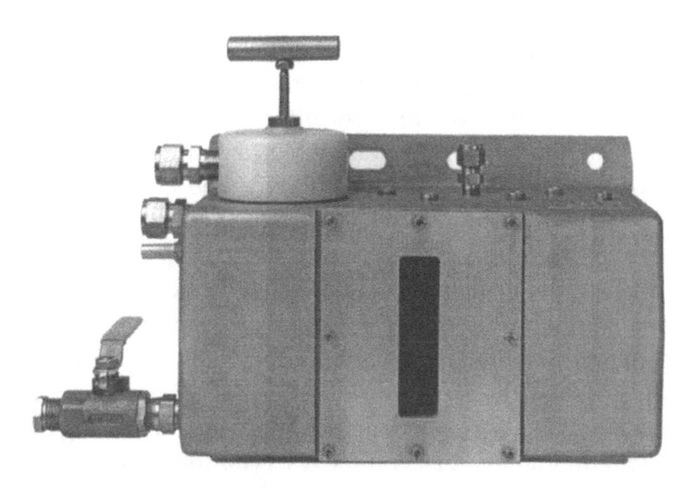

Figure 10-65. Oil collection container with manually operated pump. (Source: Lubrication Systems Company, Houston, Texas.)

Demisting Oil Supply Vessel

The 2-inch return header slopes back to and is connected into a central demisting oil supply vessel. The liquid oil drains back to this vessel. The oil mist that reaches the vessel, the very small mist particles that have traversed the entire supply piping network, gone through the bearing cavities and collection containers plus the return header system, must now be recaptured. This is accomplished by the electric motor driven, rotating filter located in the top of the demisting vessel, Figure 10-66. The rotating internal element captures the returned mist and coalesces the small particles into large droplets that then fall into the oil below. Clean, essentially hydrocarbon-free air, vents from the raised exhaust pipe.[19]

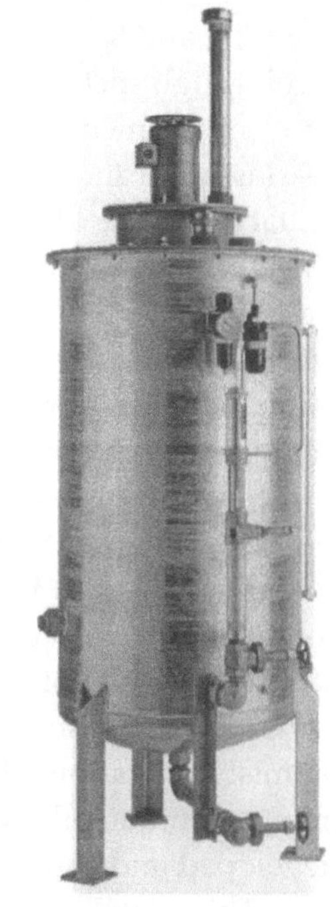

Figure 10-66. Demisting oil supply vessels as used in a closed-loop mist system. (Source: Lubrication Systems Company, Houston, Texas.)

The spinning filter has been designed so as not to "pull a vacuum" on the return header system. If it did one of the benefits of oil mist lubrication, maintaining positive pressure in bearing housings, would be negated. The filter will operate over a wide range of ambient conditions. It does not suffer the inherent problems of changing differential pressure across the filter media encountered with static filter elements. In fact, the filter media are by-passed if there is a failure of the electric motor, meaning the unit will fail in the safe mode.

The demisting vessel also acts as the oil supply tank for the central oil mist generator. It is common practice to locate the generator and demisting vessel adjacent to one another. This makes for efficient connection to the common utilities, compressed air source and electrical supply. Also, the oil supply line is kept short. The oil from the demisting vessel is pumped, on demand, through a filter and into the mist console reservoir for reuse.

Since it is expected that over 95% of the oil is recycled, superior performing, higher cost synthetic lubricants can now be easily justified for use in mist systems. The oil in the demisting vessel should be analyzed periodically to assess its quality. The design of the demisting vessel allows for piping to an oil purifier for on-line reconditioning of the mist oil.

Figure 10-67. Demisting system (closed loop) for the textile industry. (Source: Lubrication Systems Company, Houston, Texas.)

Demisting System for the Textile Industry

The manufacture of man-made synthetic fiber from polymers such as nylon and PET involves the melting, extruding, cooling, orienting, winding and sometimes crimping and/or chopping of the fibers. These processes involve high speed rotating equipment and hot, often humid, environments. There is also a strict requirement that bearing lubricants do not leak, drip or in any way come in contact with the fiber. The aggressive environment combined with the high speeds and need for reliability have led fiber manufacturers to investigate and apply oil mist lubrication decades ago.

The main hurdle that kept oil mist from penetrating this market was the potential for escaping stray mist while avoiding recycling systems that depended on negative pressure to promote the return of the oil mist. Negative pressure in bearing housings brings external contaminants such as water vapor into the system, thus offsetting many of the benefits delivered by positive pressure mist systems. In response to the desire to use oil mist lubrication while meeting the operating requirements a unique oil mist demisting filter technology was developed and commercialized.

This unit, Figure 10-67, utilizes an air blower to circulate returned oil mist through a coalescing filter. The design insures that the blower does not create a negative pressure in the return oil mist header. That header operates at atmospheric pressure. The volume of air that exhausts from the system is equal to the volume of air introduced by the mist generator.

Oil mist that has coalesced to liquid drains by gravity flow back to the reservoir of the demisting vessel. The oil mist that flows back to the unit is coalesced in the filter and the droplets created fall to the oil in the reservoir. The returned and captured oil is pumped on demand into the mist generator for reuse. This re-circulation technology has proven to be extremely effective. The exhaust is clean and these systems are operating in closed-room environments. Systems are in place lubricating crimper bearings and bearings on heated spindle rollers. The inventor has applied for patents on this technology.[20]

New Applications for Oil Mist

Rotary Lobe Blowers

This type of blower is often used in air systems for conveying solid materials in flake and pellet form. It is not uncommon to find these blowers in polymer plants producing plastics such as polyethylene, polypropylene, PVC and polystyrene. The blowers are part of the pellet conveying systems moving product from one step in the manufacturing process to another and to final load out and packaging. In this service the blowers are subject to intermittent operation. It has been reported that many users operate blowers at maximum speed and temperature. A reliability study by a major end-user of this type blower found two of the biggest contributors to premature bearing failure to be loss of lubrication and contaminated lubricant.[21] One of the steps taken to improve the reliability of their blowers was to convert from splash lubrication to oil mist.

Both purge and pure mist application are used; purge on the gear end of the machine, pure (dry sump) on the drive end. Closed-loop oil mist systems are being utilized to ensure that no stray mist escapes so that product contamination is avoided. The user now states, "It is recommended that oil mist be the first choice for bearing lubrication because it provides the bearings with a continuous supply of cool, clean oil even with the equipment in stand-by mode."[22]

In 1996, one major polyethylene plant in the United States reported 56 rotary lobe blowers being lubricated with oil mist. At this facility, failure rates have been reduced by over 95%.

Defibrator Displacement Press

This press is a device used in the pulp and paper industry. The equipment removes water from pulp by mechanical means. The operation requires large amounts of applied energy to separate the water from the fibers.[23] The unit contains an auger to feed pulp between rotating press rolls. Under heat and pressure the pulp is squeezed and the de-watered pulp is ready for the next stage of the process.

The rolls are supported by large bearings that turn at relatively low speeds. The process exerts much force to the bearings and this along with the high temperature and aggressive environment make the achievement of extended bearing life a challenge. Typical lubrication has included special greases to withstand the temperature and humidity. Also, oil bath and circulating oil systems are used. Recently a progressive pulp mill in Canada converted their oil lubricated bearings to oil mist and has found remarkable success.

The oil used in the application is a heavy grade ISO VG 460. The mist generator is equipped with both air and oil heaters to facilitate the production of oil mist. Also, because of the criticality of the application the system uses a mist monitor to ensure mist density stays within design limits.

On each of the large bearings five spray reclassifiers provide the mist lubrication. A bottom drain in the bearing housing is used to channel the mist away from the equipment thus ensuring a clean application. One of the first observable effects after the conversion to oil mist lubrication was a 20% reduction in bearing operating temperature. The application of mist to these bearings has resulted in extended bearing life while proving to be extremely reliable and trouble free.

Portable Mist Density Monitor

There is a desire by many users of oil mist systems to have the capability to test for the quality of the oil mist at various points throughout the mist system. They are confident that the fully monitored central oil mist generator is operating properly. However, having the ability to cost-effectively and quantifiably measure mist quality downstream of the generator would add to system effectiveness and reliability.

One particular model of a portable mist density measuring device is designed to utilize the push button drain valve of the mist manifold described above as a sample point. The sample of oil mist is captured in a test chamber and the relative density of that mist is then measured on a 0 to 100% scale. Optimization plans for the unit included computerization for storage and downloading of the sample information for trend analysis and statistical evaluation. Commercialization was achieved in 1999.

Miniaturized, Closed-loop Lubricator

Marketplace feedback indicates there is a need for a reliable, small lubricator that is environmentally clean (no emissions) and delivers improved machinery reliability. Such a device (Figure 10-68) is based on proprietary, circulating, closed-loop mist technology.

Operating cost of the unit is low and reliability is high because the lubrication and re-circulation is achieved with the use of compressed air. No other util-

Figure 10-68. Miniaturized closed-loop oil mist system, or "single-machine lubricator." (Source: Lubrication Systems Company, Houston, Texas.)

ities are required. The unit is safe and can be used in hazardous areas. The unit is designed for the lubrication of from one to five lubrication points. It is equipped with an air pressure regulator and filter separator. A mist pressure gauge monitors pressure in the lubrication delivery piping and tubing. An oil level viewing window allows operations personnel to conveniently see the oil level in the reservoir. Depending on losses from the bearing housings receiving lubricant, the unit is capable of re-circulating over 95% of the oil lubrication.

COST-JUSTIYING OIL MIST IN MODERN OIL REFINERIES*

Is oil mist really cost-justified? Why do so few European oil refineries use plant-wide oil mist systems? These questions were addressed in an article published in 1990, where the author noted "together with an appropriate amount of a suitable state-of-the-art synthetic lubricant, this low-cost retrofit [referring to a modern magnetic seal and a plugged vent instead of the customarily open-to-atmosphere bearing housing vent port] may extend bearing life to the point where oil mist lubrication is no longer economically attractive."

This statement is as true in 2008 as it was in 1990. But the context is of great importance, because it referred to the small but diligent group of equipment users that insist on correct pump installation, operation, and maintenance. For them, oil mist lubrication may indeed not be justifiable. These are the relatively few facilities that expertly apply the right lubricant to a particular bearing and change the oil periodically—sometimes as often as four times a year! Using a highly trained workforce and holding them accountable, industry in many European countries apparently does not experience enough bearing failures to justify additional (incremental) failure avoidance through the use of plant-wide oil mist systems.

In stark contrast, virtually all U.S. facilities will likely benefit from oil mist lubrication. There are many elements that contribute to this remarkable difference. The European mindset appears more oriented towards taking the necessary time to do things right, whereas on this side of the Atlantic the mere speed with which a repair is effected is often given more weight. In Europe, the administrative person in charge will not (usually) interfere with the experience-based judgment of a highly qualified craftsperson. If periodic oil changes are needed, they will be performed. If bearing installation tools are needed, these will be procured and properly used. Piping will be installed with proper fits, and the list just goes on.

Regrettably, the same approach is rarely practiced in the United States. All too often the person in charge may insist on quick work, or will not allocate the time it takes to understand and remedy the underlying causes of failure. When our typical person in charge manages to quickly restore equipment to running condition, he or she will be elevated to higher status. If the quick fix attempt fails, blame for having guessed wrong can usually be shifted to others. Deviations from the original quality norm become the new norm and repeat failures are experienced. Repeat failures are the precursors of extreme failures. When extreme failures occur on lubricated equipment, there is usually a fire.

Whenever truly pertinent training and accountability are lacking, the cycle repeats itself. As just one example, in many plants no one can explain why and how a constant level lubricator works and why the widely used non-balanced versions no longer represent state-of-art accessories. Lube replenishing duties are often overlooked, or carried out wrongly. Inadequate slinger rings (see "black oil," elsewhere in this text) are used in many thousands of bearing housings, and defensive bickering is often preferred over sitting down and listening to solidly science and fact-based explanations. Again, it is in those widely prevailing circumstances that upgrading to oil mist would prove highly valuable and will be quite easy to cost-justify.

Oil mist provides more than just lubrication. Oil mist *lubrication* should always be mentioned together with the term oil mist *preservation*. Because oil mist inevitably preserves stand-by equipment, the resulting reliability increase deserves to be reflected in the cost justification, as should failure avoidance and the ensuing reduction in pump-related fires. Needless to say, this type of lubrication and preservation is even more easily justified in geographic regions with high humidity, or regions with blowing sand. Additional credits are derived from oil mist lubrication for equipment drivers. Indeed, every experienced user plant applies oil mist to electric motor bearings, covered later under electric motor lubrication (Figure 10-69). As was mentioned earlier in this chapter, oil mist is routed through all bearings in accordance with the guidelines set forth in the 8th and later editions of the API-610 Standard and shown earlier, in Figure 10-51.

These API standard clearly depict the optimized

*Based on Bloch-Ehlert ""Get the Facts on Oil Mist Lubrication," *Hydrocarbon Processing*, August 2008

Figure 10-69: Oil mist application on a pump-and-motor set, with routing per API-610. (Source: Lubrication Systems Company, Inc., Houston, Texas.)

through-flow method that has now been in use at some of the world's most profitable facilities since the late 1960s. Numerous papers and articles have documented this fact. Engineers from user companies, among them some of the largest multi-national refiners and petrochemical companies, have freely shared their highly favorable experiences. Reliability professionals at these facilities are in the business of keeping plants running. At the same time, they have been tasked with finding cost-effective ways of extending and optimizing equipment uptime. Optimizing uptime does not mean adding maintenance cost and, in fact, implies maintenance cost reductions. The final outcome and ultimate test of a best-of-class facility has been, and will continue to be, lowest possible life cycle cost of all assets. In many places, oil mist lubrication has aided in meeting these test criteria.

Recent statistics are useful. Major suppliers of oil mist systems have furnished monetary data on the estimated overall economic performance for various plants. These suppliers can provide the data expressed as DCF return and payback period. One supplier divided the information into broad categories of user plants, including refineries, petrochemicals and polymers or, perhaps, metals processing. Knowledgeable vendors can also provide details on the additional benefits of the technology. There are, for instance, significant benefits derived from using oil mist for both indoor as well as outdoor storage protection (Figure 10-70), and even the "mothballing" of entire plants. (See also Chapter 19—Lube Oil Contamination, Filtration and On-stream Purification Coontrol).

Figure 10-70. Outdoor storage of equipment before its installation at a grass-rootsfacility in Thailand (Source: Lubrication Systems Company, Houston, TX

While oil mist providers are generally not allowed to give the names and locations of plants that supplied relevant data, the reader should be alerted to the benefits calculated by some of these oil mist users. It should be noted that these data include numbers other can use in calculating cost justification and payback. Keep in mind that the oil mist provider may find ways to be more specific. That said, here is what the author and publishers are at liberty to share:

- As of 2008. plant-wide oil mist lubrication has been applied in over 100 refineries and chemical plants in dozens of different countries.
- Satisfied users include major big and small names.
- The investment made by these plants generates attractive returns and short payout periods, based upon improved reliability and reduced cost.

The demonstrated areas of improvement include, for example:

1. Reduced pump and electric motor bearing failures:
   — 80-90% reduction in pump bearing failures is typical
   — Electric motor bearing failures are often lowered by over 90%
   — Competent oil mist suppliers can provide data on:
      a. Bearing failures at a major refinery in Thailand
      b. A California refinery sharing its electric motor failure history
      c. Major olefins plant failure history

2. Reduced seal failure events:
   — Reduction of seal failures in the range of 30-50%.
   — One user reported average seal life doubled to 8 years.
   — Examples include:
      a. Bearing and seal experience of an oil mist user in California
      b. Seal life comparison from an offshore facility

3. Reduced failures rates of specialty equipment:
   — Oil mist application has shown excellent results in a variety of other equipment applications.
   — Rotary lobe blowers, chemical mixers, and cooling tower fan gearboxes are examples of more specialized applications with big payouts.
   — Examples include:
      a. A polymer processing equipment failure history
      b. Applicable experience with rotary blowers
      c. Highly favorable refinery cooling tower gearbox history

4. Results expressed as MTBR improvement for pumps, drivers, and other equipment:
   — One user reports improvement from three years before oil mist, to now nine years after oil mist.
   — Another user went from four years (before) to now almost eight years (after oil mist was introduced).
   — Detailed examples are available for:
      a. A refinery in a Pacific Rim country; it published seal life comparisons
      b. Highly favorable pump MTBR experience
      c. Similarly favorable steam turbine MTBR experience

5. Disclosure of pump maintenance costs, showing significant reductions. Percentages are given
   — One user reports a 40% reduction in all work orders for pump maintenance.
   — Others reported pump repair cost reduced by 60-80%. These included:
      a. An asphalt plant
      b. Experience at several refineries
      c. Dollar-cost reduction numbers provided by one refinery

6. Operations manpower to carry out lubrication tasks was reduced
   — User reporting 47% reduction in hours needed to complete lubrication-related tasks
   — User feedback from a Pacific Rim country supports the data

7. Lubricant consumption was reduced
   — 40% typical reduction due to more efficient application
   — One user reduced consumption 70% by applying recommended recovery steps.
   — Comparison of oil usage in several affiliated Pacific refineries is available

8. Reduced energy consumption is a fact
   a. 1-2% lower energy use demonstrated in several controlled tests
   b. Energy consumption study in a South American location has been published

9. Eliminated lost production incidents
   a. Specialty polymer producer estimated 7-8% runtime improvement.
   b. Refiner eliminated value of reduction of lost production incidents on crude oil unit.

The overall economic results for five refining applications have been published:

A. Western State Refinery
   c. Applied oil mist to Crude Unit, FCCUs and steam boiler area in 1999
   d. Experienced sharp reduction in pump maintenance cost
   e. Eliminated lost production incidents on Crude Unit
   f. DCF returns exceed 200%
   g. Payback achieved in less than one year.

B. Southern Plains (USA) Asphalt Plant
   h. One system for entire plant installed in 1997.
   i. Pump repair costs dropped 72%
   j. DCF return of 150%
   k. Payback in less than one year.

C. Overseas Refinery
   a. Installed systems throughout one plant in mid-1990s
   b. Compared performance with sister plant without oil mist.
   c. Results include doubling of MTBR for pumps and seals, cutting operating manpower in half and reducing lubricant consumption
   d. Estimated DCF return for converting other refinery to mist is 54%.
   e. Estimated payback is only 1.9 years.

D. Mid-coast U.S. Refinery
   a. Two systems installed in 1996 on crude processing units.
   b. Pump bearing repairs costs dropped 88%.
   c. DCF return of 70% based only on lower repair costs.
   d. Payback was achieved in 1.5 years.

E. Southern United States Refinery
   a. Three systems were installed in 1989.
   b. Pump repair costs reduced by 65-70%
   c. DCF returns of 75% achieved, based only on pump repair savings.
   d. Payback in 1.5 years.

Similarly, the overall economic results for three petrochemical applications are available:

A. Specialty Polymer Plant in a Western State in the U.S.
   a. Failure rate on rotating equipment was about every 6 months before oil mist.
   b. Failure rate dropped 98% after mist applied.
   c. Plant availability to manufacture polymer increased 5-7%.
   d. DCF return without including increased production, exceeded 400%
   e. Payout in less than six months.

B. Commodity Polymer Plant in Mid-south Region of the U.S.
   a. High rate of rotary lobe blower failures motivated mist investment.
   b. Blower maintenance cost reduced by 90% within two years.
   c. Resulting DCF return of 45% and payback period of two years.

C. Central Gulf Coast Olefin plant
   a. Compared pump failures between plant built in early 1980s with mist and one without built 10 years earlier.
   b. Pump bearing failures 90% lower with oil mist lubricated plant.
   c. DCF return, based only on lower bearing failures, of 75%.
   d. Payback in 1.5 years.

A cost justification example involving a plant with 600 pumps. Oil mist systems vary in size and scope; they range from small oil mist generators (OMGs) serving a single application to a pump, motor or gearbox, to large console-based OMGs. Such systems often include instruments and monitoring devices associated with expansive piping networks that serve 60 to 80 applications. As of 2008, a small OMG that provides mist to a single machine often costs less than $4,000. In comparison, a large OMG representing an average size system in HPI facilities serves 30 to 50 pumps and their drivers. A pump and its driver are called a "train." For a 40-train system, the (2008) cost was typically in the $250,000 range. A proposed plant-wide system serving 600 pump sets was offered at an installed ("turn-key") cost of $4,000,000.

The following illustrations show the justification compiled for a complete HPI facility, with a 600 pump & motor count. To cover the area and number of process units and equipment in such a plant would typically require installation of 12 to 14 large-scale oil mist systems. While the example cost justification is for the entire complex, it should be noted that the cost justification approach for a single operating unit would be virtually identical. The key input would reduce the number of equipment items and dollar amounts for each line item. All percentages, including the ultimate returns on investment, will remain largely unaffected.

Of course, the actual example used in this segment of the text does not imply that the cost and frequency of incident occurrences will be identical at every plant. We have seen some that were higher and some that were lower. On the other hand, experience shows our listings as items that are commonly used by HPI facilities in justifying oil mist systems, whether for an entire complex or a single operating unit. A large Gulf Coast HPI facility recently used these items to justify installation of 14 oil mist systems for less than a one year ROI (Return On Investment).

Overall failure reductions due to oil mist have been published by Shell Oil Company (in Figure 10-71) and, since mechanical seals benefit from fewer bearing failures, Figure 10-72 illustrates applicable credits.

Figure 10-73 reflects the published electric motor bearing failures at two different Shell Refineries in the United States. The number of failure events was drastically reduced at the facility with oil mist lubrication. Similarly, there were manpower efficiencies gained with dry sump ("pure") oil mist lubrication (Figure 10-74). Also, Figure 10-75 shows monetary savings due to cooling water not being needed with dry sump oil mist. A comparison is made with conventional sump lubrication and it was assumed here that conventionally lubricated pumps would require cooling water. However, while cooled bearing housings are still used occasionally, we should point out that no cooling water is needed with rolling element bearings and suitable grades of synthetic lubricants.

Fewer pump failures clearly equate to reductions in fire incidents (Figure 10-76). Moreover, there is value in protecting a plant's standby equipment and Figure 10-77 establishes an average value reported by competent users of oil mist for equipment preservation purposes. Figure 10-78 deals with lost opportunity costs related to production interruptions at a refinery and Figure 10-79 shows the total projected savings for a refinery with 600 pump sets. The calculation is based on informal feedback from a number of reliability professionals in the United States and some overseas locations. Regrettably and as is customary in the industry, these individuals and their respective employers do not want their names publicized. A good provider of large-scale oil mist systems will discuss or informally disclose who these customers are.

In essence, many companies have found investment in oil mist highly cost-justified, although their calculations were often based on lower pump maintenance costs alone. While that is noteworthy, savings in driver maintenance, lower operating manpower, reduced lubricant consumption and energy savings (see electric motor lubrication, Chapter 13) merit consideration as well. Once these are included, the results will make the case for oil mist lubrication even more strongly. The decision to include production credits is supported by industry experience and even the obvious value of reducing the frequency of fire incidents has monetary value. For it not to be factored into cost avoidance calculations

PUMP BEARING EXPERIENCE

*Two similar olefin plants each with 200 pumps

Without Oil Mist: Unit with 200 pumps experiences 45 failures/yr;
= 54 Month MTBR

Unit with 600 pumps experiences 135 failures/yr.
Assume 10% of all failures are lubrication-related

Savings with Oil Mist: Oil Mist eliminates 90% of all lubrication failures;
Average cost for API pump repair is $14,000.

10% of 135 = 13; 13 x 90% = 11; 11 @ $14,000 ea. = $154,000

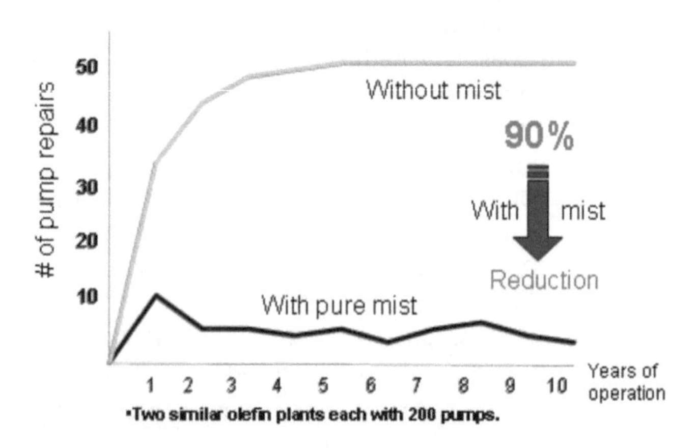

Figure 10-71. Failure Reductions Published by Shell Oil Company

would be an oversight.

Unless proven otherwise (and there are such rare instances in markets with low compensation rates for labor), plant-wide oil mist is cost-justified for both new plants and as a retrofit opportunity at existing plants. An all-inclusive and sound strategic technology package must be aimed at maximizing reliability and minimizing life cycle cost for rotating equipment. Such an analysis might quickly prove the true value of a plant-wide oil mist project.

PUMP SEAL EXPERIENCE

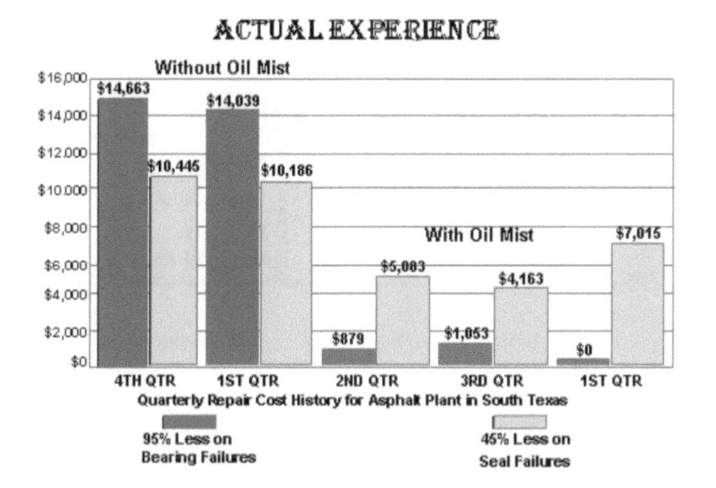

ACTUAL EXPERIENCE

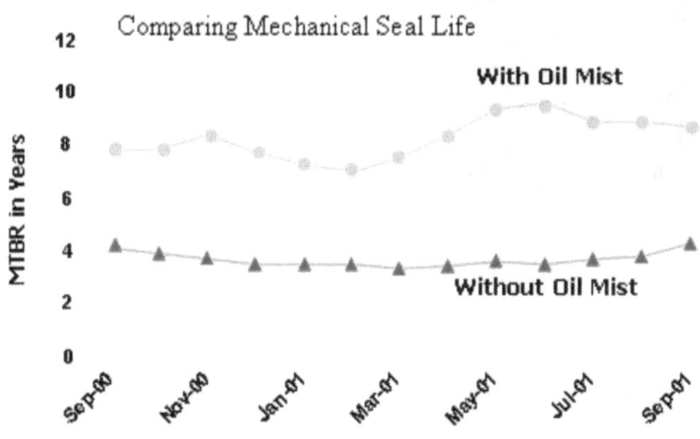

Two Refineries w/ 600 Pumps Each

Savings with oil mist on 600 pump & driver sets

60% of all pump failures are mechanical seal related.
Oil mist reduces these seal failures by 40%.
Assume $8,000 per seal and bearing repair.

135 x 60% = 81; 81 x 40% = 32 @ $8,000 ea. = $256,000

Figure 10-72. How Mechanical Seals Benefit from Fewer Bearing Failures

MOTOR BEARING EXPERIENCE

Motor Issues Cause Stress on Pump Seals and Bearings

Without Oil Mist
150 motors with 75 failures/3.5 years = 21 failures/yr.
600 motors equates to 84 failures/yr.
90% of all motor failures are lubrication related.

Savings with Oil Mist
Oil mist eliminates 80% of lubrication related failures.
Average motor repair cost is $4,000.

90% of 84 = 75; 75 x 80% = 60; 60 @ $4,000 ea.
= (Motors Only) $240,000

20% of the 84 motor failures cause pump seals to fail:

20% of 84 = 16 pumps; 16 @ $8,000 ea.
= (Stressed Pumps) $128,000

ACTUAL EXPERIENCE

3 ½ Year Study Comparing Electric Motor Bearing Life Two Shell Refineries

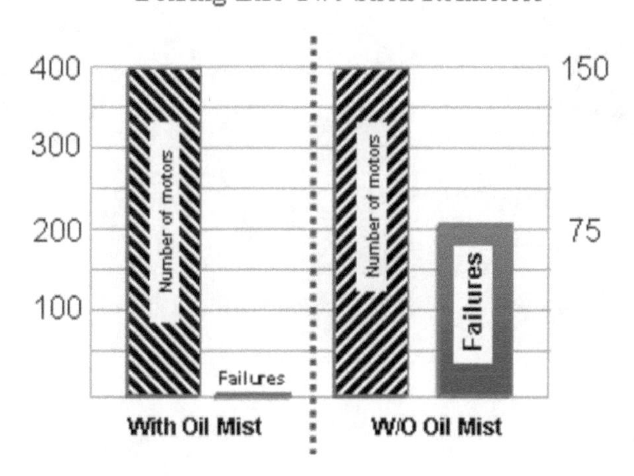

Figure 10-73. Published Findings from two Different Shell Oil Refineries in the U.S.

MANPOWER EFFICIENCY GAINS

Oil mist eliminates changing oil, checking levels, and re-greasing

Actual Experience

- 180 Oil Sumps
- Oil Changes: 2 per year
- Oil Fills/Checks: 4 per month
- Time Per Sump: 4.5 hours/year
- Cost: $67/hour
- Manpower Credit: $55,000/year
- .25 Hrs to re-grease mtrs. semi annual

Maintenance staff no longer changes oil or re-greases motors.

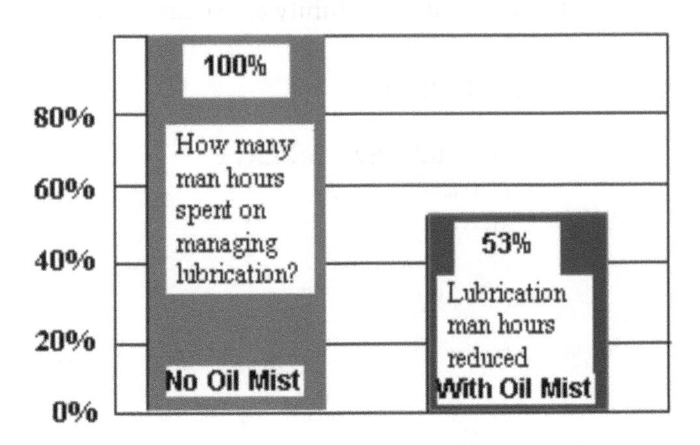

West Coast Refinery
Oil mist makes it possible to reallocte 3,000 man-hours per year to proactive failure avoidance and realiability improvement tasks.

Without oil mist, man-hours required are base-100%
800 oil sumps at 4.5 hrs/yr = 2,700 man-hours per year
600 greasing of motors at .25 hrs twice per year = 300 man-hours per year
Savings with Oil Mist
Oil PM avoided for pumps: 2.700 hrs x $67 x 40% = $85,023
Grease PM avoided for pumps: 300 hrs x $67/hr = $20,100

Figure 10-74. Manpower Efficiencies Gained with Oil Mist Lubrication

REMOVAL OF COOLING WATER

Cooling water on rolling element bearing housings is not needed with pure oil mist.

Without Oil Mist
Water treatment, consumption and system maintenance throughout the refinery
Removal from hundreds of pumps is estimated at $120,000/yr
Expect three pump failures/yr due to cooling water usage (!)
...Water causes corrosion of water jackets, bearing housing & base plate pedestal

Expenses Avoided with Oil Mist
Water treatment, consumption and associated maintenance = $120,000
Elimination of three pump failures @ $14,000 per pump = $42,000

Total Savings = $162,000

Figure 10-75. Cooling Water-related Savings with Pure Oil Mist Systems

AVOIDING POTENTIAL FIRE DAMAGE

It is routine for many oil companies to instasll oil mist systems in off-side bulk loading areas to prevent fires.

Without Oil Mist
Counting bearing & seal-related failures only
Assume for every 1000 pump failures there will be one significant fire
Assume $10 million for fire damage.

Assume $1,000,000 per day of lost production
Assume 12 days to return to full operation

Savings with Oil Mist
25 bearing related pump failures;
34 motor-related pump failures; together avoided 109 potential fires.
1000 failures/109 failures/yr = 1 fire possibilitiy every 9.2 years

$10 million/9.2 years =	Yearly Potential Fire Damage	$1,087,000
$12M/9.2 years =	Yearly Potential Value of Lost Production	$1,304,000

Figure 10-76.Avoided Pump Failures Equate to Avoided Fires

Without Oil Mist:

- Spared equipment failures of pumps, motors, turbines that caused minor production upset; assume $1,000,000/day loss for any given process unit
- Assume 5 days lead time for repair; eliminating 109 failures equates to 545 additional days/yr for production; 4.5 yr/MTBR leads to a failure every 1606 days for any given service

Lost Production Avoided With Oil Mist:
545 days / 1606 days = .33 additional production days/year

$1,000,000 x .33 days = $330,000

Items Not Included:
Spare Parts Inventory
Purchasing Cost
Warehousing Expense

Figure 10-77. Protecting Non-running Equipment with Oil Mist Saves Money

SUMMARY AND CONSIDERATION OF OIL MIST LUBRICANTS

Centralized oil-mist systems continue, for many leading process industries, to be the technically preferred approach for lubricating rotating equipment. Oil mist technology has kept pace with other developments in these industries, especially in the area of microprocessor controls and process monitoring. In addition, system design specifications have expanded, allowing oil mist systems to reach further. Moreover, today's distribution systems are more efficient because they allow coalesced oil to return to the central mist generator for reuse.

Components such as mist manifolds have been redesigned and improved and new devices such as vent/fill assemblies and automated drain legs have been developed, making today's systems both better and cleaner than defined and delivered by prior technology and installation practices. The invention and commercializa-

NON SPARED EQUIPMENT SAVINGS

When double pumping 100% to maximize production, equipment outage losses cannot be recovered.

Items Not Included:
Overtime Hours
Purchasing Cost
Inventory Cost
Warehouse Cost

Without Oil Mist
During peak production seasons each year some refiners operate the spare pumps in parallel with the main pumps so as to maximize production.

—Assume $1,000,000/day loss for any given process unit
—Assume 2.5 days required for emergency pump repair; 90% lube-related issues

Saved Production with Oil Mist
Throughput with spared pump in operation does not double unit throughput, but assume 20% increase in throughput adds $200,000/day in extra revenue
$200,000 x 2.5 days.. $500,000 x 90% = $ 450,000

Figure 10-78. Opportunity Costs Related to Production Interruptions at a Refinery

OPERATING EXPENSE SAVINGS	
• Pump Bearings	$ 154,000
• Pump Seals	$ 256,000
• Motors	$ 368,000
• Manpower	$ 105,123
• Oil Savings	$ 36,800
• Cooling Water	$ 162,000
Reliability Saving—Cost & Risk Avoidance	
• Fire Potential	$2,391,000
• Production	$ 330,000
• Non-Spared	$ 450,000
Total Savings:	$4,252,923

Figure 10-79. Summary of Savings Calculated for Pure Oil Mist Applied to 600 Refinery Pumps and Drivers

tion of closed-loop, circulating oil mist systems and related demisting equipment have positioned oil mist for greater use by process industries in an environmentally conscious world. Oil mist now meets the requirements for clean, emission-free operation while still delivering the improved reliability results expected of oil mist.

Recent advances in the areas of portable mist density measurement equipment and the introduction of miniaturized, closed-loop lubricators indicate that oil mist technology will continue to adapt to the needs of industry.

As regards lubricants, many formulations are suitable for use in modern oil mist systems. Plain diester base stocks as well as diester/PAO and mineral oil blends are available to the user. Special oil mist lubricants include Exxon's ENMIST EP, Table 10-2, which is formulated to resist premature condensation of oil ahead of the reclassifier and to minimize stray mist at the point of application. Laboratory tests with conventional oils have shown that, of the total mist delivered to the point of application, only about 75% is condensed as liquid oil, while the remainder escapes as stray mist. By contrast, tests with ENMIST EP have shown that 90% of the total mist delivered typically is condensed as liquid oil. This high delivered/stray mist ratio is achieved with special additives that reduce the number of oil particles too small to be condensed, or reclassified, at the point of application. The formulation of ENMIST EP is carefully balanced to ensure that the additives do not severely reduce the oil-to-air ratio in the oil mist system, which

Table 10-2. Typical inspections for Exxon's ENMIST *EP oil mist lubricant.*

Typical Inspections Grade	100	150	220	320	460
ISO viscosity grade	100	150	220	320	460
Gravity, °API	26.5	25.1	23.6	22.6	21.3
specific @ 15.6°C (60°F)	.8956	.9036	.9123	.9182	.9260
Viscosity,					
cSt @ 40°C	100	151	225	322	442
cSt @ 100°C	10.6	13.4	16.6	21.0	25.8
SUS @ 100°F	524	799	1203	1733	2393
SUS @ 210°F	62.4	73.3	86.7	105.9	127.9
Viscosity index	86	80	72	74	76
Pour point, °C	-24	-24	-24	-22	-21
°F	-11	-11	-11	-8	-6
Flash point, °C	212	217	222	227	232
°F	414	423	432	441	450
Rust test, ASTM D 665 A & B	Pass	Pass	Pass	Pass	Pass
Timken OK load,					
kg	27	27	27	27	27
lb	60	60	60	60	60
Four-ball wear test ASTM D 2266, 40 kg, 1200 rpm, 75°C, 1 hr, scar diameter, mm	0.48	0.46	0.45	0.44	0.44
Copper strip corrosion ASTM D 130 3 hrs@ 100°C (212°F), rating	1B	1B	1B	1B	1B
Oxidation stability, USS S-200 312 hrs @ 121°C (250°F), viscosity @ 99°C (210°F), % increase	3.4	4.3	5.4	9.5	13.6

might prevent adequate lubrication of the machine elements.

ENMIST EP has outstanding oxidation stability for better deposit control under high-temperature operating conditions, which helps reduce wear rates and extend equipment life. Its low pour points reduce the risk of wax plugging at the reclassifier, which could otherwise cause premature equipment failure, particularly in dry-sump applications.

ENMIST EP is fortified with anti-wear and EP agents that provide highly effective protection, especially for heavily loaded bearings and gears, such as those used in the metal rolling industry.

ENMIST EP provides dependable performance in the oil mist lubrication systems of major manufacturers such as Alemite, Bijur, Lubrication Systems Company, Norgren, and Trabon. It is also suitable for oil/air systems. Care should be taken to use ENMIST EP in accordance with the viscosity-temperature recommendations of the equipment manufacturer.

LUBRICATION DELIVERY SYSTEM PUMPS—MECHANICAL AND PNEUMATIC ACTIVATED TYPES

In this chapter we have addressed all variations of lubricant distribution delivery systems available. With exception to the mist, air/oil, and single point lubricating devices—that all utilize a unique integral lubricant pumping unit—centralized lubrication system designers have a variety of lubricant pumping options to choose from.

With the distribution delivery system design finalized, the next step for the system designer is to review the lubricated machine design and customer cost constraints to select a suitable lubricant pump. To narrow the selection process, the designer must first answer some basic questions:

1. Does the design budget allow for the cost of a fully automated pumping system, or is the design

restricted to much less expensive manual actuated pump that be upgraded or automated at a later date?

2. Does the machine have an integral lubricant reservoir the pump can be mounted in or on, or does the pump require its own lubricant reservoir?
3. Does the equipment have a mechanical power takeoff point, hydraulic or pneumatic power source available?
4. Is an electrical power source available?

Mechanical Powered Pump Units

If no electrical power source is available, the designer has no choice but to use a mechanically actuated pump. The pump design will usually employ a positive displacement piston whose output delivery can be adjusted by restricting the length of the piston stroke. For most manual operated pumps, a lever arm is mechanically connected to a cam that moves a single acting piston pump back and forth (some designs use a spring returned pump) with lubricant fed from an attached reservoir. The pump is actuated by manually moving the lever arm in a back and forth arc motion drawing lubricant into the piston chamber that is in turn pumped into the distribution system through an internal check valve.

If reciprocating or rotary machine motion is available, the lever arm of the manual pump can be replaced with a power takeoff pitman arm linkage attached to the motion device. Figure 10-80 shows a series progressive distribution system with a mechanical pump attached to a pitman arm arrangement attached to the end of the large diameter rotating machine shaft. The shaft attachment point is offset from the center to produce a reciprocating (up and down) motion of the arm that produces a rocking motion at the pump shaft, emulating the back and forth motion of the manual lever arm. By changing the length relationship of the pitman arm attachment point and arm length, the degree of arc will change and speed up, or slow down, the amount of pump strokes per hour. In Figure 10-81 you can see that the pump setting is incorrect, evident by the excessive grease being pumped out of the bearing seal.

Figure 10-80. Mechanically actuated grease pump

In the smaller single line resistance "oil only" type systems, a spring return piston is employed. A single push of the lever pushes lubricant out through the meter valves to the lube points. As the lubricant is apportioned, the line pressure dissipates and the spring return piston draws in the next lubricant charge.

Pneumatic Powered Pumps

Pumping lubricant to many points, over large distances, through large diameter lines is typical of large progressive and dual line type systems in heavy industries. These systems will typically employ a high-pressure pneumatic pump as shown in Figure 10-79.

Pneumatic barrel or drum pumps are unique in that they are designed to sit straight on top of a standard grease or oil drum, eliminating the need for a reservoir. The pump can deliver an output pump pressure of up to 70:1 airline input pressure. Pump design is again piston style, and is usually controlled by a stroke counter or by line pressure depending on the distribution system requirements.

In the advent years of centralized lubrication systems the automotive industry utilized "on board" vacuum operated lubricating oil pumps to automatically lubricate the suspension and steering components of luxury cars while they moved.

LUBRICATION CONTROLLERS AND SIGNAL DEVICES

To complete a centralized lubrication system design, the designer must tie the pump and delivery system together, and synchronize its operation with a combination of control and signal devices. The type of controller and signal devices used will depend on budget, level of system protection required, the type of pump and distribution system employed, and the ability of the host machine to interpret and act upon the control signals.

Controllers

A lubrication system controller is a device often described as the lubrication system's "brain." Most controllers are multi-function stand-alone devices housed in a control panel. The exceptions to this can be found in the single point lubricators (SPLs) and some of the smaller electro mechanical oil delivery pumps that have built-in circuit board style controllers that can be programmed through an LCD touch screen or a series of mechanical switches. These are usually simple control devices that turn the lubricating pump on or off, and control its rate of operation to speed up or slow down the rate of lubricant flow being pumped.

A controller's primary function is to turn a pump's power source on or off. In the case of a pneumatically powered pump the controller opens an electrically operated air solenoid valve to allow air into the piston and fire the pump. With electrically driven piston and gear pumps, the motor is electrically energized allowing the pumping action to commence. Once the controller receives a signal informing it that the pump has fired, a given time has elapsed, a lubricant line pressure has been achieved, or a distribution block cycle has been completed, the pump's power source is shut down until the next lubrication cycle commences. (The only exception to this is a recirculating oil system that is powered up on machine startup and runs continually until it is turned off when the machine is idle or shut down.) Lubrication cycles can be controlled by a counter that counts the number of machine or production operations, a programmed or set timer, or by a condition signal. E.g. amperage draw meter indicating energy draw increases on the machine system motor(s) due to a mechanical friction rise caused by lack of lubrication. (This is a popular control mechanism that measures the amperage of the conveyor drive and take-up motors to activate and deactivate the "power and free" conveyor chain and pin lubricators used in automotive assembly plants.).

A controller's secondary function is to take an emergency signal, shutdown the system and activate an alarm. This alarm can be a simple light or buzzer wired directly to a solenoid in the control panel, to the activation of an alarm email or work order in the CMMS/EAM maintenance management software—or both.

A controller's level of sophistication can range from a manual on/off device, to a simple count driven on/off device, all the way up to a very sophisticated programmable PLC/computerized PC device. The controller sophistication is usually underwritten by the lubrication system's consequence of failure where public safety is a concern. E.g. nuclear industry, chemical industry, etc., or where production losses are a major concern when the lubricated machinery is a production constraint, or machinery failure results in high downtime costs.

Signal Devices

Signal devices are used for both control and system protection and can be mechanical, hydraulic, pneumatic or electrical in design, and active or passive in operation.

Different delivery system designs will use controls differently. E.g. in most single or dual line system designs, the pump must continue to pump until the line pressure has reached an end-of-line line pressure of at least 800 psi allowing the injectors to fire. Once attained, a pressure signal switch informs the controller to shut off the pump, which in turn also reverses a flow valve allowing the lubricant to relieve back to the reservoir and allow the injectors to reset. A timer then counts operations or elapsed clock time and tells the controller to start the process all over again. In this system type, the pressure switch can also be coupled to a time out switch set to signal an alarm state if the system doesn't achieve its line pressure (due to a broken line or no lubricant) in a set time period.

Progressive divider systems can employ simple counters attached to the top piston in a primary delivery block. Once every outlet in the block has fired lubricant the top pin will have moved in and out the block once signaling one complete operation of the block. The counter is then linked up to the controller that actuates the pump based on the number of block cycles called for. Progressive blocks also utilize passive hydraulic blocked line indicators that actuate when hydraulic lockup occurs in a line blockage situation causing a pin to "pop" out visually indicating which line is blocked, speeding up the troubleshooting process what line is blocked. (see Figure 10-81.)

These are many control and protection devices available to the customer, question your suppliers to ensure you have the right system and right level of control you need.

MAINTAINING YOUR CENTRALIZED LUBRICATION DELIVERY SYSTEM

A centralized lubrication system is an integral and critical part of the host machine it lubricates and as such, must be maintained accordingly.

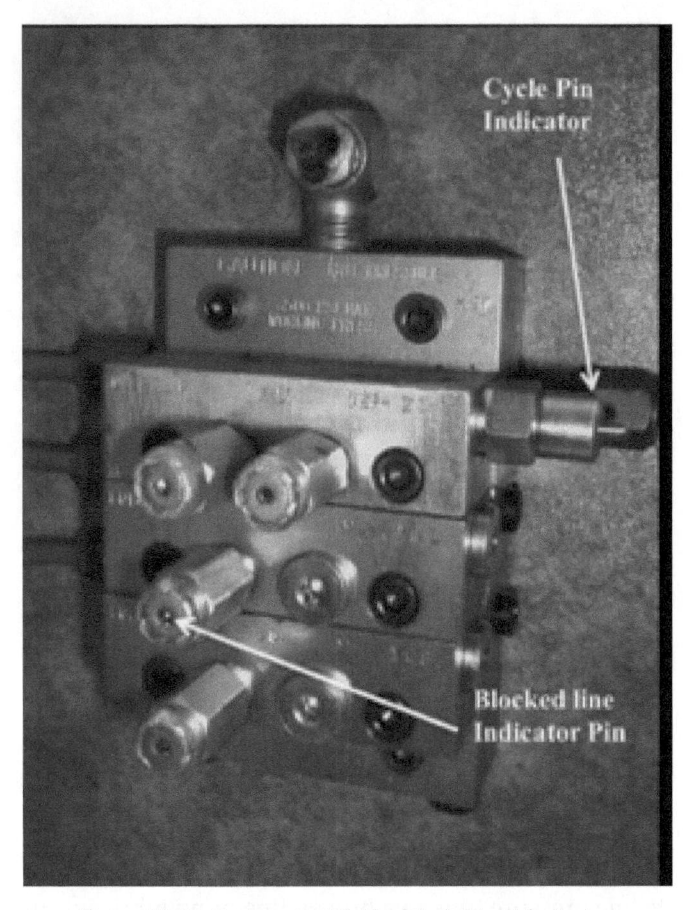

Figure 10-81. Progressive Divider Block Signal Indicators

Cleanliness and Contamination Control

The old adage "cleanliness is godliness" is a mantra to live by when dealing with lubricants and lubrication systems. Induced system contamination is a major causal factor for premature bearing and lubrication system wear. Great care must be taken when transferring lubricants into lubrication system reservoirs so as peripheral dirt is not introduced into the system and passed through to the bearing points. By their very design and nature, lubrication system components are not dirt tolerant as many of the systems employ fine tolerance pistons and spool valves in their pumps and delivery blocks and injectors similar to those found in the bearings they are to lubricate. Utilizing the following simple maintenance and set up tips will alleviate most contamination problems:

New Installation

- To avoid cross lubricant contamination, ensure the lubricant reservoir tag identifying the correct lubricant is the same lubricant about to be dispensed into the reservoir
- Clean the reservoir fill area and fill the pump reservoir with clean lubricant using a filter cart with a cleaned dispense nozzle and a clean dedicated transfer funnel in the case of oil, or a fully cleaned positive coupled air powered grease barrel pump when filling a grease reservoir
- Start up pump and purge lubricant through the pump before connecting the main delivery lines
- Clean all lube lines of swarf and debris before connecting the lines. Use a air powered "wad" cleaning system to shoot wadding through the lines to ensure no dirt is in the lines prior to start up
- Connect lines to cleaned dispensing blocks and purge with lubricant before connecting and purging the secondary lines prior to connecting top the bearing points.
- Check for system leaks and repair any leak immediately cleaning up all traces of leaked lubricant
- After system had run for a number of hours, perform a second leak check
- After system has run for a number of days, perform a lubrication check at each bearing point to ensure no lubricant has purged through the bearing. If lubricant is evident the system will require further calibration.

Existing Installation

- Set up a PM task to clean the lubrication pump and reservoir regularly
- To avoid cross lubricant contamination, ensure the lubricant reservoir tag identifying the correct lubricant is the same lubricant about to be dispensed into the reservoir
- Clean the reservoir fill area and fill the pump reservoir with clean lubricant using a filter cart with a cleaned dispense nozzle and a clean dedicated transfer funnel in the case of oil, or a fully cleaned positive coupled air powered grease barrel pump when filling a grease reservoir
- Perform a system leak check

Regular PM/Operator Maintenance

Daily Lubrication system checks are essential for ensuring the lubrication system is operating as designed and that there is lubricant in the system. This is often best performed on a daily basis by the equipment operator who visually checks the entire system in a quick system walk-around and only notifies the maintenance department when an exception is found. Check functions can include:

- Check reservoir fill level—is the level between the Lo and Hi mark on the reservoirs see Figure 10-82.

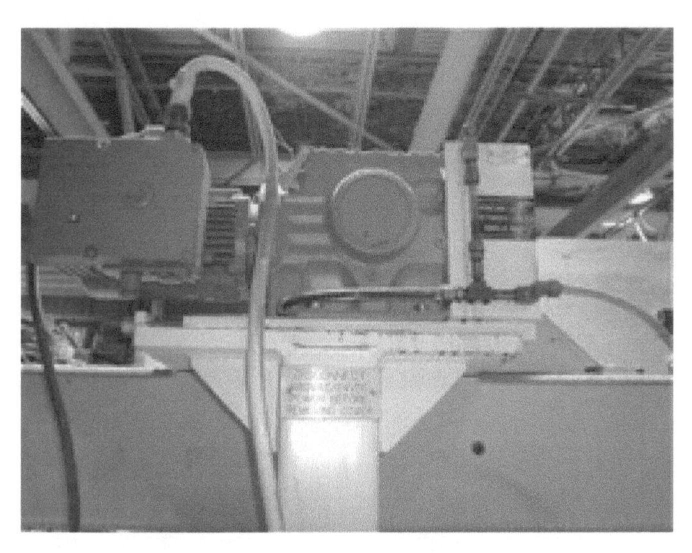

Figure 10-82. Gearbox reservoir Ho-Lo arrow indicators

- Check for and notify immediately of any system leaks
- Check for apparent system damage including line crush and any overpressure indicator signal denoting back pressure in the system caused by a damaged or blocked bearing or line
- Check for controller warning signals/lights and report immediately
- Check any pressure filter (used in recirculating oil and hydraulic systems) is not showing a red flag signal indicating the filter is full and in bypass mode

Of course, the type of lubrication system and lubricant will dictate the level of checking required. For example, recirculating oil systems are prime candidates for oil analysis allowing the lubricant to be changed only when needed, based on its condition.

References

1. Bloch, H.P., "Large Scale Application of Pure Oil Mist Lubrication in Petrochemical Plants," ASME Paper 80-C2/Lub-25 (1980).
2. Miannay, C.R., "Improve Bearing Life with Oil Mist Lubrication," *Hydrocarbon Processing Magazine*, (May 1974), 113-115.
3. Ward, T.K., "1995 Refinery Oil Mist Usage Survey," Lubrication Systems Company Internal Memo, (February 1996).
4. Towne, C.A., "Oil Mist Lubrication for the Petrochemical Industry," Proceedings of the 11th International Pump Users Symposium, The Turbomachinery Laboratory, Texas A & M University, College Station, Texas, (1994), p. 107.
5. Shamim A., Kettleborough, C.F., "Tribological Performance Evaluation of Oil Mist Lubrication," Texas A&M University Research Paper, (December, 1994).
6. SKF USA Inc., Bearing and Installation Guide; SKF Publication 140170, (February, 1992), p. 35.
7. Bloch, "Oil Mist Lubrication Handbook," Gulf Publishing Company, Houston, Texas, USA (1987).
8. Bloch, H.P., and Shamim, A., *Oil Mist Lubrication, Practical Applications*, The Fairmont Press, Lilburn, Georgia, USA (1998).
9. Ehlert, C.W., Inventor, United States Patent 5,125,480, (Issued June 30, 1992).
10. Lubrication Systems Company of Texas, Inc., Design and Installation of LubriMist® Oil Mist Lubrication Systems, (September 1996).
11. Stewart-Warner Corporation, Oil Mist Lubrication Systems for the Hydrocarbon Processing Industry, (1982)
12. Lubrication Systems Company of Texas, Inc., LubriMist® Accessories Brochure, (March 1996).
13. Ibid.
14. Ehlert, C. W., Inventor, United States Patent 5.318.152, (Issued June 7, 1994)
15. Lubrication Systems Company of Texas, Inc., LubriMist® Accessories Brochure.
16. Ibid.
17. Ibid.
18. United States Patent 5.318.152.
19. Ibid.
20. Ehlert, C.W., Inventor, Patent Application, (January 30, 1996)
21. Arnold, D.R., "Improving Rotary Lobe Blower Reliability," Blower Reliability Conference, (April 26, 1995).
22. Ibid.
23. Smook, G.A., Handbook for Pulp & Paper Technologists, Joint Textbook Committee of the Paper Industry, Atlanta, Georgia, (1987).
24. Bradshaw, Simon; "Investigations into the Contamination of Lubricating Oil in Rolling Element Pump Bearing Assemblies," *Proceedings of the Texas A&M Pump Users Symposium* (2000)

Bibliography

Bannister, Kenneth E., Manual Lubrication Delivery Devices, *Lubrication and Fluid Power* magazine, Feb 2006

Bannister, Kenneth E., Anatomy of a Centralized Lubrication System article series, *Lubrication Management Technology* magazine, 2011-2013

Bannister, Kenneth E., Industrial Lubrication Fundamentals Article series, *Lubrication Management & Technology* magazine, 2013-2015

Bannister, Kenneth E., Liquid Gold—Lubrication Principles Workshop Training Manual, *ENGTECH Industries Inc.*, 2008-2011

Bannister, Kenneth E., Industrial Lubrication Fundamentals—Certification Preparatory Training—Level 1 Workshop Manual, *ENGTECH Industries Inc.*, 2012-2016

Bloch, Heinz P., "Dry Sump Oil Mist Lubrication for Electric Motors," *Hydrocarbon Processing*, March 1977

Bloch, Heinz P., "Oil Mist Lubrication Cuts Bearing Maintenance" (*Plant Services Magazine*, November 1983)

Bloch, Heinz P.; "Benefits of Oil Mist Lubrication for Electric Motor Bearings" (*Plant Services Magazine*, April 1986)

Bloch, Heinz P.; "Preservation by Oil Mist Application," (*Plant Services Magazine*, November 1987)

Bloch, Heinz P.; "Oil Mist Lubrication: Is it Justified and How Should it be Executed in the 90s?" *Hydrocarbon Processing*, October 1990, pp. 25

Bloch, Heinz P.; "Best-of-Class" Lubrication for Pumps and Drivers," *Pumps & Systems*, April 1997

Bloch, Heinz P.; "Oil Mist Lubrication for Electric Motors," *Hydrocarbon Processing*, August 2005

Bloch, Heinz P.; "Applying Oil Mist," *Lubrication & Fluid Power*, February 2005

Bloch, Heinz P. and Fred Geitner; "Machinery Uptime Improvement," (2006) Elsevier-Butterworth-Heinemann, Stoneham, MA (ISBN 0-7506-7725-2)

Bloch, Heinz P.; *Improving Machinery Reliability*, (1982 and all later Editions), Gulf Publishing Company, Houston, TX, ISBN 0-87201-376-6; ISBN 0-87201-455-X; ISBN 0-88415-661-3

Bloch, Heinz P. and Allan Burris; *Pump User's Handbook—Life Extension*, (2006) 2nd Edition, The Fairmont Press, Lilburn, GA 30047, ISBN 0-88173-517-5

SKF USA Inc, "Bearings in Centrifugal Pumps, Application Handbook" *Publication 100-955* (1995), pp. 20

Wilcock, Donald F., and E.R. Booser, 1957, *Bearing Design and Application*, McGraw-Hill Book Company, New York, NY 10121

Chapter 11

Lubricating Bearings and Other Machine Elements

If rolling bearings are to operate reliably they have to be adequately lubricated to prevent direct metallic contact between the rolling elements, raceways and cages, to prevent wear and to protect the bearing surfaces against corrosion. The choice of a suitable lubricant and method of lubrication for each individual bearing application is therefore important, as is correct maintenance.*

The following information and recommendations relate to bearings without integral seals or shields. In general, bearings and bearing units with integral seals (shields) are supplied pre-greased. The standard greases used by competent bearing manufacturers for these products have operating temperature ranges and other properties to suit the intended application areas and filling grades appropriate to bearing size. The service life of the grease often exceeds bearing life so that, with some exceptions, no provision is made for relubrication.

As shown in this text, a wide selection of greases and oils is available for the lubrication of rolling bearings, and there are also solid lubricants, e.g., for extreme temperature conditions. The actual choice of lubricant depends primarily on the operating conditions, i.e., the temperature range and speeds as well as the influence of the surroundings. The most favorable operating temperatures will be obtained when the bearing is supplied with the minimum quantity of lubricant needed to provide reliable lubrication. However, when the lubricant has additional tasks, such as sealing or the removal of heat, larger quantities are required.

The lubricant fill in a bearing arrangement gradually loses its lubricating properties during operation as a result of mechanical work, aging and build-up of contamination. It is therefore necessary for grease to be replenished or renewed from time to time and for oil to be filtered and also changed at certain intervals (see "Relubrication" and "Oil Change," later in this segment).

Because of the large number of different lubricants which are available and, particularly where greases are concerned, because there may be differences in the lubricating properties of seemingly identical greases produced at different locations, **a bearing manufacturer cannot accept liability for the lubricant or its performance**. The user is therefore advised to specify the required lubricant properties in detail and to obtain a guarantee from the lubricant supplier that the particular lubricant will satisfy these demands.

Grease Lubrication

Grease can be used to lubricate rolling bearings under normal operating conditions in the majority of applications. Where grease lubrication of spherical roller thrust bearings is concerned, please refer to a later page.

Grease has the advantage over oil that it is more easily retained in the bearing arrangement, particularly where shafts are inclined or vertical, and it also contributes to sealing the arrangement against contaminants, moisture or water.

An excess of lubricant will cause the operating temperature to rise rapidly, particularly when running at high speeds. As a general rule, therefore, only the bearing should be completely filled, while the free space in the housing should be partly (between 30 and 50 %) filled with grease. Recommended grease quantities for the initial fill of bearing housings will be found in bearing manufacturers' housing tables.

Where bearings are to operate at very low speeds and must be well protected against corrosion, it is advisable to completely fill the housing with grease.

A speed rating for grease lubrication is quoted in the bearing manufacturers' literature. Suffice it to say that the values are lower than corresponding speed ratings for oil lubrication to take account of the initial temperature peak which occurs when starting up a bearing which has been filled with grease during mounting or which has just been relubricated. The operating temperature will sink to a much lower level once the grease has been distributed in the bearing arrangement. The pumping action inherent in certain bearing designs, e.g.,

*Source: SKF America, Kulpsville, Pennsylvania. Adapted, by permission, from General Catalog 4000 US, 1991.

in angular contact ball bearings and taper roller bearings, and which becomes more accentuated as speeds increase, or the pronounced working of the grease which occurs, for example, in full complement cylindrical roller bearings, also make it necessary for the speed ratings for grease lubrication to be lower than those for oil lubrication.

Lubricating Greases

Lubricating greases are thickened mineral or synthetic oils, the thickeners usually being metallic soaps. Additives can also be included to enhance certain properties of the grease. The consistency of the grease depends largely on the type and concentration of the thickener used. When selecting a grease, the viscosity of the base oil, the consistency, operating temperature range, rust inhibiting properties and the load carrying ability are the most important factors to be considered.

The *base oil viscosity* of the greases normally used for rolling bearings lies between 15 and 500 mm^2/s at 40°C. Greases based on oils having viscosities in excess of this range will bleed oil so slowly that the bearing will not be adequately lubricated. Therefore, if a very high viscosity is required because of low speeds, oil lubrication will generally be found more reliable.

The base oil viscosity also governs the maximum permissible speed at which a given grease can be used for bearing lubrication. For applications operating at very high speeds, the most suitable greases are those incorporating diester oils of low viscosity. The permissible operating speed for a grease is also influenced by the shear strength of the grease, which is determined by the thickener. A speed factor nd_m is often quoted by grease manufacturers to indicate the speed capability; n is the operating speed and d_m the mean diameter (mm) of the bearing, $d_m = 0.5 (d + D)$.

Greases are divided into various *consistency classes* according to the National Lubricating Grease Institute (NLGI) Scale. Please refer to the entry "Grease Classification" in Chapter 3 of this text.

The consistency of greases used for bearing lubrication should not change unduly with temperature within the operating temperature range or with mechanical working. Greases which soften at elevated temperatures may leak from the bearing arrangement. Those which stiffen at low temperatures may restrict rotation of the bearing.

Metallic soap thickened greases of consistency 1, 2 or 3 are those normally used for rolling bearings. The consistency 3 greases are usually recommended for bearing arrangements with vertical shaft, where a baffle plate should be arranged beneath the bearing to prevent the grease from leaving the bearing. In applications subjected to vibration, the grease is heavily worked as it is continuously thrown back into the bearing by vibration. Stiffness alone does not guarantee adequate lubrication; mechanically stable greases should be used for such applications.

Greases thickened with polyurea can soften and harden reversibly depending on the shear rate in the application, i.e., they are relatively stiff at low speeds and are soft or semifluid above a given speed. In applications with vertical shafts there is consequently a danger that a polyurea grease will leak when it is in the semi-fluid state.

The *temperature range* over which a grease can be used depends largely on the type of base oil and thickener as well as the additives. The lower temperature limit, i.e., the lowest temperature at which the grease will allow the bearing to be started up without difficulty, is largely determined by the type of base oil and its viscosity. The upper temperature limit is governed by the type of thickener and indicates the maximum temperature at which the grease will provide lubrication for a bearing. It should be remembered that a grease will age and oxidize with increasing rapidity as the temperature increases and that the oxidation products have a detrimental effect on lubrication. The upper temperature limit should not be confused with the "dropping point" which is quoted by lubricant manufacturers. The dropping point only indicates the temperature at which the grease loses its consistency and becomes fluid.

Table 11-1 gives the operating temperature ranges for the types of grease normally used for rolling bearings. These values are based on extensive testing carried out by SKF laboratories and may differ from those quoted by lubricant manufacturers.

They are valid for commonly available greases having a mineral oil base and with no EP additives. Of the grease types listed, lithium and more particularly lithium 12-hydroxystearate base greases are those most often used for bearing lubrication.

Greases based on synthetic oils, e.g., ester oils, synthetic hydrocarbons or silicone oils, may be used at temperatures above and below the operating temperature range of mineral oil based greases.

If bearings are to operate at temperatures above or below the ranges quoted in the table and are to be grease lubricated, the bearing manufacturer should be contacted for advice.

The *rust inhibiting properties* of a grease are mainly determined by the rust inhibitors which are added to

Table 11-1. Operating temperature ranges for greases used in rolling element bearings.

Grease type (thickener)	Recommended operating temperature range			
	from	to	from	to
—	°C		°F	
Lithium base	–30	+110	– 22	+ 230
Lithium complex	–20	+140	–4	+284
Sodium base	–30	+80	–22	+176
Sodium complex	–20	+140	– 4	+284
Calcium (lime) base	–10	+ 60	+14	+140
Calcium complex	–20	+130	– 4	+266
Barium complex	–20	+130	– 4	+266
Aluminum complex	–30	+110	– 22	+230
Inorganic thickeners (bentonite, silica gel etc.)	– 30	+130	–22	+266
Polyurea	–30	+140	–22	+284

the grease and its thickener.

A grease should provide protection to the bearing against corrosion and should not be washed out of the bearing in cases of water penetration. Ordinary sodium base greases emulsify in the presence of water and can be washed out of a bearing. Very good resistance to water and protection against corrosion is offered by lithium and calcium base greases containing lead-base additives. However, because of environmental and health reasons such additives are being replaced by other combinations of additives which do not always offer the same protection.

For *heavily loaded bearings*, e.g., rolling mill bearings, it has been customary to recommend the use of greases containing EP additives, since these additives increase the load carrying ability of the lubricant film. Originally, most EP additives were lead-based compounds and there was evidence to suggest that these were beneficial in extending bearing life where lubrication was otherwise poor, e.g., when κ (calculated as explained in conjunction with Figures 11-7 and 11-8 later), is less than 1. However, for the reasons cited above, many lubricant manufacturers have replaced the lead-based additives by other compounds, some of which have been found to be aggressive to bearing steels. Drastic reductions in bearing life have been recorded in some instances.

The utmost care should therefore be taken when selecting an EP grease and assurances should be obtained from the lubricant manufacturer that the EP additives incorporated are not of the damaging type, or in cases where the grease is known to perform well a check should be made to see that its formulation has not been changed.

It is important to consider the *miscibility* of greases when, for whatever reason, it is necessary to change from one grease to another. If greases which are incompatible are mixed, the consistency can change dramatically and the maximum operating temperature of the grease mix be so low, compared with that of the original grease, that bearing damage cannot be ruled out.

Greases having the same thickener and similar base oils can generally be mixed without any detrimental consequences, e.g., a sodium base grease can be mixed with another sodium base grease. Calcium and lithium base greases are generally miscible with each other but not with sodium base greases. (Refer to Table 9-3, earlier.) However, mixtures of compatible greases may have a consistency which is less than either of the component greases, although the lubricating properties are not necessarily impaired.

In bearing arrangements where a low consistency might lead to grease escaping from the arrangement, the next relubrication should involve complete replacement of the grease rather than replenishment (see segment "Relubrication"). As of this writing, the preservative with which SKF bearings are treated is compatible with the majority of rolling bearing greases but not with polyurea greases. Other manufacturers may have similar guidelines. The user will have to explore these issues with a particular supplier.

Many bearing manufacturers are able to supply suitable greases for their bearings. For example, the SKF range of lubricating greases for rolling bearings comprises six different greases and covers virtually all application requirements. As can be expected, these greases have been developed based on the latest know-how regarding rolling bearing lubrication and have been thoroughly tested both in the laboratory and in the field.

The most important technical data on SKF greases are given in Table 11-2.

This vendor/manufacturer is able to provide further information to users requesting additional details.

Relubrication

Rolling bearings have to be relubricated if the service life of the grease used is shorter than the expected service life of the bearing. Relubrication should always be undertaken at a time when the lubrication of the bearing is still satisfactory.

The time at which relubrication should be undertaken depends on many factors which are related in a complex manner. These include bearing type and size, speed, operating temperature, grease type, space around the bearing and the bearing environment. It is only possible to base recommendations on statistical rules; the SKF relubrication intervals are defined as the time period, at the end of which 99% of the bearings are still reliably lubricated, and represent L_1 grease lives. The L_{10} grease lives are approximately twice the L_1 lives.

The information given in the following is based on long-term tests in various applications but does not pertain in applications where water and/or solid contaminants can penetrate the bearing arrangement. In such cases it is recommended that the grease is frequently renewed in order to remove contaminants from the bearing.

The *relubrication intervals* t_f for normal operating conditions can be read off as a function of bearing speed n and bore diameter d of a certain bearing type from Figure 11-1. The diagram is valid for bearings on horizontal shafts in stationary machines under normal loads. It applies to good quality lithium base greases at a temperature not exceeding 70°C. To take account of the accelerated aging of the grease with increasing temperature it is recommended that the intervals obtained from the diagram are halved for every 15° increase in bearing temperature above 70°C, remembering that the maximum operating temperature for the grease given in Tables 11-1 and 11-2 should not be exceeded. The intervals may be extended at temperatures lower than 70°C but as operating temperatures decrease the grease will bleed oil less readily and at low temperatures an extension of the intervals by more than two times is not recommended. It is not advisable to use relubrication intervals in excess of 30,000 hours. For bearings on vertical shafts the intervals obtained from the diagram should be halved.

For large roller bearings having a bore diameter of 300 mm and above, the high specific loads in the bearing mean that adequate lubrication will be obtained only if the bearing is more frequently relubricated than indicated by the diagram, and the lines are therefore broken. It is recommended in such cases that continuous lubrication is practiced for technical and economic reasons. The grease quantity to be supplied can be obtained from the following equation for applications where conditions are

Table 11-2. Lubricating greases marketed by SKF

Properties	Lubricating greases (designations)					
	LGMT 2	LGMT 3	LGEP 2	LGEM 2	LGLT 2	LGHT 3
Thickener	Lithium soap	Lithium soap	Lithium soap	Lithium soap	Lithium soap	Complex lithium soap
Base oil	Mineral oil	Mineral oil	Mineral oil	Mineral oil	Diester	Mineral oil
Temperature range, °C (continuous running)	–30 to +120	–30 to +120	–30 to +110	–20 to +120	–55 to +110	–30 to +150
Kinematic viscosity of base oil, mm^2/s	91	120	195	510	116	110
Consistency (NLGI scale)	2	3	2	2	2	3

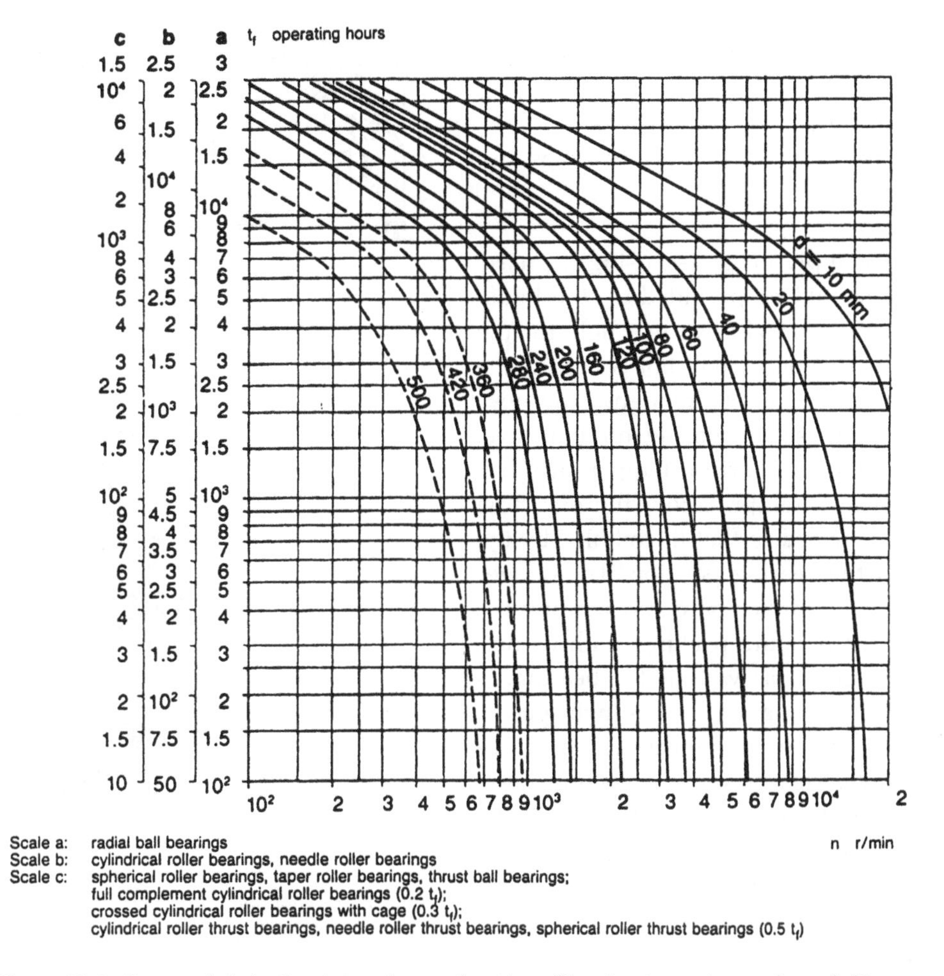

Figure 11-1. Grease relubrication intervals as a function of bearing type, size, and speed. (Source: SKF America, Kulpsville, Pennsylvania.)

otherwise normal, i.e., where external heat is not applied (recommendations for grease quantities for periodic relubrication are given in the following section).

$$G_k = (0.3 \ldots 0.5)\, D\, B \times 10^{-4}$$

where

- G_k = grease quantity to be continuously supplied, g/h
- D = bearing outside diameter, mm
- B = total bearing width (for thrust bearings use total height H), mm

One of the two *relubrication procedures* described below should be used, depending on the relubrication interval t_f obtained:

— if the relubrication interval is shorter than 6 months, then it is recommended that the grease fill in the bearing arrangement be replenished (topped up) at intervals corresponding to 0.5_{t_f}; the complete grease fill should be replaced after three replenishments, at the latest;

— when relubrication intervals are longer than 6 months it is recommended that all used grease be removed from the bearing arrangement and replaced by fresh grease.

The six-month limit represents a very rough guideline recommendation and may be adapted to fall in line with lubrication and maintenance recommendations applying to the particular machine or plant.

By adding small quantities of fresh *replenishment grease* at regular intervals the used grease in the bearing arrangement will only be partially replaced. Suitable quantities to be added can be obtained from

$$G_p = 0.005\, D\, B$$

where

G_p = grease quantity to be added when replenishing, grams
D = bearing outside diameter, mm
B = total bearing width (for thrust bearings use total height H), mm

To facilitate the supply of grease using a grease gun, a grease nipple should be provided on the housing. It is also necessary to provide an exit hole for the grease so that excessive amounts will not collect in the space surrounding the bearing. This might otherwise cause a permanent increase in bearing temperature. However, as soon as the equilibrium temperature has been reached following a relubrication, the exit hole should be plugged or covered so that the oil bled by the grease will remain at the bearing position. The danger of excess grease collecting in the space surrounding the bearing and causing temperature peaking, with its detrimental effect on the grease as well as the bearing, is most pronounced when bearings operate at high speeds. In such cases it is advisable to use a grease escape valve rather than an exit hole. This prevents over-lubrication and allows relubrication to be carried out, without the machine having to be stopped. A grease escape valve, Figure 11-2, consists basically of a disc which rotates with the shaft and which forms a narrow gap together with the housing end cover. Excess and used grease is thrown out by the disc into an annular cavity and leaves the housing through an opening on the underside of the end cover. The use of small check valves that are expected to perform as grease escape valves has proven problematic and should be discouraged.

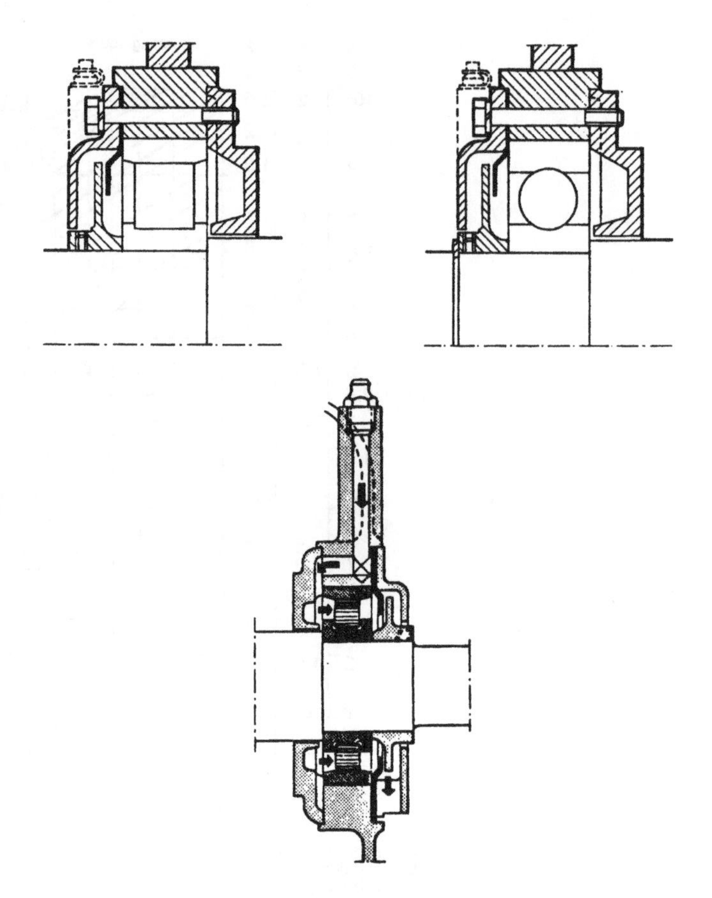

Figure 11-2. Advantageous, simple grease escape "valve." (Arrangements found in ASEA Electric Motors.)

To ensure that fresh grease actually reaches the bearing and replaces the old grease, the lubrication duct in the housing should either feed the grease adjacent to the outer ring side face or, better still, into the bearing which is possible, for example, with spherical roller bearings and double row full complement cylindrical roller bearings.

Where centralized lubrication equipment is used, care must be taken to see that the grease has adequate pumpability over the range of ambient temperatures.

If, for some reason, it is necessary to change from one grease to another, a check should be made to see that the new and old greases are compatible (see under "Miscibility," earlier in this segment.

When the end of the relubrication interval t_f has been reached the used grease in the bearing arrangement should be completely removed and replaced by fresh grease. As stated earlier, under normal conditions, the free space in the bearing should be completely filled and the free space in the housing filled to between 30 and 50% with fresh grease. The requisite quantities of grease to be used for a particular housing are usually stipulated in the manufacturer's literature.

In order to be able to *renew the grease fill* it is essential that the bearing housing is easily accessible and easily opened. The cap of split housings and the cover of one-piece housings can usually be taken off to expose the bearing. After removing the used grease, fresh grease should first be packed between the rolling elements. Great care should be taken to see that contaminants are not introduced into the bearing or housing when relubricating, and the grease itself should be protected. Where the housings are less accessible but are provided with grease nipples and exit holes or grease valves it is possible to completely renew the grease fill by relubricating several times in close succession until it can be assumed that all old grease has been pressed out of the housing. This procedure requires much more grease than is needed for manual renewal of the grease fill.

Oil Lubrication

Oil is generally used for rolling bearing lubrication only when high speeds or operating temperatures preclude the use of grease, when frictional or applied heat has to be removed from the bearing position, or when adjacent components (gears, etc.) are lubricated with oil.

Methods of Oil Lubrication

The most simple method of oil lubrication is the oil bath, Figure 11-3. The oil, which is picked up by

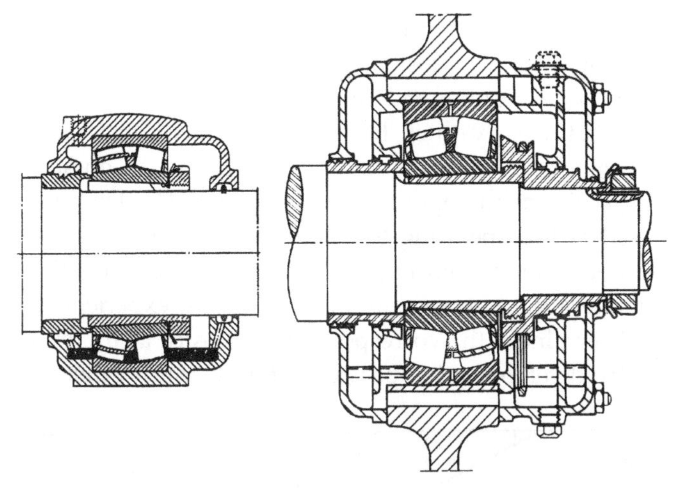

Figure 11-3. Oil bath/oil spray on double row spherical roller bearings. In the oil bath configuration (left), the oil level reaches the center of the rollers at the bottom. In the spray configuration (right), the oil is conveyed to the tapered flinger by an oil ring which dips into the oil bath. An important feature of this application is that the air pressure on both sides of the bearing and enclosures is equalized by connecting ducts. This prevents leakage of the lubricant when the housings are located in an air stream.

the rotating components of the bearing or by a flinger ring, is distributed within the bearing and then flows back to the oil bath. The oil level should be such that it almost reaches the center of the lowest rolling element when the bearing is stationary. Speed ratings for oil lubrication given in manufacturers' bearing tables normally apply to oil bath lubrication.

Operating at higher speeds will cause the operating temperature to increase and will accelerate aging of the oil. To avoid frequent oil changes, circulating oil lubrication, Figure 11-4, is generally preferred; the circulation is usually produced with the aid of a pump. After the oil has passed through the bearing it is filtered and, if required, cooled before being returned to the bearing. Cooling of the oil enables the operating temperature of the bearing to be kept at low level.

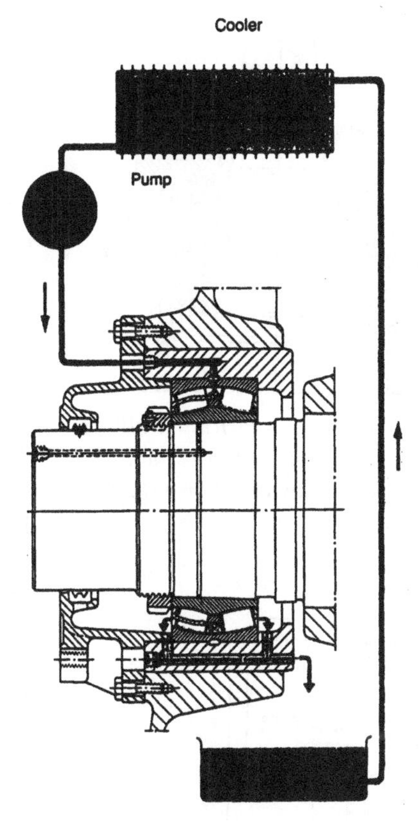

Figure 11-4. Circulating oil lubrication.

For very high-speed operation it is necessary that a sufficient but not excessive quantity of oil penetrates the bearing to provide adequate lubrication without increasing the operating temperature more than necessary. One particularly efficient method of achieving this is the oil jet method, Figure 11-5, where a jet of oil under high pressure is directed at the side of the bearing. The velocity of the oil jet must be high enough (at least 15 m/s), so that at least some of the oil will penetrate the turbulence surrounding the rotating bearing.

With the air-oil method, Figure 11-6 (also earlier, Chapter 10), very small, accurately metered quantities of oil are directed at each individual bearing by compressed air. This minimum quantity enables bearings to operate at lower temperatures or at higher speeds than any other method of lubrication. The oil is supplied to the points of application by a metering unit at given intervals. The oil is transported by compressed air; it coats the inside of applicator tubing or wires, and "creeps" along them. It is injected to the bearing via a nozzle. The compressed air serves to cool the bearing and also produces an excess pressure in the bearing arrangement which prevents contaminants from entering.

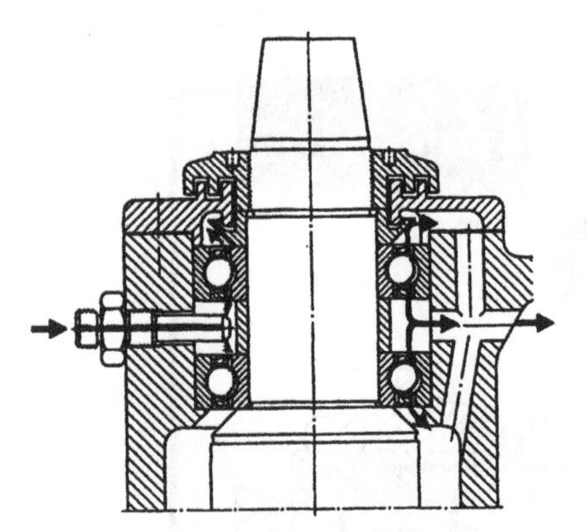

Figure 11-5. Oil-jet lubrication allows bearings to operate at lower temperature or at higher speeds than any other method of lubrication.

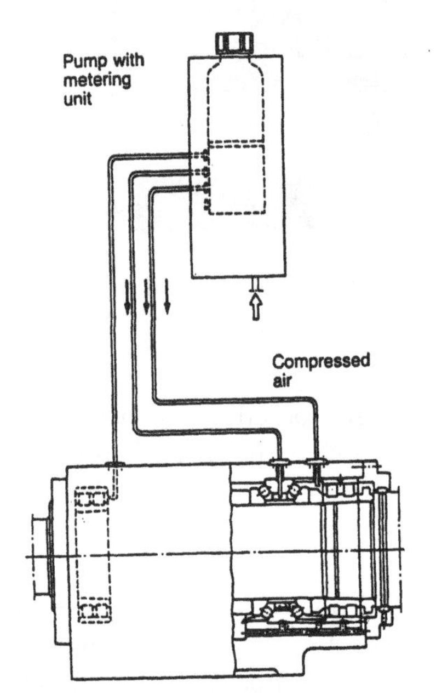

Figure 11-6. Air-oil lubrication schematic.

When using the circulating oil, oil jet and air-oil methods, it is necessary to ensure that the oil flowing from the bearing can leave the arrangement by adequately dimensioned ducts.

Straight mineral oils without additives are generally favored for rolling bearing lubrication. Oils containing additives for the improvement of certain lubricant properties such as extreme pressure behavior, aging resistance etc. are generally only used in special cases. *Synthetic oils* are generally only considered for bearing lubrication in extreme cases, e.g., at high loads, and very low or very high operating temperatures. It should be remembered that the lubricant film formation when using a synthetic oil may differ from that of a mineral oil having the same viscosity.

The remarks covering EP additives in the earlier segment on greases entitled "Load carrying ability," also apply to EP additives in oils.

As was brought out in our earlier chapters, the *selection of an oil* is primarily based on the viscosity required to provide adequate lubrication for the bearing at the operating temperature.

The viscosity of an oil is temperature dependent, becoming lower as the temperature rises. The viscosity/temperature relationship of an oil is characterized by the viscosity index, VI. For rolling bearing lubrication, oils having a high viscosity index (little change with temperature) of at least 85 are recommended.

In order for a sufficiently thick film of oil to be formed in the contact area between rolling elements and raceways, the oil must retain a minimum viscosity at the operating temperature. The kinematic viscosity υ_1 required at the operating temperature to ensure adequate lubrication can be determined from Figure 11-7 provided a mineral oil is used. When the operating temperature is known from experience or can otherwise be determined, the corresponding viscosity at the internationally standardized reference temperature of 40°C, or at other test

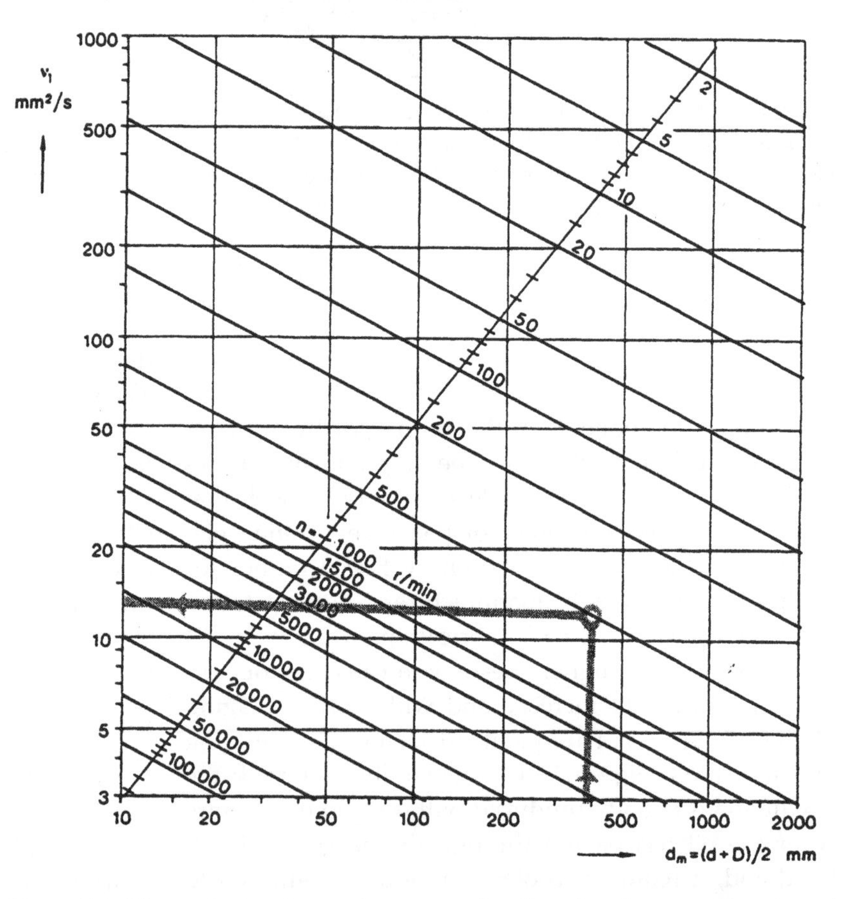

Figure 11-7. Kinematic viscosity requirement as a function of bearing mean diameter and speed.

temperatures (e.g., 20 or 50°C) can be obtained from Figure 11-8 which is compiled for a viscosity index of 85. Certain bearing types, e.g., spherical roller bearings, taper roller bearings, and spherical roller thrust bearings, normally have a higher operating temperature than other bearing types e.g., deep groove ball bearings and cylindrical roller bearings, under comparable operating conditions.

When selecting the oil the following aspects should be considered.

Bearing life may be extended by selecting an oil whose viscosity v at the operating temperature is somewhat higher than υ_1. However, since increased viscosity raises the bearing operating temperature there is frequently a practical limit to the lubrication improvement which can be obtained by this means.

If the viscosity ratio $\kappa = \upsilon/\upsilon_1$ is less than 1 an oil containing EP additives is recommended and if κ is less than 0.4 an oil with such additives must be used. An oil with EP additives may also enhance operational reliability in cases where κ is greater than 1 and medium and large-sized roller bearings are concerned. It should be remembered that only some EP additives are beneficial, however (see also under "Load carrying ability," earlier in this segment.

For exceptionally low or high speeds, for critical loading conditions or for unusual lubricating conditions please consider discussions with the applications engineering staff of major bearing manufacturers.

Example

A bearing having a bore diameter d = 340 mm and outside diameter D = 420 mm is required to operate at a speed n = 500 r/min. Since d_m = 0.5 (d + D), d_m = 380 mm. From Figure 11-7, the minimum kinematic viscosity υ_1 required to give adequate lubrication at the operating temperature is 13 mm^2/s. From Figure 11-8, assuming that the operating temperature of the bearing is 70°C, an oil having a viscosity υ at the reference temperature of 40°C of at least 39 mm^2/s will be required.

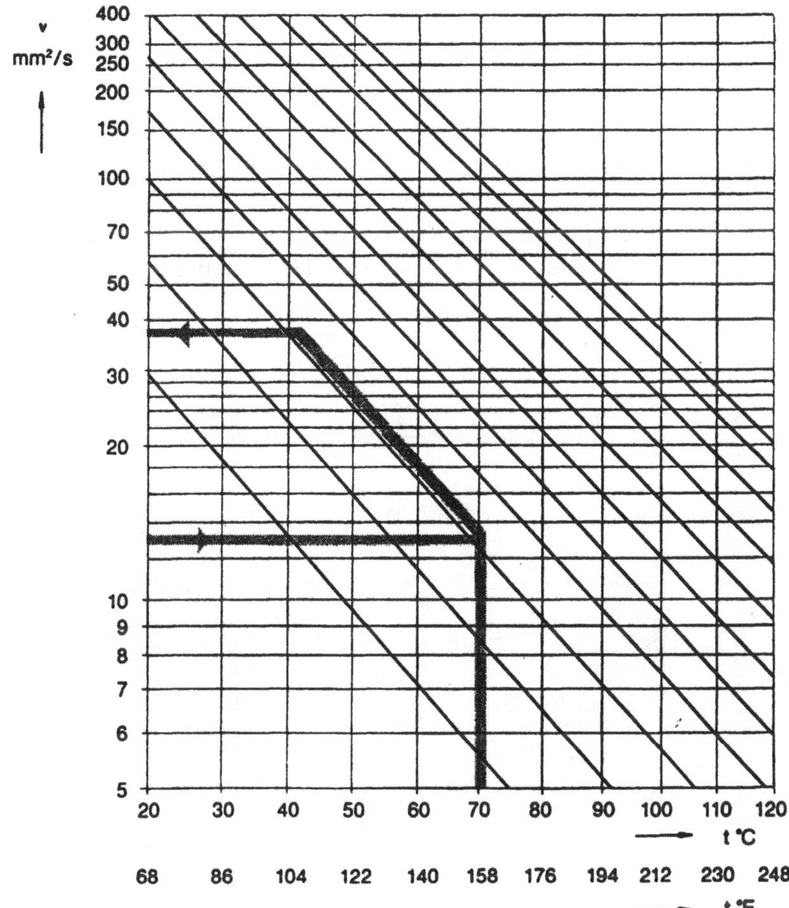

Figure 11-8. The required ISO-grade of a lubricant can be obtained from this graph.

The *frequency with which it is necessary to change the oil* depends mainly on the operating conditions and the quantity of oil.

With oil bath lubrication it is generally sufficient to change the oil once a year, provided the operating temperature does not exceed 50°C and there is little risk of contamination. Higher temperatures call for more frequent oil changes, e.g., for operating temperatures around 100°C, the oil should be changed every three months. Frequent oil changes are also needed if other operating conditions are arduous.

With circulating oil lubrication, the period between two oil changes is also determined by how frequently the total oil quantity is circulated and whether or not the oil is cooled. It is generally only possible to determine a suitable interval by test runs and by regular inspection of the condition of the oil to see that it is not contaminated and is not excessively oxidized. The same applies for oil jet lubrication.

With air-oil lubrication the oil only passes through the bearing once and is not recirculated. The same is generally true of oil mist lubrication, Figure 11-9, a superior means of conveying and applying liquid lubricants (see Chapter 11).

Spherical roller bearings present a special lubrication challenge. It is gener-

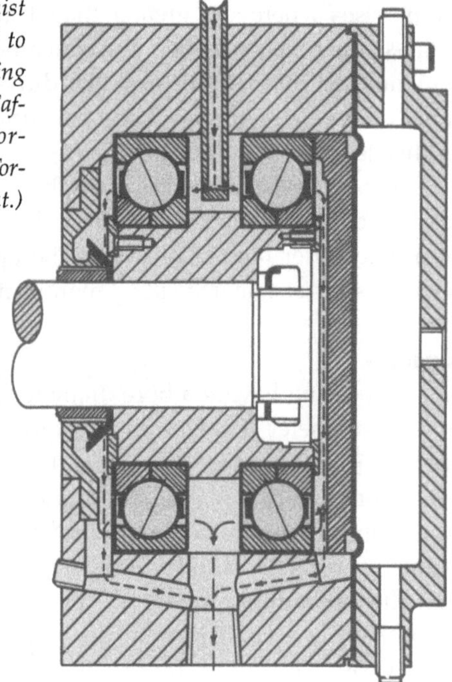

Figure 11-9. Oil mist lubrication applied to a set of split inner ring bearings. (Source: Fafnir Division of Torrington Company, Torrington, Connecticut.)

ally recommended that spherical roller thrust bearings should be oil lubricated. Grease lubrication can be used in special cases, for example, under light loads and at low speeds, particularly where bearings incorporating a pressed steel cage are concerned.

When using grease as the lubricant it is necessary to ensure that the roller end/flange contacts are adequately supplied with grease. Depending on the actual application, this can best be done by completely filling the bearing and its housing with grease or by regular relubrication.

The speed ratings quoted in the bearing tables for grease lubricated bearings fitted with pressed steel cages are valid for bearing arrangements where the shaft is horizontal. For arrangements with vertical shafts, the values should be approximately halved.

Because of their internal design, spherical roller thrust bearings have a pumping action which may be exploited under certain conditions and should be taken into consideration when selecting lubrication method and seals.

More detailed information regarding the lubrication of spherical roller thrust bearings can be provided by the application engineering service groups of competent manufacturers.

In order to assure the satisfactory operation of all ball and roller bearings they must always be subjected to a given *minimum load.* This is also true of spherical roller thrust bearings, particularly if they run at high speeds where the inertia forces of the rollers and cage, and the friction in the lubricant can have a detrimental influence on the rolling conditions in the bearing and may cause damaging sliding movements to occur between the rollers and the raceways.

The requisite minimum axial load to be applied in such cases can be estimated from

$$F_{am} = 1.8\,F_r + A\left(\frac{n}{1000}\right)^2$$

(If $1.8\,F_r < 0.0005\,C_0$, then $0.0005\,C_0$ should be used in the above equation instead of $1.8\,F_r$)

where

- F_{am} = minimum axial load, N
- F_r = radial component of load for bearings subjected to combined load, N
- C_0 = basic static load rating, N
- A = minimum load factor, see manufacturer's bearing tables
- n = speed, r/min

The weight of the components supported by the bearing, together with the external forces, often exceeds the requisite minimum load. If this is not the case, the bearing must be preloaded (e.g., by springs).

Finally, the reader may wish to use Figure 11-10, a simplified oil viscosity selection chart devised by the Fafnir Bearings Division of The Torrington Company,

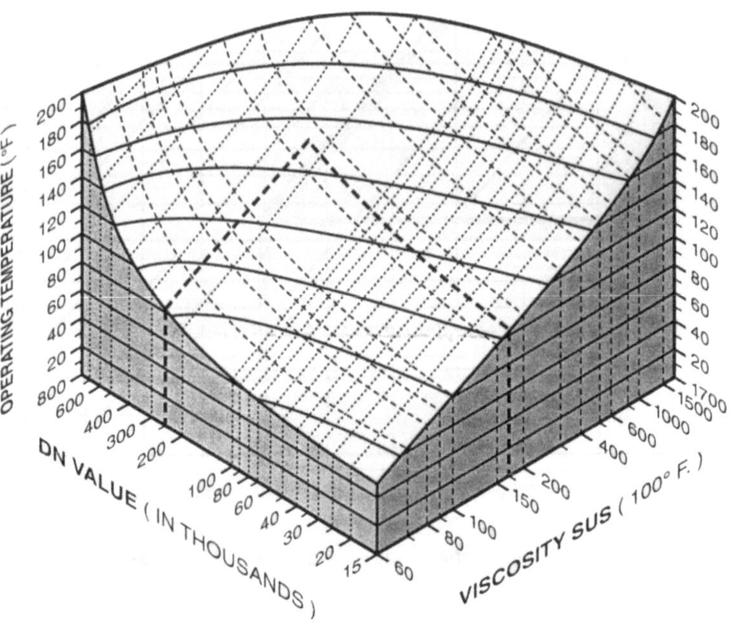

Figure 11-10. Simplified oil viscosity selection chart.

Torrington, Connecticut.

This chart may be used to approximate the proper oil viscosity for all bearing applications.

To use the chart proceed as follows:

1. Determine the DN value—Multiply the bore diameter of the bearing, measured in millimeters, by the speed of the shaft, measured in revolutions per minute.

2. Select the proper temperature—The operating temperature of the bearing may run several degrees higher than the ambient temperature depending upon the application. The temperature scale of this chart reflects the operating temperature of the bearing.

3. Enter the DN value in the DN scale on the chart.

4. Follow or parallel the "dotted" line to the point where it intersects the selected "solid" temperature line.

5. At this point follow or parallel the nearest "dashed" line downward and to the right to the viscosity scale.

Figure 11-11. 20MW naval gearbox fitted with Glacier Series II flooded lubrication standard bearings for medium speed duties. (Photo courtesy of Maag Gear Company, Zurich, Switzerland.)

6. Read off the approximate viscosity value-expressed in Saybolt Universal Seconds at 100°F.

Typical Example

PROBLEM:

Determine the proper oil viscosity required for a 50 mm ball bearing operating at a speed of 5000 RPM at a temperature of 150°F.

SOLUTION:

Determine the DN value-the bore diameter of the ball bearing is 50 mm. Multiply this by the shaft speed in RPM; 50 mm × 5000 RPM = 250,000 DN.

Enter this value on the DN scale. Parallel the "dotted" lines to the point of intersection with the projected "solid" 150°F temperature line. At this junction, parallel the nearest "dashed" line downward and to the right to the viscosity scale. Read off the approximate viscosity of 170 SUS at 100°F.

TILTING PAD THRUST BEARINGS*

Tilting pad thrust bearings are designed to transfer high axial loads from rotating shafts with minimum power loss, while simplifying installation and maintenance. The shaft diameters for which the bearings are designed typically range from 20mm to over 1000mm. The maximum loads for the various bearing types range from 0.5 to 500 tons. Bearings of larger size and load capacity are considered non-standard, but can and have been made to special order.

Each bearing consists of a series of pads supported in a carrier ring; each pad is free to tilt so as to create a self-sustaining hydrodynamic film. The carrier ring may be in one piece or in halves, and there are various location arrangements.

Two options exist for lubrication. One is by fully flooding the bearing housing, the other, which is more suitable for higher speed applications, directs oil to the thrust face; this oil is then allowed to drain freely from the bearing housing.

Similarly, two geometric options exist. The first option is shown in Figures 11-11 and 11-12; it does not use equalizing or leveling links. This option is used in many gear units and other shaft systems where perpen-

*Sources: The Glacier Metal Company, London, England, and Mystic, Connecticut; also, Kingsbury Inc., Philadelphia, PA, and Waukesha Bearings, Waukesha, WI.

dicularity between shaft centerline and bearing faces is assured.

This design, Figure 11-13, is intended for machines where an equalized thrust bearing is specified by API requirements or where this bearing may be required for other reasons.

Flooded Lubrication vs. Directed Lubrication

The conventional method of lubricating tilting pad thrust bearings is to flood the housing with oil, using an orifice on the outlet to regulate the flow and maintain pressure. A typical double thrust bearing of this type is illustrated in Figure 11-14. A housing pressure of 0.7 -1.0 bar is usual and, to minimize leakage, seal rings are required where the shaft passes through the housing.

Although flooded lubrication is simple, it results in high parasitic power loss due to turbulence at high speed. Where mean sliding speeds in excess of 50 m/s are expected, these losses may be largely eliminated by employing the system of directed lubrication. As well as reducing power loss by (typically) 50%, directed lubrication also reduces the bearing temperature and in most cases oil flow.

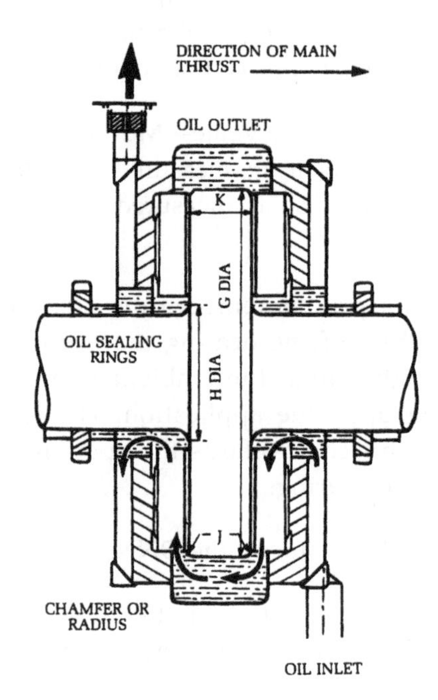

Figure 11-12. Flooded lubrication: typical double thrust arrangement.

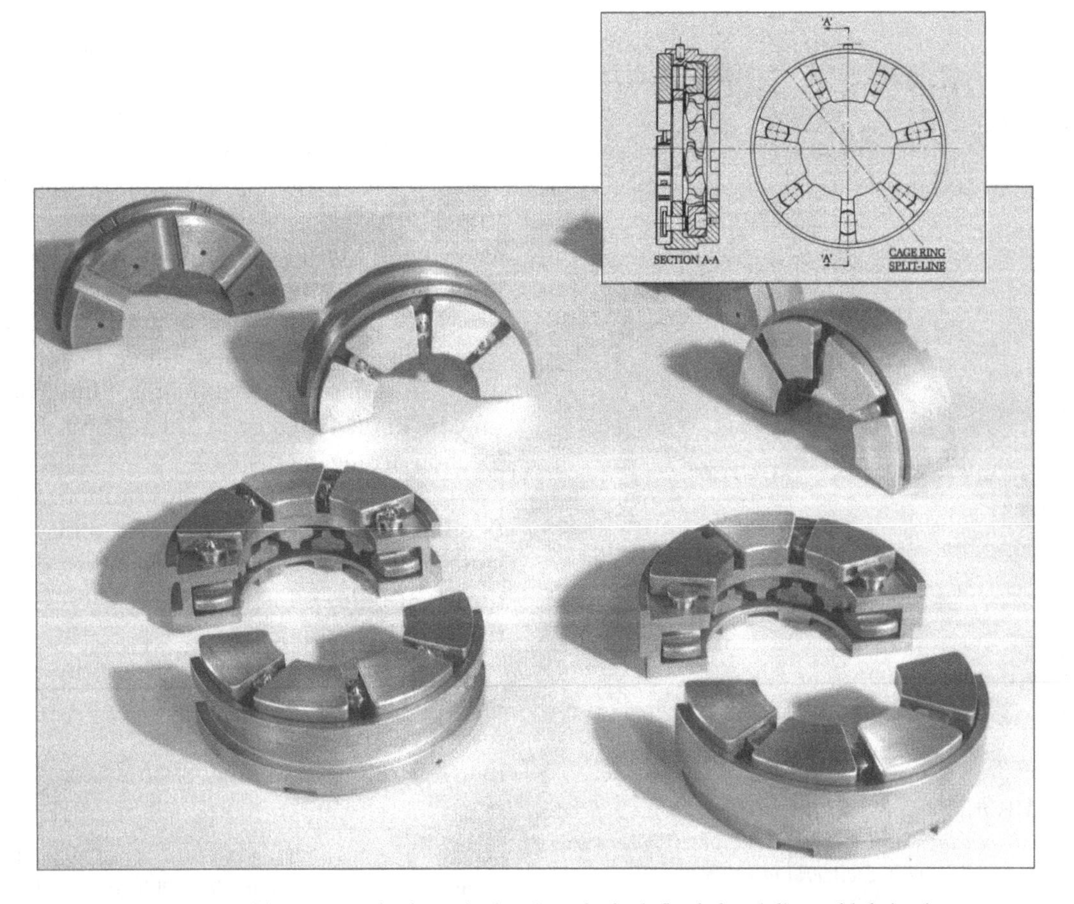

Figure 11-13. Glacier's standard 7 series bearings for both flooded and directed lubrication.

Figure 11-14. Glacier double thrust bearing size 10293 with directed lubrication installed in an ABB gas turbine. (Photo: ABB, Baden, Switzerland.)

Some typical double thrust bearing arrangements using directed lubrication are shown in Figure 11-15.

It should be noted that

- Directed and flooded bearings have the same basic sizes, and use identical thrust pads.

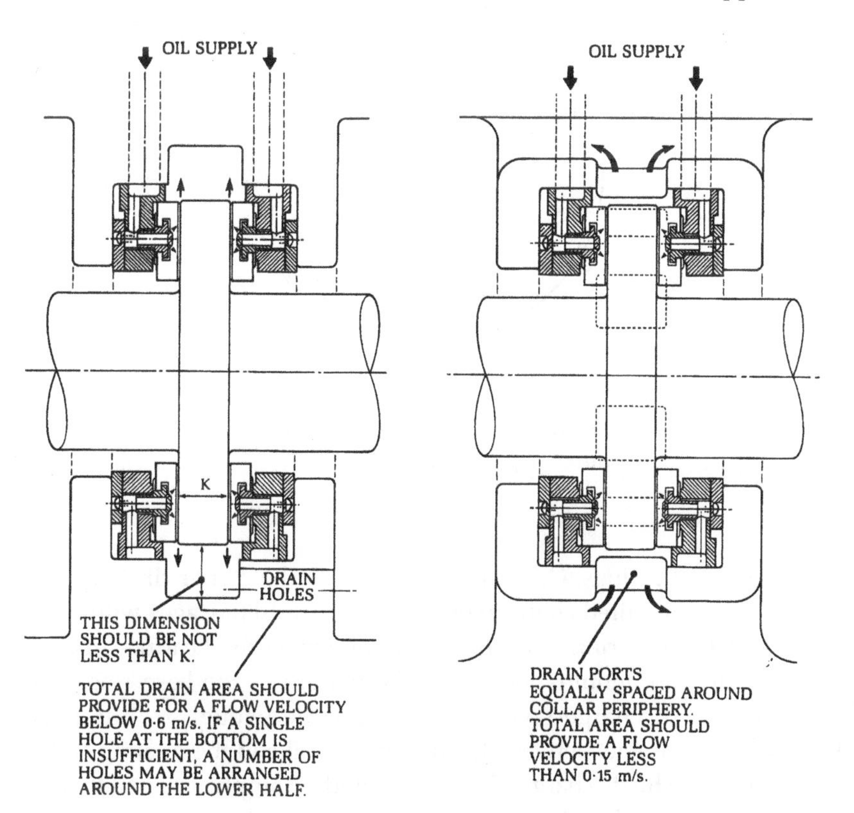

Figure 11-15. Directed lubrication: typical double thrust arrangements designed to prevent bulk oil from contacting the collar.

- Preferred oil supply pressure for directed lubrication is 1.4 bar.
- Oil velocity in the supply passages

Experienced manufacturers can offer a variety of pad materials. Some polymeric materials are capable of operating at temperatures up to 120°C higher than conventional white metal or babbitt. Also, pad pivot position can have an effect on thrust pad temperature (Figure 11-16).

All pads can be supplied with offset pivots, but center-pivoted pads are preferred for bi-directional running, foolproof assembly and minimum stocks. At moderate speeds the pivot position does not affect load capacity but where mean sliding speeds exceed 70 m/s offset pivots can reduce bearing surface temperatures and thus increase load capacity under running conditions

Thrust bearings can be fitted with temperature sensors (Figures 11-17 and 11-18), proximity probes (Figure 11-19) and load cells (Figure 11-20).

In hydraulic thrust metering systems, a hydraulic piston is located behind each thrust pad, these pistons being interconnected to a high pressure oil supply. The pressure in the system then gives a measure of the applied thrust load. Figure 11-21 shows a typical installation of this type complete with system control panel which incorporates the high pressure oil pump and system pressure gauge calibrated to read thrust load.

For systems incorporating load cells or hydraulic pistons, it will normally be necessary to increase the overall axial thickness of the thrust ring.

Finally, there are thrust bearings that incorporate hydraulic jacking provisions. These provisions ensure that an appropriate oil film exists between thrust runner and bearing pads while operating at low speeds.

At the instant of start up, the load carrying capacity of tilting pad thrust bearings is restricted to approximately 60% of the maximum permissible operating load. If the start up load on a bearing exceeds this figure and a larger bearing is not a feasible option, the manufacturer can supply thrust bearings fitted with a hydrostatic jacking system to allow the bearing to operate with heavy loads at low speeds. This system introduces oil at high pressure (typically 100-150 bar) between the bearing

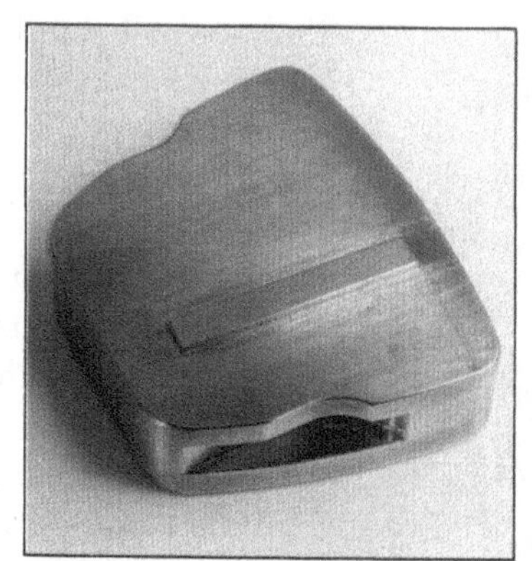

Offset pivot thrust pad.

Figure 11-17. Glacier bearing fitted with thermocouples ready for high-speed petrochemical application. Note half pad stops which avoid handling loose pads on assembly.

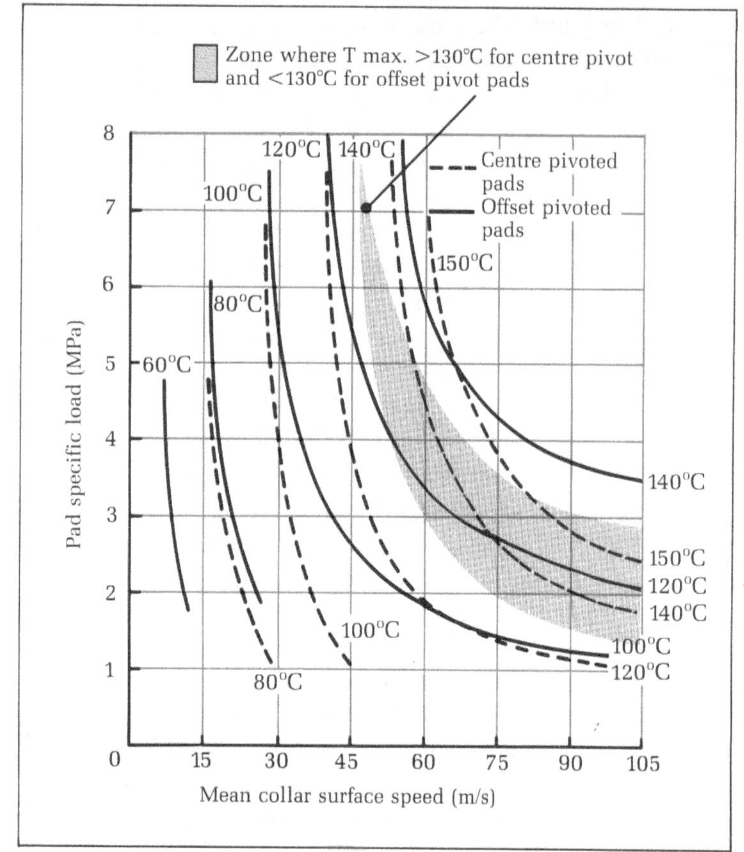

Figure 11-16. Offset pivots: effect on thrust pad temperature.

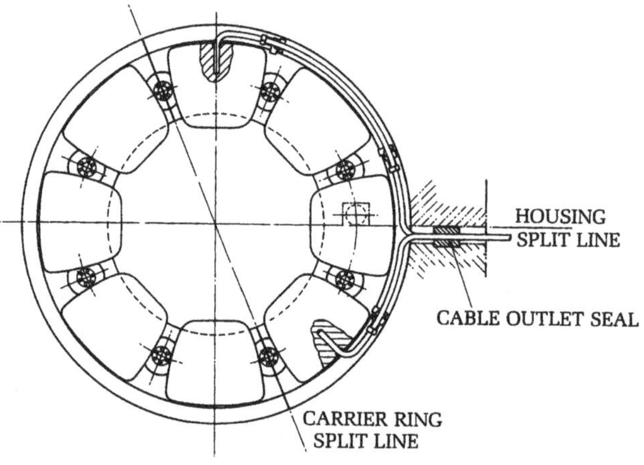

Figure 11-18. Temperature sensors: typical method of fitting to thrust rings.

surfaces to form a hydrostatic oil film; Figures 11-22 and 11-23 show typical bearings of this type.

It should be noted that a very similar approach is taken when making hydraulic jacking provisions for radial bearings. A "hybrid" thrust bearing is offered by Kingsbury and UK-based Colherne Company under the name KingCole. This pivoting pad leading edge groove (LEG) bearing is illustrated in Figure 11-24.

The bearing housing requirements for the KingCole "LEG" bearing are similar to those of standard thrust bearings. Oil seals at the back of the carrier rings are not required as the inlet oil is confined to passages within the base ring assembly. Fresh oil enters the bearing through an annulus located at the bottom of the base ring. The discharge space should be large enough to minimize contact between the discharged oil and the rotating collar. The discharge oil outlet should be amply sized so that oil can flow freely from the bearing cavity.

The manufacturer recommends a tangential discharge opening, equal in diameter to 80% of the recom-

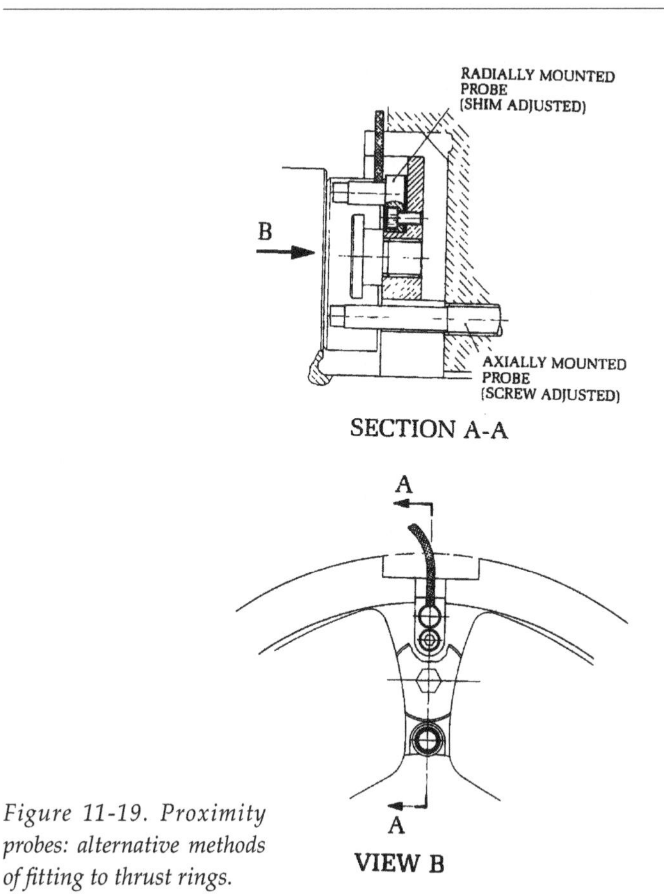

Figure 11-19. Proximity probes: alternative methods of fitting to thrust rings.

mended collar thickness. If possible, the discharge outlet should be located in the bottom of the bearing housing. Alternatively, it should be located tangential to the collar rotation.

The bearing pads and carrier ring are constructed so that cool undiluted inlet oil flows from the leading edge groove in the bearing pad directly into the oil film. The cool oil in the oil film wedge insulates the white metal face from the hot oil carryover that adheres to the rotating collar.

In contrast to the KingCole "LEG" bearing, the oil for spray-fed bearings is injected not directly onto the bearing surfaces but between them. This can result in uneven bearing lubrication and the need to supply impractically high pressure to get true effective scouring of the hot oil carryover adhering to the thrust collar. There is also a possibility for the small jet holes to clog with foreign material.

Friction power loss is claimed to be lower than both flooded and spray feed bearings due to the reduced oil flow. The flow of cool oil over the leading edge lowers pad surface temperatures, apparently increasing the KingCole's capacity.

The resulting performance improvements are shown in Figure 11-25.

Assuming an oil inlet temperature of 50°C, it is possible to estimate the white metal temperature of KingCole leading edge bearings from Figure 11-26. These temperatures are a function of surface speed and contact pressure.

Bearing Selection

Thrust load, shaft RPM, oil viscosity and shaft diameter through the bearing determine the bearing size to be selected.

Leading edge bearings are sized for normal load and speed when transient load and speed are within 20% of normal conditions.

Although the graphs in Figures 11-27 through 11-29 pertain only to 8-pad bearings by KingCole, they will convey ball-park data for thrust bearings in the size range given in Table 11-3.

Friction losses are based on recommended flow rates and an evacuated drain cavity. To calculate friction losses for double element bearings, add 10% to the values in these graphs to accommodate the slack-side bearing.

To calculate lubricant supply for double element bearings, add 20% to the values in these graphs.

All curves are based on an oil viscosity of ISO VG32, with an inlet oil temperature of 50°C. The manufacturer recommends ISO VG32 oil viscosity for moderate through high-speed applications.

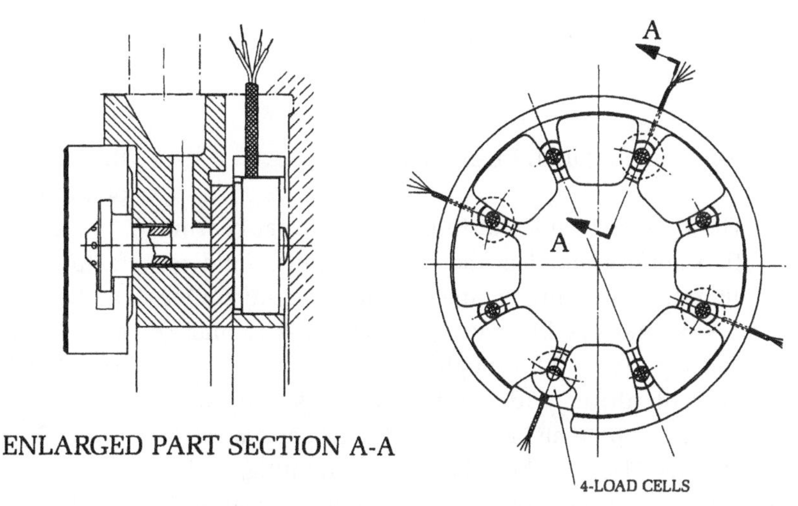

Figure 11-20. Load cells: installation in directed lubrication carrier ring.

Table 11-3. Thrust bearing designation numbers and bearing area (KingCole 8-pad thrust bearings).

Bearing Designation	Bearing Area in²	mm²
103	8	5142
112	9.5	6155
123	11.2	7265
134	13.7	8846
146	16.2	10446
159	20	12859
174	23.5	15188
190	29.2	18833
207	35	22530
225	42.5	27428
246	49	31666
269	57.6	37176
293	69.2	44616
320	82	52835

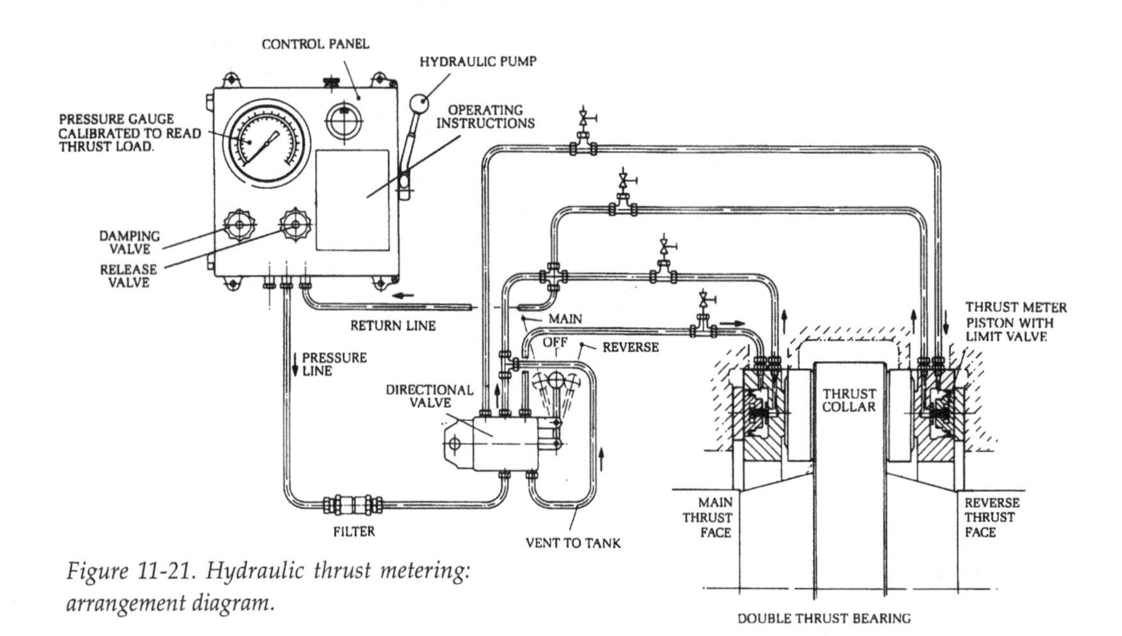

Figure 11-21. Hydraulic thrust metering: arrangement diagram.

TILTING PAD RADIAL BEARINGS

The basic principles of tilting pad journal bearing operation are explained in the selection guides and related literature of many competent manufacturers. One of these, Waukesha Bearings (Waukesha, Wisconsin) provided Figures 11-30 through 11-32.

Typical tilting pad journal bearings consist of three basic components: a shell, end plates and a set of pads (Figure 11-30). When the shaft is rotating, radial forces are transferred from the shaft to the journal pads through a film of oil that is self-generated between the shaft and the pads. The radial force then passes from the pads, through the bearing shell to the foundation or machine support.

To develop a hydrodynamic film in a bearing, three factors are required:

a) Viscous fluid
b) Relative motion
c) Converging geometry

The first factor, viscous fluid, is available since the bearings under consideration are fluid-lubricated (primarily with oil). Relative motion is provided by the rotation of the shaft relative to the surface of the tilting

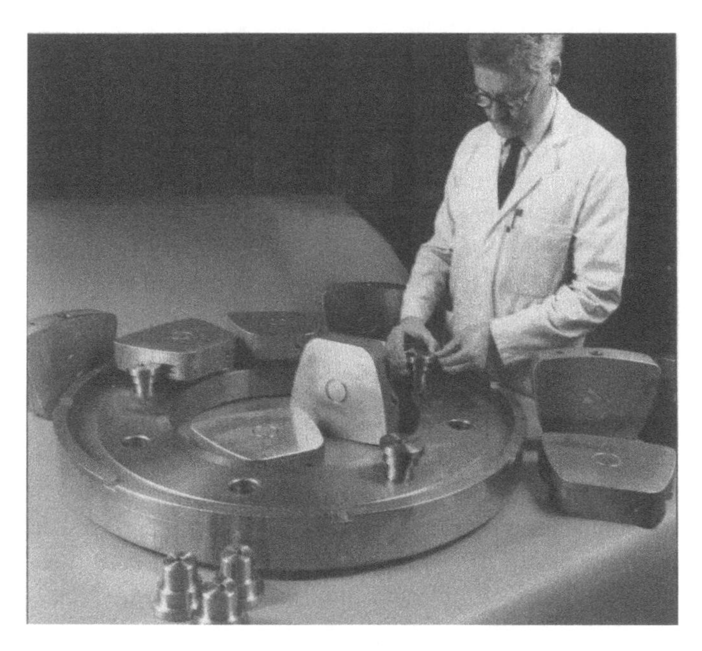

Figure 11-22. Glacier bearing featuring high pressure jacking oil for start up and run down. Jacking oil annulus can be seen on the surface of each thrust pad.

pads. Converging geometry is provided by the slight difference in the diameter of the shaft and the bore of the bearing pads (Figure 11-31). Clearances are exaggerated in Figure 11-32 for illustrative purposes.

The principle of pressure build-up in the oil film from the three factors outlined above is shown in Figure 11-32. Oil adheres to the moving (and stationary) surfaces, and thus there is flow into the converging volume. Since oil is basically incompressible, pressure builds within the converging oil film. This pressure provides a means for the oil film to transfer the load from the shaft to the pad.

The thickness of the film is of prime importance in the design and operation of a hydrodynamic oil film bearing. For the bearing and associated machinery to operate satisfactorily, it is important that the oil film fully separates the shaft and the journal bearing pad surfaces. However, during start-up and shut-down there are momentary periods when the combination of relative speed and load does not generate a full film, and at least some metal-to-metal contact results. Operating conditions such as these dictate the use of combinations of materials, such as babbitt faced journal pads operating against a steel shaft, that allow these contacts to occur without surface damage.

Modern bearings typically use tin base babbitt as the standard bearing material. Though other types of bearing material are available, each with its own advantages and limitations, tin base babbitt meets desired bearing properties such as compatibility, corrosion resistance, conformability and embeddability to such a high degree that it is widely used and accepted (Table 11-4).

Bearing instability is often a factor in selecting journal bearings. Instability refers to the problem of half-frequency whirl of the shaft within the bearing. The most serious form of this condition may occur when the operating speed is near and above twice the first critical speed of the shaft.

On simple journal bearings the displacement of the shaft within the bearing clearance in response to a radial force is not, in general, in line with the direction of that force. This lateral component of the movement of the shaft within the bearing clearance can lead to instability. This problem must be considered on lightly loaded, high speed bearing applications.

Tilting pad journal bearings are widely used because of their stability characteristics. If the load is either directly in line with a pad pivot or directly between two pivots, the displacement of a shaft operating hydrodynamically within that bearing will be directly in line with the direction of load on that shaft. Thus there is no component of motion at right angles to the direction of force. Such a bearing is inherently resistant to half-frequency bearing instability.

The potentially huge size of tilting pad radial bearings is shown in Figure 11-33.

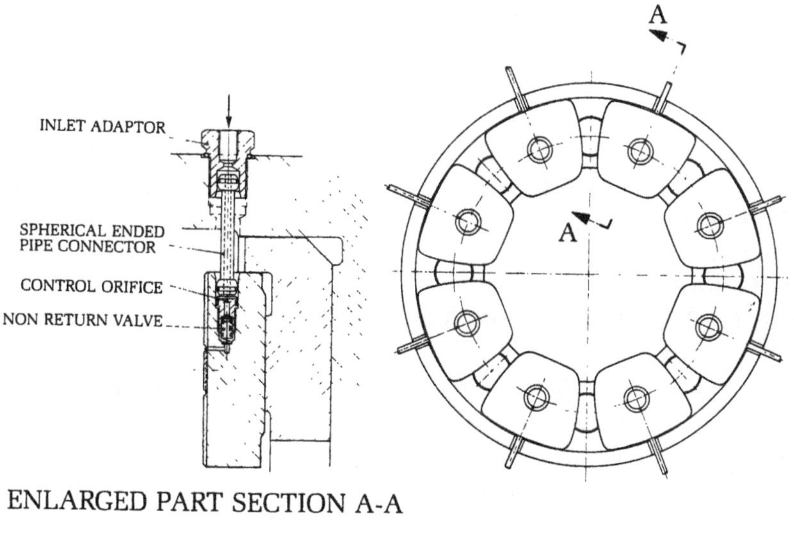

INLET ADAPTOR
SPHERICAL ENDED PIPE CONNECTOR
CONTROL ORIFICE
NON RETURN VALVE
ENLARGED PART SECTION A-A
A
A

Figure 11-23. Jacking oil system in thrust bearing.

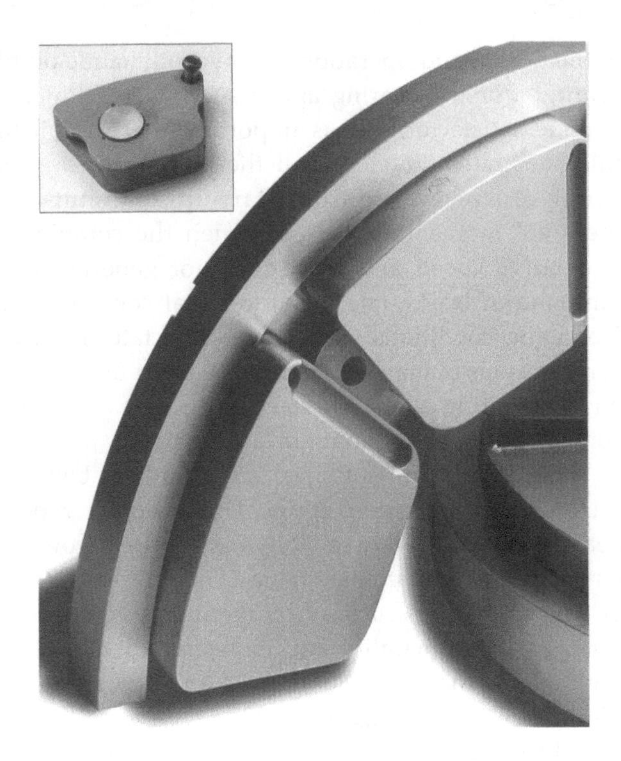

Figure 11-24. Leading Edge Groove (LEG) bearing. (Source: KingCole/Kingsbury, Inc., Philadelphia, Pennsylvania, and Newton, Cheshire, UK.)

Instrumentation

Temperature Measurement

Changes in load, shaft speed, oil flow, oil inlet temperature, or bearing surface finish can affect bearing surface temperatures. At excessively high temperatures, the pad white metal is subject to wiping, which causes bearing failure. While computer predictions of operating temperature are typically based on extensive empirical data, the algorithms used do include assumptions about the nature of the oil film shape, amount of hot oil carryover, and average viscosity. Consequently, for critical applications, onen often uses pads with built-in temperature to allow monitoring of actual metal temperatures under all operating conditions. Either thermocouples or resistance temperature detectors (RTDs) can be installed in contact with the white metal or in the pad body near the pad body/white metal interface.

Thrust Measurement

For bearings subject to critically high loads, continual thrust measurement can provide a vital indication of machine and bearing condition. On many bearing configurations, it is possible to install a strain gauge load cell in one or more places in the bearing.

Table 11-4. Properties of bearing alloys.

Bearing Material	Hardness Room Temp. Brinell	Min. Shaft Hardness Brinell	Compatibility*	Conformability and Embeddability*	Corrosion Resistance*
Tin-Base Babbitt	20-30	150 or less	1	1	1
Lead-Base Babbitt	15-20	150 or less	1	1	3
Copper-Lead	20-30	300	2	2	5
Tin Bronze	60-80	300-400	3	5	2
Lead Bronze	40-30	300	3	4	4
Aluminum Alloy	45-50	300	5	3	1

*Arbitrary scale with 1 being the best material and 5 the worst.

COMBINATION THRUST AND RADIAL BEARINGS

Several manufacturers produce combination bearings similar to the Glacier model depicted in Figure 11-34. Others are able to supply Leading Edge Groove (LEG) technology to both thrust and radial pads (11-35 and 11-36).

LEG journal bearings, Figure 11-37, oil than standard journal bearings, reducing friction power loss and oil system requirements. They also operate with significantly lower white metal temperatures.

Load cells can be installed in conventional and "LEG" bearings in place of the pad support, Figure 11-38.

PLAIN BEARINGS

Plain bearings are machine elements transmitting forces between machine elements that move relative to each other.

A distinction is made between

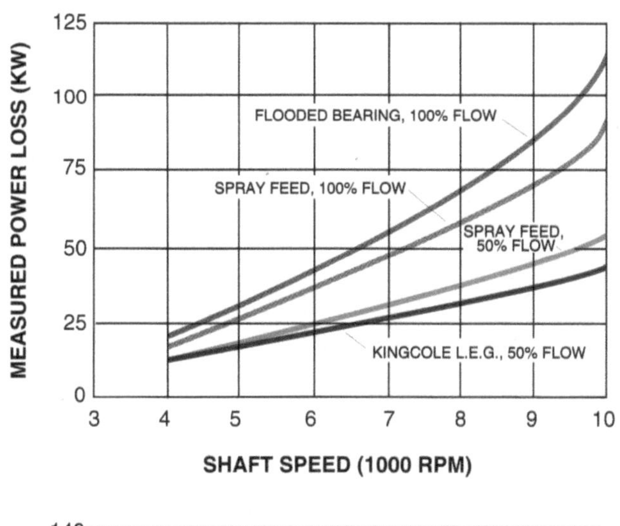

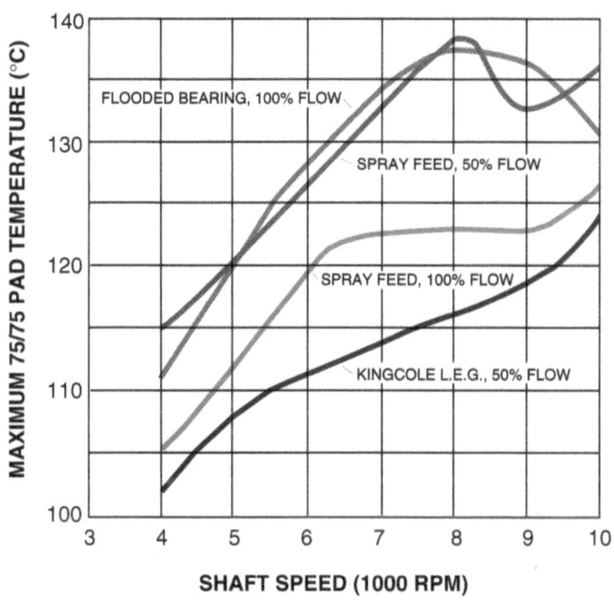

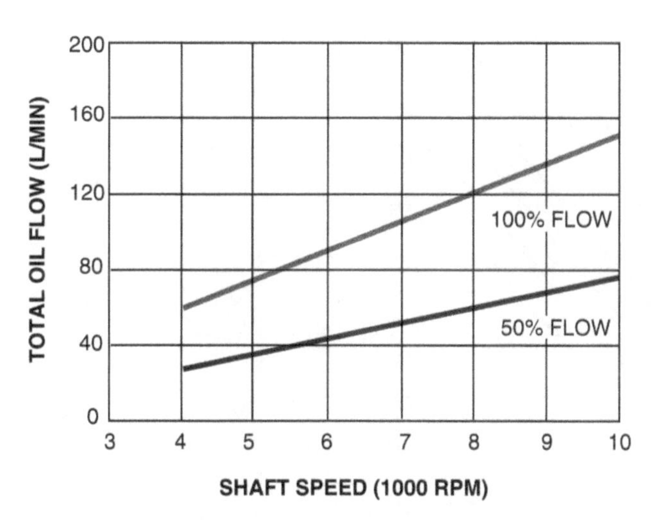

Figure 11-25. "LEG" bearings vs. standard flooded bearings and spray-fed bearings.

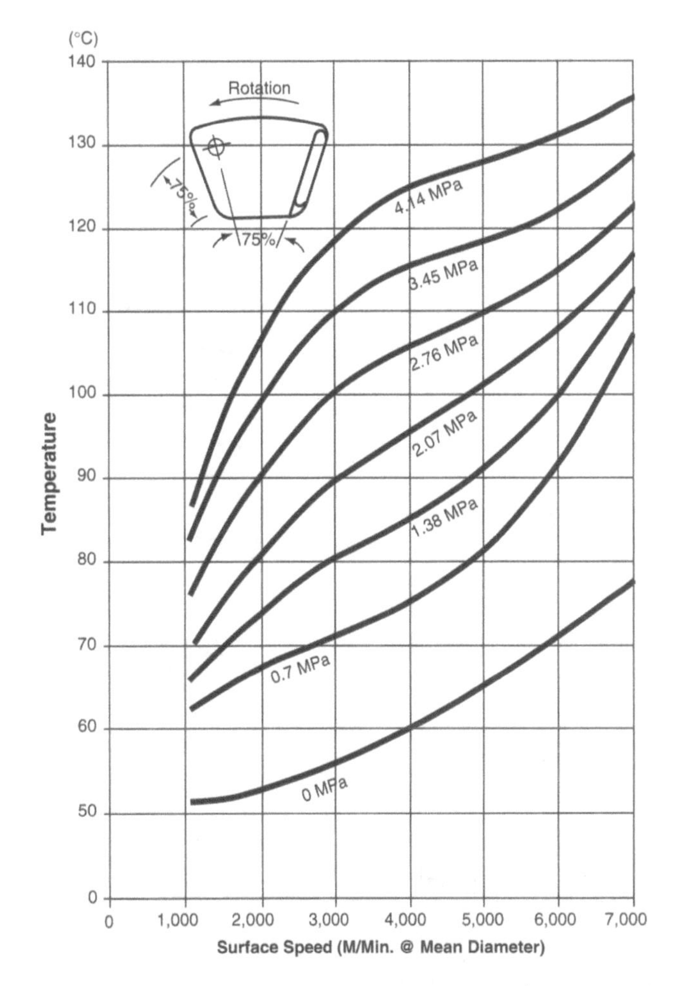

Figure 11-26. "LEG" white metal temperatures at 75/75 position (6 and 8-pad series, steel pads).

- **hydrodynamic sliding bearings**
  where pressure is built up in a converging lubricating gap

- **hydrostatic sliding bearings**
  where pressure is built up outside the lubricating gap.

- **dry sliding bearings**
  made of non-metallic or metallic sliding materials

- **sintered bearings**
  made of porous sliding materials

The service conditions may vary depending on the load and stress factors.

Sliding bearings can only function properly when they are adequately lubricated.

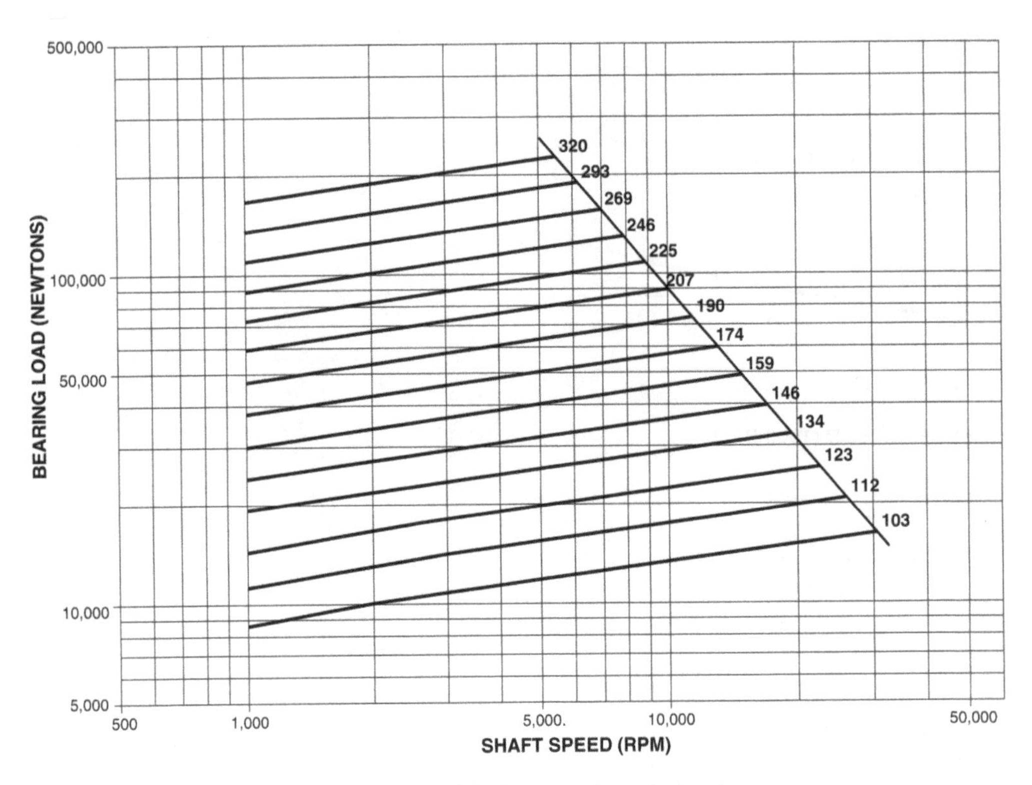

Figure 11-27. Rated load for 8-pad "LEG" bearings

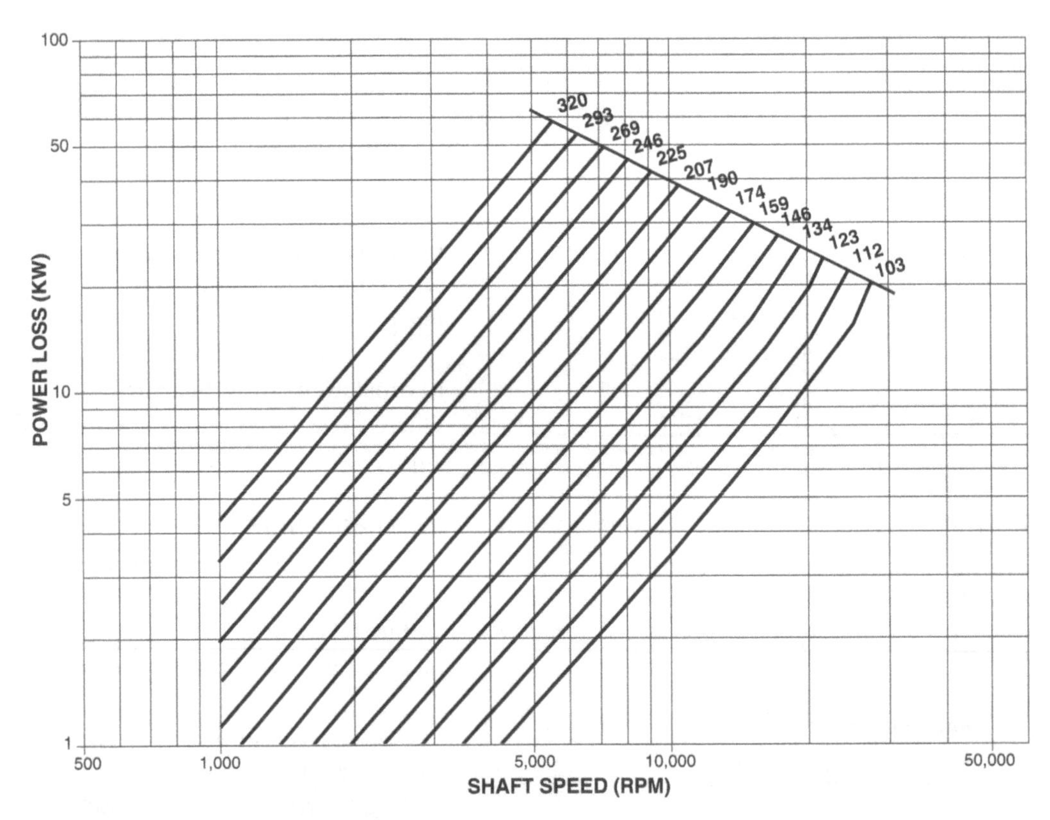

Figure 11-28. Frictional loss for single element 8-pad "LEG" bearings.

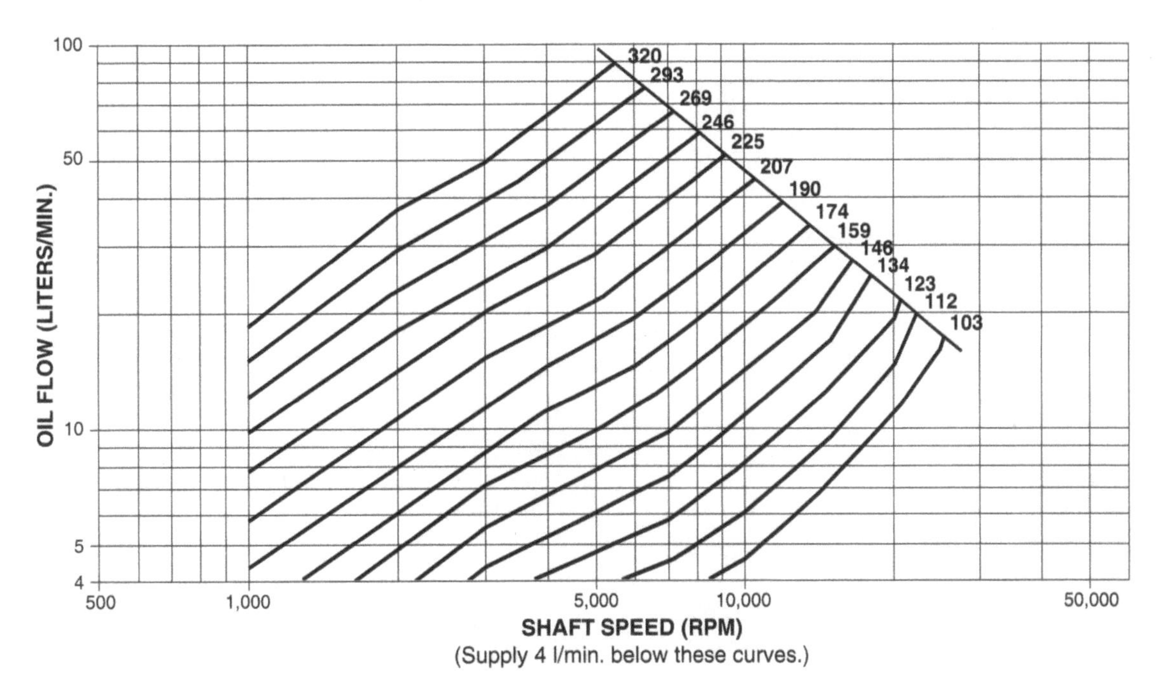

Figure 11-29. Recommended lubricant supply for single element 8-pad "LEG" bearings.

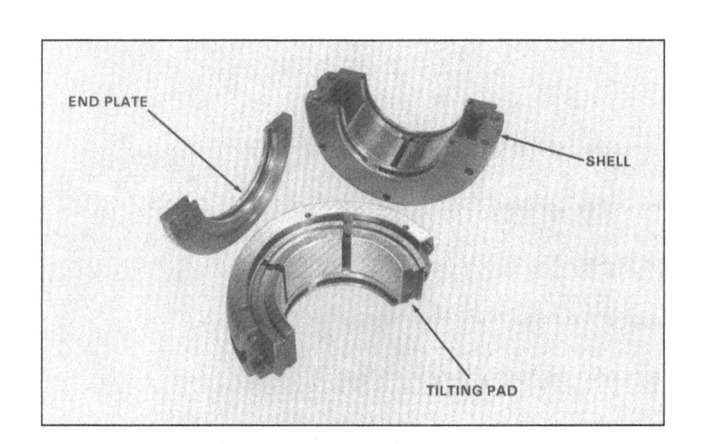

Figure 11-30. Tilting pad bearing components.

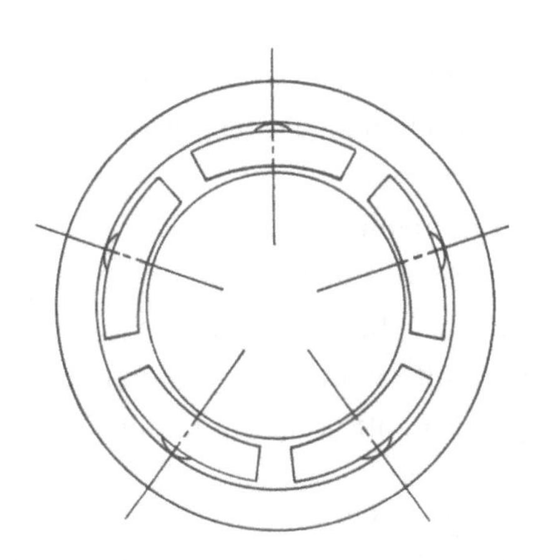

Figure 11-31. Tilting-pad journal bearing, converging geometry.

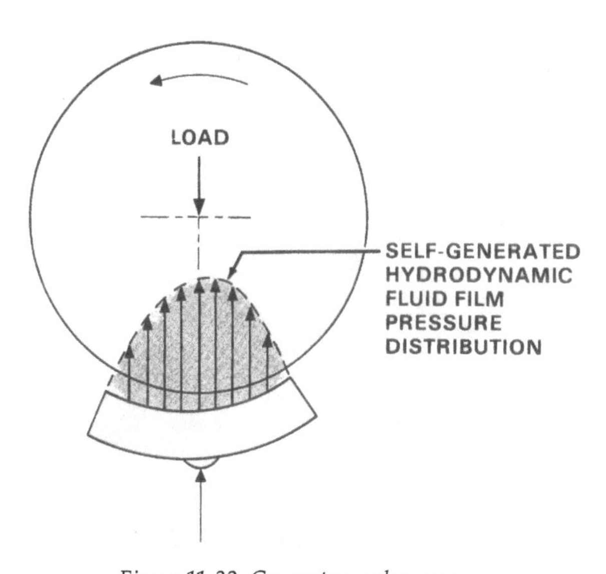

Figure 11-32. Geo-metry and pressure.

Lubrication of Hydrodynamic Sliding Bearings

Hydrodynamic sliding bearings are used to transfer power from a shaft to a housing via oil-lubricated bearing shells. Apart from an appropriate design, the main prerequisite for the reliable operation of a hydrodynamic sliding bearing is that its lubrication must be tailored to suit the operating conditions. Except during starting and stopping and during slow-turn operations, a hydrodynamic sliding bearing is subject to the laws of fluid friction described in our introductory chapter.

The selection of a proper lubricant and an adequate

Figure 11-33. Glacier bearing for 400mm shaft at 3000 rpm. This application is on a 130MW gas generator.

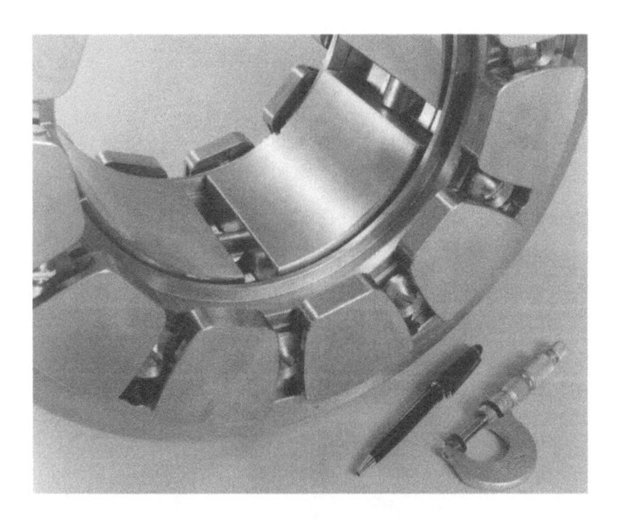

Figure 11-34. Glacier combination thrust/radial tilt pad bearing.

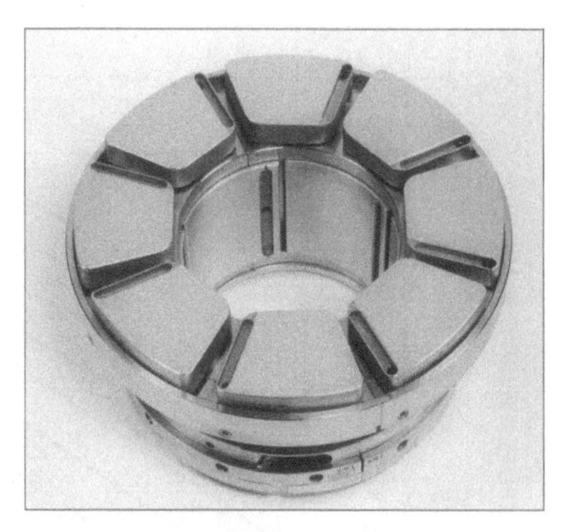

Figure 11-35. Kingsbury ("KingCole") combination "LEG" thrust and "LEG" journal bearing.

Figure 11-36. KingCole's unique split-ring design makes installation easier than standard bearings.

viscosity depends on special or additional requirements, such as

- good adhesion
- good corrosion protection
- self-lubrication with oil
- special bearing shell material
- high speed mixed friction conditions
- high and/or low temperatures
- compatibility with coatings
- compatibility with plastics/elastomers
- extended oil change intervals
- long service life
- small oil quantities
- high loads
- special steel shafts
- compatibility with the environment
- compliance with food regulations

In addition, there is the possibility to optimize hydrodynamic sliding bearings by means of tribo-system coatings or tribo-system materials.

Table 11-5 represents a broad-brush overview of

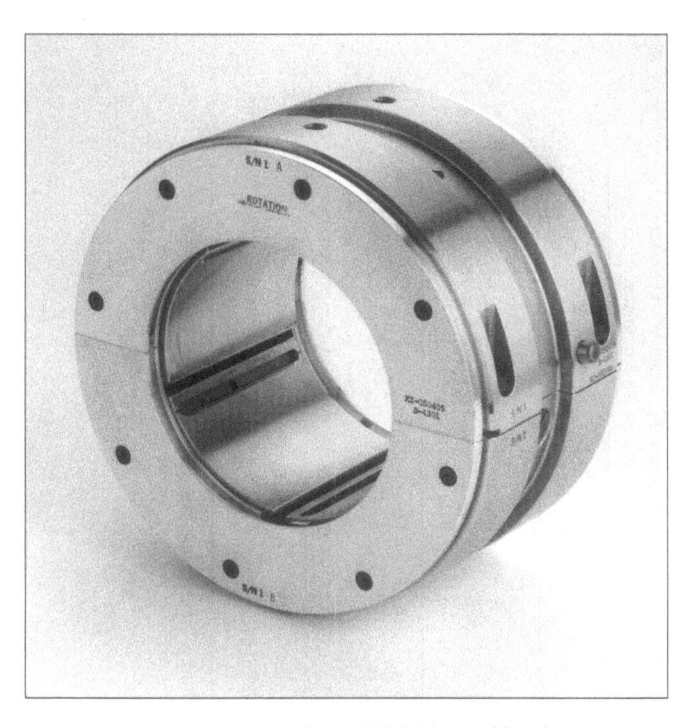

Figure 11-37. Kingsbury "LEG" journal bearing.

eight different lubricants recommended by Klüber for a variety of applications where hydrodynamic sliding bearings are used.

A typical hydrodynamic equipment bearing is shown in Figure 11-39. Here, a medium wall thickness, babbitted liner is fitted to a gear unit. At high speeds and light loads stability becomes a problem. Special bore profiles such as lemon bore, or lobes can give better shaft control and avoid oil film whirl.

Where machines approach or run through critical speeds bearing oil films are often the major source of damping. The right choice of oil viscosity and bearing bore profile (Table 11-6) can significantly reduce vibration amplitude. Also, bearing clearances (Figure 11-40) play an important role.

Sliding Bearings in the Mixed Friction Regime

It is one of the most difficult tasks in terms of tribo-engineering to lubricate sliding bearings operating in the mixed friction regime. A lubricating wedge cannot form due to the low speed and the oscillating or intermittent movements.

This is where the user may have to seek guidance from manufacturers with experience encompassing lubricating greases, waxes and wax emulsions used in the mixed and partial friction regimes. A lubricating grease is recommended for temperatures above 60°C and for friction points that are relubricated via a lubrication system. A lubricating wax is preferred for small bearings to ensure a non-tacky lubricant film suitable for lifetime lubrication. Tribo-system materials and dry lubricants are an alternative to lubricating greases and waxes.

One such tribo-system material, Klüberdur, is suitable to fill the lubricating holes in metallic bushings, thus making them dry-running bearings. Pre-start lubrication with a lubricating grease or wax could be essential to adequate running-in.

Dry-running bearings (Figure 11-41) can also be manufactured from semi-finished tubes for bushings. A fluid tribo-system material is poured in the tube and the tube is rotated to ensure that the lubricant distributes evenly. The bushings are then cut in length and subject to the finishing process.

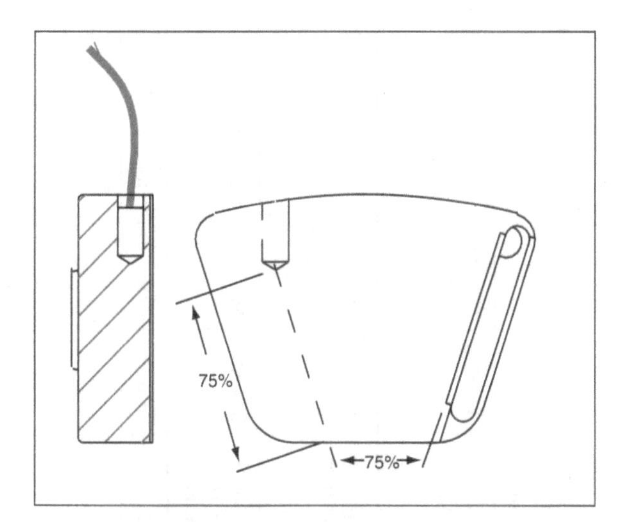

Recommended Sensor Locations

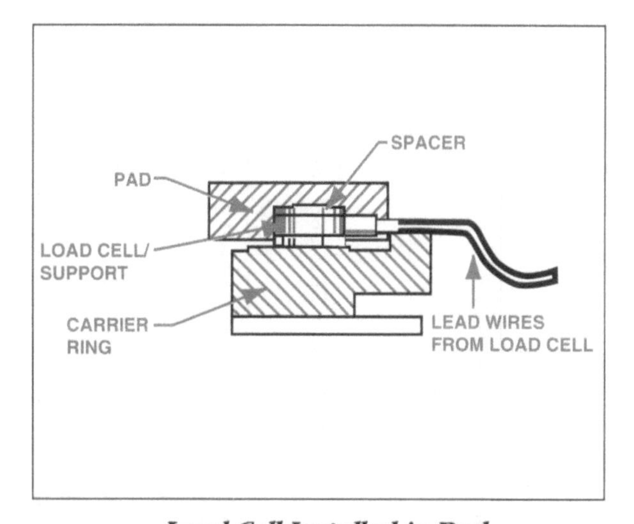

Load Cell Installed in Pad

Figure 11-38. Temperature sensor (left) and load cell locations (right) in KingCole "LEG" bearing pads.

Table 11-5. Selection criteria for lubricant used with hydrodynamic sliding bearings.

Selection criteria	Product name	Service tempera-ture-range (°C) ≈	ISO VG DIN 51 519	Density at 20°C (g/ml) DIN 51757	Kinematic viscosity DIN 51561, (mm²/s) at ≈ 40°C	Kinematic viscosity DIN 51561, (mm²/s) at ≈ 100°C	Viscosity index DIN ISO 2909 (VI)	Pour point DIN ISO 3016 (°C) ≈	Flash point DIN ISO 2592 (°C) ≈	Notes
For normal temperatures	Klübersynth GEM 1-68	–15 to 100	68	0.88	68	8	100	<–20	<200	For bearing temperatures up to approx. 80°C. From ISO VG 46 to 460, depending on the requirement.
For high temperatures	Klübersynth GH 6-80	–35 to 160	—	1.05	80	16	>200	<–35	>280	Not for aluminum sliding surfaces. Observe compatibility with some plastics and elastomers. From ISO VG 32 to 460, depending on the requirements.
For very low temperatures	ISOFLEX PDP 38	-65 to 100	12	0.93	12	3,2	125	-70	205	Observe compatibility with some plastics and elastomers. Various viscosity grades, depending on the requirements.
For the food processing industry	Klüberoil 4 UH 1-68	-35 to 110	68	0.84	68	12	>160	≤-35	>200	Especially suitable for applications in the food industry. USDA-H1 authorized. From ISO VG 32 to 460, depending on the requirements.
For high temperatures	Klübersynth GEM 4-68									From ISO VG 32 to 460, depending on the requirements
For special ambient conditions	Klüberbio CA 2-100	-25 to 120	100	0.94	100	16	>160	<-30	>200	Rapidly biodegradable, good anti-wear and anti-oxidation properties. Various viscosity grades, depending on the requirements.
For very high temperatures	Klüberalfa DH 3-100	-25 to 180	100	1.89	100	12	114	-30	without	For lubrication at very high temperatures. Excellent resistance to aging and oxidation. Various viscosity grades, depending on the requirements
Tribo-system coating / tribo-system material	Klüberdur KM 01-854	-40 to 180	—	—	—	—	—	—	—	Reduction of wear in the mixed friction regime. Available in a fluid or solid state, depending on the requirements

*Service temperature ranges refer to the individual type of oil and depend on the special case of application and the operating conditions

Table 11-6. Bearing bore options.

	Bearing type	Resistance to Half Speed Whirl	Load Capacity	Stiffness	Damping	Acceptance of Rotating Load	Relative Cost	Advantages	Disadvantages	Applications	Usage
1	Plain Bore Two Axial Groove	●	●●●●●	●●	●●	●●	○	Simple to install and line up machine	Prone to half speed whirl with light load and may require very low length/diameter ratio	All rotating machines, gearboxes and screw compressors	Wide
2	Plain Bore Circumferential Single Groove	●	●●●	●●	●●	●●●●●	○	Best for rotating load. Simple to install and line up machine	Requires high oil supply pressure. Prone to half speed whirl with light load	Blowers. Quill shaft steady bearings, some gearbox bearings	Moderate
3	Lemon Bore	●●	●●●●	●●●	●●●	●●	○○	Simple to install and line up machine	Low stiffness in horizonal direction	Large turbines and electrical machines. Pumps, gearboxes	Wide
4	Lemon Bore with bedded arc	●●	●●●●	●●●	●●●	●●	○○	Simple to install and line up machine	Low stiffness in horizontal direction	Large turbines and electrical machines	Moderate
5	Offset Halves	●●●●	●●●●	●●●●●	●●●●●	●●	○○	High lubricant pumping action gives low operating temperatures	Not suitable for reverse rotation	Turbines, pumps, electric machines, compressors, gearboxes	Increasing
6	Tilting Pad	●●●●●	●●●	●●●●	●●●●	●●	○○○	Inherently most stable of all bearings. Adapts to changing load conditions	Pivots liable to fretting with rotating or dynamic load	All rotating machines, horizontal or vertical	Increasing
7	Four Lobe	●●	●●	●●●	●●●	●●	○○	Used mainly in small sizes as a thick wall flanged bearing with taper land thrust faces	Low load capacity due to four grooves, and large lobe clearances	Small turbines Compressors Blowers Pumps	Wide

Performance Rating ●●●●● Very High ● Low Relative Cost ○ Low ○○○ High

Figure 11-39. Lowering a large Glacier medium wall bearing into a marine propulsion gearbox. (Photo: GEC Alsthom Gears Ltd.)

Metallic bushings can be coated with a "Klüberplast" sheet (see Table 11-7 for typical properties), thus making them dry-running bearings. Metallic bearings with damaged surfaces can be repaired easily with such a coating. Except for a lateral rib of approx. 0.3 mm to guide the sheet, the bearing material is cut off in a layer over the entire width.

Dry lubricants for tribo-systems can be used to impart a running layer to smooth metal or plastic bushings.

Lubrication of Sintered Metal Sliding Bearings

Sintered bearings are made of powder composites subject to pressure and heat. Depending on the material composition (sintered iron, steel, bronze), sintered metal sliding bearings have a different porosity. This is illustrated in Table 11-8.

Sintered metal sliding bearings (Figure 11-42) have open pores which are filled with a lubricant in an immersion process. They are not operational without a lubricant and are therefore generally lubricated for life. The better the lubricant fulfills its task, the longer the bearing's service life.

The high requirements, in terms of temperature stability, corrosion and wear protection as well as oxidation resistance, are met even under difficult operating conditions, such as

- low and/or high temperatures
- low and/or high speeds
- low noise
- low starting and running torque

Table 11-7. Lubricant selection for small sliding bearings.

Surface opposing steel / Type of lubricant	Metals			Pivoting bearings	Plastics	
	Al alloys	Cu alloys	Steel		PA, PBTP, PC, POM	PTFE
Lubricating greases	POLYLUB GLY 791	Klüberplex BE 31-502	Klüberplex BE 31-502	COSTRAC GL 1501 MG(pivoting angle, 15°) ISOFLEX TOPAS MB 5-51 (pivoting angle 15°)	POLYLUB GLY 801	ISOFLEX TOPAS NB 5051
Lubricating wax	Klüberplus SK 04-295		Klüberplus SK 04-295	Klüberplus SK 02-295		—
Tribo-system materials	Klüberdur KM 01-854/s 50[2] Klüberplast LD oder W2[4]	Klüberdur KM 01-854[1] Klüberplast LD oder W2[4]	Klüberdur KM 01-854/S 50[2] Klüberplast LD oder W2[4]	Klüberdur KM 01-854/S 50	Consulting required[3]	—
Dry lubricants tribo-systems	UNIMOLY CP Klübertop TG 05-371[5]		UNIMOLY C 220 Klübertop TG 05-371[5]	UNIMOLY C 220		—

[1]To fill lubricating holes, pockets and grooves.
[2]To fill into the tube before cutting the bushings.
[3]To determine additives for incorporated lubrication.
[4]To insert the sheet into the running surface.
[5]For emergency lubrication of hydrodynamic bearings.

Table 11-8. Porosity of sintered metal sliding bearings.

Material category	Porosity %	Preferred application
Sint A	25 ± 2.5	Sliding bearings
Sint B	20 ± 2.5	Sliding bearings, molded parts with sliding properties
Sint C	15 ± 2.5	Sliding bearings, molded parts

- uniform operation
- very long service life
- high or low specific surface pressure

Klüber lubricants for sintered metal sliding bearings have proven effective in practical applications. Their special properties, such as low-noise behavior, high load-carrying capacity, low friction moments, constant friction values during speed changes, etc. are characteristic of impregnating fluids.

If the service life of a sintered metal sliding bearing should be increased considerably, additional lubrication with an appropriately selected specialty lubricant provides substantial advantages as compared to felt or depot grease lubrication. This is shown in Figure 11-43, contrasting felt (oil-soaked film lubrication) and Klüber's "Mikrozella."

LUBRICATION OF MACHINE ELEMENTS*

Lubrication of Screws

Screws are the most frequently used detachable fastening elements. It is therefore quite astonishing that

Source: Klüber Lubrication North America, Inc., Londonderry, New Hampshire. Adapted by permission.

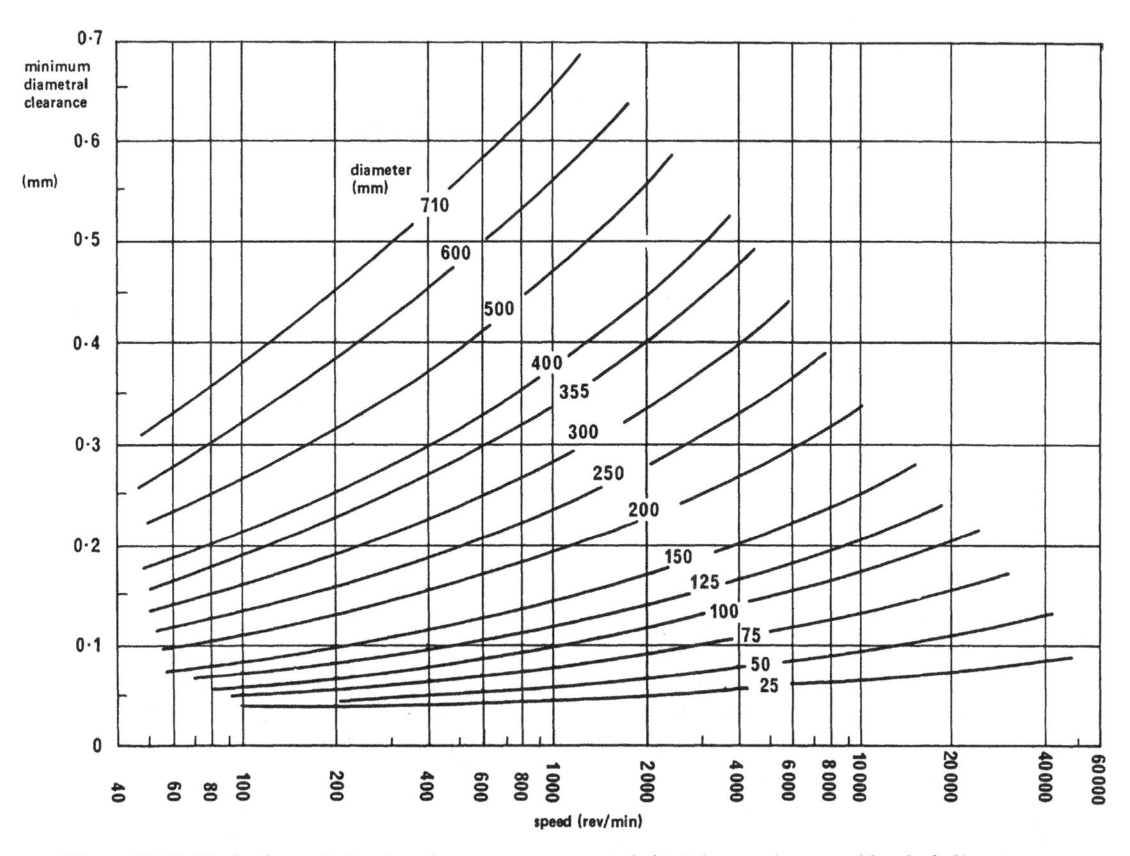

Figure 11-40. Hydrodynamic bearing clearances: recommended minima against speed by shaft diameter.

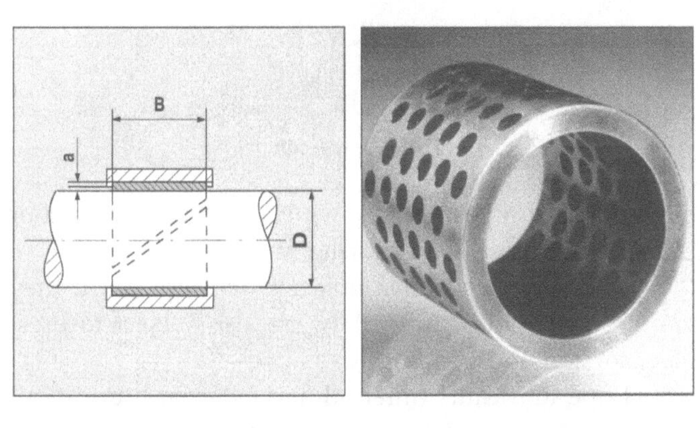

Figure 11-41. Dry-running bearings: sheet bearing (left), having lubricating holes filled (right) with "Klüberdur" tribo-system material.

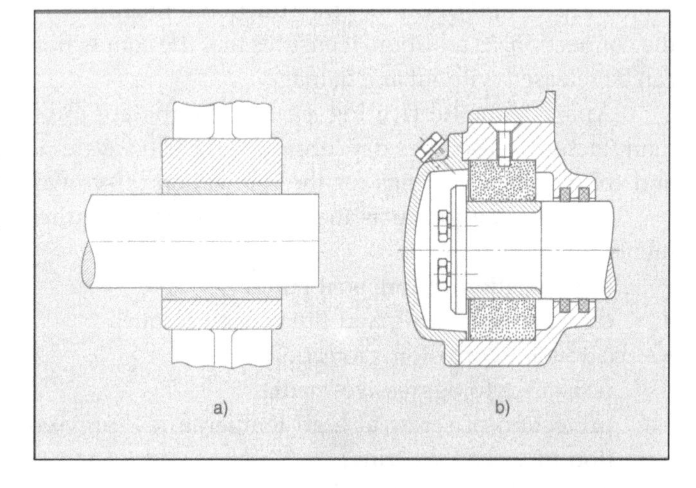

Figure 11-42. Small bearings; a) dry sliding bearing; b) sintered bearing.

their lubrication is often neglected. Damage due to insufficient or inadequate lubrication may lead to component failure, resulting in expensive maintenance or production losses.

A screw connection is a power locking connection. Therefore, the criterion for the quality of a screw connection is the required preloading force which determines the extent to which the joined components are pressed together. Screw connections that are too highly preloaded tend to fail during assembly due to elongation or breaking (Figure 11-44). If the preload is too low, the connections will fail during operation due to fatigue fracture or unintentional releasing.

The torque-controlled tightening methods currently used most frequently generate the preloading force via the tightening torque.

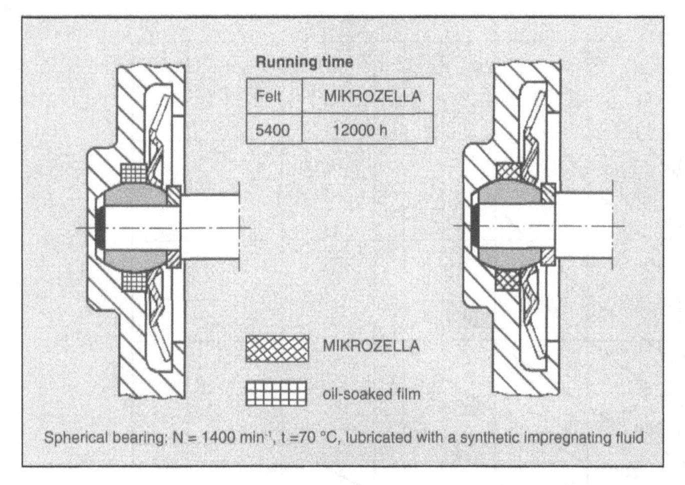

Figure 11-43. Plastic oil depot lubrication of a spherical bearing.

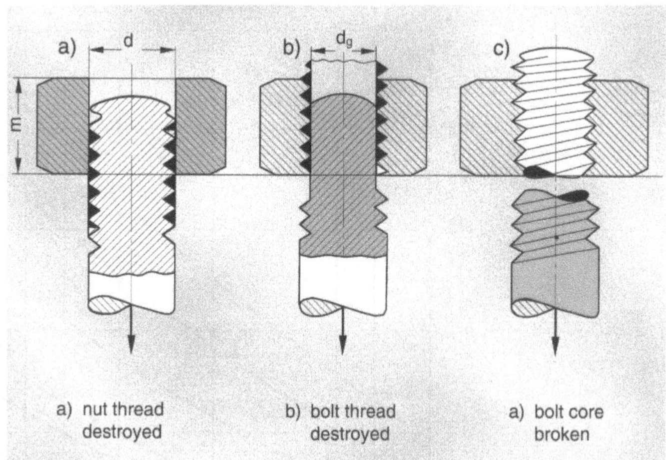

Figure 11-44. Screw connections that are too tightly preloaded tend to fail.

Apart from the assembly method, the friction behavior has a substantial impact on the clamping effect. For example, up to 90 % of the applied torque is consumed in the form of friction related to the screw head and thread. Only 10 % is available to build up the required preload.

Adequate lubrication (Figure 11-45) reduces head and thread-related friction, increases the available screw preload force and optimizes the functional reliability of the connection. In addition, it ensures that the connection can be released without any damage.

Apart from lubricating pastes, competent lube manufacturers also offer dry lubricants for tribo-systems and tribo-system coatings for the lubrication of screws.

Appropriate products meet the following requirements:

- minimization of torsional stress
- constant tightening and breakaway torques
- extended corrosion protection
- resistance to aggressive media
- protection of screw at high temperatures (prevention of thread welding)
- easy and clean application

Table 11-9 gives an overview of selection criteria, operating range, and important characteristics of screw lubricants.

Lubrication of Ropes

Ropes are elements used for materials handling. They are made of stranded wires that are combined to form ropes (Figure 11-46).

Depending on the application, they are classified

Figure 11-45. Screw compound.

into conveyor ropes (cranes, winches, elevators), anchor ropes (guy wires), load-bearing and sling ropes.

Ropes are generally subject to tensile load. If they are led over return units, they are also subject to pressure, torsion and bending loads.

Rope lubricants (internal and external lubrication) have to meet the following requirements:

- protect against wear
- protect against corrosion
- ensure required friction moments (friction pulley conveyor ropes)

Quality lubricants for ropes ensure

- compatibility with rope materials
- long service life
- weather resistance
- no dripping
- good pumpability in lubrication systems

Table 11-9. Selection criteria for screw lubricants.

Selection criteria	Product name	Base oil/ thickener	Service temperature range (°C) ≈	Density at 20°C (g/cm^3) DIN 51757 ≈	Base oil viscosity DIN 51561 (mm^2/s) ≈	Color	Drop point DIN ISO 2176 (°C)	Speed factor ($n \bullet d_m$) $mm \bullet min^{-1}$ ≈	Worked penetration DIN ISO 2137 (0,1mm)	Consistency NLGI grade DIN 51 818	Apparent viscosity KL viscosity grade	Notes
High-temperature	WOLFRACOAT C	Synthetic hydro-oil/silicate, pigments (copper...)	-30 to 200 above that dry lubrication up to 1200	1.01	—	gray, copper	without	—	270 to 310	—	M/S	Apply a thin layer to the screw or the bolt thread. Effective against tribo corrosion
White multi-purpose paste	Klüberpaste 46 MR 401	Synthetic oil/ inorganic solid lubricant	-45 to 150	1.23	380	ivory	>185	—	300 to 340	—	M	For refined steel screws

Selection criteria	Product name	Service temperature range (°C) ≈	Color	Dry to handle at ...(°C) after ...(min) ≈	Yield at a film thickness of 10μm, (m^2/l)	Burning-in temperature (°C) hardening time (minimum)	Friction coefficient (Tannert, 20°C, V_{max} = 0,243 mm/s F = 300 N)	Notes
Dry lubricant for tribo-systems	UNIMOLY * C 220	-180 to 450	gray	20/5	8	20/30 min	approx. 0.10	MoS_2 base dry lubricant, ensures highly efficient coating at a competitive price

Selection criteria	Product name	Service temperature range (°C) ≈	Density at 20°C g/ml DIN 51757 ≈	Color	Ph value ≈	Viscosity DIN 53 211 at 20°C 4mm nozzle	Four-ball tester, welding force DIN 51 350 Pt 4	Corrosion protection DIN 50 017 KFW towards ST 14	Reichert friction and wear tester, specific surface pressure N/cm^2 ≈	Notes
Lubricating wax emulsion	Klüberplus SK 01-205	-40 to 120*	0.98	ivory	8	20 to 30 s	≤ 1,200 N	≤ 20 cycles no corrosion	2,000	Ready-to-use product; dilution with water between 1:1 and 1:5 is possible if the resulting lubricant film is sufficiently thick for the intended use

*As a dry, water-free film, otherwise 5 to 95°C

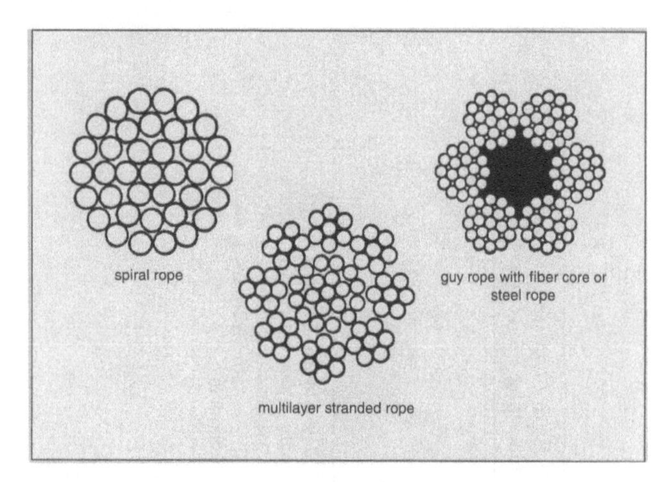

Open ropes

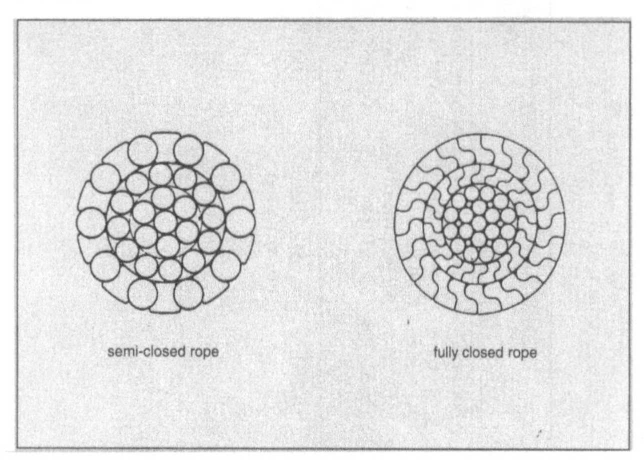

Closed ropes

Figure 11-46. Open and closed ropes.

- availability all over the world

Refer to Table 11-10 for typical selection criteria.

Lubrication of Seals

Seals are machine elements that separate spaces containing different substances and/or are subject to different pressures. For example, a seal ensures the long service life of a rolling bearing by preventing lubricant leaks and the ingress of foreign matter into the bearing. Simple seals are depicted in Figure 11-47.

A suitable lubricant is required in order for a seal to ensure a component's operational reliability over a predetermined service life. The lubricant has to meet the following requirements:

- permit damage-free installation of the seal
- dissipate frictional heat
- increase the sealing effect
- prevent adhesion of the seal even after a long standstill
- permit easy disassembly
- be compatible with the sealing material and resistant to ambient media

The product range offered by world-class lubricant manufacturers includes suitable lubricants for all types of seals and applications: for seals subject to static or dynamic loads, seals operating under extreme temperatures, in the presence of aggressive media or oxygen, or seals used in the food processing industry.

Table 11-11 gives an overview of products offered by one manufacturer, Klüber, together with relevant physical characteristics.

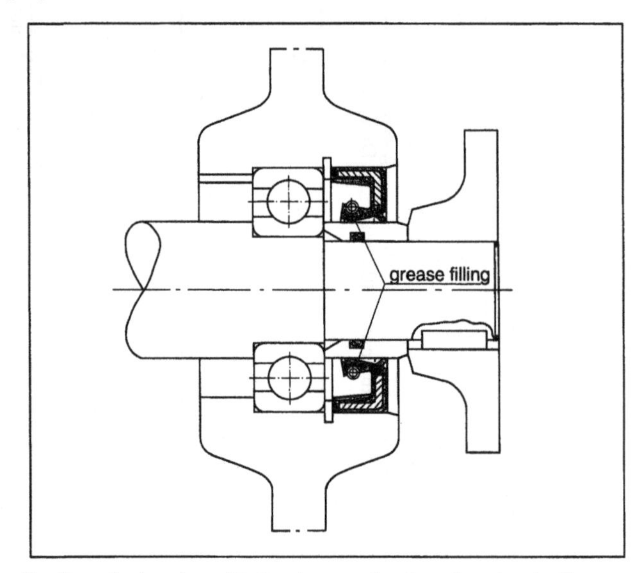

Sealing of a bearing with the danger of external contamination

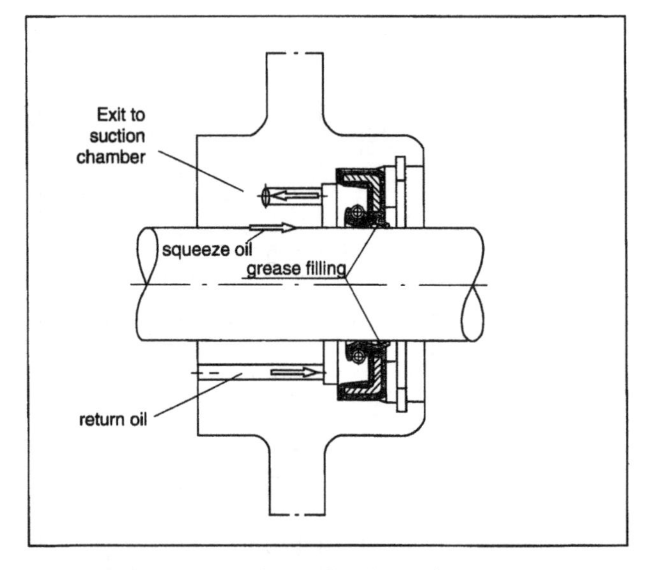

Sealing against pressure (example: oil pump)

Figure 11-47. Simple elastomeric seals benefit from lubrication.

Table 11-10. Selection criteria for rope lubricants.

Selection criteria	Product name	Base oil/ thickener	Service temperature range (°C) ≈	Density at 20°C (g/cm³) DIN 51757 ≈	Base oil viscosity DIN 51561 (mm²/s) ≈ 40°	100°C	Color	Drop point DIN ISO 2176 (°C)	Consistency NLGI grade DIN 51 818	Speed factor (n • d_m) mm • min⁻¹ ≈	Worked penetration DIN ISO 2137 (0,1mm)	Apparent viscosity KL viscosity grade DIN	Notes
External and internal rope lubrication (except friction pulley ropes)	GRAFLOSCON C-SG 2000 PLUS	Mineral hydrocarbon oil Al soap, solid lubricant	-30 to 200	0.98	—	—	black-gray	>70	00	—	400 to 430	M	Multi-purpose lubricant, good corrosion protection, resistant to ambient media
Grease for window lifter and sunroof ropes	ISOFLEX TOPAS L 32	Synthetic hydrocarbon oil/Li soap	-60 to 130	0.86	16 -19	3,5 -4,2	beige	>180	2	—	265 to 295	EL/L	Low starting torque also at low temperatures; neutral towards thermoplastics and duroplastics
Grease for clutch, accelerator and brake system ropes	UNISILKON GLK 112	Silicone oil/ spec. Li soap	-55 to 180	0.97	90 -110	53	white	>200	2	—	265 to 295	M	Especially low starting torque, low variation throughout the service temperature range
	ISOFLEX TOPAS AK 50	Synthetic hydrocarbon oil, Al complex soap	-50 to 150	0.87	3	5,5	whitich, almost transparent	>200	—	—	355 to 385	EL	For the same application range, but if a silicone-free product is required
Lubricant for friction pulley ropes	Klüberplus SK 03-498	—	—	—	—	—	—	—	—	—	—	—	Meets Austrian requirement for wire ropes (DSB 80), test certificate by the Techniche Versuchs- und Forschugsanstalt TU Wien (1993)

Selection criteria	Product name	Base oil/ thickener	Density at 20°C (g/ml) DIN 51757 ≈	Kinematic viscosity DIN 51561, (mm²/s) at ≈ 40°C	100°C	ISO VG DIN 51 519	Viscosity index DIN ISO 2909 (VI)	Pour point DIN ISO 3016 (°C) ≈	Flash point DIN ISO 2592 (°C) ≈	Notes
Relubrication of ropes (except friction pulley ropes)	Klüberbio C 2-46	Ester oil/ fatty oil	0.93	46	10	46	—	<-30	>200	Rapidly biodegradable oil
Relubrication of friction pulley ropes	Klüberplus SK 06-418	Synthetic hydrocarbon oil. Al complex soap	0.80	—	—	—	—	—	>55	Very adhesive oil for the relubrication of friction pulley ropes

Table 11-11. Overview of lubricants for elastomeric seals.

Selection criteria	Product name	Base oil/ thickener	Service tempera- ture range (°C) ≈	Density at 20°C (g/cm^3) DIN 51757 ≈	Color	Drop point DIN ISO 2176 ≈	Worked penetration DIN ISO 2137 (0,1mm) ≈	Consis- tency NLGI grade DIN 51 818	Apparent viscosity KL viscosity grade	Compati- bility with elastomers*	Notes
Assembly and sealing grease	MICRO- LUBE GL 261	Min. hydro- carbon oil/ spec. Li	-30 to 140	0.89	yellowish brown, almost transparent	>220	310 to 340	1	L/M	NBR, MVQ, FPM, AU	Particularly suitable for shaft seals, O-ring seals, flat seals; wide service temperature range
Starting support for rotary shaft lip seals	Klüber- plus SK 01-205	Synthetic hydrocar- bons, sur- factants, emulsifiers	-40 to 12-**	0.98	ivory	—	—	—	Viscosity, DIN 53211 at 20°C 4 mm	NBR, EPDM, MVQ, FPM	Ready-to-use lubricant emulsion, water-miscible, free from solvents and silicone; can be diluted with tap water as required; once dry, there is a thin lubricant film that is fast to handle and protects against corrosion
High-tem- perature metal paste for flat seals	WOFERA- COAT C	Synth. hydro- carbon oil/silicate pigments	-30 to 200 above that dry lubrication up to 1200	1.01	gray, copper	without	270 to 310	—	M/S	NBR, MVQ FPM	Prevents adhesion of the seal to the opposing surface, permits easy dis- assembly; especially suitable for metal seals subject to very high temperatures
Sealing grease for food applications	PARALIQ GB 363	Paraffinic mineral hydrocarbon oil/silicate	-30 to 140	0.91	transparent to yellowish	without	215 to 245	—	S	NBR, MVQ,	USD-H1 authorized; resistant to dis- infectants and cleaning agents, does not impair the formation of beer froth, no smell or taste
	PARALIQ GTE 703	Silicone oil/PTFE	-50 to 150	1.31	white, cream	>250	220 to 250	3	S	NBR, EPDM, FPM	
Sliding agents for seals in oxygen installations	OXIGENO- EX FF 250	Fluorinated polyether oil/PTFE, silicate	-50 to 200	2.0	light gray	without	240 to 270	—	S	NBR, EPDM,	For applications in contact with gaseous oxygen up to the following oxygen pressure limits: to 60°C:260 bar between 60°C and 150°C: 220 bar between 150°C and 200°C: 210 bar
Dry lubri- cant for tribo- systems	FLUORO- PAN CO 60	Organic binder/ solvents (ester, alco- hol), solid lubricants (e.g., PTFE	-15 to 80	0.9	black	—	—	—	Viscosity DIN 53211 at 20°C 2 mm nozzle(s) 50 - 64	NBR, EPDM,	Organic air-drying bonded coating, PTFE base, excellent adhesion to metals and plastics

*Data in accordance with current experience. No claim for completeness. Owing to the many possible combinations of elastomer materials we recommend having the lubricant tested for compatibility by the elastomer manufacturer before using it in series applications.
**As a dry, water-free film; otherwise 5 to 95°C

Lubrication of Chains

Chains (Figure 11-48) are multi-purpose design elements used to transfer power. They consists of many links, mostly metallic ones. They are used, among others, as

- drive chains (e.g., bicycle)
- control chains (e.g., automotive engine)
- lifting chains (e.g., sluice gate)
- transport chains (e.g., conveyor system)

There are various types of chains suiting most different requirements, for example roller, bushing, pin and inverted tooth chains. Owing to its versatility, the roller chain is widely used.

As a chain performs a very complex movement, the tribo-system needs a special lubricant to meet all requirements.

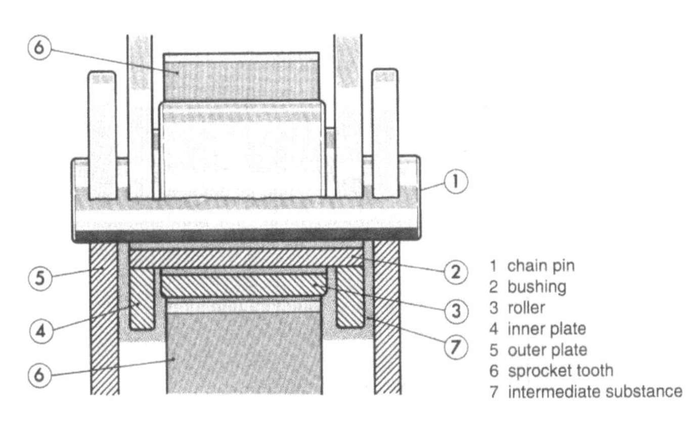

Figure 11-48. Chain components.

Figure 11-49. Chain test rig at Klüber laboratories.

- The oscillating movements of the friction components result in a permanent state of mixed friction.

- The line contact of pins, bushings and rollers results in very high specific surface pressures.

- The intermeshing of a chain link and a gear tooth results in high shock loads.

Chain lubricants must be able to absorb high pressures and to reliably protect against wear in order to ensure that only a minimum of so-called "permissible wear" takes place even though the whole system operates in the mixed friction regime.

Depending on the individual application, chain lubricants must also meet the following requirements:

- corrosion protection
- wetting/spreading properties
- adhesiveness
- high-temperature stability
- dissolution of used lubricant
- low coking tendency
- suitability for low temperatures
- resistance to ambient media

Other selection criteria may include:

- compliance with food regulations
- environmental aspects (rapidly biodegradable)
- noise damping

Again, competent manufacturers have developed and manufactured efficient chain lubricants for years. Some have built their own chain test rigs to test, among others, a product's antiwear properties, lubricating efficiency and suitability for high-temperature applications in practice-oriented tests (Figure 11-49). Many offer fully synthetic chain oils for high-temperature applications tailored to suit specific requirements.

Refer to Table 11-12 for selection criteria and physical properties of typical chain lubricants.

Shaft/hub Connections

Shaft/hub connections are positive or friction/power-locking connections to transfer torques. Positive-locking connections make it possible to move the hub and the shaft in an axial direction.

Figure 11-12. Selection criteria and typical properties of chain lubricants.

Selection criteria	Product name	Service temperature range (°C) ≈	ISO VG DIN 51 519	Density at 20°C (g/ml) DIN 51757	Kinematic viscosity DIN 51561, (mm^2/s) at ≈ 40°C	100°C	Viscosity index DIN ISO 2909 (VI)	Pour point DIN ISO 3016 (°C) ≈	Flash point DIN ISO 2592 (°C) ≈	Notes
Synthetic high-temperature chain oils	HOTEMP OY 95	0 to 250	—	0.95	95	13	120	-35	>280	Main application: conveyor systems comprising a drying section; polymerization furnaces for glass/rock wool
	HOTEMP SUPER	0 to 250	—	0.91	260	30	150	<-30	230	For continuous particle board or laminate presses
	HOTEMP PLUS	0 to 250	320	0.95	320	27	110	>-25	>250	Especially for fabric conveyor chains in textile finishing machines
	PRIMIUM SUPER	0 to 250	—	0.95	300	26	>100	-20	>230	For conveyor chains in film stretching machines
Synthetic lubricating oils for the food processing industry for low and normal temperatures	Klüberoil 4 UH 1 68	-40 to 130	68	0.84	68	12	>160	≤-35	>200	Food grade lubricant with USDA-H1 authorization; for the lubrication of drive and conveyor chains operating at low ambient temperatures
	Klüberoil 4 UH 1 460	-30 to 130	460	0.85	460	52	>170	≤35	>200	Food grade lubricant with USDA-H1 authorization; for the lubrication of drive and conveyor chains operating at low ambient temperatures
Mineral hydrocarbon chain lubricants for normal temperatures and for the wet section	STRUCTOVIS EHD	-5 to 150	460	0.89	460	36	>100	-10	>220	Hydrokapilla effect (spreads under humid spots/water); for conveyor chains operating in a wet section
Rapidly biodegradable chain oil for the normal temperature range	Klüberbio C 2-46	-40 to 80	46	0.93	46	10	—	<-30	>200	Rapidly biodegradable chain oil, water hazard category 0; for escalator chains or chains in construction and agricultural machines
Dry lubricant suspension	WOLFRASYN UL 129 G 10	-30 to 500	—	1.06	360	55	—	—	—	Plate carrier chain in baking ovens; dry lubrication above approx. 200°C
Solid lubricant suspension	WOLFRAKOTE TOP FLUID	-25 to 1000	—	1.10	310	20	—	—	—	For chains used e.g. in burning, annealing or melting furnaces; dry lubrication above approx. 200°C
Dry lubricant for tribo-systems	UNIMOLY C 220* MoS_2-bonded coating	-180 to 450	Further data see machine elements "screws"							For special cases in which the application of an oil or grease is not possible, or to improve the emergency- running properties

*also available as a spray (CFC-free)

A lubricant used in shaft/hub connections has to meet various requirements. The main task, however, is to prevent fretting and tribo-corrosion which would have a negative effect on the surface of the friction components. Tribo-corrosion often occurs in positive and powerlocking machine elements.

Tribo-corrosion is a generic term describing the physical and chemical influences on materials. Small relative movements (microsliding) in the contact zone mechanically excite the surface layers, resulting in a strong reaction of the component material and the atmospheric oxygen. Oxidation products (wear particles) accumulate in the joints which, unless removed, will lead to malfunctions and have an impact on the axial sliding movements (Figure 11-50).

Tribo-corrosion is caused by the following load factors:

- vibrations
- micro-sliding
- oscillations
- condensation water
- atmospheric oxygen
- torque changes

Specialty products with the characteristics shown in Table 11-13 control these factors to such an extent that functional reliability is ensured. They also provide advantages during assembly and disassembly by facilitating

- pressing-in,
- sliding-on,
- pressing-out, and
- sliding off of the components.

For the selection of an adequate lubricant it is important to take into consideration the type of shaft/hub connection and the bearing fit (interference, transition or loose fit). As can be seen on Figure 11-51, the friction-locking conditions of interference fits require a different lubricant than the transition or loose fit in case of axial sliding movements. Again, refer to Table 11-13 for selection criteria.

Lubrication of Valves and Fittings

Valves and fittings are integral elements of pipe systems fulfilling a

- control (open/close) and
- adjustment (mixing, etc.) function.

They are used in pipes transporting solids, fluids and gases.

An adequate lubricant on the individual valve components (e.g., seal, stuffing box, spindle, ceramic disks, plug, as illustrated in Figure 11-52), provides smooth operation and sealing, and minimizes wear.

To ensure long-term operational reliability of valves, lubricants should meet the following requirements:

- Resistance to ambient media
- Securing effect
- Neutrality towards other materials (metals, plastics, elastomers)
- Compliance with food regulations.

These regulations, or their appropriate US or other national codes, should be the equivalent of:

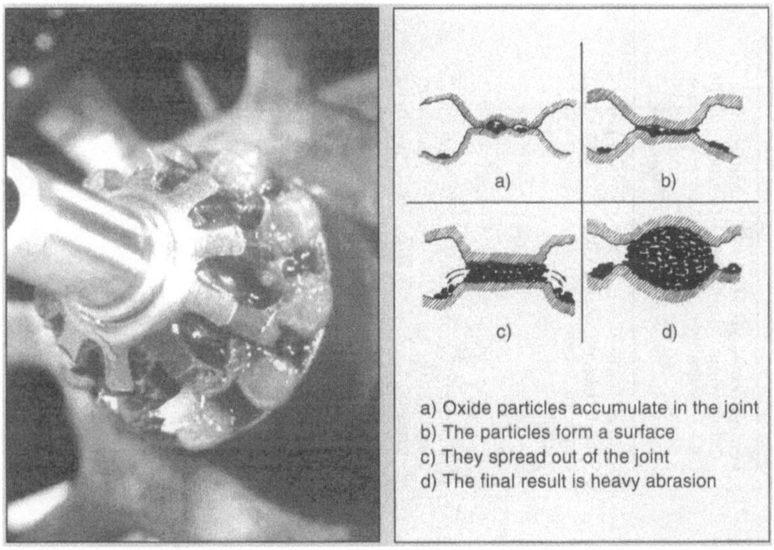

a) b) c) d)

a) Oxide particles accumulate in the joint
b) The particles form a surface
c) They spread out of the joint
d) The final result is heavy abrasion

Figure 11-50. Lubrication of a multiple spline shaft and formation of fretting corrosion.

Table 11-13. Selection criteria for assembly lubricants.

Selection criteria	Product name	Base oil/ thickener	Service temperature range (°C) ≈	Density at 20°C (g/cm^3) DIN 51757 ≈	Color	Drop point DIN ISO 2176 ≈	Speed factor ($n \bullet d_m$) $mm \bullet min^{-1}$ ≈	Worked penetration DIN ISO 2137 (0,1mm) ≈	Consistency NLGI grade DIN 51 818	Apparent viscosity KL viscosity grade	Notes
Bearing fits Lubricating and assembly paste for interference and transition fits	ALTEMP Q NB 50	Mineral hydrocarbon oil/Ba complex soap and solid lubricants	-15 to 150	1.4	beige	>170	—	250 to 270	3/2	M/S	Lubricating and assembly paste for universal assembly application in mechanical engineering. Apply in a uniformly thin layer by hand, cloth or brush. Do not rub in.
Assembly paste for loose fits	STABURAGS NBU 30 PTM	Mineral hydrocarbon oil/Ba complex soap and solid lubricants	-10 to 160	1.1	light gray	>220	—	245 to 275	3/2	S	Assembly paste for humid outdoor applications. Apply a thin layer by hand, cloth or brush. Do not rub in.
Feather keys lubricating and assembly paste	ALTEMP Q NB 50	Min. hydrocarbon oil/Ba complex soap and solid lubricants	-15 to 150	1.4	beige	>170	—	250 to 270	3/2	M/S	Lubricating and assembly paste for loose and interference fits and for general assembly applications in mechanical engineering
Spline shafts Serrated teeth Special lubricating grease	MICROLUBE GL 261	Mineral hydrocarbon oil/special Li soap	-30 to 140	0.89	yellowish brown	>220	300 000	310 to 340	1	L/M	Special lubricating grease suitable for rolling and plain bearings, girth gears, small gears, adjusting gears, winches, roller guides. Also suitable as a sealing and sliding grease for rubber seals resistant to oil and grease
Special lubricating grease	MICROLUBE GNY 202	Mineral hydrocarbon oil, synthetic oil/Na complex soap	-20 to 150	0.9	brownish	>220	400 000	245 to 275	3/2	M	Special lubricating grease for the lifetime lubrication of rolling bearings, e.g., in electric motors, fans, vibrators
Dry lubricant for tribo-systems (in combination with special lubricating grease = sandwich lubrication	UNIMOLY C 220 Spray UNIMOLY C 220	Solid lubricant/ inorganic binder	-180 to 450	1.06	gray	—	—	—	—	—	MoS_2 base dry lubricant for tribo-systems. Particularly suitable for dry and sandwich lubrication. Spray a very thin layer of UNIMOLY C 220 to the clean, dry surfaces and grease for example with MICROLUBE GL 261 or MICROLUBE GNY 202

- BAM* test for an application in oxygen installations
- DVGW† approvals in accordance with the pertinent drinking water regulations for an application in sanitary and drinking water valves
  DVGW approval in accordance with DIN 3536 for an application in gas installations
- WRC¶ approval for use in potable water supplies

Refer to Table 11-14 for selection criteria and properties.

---

*BAM = German Federal Institute for Materials Research and Testing
†DVGW = German Association of Plumbers—Regulations pertaining to synthetic materials in drinking water installations
¶WRC = Water Research Council

Lubrication of Electrical Switches and Contacts

Switches are components consisting of one or more electrical contacts. They are actuated mechanically, thermally, electromagnetically, hydraulically or pneumatically.

Their function is to separate or close electric circuits even when subjected to high loads (Figure 11-53).

Lubricants applied in switches ensure the following advantages:

- increased life cycle
- protection against wear
- reduced switching pressure
- reduced switching noise
- reliable contacts
- protection against corrosion
- prevention of fretting
- reduction of friction forces

Competent lubricant manufacturers can offer special formulations that meet all of the following requirements and even surpass them in many respects:

- high affinity towards metals
- compatibility with plastics
- thermal stability
- purity
- constant high quality
- excellent aging resistance
- easy application

Table 11-15 gives an overview of selection parameters and pertinent characteristics of these lubrication products.

Detachable and/or movable contacts such as switches, sliding and plug-in contacts are interesting from a tribological point of view. They are normally made of metal alloys plus a coating depending on the application (Figure 11-54 and 11-55). Their geometries vary according to the intended use.

All contacts must be able to conduct electric power. In addition, switches must interrupt, close and insulate an electric circuit.

The resulting loads and requirements can only be met with special lubricants:

- reduction of plug-in forces
- avoidance of tribo-corrosion
- protection against oxidation

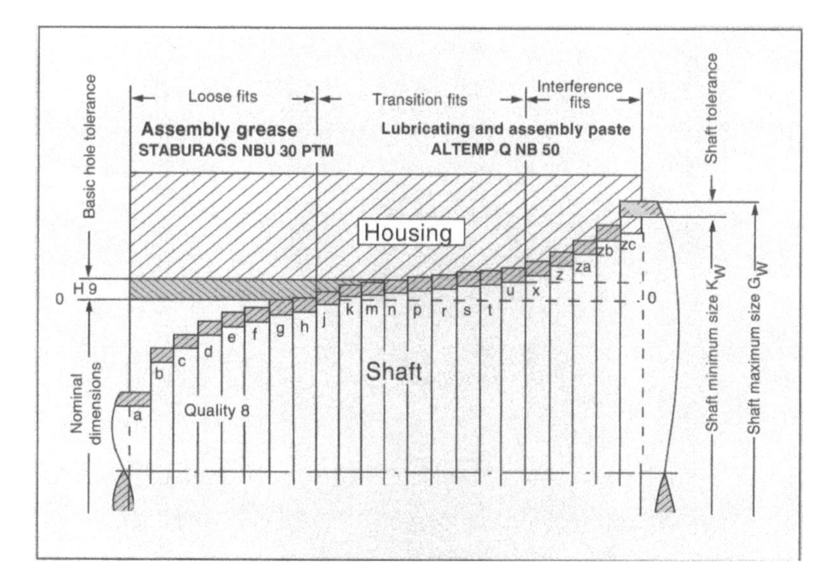

Figure 11-51. Loose fits require different lubricants than interference fits. (Source: Klüber Lubrication North America, Londonderry, New Hampshire.)

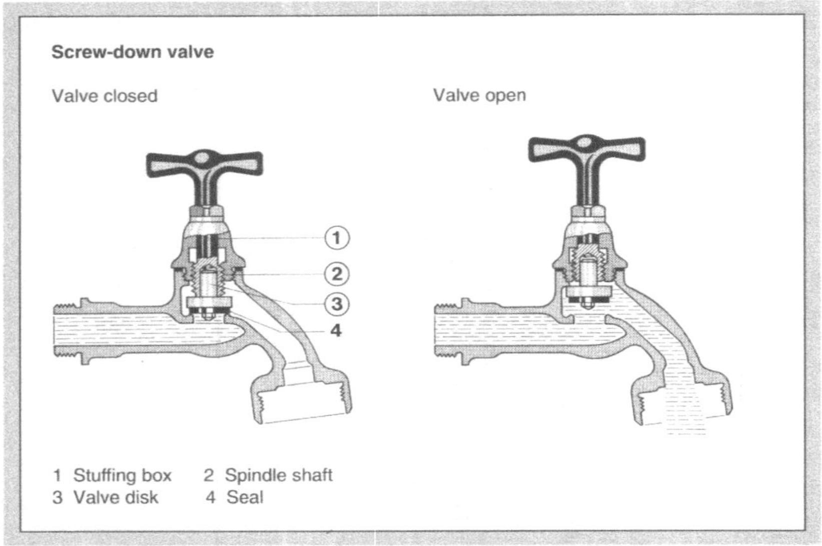

Screw-down valve
Valve closed
Valve open
1
2
3
4
1 Stuffing box
2 Spindle shaft
3 Valve disk
4 Seal

Valves (① ② ③ = lubricated)

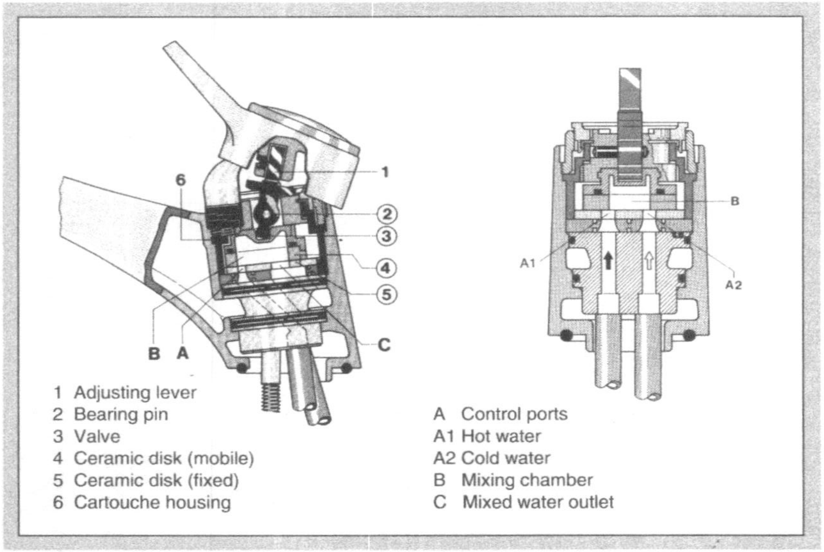

6
1
2
3
4
5
C
B
A
A1
A2
B
1 Adjusting lever
2 Bearing pin
3 Valve
4 Ceramic disk (mobile)
5 Ceramic disk (fixed)
6 Cartouche housing
A Control ports
A1 Hot water
A2 Cold water
B Mixing chamber
C Mixed water outlet

Single-lever mixer tap with ceramic disks (② - ⑤ = lubricated)

Figure 11-52. Valve components may require lubrication.

DIL EM-10
MOELLER

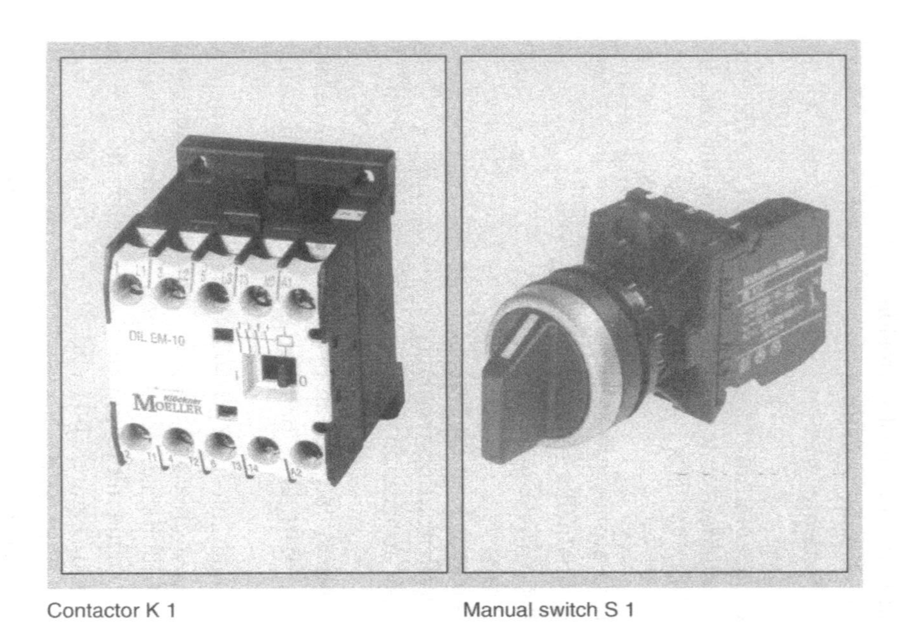

Contactor K 1

Manual switch S 1

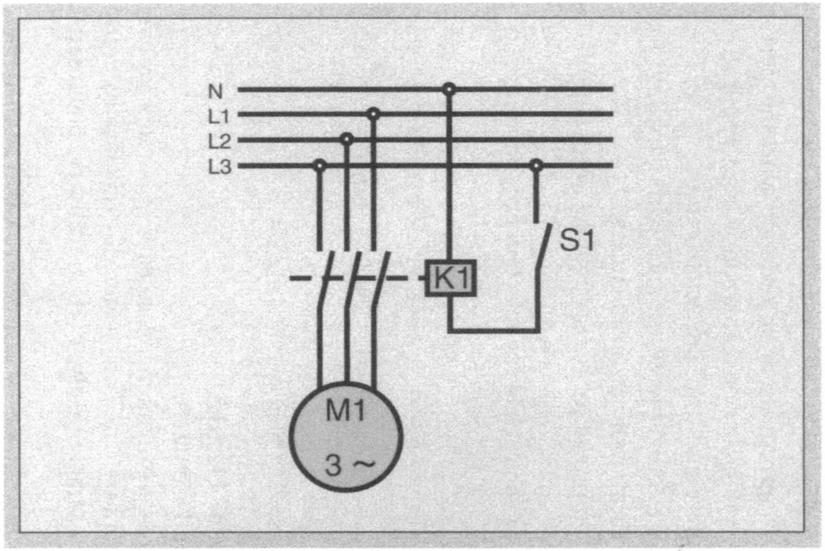

N
L1
L2
L3
S1
K1
M1
3 ~

Simplified connection diagram depicting a three-phase motor switched on/off by a contactor K 1 via a manual switch S 1

Figure 11-53. Some electrical switches and contactors must be lubricated.

Table 11-14. Selection criteria for valve lubricants.

Selection criteria	Product name	Base oil/ thickener	Service temperature range (°C) ≈	Density at 20°C (g/cm^3) DIN 51757 ≈	Color	Drop point DIN ISO 2176 ≈	Worked penetration DIN ISO 2137 (0,1mm) ≈	Consistency NLGI grade DIN 51 818	Apparent viscosity, KL viscosity grade	Test certificates, approvals, recommendations	Notes
Valve grease and drinking water applications	UNI-SILKON L 641	Methyl silicone oil/PTFE	-40 to 160	1.25	white cream	>230	300 to 320	—	—	DIN DVGW-KTW, WRC, USDA H 1	Single-lever and thermostatic mixer taps, drain valves
	Klübersynth VR 69-252	Fatty/oil/ silicate	-10 to 100	1.0	light olive	without	280 to 310	—	M/S	DIN DVGW-KTW	Spindle heads with EPDM/NBR elastomers
Sealing disk (sanitary and drinking water applications)	Klüberplast W 2*	—	-240 to 260	—	black	—	—	—	—	DIN DVGW-KTW, WRC USDA	Tribo-system material for sealing disks and stuffing boxes used in valves
Valve grease (beverage and food installations)	Klübersynth UH 1 14-31	Synthetic hydrocarbon oil, ester oil/ Al complex soap	-50 to 120	0.9	white	>220	310 to 340	1	—	USDA H 1	High resistance to aging and oxidation
	PARALIQ GTE 703	Silicone oil/ PTFE	-50 to 150	1.31	white cream	>250	220-250	3	S	USDA H 1	Special grease for the food processing industry. Neutral towards beer froth, resistant to hot and cold water, steam, disinfectant solutions. Well-proven in filling valves with EPDM seals. Universal application in plug and spindle valves
Gas valve grease	STABURAGS N 32	Mineral hydrocarbon oil/Na complex soap, solid lubricant (MoS_2)	0 to 150	1.22	black metallic	>220	185 to 215	4	S	DIN-Bez 3536-0-150G DVGW Reg. Nr. 91.03 e 47	High temperature grease for gas valves carrying propane, natural and town gas; resistant to oxidation and high temperatures; emergency running properties
Gas valve grease	BARRIERTA LP	Fluorinated polyether oil/ PTFE	DIN 3536 -20 to 150; min. and max. service temperature limit -25 to 250	2.0	whitish, cream	without	205 to 235	—	ES	DIN-Bez/ 3536-20-150 L DVGW Reg. Nr. 91.02e47	Fully synthetic grease for gas valves, also for LPG lines. Very resistant to oxidation and high temperatures, resistant to gaseous and fluid hydrocarbons, neutral towards elastomers and plastics
Sliding agents (oxygen installations)	OXIGEN-OEX FLUID C3	Silicone oil/ chlorinated paraffin oil	-40 to 60	1.1	yellowish clear	—	-	—	—	BAM-Tgb.-Nr. 4705/91 4-1897	Fluid safety sliding agent for gaseous oxygen in installations with an oxygen pressure limit of 30 bar at 60°C
Sliding agents (oxygen installations)	OXIGEN-OEX FF 250	Fluorinated polyether oil/ PTFE, silicate	-50 to 200	2.0	light gray	without	240 to 270	—	S	BAM-Tgb.-Nr. 7002/93 4-3936	Pasty safety sliding agent for gaseous oxygen in installations with an oxygen pressure limit of 260 bar at 60°C

*For further information refer to tribo-system materials.

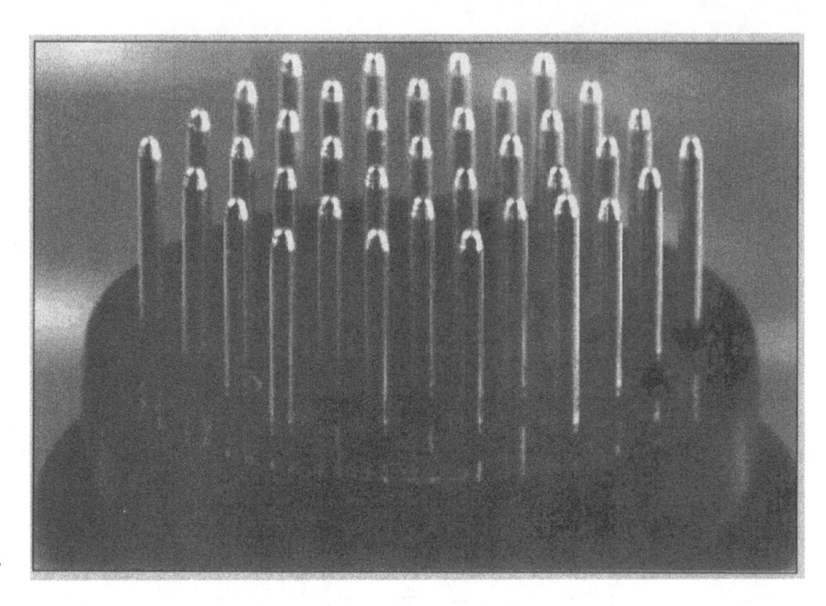

Figure 11-54. Plug contact.

Figure 11-55. Advantages of a lubricated, gold-plated plug contact as a function of the number of connections.

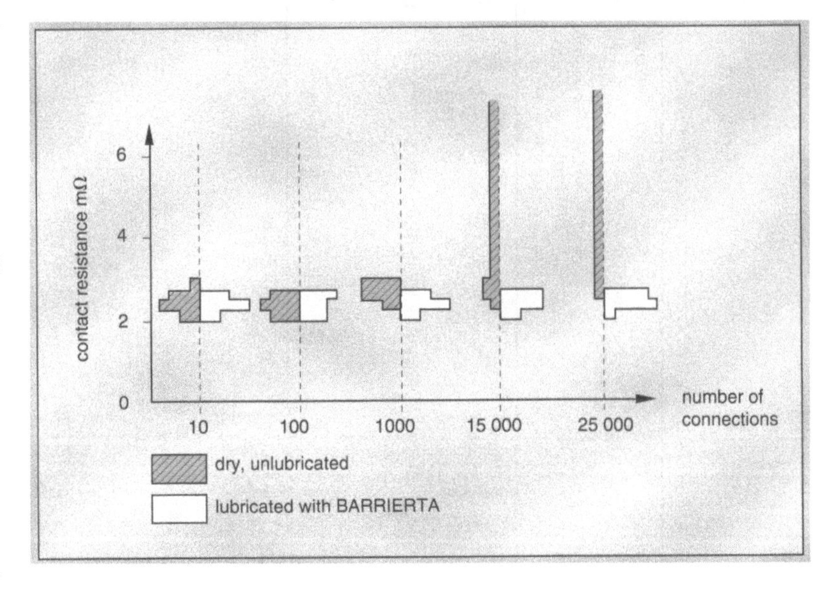

- protection against wear
- high number of actuations

Moreover, these products must also meet other important requirements such as

- excellent aging stability
- high affinity towards metals
- thermal stability
- compatibility with plastics
- purity

Refer to Table 11-16 for typical characteristics and selection criteria.

Lubrication of Industrial Spring

All units made of an elastic material are resilient. This capacity can be utilized by imparting a special shape.

Springs are used to store energy, restore components to their former position, absorb shocks, distribute, limit or measure power, maintain powerlocking connections, and function as a vibration suppression element. Coil springs and, especially, annular plate springs or Belleville washers (Figures 11-56), have to meet certain requirements in terms of material and shape. Their performance can be improved by applying special lubricants and/or tribosystem coatings.

Depending on the operating conditions, an adequate lubricant will optimize the work consumed by friction.

The lubricant has to meet the following requirements:

- reduction of wear
- good adhesion in case of vibration and shock
- low friction even at extremely low or high temperatures
- excellent protection against corrosion
- good behavior if used in conjunction with plastics and elastomers
- efficient protection against tribocorrosion
- favorable behavior towards non-ferrous metals
- compliance with food and water regulations
- uniform transmission of power
- protection against aggressive media
- prevention of stick-slip

Table 11-15. Typical selection parameters for switch and contact lubricants.

Selection criteria	Product name	Base oil/ thickener	Service temperature range (°C) ≈	Density at 20°C (g/cm³) DIN 51757 ≈	Base oil viscosity DIN 51561 (mm²/s) ≈ 40°C	Base oil viscosity DIN 51561 (mm²/s) ≈ 100°C	Color ≈	Drop point DIN ISO 2176	Specific electric resistance Ω•cm	Worked penetration DIN ISO 2137 (0.1 mm)	Apparent viscosity, KL viscosity grade	Notes
Lubricating greases for actuating elements	SYNTHESO GLEP 1	Polyglycol oil/ special Li soap	-50 to 150	0.97	370	55	beige, almost transparent	>220	$2.6 \bullet 10^{9}$	280 to 310	M	Excellent antiwear and contacting properties at temperatures down to -10°C in actuating elements and sliding contacts (down to -40°C in cases of high contact pressure)
	ISOFLEX TOPAS L 32	Synthetic hydrocarbon oil/Li soap	-60 to 130	0.86	19	4.2	beige	>180	3.7 to 10^{14}	265 to 295	EL/L	Dynamically light grease, suitable for low contact forces, compatible with plastics
	ISOFLEX TOPAS AK 50	Synthetic hydrocarbon oil/Al	-50 to 150	0.87	30	5.5	whitish, almost transparent	>180	3.7 to 10^{14}	355 to 385	EL	Especially suitable for the lubrication of plastic actuating elements and Al and Zn contact materials
	ISOFLEX TOPAS NB 52	Synthetic hydrocarbon oil/Ba complex soap	-50 to 150	0.96	30	5.5	light beige cream	>240	1.3 to 10^{14}	265 to 295	M	For switching actuators and contacts, especially for high-tension switches
Low-temperature contact grease	SYNTHESIN PLD 250/01	Ester oil/ Li soap	-60 to 125	0.92	12	3	beige	>190	$8.8 \bullet 10^{9}$	260 to 280	L	Suitable for extremely low temperatures (-40°C) and very low contact forces
	ISOFLEX TOPAS NCA 5051	Synthetic hydrocarbon oil/Ca complex soap	-50 to 140	0.85	30	6	beige	>180	$4.2 \bullet 10^{15}$	385 to 415	EL	Excellent compatibility with plastics, particularly suitable for mechanical elements
Damping greases	POLYLUB GLY 801	Synth. hydrocarbon oil, min. hydro-carbon/ oil sp. Li soap	-40 to 150	0.88	660 to 800	55 to 65	beige	>250	$2.2 \bullet 10^{15}$	310 to 340	M	Suitable for mechanical elements only (e.g., steering column shifts), excellent damping properties and compatibility with elastomers
Aging-resistant lubricants for contacts and actuating elements	BARRIERTA I EL	Fluorinated polyether oil, PTFE	-50 to 180	1.95	100	12	whitish, cream	not measurable	$3.0 \bullet 10^{16}$	265 to 295	M	Extremely resistant to aging at increased operating temperatures, neutral towards plastics
Lubricating grease for gold contacts	BARRIERTA L 25 DL	Fluorinated polyether oil, PTFE	-35 to 150	1.95	90	11	whitish cream	not measurable	$3.0 \bullet 10^{16}$	270 to 300	M	High affinity towards gold, excellent wear protection, wide temperature range
Switchgear installations	BARRIERTA L 55/2	Fluorinated polyether oil, PTFE	-40 to 260	1.96	380 to 420	36.5 to 39	whitish cream	not measurable	$3.0 \bullet 10^{16}$	265 to 295	S	Suitable for switching units and mechanical parts in the switching chamber in case of an SF_6 gas atmosphere

Table 11-16. Lubricants for electrical contacts—selection criteria and characteristics.

Selection criteria	Product name	Base oil/ thickener	Service tempera-ture range (°C) ≈	Density at 20°C (g/cm^3) DIN 51757 ≈	Base oil viscosity DIN 51561 (mm^2/s) 40°C	Base oil viscosity 100°C	Color	Drop point DIN ISO 2176 (°C)	Specific electric resistance Ω•cm	Worked penetra-tion DIN ISO 2137 (0.1 mm)	Apparent viscosity, KL viscosity grade	Notes
Lubricating greases for contacts	SYNTHESO GLEP 1	Polyglycol oil/ spec. Li	-50 to 150	0.97	370	55	beige, almost trans-parent	>220	2.6 • 10^9	280 to 310	M	Excellent antiwear behavior, ensures contact down to -10°C in sliding contact switches (down to -40°C in case of high contact forces)
	ISOFLEX TOPAS L 32	Synthetic hydrocarbon oil/Li	-60 to 130	0.86	19	4.2	beige	>180	3.7 × 10^{14}	265 to 295	EL/L	Dynamically light grease, suitable in case of low contact forces, compat-ible with plastic materials
	ISOFLEX TOPAS AK 50	Synthetic hydrocarbon oil/Al compl.	-50 to 150	0.87	30	5.5	whitish, almost transparent	>200	9.8 × 10^9	355 to 385	EL	Particularly suitable for the lubrica-tion of Al and Zn contact materials
Aging-resistant lubricants	BARRIERTA I EL	Fluorinated polyether oil PTFE	-50 to 180	1.95	100	12	whitish, cream	not mea-surable	3.0 ¥ 10^{16}	265 to 295	M	High resistance to aging in case of increased permanent temperature, neutral towards plastics
Lubricating grease for gold contacts	BERRIERTA L 25 DL	Fluorinated polyether oil PTFE	-35 to 150	1.95	90	11	whitish cream	not mea-able	3.0 ¥ 10^{16}	270 to 300	M	High affinity towards gold, excellent wear protection, wide service temperature range
Grease for tin-plated contacts	BARRIERTA GTE 403	Fluorinated polyether oil PTFE	-35 to 260	1.95	400	35	whitish, cream	not mea-surable	3.0 ¥ 10^{16}	220 to 250	S	Particularly suitable for plug-in contacts in automotive applications
Switcher installations	BARRIERTA	Fluorinated polyether PTFE	-40 to 260	1.96	380 to 420	36.5 to 39	whit-ish, cream	not mea-surable	3.0 ¥ 10^{16}	265 to 295	S	Suitable for switching units and actuating mechanisms in switching chambers of an SF_6 gas atmosphere

Selection criteria	Product name	Service tempera-ture range (°C) ≈	Base oil/ thickener	Density at 20°C (g/cm^3) DIN 51757 ≈	Kinematic viscosity DIN 51561, (mm^2/s) at ≈ 40°C	Kinematic viscosity 100°C	Viscosity index DIN ISO 2909 (VI) ≈	Pour point DIN ISO 3016 (°C) ≈	Flash point DIN ISO 2592 (°C) ≈	Specific electric resistance Ω • cm	Notes
Aging-resistant oil for contacts and actuating mechanisms	BARRIERTA I EL Fluid	-35 to 180	Fluorinated polyether oil	1.90	100	12	120	-40	not flammable	3.9 × 10^{13}	High resistance to aging in case of an increased permanent temperature, neutral towards plastics, especially suitable to provide wear protection for electroplated snap contacts

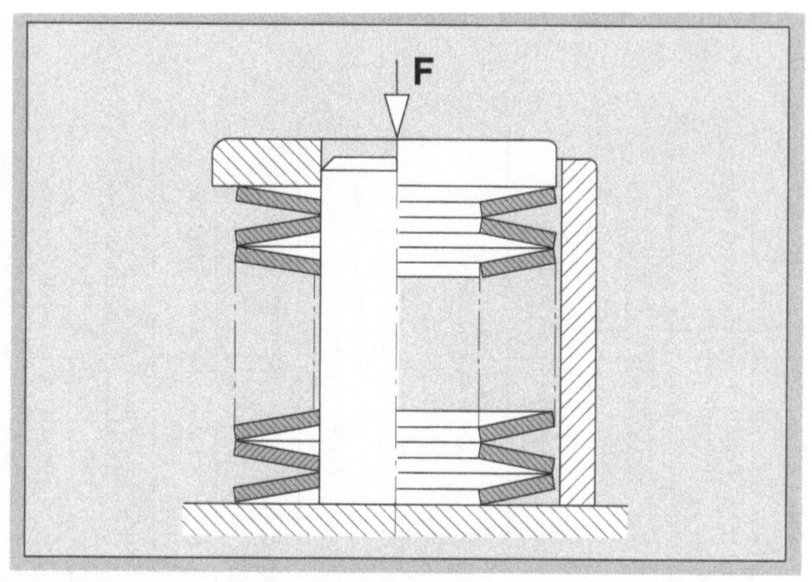

Plate spring assembly internally, externally guided

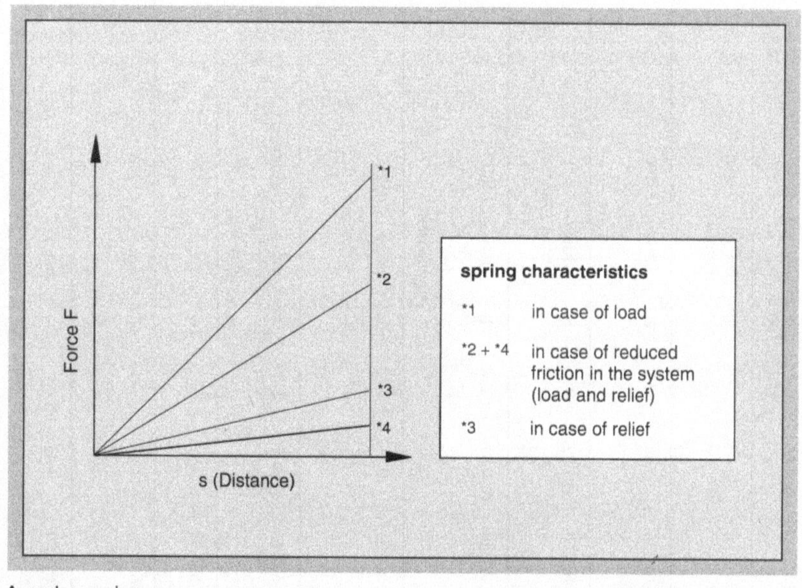

Annular spring

Figure 11-56. The performance of plate and annular springs can be improved with appropriate lubrication.

By utilizing selected lubricants it is possible to optimize individual components and even entire systems.

Selection criteria and physical characteristics of suitable lubricants are given in Table 11-17.

Lubrication of Pneumatic Components

Pneumatic cylinders (Figure 11-57), and valves (Figure 11-58) are components used in pneumatic systems.

Pneumatic cylinders convert pneumatic into mechanical energy which is subsequently used to perform linear movements to move, lift or return workpieces or tools.

By controlling starts, stops, directions, pressure and throughput, pneumatic valves ensure that the pressurized air carrying the energy follows the "right paths."

To ensure functional reliability it is indispensable

Table 11-17. Selection criteria for lubricants used with mechanical springs.

Selection criteria	Product name	Base oil/ thickener	Service tempera-ture range (°C) ≈	Density at 20°C (g/cm^3) DIN 51757 ≈	Base oil viscosity DIN 51561 (mm^2/s)		Color	Drop point DIN ISO 2176 (°C)	Consis-tency NLGI grade DIN 51 818	Worked penetra-tion DIN ISO 2137 (0.1 mm)	Apparent viscosity, KL viscosity grade	Notes
					40°C	100°C						
Impact of ambient media and humidity	Klüberplex BE 31-222	Mineral hydro-carbon oil/ special Ca soap	—15 to 140	0.86	220	19	beige, brownish	>190	2	265 to 295	M	Effective corrosion and wear protec-tion, good adhesion
Very low temperatures	ISOFLEX LDS 18 Special A	Mineral hydro-carbon oil, ester oil/soap	-50 to 120	0.88	15	2,7	yellow	>190	2	265 to 295	L	Reduction of friction and wear, suitable for all technical metal springs
Very high tempera-tures	Klüberalfa BHR 53-402	Perfluorinated polyether oil/ Na complex soap	-40 to 260	1.82	400	35	white	not measur-able	2	265 to 295	S	Reduction of friction and wear, extremely resistant to aging and oxidation
Protection against tribo-corrosion	MICROLUBE GL 262	Min. hydro-carbon, oil, synth. hydrocarbon oil/ special Li soap	-25 to 140	0.89	260	18	yellow-brown, almost transparent	>250	2	265 to 295	M	Reduction of friction and wear in case of micro-movements
Lubrication of synthetic materials	POLYLUB GLY 801	Min. hydrocarb. oil, synth. hydro-carbon oil, special Li soap	-40 to 150	0.88	730	57	beige	>250	1	310 to 340	M	Reduction of friction and wear in the mixed friction regime in case of plastics and elastomers
Food-grade lubricant	PARALIQ GB 363	Min. hydrocarb. oil, synth, hydro-carbon oil, silicate	-30 to 140	0.91	2420	156	trans-parent to yellowish	without	3	215 to 245	S	USDA-H1 authorized
Drinking water applications	UNISILKON NCA 3001	Silicone oil/ special Ca soap	-50 to 150	0.95	3700	1400	beige	>190	1	295 to 325	S	Released in accordance with KTW
Special envir-onmental requirements	Klüberbio M 32-82	Ester oil/Ca complex soap	-30 to 120	0.93	90	17	brown	>220	—	265 to 295	L/M	Water hazard category 0, rapidly bio-degradable, good protection against wear, high resistance to aging
Dry lubricant for tribo-systems	Klübertop TP 03-111	Solid lubricant (PTFE), organic binder	-40 to 180	0.05	—	—	black	—	—	—	—	Protects against corrosion, reduces friction and wear, very elastic, good adhesion to metal and plastic surfaces
	UNIMOLLY CP	Solid lubricant (MoS_2) organic binder	-40 to 220	0.12	—	—	reddish-brown	—	—	—	—	

to apply a prestart lubricant to all components performing a relative movement, for example the piston rod, pressure tube, valve elements and seals. If required, air compressor oils can be applied during operation.

Premium lubricants for pneumatic components must ensure

- an optimum sealing effect and improved efficiency
- increased performance
- operation without stick slip (e.g., in case of feed movements with low pistons speeds and long strokes)
- low breakaway moments (also after extended standstill periods)
- excellent adhesion and wetting properties on materials such as steel, refined steel, aluminum, brass, ceramic materials, plastics and elastomers

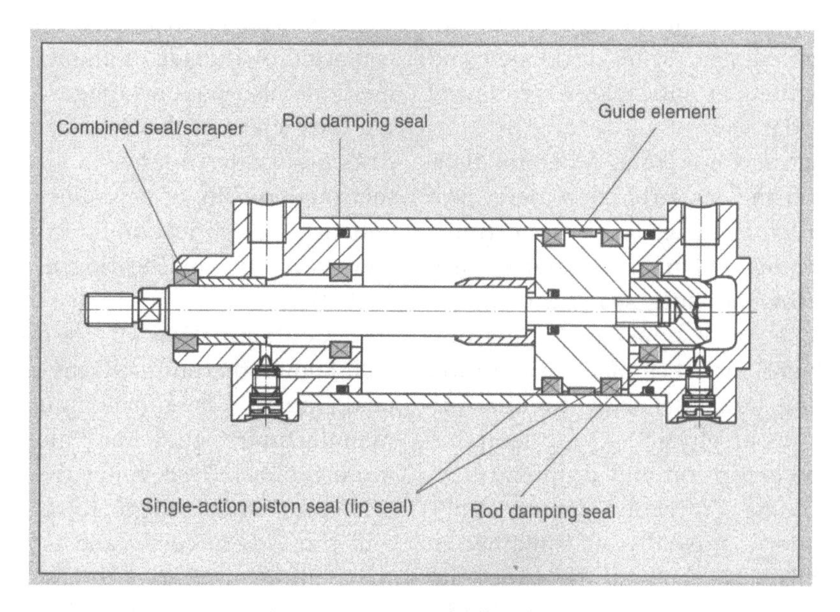

Figure 11-57. Pneumatic cylinder.

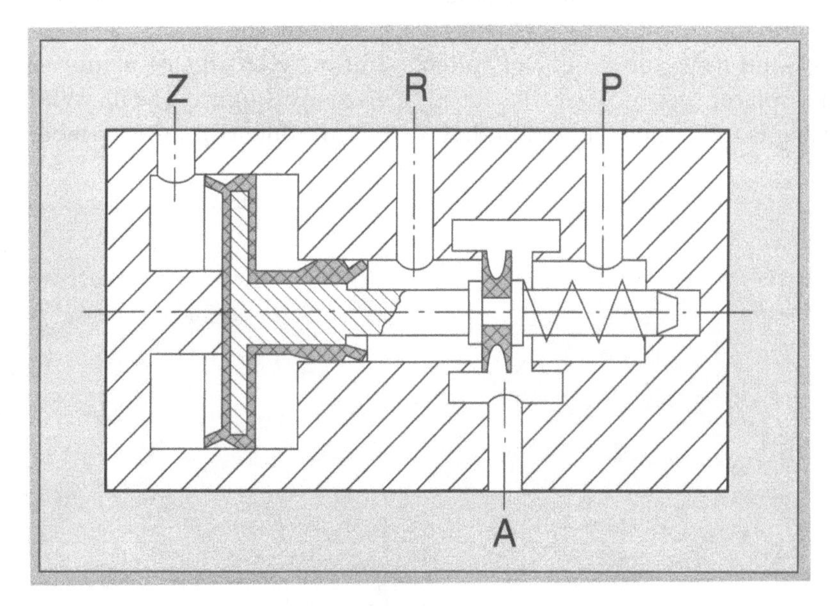

Figure 11-58. Pneumatic valves.

BEARING PROTECTOR SEALS

Much is known about the ill effects of moisture and particle contamination in process and industrial machinery lubricants. Dirt and water ingress often occur because the sealing capability of conventional lip seals and straight labyrinth seals is limited at best. Reliability-focused user companies therefore insist on better means of excluding both moisture and dust from equipment bearing housings. In the late 1990s, the top reliability-minded companies tended to accomplish this exclusion with seals that are sometimes called bearing protectors or bearing isolators. In the mid-2000s, improved bearing protector designs were developed and old notions about configurations and styles were in need of reassessment and update.

Where contaminants come from. Moisture typically enters bearing housings as airborne water vapor or a stream of water from hose-down operations. Also, and unless the equipment manufacturer or user provides suitable bearing housing seals, a contamination-inducing interchange of housing-internal and housing-external (ambient) air takes place in the form of "breathing." Bearing housings will "breathe" in accordance with the laws of physics: Rising temperatures cause gas volume expansion and dropping temperatures cause gas volume contraction (Figure 11-59). The gas volume of concern is usually air contained in the bearing housing; this air is rarely at the same temperature as the ambient surroundings. Open or inadequately sealed bearing housings thus seek and obtain pressure equalization with the surrounding ambient air and a continuous back-and-forth movement of (often contaminated) air takes place.

If a bearing housing is fully sealed, it will undergo pressure changes in proportion to temperature swings. "Fully sealed" implies the use of bearing protectors akin to face seals; it also implies the elimination of housing vents. Our text has previously alluded to it by mentioning "hermetic sealing". Of course, any particular lubricant application method must accommodate changes in pressure and temperature.

In "sealed" bearing housings, i.e. housings sealed with modern face seals and with plugged vent ports, temperature changes cause pressures to deviate from ambient. Such application methods include the balanced constant level lubricators mentioned earlier in this chapter; balanced lubricators do not allow lubricant being contacted by outside or ambient air. Modern application methods also include flinger discs (Ref. 1) and, while not the subject of this segment of our text, these flinger discs are vastly superior to oil rings. It should be noted that modern flinger discs are readily available and will enhance the lubrication effectiveness of both vented as well as non-vented bearing housings in equal measure (Ref. 1).

Contamination effects on bearing life. It is also well known that in many environments airborne dust is being ingested in sufficient quantities to exceed bearing manufacturers' guideline values by factors of 100 and more (Ref. 2). That being the case, operation without sealing or with inadequate sealing of bearing housings will prove very costly and is virtually never found in reliability-focused user companies.

In Ref. 3, European bearing manufacturer FAG emphasizes that the severity of the undesirable end effects of contamination depends on the ratio of operating viscosity of a lubricant divided by its rated viscosity (Figure 11-60). While there obviously could be an almost infinite number of combinations in the

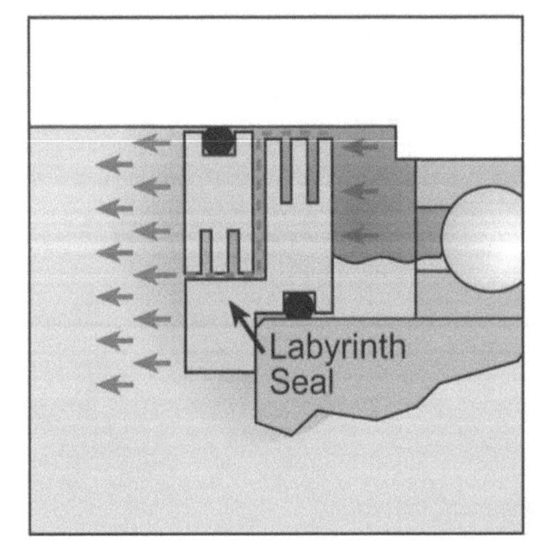

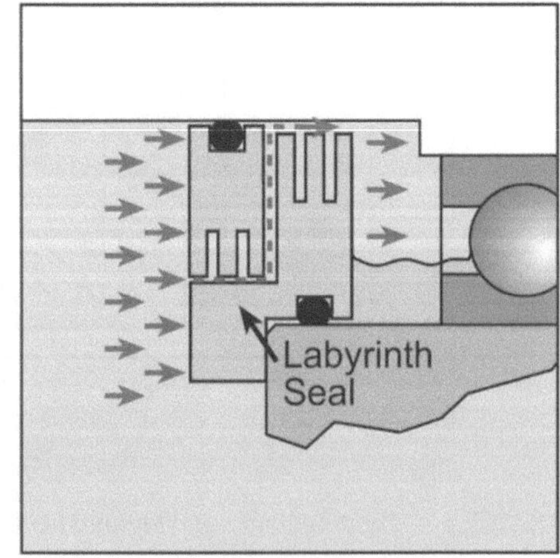

Figure 11-59. "Breathing out" warm air and "breathing in" cooler ambient air

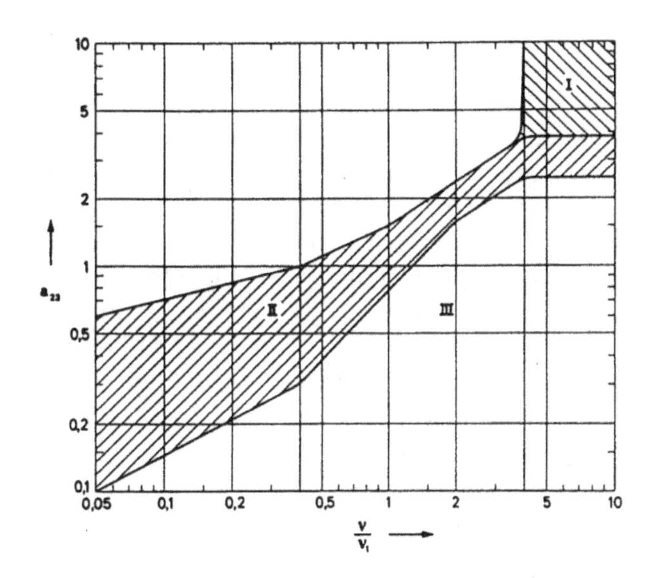

Figure 11-60. Contamination adjustment factor "a23" vs. viscosity ratio (Source: FAG Bearings, Schweinfurt, Germany, also Ref. 1)

amount of contamination and ratios of viscosity, ratios of 0.5 to perhaps 1.0 are thought rather typical. Using 0.5 for this ratio, and plotting from the mid-point of the zone labeled "contaminants in lubricant" (Zone lll) to the mid-point of the zone labeled "high degree of cleanliness in the lubricating gap," (Zone ll) we would find a four-fold increase in bearing life for the cleaner oil. At a viscosity ratio of 2:1, the projected bearing life increase traversing from "contaminated" to "clean" would be approximately seven-fold. It should be noted that we are not here considering "ultra-clean" (Zone l) oil, since it would be unrealistic to find this degree of cleanliness in field-installed process pump bearing housings.

Having understood the life-limiting effects of lube oil contamination, reliability-focused equipment users found it easy to cost-justify and/or retrofit superior means of bearing protection (Ref. 4). In lieu of inexpensive lip seals, either magnetic face seals or thoughtfully engineered, well-tested rotating labyrinth seals are now applied by bottom-line cost conscious users. These modern bearing housing seals are found in thousands of process machines; these include centrifugal pumps, fans and gear units where a variety of other, demonstrably inferior sealing strategies were previously utilized.

But the most authoritative data on the effects of lubricant contamination might perhaps be gleaned from the General Catalog of another leading bearing manufacturer, SKF (Ref. 5). For the example shown in their catalog, SKF applied its New Life Theory to an oil-lubricated 45 mm radial bearing running at constant load and speed. Under ultra-clean conditions (nc = 1), this example bearing was calculated to reach 15,250 operating hours. The SKF catalog text goes on to explain that, if the example were to be calculated for contaminated conditions such that nc =0.02, bearing life would be only 287 operating hours. We believe this is powerful analytical evidence in support of industry moves towards better bearing housing protection.

Rotating Labyrinth Seals: How they work and how they often differ. It can be shown that in process machinery such as the literally millions of centrifugal pumps operating today, modern housing seals make much economic sense. Findings of rapid payback and quantifiable failure reductions are supported both by industry statistics and the failure rate plots issued by several lip seal manufacturers. For decades, lip seals (Figure 11-61, upper portion) have been out of compliance with the minimum requirements stated in the widely accepted API-610 industry standard for centrifugal pumps (Ref. 6). Indeed, most rotating labyrinth seals are a good choice for fluid machine bearing protection and the one shown in Figure 11-61 (lower portion) will generally outperform lip seals by a wide margin (Ref. 7).

It must be realized, however, that there are many types, configurations or versions of rotating labyrinth seals. Different configurations will allow anything from

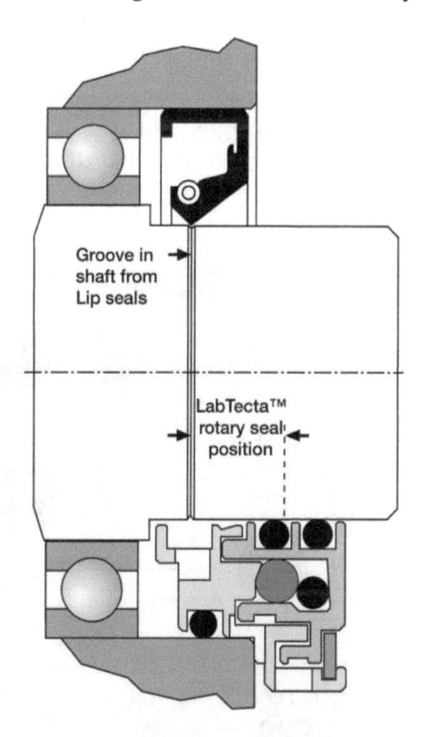

Figure 11-61. Relative location of lip seal (upper view) v. advanced, field-repairable rotating labyrinth seal (lower view). Two O-rings clamp the seal to the shaft in a region not previously damaged by the lip seal (Source: AESSEAL plc, Rotherham, UK and Knoxville, TN)

a truly minimal amount of "breathing" and virtually zero leakage, to a rather significant amount of breathing and worrisome leakage. The amounts of breathing and leakage depend very much on the design and construction features of a given make or brand and must be compared against the configuration and/or construction features of another make or brand.

A detailed, albeit brief, review of their relevant drawings or patent applications will often prove revealing. For instance, it is evident that some rotating labyrinth designs and configurations are decidedly not field-repairable whereas other manufacturers have made it their business to design this "in-place repairability" into rotating labyrinth seals. Also, certain hybrid designs incorporate sliding lip seals between rotating and stationary components while others are fitted with an O-ring that moves radially in and out of a groove. Unfortunately, any design based on a dynamic O-ring being located part-way in a groove machined into a stationary part and part-way in a groove machined in a rotating part (Figure 11-62) tends to encounter the sharp edges of either of the two grooves. Envision the difficulty of maintaining close manufacturing tolerances in bearing protectors whose effectiveness depends on sliding motion of a dynamic (or moving) O-ring against the edges of a discontinuity such as a groove. In contrast, visualize another, more modern-design that incorporates a dynamic O-ring making contact with a relatively large area (Figure 11-63). Since pressure equals force divided by unit area, the two designs will expose their respective dynamic O-rings to pressures that differ by perhaps two orders of magnitude. Because this would suggest trouble for equipment being slow-rolled during, say, the warm-up period for turbine-driven pumps, one would choose the one with lower contact pressure because its elastomeric O-ring life is likely to be greater.

In any event, some bearing protector seals will not run without risk under slow rolling ("warm-up") conditions. Designs with dynamic O-rings may not be appropriate for many applications at less than 500 rpm. At low peripheral speeds, dynamic O-rings may drag and create high frictional heat. Likewise, and in extreme cases at high speeds, a single O-ring used for clamping the rotating component to the shaft might cause the rotor to lose perpendicularity relative to the shaft. In that case, contact between the rotor and stator, or shaft and rotor has resulted in frictional temperatures high

Figure 11-62. O-ring degradation risk with sharp-edged components (Ref. 7)

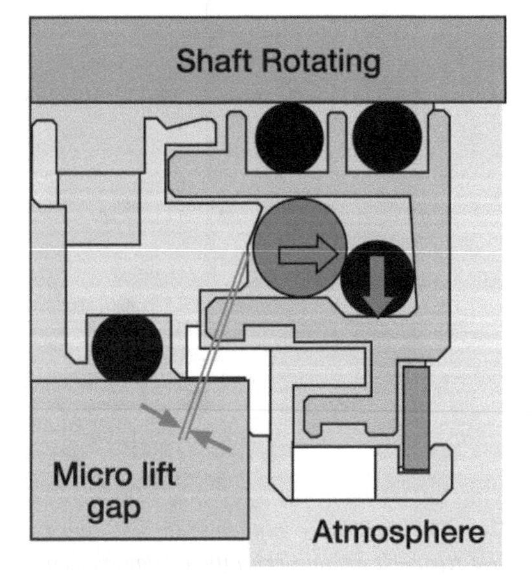

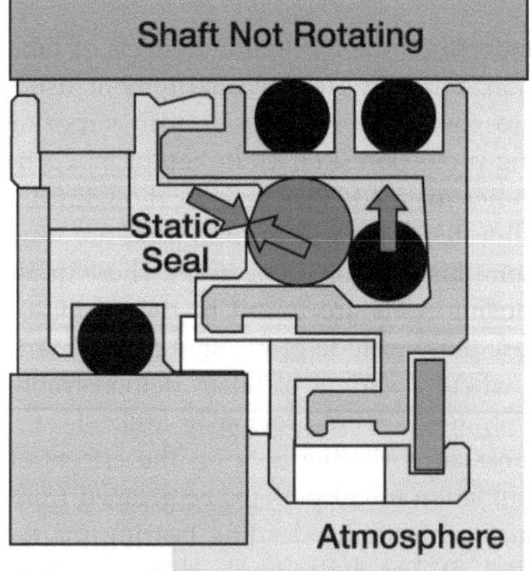

Figure 11-63. Advanced rotating labyrinth-type bearing housing seal (Source: AESSEAL plc, Rotherham, UK and Knoxville, TN)

enough to turn bronze blue.

Back to the advanced design of Figure 11-63. In the equipment idle condition (right portion illustration), the O-ring with the largest circumference ("energizing member") applies a radial load on the upper quadrant of the two dynamic O-rings which, together, constitute a quasi "shut-off valve". When the equipment is operating (left portion of Figure 11-63), the energizing member is subjected to centrifugal forces that cause it to move radially outward and disengage it from the primary (i.e. area-contacting) O-ring. The primary O-ring acts as a shut-off valve. Radial outward movement of the energizing O-ring allows the primary shut-off valve to move away from the primary O-ring. The resulting micro-gap allows the bearing housing to breathe. As the equipment stops, the energizing member again takes up its idle position and axially moves the primary shut-off valve to make a static seal with the stator (or stationary ring).

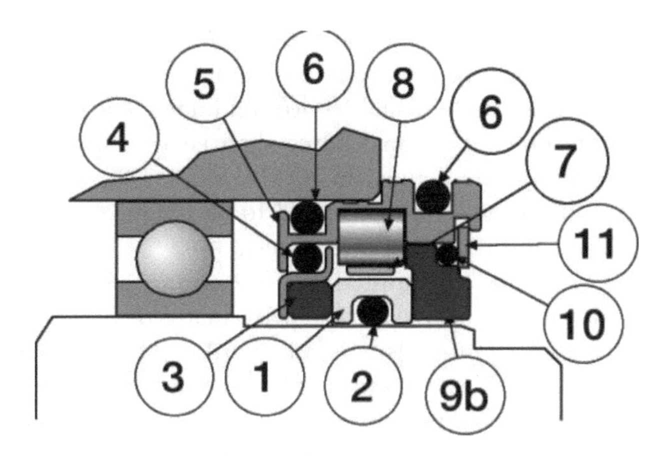

Figure 11-64. Dual-face magnetic seal and O-rings (2, 4, 6, 10), snap ring (11), rotating face (1), stationary faces (3, 9b), stationary magnets (8), magnet carrier (7), and outer body (5). (Source: AESSEAL plc, Rotherham, UK, and Knoxville, Tennessee)

The second moisture ingress prevention feature of the seal shown in Figure 11-63 is a multi-tiered terrace system comprised of a complex tortuous path between the counter-rotating members together with a series of orifice outlets. The system creates a cascading effect and extensive testing confirmed that water intrusion is virtually non-existent. In advanced bearing protector seals, two O-rings are used to clamp the rotating component to the equipment shaft.

Magnetic seals and application criteria. For "hermetic" sealing, however, face-type seals will be needed. Properly applied, the latest iteration of dual-face magnetic seals (Figure 11-64) has had an outstanding performance record and hundreds of reliability-focused users have contributed to an impressive reference list. Precursor-style single face magnetic seals have been applied in thousands of aircraft, including decades-old Boeing 707, 747 and the more recent Boeing 777 commercial aircraft. Time and space limitations don't allow us to explain the numerous aerospace case histories where precursor magnetic seals have been in service for over 35,000 operating hours and have demonstrated low coefficients of friction. When a hydrodynamic film is maintained, the coefficient of friction can be less than 0.05. The coefficient is normally higher during early periods of operation when the film and interface are being developed. Once developed, the film helps control wear, reduces frictional heat, and stops bounce, or chatter.

For maximum life, face loads are usually kept low and certain peripheral speed and lubrication guidelines apply. Proper cooling and/or lubrication must be provided to remove frictional heat. Lubrication, of course, is present in the overwhelming majority of oil-lubricated machines operating in process plants. In the very few instances where bearing lubrication would not reach the adjacent magnetic bearing housing protector seal, it might be wise to question the adequacy of bearing lubrication as well.

Applicable Industry Standards. Probably all rotating labyrinth seals (but no lip seals) marketed in the United States comply with the relevant stipulations of API-610 9th and later Editions. Recognizing that single-acting lip seals cannot prevent forced water intrusion, this universally used refinery pump standard states "lip-type seals shall not be used in pumps." Water intrusion is likely to occur past the lip seal (Figure 11-61, upper half of illustration) when hosing down equipment. Such hosing down is quite typical in the cleanup of food processing and pharmaceutical manufacturing facilities. Likewise, the shaft wear ridge shown in this illustration compromises equipment integrity.

Advanced rotating labyrinth seals—and certainly the one depicted in Figure 11-63—comply with both (UK) Ingress Protection Standard IP55 and the (U.S.) IEEE 841-2001 standard. In fact, IEEE 841-2001, the standard for severe duty TEFC (Totally Enclosed Fan Cooled) motors intended for use by the petroleum and chemical industry makes clear reference to IP55 (Ref. 8) by stating, in paragraph 4.5:

"Enclosures shall be TEFC or totally enclosed non-ventilated (TENV) and shall have a degree of protection of IP54 except as follows: For frame sizes 320 and larger, the degree of protection for bearings shall be IP55. If replaceable shaft seals are used to achieve this degree of protection, they shall be the non-contact or non-contacting-while-rotating type with a minimum expected seal life of 5 years under usual service conditions. The degree of protection for thermal boxes shall be IP55. Degrees of protection are defined in NEMA MG-1-1998, Part 5. Drain fitting holes are permitted to be plugged during the enclosure IP54 dust ingress test."

That, now, prompted us to examine what IP55 certification really means and why many of the older rotating labyrinth seals would be hard pressed to measure up to IP55 as an "Index of Protection." In the standard, IP55 indicates test results with no dust entering in harmful quantities and no risk of direct contact with rotating parts. It further indicates a machine protected against jets of water from all directions from hoses at 3 m distance with a flow rate of 12.5 l/min at 0.3 bar. These indices of protection of electrical equipment enclosures conform to IEC standard 34-5-EN 60034-5 (IP)-EN 50102 (IK).

Face-type seals can be used in electric motors as long as they are applied within their respective design envelopes. The dual-face magnetic seal shown earlier in Figure 11-64 complies with ATEX, the European testing authority for components operating in explosive atmospheres. The Health & Safety Executive also tested and certified the sparking and hazard-related performance of these dual-face magnetic seals.

User Standards. There are also user standards worthy of note. As an example, a large U.S. pulp and paper company found that 30% to 40% of their many motor failures were caused by water-related bearing damage (Ref. 8). They then decided to develop the following standard for motor bearing housing seals:

"The bearing isolators shall exclude all water (including water vapor) entry into the bearing grease cavity when the motor is subjected to the following low pressure flooding water hose test continuously for twenty-four (24) hours:

The drive end bearing shall be subjected to a water hose nozzle pressure of not less than 15 psig at a distance of eight (8) inches or less from the bearing housing The stream is to be parallel to the shaft and directed so as to impinge at the joint between the moving shaft parts and the stationary housing. The water flow shall be not less than three gallons per minute (3 gpm) and sufficient that the entire bearing environment is continuously flooded.

The fan end bearing shall likewise be subjected to water flooding through the upper air intake at a rate sufficient to flood the bearing environment but not to damage the fan.[The fan may be removed for this test].

This test shall be conducted with the motor operating on a duty cycle of one hour running and one hour off for twenty-four (24) consecutive hours.
Criteria for passing test: No water entry into the bearing grease cavity.

The above twenty-four (24) hour test shall then be repeated using wet saturated steam vapor instead of water. The quantity shall be such that each bearing is completely engulfed in steam vapor. As before, the criterion for passing the test is no water vapor entry into the bearing grease cavity."

Again, there are compelling reasons to protect electric motor bearings from the ingress of airborne dust and water. Properly designed and applied face seals (Figure 11-64) will pass this test without difficulty, as will the most effective multi-tiered rotating labyrinth seals available (Figure 11-65). Special variants are available for small steam turbines and applications that require considerable axial float capability. One of these is shown in Figure 11-66.

Consider this summary. The reliability advantages and cost justification of using bearing protection seals are very widely known; numerous articles and at least ten books cover the subject. However, marketing-driven and often misleading claims abound in a world where one rarely gets to hear the full story. Reliability-focused best-of-class companies make it their goal to investigate the whole story and become familiar with a number of issues:

1. Properly designed and applied, magnetic seals are a superior product category; thousands of successful

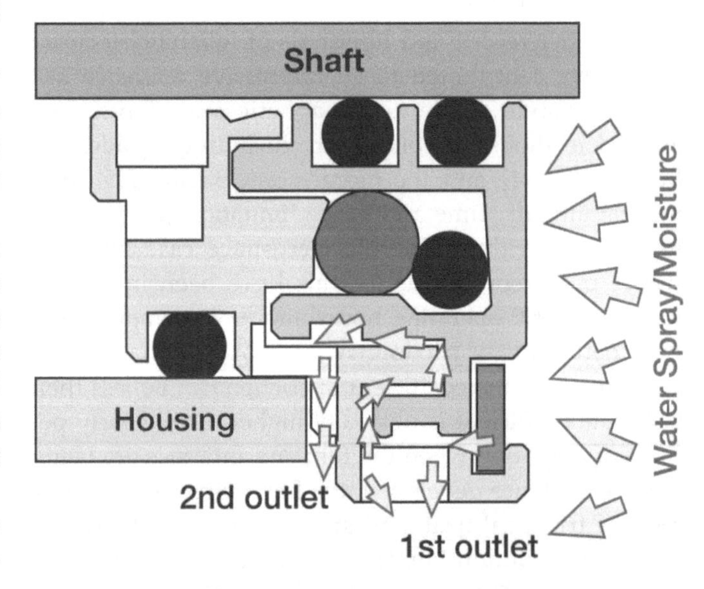

Figure 11-65. Multi-tiered arrangement (tortuous path) and multiple expulsion orifices of modern, field-repairable rotating labyrinth seal (Source: AESSEAL plc, Rotherham, UK, and Knoxville, Tennessee)

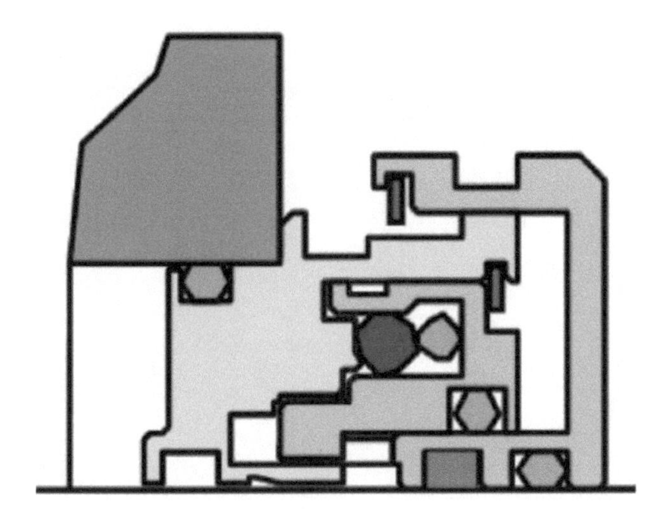

Figure 11-66. This variant of Figure 11-65 incorporates considerable axial float capability and is used on small steam turbines (Source: AESSEAL plc, Rotherham, UK and Knoxville, Tennessee)

installations are a matter of undisputable record.

2. In the hands of value-conscious engineers and technicians, dual-face magnetic seals have saved process plants hundreds of thousands of dollars. If one allows the product to be mistreated, it will fail.

3. Some established manufacturers deserve credit for entrepreneurship and designs that have rightly replaced lip seals in many process plant machinery applications. But what if a close examination and thorough testing indicate that better products are now entering the market? An unbiased examination by knowledgeable engineers will show that many old-style bearing housing seals have limitations. These limitations are rooted in the laws of physics and the concepts of lubrication and wear.

4. Reliability-focused users examine both test data and the construction features of cost-competitive products (old and new) and then draw their own, fact-based conclusions.

5. A thorough review of testing setups and test results will uncover strengths in one and weaknesses in another. One model will be field-repairable, the other will not be and this is important to many reliability-focused users. Models that use a single O-ring to hold the rotating element on the shaft are more likely to experience an out-of-perpendicularity condition relative to the shaft centerline. At high speeds, serious rubbing contact could result.

6. Superior designs use two O-rings to hold the rotating component to the shaft. However, even the best bearing protector seals will (usually) require better installation tools than heavy hammers. In essence, no process plant should tolerate hammer blows on bearings and seals.

In observing process plants, their work processes and procedures, their goals and aspirations, we noted a strange paradox: everyone talks "reliability" but many only pay lip service to the term. Let's face it—many managers are driven by short-term goals only. Similarly, some wage personnel have voiced concern that their jobs may be phased out when there are fewer equipment failures, so why cooperate? And while allowing modern components in one's plant may run counter to some instincts, it doesn't run counter to logic and experience.

Finally, there is ample evidence that business-as-usual attitudes and procurement decisions based on advertising claims alone will not yield best-in-class performance. True reliability professionals investigate first the facts. They periodically update their knowledge base and when they have the facts, they freely share them with everyone who's teachable. There are even times when engineers and technicians "see the light" before a manager does. When that is the case, consider making your points based on evidence. Hopefully, we've equipped many readers to do so with this bearing protector knowledge update.

References

1. TRICO Mfg. Corporation, Pewaukee, Wisconsin, (www.tricomfg.com)
2. Adams, Erickson, Needelman and Smith, (1996) Proceedings of the 13th International Pump User's Symposium, Texas A&M University, Houston, TX, pp. 71-79
3. Eschmann, Hasbargen and Weigand; (1985), ISBN 0-471-26283-8), "Ball and Roller Bearings," John Wiley & Sons, New York, NY, pg. 183
4. Bloch, Heinz P. and Alan Budris; (2006) "Pump User's Handbook: Life Extension," Second Edition, The Fairmont Press, Inc., Lilburn, GA 30047, ISBN 0-88173-517-5
5. SKF Catalog 140-170, August 1988, Page 40, Figure 17
6. API-610, Standard for Refinery-Type Centrifugal Pumps, 8th and later Editions, The American Petroleum Institute, Washington, DC.
7. Bloch, Heinz P.; "Twelve Equipment Reliability Enhancements With 10:1 Payback,," NPRA Paper RMC-05-82, Presented at NPRA Reliability and Maintenance Conference, New Orleans, May 2005
8. Roddis, Alan; "21st Century Bearing Protectors," presented at 2nd International Symposium on Centrifugal Pumps, The State-of-Art and New Developments, London, U.K., September 22, 2004

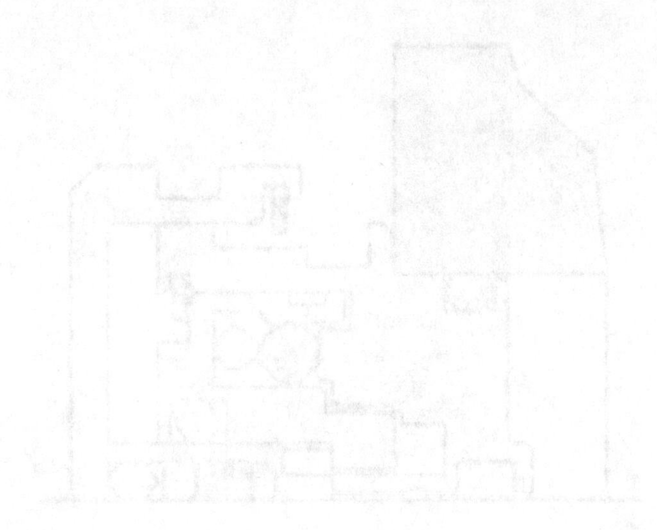

Chapter 12
Lubricating Gears

Gears are one of man's oldest mechanical devices. In the public mind, gears are one of the most well recognized kinds of machinery. They create the impression of positive action, coordinated-interlocked-precise application of effort to secure a desired result. The primary early uses of gears were for navigation, timekeeping, grinding, etc. The automobile transmission is probably the most common use of gearing for the everyday citizen.

Gears are machine elements that transmit motion by means of successively engaging teeth (Figures 12-1 and 12-2). Of two gears that run together, the one with the larger number of teeth is called the gear. A pinion is a gear with the smaller number of teeth. A rack is a gear with teeth spaced along a straight line and suitable for straight-line motion. Many kinds of gear teeth are in general use. For each application, the selection will vary depending on the factors involved. One basic rule is that to transmit the same power, more torque is required as speed is reduced. The torque is directly proportional to speed; therefore, the input and output torque for power transmissions are directly proportional to the ratio.

The gear designer must do more than just provide a mechanism that will develop the required speeds. He must make sure that his mechanism does not break or wear out prematurely because of the power being transmitted. Needless to say, gears must be lubricated, and the oil must be kept clean (see Chapter 11 and our segment on bearing housing protector seals).

LUBRICANT SELECTION FOR CLOSED GEARS*

Closed gears are used in many different arrangements, Figures 12-3 and 12-4. The factors affecting lubricant selection for the various arrangements are shown below and should be reviewed and evaluated to determine the required properties.

Figure 12-1. Single-helical, high-speed gear unit being assembled. (Source: Maag Gear Company, Zurich, Switzerland)

*James R. Partridge, Lufkin Industries, Lufkin, Texas; also Exxon Company, USA, Houston, Texas.

Factor	Lubricant Properties
Load	Viscosity EP Additives
Speed	Viscosity EP Additives
Temperature (Operating and Ambient)	Viscosity Viscosity Index Fluidity Oxidation Stability
Contamination	Demulsibility Corrosion Protection Oxidation Stability
Life	Oxidation Stability Additive Depletion
Compatibility	Synthetic (Paint and Seals) EP Additives

Also, Tables 12-1 and 12-2 taken from AGMA 250.04 show recommended viscosity ranges based on gear center distance. These recommendations should be used with caution since they are very loosely written.

Viscosity is probably the single most important factor in lubricant selection and relates to load, speed, and temperature.

Table 12-3 is a general guide based on the required viscosity in relation to the Operating "K" factor of the gears and pitch line speed. However, required viscosities can also be calculated from two empirical expressions:

$$Vg = 420\ (K/V)^{0.43}$$

Where: Vg = Viscosity, Centistokes
K = Operating "K" Facor

$$k = \frac{W}{dF}\left(\frac{m_G + 1}{m_G}\right)$$

W = Tangential Load
d = Pitch Diameter of Pinion
F = Effective Face Width
m_G = Ratio
V = Pitch Line Velocity, ft./min.

and:

$$SSU = \frac{V_g + (V_g^2 + 158.4)^{.5}}{0.44}$$

This formula was first published by Shell, and all values were converted to SSU.

It should be noted that by this formula the high speed gears (above 5000 FPM) require a heavier oil than the 150 SSU @100°F usually used, but compromises are made for bearings and sometimes seals.

We like to use a general rule that for high speed gearing, the minimum viscosity at supply temperature should be 100 SSU.

Film Thickness

Several authorities have stated that film thickness is a function of operating speed only. Based on this theory, the following formula can be used as a guide which was derived from experimental results by Crook and Arcard.

$$h_{min} = 0.33d\ \ V_g\,(n_p)\left(\frac{m_G}{m_G + 1}\right)^{.5}$$

h = Film thickness (micro-inches)

Table 12-1. Viscosity ranges for AGMA lubricants.

Rust and Oxidation Inhibited Gear Oils	Viscosity Range ASTM System	Extreme Pressure Gear Lubricants	
AGMA Lubricant No.	SSU @ 100°F.	AGMA Lubricant No.	
1		193 to 235	
2		284 to 347	2 EP
3		417 to 510	3 EP
4		626 to 765	4 EP
5		918 to 1122	5 EP
6	1335 to 1632	6 EP	
7 comp.	1919 to 2346	7 EP	
8 comp.	2837 to 3467	8 EP	
8A comp.	4171 to 5098		

Figure 12-2. Double-helical, low-speed gear unit being checked at Lufkin Gear Company, Lufkin, Texas.

d = P.D. of pinion
V_g = Viscosity, Centistokes at gear blank temperature
n_p = Pinion speed, RPM
m_G = Ratio

Using this formula, a 5″ pinion running with a 20″ gear has a film thickness of 385 and 640 microinches at 3,600 and 10,000, respectively, using 100 SSU oil at the mesh.

Due to the variables involved, film thickness calculation procedures are useful only for design comparisons and should not be used to decide that a particular gear set will not work.

Lubricant Types

Mineral Oils

Although synthesized hydrocarbons (diesters and PAOs) are rapidly gaining more wide-spread acceptance, mineral oils are still the most commonly used type of gear lubricant. Containing rust and oxidation inhibitors, these oils are less expensive, readily available, and have very long life. When gear units operate at high enough speed or low enough load intensity, the mineral oil is probably the best selection.

Extreme Pressure Additives

Extreme pressure additives of the lead-napthenate or sulphur-phosphorus type are recommended for gear drives when a higher load capacity lubricant is required. As a general rule, this type of oil should be used in low speed, highly loaded drives, with a medium operating temperature.

It should be remembered that EP oils are more expensive and must be replaced more often than straight mineral oils. Some of these oils have a very short life above 160°F temperature.

A good gear EP oil would have a Timken OK load above 60 pounds and pass a minimum of 11 stages of the so-called FZG test.

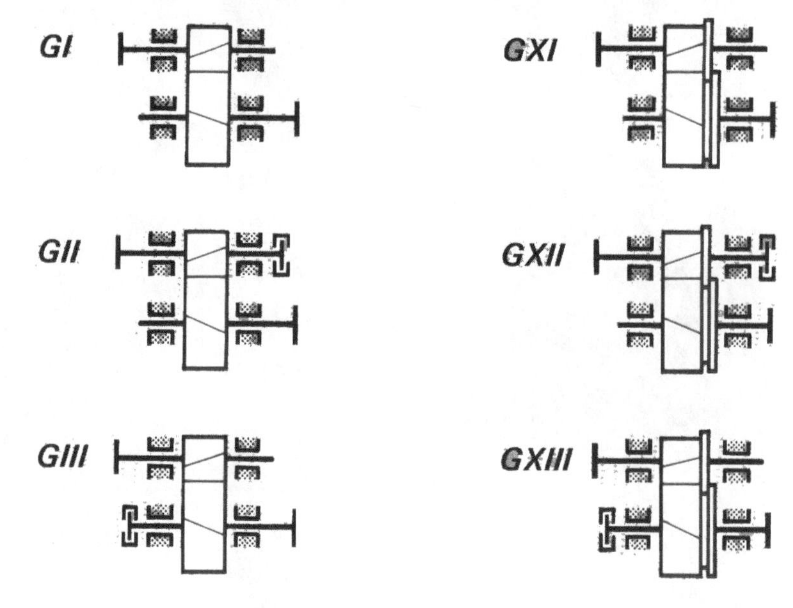

Figure 12-3. Basic gear designs. (Source: Maag Gear Company, Zurich, Switzerland)

Table 12-2. Equivalent Viscosities of other systems (for reference only)

AGMA LUBRICANT NO.	EQUIVALENT ASTM-ASLE GRADE NO.	EQUIVALENT ISO GRADE NO.	METRIC EQUIVALENT VISCOSITY RANGES cSt AT 37.8°C (100°F)
1	S215	46	41.4 to 50.6
2, 2 EP	S315	68	61.2 to 74.8
3, 3 EP	S465	100	90 to 110
4, 4 EP	S700	150	135 to 156
5, 5 EP	S1000	220	198 to 242
6, 6 EP	S1500	320	288 to 352
7 comp., 7 EP	S2150	460	414 to 506
8 comp., 8 EP	S3150	680	612 to 748
8A comp.	S4650	1000	900 to 1100

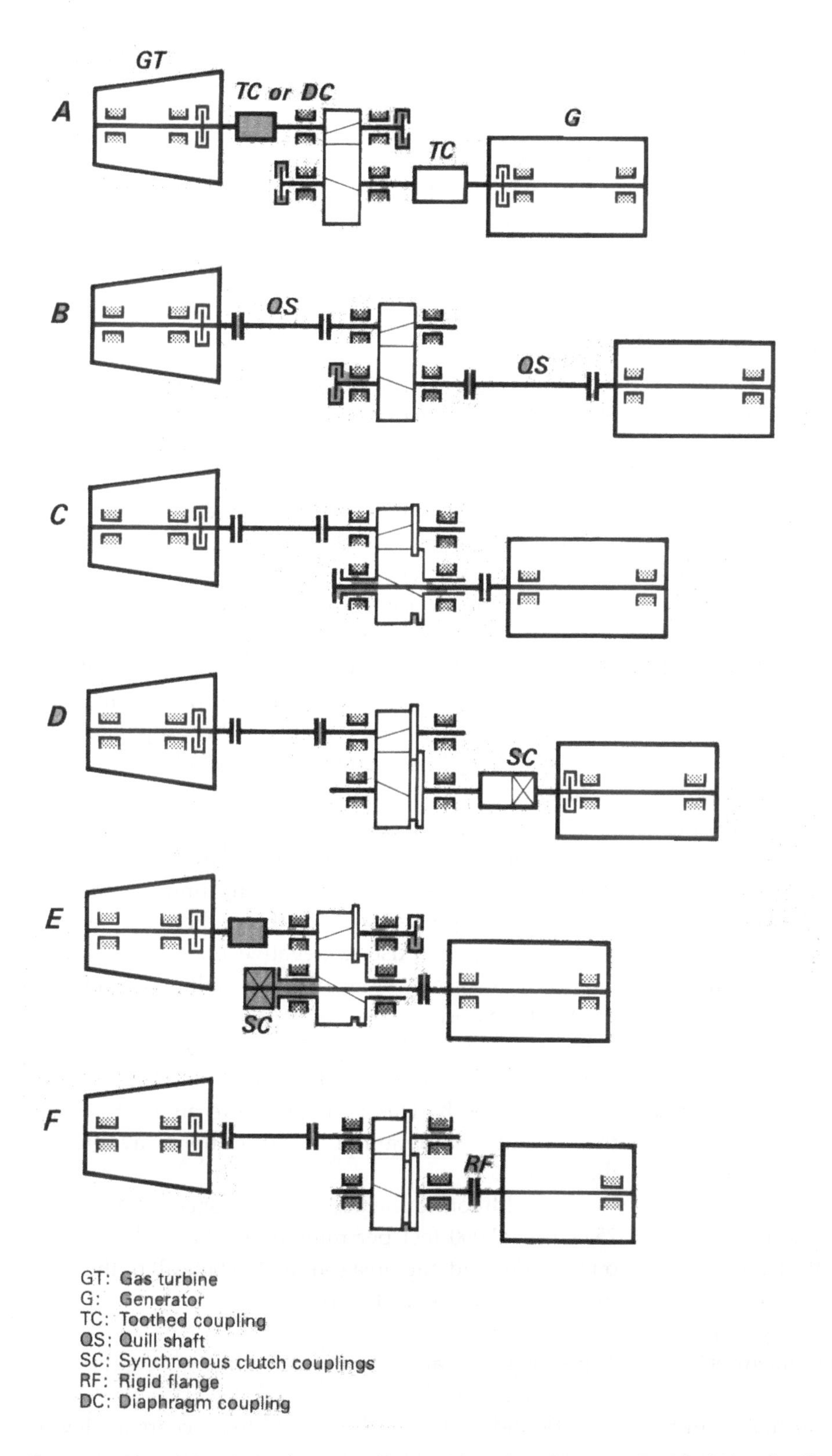

Figure 12-4. Examples of geared systems. (Source: Maag Gear Company, Zurich, Switzerland)

Boron compounds, as EP additives, are being tested; and these show promise as an extremely high load capacity lubricant. The compounds being tested show Timken OK loads greater than 100 pounds and 14 stages of the FZG test. This additive is nontoxic, highly stable, but sensitive to water.

Synthetic Lubricants

Not to be confused with the highly desirable synthesized hydrocarbons (typically diesters and PAOs), true "synthetic" lubricants are not recommended for general gear applications due to cost, availability, and lack of knowledge of their properties. In extreme applications of high or low temperature and fire protection, they are used. The user must be careful when selecting these lubricants since some of them remove paint and attack rubber seals.

The more recent synthesized hydrocarbons (SHC) have many desirable features such as compatibility with mineral oils and excellent high and low temperature properties. They should be an excellent selection when EP lubricants are required along with high temperature operation. Refer to Chapter 7 for details.

Compounded Oils

Compounded oils are available with many different additives. The most commonly available is a molydenum disulfide compound that has been successfully used in some gear applications. It is difficult for a gear manufacturer to recommend these oils since some of these additives have a tendency to separate from the base stock.

Viscosity Improvers

Viscosity improvers in gear drives should be used with great care. These polymer additives do great textbook things for the viscosity

Table 12-3. Viscosity, SSU @ 100°F.

	V = Pitch Line Velocity, FPM							
K	50	100	200	500	1000	5000	10000	15000
50	1900	1400	1050	700	500	250	200	175
100	2600	1900	1400	950	700	350	250	225
150	3100	2300	1700	1150	850	425	325	250
200	3500	2600	1900	1300	950	475	350	300
250	3800	2800	2100	1400	1050	525	400	325
300	4100	3100	2300	1500	1150	575	425	350
400	4700	3500	2600	1700	1300	650	475	400
600	5500	4100	3100	2100	1500	750	575	475

index and extend the operating temperature range of an oil. What must be remembered is that polymers are non-Newtonian fluids, and the viscosity reduces with shearing. A gear drive is a very heavy shear application; and as a result, the viscosity reduces rapidly if too much polymer is used.

Lubricants in gear units have basically two functions: (1) to separate the tooth and bearing surfaces and (2) cooling. On low speed gear units, the primary function is lubrication; on high speed units, the primary function is cooling. This does not mean that both are not important but relates to the relative quantity of oil.

On low speed units, the amount of oil is determined by what is required to keep the surfaces wetted. On high speed units, quantity is generally determined by heat loss (or inefficiency). As a general rule, one GPM must be circulated for each 100 HP transmitted which results in a temperature rise of approximately 25°F. Higher HP units use a 40°F to 50°F temperature rise and require .5 to .6 GPH per 100 HP transmitted. This is based on a 98% efficiency.

Lubrication of High Speed Units

The oil furnished to high speed gears has a dual purpose: Lubrication of the teeth and bearings, and cooling. Usually, only 10% to 30% of the oil is for lubrication and 70% to 90% is for cooling.

A turbine type oil with rust and oxidation inhibitors is preferred. This oil must be kept clean (filtered to 40 microns maximum, or preferably 25 microns), cooled, and must have the correct viscosity. Synthetic oils should not be used without the manufacturer's approval.

For some reason, the high speed gear makes all the compromises when oil viscosity for a combined lube oil system is determined. Usually a viscosity preferred for compressor seals or bearings is selected and gear life is probably reduced. The bearings in a gear unit can use the lightest oils available, but gear teeth would like a much heavier oil to increase the film thickness between the teeth.

When selecting a high speed gear unit, the possibility of using an AGMA No. 2 Oil (315 SSU @ 100°F) should be considered. In most cases, the sleeve bearings in the system can use this oil and, if not, a compromise 200 SSU at 100°F oil should be considered.

When 150 SSU at 100°F oil is necessary, inlet temperatures should be limited to 110°F to 120°F to maintain an acceptable viscosity. Oil should be supplied in the temperature and pressure range specified by the manufacturer.

Up to a pitch line speed of approximately 15,000 feet per minute, the oil should be sprayed into the out-mesh. This allows maximum cooling time for the gear blanks and applies the oil at the highest temperature area of the gears. Also, a negative pressure is formed when the teeth come out of mesh pulling the oil into the tooth spaces.

Above approximately 15,000 feet per minute, 90% of the oil should be sprayed into the out-mesh and 10% into the in-mesh. This is a safety precaution to assure the amount of oil required for lubrication is available at the mesh.

In addition to the above, in the speed ranges from 25,000 to 40,000 feet per minute, oil should be sprayed on the sides and gap area (on double helical) of the gears to minimize thermal distortion.

Types of Lubrication in Gear Teeth

Boundary Lubrication

Boundary lubrication most often occurs at slow to moderate speeds, on heavily loaded gears, or on gears subject to high shock loads. The oil film is not thick enough to prevent some metal-to-metal contact. This condition usually shows some early wear and pitting due to surface irregularities in the tooth surfaces.

When boundary lubrication is encountered, extreme pressure oils should be used to minimize wear and possible scuffing.

Hydrodynamic Lubrication

Hydrodynamic lubrication occurs when two sliding surfaces develop an oil film thick enough to prevent metal-to-metal contact. This type lubrication usually only exists on higher speed gearing with very little shock loading.

Elastohydrodynamic (EHL) Lubrication

Elastohydrodynamic theory of lubrication is now accepted as very common in gear teeth. The formation of EHL films depends on the hydrodynamic properties of the fluid and deformation of the contact zone. This flattening of the contact area under load forms a pocket that traps oil so that the oil does not have time to escape resulting in an increase in oil viscosity. This increase makes possible the use of light oils in high speed drives and usually only occurs above 12,000 FPM.

Methods of Supplying Lubricant

Splash Lubrication

Splash lubrication is the most common and foolproof method of gear lubrication. In this type system, the gear dips and in turn, distributes oil to the pinion and to the bearings. Distribution to the bearings is usually obtained by throw off to an oil gallery or is taken off by oil wipers (or scrapers) which deliver the oil to an oil trough.

Care must be taken that the operating speed is high enough to lift and throw off the oil. In the throw-off system, the minimum speed, n_p, is:

$$n_p = (70{,}440/d)^{.5}$$

Where:

d = Pitch diameter

n_p = RPM

Oil wiper systems can operate at much lower speeds which are generally determined by test. The splash system can be used up to 4000 FPM pitch line velocity. Higher speeds can be splash lubricated with special care.

Forced Feed Lubrication

Forced feed lubrication is used on almost all high speed drives and on low speed drives when splash lubrication cannot be used due to gear arrangement.

A simple forced feed system consists of a pump with suction line and supply lines to deliver the oil. However, lubrication supply systems for high speed drives include many of the components show in the table below.

Many of these systems are well designed and constructed for optimum performance.

Scoring or scuffing (adhesive wear) is caused when the oil film does not prevent contact between mating surfaces. Areas touch each other due to load which results in welding of the two surfaces. As sliding continues, these surfaces break apart. These particles adhere to the surfaces, and rapid adhesive wear occurs.

The flash temperature theory of this type failure indicates that the welding is caused by the high temperature generated locally in the contact area. Calculations can be made to determine the scoring risk for higher speed drives but data are not available for all oil types.

Pitting or surface fatigue comes from the formation of small sub-surface cracks that are developed by fatigue failure of the tooth surface under repeated load. As fatigue failure progresses, the surface begins to break up, and pits form. Pitting usually starts close to the pitch line in the dedendum area.

If a gear is operating above the basic strength of the gear material, no lubricant can prevent pitting. Some pitting is corrective in nature and up to a point, is not detrimental to the gearing. Pitting in case hardened gearing usually leads to failure. Extreme pressure oils and higher viscosity can help reduce pitting.

Abrasive wear is usually caused by a very rough surface finish on the gear teeth or foreign particles in

- Large reservoir
- Filters (Duplex or Single)
- Shaft driven pump
- Auxiliary Pumps (Motor and Steam)
- Heat exchangers (Single or Duplex)
- Accumulators
- Pressure control devices
- Safety alarms and shutdowns (Temp. and Pressure)
- Temperature Regulators
- Isolation valves
- Heaters (Steam or Electric)
- Purifier (Removes water and oxidation products)
- Flow Indicators

the oil. The foreign particles adhere temporarily to one surface and in turn, scratch a groove in the second surface. Generally, there is very little problem with abrasive wear if the lubricant is clean.

Gear lubrication, at the present time, is not a highly developed technology in general industrial applications and the ultimate capacity of gearing is partially determined by lubricant load limits. Fortunately future higher capacity lubricants (synthesized hydrocarbons) may well solve gear problems by generating thick films without the problems now associated with the extremely heavy oils.

LUBRICATION OF LARGE OPEN GEARS

Large open gears (Figure 12-5) are toothed gear systems, i.e., the gear and the pinion are not situated in a joint housing. The drive cover frequently is not oil-tight. Large open gears are mainly used in the base material industry, for example in ore and raw material processing installations, fertilizer, waste incineration and composting plants, coal-fired plants, rotary kilns, tube mills, drying, cooling and conditioning cylinders.

As the pinion and the gearwheel are supported separately, and owing to the low peripheral speed, extremely high flank load and surface roughness, such gears are mainly operated in the mixed friction regime. To ensure operational reliability it is therefore necessary to apply special adhesive lubricants having specific physical and chemical properties to form a protective layer on the tooth flanks and avoid direct contact between the metal surfaces. Competent vendors often recommend the application of running-in lubricants to rapidly reduce surface roughness when putting the gear into operation and achieve a good load distribution over the flanks and the faces.

One lubricant manufacturer, Klüber, has developed a so-called A-B-C system of lubrication to ensure optimum lubrication during all operating stages and to protect the drive against any damage right from the first assembly-related turns of the gear.

They offer special lubricants tailored to suit any of these steps as well as the various lubricant application methods. Tables 12-4 and 12-5 show the products recommended for the individual step and application method. This lubrication system comprises the following steps:

A = Priming and prestart lubrication
B = Running-in lubrication
C = Operational lubrication

Figure 12-5. Girth gear drive of a drying cylinder.

LUBRICATION OF WORM GEARS

Worm gears are of a crossed-axis geometry. They have a constant transmission ratio and are used as speed and torque converters between driving engines and machines. The high sliding percentage of worm gear engagement ensures low-noise and low-vibration operation.

As compared to other gears (e.g., bevel gears), the efficiency of worm gears is relatively poor. However, they make high transmission ratios possible in a single step.

Suitable synthetic lubricants, Table 12-6, reduce the friction and power loss in worm gears up to 30 %. The operational wear of the worm wheels, which usually consist of copper bronze, can be reduced substantially with synthetic lubricants containing suitable additives.

The use of synthetic lubricants in worm gears results in an improvement of the gears' efficiency and service life and makes them suitable for many applications.

Properly formulated special synthetic gear lubricants will not only improve gear efficiency; their anti-wear additives can optimize the gear's wear behavior. Figures 12-6 and 12-7 show the efficiency and wear curves of synthetic Klüber lubricants and mineral hydrocarbon oil. The pertinent worm gear had a center distance of 63 mm, a worm speed of 350 rpm and a

Table 12-4. Lubricant application methods.

Type of lubrication	Type of application	Klüber lubricant A: GRAFLOSCON A-G 1 PLUS	B: Klüberfluid B-F 1	B: GRAFLOSCON B-SG 00 PLUS	C: Klüberfluid C-F 1	C: Klüberfluid C-F 1	C: GRAFLOSCON C-SG 0 Plus	C: GRAFLOSCON C-SG 2000 PLUS
	Dip-feed lubrication		•		•	•		
Continuous lubrication	Lubrication with the Klübematic PA system		•		•	•		
	Dip-feed circulation lubrication		•		•	•		
	Transfer lubrication							
Long-term lubrication	with paddle wheel		•		•	•		
	with transfer pinion (plunging into the lubricant)		•		•	•		
	Manual lubrication							
Intermittent lubrication	by brush or spatula	•						
	with pressurized air spray gun		o	o	o	o	o	o
	Automatic spray lubrication			•	o	o	•	•
Total loss lubrication	Transfer lubrication with pinion (lubricant fed to the pinion by pump)			•	o	o	•	•

• Preferred method
o Possible method

drive torque of 300 Nm. (See also Chapter 7.)

The synthetic lubricants' excellent resistance to aging allows extended lubricant change intervals. These can be three to five times longer than intervals recommended with mineral hydrocarbon oils, which means lifetime lubrication in many cases.

LUBRICATION OF SMALL GEARS

Small gears comprise spur, bevel and worm gears of an open, semi-closed or closed design (often not oil-tight).

Small and miniature gears are used in adjusting and control drives in the automotive industry, office machines, household appliances and machines for do-it-yourselfers, Figures 12-8 and 12-9. Their main task is to transfer movements, sometimes also power.

Due to the construction of these gears and the materials that are used—steel/steel, steel/bronze, steel/plastic and plastic/plastic components—the lubricants have to meet various requirements, including

- lifetime lubrication
- noise damping
- low starting torque
- low and high temperature operation
- resistance to ambient media
- compatibility with the materials used

As small gears often are not oil-tight they are mainly lubricated with greases applied by dip-feed lubrication or one-time lubrication of the tooth flanks.

Dip-feed lubrication is preferred for gears in continuous operation or gears used for power transmission. One-time lubrication is suitable for gears used for the transmission of movements or gears only operating for short intervals or intermittently.

Lubricating greases of NLGI grade 000 to 0 are used for dip-feed lubrication, and of grade 0-2 for lifetime lubrication. To avoid the lubricant being thrown off the gear, pastier greases are preferred in case of increased peripheral speeds. In case of dip-feed lubrication at higher peripheral speeds, however, the grease should be softer to avoid channeling.

Table 12-5. Typical products recommended for large open gear drives.

Selection criteria	Product name	Base oil/ thickener	Service temperature range (°C) ≈	Density at 20°C (g/cm³) DIN 51757 ≈	Base oil viscosity DIN 51561 (mm²/s) ≈ 40°C	Base oil viscosity DIN 51561 (mm²/s) ≈ 100°C	Color	Drop point DIN ISO 2176 (°C)	Functional lubricant film	Worked penetration DIN ISO 2137 (0,1mm) 51 818	Consistency NLGI grade DIN	Apparent viscosity KL viscosity grade	Notes
Prestart lubrication priming	GRAFLOSCON A-G 1 PLUS	Min. hydrocarbon oil/Al complex soap, fine graphite	-15 to 90	1.07	500	31	gray-black	>220	-30 to +200*	310 to 340	1	S	Adhesive lubricant, free from solvents, bitumen and lead
Running-in lubrication, spray lubrication	GRAFLASCON B-SG 00 PLUS	Min. hydrocarbon oil/ silicate, graphite	-15 to 90	0.99	470	92	gray-black	—	-30 to +200*	400 to 430	00	—	Sprayable adhesive lubricant, free from solvents, bitumen and lead. Scuffing load >12 in the special FZG test A/2/76/50, change in specific weight approx. 12 mg/kWh
Running-in lubrication, dip-feed and circulation lubrication	Klüberfluid BF-1	Min. hydrocarbon oil/ Al soap, fine graphite	0 to 80	0.99	1300	60	dark brown gray-black	—	-30 to +200*	—	—	—	Running-in lubricant for dip-feed and circulation lubrication, free from solvents, bitumen and lead. Scuffing load >12 in the special FZG test A/2.76/50, change in specific weight approx. 1 mg/kWh
Operational lubrication, spray lubrication	GRAFLOSCON C-SG 0 PLUS	Min. hydrocarbon oil/Al soap, fine graphite	0 to 90	0.96	680	40	gray-black	>90	-30 to +200*	355 to 385	—	M	Sprayable adhesive lubricant, free from solvents, bitumen and lead. Scuffing load >12 in the special FZG test A/2,76/50, change in specific weight approx. 0.2 mg/kWh.
	GRAFLOSCON C-SG 2000 Plus	Min. hydrocarbon oil/ Al soap, fine graphite	10 to 90	0.98	1500	60	gray-black	>70	-30 to +200*	400 to 430	00	-	GRAFLOSCON C-SG 2000 plus is particularly suitable for hot climatic conditions
Operational lubrication, dip-feed and circulation	Klüberfluid CF-1	Min. hydrocarbon oil/ Al soap, fine graphite	-15 to 35	0.96	250	20	dark brown, gray-black	—	-40 to +150*	—	—	—	GRAFLOSCON C-SG 2000 plus Operational lubricant for dip-feed and circulation lubrication, free from solvents, bitumen and lead. Scuffing load >12 in the special FZG test A/2.76/50, change in specific weight approx. 0.2 mg/kWh
	Klüberfluid CF-2	Min. hydrocarbon oil/Al soap, fine graphite	5 to 100	0.96	3200	94	dark brown, gray-black	—	-30 to +200*	—	—	—	

*Depending on relubrication

Table 12-6. Typical worm gear lubricants.

Selection criteria	Product name	Base oil/ thickener	Service temperature range (°C)	ISO VG DIN 51 519	Density at 20°C (g/ml) DIN 51757	Kinematic viscosity DIN 51561, (mm^2/s) at ≈(VI) 20°C	40°C	100°C	Viscosity index DIN ISO 2909 (°C)	Pour point DIN ISO 3016 (°C) ≈	Flash point DIN ISO 2592 (°C) ≈	Notes
Synthetic gear oils for normal and high temperatures	Klübersynth GH 6-220	Poly-glycol oil	-30 to 160	220	1.05	630	220	41	>220	<-30	>300	Scuffing load stage >12 in the special FZG test A/16.6/140; particularly suitable for steel/bronze pairings/ check compatibility with sealing materials and paint coatings
	Klübersynth GH 6-460	Poly-glycol oil	-25 to 160	460	1.05	1240	460	79	>240	<-25	>300	
Synthetic gear oils for normal temperatures	Klübersynth GEM 4-220	Synthetic hydro-carbon oil	-40 to 140	220	0.86	627	220	26	>150	<-40	>200	Scuffing load stage >12 in the special FZG test A/16.6/140; also suitable for worm wheels made of aluminum bronze
	Klübersynth GEM 4-460	Synthetic hydro-carbon oil	-30 to 140	460	0.86	1440	460	48	>160	<-35	>200	
Synthetic gear oils for the food-processing industry	Klüber-oil 4 UH 1-220	Synthetic hydro-carbon oil	-30 to 110	220	0.85	670	220	33	>170	-35	>200	Scuffing load stage 12 in the FZG test A/8,3/90, DIN 51354 Pt 2; USDA-H1 authorized
	Klüberoil 4 UH 1-460	Synthetic hydro-carbon oil	-30 to 110	460	0.85	1370	460	52	>170	-35	>200	

Selection criteria	Product name	Base oil/ thickener	Service temperature range (°C) ≈	Density at 20°C (g/cm^3) DIN 51757 ≈	Base oil viscosity DIN 51561 (mm^2/s) ≈ 40°C	100°C	Color	Drop point DIN ISO 2176 (°C)	Speed factor ($n \bullet d_m$) $mm \bullet min^{-1}$ ≈	Worked penetration DIN ISO 2137 (0.1mm)	Consistency NLGI grade DIN 51 818	Apparent viscosity KL viscosity grade	Notes
Fluid synthetic grease for gears that are <u>not</u> oiltight	Klüber-synth GE 46-1200	Polyalkyleneglycol oil Li soap	-50 to 120	0.99	120	20	brown	>160	—	400 to 430	00	EL	Scuffing load stage >12 in the FZG test A/8,3/90, DIN 51354 Pt 2

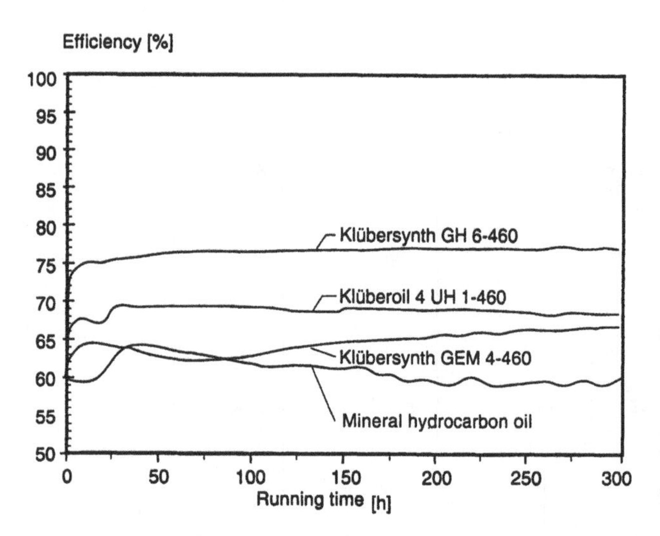

Figure 12-6. Efficiency of mineral hydrocarbon oil compared to synthetic oils.

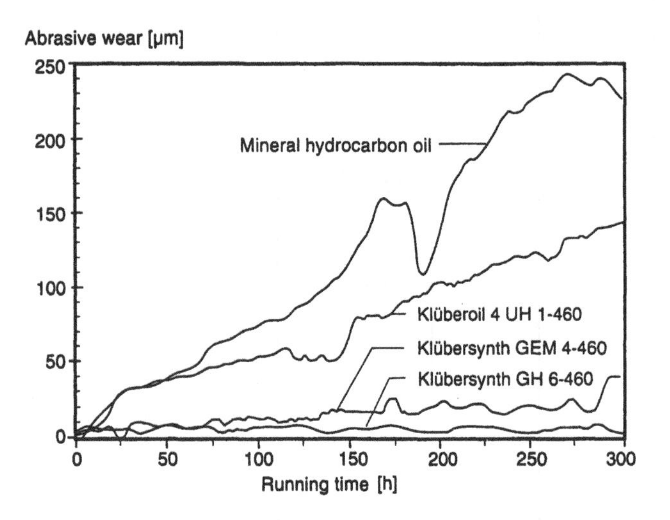

Figure 12-7. Wear curve of mineral hydrocarbon oil compared to synthetic oils.

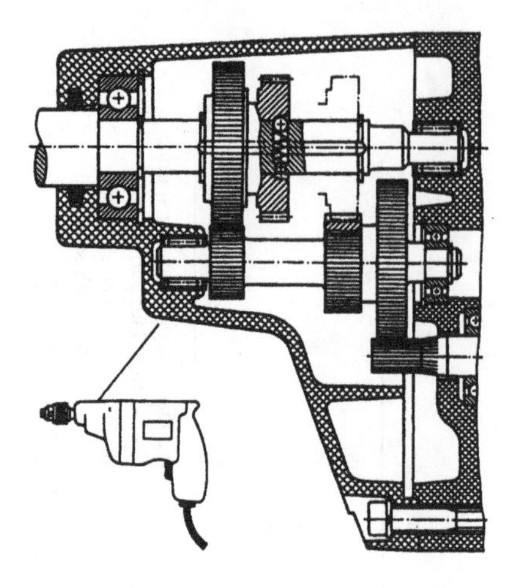

Figure 12-8. Manual drill, double-stage gear.

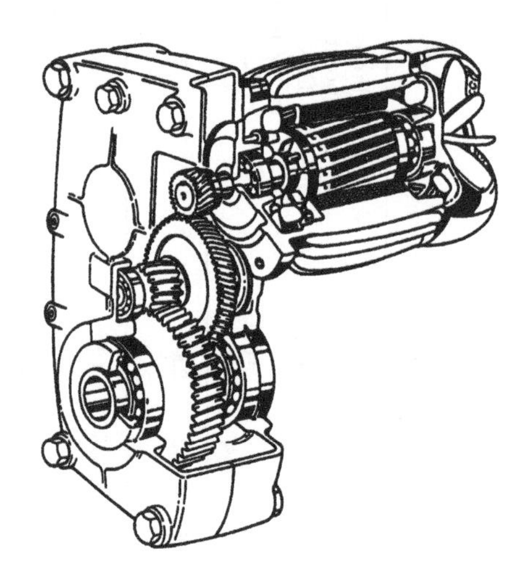

Figure 12-9. Gear motor, double-stage spur gear.

Greases with a synthetic base oil are particularly suitable for applications where high resistance to high and low temperatures and aging are required and where the gear friction has to be low. Refer to Table 12-7.

TESTING THE PERFORMANCE OF GEAR OILS

There are three major characteristics a superior EP gear oil should exhibit: extreme pressure capability, cleanliness and demulsibility. In other words, the oil must perform under pressure for extensive time periods, must keep machine systems (and reservoirs) clean and must separate rapidly from water. In developing their SPARTAN EP gear lubricants, Exxon extensively tested for these and other characteristics for two years in order to ensure that the formulation is the best, most dependable lubricant the company could produce with today's leading-edge technology. Such dependability is vital, given the high cost involved in equipment failure and downtime.

Five primary tests run during this two-year period were: EP retention, oxidation, panel coker, copper strip corrosion and demulsibility. The tests measure EP capability, the degree of cleanliness and the demulsibility a lubricant exhibits.

Performance under Pressure

Superior EP lubricants last. They hold up under pressure and over time. They don't have to be replaced as often. That saves money and helps gear units last longer.

Table 12-7. Typical small gear lubricants.

Selection criteria	Product name	Base oil/ thickener	Service temperature range (°C) ≈	Density at 20°C (g/cm³) DIN 51757 ≈	Base oil viscosity DIN 51561 (mm²/s) ≈ 40°C	100°C	Color	Drop point DIN ISO 2176 (°C)	Speed factor ($n \bullet d_m$) mm•min⁻¹ ≈	Worked penetration DIN ISO 2137 (0.1mm)	Consistency NLGI grade DIN 51 818	Apparent viscosity KL viscosity grade	Notes
Dip-feed lubrication, normal temperature range	MICRO-LUBE GB 00	Min. hydrocarbon oil/ silicate	0 to 100	0.90	600	38	reddish black	>200	— to +200*	430 to 475	00/000	EL	For gears subject to high loads, scuffing load stage >12 in the special FZG test A/2,76/50
	CENTO-PLEX HO	Min. hydrocarbon oil/Li	0 to 100	0.88	110	11.5	beige to brownish almost transparent	>170	5 x 10^5	370 to 380	0	EL	For low-power small gears
Dip-feed lubrication, low and high temperatures	Klübersynth GE 46-1200	Poly-alkylene glycol/Li	-50 to 120	0.99	120	20	brown	>160	—	400 to 430	00	EL	For extreme loads, long-term and lifetime lubrication; scuffing load stage >12 in the FZG test A/8,3/90, Din 51354 Pt 2
	STRUC-TOVIS P LIQUID	Poly-alkylene glycol/Li	-35 to 130	1.00	365	57	greenish brownish	—	—	600 to 700 (unworked penetration in acc. with Klein at 25°C)	—	EL	Fluid grease for worm gears, long-term and lifetime lubrication
Tooth flank or dip-feed lubrication, normal temperature range	MICRO-LUBE GB 0	Min. hydrocarbon silicate	-0 to 100	0.90	400	25	reddish brown	>180	—	355 to 385	0	L	For gears subject to high loads, scuffing load stage >12 in the special FZG test A/2,76/50
	Klüber-synth GE 11-680	Min. hydrocarbon oil/ Al complex	0 to 100	0.94	680	35	brown	>160	—	380 to 420	0/00	L/M	Adhesive lubricant, noise damping, scuffing load stage >12 in the special FZG test A/2.76/50
Tooth flank or dip-feed lubrication, low and high temperatures	Klüber-synth G 34-130	Syn. hydrocarb. oil, mineral hydrocarb. oil/special Ca	-30 to 150	0.87	130	16	beige/ brownish	>180	—	355 to 385	0	EL	Lubricant for small gears, e.g., in machines for do-it-yourselfers
	ISOFLEX TOPAS NCA 5051	Synth. hydrocarb. oil/special Ca	-5- to 140	0.80	30	6	beige	>180	1 x 10^6	385 to 415	0/00	El	Low starting torques, noise damp-with plastic friction components
Tooth flank lubrication	POLY-LUB GLY 801	Synth. hydrocarb. oil, mineral hydrocarb. oil/special Li	-40 to 150	0.88	730	57	beige/	>250	—	310 to 340	1	M	For flank lubrication in case of non-metallic materials, good adhesionand noise damping
	ISOFLEX TOPAS NCA 5051	Synthetic hydrocarbon oil/special Ca	-60 to 130	0.86	18	4	beige	>180	1 x 10^6	265 to 295	2	L	For plastic/plastic and plastic/ steel components, low starting torque
	ISOFLEX TOPAS NCA 52	Synth. hydrocarb. oil/special Ca	-50 to 150	0.80	30	5.5	beige	>220	1 x 10^6	265 to 295	2	L	For plastic/plastic and steel/steel components, low starting torque

*Gear relubrication

The Reserve EP Capability Test measures a lubricant's load-carrying ability over time. EP—extreme pressure—oils are specially formulated to lubricate under heavy-load conditions. The longer a lubricant can maintain its load-carrying ability, the less often it must be replaced...which reduces operating costs.

Before starting this test the specific amount of a key load-carrying ingredient—phosphorus—is measured. Then the lubricant is run through the modified U.S. Steel S-200 Oxidation Test called for in U.S. Steel specification 224. (See the following section on "Superior Oxidation Stability.") Once the test is finished the amount of phosphorus is remeasured, to check for any depletion. The more that's left, the more reserve EP capability the lubricant has.

SPARTAN EP gear oil's keep-clean ability helps hold the EP additive in the oil, where the gear can use it—not trapped in the sludge at the bottom of a gear case where it's lost to the gears.

SPARTAN EP not only tested better than the competition initially, it tested better in the long run. SPARTAN EP showed very little phosphorus depletion after the modified U.S. Steel S-200 Oxidation Test called for in U.S. Steel specification 224. Table 12-8 presents an overview of a competitive product survey while Table 12-9 lists typical inspections for SPARTAN EP Gear Oils.

Keeping Machine Systems Clean (Figure 12-10)

If a gear oil leaves dirt and sludge in machine systems, it can eventually affect an entire plant's operation. Deposits build in the system, reducing the lubricant's effectiveness, leading to equipment failure... breakdown...or even plant shutdown. SPARTAN EP is designed to minimize these deposits.

When changing to SPARTAN EP it is recommended that contaminants be flushed from the internal gear oil system. Built-up sludge should be removed from centralized system sumps, lines flushed and filters changed. System filters and screens should also be checked during startup and changed if required.

The following tests measured the ability of SPARTAN EP to keep systems clean.

Superior Oxidation Stability (Figure 12-11)

All oils in service are exposed to oxidation—a form of deterioration that occurs when oxygen interacts with the oil. Oxidation raises the viscosity of the oil. It also leaves acidic materials which cause the soft sludge deposits or hard, varnish-like coatings. And that can lead to equipment failure. Oxidation stability is particularly

Table 12-8. Gear oil competitive product survey.

Grade	SPARTAN EP 220	A	B	C	Competitor D	E	F	G	H
Kinematic viscosity									
cSt @ 40°C, ASTM D 445	217.5	200.8	206.3	211.0	228.8	208.6	205.1	215.0	215.3
cSt @100°C	18.62	18.11	19.00	18.35	19.39	17.69	18.29	18.19	18.18
Viscosity index	95	98	95	96	96	92	98	92	93
Neutralization number ASTM D974	0.44	0.53	0.24	0.83	0.59	0.52	0.48	0.91	0.22
Oxidation stability, USS S-200*									
Viscosity increase @ 100°C, %¶	2.3	5.7	1.2	4.9	4.5	7.1	5.8	4.0	7.0
Precipitation number									
Before	nil	nil	nil	nil	nil	nil	nil	nil	nil
After	trace	trace	trace	trace	trace	trace	trace	trace	trace
ER&E analysis of oils after USS S-200†									
Viscosity increase @100°C¶	2.9	5.7	0.9	6.2	5.0	6.9	6.8	4.4	7 3
Phosphorous depletion loss %¶	2.1	56.0	15.9	3.7	19.2	20.1	15.9	53.4	54.8

*U.S. Steel Lab results.
¶See bar graphs, page 386.
†Exxon Research & Engineering Lab results.

Table 12-9. Spartan EP typical inspections.

Grade	68	100	150	220	320	460	680	USS	AGMA
ISO viscosity grade	68	100	150	220	320	460	680	—	—
AGMA lubricant number	2EP	3EP	4EP	5EP	6EP	7EP	8EP	—	—
Viscosity:									
cSt @ 40°C, ASTM D 445	68 0	100.0	150.0	220.0	320.0	460.0	680.0	—	—
cSt @ 100°C	8.6	11.1	14.5	18.8	24.0	30.9	41.2	—	—
SSU @ 100°F, ASTM D 2161	353.0	522.0	789.0	1165.0	1705.0	2465.0	3660.0	—	—
SSU @ 210°F	55.0	64.0	78.0	96.0	120.0	151.0	201.0	—	—
Viscosity index	95	95	95	95	95	97	100	95[1]	90[1]
Gravity, ASTM D 287, °API	30.0	29.5	27.8	27.2	26 6	25.7	25 0	25[1]	—
Flash point, ASTM D 92, °C(°F)[2]	210(410)	221(430)	238(460)	243(469)	249(480)	257(495)	257(495)	232(450)	—
Pour point, ASTM D 97, °C(°F)	-21(-6)	-18(0)	-18(0)	-18(0)	-9(16)	-9(16)	-9(16)	-9(16)	—
Timken OK load, ASTM D 2782 kg (lb)	27(60)	27(60)	27(60)	27(60)	27(60)	27(60)	27(60)	27(60)[1]	27(60)[1]
Four-ball EP test, ASTM D 2783									
Load wear index, kg	47	47	47	47	47	47	47	45[1]	—
Weld point,	250	250	250	250	250	250	250	250[1]	—
Four-ball wear test, ASTM D 2266									
Scar diem, mm 20kg, 1800 rpm 54°C(130°F) 1 hr		0.30	0.30	0.30	0.30	0.30	0.30	0.30	0.35[3]
FZG gear test, minimum stages	12	12	12	12	12	12	12	11	11
Oxidation stability, USS S-200									
312 hr @ 121°C(250°F) viscosity @ 99'C(210.F), % increase	2.8	2.7	2.8	2.3	4.0	4.4	4.9	6[3]	10[3,4]
Demulsibility ASTM D 2711									
Water in oil, %	0.3	0,4	0.5	0.8	0.8	1.0	0.9	2.0[3]	1.0[3]
Total free water, ml	85	85	85	85	85	85	85	80[3]	50-60[1]
Emulsion, ml	nil	nil	nil	nil	nil	nil	nil	1[3]	2-4[3]
Foam test, ASTM D 892, ml									
Sequence 1	Tr/0	Tr/0	Tr/0	Tr/0	Tr/0	Tr/0	Tr/0	—	75/10
Sequence 2	Tr/0	Tr/0	Tr/0	Tr/0	Tr/0	Tr/0	Tr/0	—	75/10
Sequence 3	Tr/0	Tr/0	Tr/0	Tr/0	Tr/0	Tr/0	Tr/0	—	75/10
Rust test, ASTM D 665 A/B									
Distilled water	pass	pass	pass	pass	pass	pass	pass	no rust	no rust
Synthetic sea water	pass	pass	pass	pass	pass	pass	pass	no rust	no rust
Copper strip corrosion, ASTM D 130									
3 hr @ 100°C(212°F), rating	1	1	1	1	1	1	1	1B[3]	1B[3]

[1]Minimum
[2]For ISO Grade 68 and 100, 200°C(400°F) is the minimum flash point; ISO 150 and above, 232°C(450°F) is the minimum.
[3]Maximum
[4]ASTM D 2893 95.C(203°F)

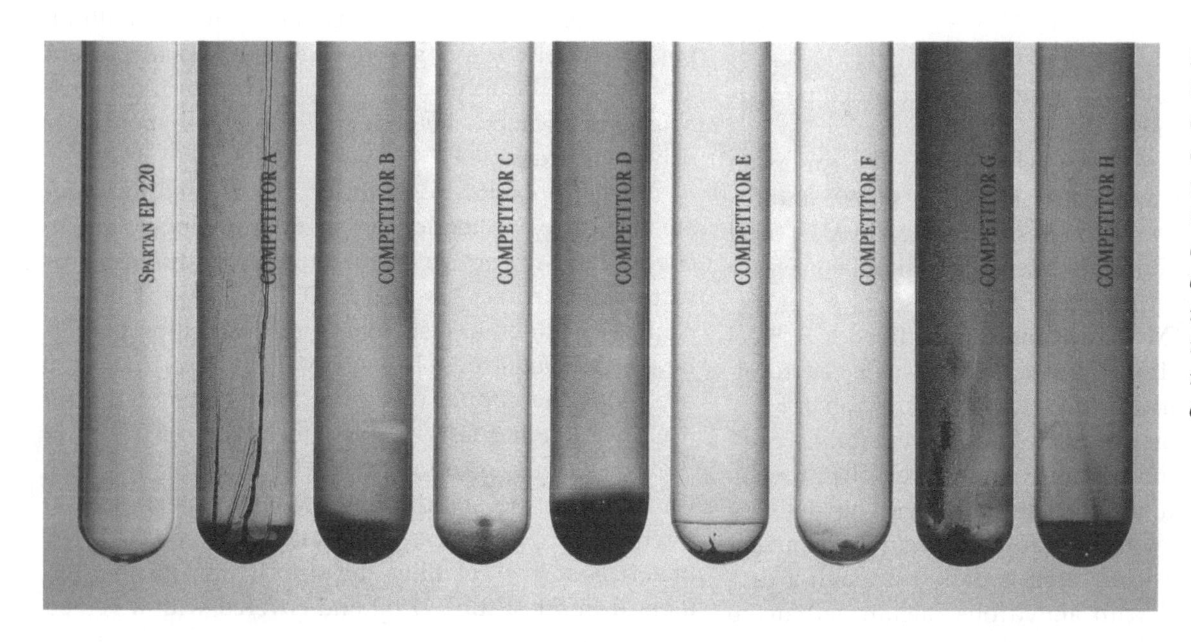

Figure 12-10. SPARTAN EP clearly beats the competition. If most other test tube samples from competitive gear oils look this bad, imagine what their reservoirs must look like after extended service.

Phosphorus Depletion

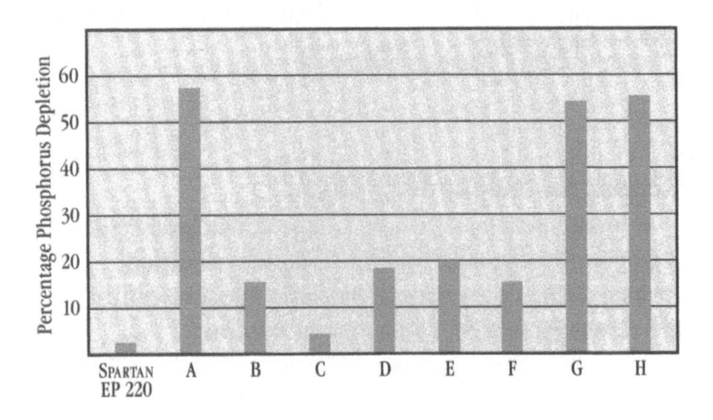

Oxidation Stability

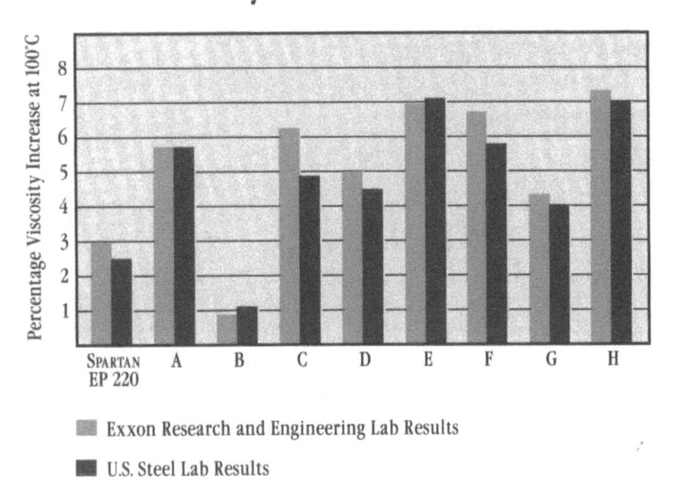

Figure 12-11. SPARTAN EP: low additive depletion results in extended life.

important in gear oils that circulate for extended periods at high temperatures.

The primary tests for oxidation are the American Society for Testing and Materials (ASTM) D 2893 and the U.S. Steel S-200 Oxidation Test called for in U.S. Steel specification 224. Both are intended to simulate service conditions on an accelerated basis.

In both ASTM D 2893 and the modified U.S. Steel S-200 Oxidation Test, oil samples are subjected to elevated temperatures in the presence of dry air for 312 hours (13 days). First, oil samples are tested for viscosity at 100°C (212°F) and precipitation number. Then they're poured into test tubes specially fitted with air delivery tubes and flowmeters to ensure an accurate, constant flow of dry air. (The air itself is passed through a drying tower packed with anhydrous calcium sulfate to ensure that it is moisture-free.) The test tubes are immersed in a heating bath and air is bubbled through the oil. The samples are kept at a constant temperature of 95°C(203°F) (in the ASTM D 2893 test) or 121°C(250°F) (in the USS S-200 test) for 312 hours. Test samples are then removed from the bath, mixed thoroughly and tested again for viscosity and precipitation number.

SPARTAN EP controlled the percentage of viscosity increase and—as shown in Figures 12-11 through 12-14—controlled deposits, passing both the ASTM and USS tests with nearly spotless results.

Panels Sparkle with SPARTAN EP (Figure 12-12)

Deposits build up...leading to friction...increased heat...more deposits...increased wear and eventual equipment breakdown. The Panel Coker Test measures the effectiveness of a lubricant in keeping the system clean and free from deposits.

This test, which simulates extreme environments, compared SPARTAN EP oil's "keep clean" capability with that of its major competitors. The procedure oil was continuously splashed against metal panels heated to 246°C(475°F) for four hours.

Results are read as the amount of deposits left on the panels. The darker the panels, the more deposits that have been left. The cleaner the panel, the fewer deposits, so the cleaner the lubricant keeps the machine system.

SPARTAN EP kept the panel clean throughout the test.

No Corrosion Here

Many types of industrial equipment have parts made of copper or bronze. So any oil that comes in contact with these parts must be non-corrosive. With EP oils, chemically active additives are needed to prevent steel-on-steel scoring and seizure. They're indispensable to applications involving steel parts—hypoid gear drives, for example.

The Copper Strip Corrosion Test is used to evaluate the corrosive tendencies of oils to copper and to check them for active sulfur-type EP additives. This test—standardized as ASTM D 130—should not be confused with other tests for the rust-inhibiting properties of petroleum oils, like ASTM D 665. ASTM D 130 evaluates the copper-corrosive tendencies of the oil itself, while other tests evaluate the ability of the oil to prevent ferrous corrosion.

During the standard ASTM D 130 test, special three-inch copper strips are cleaned and polished. For materials of low volatility, the strip is immediately immersed in a test tube of oil and covered with a vented

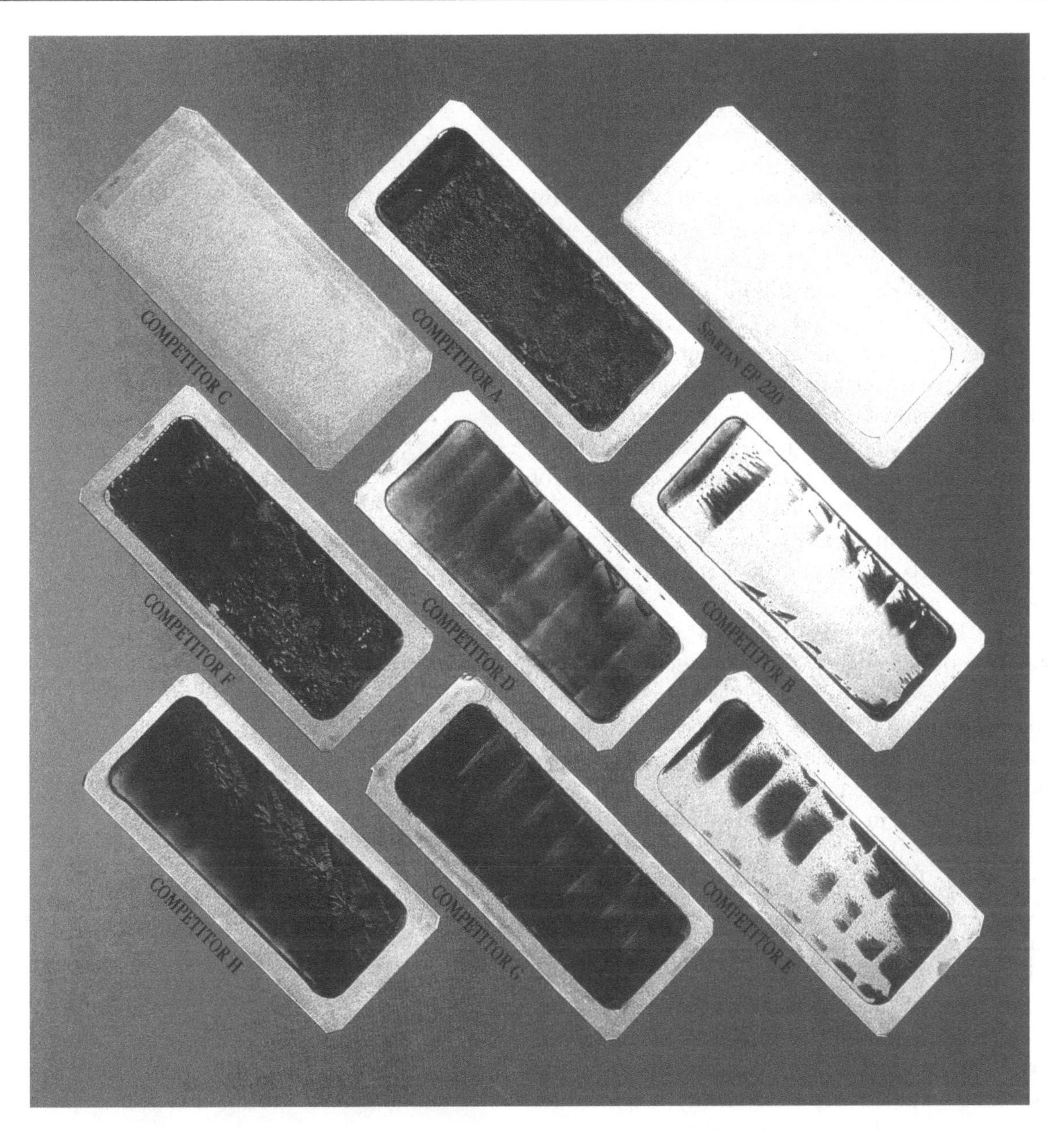

Figure 12-12. SPARTAN EP kept the panel clean, while the competition coked.

stopper. The tube is held in a water bath for three hours at a temperature of 100°C (212°F). At the end of the exposure period, the strip is removed and the oil wiped off. It's compared to a specially prepared set of standardized reference strips that illustrate from Class 1 (slight tarnish) to Class 4 (heavy tarnish). Results are reported as an ASTM D 130 rating. All seven of the lighter ISO grades of SPARTAN EP passed the copper corrosion test with a Class 1 rating.

Another test that measures a lubricant's corrosiveness tendencies to copper-containing compounds is the Radicon Worm Gear Test. In this test, the lubricant is applied to Radicon worm and wheel gears and the clearance between them is measured. The gears are then run for 250 hours at 90°C(194°F), and the end clearance is measured. SPARTAN EP measured 0.011 inches at the beginning and 0.0123 inches at the end, which translates as excellent results.

Excellent Demulsibility

Gear oils are frequently exposed to water, from damaged coolers, lines, atmospheric moisture or occasional steam sources. Water speeds the rusting of ferrous machine parts and accelerates the oil's oxidation. So a gear oil must have good demulsibility characteristics for quick, effective water removal.

SPARTAN EP and Water Don't Mix (Figure 12-13)

A high-quality gear oil resists emulsification and will separate rapidly and thoroughly from water. The test for measuring demulsibility characteristics for EP oils has been standardized as ASTM D 2711.

In this test a 360-ml sample of oil and a 90-ml sample of distilled water are vigorously stirred together for five minutes in a special graduated separating funnel, with the temperature maintained at 82°C(180°F). After a 5-hour settling period, a 50-ml sample drawn from near the top of the oil layer is centrifuged to determine the "percent water in the oil." 'Milliliters of free water" is also measured and reported. Then the mixture is siphoned off until only 100 ml remain in the bottom of the funnel. This is centrifuged and the milliliters of water and emulsion are reported. This amount of water is added to the amount of free water determined in the first separation step above, and the sum is reported as "total water."

SPARTAN EP satisfied both AGMA and U.S. Steel specification requirements in all three steps. These oils represent a multi-industry tool, effective in operating temperatures up to 93°C(200°F). From docks to oil fields, steel mills to paper mills, mines to rolling mills, textile plants to sugar mills, SPARTAN EP keeps the gears turning and plants operating in a wide variety of industries.

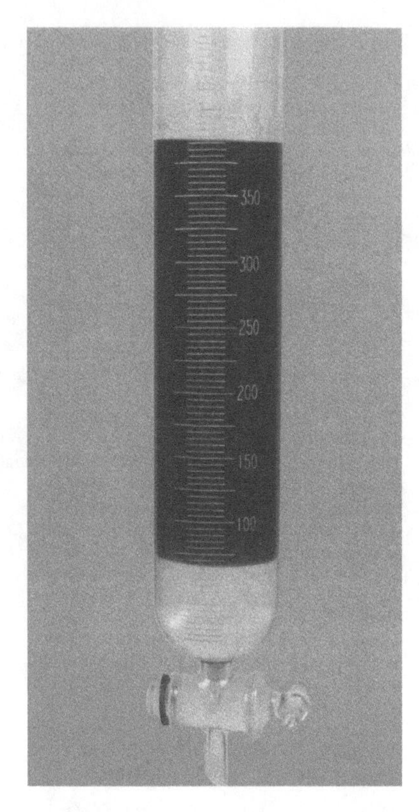

ABOVE: Figure 12-13. SPARTAN EP separates rapidly from water.

High-Temperature, Heavy-Duty Power for the Steel Mills

Modern steelmakers need a lubricant that performs under high temperatures and exhibits excellent demulsibility. SPARTAN EP is designed for the heavy-duty, high-temperature applications found in steel mills. It demonstrates a high level of load-carrying performance— determined in both extreme-pressure and anti-wear tests—and its keep-clean ability helps reduce deposits. It also protects against rust and oxidation, and provides a high level of water shedding and anti-foaming characteristics.

LEFT: Figure 12-14. SPARTAN EP has the high viscosity index needed for demanding marine applications.

High Viscosity Index Means Effective Marine Use (Figure 12-14)

A gear oil used in a marine environment must perform in a wide range of temperatures and conditions. Without this capability, at high temperatures the oil's viscosity may drop to a point where the lubricating film is broken...and that means metal-to-metal contact and severe wear. At the other extreme, the oil may become too viscous for proper circulation, or may set up such high viscous forces that proper operation of machinery is difficult. SPARTAN EP has a high viscosity index—ideal for marine applications.

And of course, lubes used in a marine environment will be exposed to water. As the Demulsibility Test demonstrates, SPARTAN EP separates rapidly and thoroughly from water. Its superior demulsibility characteristics mean quick, effective water removal. Additionally, SPARTAN EP exhibits excellent anti-rust properties in the presence of sea water.

SPARTAN EP is also recommended for use in the deck equipment of motorships, including capstans, winches and windlasses.

Protection for Pulp and Paper Applications

From harvesting and transporting wood to producing lumber, pulp and paper, the forest products industry needs a lubricant designed to minimize equipment downtime and ensure maximum productivity.

Like steel mill machinery, pulp and paper mill machinery requires a lubricant that performs well under high temperatures and exhibits excellent demulsibility. Good protection under severe load conditions and protection against welding and scoring of gear components are needed.

Spartan EP oil's excellent rust protection is a big plus in the humid paper mill environment. It represents an effective lubricant for forest and woodyard equipment such as conveyors, and for pulp and paper mill equipment such as paper machines, gear drives, vacuum pump gear cases and vibrating screen gear cases. The high cost of this machinery necessitates a lubricant that maximizes yield and helps ensure reliable equipment performance.

Applications Above and Below Ground in Mining Use

Mining costs continue to rise. Mining machines are being pushed to the limits of their capabilities in order to maximize tonnage and increase efficiencies. Productivity is a must. As a result, typical operating temperatures and pressures are significantly higher than in the past. These demanding circumstances require a proven performer under severe operating conditions.

SPARTAN EP provides the superior performance properties demanded by increasingly sophisticated mining machinery and techniques.

Maximum Equipment Life for Aluminum Rolling Industry (Figure 12-15)

The use of aluminum containers for beverages continues to grow. It's the preferred container among brewers, bottlers and distributors—compact, easily adaptable to high-speed filling lines and lightweight. And that makes handling and transportation easier. Its high recycling value is also a major factor in its favor. Cans are currently the most popular single-service soft drink containers, while beer manufacturers continue to switch from glass containers to aluminum.

Nonetheless, this is a highly competitive market and every aluminum manufacturer must continually focus on mill reliability and productivity.

That's where an oil with superior EP capabilities and high viscosity index can save mills money in equipment maintenance and turnover. SPARTAN EP is recommended in roll stands and gear boxes throughout aluminum cold rolling mills.

Figure 12-15. Saving money in equipment maintenance and turnover is a high priority for today's aluminum rollers—so they can rely on SPARTAN EP.

GEAR COUPLING LUBRICATION

If a user elects to use grid or gear couplings instead of nonlubricated coupling types, he should be made aware of their vulnerability. The gear coupling, Figures 12-16 and 12-17, is one of the most critical components in a turbomachine and requires special consideration from the standpoint of lubrication.

There are two basic methods of gear coupling lubrication: batch and continuous flow. In the batch method the coupling is either filled with grease or oil; the continuous flow type (Figure 12-17) uses only oil, generally light oil from the circulating oil system.

The grease-filled coupling requires special quality grease. The importance of selecting the best quality grease cannot be overemphasized. A good coupling grease must prevent wear of the mating teeth in a sliding load environment and resist separation at high speeds. It is not uncommon for centrifugal forces on the grease in the coupling to exceed 8,000 Gs.

Testing of many greases in high speed laboratory centrifuges proved a decided difference existed between good quality grease and inferior quality grease for coupling service. Testing also showed separation of oil and soap to be a function of G level and time. In other words, oil separation can occur at a lower centrifugal force if given enough time. The characteristics of grease that allow the grease to resist separation are high viscosity oil (Figure 12-18), low soap content, and soap thickener and base oil as near the same density as possible. In the late 1970s, a number of greases were tested for separation characteristics in a Sharples high speed centrifuge and for wear resistance on a Shell 4

Grid Type

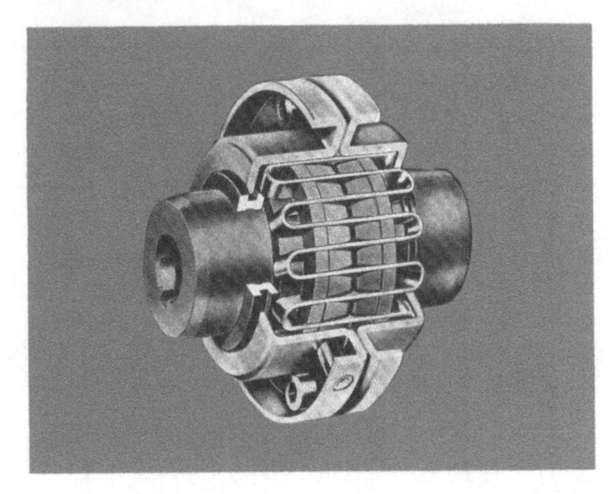

Gear Type

Figure 12-16. Typical couplings requiring grease lubrication.

Ball Extreme Pressure Tester. It was found (Table 12-10), that Grease B exceeded all other greases tested in separation characteristics. Zero separation was recorded at all speeds up to and including 60,000 Gs. Greases A, C, and D were rated poor in separation characteristics at all speeds tested.

Table 12-11 illustrates how these four greases performed on the Shell 4 Ball Extreme Pressure Tester in comparison with a typical Extreme Pressure gear oil. Based on these data, Greases A and B should provide excellent wear protection in severely loaded service. However, only Grease B passes the oil separation test and would qualify for long-term service at a modern facility.

Refer to Table 12-12 for typical inspections on Exxon's HIGH SPEED COUPLING GREASE, reflecting the testing practices of 1997 and 1998. This high-quality grease was formulated to lubricate all flexible couplings. In particular, it meets the lubrication needs of couplings operating at high speeds and with high centrifugal forces. It provides extreme-pressure protection and is suitable for operating temperatures between -40°C(-40°F) and 149°C(300°F). EXXON HI-SPEED COUPLING GREASE offers excellent resistance to oil separation, as indicated by ASTM D 4425 test results (K36 typically <4/24).

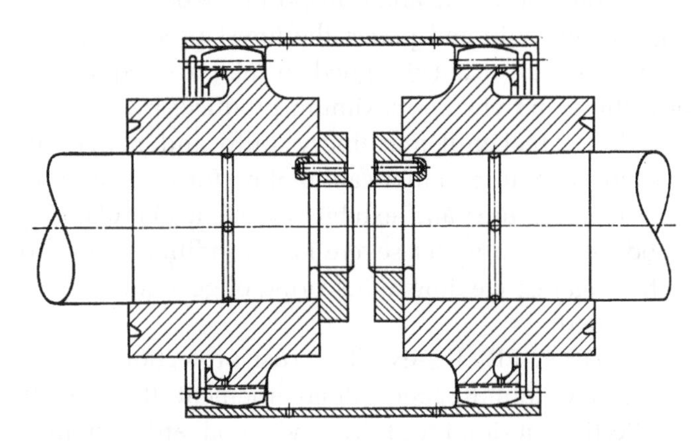

Figure 12-17. Section through a tooth-type coupling. (Source: ASEA Brown-Boveri, Baden, Switzerland)

HIGH SPEED COUPLING GREASE utilizes a state-of-the-art calcium sulfonate thickener system. This thickener system has several unique performance advantages over conventional lithium-polymer thickener systems used in many competitive coupling greases. Some of the performance benefits of the calcium sulfonate thickener are:

- Excellent corrosion prevention, even in the presence of salt mist
- Inherent EP and anti-wear protection
- High dropping points
- Superior oxidation resistance
- Excellent shear stability

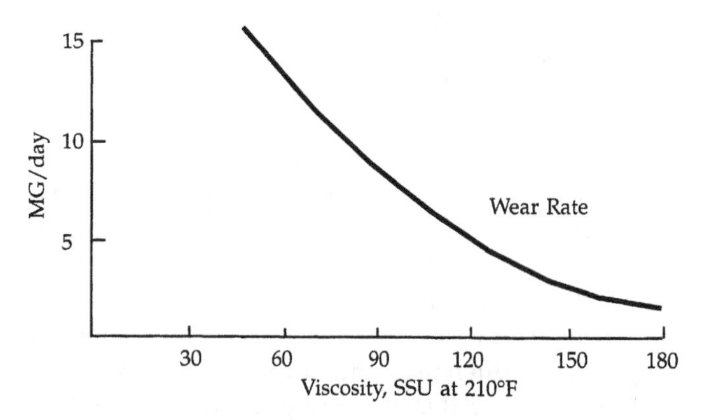

Figure 12-18. Viscosity vs. wear plot.

Table 12-10. Oil separation (%) observed on four coupling greases.

	"G" Level		
GREASE	60,000 2.5 hr	32,000 6 hr	9,700 6 hr
GREASE "A"	53	35	—
GREASE "B"	0	0	0
GREASE "C "	79	75	40
GREASE "D"	51	—	—

Table 12-11. Shell 4 ball test—one minute wear load performance of four coupling greases.

GREASE	KILOGRAM LOAD PASSED
GREASE "A"	80
GREASE "B"	90
GREASE "C "	50
GREASE "D"	20
TYPICAL E.P. 140 OIL	90

Table 12-12. Typical inspections for Exxon's HIGH-SPEED COUPLING GREASE.

Soap type	Calcium Sulfonate
NLGI grade	1
Appearance	Light brown or tan
Dropping point, ASTM D 2265, °C (°F)	>260 (>500)
Penetration ASTM D 217, mm/10, worked 60 ×	320
Timken OK load, ASTM D 2509, lb	60
Four Ball EP, ASTM D 2596	
LWI, kg	62
Weld point, kg	500
Base oil viscosity, ASTM D 445	
cSt @ 40°C	329
cSt @ 100°C	22.5
Rust test, ASTM D 1743	Pass
Water washout, ASTM D 1264, 79°C, % loss	2.5
Centrifugal separation,	
K36 = % oil separated/test hours, ASTM D 4425	K36 <4/24

LUBRICATION OF SMALL GEARED BLOWERS

Large blowers often use lubricant application methods that are virtually identical to those found in rotating machinery such as compressors (see Chapter 17). That may not be the case with small positive displacement blowers that often incorporate gears and rolling element bearings. Gears and rolling element bearings may each have their own distinct lubrication requirements and oil mist (Ref. 1) may be an advantage for these machines.

Positive displacement blowers (Fig. 12-19) are often used to move air ranging in pressure from slightly negative to about 15 psig (~ 1 bar) positive. These rotary lobe machines are often used for polymer powder or pellet transfer and they are manufactured in different sizes. A number of models with 10, 12, and 12-inch shaft center to shaft centers are among the most widely used versions. Drives for these blowers include direct-con-

Fig. 12-19: Roots-type positive displacement blower at a petrochemical plant. (Source: Lubrication Systems Company, Houston, Texas)

nected motors as well as a number of gearbox and belt arrangements. Many positive displacement blowers are oil splash lubricated, while some of the larger ones are forced-feed lubricated (Ref. 2).

Case Histories

When some of these blowers were installed at one U.S. facility in the early 1990s there was—at that time—an understanding between the oil mist provider and user-owner that pure mist (dry sump) would be used for the blower bearings at a later date. A wet sump (purge mist) arrangement was considered appropriate on the timing gear/oil sump side of the splash-lubricated units. Years later, the user-owner wanted to implement an upgrade project and researched if it was acceptable to eliminate forced-feed lubrication from the larger blower models in his plant. Although not disallowing oil mist, the original equipment manufacturer (OEM) had indicated insufficient experience converting forced-feed units to best-practice oil mist lubrication.

Suffice it to say that, regardless of the OEM's position, oil mist lubrication has served Roots-type blowers for many years and has been quite successful on many sizes. Competent oil mist providers can certainly point to case histories and would be pleased to share these with any prospective client. More typically, many Roots-type blowers use a combination of dry sump and wet sump, as shown in the cross-sectional view of Fig. 12-20, and conversions from forced-feed are generally feasible. However, first and foremost, the dry sump oil mist is intended for antifriction bearings, as shown on the right side of Fig. 12-20. Wet sump (purge mist) is shown on the left side, where the gears pick up oil and fling the lubricant into the two spherical roller bearings. The exact point where the mist is applied is of critical importance. If routed per API-610 (8th through 10th Editions), pure oil mist will protect bearings better than anything short of an oil jet impinging on the rolling elements.

On the other hand, oil mist can work even if not applied per API-610. Yet, a facility doing it the way it used to be done in the 1960s cannot, in 2009, claim to be reliability-focused. As was brought out in Chapter 11, for best effectiveness the dry sump oil mist connections must be arranged for mist flow through the bearings. The wet sump (purge mist) side requires no such porting since purge mist must merely represent a region of elevated pressure. Recall that the only purpose for purge mist is to keep out atmospheric contaminants. Although oil jet lubrication would be the one superior form of lubricant application to bearings, jet lube would be difficult to cost-justify on small positive displacement blowers. But dry sump (pure) oil mist—properly applied—is an attractive lubrication method for rolling element bearings in virtually all industries and for rolling element bearings in all types of rotating machinery. It also serves as a protective environment for stored or non-running standby machines.

When utilizing a suitable synthetic lubricant either as a liquid or an oil mist, the maximum allowable temperature is usually set by bearing-internal clearance and bearing metal considerations (generally limited to 230 degrees F maximum). Nevertheless, clearances are rarely an issue as long as bearing inner and outer rings are within 60 or 70 degrees F of each other, and neither are

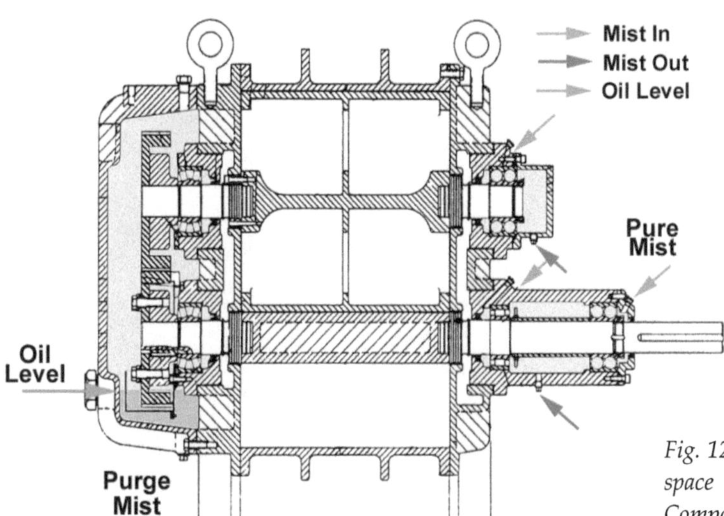

Fig. 12-20: Roots-type blower cross-section with oil mist purge in gear space (left), and pure oil mist (right). (Source: Lubrication Systems Company, Houston, Texas)

lubricant temperatures, since the temperature permitted for many premium grade synthetic lubricants certainly exceeds 300 degrees F.

Some potential users of oil mist wish to engage in experience surveys and that's fine. In that case, try not to put blind faith in what someone said, regardless of whether they claim good, or bad, or "just so-so" experiences. Sometimes, a non-expert's word-of-mouth experience is not relevant at all. Since the non-expert rarely does root cause failure analysis (RCFA), any feedback or opinion would have to be linked to several variables and would mandate that we knew these variables. In any event, there is never a substitute for understanding how parts work, and how they fail.

So, a simple review of the cross-sectional configuration of a given positive displacement blower would be helpful. Anything short of such a review is just guesswork; it rarely adds value but always adds some risk. As long as this diligent review includes all components and fully explains the machine's inner workings, an owner-operator will prosper. Such a review is as important for oil mist on blowers as it is for the bearing housing seals on these machines. Examine these seals, using material in Chapter 11 as your guide. If, in a seal, half the O-ring is contacted by a groove in the stationary part and the other half is contacted by a groove in the rotating part, ask what happens to the O-ring at slow-roll, or when there is axial movement of the two parts relative to each other. As you think about it, you may understand that you have not, perhaps, purchased the best available bearing protector seal (Refs. 3 and 4), or the most cost-effective bearing, or the most suitable lubricant. Understanding how components work and then making smart buying decisions will surely contribute to downtime avoidance.

References

1. Bloch, Heinz P. and Abdus Shamim; "Oil-Mist Lubrication Handbook"—Practical Applications," (1998), Fairmont Press, Lilburn, GA, 30047 (ISBN 0-88173-256-7)
2. Bloch, Heinz P.; "Improving Machinery Reliability," (1998), Third Edition, Gulf Publishing Company, Houston, TX, 77520 (ISBN 0-88415-661-3)
3. Bloch, Heinz P.; "Consider Dual Magnetic Hermetic Sealing Devices for Equipment In Modern Refineries," (Pumps & Systems, September 2004)
4. Bloch, Heinz P.; "Counting Interventions Instead of MTBF," (Hydrocarbon Processing, October 2007)

Chapter 13

Lubricating Electrical Motors

In the petrochemical industry, approximately 60 percent of all motor difficulties are thought to originate with bearing troubles. One plant, which had computerized its failure records, showed bearing problems in 70 percent of all repair events. This figure climbs to 80 percent in household appliances with "life-time" lubrication. If a bearing defect is allowed to progress to the point of failure, far more costly motor rewinding and extensive downtime will often result. Improvements in bearing life should not be difficult to justify under these circumstances, especially if it can be readily established that most incidents of bearing distress are caused by lubrication deficiencies.

There is some disagreement among electric motor manufacturers as to the best bearing arrangement for horizontal-type, grease-lubricated, ball bearing motors. There is disagreement also on the best technique for replenishing the grease supply in the bearing cartridge. If the user of these motors wishes to follow the recommendations of all these manufacturers for their specific motors, he must stock or have available ball bearings in a given size with no shield, single-shield, and double-shield. He should also try to train personnel in the relubrication techniques to be followed for each make of motor. The confusion thus created in the mind of maintenance personnel may indeed bring about a less than satisfactory method of maintaining expensive, important equipment.

The users, too, disagree on such matters as lubrication method, bearing type, and relubrication frequency in seemingly similar plants (Table 13-1). A 1980 study of 12 petrochemical facilities showed that lubrication practices for electrical motors varied from the extreme of having no program to the opposite, and certainly laudable extreme of continuous lubrication via oil mist. Four plants stated they had no lubrication program for motors and ran motors to failure. These plants specified sealed bearings for motors. Two plants were apparently trying oil mist on some motors and another plant had all (which is to say several thousand) electric motors with anti-friction bearings on oil mist. As of 1998, the plant that used oil mist lubrication was able to point to decades of highly satisfactory experience.* Another plant submitted their computerized failure history and demonstrated that no more than ten bearing failures per year is an achievable goal for a facility with 540 electric motors hooked up to oil mist lube systems! For more

*The most comprehensive treatment of the topic can be found in the Bloch/Shamim book *Oil Mist Lubrication: Practical Applications*, Fairmont Press, Lilburn, GA 30247 (1998), ISBN 0-88173-256-7

Table 13-1. Lubrication strategies can vary widely from plant to plant.

Lubrication Strategies Used By Six Plants in 1980

Facility	Oil	Grease	By Whom	Scheduling
1	As required	3 - 6 months	Oil: Operators Grease: Electricians	Manual now, computer-based anticipated in future
2	As required	1800 RPM/2-3 years 3600 RPM/1 year 3600 VERT/6 months	Mechanics	Manual
3	1 year	1 year	Electricians	Manual
4	2 years	1 year	Electricians	Computer
5	—	1 year	Electricians	Manual
6	1 year	1 year	Electricians	Manual

general information on oil mist lubrication, for electric motors, please refer to Chapter 10 of this text.

The absolute superiority of oil mist lubrication can also be gleaned from Table 13-2, which shows the influence of lubrication on the service life of rolling element bearings.

However, this segment of our text will focus on the more conventional and most frequently used grease lubrication methods for electric motor bearing bearings. All too often, an industrial user will employ less-than-ideal lubrication strategies, or vulnerable bearing housing configurations. These are the issues we will address first.

How Grease-lubricated Bearings Function in Electric Motors

A shielded, grease-lubricated ball bearing (Figure 13-1) can be compared to a centrifugal pump having the ball-and-cage assembly as its impeller and having the annulus between the stationary shield and the rotating inner race as the eye of the pump. Shielded bearings are not sealed bearings. With the *shielded* type of bearing, grease may readily enter the bearing, but dirt is restricted by the close fitting shields. Bearings of the *sealed* design will not permit entry of new grease, whereas with shielded bearings grease will be drawn in by capillary action in as the bearing cage assembly rotates. The grease will then be discharged by centrifugal force into the ball track of the outer race. If there is no shield on the back side of this bearing, the excess grease can escape into the inner bearing cap of the motor bearing housing.

Single-Shield Bearings

A large petrochemical complex in the U.S. Gulf Coast area considers the regular single-shield bearing with the shield facing the grease supply (Figure 13-2) to be the best arrangement. Their experience indicates this simple arrangement will extend bearing life. It will also permit an extremely simple lubrication and relubrication technique if so installed. This technique makes it unnecessary to know the volume of grease already in the bearing cartridge. The shield serves as a baffle against agitation. The shield-to-inner-race annulus serves as a metering device to control grease flow. These features prevent premature ball bearing failures caused by contaminated grease and heat buildup due to excess grease. Further, warehouse inventories of ball bearings can be reduced to one type of bearing for the great bulk of existing grease-lubricated ball bearing requirements. For other services, where an open bearing is a "must," as in some flush-through arrangements, the shield can be removed in the field.

Table 13-2. Influence of lubrication on service life. (Source: FAG Bearing Corporation)

Decreasing service life** ↓

Oil	Oil	Grease	Dry Lubricant
Rolling bearing alone	Rolling bearing with gearwheels and other wearing parts	Rolling bearing alone	Rolling bearing alone
Circulation with filter, automatic oiler	Circulation with filter	Automatic feed	
Oil-air	Oil-air		
Oil-mist	Oil-mist		
Circulation without filter *			
Sump, regular renewal	Circulation without filter*	Regular regreasing of cleaned bearing	
	Sump, regular renewal *		
Sump, occasional renewal	Rolling bearing a) in oil vapor b) in sump (c) oil circulation	Regular grease replenishment	Regular renewal
	Sump, occasional Renewal (a) in oil vapor (b) in sump (c) oil circulation	Occasional renewal Occasional replenishment Lubrication for-life	
			Lubrication for life

*By feed cones, bevel wheels, asymmetric rolling bearings.
** Condition: Lubricant service life < fatigue life.

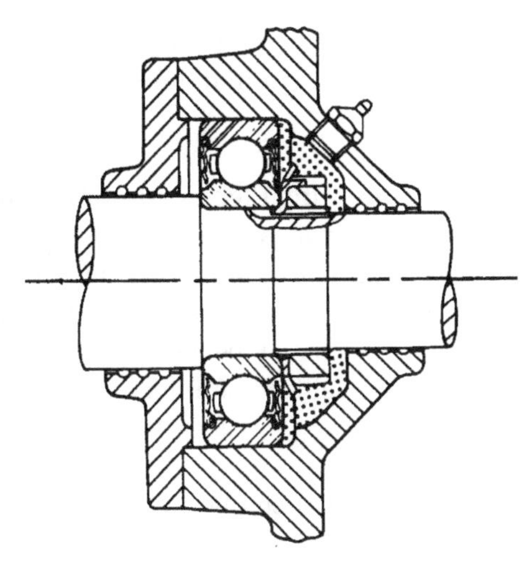

Figure 13-1. Shielded, grease-lubricated bearing.

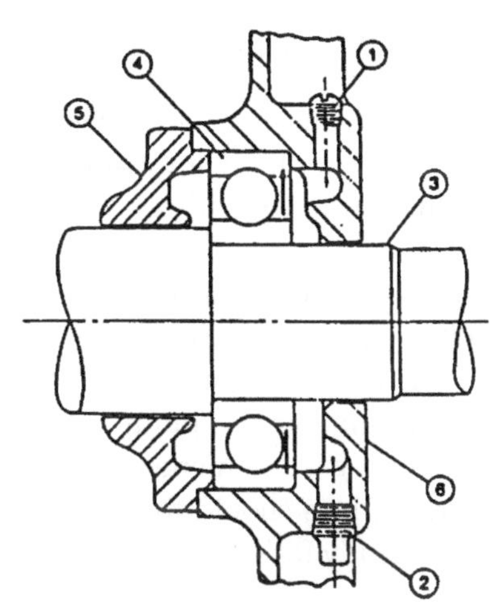

1. Lubrication Entry
2. Drain
3. Shaft
4. Bearing
5. Inner Cap
6. Bracket

Figure 13-2. Single-shield motor bearing, with shield facing the grease cavity.

Double-shielded Bearings

Some motor manufacturers subscribe to a different approach, having decided in favor of double-shielded bearings. These are usually arranged as shown in Figure 13-3. The housings serve as a lubricant reservoir and are filled with grease. By regulating the flow of grease into the bearing, the shields act to prevent excessive amounts from being forced into the bearing. A grease retainer labyrinth is designed to prevent grease from reaching the motor windings on the inner side of the bearing.

On motors furnished with this bearing configuration and mounting arrangement, it is not necessary to pack the housing next to the bearing full of grease for proper bearing lubrication. However, packing with grease helps to prevent dirt and moisture from entering.

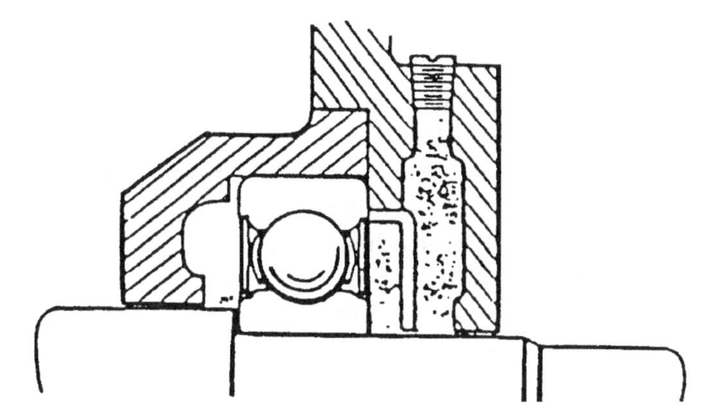

Figure 13-3. Double-shielded bearing with grease metering plate facing grease reservoir.

Oil from this grease reservoir can and does, over a long period, enter the bearing to revitalize the grease within the shields. Grease in the housing outside the stationary shields is not agitated or churned by the rotation of the bearing and consequently, is less subject to oxidation. Furthermore, if foreign matter is present, the fact that the grease in the chamber is not being churned reduces the probability of the debris contacting the rolling elements of the bearing.

On many motors furnished with grease-lubricated double-shielded bearings, the bearing housings are not usually provided with a drain plug. When grease is added and the housing becomes filled, some grease will be forced into the bearing, and any surplus grease will be squeezed out along the close clearance between the shaft and the outer cap because the resistance of this path is less than the resistance presented by the bearing shields, metering plate, and the labyrinth seal.

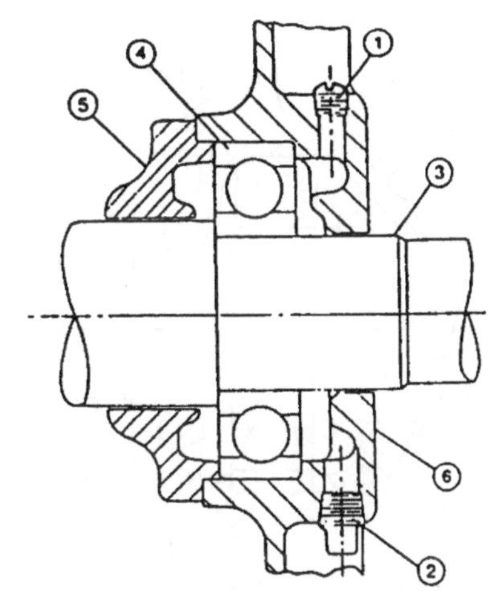

1. Lubrication Entry
2. Drain
3. Shaft
4. Bearing
5. Inner Cap
6. Bracket

Figure 13-4. High load and/or high speed bearings are often supplied without shield, as shown.

Open Bearings

High-load and/or high speed bearings are often supplied without shields to allow cooler operating temperature and longer life. One such bearing is illustrated in Figure 13-4. If grease inlet and outlet ports are located on the same side, this bearing is commonly referred to as "conventionally grease lubricated." If grease inlet and outlet ports are located at opposite sides, we refer to it as "cross-flow, or "cross-lubrication." Figure 13-5 shows a cross-flow lubricated bearing.

Life-time Lubricated, "Sealed" Bearings

Lubed-for-life bearings incorporate close-fitting seals in place of, or in addition to shields. These bearings are customarily found on low horsepower motors or on appliances which operate intermittently. Although it has been claimed that sealed ball bearings in electric motors will survive as long as bearings operating temperatures remained below 150°C (302°F) and speed factors DN (mm bearing bore times revolutions per minute) did not exceed 300,000, other studies showed that close-fitting seals can cause high frictional heat and that loose fitting seals cannot effectively exclude atmospheric air and moisture which will cause grease deterioration. These facts preclude the use of lubed-for-life bearings in installations which expect "life" to last more than three years in the typical plant environment. Moreover, we believe this to be the reason why bearing manufacturers advise against the use of sealed bearings larger than size 306 at speeds exceeding 3600 RPM. This would generally exclude sealed bearings from 3600 RPM motors of 10 or more horsepower.

A 1989 guideline issued by a major bearing manufacturer gives a DN value of 108,000 as the economic, although not technically required, limit for "life-time-lubrication."

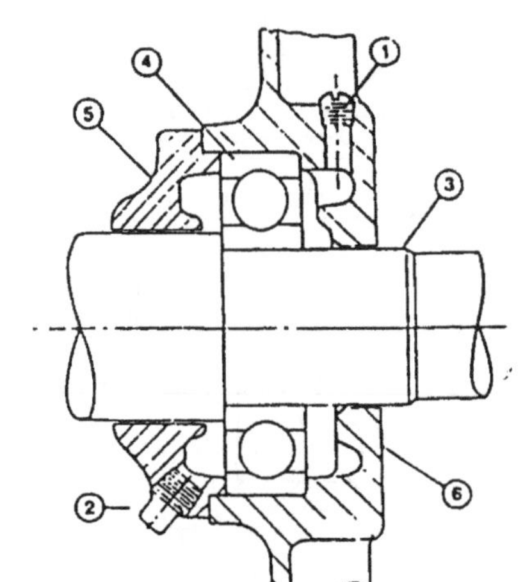

Figure 13-5. Open bearing with cross-flow grease lubrication.

Procedures for Re-greasing Electric Motor Bearings

Electric motor bearings should be re-greased with a grease which is compatible with the original charge. It should be noted that the polyurea greases often used by the motor manufacturers may be incompatible with lithium-base greases. (See Table 9-3 for details.)

Single-Shielded Bearings

To take advantage of single-shielded arrangements in electric motors, competent users have developed three simple recommendations which differ, somewhat, from the manufacturers' idealized guidelines.

1. Install a single-shield ball bearing with the shield *facing* the grease supply in motors having the grease fill-and-drain ports on that same side of the bearing. Add a finger full of grease to the ball track on the back side of the bearing during assembly.

2. After assembly, the balance of the *initial* lubrication of this single-shielded bearing should be done with the motor idle. Remove the drain plug and pipe. With a grease gun or high volume grease pump, fill the grease reservoir until fresh grease emerges from the drain. The fill and drain plugs should then be reinstalled and the motor is ready for service.

It is essential that this initial lubrication not be attempted while the motor is running. It was observed that to do so will cause, by pumping action, a continuing flow of grease through the shield annulus until the overflow space in the inner cartridge cap is full. Grease will then flow down the shaft and into the winding of the motor where it is not wanted. This will take place before the grease can emerge at the drain.

3. Relubrication may be done while the motor is either running or idle. (It should be limited in quantity to a volume approximately one-fourth the bearing bore volume.) Test results showed that fresh grease takes a wedge-like path straight through the old grease, around the shaft, and into the ball track. Thus, the overflow of grease into the inner reservoir space is quite small even after several relubrications. Potentially damaging grease is thus kept from the stator winding. Further, since the ball and cage assembly of this arrangement does not have

to force its way through a solid fill of grease, bearing heating is kept to a minimum. In fact, it was observed that a maximum temperature rise of only 20°F occurred 20 minutes after the grease reservoir was filled. It returned to 5°F two hours later. In contrast, the double-shield arrangement caused a temperature rise of over 100°F (at 90°F ambient temperature the resulting temperature was 190°F) and maintained this 100°F rise for over a week.

Double-shielded Bearings

A. Ball Bearings
   1. Pack (completely fill) the cavity adjacent to the bearing. Use the necessary precautions to prevent contaminating this grease before the motor is assembled.
   2. After assembly, lubricate stationary motor until a full ring of grease appears around the shaft at the relief opening in the bracket.

B. Cylindrical Roller Bearings
   1. Hand pack bearing before assembly
   2. Proceed as outlined in (1) and (2) for double-shielded ball bearings.

If under-lubricated after installation, the double-shielded bearing is thought to last longer than an open (non-shielded) bearing given the same treatment, because of grease retained within the shields (plus grease remaining in the housing from its initial filling).

If over-greased after installation, the double-shielded bearing can be expected to operate satisfactorily without overheating as long as the excess grease is allowed to escape through the clearance between the shield and inner race, and the grease in the housing adjacent to the bearing is not churned, agitated and caused to overheat.

It is not necessary to disassemble motors at the end of fixed periods to grease bearings. Bearing shields do not require replacement.

Double-shielded ball bearings should not be flushed for cleaning. If water and dirt are known to be present inside the shields of a bearing because of a flood or other circumstances, the bearing should be removed from service. All leading ball-bearing manufacturers are providing reconditioning service at a nominal cost when bearings are returned to their factories. As an aside, reconditioned ball bearings are generally *less prone* to fail than are brand new bearings. This is because grinding marks and other asperities are now burnished to the point where smoother running and less heat generation are likely.

Open Bearings

Motors with open, conventionally greased bearings are generally lubricated with slightly different procedures for drive-end and opposite end bearings.

Lubrication procedures for drive-end bearings:

1. Relubrication with the shaft stationary is recommended. If possible, the motor should be warm.

2. Remove plug and replace with grease fitting.

3. Remove large drain plug when furnished with motor.

4. Using a low pressure, hand operated grease gun, pump in the recommended amount of grease, or use 1/4 of bore volume.

5. If purging of system is desired, continue pumping until new grease appears either around the shaft or at the drain opening. Stop after new grease appears.

6. On large motors provisions have usually been made to remove the outer cap for inspection and cleaning. Remove both rows of cap bolts. Remove, inspect and clean cap. Replace cap, being careful to prevent dirt from getting into bearing cavity.

7. After lubrication allow motor to run for fifteen minutes before replacing plugs.

8. If the motor has a special grease relief fitting, pump in the recommended volume of grease or until a one inch long string of grease appears in any one of the relief holes. Replace plugs.

9. Wipe away any excess grease which has appeared at the grease relief port.

Lubrication procedure for bearing opposite drive end:

1. If bearing hub is accessible, as in drip-proof motors, follow the same procedure as for the drive-end bearing.

2. For fan-cooled motors note the amount of grease used to lubricate shaft end bearing and use the same amount for commutator-end bearing.

Motor bearings arranged with housings provisions as shown in Figure 13-5, with grease inlet and outlet ports on opposite sides, are called cross-flow lubricated. Regreasing is accomplished with the motor running. The following procedure should be observed:

1. Start motor and allow to operate until normal motor temperature is obtained.

2. Inboard bearing (coupling end)
   a. Remove grease inlet plug or fitting.
   b. Remove outlet plug. Some motor designs are equipped with excess grease cups located directly below the bearing. Remove the cups and clean out the old grease.
   c. Remove hardened grease from the inlet and outlet ports with a clean probe.
   d. Inspect the grease removed from the inlet port. If rust or other abrasives are observed, *do not grease the bearing*. Tag motor for overhaul.
   e. Bearing housing with outlet ports:
      (1) Insert probe in the outlet port to a depth equivalent to the bottom balls of the bearing.
      (2) Replace grease fitting and add grease slowly with a hand gun. *Count strokes of gun* as grease is added.
      (3) Stop pumping when the probe in the outlet port begin to move. This indicates that the grease cavity is full.
   f. Bearing housings with excess grease cups:
      (1) Replace grease fitting and add grease slowly with a hand gun. *Count strokes of gun* as grease is added.
      (2) Stop pumping when grease appears in the excess grease cup. This indicates that the grease cavity is full.
      (3) Outboard bearing (fan end)
         a. Follow inboard bearing procedure provided the outlet grease ports or excess grease cups are accessible,
         b. If grease outlet port or excess grease cup is not accessible, add 2/3 of the amount of grease required for the inboard bearing.

4. Leave grease outlet ports open—*do not replace* the plugs. Excess grease will be expelled through the port. Consider using a short section of open pipe in lieu of the plug.

5. If bearings are equipped with excess grease cups, replace the cups. Excess grease will expel into the cups.

APPLICATION LIMITS FOR GREASES USED IN ELECTRIC MOTOR BEARINGS

Bearings and bearing lubricants are subject to four prime operating influences: speed, load, temperature, and environmental factors. The optimal operating speeds for ball and roller type bearings—as related to lubrication—are functions of what is termed the DN factor. To establish the DN factor for a particular bearing, the bore of the bearing (in millimeters) is multiplied by the revolutions per minute, i.e.:

$$75 \text{ mm} \times 1000 \text{ rpm} = 75{,}000 \text{ DN value}$$

Speed limits for conventional greases have been established to range from 100,000 to 150,000 DN for most spherical roller type bearings and 200,000 to 300,000 DN values for most conventional ball bearings. Higher DN limits can sometimes be achieved for both ball and roller type bearings, but require close consultation with the bearing manufacturer. When operating at DN values higher than those indicated above, use either special greases incorporating good channeling characteristics or circulating oil.

RELUBRICATION FREQUENCY (FOR GREASE) RECOMMENDED BY MOTOR MANUFACTURERS

Correct seal design is the prime factor in preventing contaminants from entering a bearing, but relubrication at proper pre-scheduled intervals offers the advantage of purging out any extraneous material from the seals before they have had an opportunity to gain access to the bearings or the housing cavity. Adherence to proper scheduled regreasing intervals will also ensure that the bearing has a sufficient amount of grease at all times, and will aid in protecting the bearing component parts against any damaging effects from corrosion.

The frequency of relubrication to avoid corrosion and to aid in purging out any solid or liquid contaminants is difficult to establish since relubrication requirements vary with different types of applications.

Anticipating a not-quite-clean to moderately dirty environment as can be assumed to be present in refineries and petrochemical plants, one authority suggests greasing intervals ranging from 1 to 8 weeks. Noting that the period during which a grease lubricated bearing will function satisfactorily without relubrication is dependent on the bearing type, size, speed, operating temperature and the grease used, a major bearing manufacturer suggests use of the graph shown in Figure 13-6. However, Figure 13-6 was developed for an age-resistant, average quality grease and for bearing operating temperatures up to +70°C (+158°F) measured at the outer ring. The intervals should be halved for every 15°C (27°F) increase in temperature above +70°C (158°F), but the maximum permissible operating temperature for the grease must not be exceeded. On the other hand, SKF has published lube interval data for motor bearings in very clean locations which exceed those shown in Figure 13-6 by a factor of 3.

SKF believes that if there is a definite risk of the grease becoming contaminated the above relubrication intervals should be reduced. This reduction also applies to applications where the grease is required to seal against moisture, e.g., bearings in paper making machines (where water runs over the bearing housing) should be relubricated once a week. (See also Figure 12-1, page 287.)

The FAG Bearing Company also opted for a graphical representation showing recommended relubrication intervals, Figure 13-7. Here, the horizontal scale depicts the ratio of running speed over the maximum allowable running speed for grease lubrication of a given bearing. This is basically similar to actual DN over the limiting DN of, say, 200,000. Most ball bearing-equipped motors are supplied with bearings operating at n/n_g approximately equal to 0.5.

FAG recognizes that the relubrication interval depends on the stressing of the grease by friction, speed, and environmental conditions. Figure 13-7 demonstrates the lubrication interval T that can be achieved in the various bearing types, under favorable ambient conditions, as a function of the speed ratio n/n_g.

The lubrication interval shortens when high temperatures or vibrations subject the grease to higher stressing or when the lubricity is impaired by dust and humidity. The relubrication interval applicable to poor operating conditions T_N is

$$T_N = T \cdot q \text{ hours}$$

The reduction factor q consists of three components: f_1 covering the influence of dust and moisture, f_2 accounting for shock and vibration, and f_3 accounting for higher temperatures.

Indicative values for f_1, f_2, and f_3 are given in Table 13-3. Table 13-4 lists the reduction factor q for a number

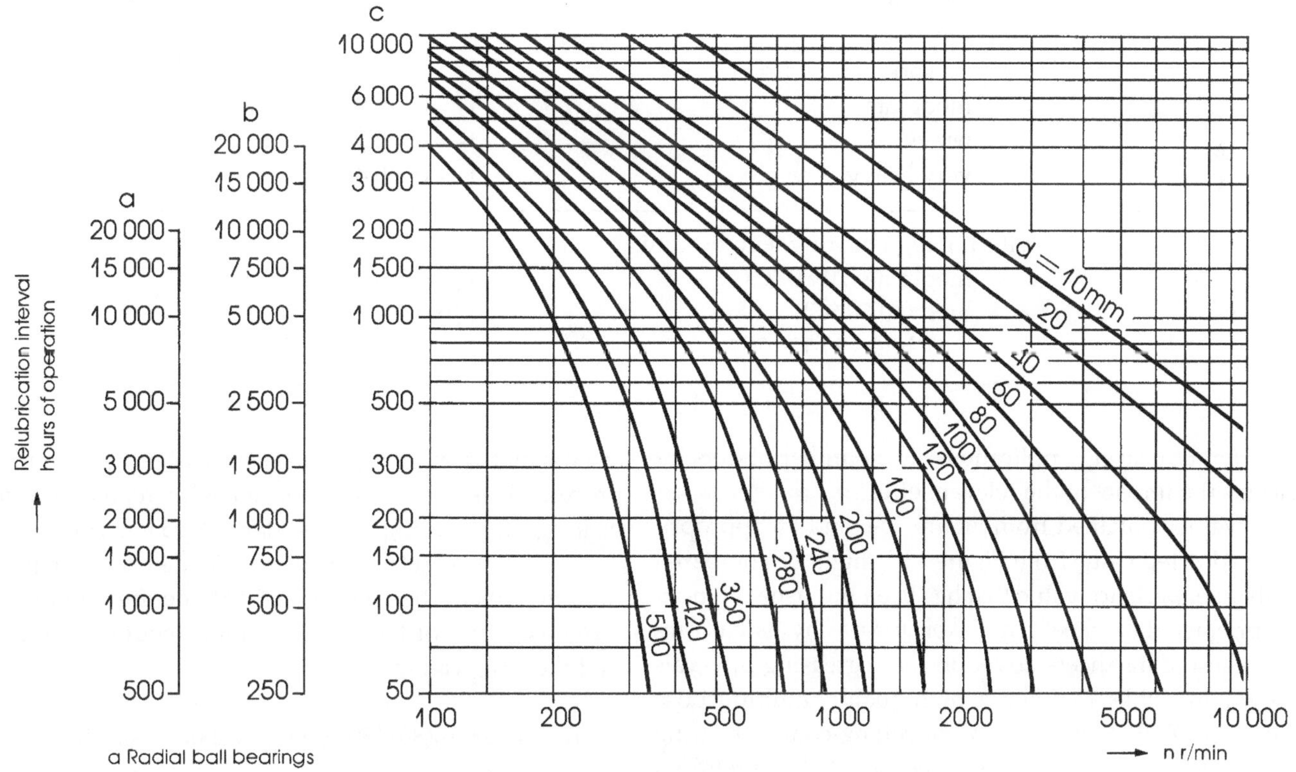

Figure 13-6. Relubrication intervals recommended by SKF. USA, Kulpsville, Pennsylvania.

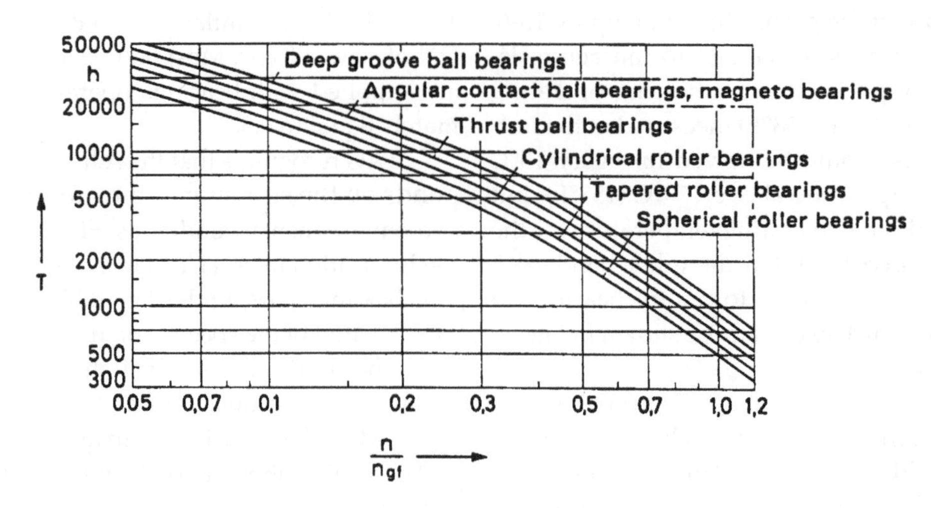

Figure 13-7. Lubrication interval T for grease-lubricated rolling bearings under favorable ambient conditions (q = 1; lithium soap grease). Source: FAG Bearing Company.

Table 13-3. Reduction factors f_1, f_2, and f_3. (Source: FAG, Schweinfurt, Germany)

Effect of dust and moisture at the bearing contact faces	
moderate	$f_1 = 0.7\text{-}0.9$
heavy	$f_1 = 0.4\text{-}0.7$
very heavy	$f_1 = 0.1\text{-}0.4$
Effect of shock and vibration	
moderate	$f_2 = 0.7\text{-}0.9$
heavy	$f_2 = 0.4\text{-}0.7$
very heavy	$f_2 = 0.1\text{-}0.4$
Effect of high bearing temperature	
moderate (70-80 °C)	$f_3 = 0.7\text{-}0.9$
high (80-90 °C)	$f_3 = 0.4\text{-}0.7$
very high (90-120 °C)	$f_3 = 0.1\text{-}0.4$

of different bearing applications. The number of dots indicates the impact of the relevant effect.

Values obtained from Figure 13-6 and an appropriately adjusted value from Figure 13-7 might be compared with an experience value that has been published in non-proprietary data sheets by Exxon for use by its customers. These data sheets advocate the regreasing intervals shown in Table 13-5 for a high-quality multipurpose grease (RONEX) and a premium rolling-contact bearing grease (UNIREX N). It should be noted that premium greases are generally recommended for motor bearing lubrication.

Here's an example dealing with a 200 hp, 1800 rpm electric motor operating in a severe, 24-hour (continuous) process. This motor is furnished with regreasable ball bearings; the bearing bore is 80mm (3.15 inches).

From Figure 13-6 we would determine regreasing intervals in the vicinity of 5000 operating hours. Alternatively, use of Figure 13-7 would require calculating first the n/n_g value:

$$n/n_g = (1800)(80)/300{,}000 = 144{,}000/300{,}000 = 0.48$$

Since Table 13-4 recommends a factor q = 1 for electric motors, the time value (regreasing interval) corresponding to $n/n_g = 0.48$ could be directly obtained from

Table 13-4. Reduction factor q for various machines. (Source: FAG, Schweinfurt, Germany)

Application	Dust, moisture	Shocks, vibrations	High temperatures	Reduction factor q
Stationary electric motor				1
Lathe spindle				1
Grinding spindle				1
Face grinder				1
Circular saw shaft	•			0.8
Flywheel of a car body press	•			0.8
Hammer mill	•			0.8
Dynamometer			•	0.7
Journal roller bearing of locomotive	•	•	•	0.7
Rope return sheaves of aerial ropeway	••			0.6
Passenger car front wheel	•	•		0.6
Textile spindle		•••		0.3
Jaw crusher	••	••	•	0.2
Vibratory motor	•			0.2
Wire section roll (paper-making machine)	•••			
Wet section press roll (paper-making machine)	•••			0.2
Work roll (steel mill)	•••		•	0.2
Bucket wheel reclaimer	•••			0.1
Double crank saw frame	•	•••		<0.1
Vibrator roll	•	•••	•••	<0.1
Vibrating screen	•	•••		<0.1

• = Moderate effect.
•• = Heavy effect.
••• = Very heavy effect.

Figure 13-7. It would be approximately 4500 hours. Likewise, from Table 13-5, we would obtain Exxon's conservative, experience-based value for motor bearings lubricated with a premium-grade grease: 3 months. A reasonable maintenance approach would thus call for relubrication every 3 to 6 months.

Finally, we could consult the relubrication guidelines issued by knowledgeable electric motor manufacturers. Not unlike the Exxon recommendations, we find these similarly conservative and aimed at a reliability-conscious user. Arkansas-based Baldor Electric prefaces their guidelines by stating that regreasing intervals are assuming "average use." Service conditions, lubrication interval multiplier, and volume of grease to be added are given in Table 13-6.

What It Costs to Lubricate Electric Motors

In 1976, a major petrochemical company in West Virginia calculated the cost of their manual greasing program at $2.08 per pump per year. These data referred to costs incurred in 1974 and were based on a three-month relubrication schedule.

If we assume motor lubrication to have cost the same amount and use a reasonable average inflation escalator, we arrive at the equivalent present-day cost.

From personal observation, we believe a process worker, electrician or contractor can lubricate 6 motors per hour. Also, let us assume a medium-size petrochemical complex has 1200 motors. To lubricate these motors four times per year, we expend 800 man-hours at a total cost of perhaps $40,000, which includes a few pounds of premium grease. With this adequate lubrication program we anticipate 3% of 1200, or 36 motor bearing failures per year. Without this program, we might expect at least 12% of 1200, or 144 motor bearing failures per year. The cost of each bearing-related motor failure is at least $1,800 for material and labor. Therefore, an expenditure of $40,000 has bought us motor repair cost credits of $(144-36)(1,800) = $194,400. The actual credits are probably much greater because production loss credits,

Table 13-5. Maximum relubrication intervals for motors lubricated with RONEX MP and UNIREX N2 (months).

Motor Size, Horsepower	1/4 to 7-1/2		10 to 40		50 to 150		Over 150	
Type of Grease	Ronex	Unirex	Ronex	Unirex	Ronex	Unirex	Ronex	Unirex
Type of Service								
I. Easy, infrequent operation (1hr/day). Valves, door openers, and portable tools.	60	120	60	84	48	48	12	12
II. Standard, 1 or 2-shift operation, Machine tools, air-conditioners, conveyors, refrigeration equipment, laundry and textile machinery, woodworking machinery, light-duty compressors and pumps.	60	84	48	48	12	18	6	6
III. Severe, continuous running (24 hr/day). Motors, fans, motor-generator sets, coal and other mining machinery, steelmill machinery and processing equipment.	36	48	12	18	6	9	2	3
IV. Very severe. Dirty, wet, or corrosive environment, vibrating applications, high ambient temperatures (over 40°C, 100°F), hot pumps and fans.								

Note: Relubrication interval for Class F motors in Service Types III and IV should not exceed 12 months.

reduction of fire incidents, and less frequent damage to motor windings as a consequence of bearing damage have been achieved as well.

What about automatic single-point grease lubricators (ASPGLs)? The performance of single-point automatic grease lubricators has not been totally flawless. Depending on the type of grease, ambient conditions and bearing configuration, the user may be faced with such phenomena as separation of grease into its oil and soap constituents. Due to this experience, plants have generally found it necessary to discard the ASPGLs after about 6 months of operation. Assume 1200 motors that each have two grease inlet ports; hence, 2400 ASPGLs would be installed, and 4800 be purchased each year at a cost roughly from $15 to as much as $50 each. Using $50 as the average cost of labor and materials, the ASPGLs would cost $240,000 per year. Manual lubrication with grease guns would cost considerably less!

Grease lubrication should not be left to chance. The most desirable prerequisite to the establishment of a program would be a thorough knowledge of the bearing and bearing housing configuration in your motors. Procurement specifications should address this requirement.

Automatic single point grease lubricators (Figure 13-8) should be used judiciously. These lubricators have their place but cannot be applied indiscriminately. They are quite useful in keeping bearing housing grease cavities full, but this is an advantage only if the bearing is constructed and installed so as to avoid detrimental overgreasing. Although automatic single-point grease lubricators are attractive in inaccessible locations, there may be no acceptable solution to the grease separation problems which are frequently observed in plants. One ASPGL manufacturer suggested use of low-temperature or extreme-pressure greases instead of the premium high-temperature greases recommended by most electric motor manufacturers. Also, it would seem prudent to look at cost justifications before using ASPGLs for every lubrication point in the plant. Moreover, there is some concern that field-refillable ASPGLs may be refilled with the wrong type of grease unless special precautions are taken to ward off this possibility.

Lubed-for-life bearings have serious limitations. Indeed, there is much evidence pointing to limitations of lubed-for-life, or sealed bearings in installations of 10 or more horsepower at speeds over 3600 rpm, and even more evidence against the "run until it fails" philosophy occasionally practiced in motor bearing lubrication.

Table 13-6. Baldor Electric Company guidelines for motor relubrication.

NEMA/(IEC) Frame Size	Rated Speed - RPM			
	3600	**1800**	**1200**	**900**
Up to 210 incl. (132)	5500 Hrs.	12000 Hrs.	18000 Hrs.	22000 Hrs.
Over 210 to 280 incl. (180)	3600 Hrs.	9500 Hrs.	15000 Hrs.	18000 Hrs.
Over 280 to 360 incl. (225)	*2200 Hrs.	7400 Hrs.	12000 Hrs.	15000 Hrs.
Over 260 to 5000 incl. (300)	*2200 Hrs.	3500 Hrs.	7400 Hrs.	10500 Hrs.

*Bearings in 360 through 5000 frame, 2 pole motors are either 6313 or 6314 bearings and lubrication interval is shown in the table. If roller bearings are used, and the bearings must be lubricated more frequently, divide the listed lubrication interval by 2.

Service Conditions

Severity of Service	Ambient Temperature Maximum	Atmospheric Contamination	Type of Bearing
Standard	40°C	Clean, Little Corrosion	Deep Groove Ball Bearing
Severe	50°C	Moderate dirt, Corrosion	Ball Thrust, Roller
Extreme	>50°C* or Class H Insulation	Severe dirt, Abrasive dust, Corrosion	All Bearings
Low Temperature	<-30°C**		

* Special high temperature grease is recommended
** Special low temperature grease is recommended

Lubrication Interval Multiplier

Severity of Service	Multiplier
Standard	1.0
Severe	0.5
Extreme	0.1

Bearings Sizes and Types

Frame Size NEMA (IEC)	Bearing Description (These are the "Large" bearings (Shaft End) in each frame size)					
	Bearing	OD D mm	Width B mm	Weight of grease to add oz (grams)	Volume of grease to be added in^3	teaspoon
Up to 210 incl. (132)	6307	80	21	0.31 (8.4)	0.6	2.0
Over 210 to 280 incl. (180)	6311	120	29	0.61 (17.4)	1.2	3.9
Over 280 to 360 incl. (225)	6313	140	33	0.81 (23.1)	1.5	5.2
Over 360 to 5000 incl. (300)	NU322	240	50	2.12 (60.0)	4.1	13.4

Weight in grams = .005 DB

STAR

PLASTIC HOUSING
On/Off Switch Location
BATTERY PACK
DRIVE UNIT
Dip Switch Location
Number Sticker Placement
RESERVOIR/LC UNIT
Reservoir Threads
1/4" NPT Outlet
Outlet Plug

A system which consists of a drive unit, microprocessor including batteries and a replaceable lubrication canister (LC/reservoir). The STAR can be set for discharge periods up to 12 months, and turned "ON" and "OFF" with a switch. When the STAR is turned "ON" the motor will begin to run and the lubricant will be discharged. The STAR builds up to 5 bar (75 psi) of pressure at the initial discharge. With the discharge period set, the STAR will dispense .75cc of lubricant every time it cycles. The discharge periods can be changed during the lubrication period, if required. At the end of the discharge period, simply replace the lubrication canister (LC/reservoir) and batteries to begin a new lubrication cycle.

Classic • Futura • Frost

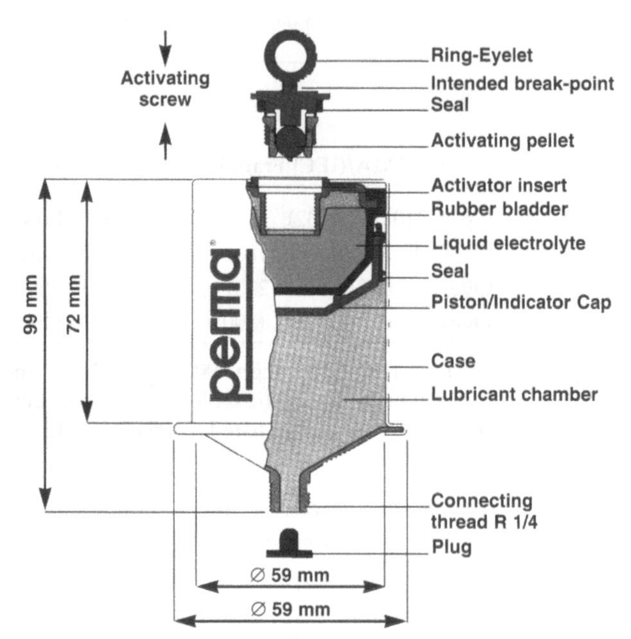

By tightening the plastic activating screw, the gas generator falls into the liquid electrolyte. The chemical reaction builds up pressure which causes the piston to move forward. The lubricant is continuously injected into the lubrication point. At the end of the lubrication period the discharge indicator cap becomes clearly visible, indicating that the lubricant has been fully discharged except for the Futura. Due to its transparent casing the movement of the piston can be seen during the lubrication cycle. The delay between activation and lubrication depends on the Perma type.

Figure 13-8. Two models of automatic single-point grease lubricators (ASPGLs) manufactured by Perma USA (www.permausa.com).

That leaves us with a consideration of decade-long, highly successful experience with oil mist lubrication of electric motors.

OIL MIST FOR ELECTRIC MOTORS

Plant-wide oil mist systems were explained in Chapter 10 of this text. Oil mist consists of a mixture of 200,000 volume parts of clean and dry plant or instrument air and one part of lubricating oil. Since the 1970s, dry sump ("pure") oil mist has provided ideal lubrication for thousands of rolling element bearings in electric motors. In the intervening decades, this lubricant application method has gained further acceptance at many reliability-focused process plants in the United States and overseas.

Wide application range documented for electric motors. For the past 40 years, empirical data have been employed to screen the applicability of oil mist. The influences of bearing size, speed, and load have been recognized in an empirical oil mist applicability formula, limiting the parameter "DNL" (D= bearing bore, mm; N= inner ring rpm; and L= load, lbs) to values below 10E9, or 1,000,000,000. An 80 mm electric motor bearing, operating at 3,600 rpm and a load of 600 lbs, would thus have a DNL of 172,000,000—less than 18% of the allowable threshold value.

Major grass-roots olefins plants commenced using oil mist on motors as small as one hp (0.75 kW) in 1975. Although the largest electric motors using pure oil mist in refineries and petrochemical plants exceed 2,000 hp in size, the more typically prevailing practice among reliability-focused users is to apply oil mist on horizontal motors, 15 hp and larger, and on vertical motors of 3 hp and larger. In all cases, these electric motors are fitted with rolling element bearings.

API-610, the most widely used pump standard in

the petrochemical and refining industries, asks for oil mist to be routed *through* the bearings (which was done in Figure 13-9) instead of *past* the bearings, Figure 13-10. Although intended for pumps, this recommendation will work equally well for electric motor rolling element bearings. The resulting *diagonal through-flow* route guarantees adequate lubrication, whereas oil mist entering and exiting on the same side might allow some of the mist to leave without first wetting the rolling elements. Through-flow is thus one of the keys to a successful installation.

Figure 13-9. Oil mist routed through electric motor bearings

Figure 13-10. Oil mist applied to the same side of a bearing is not providing optimal lubrication; much of the mist is simply flowing from entry to drain.

Major electric motor manufacturer Siemens A.G. has published technical bulletins showing oil mist as a superior technique for electric motors ranging in size from 18 to 3,000 kW.

Mist flow quantified. The required volume of oil mist is often expressed in "bearing-inches," or "BIs." A bearing-inch is the volume of oil mist needed to satisfy the demands of a row of rolling elements in a one-inch (~25 mm) bore diameter bearing. One BI assumes a rate of mist containing 0.01 fl. oz., or 0.3 ml, of oil per hour. Certain other factors may have to be considered to determine the needed oil mist flow and these are known to oil mist providers and bearing manufacturers. The various factors are also extensively documented in several references; they are readily summarized as:

a. *Type of bearing*. The different internal geometries of different types of contact (point contact at ball bearings and linear contacts at roller bearings), amount of sliding contacts (between rolling elements and raceways, cages, flanges or guide rings), angle of contact between rolling elements and raceways, and prevailing load on rolling elements. The most common bearing types in electrical motors are deep groove ball bearings, cylindrical roller bearings and angular contact ball bearings.

b. *Number of rows of rolling elements*. Multiple row bearing or paired bearing arrangements require a simple multiplier to quantify the volume of mist flow.

c. *Size of the bearings*, related to the shaft diameter—inherent in the expression "bearing-inches."

d. *The rotating speed*. The influence of the rotating speed should not be considered as a linear function. It can be linear for a certain intermediate speed range, but at lower and higher speeds the oil requirements in the contact regions may differ from straight linearity.

g. *Bearing load conditions* (preload, minimum or even less than minimum load, heavy axial loads, etc.)

f. *Cage design*. Different cage designs may affect mist flow in different ways. It has been reasoned that stamped (pressed) metal cages, polyamide cages, or machined metal cages might produce different degrees of turbulence. While different rates of turbulence may cause different amounts of oil to "plate out" on the various bearing components, the concern vanishes when oil mist is applied in through-flow mode.

Through-flow oil mist will accommodate all of the arrangements listed above. Major electric motor manufacturers, such as Reliance Electric (Cleveland, Ohio), are aware of this fact. Representing "Best Technology," their bearing housings are arranged for through-flow.

Decades of experience confirm that no further investigation is needed for bearings in the operating speed and size ranges encountered by motors driving process pumps. As of 2009, tens of thousands of oil mist lubricated electric motors continue to operate flawlessly in reliability-focused user plants. Capitalizing on this favorable experience, the procurement specifications for both new projects and replacement motors at many of these plants require oil mist lubrication in motor sizes 15 hp and larger.

Sealing and drainage issues. Although oil mist will not attack or degrade the *winding* insulation found on electric motors made since the mid-1960s, mist entry and related sealing issues must be understood and merit being included in this overview.

Regardless of motor type, i.e. TEFC, X-Proof or WPII, *cable* terminations should not be made with conventional electrician's tape. The adhesive in this tape will last but a few days and then become tacky to the point of unraveling. Instead of inferior products, competent motor manufacturers use a modified silicone system ("Radix") that is highly resistant to oil mist. Radix has consistently outperformed the many other "almost equivalent" systems.

Similarly, and while it must always be pointed out that oil mist is neither a flammable nor explosive mixture, it would be unsightly to allow a visible plume of mist to escape from the junction box cover. The wire passage from the motor interior to the junction box should, therefore, be sealed with 3M Scotch-Cast Two-Part Epoxy potting compound to exclude oil mist from the junction box.

Finally, it is always good practice to verify that *all* electric motors have a small (3 mm) weep hole and that XP-motor drains are given closer attention. The latter are furnished with either an explosion-proof rated vent or a suitably routed weep hole passage at the bottom of the motor casing or lower edge of the motor end cover. Intended to drain accumulated moisture condensation, the vent or weep hole passage will allow coalesced or atomized oil mist to escape. Note, however, that explosion-proof motors are still "explosion-proof" with this passage. Reasoning on the issue should convince us that a motor with its interior slightly pressurized by non-explosive oil mist cannot ingest explosive vapors from a surrounding atmosphere. The suitability of oil mist for Class 1, Group C and D locations was specifically re-affirmed by Reliance Electric in July of 2004.

TEFC vs. WPII Construction. On TEFC (totally enclosed, fan-cooled) motors, there are documented events of liquid oil filling the motor housing to the point of near contact with the spinning rotor. Conventional wisdom to the contrary, there neither were nor will there be detrimental effects with the oils used in normal industry. The motor could have run indefinitely! TEFC motors are suitable for oil mist lubrication by simply routing the oil mist *through* the bearing, as has been explained in documents cited in the references and/or bibliographies in Chapters 10 and later in this chapter. There are numerous other references, including API-610. No special internal sealing provisions are needed with pure oil mist filling a TEFC motor as long as the pressurized mist keeps dirty atmospheric air from entering.

On weather-protected (WPII) motors, merely adding oil mist has often been done and has generally worked surprisingly well. In this instance, however, it was found important to lead the oil mist vent tubing away from regions influenced by the motor fan. Still, weather-protected (WPII) electric motors do receive additional attention from reliability-focused users and knowledgeable motor manufacturers.

Air is constantly being forced through the windings and an oil film deposited on the windings could invite dirt accumulation. To reduce the risk of dirt accumulation, suitable means of sealing should be provided between the motor bearings and the motor interior. Since V-rings and other elastomeric contact seals are subject to wear, low-friction face seals are considered technically superior. The axial closing force on these seals could be provided either by springs or small permanent magnets (see Chapter 11, segment on bearing protector seals).

As is so often the case, the user has to make choices. Some low friction axial seals (face seals) may require machining of the cap, but long motor life and the avoidance of maintenance costs will make up for the added expense. Nevertheless, double V-rings using nitrile or Viton elastomeric material should not be ruled out since they are considerably less expensive than face seals.

Sealing to avoid stray mist stressing the environment. Even when still allowed under prevailing regulatory environmental regulations (e.g. OSHA or EPA), air quality and greenhouse concerns make it desirable to minimize stray oil mist emissions. It is helpful to recall that state-of-art oil mist systems are fully closed, i.e. are configured so as *not* to permit any mist to escape.

Combining effective seals and a closed oil mist lubrication system has, for many decades, represented a well-proven solution. The combination not only eliminates virtually all stray mist and oil leakage, but makes possible the recovery, subsequent purification, and re-use of perhaps 97% of the oil. These recovery rates make the use of more expensive, superior quality synthetic lubricants economically attractive.

Closed systems and oil mist-lubricated electric motors give reliability-focused users several important advantages:

- Compliance with actual and future environmental regulations
- Convincing proof that oil mist lubrication benefits electric motors and the maintenance budget
- The technical and economic justification to apply high-performance synthetic oils

PAO and diester-based "synthetic" lubricants embody most of the properties needed for extended bearing life and greatest operating efficiency. These oils excel in the areas of bearing temperature and friction energy reduction. It is not difficult to show relatively rapid returns on investment for these lubricants, providing, of course, the system is closed, and the lubricant re-used after filtration (see Chapter 10).

Needed: The right bearing and a correct installation procedure. Very significant increases in bearing life and overall electric motor reliability have been repeatedly documented in the decades since 1960. Oil mist cannot eliminate basic bearing problems; it can, however, provide one of the best and most reliable means of lubricant application. Bearings must be:

- Adequate for the application, i.e. deep groove ball bearings for coupled drives, cylindrical roller bearing to support high radial loads in certain belt drives, or angular contact ball bearings to support the axial (constant) loads in vertical motor applications
- Incorporating correct bearing-internal clearances
- Mounted with correct shaft and housing fits
- Carefully and correctly handled, using tools that will avoid damage
- Correctly assembled and fitted to the motor caps, carefully avoiding misalignment or skewing
- Part of a correctly installed motor, avoiding shaft misalignment and soft foot, or bearing damage incurred while mounting either the coupling or drive pulley
- Subjected to a vibration spectrum analysis. This will indicate the lubrication condition as regards lubricating film, bearing condition (possible bearing damage) and general equipment condition, including misalignment, lack of support (soft foot), unbalance, etc.

Additional considerations for converting electric motors already in use. When converting operating motors from grease lubrication to oil mist lubrication, consider the following measures in addition to the above:

a. Perform a complete vibration analysis. This will confirm or rule out pre-existing bearing distress and will indicate if such work as re-alignment or base plate stiffening is needed to avert incipient bearing failure
b. Measure the actual efficiency of the motor. If the motor is inefficient, consider replacing it with a modern high efficiency motor, using oil mist lubrication in line with the above recommendations. This will allow capture of all benefits and will result in greatly enhanced return on investment.
c. Last, but not least, evaluate if the capacity of the motor is the most suitable for the application. "Most suitable" typically implies driven loads that represent 75% to 95% of nominal motor capacity. The result: Operation at best efficiency. Note that converting an overloaded, hot-running electric motor to oil mist lubrication will lead to marginal improvement at best.

ENERGY COST ADVANTAGES WITH OIL MIST LUBRICATED ELECTRIC MOTORS*

In 1980, an important presentation on the "Effects of Synthetic Industrial Fluids on Ball Bearing Performance" (ASME Paper 80-Pet-3) indicated that a readily available synthetic lubricant, having a viscosity of 32 cSt at a temperature of 40C (98F), offered long-term contact surface protection. Although "only" 32 cSt, the protective effect of this *synthetic* lubricant was found to be equivalent to that of a base line *mineral oil* with the higher viscosity of 68 cSt. Equivalent protection was verified for the 32 cSt *synthetic* without reducing bearing service life below the theoretically predicted levels. The same good wear protection could not be achieved with a reduced viscosity *mineral* oil. In the late 1970s, tests showed the lower vis-

*Based on Bloch-Bonnin; "Reduce Both Energy and Maintenance Costs," *Hydrocarbon Processing*, May 2004

cosity synthetic lubricant providing energy savings that are even harder to ignore in the cost and reliability-focused plant environment of 2009.

Since 1980, additional means of achieving energy savings have become available in the form of superior additives technology. While the important and ever-relevant ASME Paper 80-Pet-3 focused on a diester-base oil and viscosity effects. Since then, modern additives technology has further strengthened wear protection and offers reduced energy consumption with other synthetic base oils and without requiring reductions in viscosity. This segment of our text recaps and incorporates the findings of this very important ASME paper. Also, it condenses more recent findings regarding the equivalent effectiveness of PAO-based synthetic lubes formulated with advanced additives.

Although it was well known that synthetic lubes reduce friction, little quantitative work had been done before 1980. It was then that Morrison, Zielinski and James rigorously documented the beneficial effects of diester fluids on the frictional power losses of industrial equipment. Because synthetic fluids are chemically different from mineral oils, one might expect effects that go beyond those attributable to viscosity relationships alone. Indeed, lubricant properties and application methods also affect lubrication effectiveness and the frictional torque to be overcome. The potential cost savings through power loss reduction appear to be quite substantial. It has been estimated that industrial machines consume 31% percent of the total energy in the United States. It has also been estimated that as much as 5% of the mechanical losses[26] of these machines could be avoided through a combination of improved equipment design and lubricant optimization.

Quantifying the energy savings potential. The experimental investigations of 80-Pet-3 established that power losses in rolling element bearings could be reduced as much as 37%. The test rig instrumentation used by the three researchers established that, on 65 mm bearings in the typical size range used in 15 hp (11.3 kW) process pumps, 0.11 kW could be saved. Understandably, the small absolute value of 0.11 KW per bearing tends to make the savings appear insignificant. However, petrochemical process pump rotors are typically supported by a double row radial ball bearing and two angular contact ball thrust bearings. Most of these pumps operate at 3,600 rpm, have a relatively high self-induced axial load, and are oil mist lubricated. The average pump is driven by an electric motor with an estimated 15 hp; the motor rotor is typically supported by two grease-lubricated ball bearings.

The authors of 80-Pet-3 correctly pointed out the total number of bearings in a typical electric motor driver and pump set. In actual fact, the number of bearings on a pump-and-driver combination (or "pump set") is comprised of two motor bearings, two rows of radial pump bearings and two thrust pump bearings. These then represent a total of 4.8 test-equivalent bearings $[(4 \times .7) + (2 \times 1) = 2.8 + 2 = 4.8]$. Therefore, the total savings available from an actual motor-driven pump set are 4.8 times the single test bearing energy savings of 0.11 kW. It can thus be shown that with the above-mentioned change in lubricants and using oil mist as the application method, power loss reductions of approximately 0.53 kW could be realized in the combined 15 hp (11.2 kW) pump and driver system. This 0.53 kW reduction represents a power saving of 4.7% and Tables 13-7 through 13-9, and Figures 13-11 through 13-14, attest to the details of these diligent investigations. These illustrations start with temperature rise observations (Table 13-7 and Figure 13-11), move to cataloging percentage changes (Table 13-8 and Figure 13-12), quantification of power loss (Table 13-9 and Figure 13-13), and the energy loss percentage effects of different oils and application methods (Table 13-10 and Figure 13-14).

Assuming that the average pump operates 90 percent of the time, and rounding off the numbers, this difference amounts to energy savings of 4,180 kW-hrs per year. At $0.10 per kWh, yearly savings of $418 should be expected. It should be noted that these are realistic expectations in spite of the higher cost of synthetic lubricating fluids. While making the case for oil mist lubrication, it should be pointed out that in conventional sump-lubricated pumps, longer drainage intervals are entirely feasible. Oil replacement schedules are typically extended four-fold; the extended drain intervals more than compensate for the higher cost of synthetic lubricants.

When closed-loop oil mist systems are used, the bearings run cooler and the lubricant remains cleaner and dryer for longer periods than those typically found in conventional oil sump applications. While open oil mist systems typically consume 12-22 liters (3.1-5.7 gallons) per pump set per year, closed oil mist systems consume no more than 10% of these yearly amounts. Again, at these extremely low make-up or consumption rates and compared to the cost of mineral oils, the incremental cost of synthetic lubricants is relatively insignificant.

Considering annual energy saving per 15 hp pump and driver set to be worth $418, we realize that these savings should be multiplied by the number of pumps actually operating in large refineries—850 to 1,200. Again using $0.10 per kWh, annual savings in the vicinity of $

Table 13-7. Average temperature rise in a ball bearing test. Two different oils are used at two different viscosities (Ref. 1)

Average Temperature Rise (degrees C)		
Load = 8.9 KN (2000lbf)	**Oil Sump**	**Oil Mist**
MIN 68	66	48
SYN 32	52	43

Table 13-8. Overview of different changes (oil type and application method) and resulting temperature reductions

Change	Δ *T*	*Total reduction*
Sump: MINERAL 68 to SYN 32	-14	21%
Mist: MINERAL 68 to SYN 32	-5	10%
Sump MIN 68 to Mist MIN 68	-18	27%
Sump SYN 32 to Mist SYN 32	-9	14%
Sump MIN 68 to Mist SYN 32—Represents greatest reduction, and therefore best choice	**-23**	**35%**

Table 13-9. Overview of power loss with different oils and application methods (Ref. 1)

Power loss per bearing (KW)		
L = 8.9 KN (2000 lbf)	**Oil Sump**	**Oil Mist**
MIN 68	0.271	0.192
SYN 32	0.254	0.169

450,000 would not be unusual. A detailed calculation will prove the point:

- Total pump hp installed at the plant = 15 hpm × 1,000 = 15,000 hp
- Total pump kW installed at the plant = 15,000 × .746 = 11,190 kW
- Total consumption kWh per year, considering 90% of 8,760 h/yr = 8,760 h × .90 = 7,884 h/yr = 7,884 × 11,190 kW = 88,220,000 kWh/yr
- Total USD value of yearly energy consumption at $0.10/kWh: =$ 8,822,000

Total energy savings for 1,000 average-sized pump sets would thus equal 0.047 × 8,822,000 = $414,600. That is an amount that cannot be overlooked.

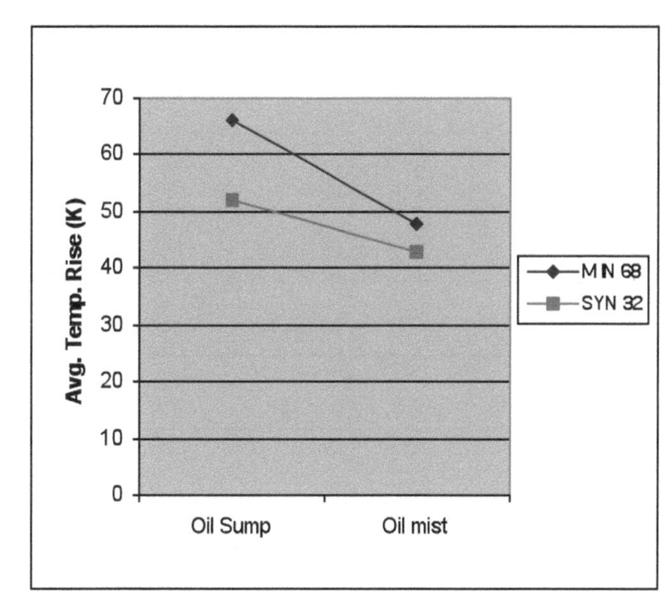

Figure 13-11. Average temperature rise plot for the ball bearing test of Table 13-7. Note that two different oils are used at two different viscosities (Ref. 1)

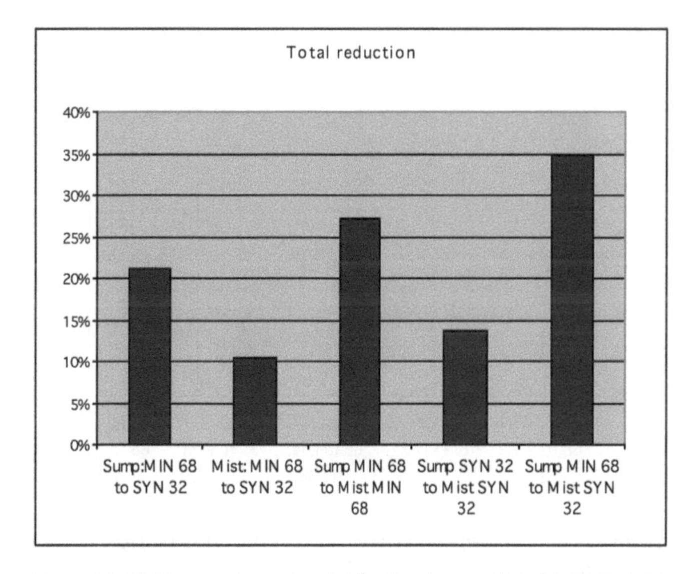

Figure 13-12. Temperature rise plot for the changes listed in Table 13-8. Note that two different oils are used at two different viscosities (Ref. 1)

Bibliography

1. Miannay, Charles; "Grease Life Estimation In Rolling Bearings," Engineering Sciences Data (U.K.) Number 78032, November 1978.
2. Anonymous; "Oil Mist Arrests Bearing Failure In Aruba," *Oil and Gas Journal*, September 16, 1974.
3. Autenrieth, J.R.; "Motor Lubrication Experience at Phillips Petroleum, Sweeney, Texas." Documentation prepared for earlier NPRA meetings.
4. Aviste, M.; "Lubrication And Preventive Maintenance," *Lubrication Engineering*, Volume 37,2, February, 1981, pp. 72-81.
5. Bloch, H.P.; "Dry Sump Oil Mist Lubrication for Electric Motors" *Hydrocarbon Processing Magazine*, March 1977.

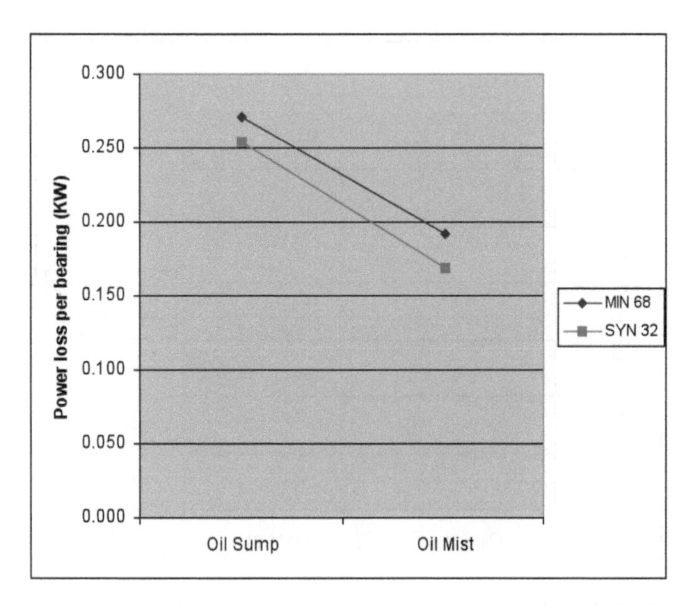

Figure 13-13. Power loss plot for the ball bearing test of Table 13-9. Note that two different oils are used at two different viscosities (Ref. 1)

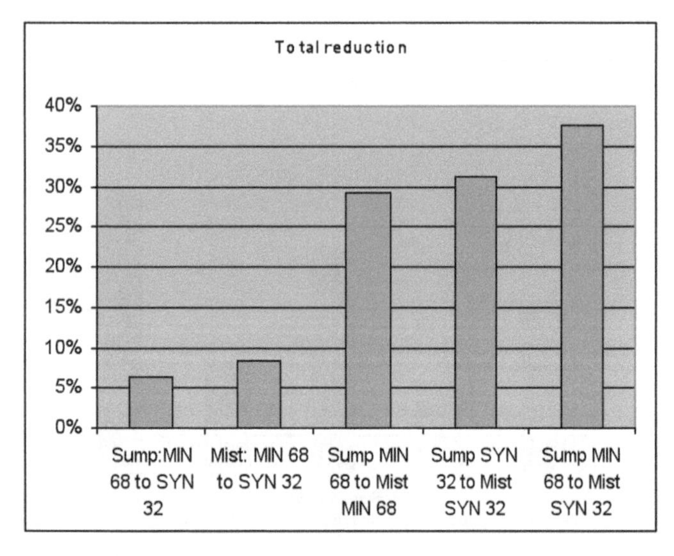

Figure 13-14. Plot of different changes and power reduction percentages that resulted (Ref. 1)

Table 13-10. Overview of power loss and loss reduction percentages with different oils and application methods (Ref. 1)

Change	Δ Power loss per bearing	Total reduction
Sump:MIN 68 to SYN 32	0.017	6%
Mist: MIN 68 to SYN 32	0.022	8%
Sump MIN 68 to Mist MIN 68	0.080	29%
Sump SYN 32 to Mist SYN 32	0.085	31%
Sump MIN 68 to Mist SYN 32	0.11	38%

6. Bloch, H.P.; "Large Scale Application of Pure Oil Mist Lubrication in Petrochemical Plants," ASME Paper No. 80-C12/ Lub-25, August 1980.
7. Bloch, H.P.; "Optimized Lubrication of Antifriction Bearings for Centrifugal Pumps," ASLE Paper No. 78-AM-1D-2, April 1978.
8. Booser, E.R.; "When To Grease Bearings," *Machine Design*, August 21, 1975, pp. 70-73.
9. Brozek, R.J., and Bonner, J.J.; "The Advantages of Ball Bearings and Their Application On Large-Horsepower High-Speed Horizontal Induction Motors," IEEE Transactions, Vol. IGA-7, No. 2, March/April 1971.
10. Clapp, A.M.; "Plant Lubrication," Proceedings of the Seventh Texas A&M University Turbomachinery Symposium, December 1978.
11. Electrolube Automatic Electronic Lube Dispensing Systems, Technical Bulletin, A.T.S. Electro-Lube, LTD., Delta, B.C., Canada V4G 1CB.
12. Eschmann, Hasbargen & Weigand, "Ball and Roller Bearings—Theory, Design and Application," John Wiley & Sons, New York, 1985 (ISBN 0-471-26283-8)
13. Hafner, E.R.; "Proper Lubrication, The Key To Better Bearing Life," *Mechanical Engineering*, November 1977, pp. 46-49.
14. Kugelfischer Georg Schaefer and Company, (FAG); "The Lubrication of Rolling Bearings," Publication No. 81 103EA, Schweinfurt, 1977.
15. Miannay, C.R.; "Improve Bearing Life With Oil Mist Lubrication," *Hydrocarbon Processing*, May 1974, pp. 113-115
16. Miller, N.H., and Pattison, D.A.; "How To Select The Right Lubricant," *Chemical Engineering*, March 11, 1968, pp. 193-198.
17. PERMALUBE, Technical Data Bulletin 10473-2-2, Quincy, Illinois, 62301.
18. PETROMATIC, Technical Data Bulletin (Jemalee Industries, Inc.) Grand Prairie, Texas 75051
19. Reliance Electric, Cleveland, OH.; Instruction Manual B-3620-14
20. Siemens Corporation, E & C Newsletter, April 1982.
21. SKF Industries, Bulletin 144-110, "A Guide To Better Bearing Lubrication," July 1981.
22. Smeaton, R.W.; "Motor Application and Maintenance Handbook," McGraw-Hill Book Company, New York, 1981.
23. Smith, R.L., and Wilson, D.S.; "Reliability of Grease-Packed Ball Bearings for Fractional Horsepower Motors," *Lubrication Engineering*, Volume 36, July, 1980, pp. 411-416.
24. Towne, C.A.; "Practical Experience With Oil Mist Lubrication," ASME Paper 82-AM-4C1, April 1982.
25. UNILUBE Single Point Lubricator Bulletin, TM Industries, Inc., Westwood, New Jersey 07675.
26. Morrison, F.R., Zielinsky, J., James, R., "Effects of synthetic fluids on ball bearing performance," ASME Publication, February, 1980.
27. Pinkus, O., Decker, O., and Wilcock. D.F. "How to save 5% of our energy," *Mechanical Engineering*, September 1997.

Chapter 14

Lubricating Pumps

The substantive and authoritative text "Ball and Roller Bearings" (Ref. 1) recommends the relative rankings per Table 14-1 for general guidance addressing bearings and lubrication applied to a wide variety of machines, whereas this chapter only makes observations relating to process pumps and their electric motor drivers.

Published in 1985, (Ref. 1) represents a general consensus. Its relative accuracy has been reaffirmed and largely corroborated by an additional 30 years of experience. We find it particularly reassuring that Ref. 1 ranked circulating filtered oil and/or oil spray into the cages of rolling element bearings at the very top of the list. As of 2015, all world-scale manufacturers of rolling element bearings continue to support oil spray as their best lubricating choice. They note that, after first filtering, oil is sprayed at a precise rate needed for an oil film of proper thickness. No undue amounts of frictional energy are created and Reference 1 gave oil spray 10 out of 10 points. The information and guidance in this chapter will revert back to these recommendations and will harmonize with the bearing manufacturers. In other words, it will answer the question "so what?"—a question of importance and interest to many conscientious reliability professionals [Refs. 2 and 3]

The chapter does not infringe on, or limit, the choices made by manufacturers and/or purchasers. All kinds of lubrication are presently available. However, the authors set out to address the needs of individuals and groups searching for better solutions. There are many ways to reduce pump bearing failure risks and frequencies; they are the subject of this chapter.

Table 14-1. Relative ranking of lubrication methods dating back to early 1980s (Ref. 1). (Note Oil-Air and Oil Mist ranked near top)

(Decreasing service life** ↓)

Oil	Grease
Rolling bearing alone	Rolling bearing alone
Circulation with Filter, automatic oiler Oil-air Oil-mist	Automatic feed
Circulation without filter*	
Sump, regular renewal	Regular regreasing of cleaned bearing
	Regular grease replenishment
Sump, occasional renewal	
	Occasional renewal Occasional replenishment
	Lubrication for-life

*By feed cones, bevel wheels, asymmetric rolling bearings.
**Condition: Lubricant service live < Fatigue life.
Source: Eschmann, Hasbargen and Weigand, *Ball and Roller Bearings*, John Wiley & Sons, Ltd., New York, 1985.

OIL VERSUS GREASE

The primary purpose of lubrication is to separate stationary from rotating parts by placing lubricant molecules between the components. A flow of lubricant also serves to carry away heat. Oil has advantages over grease because it removes more heat. Grease has an advantage because it is more easily confined. Both oil and grease can be applied in many different ways.

Grease is normally used in electric motor drivers ranging from fractional HP through approximately 500 kW; at over 500 kW oil often represents an overall cost advantage. This is because grease can be readily introduced in the small-to-medium electric motor sizes where motor end caps readily accommodate grease. However, sleeve bearing-equipped motors require limited end float couplings and other specifics. These are not the subject of this text.

With occasional exceptions, experienced petrochemical plants prefer oil over grease in one of its feasible application modes once 500 kW is exceeded. However, there are many different oil application details; likewise, there are many different grease application details. Each merits further elaboration and will be discussed later.

OIL LUBRICATION EXPERIENCE

Oil applied as a static sump is often called an "oil bath." Static sumps—oil baths—are acceptable for relatively low bearing velocities. With a static sump, the oil level would be at or near the center of whichever rolling element passes through the 6 o'clock (bottom) position. Oil bath lube is feasible for low-to-moderate shaft velocities. Once bearing elements plough through an oil bath at "high" velocities, heat generation will be of concern. Elevated bearing temperatures can degrade lubricant oxidation stability.

A widely used approximation suggests a "DN-value" (inches of shaft diameter multiplied by revolution per minute) of 6,000 as the threshold where bearing elements should no longer move through the oil bath and where, instead, lube oil is introduced into the bearings by other means. Traditionally, these other means have included oil rings (Figure 14-1), flinger discs (Figure 14-2), "jet oil spray" (Figure 14-3), and oil mist (Figure 14-4).

Flinger discs must be carefully engineered for the intended duty and must be securely fastened to shafts.

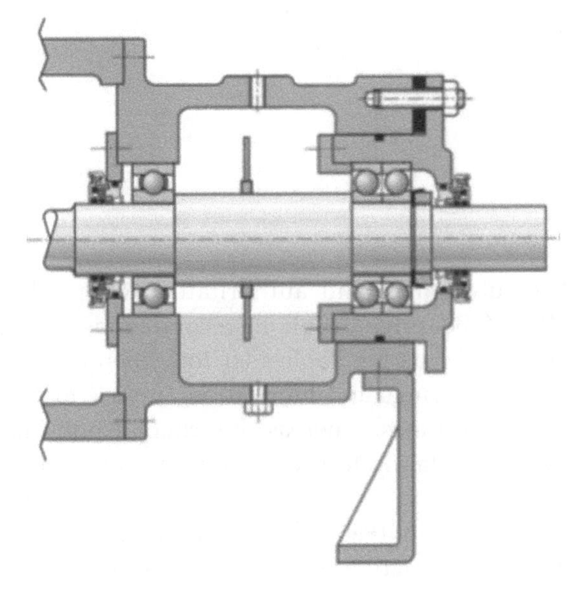

Figure 14-2. Flinger discs (arrow) avoid issues with oil rings; they can be accommodated in bearing housings fitted with cartridges designed to allow access and insertion

Experienced European manufacturers often offer them as standard components. The discs allow moderate deviation from precise horizontality of shafts systems; they make contact with the oil level or are partially immersed in the bearing housing oil sump.

As of 2016, oil mist is used on ~160,000 process pumps and ~27,000 electric motors world-wide. API-610 gives application details very similar to Fig 14-3, as do Refs. 4 and 5. The key point is that oil mist is introduced between a long-life bearing housing protector seal (Figure 14-5) and a vent location [Ref. 6]. As the mist flows through the bearing and while shaft rotation creates turbulence, atomized oil globules combine and form larger oil droplets. The coalesced oil then coats and cools the bearing. Because the bearing housing is at slightly higher than atmospheric pressure, inward migration of atmospheric contaminants is avoided.

Figure 14-1. An unrestrained oil ring can touch portions of the inside of the bearing housings and suffer abrasive damage

Using Constant Level Lubricators

Traditional lowest first-cost application of oil involves using one of many available constant level lubricators. Two widely used versions are shown in Figure 14-6. Side-mounted constant level lubricators or "oilers" are unidirectional. They should be mounted on the up-arrow side of the bearing

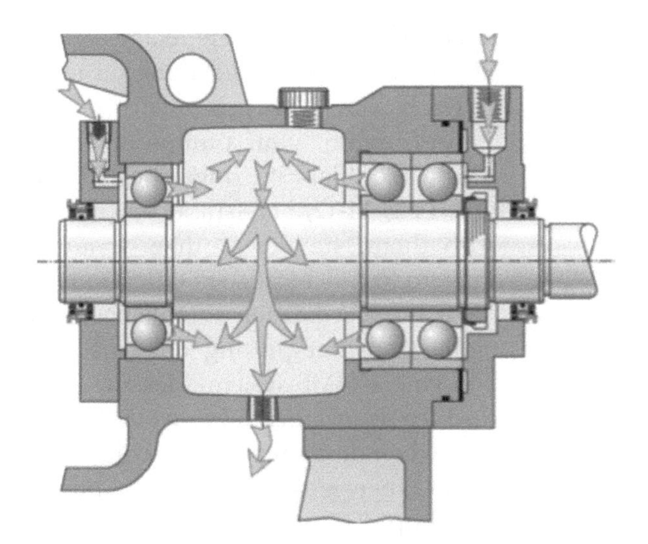

Figure 14-3. Oil spray (or, in similar fashion, an oil mist) directed into the bearing cage overcomes the "fan effect" (or windage) of inclined angular contact cages and provides an optimum thickness oil film for lubrication and heat removal at any bearing orientation

housing. Therefore, the constant level lubricator shown on the left side of Figure 14-6 should be removed.

The oval in Figure 14-6 points to caulking. Caulking has a finite life, requiring oilers to be replaced every few years.

Visualize how air will be sucked in unless the lubricator is properly mounted, or how a small lowering of an oil level may deprive a bearing of lubrication. If the pressure in a closed bearing housing increases due to a slight temperature increase, pressure will cause the oil level to go down and oil will suddenly no longer flow into the bearing. Black oil will form and the bearing will start to fail. Pressure-balanced lubricators (Figure 14-7) are preferred over unbalanced types.

Vulnerability of Oil Rings

Oil rings will work (Refs. 7 and 8); they are found in many machines. They have to be installed on a truly horizontal shaft system and are not allowed to make contact with housing-internal surfaces. Prevent them from wedging under the long limiter screw in Figure 14-8. Maintain depth of immersion and lube oil viscosity within acceptable ranges. Also, ascertain that bore eccentricity stays within the 0.002 or 0.003 inches recommended in Ref. 8.

There are other issues with Figure 14-8; oil can get trapped and overheated behind the thrust bearing because no oil slot was provided, although an oil return slot is shown below the radial bearing. Beware of adding just any bearing protector seal; doing so may result in somewhat higher pressure to the right of the thrust bearing compared to the pressure in the large, often well-vented, space near the center of the bearing housing.

Figure 14-4. Pure oil mist on an API pump. A small transparent container (circle) with vent tubing is located at the bottom

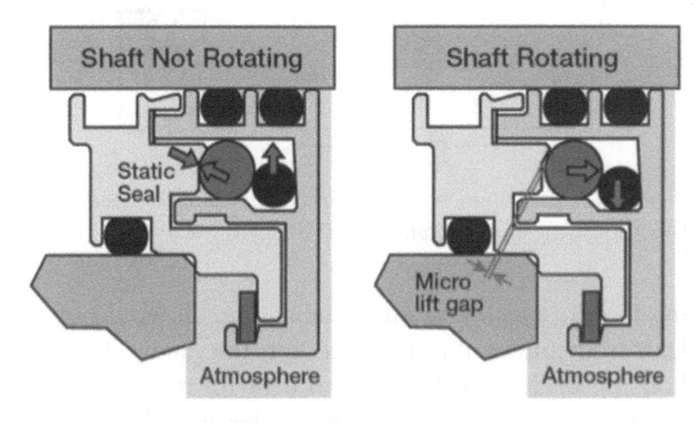

Figure 14-5. Bearing housing protector seal in stopped (left and operating (right) condition (Source: AESSEAL Inc.)

The prudent engineers at Igor Karrassik's Worthington Pump Company in the 1990s wisely placed a series of pressure equalization holes around each bear-

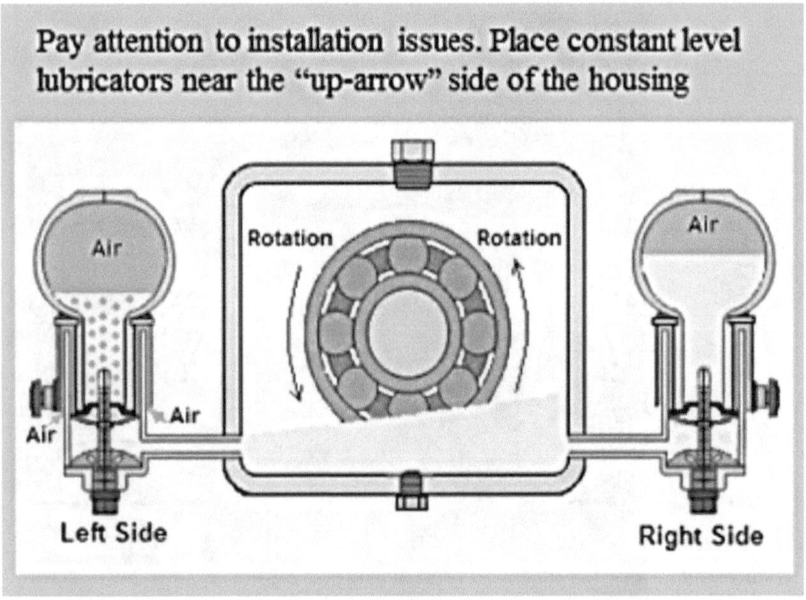

Pay attention to installation issues. Place constant level lubricators near the "up-arrow" side of the housing

Figure 14-6. Traditional liquid oil application with static sump (oil bath). The lubricator on the left side should be removed

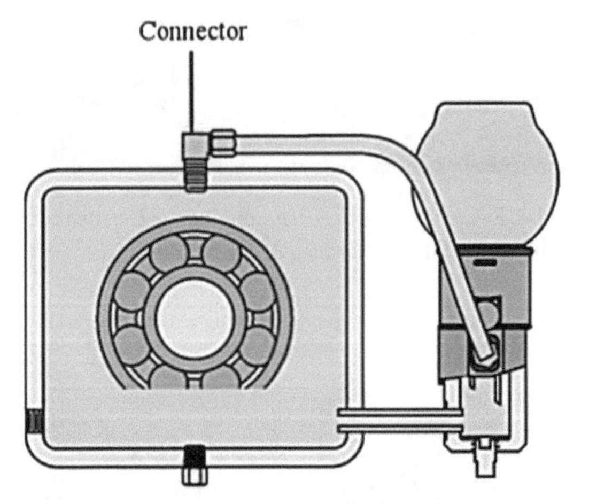

Figure 14-7. Constant level lubricator with pressure balance between bearing housing and lubricator body (TRICO Mfg. Co)

ing (Figure 14-9). Without equalization, oil would often leak from under the lip seals. Yet, the "new" bearing housing of the late 1990s (Figure 14-10) does not have equalization holes. That's probably because these holes, admittedly, are not always needed. A problem occurs if they are needed and they are not there.

Oil Ring Cross-section and Oil Gallery

In a series of rigorous tests, a pump manufacturer established that oil rings with cross-sections resembling the mirror-image trapezoids in Figure 14-11 may tolerate wave motion on seaborne vessels (Ref. 7). The trapezoidal oil rings were retained on contoured ring carriers and flung oil into oil galleries.

But, again, your pumps may incorporate the "flat" oil rings shown in Figures 14-1, 14-12, or 14-13. If you're happy with them, stay the course. If not, it's time to write better specifications and insist on specification compliance. Meanwhile, consider putting relevant details into your CMMS system. Measure and record the as-installed oil ring out-of-roundness when doing repairs. Measure oil ring width when next dismantling a pump in the shop. The difference between these measurements is evident to the naked eye, as seen in Figure 14-14. This difference ended up as abrasion product and caused premature failure of the bearings.

These are among the questions for which competent reliability engineers seek answers. If no answers are found, the facility may find that random repeat failures occur from then on.

GREASE LUBRICATION

There are isolated instances when bearings should be fully packed with grease. A boat trailer is such an isolated instance. As he backs the trailer onto the boat ramp and launches his small boat, its owner wishes

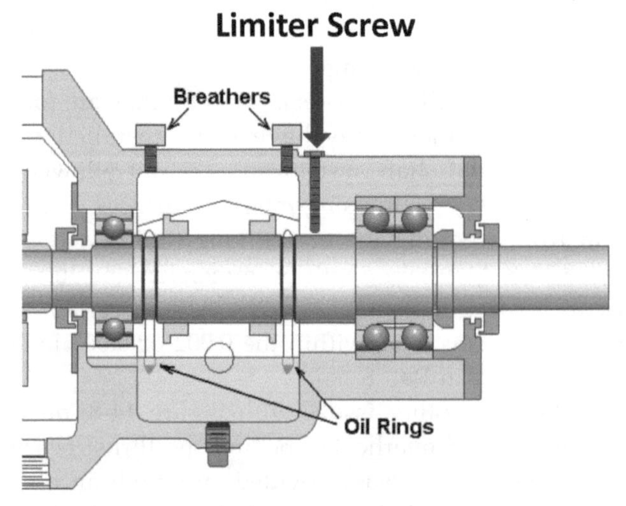

Figure 14-8. Oil return slot is (appropriately) shown below the radial bearing. But oil can get trapped, overheat and oxidize behind the thrust bearing unless an oil return slot is provided

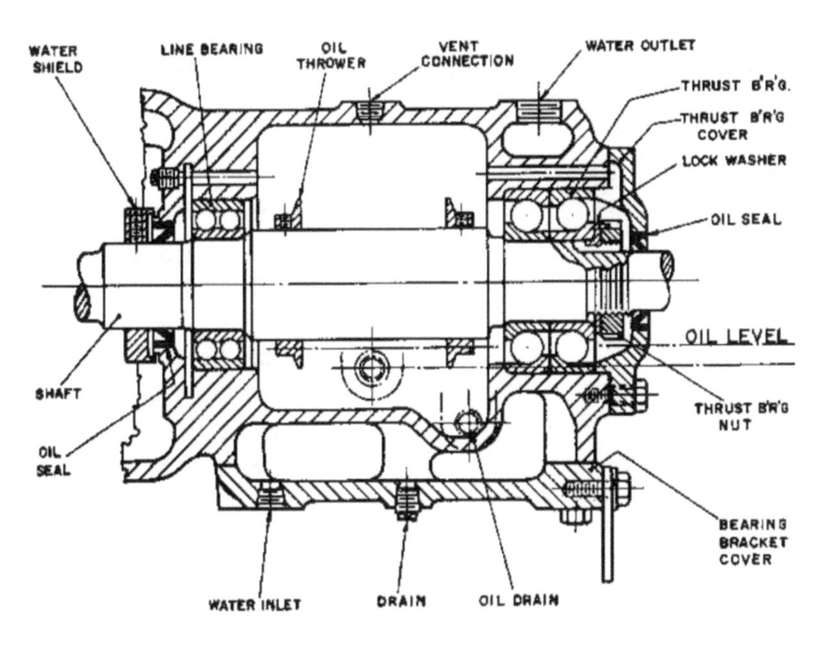

Figure 14-9. A Worthington Pump Company bearing housing from the 1960s. (Note pressure equalization holes surround bearings)

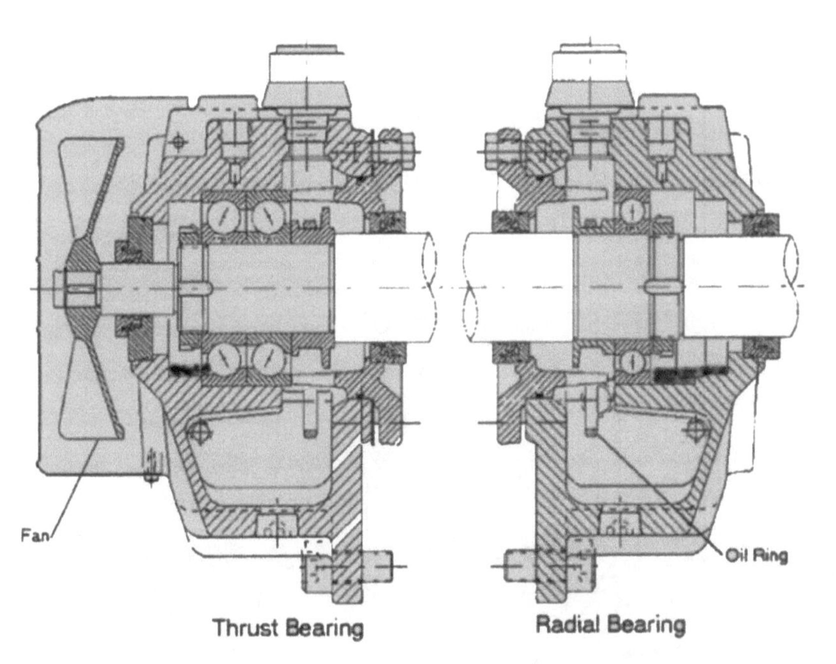

Figure 14-10. Oil trapped behind bearings is likely to overheat, oxidize and turn black (illustration from Ref. 2). No oil equalization holes are shown.

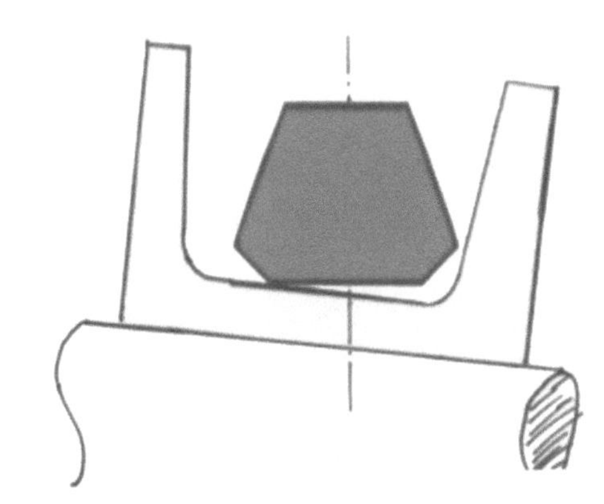

Figure 14-11. Mirror-image trapezoidal oil ring on a contoured carrier sleeve (Ref. 6)

to keep water away from the trailer's wheel bearings. When he tows the boat on a highway, the bearings rotate at usually no more than 900 rpm. We will assume that, on average, the owner tows the trailer 200 hours per year.

Compare this to the average electric motor bearing. Its shaft diameter is twice that of the boat trailer's axle. The electric motor bearing rotates at twice or four times the speed and we hope it lasts 24,000 hours—three or more years. That's why grease in electric motor bearings should take up only 30-40% of the space between bearing rolling elements. Packing the bearing full of grease would create excess heat and reduce bearing life.

Lubrication Charts

Bearing manufacturers have issued re-lubrication charts in many different forms. On the one shown in Figure 14-15, a particular bearing style is indicated by the letter "C" on top of the vertical grid axis. Depending on shaft size and speed the recommended intervals can be read off on the vertical axis as hours between re-lubrication. While these intervals were conservative and pertained to standard greases, Figure 14-15 was often used to envision where the use of life-time lubrication (meaning fully sealed, non-regreasable bearings) should be discouraged. When a bearing cannot be re-greased as with sealed bearings, the indicated interval will at least provide a general guide on expected bearing life. The advent of entirely different non-hydrocarbon greases means that Figure 14-15 will have to be modified for PFPE-PTFE greases (Figure 14-16).

PFPE greases (perfluoropolyether, including Teflon®-based products, Ref. 11) present an interesting lubrication alternative which was studied and fully validated at smaller and/or non-HPI facilities in recent years. It was found that developments in grease tech-

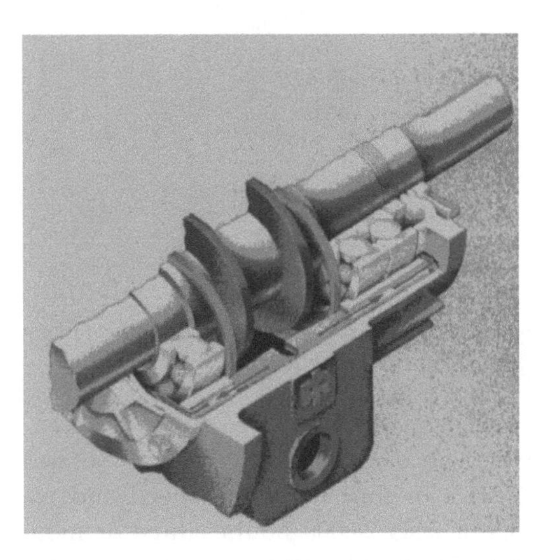

Figure 14-12. Two oil rings deposit oil in the slanted oil gallery of this Ingersoll-Rand pump bearing housing. Plastic discs limit oil ring travel. Note oil flow in gallery denoted by arrows

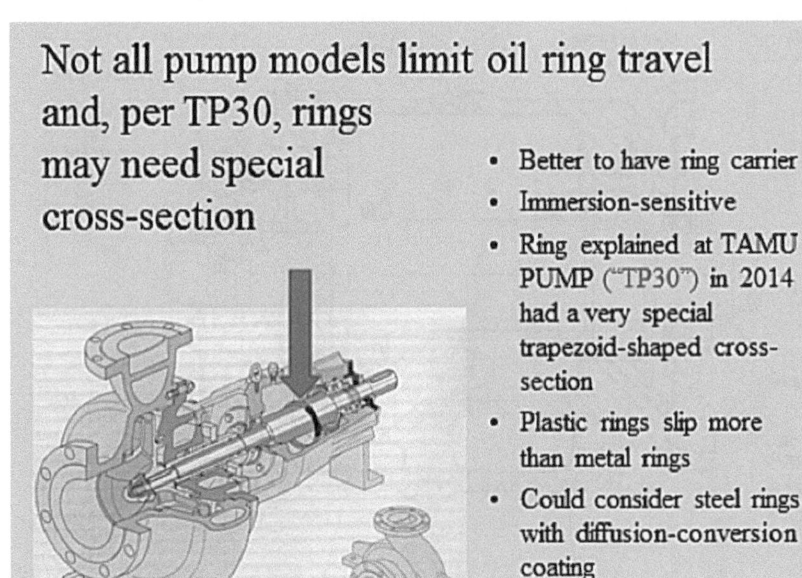

Figure 14-13. Many API-compliant pumps incorporate neither oil ring carriers nor oil galleries. The oil ring travel or skips around (Obtained from a manufacturer's marketing bulletin)

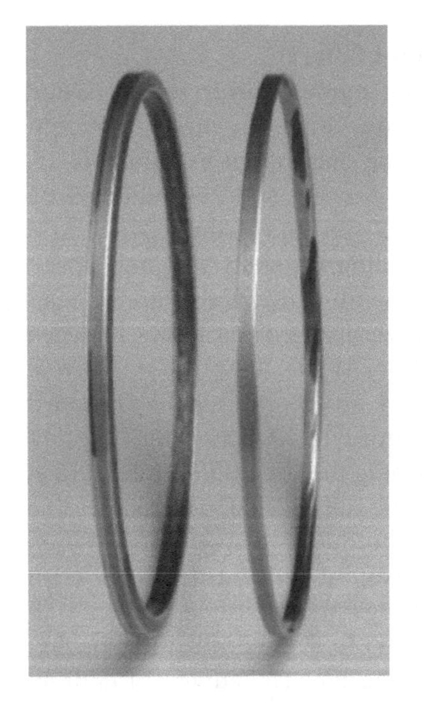

Figure 14-14. The wide oil ring on the left is new; the narrow oil ring on the right is badly worn. The difference between the two measurements is abraded material which contaminated the oil and caused premature bearing failure (Source: TRICO Mfg. Co)

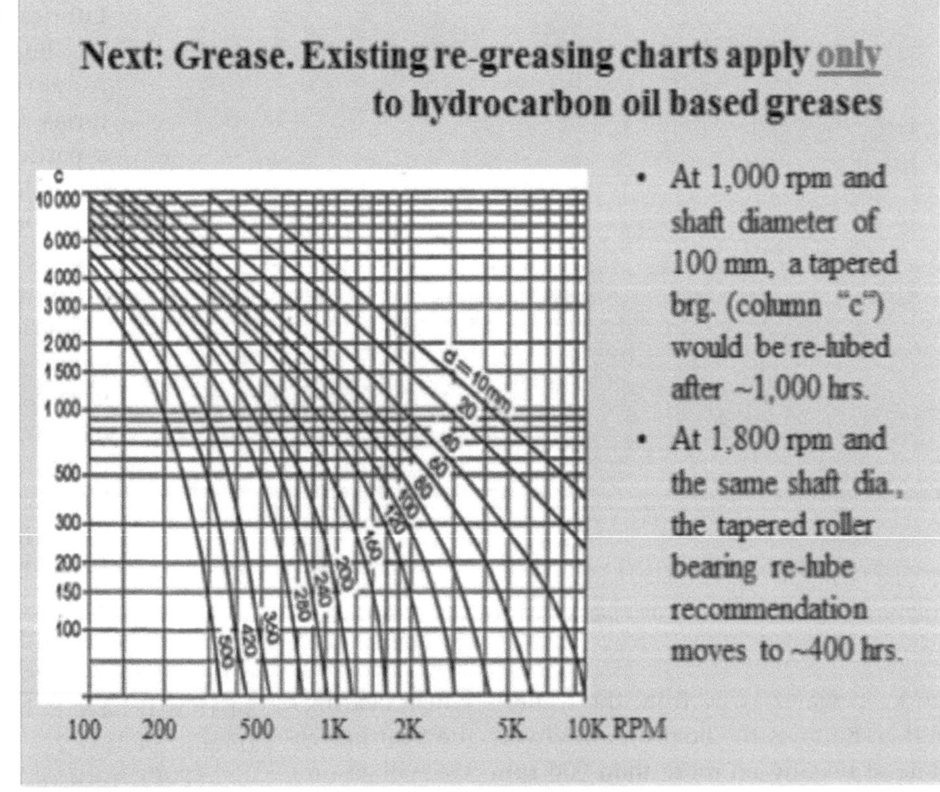

Figure 14-15. Bearing re-lubrication chart intended for a particular bearing style (e.g. tapered roller bearing) as a function of shaft speed and shaft size (Source: Ref. 8)

nology can greatly extend the life of sealed bearings in the application range traditionally associated with "life-time" lubrication of electric motor bearings.

In one study case, traditional motor bearings were supplied with sealed-in PFPE grease of the proper consistency. DuPont-Chemours' trade name is "Krytox®," the PTFE ingredient is more commonly known as Teflon®, and as of 2016 the spin-off now involved with this grease is now named Chemours®. The life extensions resulting from this grease have been tracked and explained on the Weibull comparison plots in Figure 14-16. A detailed cost study (at a Canadian paper mill) showed benefits over periodic re-greasing in certain industries and environments.

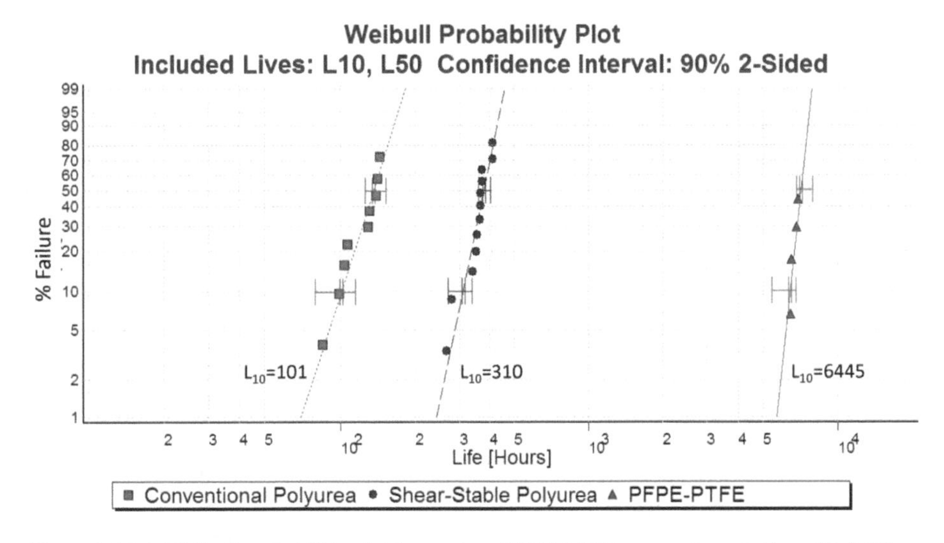

Figure 14-16. Weibull probability plot for modern PFPE-PTFE grease formulations (Ref. 11)

However, the experience with PFPE-PTFE may not apply to every situation and careful follow-up is always recommended. Also, these greases cannot be mixed with even trace quantities of traditional grease types.

Shields versus no Shields in Pump Electric Motor Bearings

Through-path, unshielded, self-relieving bearings are shown in Figure 14-17, which is somewhat typical of motors produced in Europe. In Figure 14-17, the motor manufacturer and designer ascertain that over-greasing is simply not possible; excess grease flows into a spent grease cavity or to the atmosphere. No shields are used in these bearings.

Chances are the through-flow design in Figure 14-18 saves pennies over Figure 14-17; however, spent grease will be expelled from the arrangement in Figure 14-18 only if the drain plug is first removed from the bearing housing. The human elements of (a) training and (b) conscientiously carrying out proper work procedures are then getting very important.

Instead of through-flow, shielded bearings were soon adopted; a single shield is shown in Figure 14-19 and the grease reservoir is adjacent to the shield. There is an annular gap of between 0.05 and 0.10 mm (radial measurement) between the shaft surface and the shield's inner diameter. The design intent is for a small amount of oil to migrate from the grease reservoir into the bearing by capillary action. Grease replenishment would require removal of the drain plug and orienting the shield as shown here would prevent packing the bearing with grease.

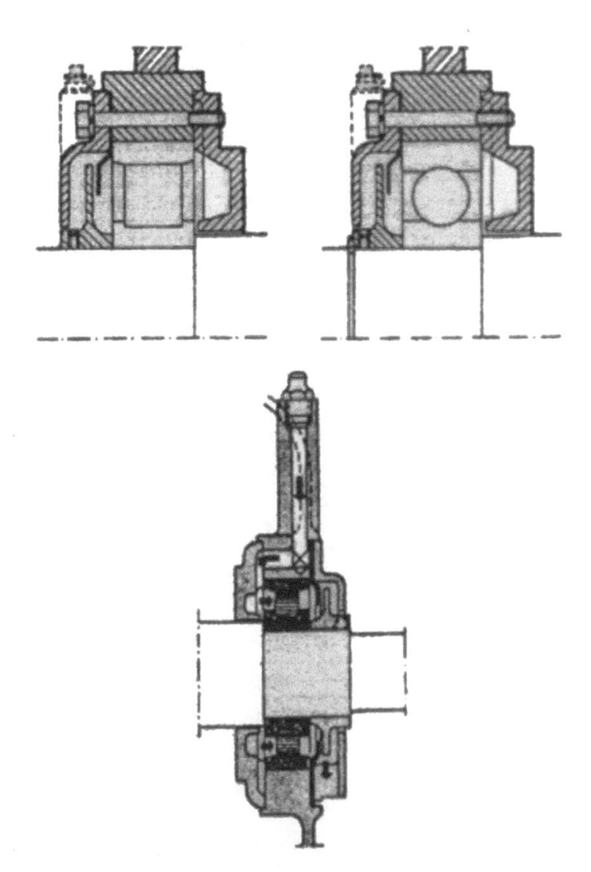

Figure 14-17. Through-path, unshielded, self-relief bearings

Unfortunately, inadequate training or personnel hoping to take shortcuts have often left drain plugs in place. In that case, excessive grease gun pressures forced shields into rubbing and scraping contact with rolling elements upon which bearing failure is inevitable.

1 Lubrication Entry
2 Drain
3 Shaft
4 Bearing
5 Inner Cap
6 Bracket
Open bearing with cross-flow grease lubrication

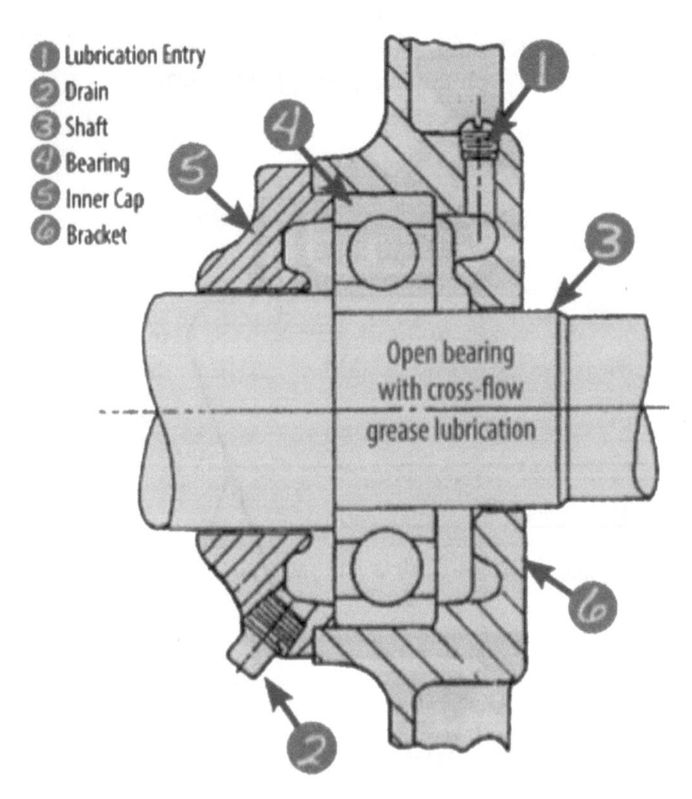

Figure 14-18. Through-flow grease; the drain plug must be removed during re-greasing

Double-shielded bearings (Figure 14-20) at least eliminated questions as to which side should face the reservoir, but leaving drain plugs in place still killed bearings. Industry virtually forgot that all we're asking for is to put a new charge of unpressurized grease into the reservoir every few years and to then let capillary action slowly move exceedingly small quantities of oil into the ball path. Reliance Electric added a metering plate (red part, Figure 14-21) in a valiant effort to ward of the over-greasing calamity. The results were mixed; mechanics often took one look at metering plates and threw them away.

Decades ago, ARCO Alaska mitigated over-pressuring with a drain pipe, Figure 14-22, left open on purpose. A small volume of spent grease formed a "natural plug." The plug advances each time new grease is added to the reservoir. Again, we will spare the reader a description of what a heap of grease looks like when it is expelled onto the pump base. After a year or so it's more than just a housekeeping problem.

Still, Figure 14-22 shows the best solution from technical acceptability and from "not wanting to argue with my workers" points of view.

Alternatively, we might try accountability and insist on staffers following instructions. We found out how well this admirable approach worked in the

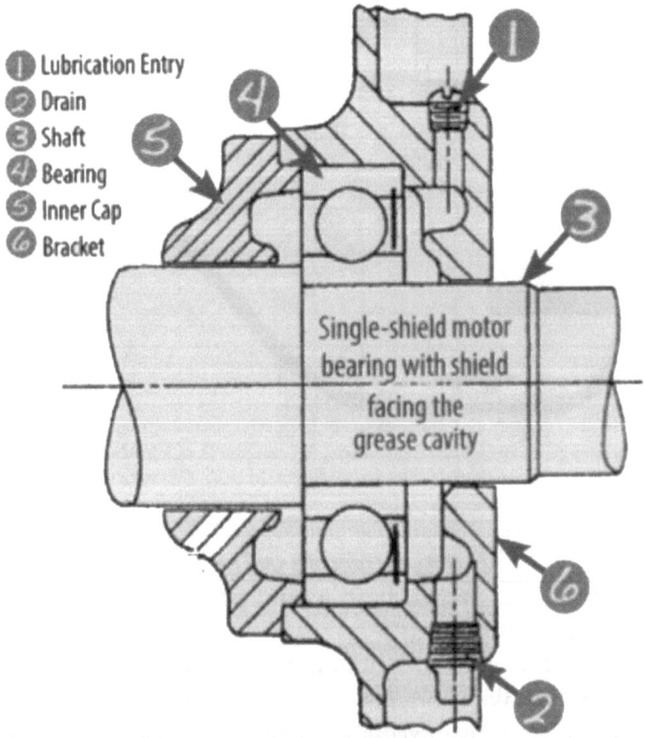

1 Lubrication Entry
2 Drain
3 Shaft
4 Bearing
5 Inner Cap
6 Bracket
Single-shield motor bearing with shield facing the grease cavity

Figure 14-19. Single-shielded bearing with removable plug (2)

1 Lubrication Entry
2 Drain
3 Shaft
4 Bearing
5 Inner Cap
6 Bracket

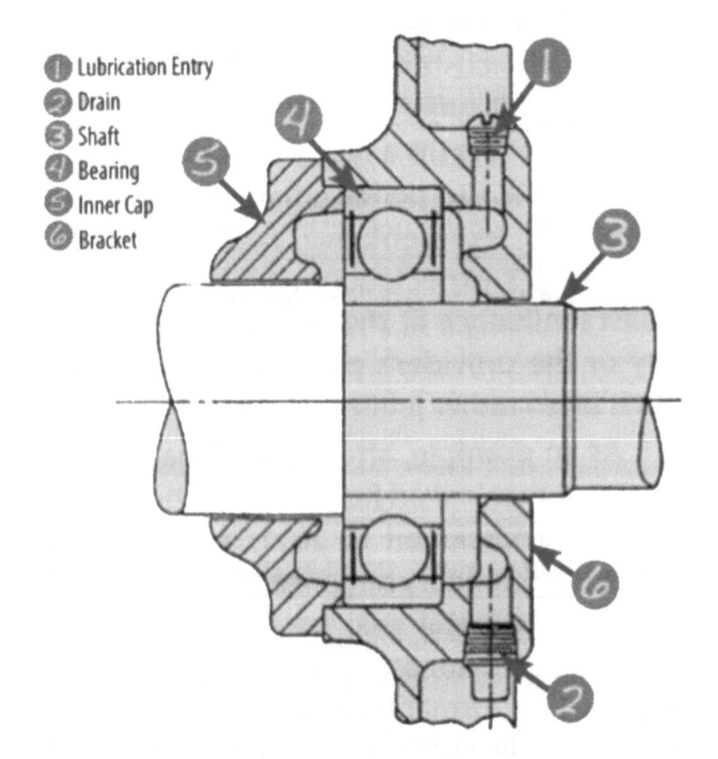

Figure 14-20. Double-shielded bearings are sometimes used because their symmetry allows installation in either direction

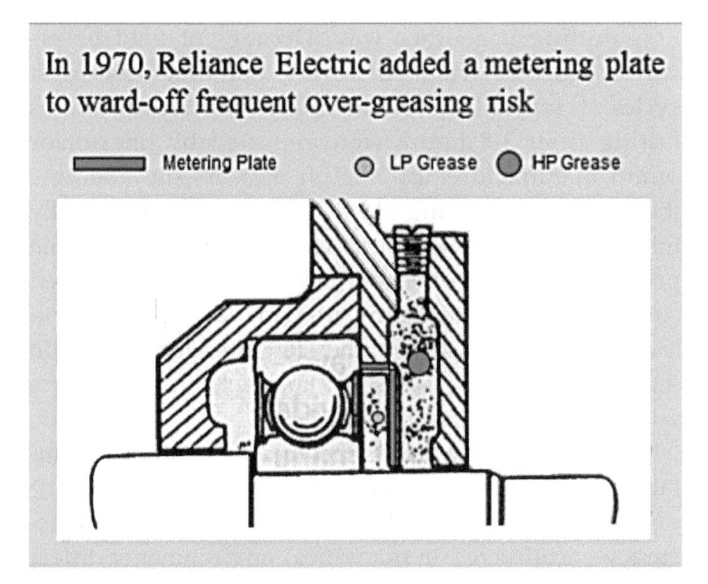

Figure 14-21. Metering plate as a first line of defense against over-pressuring the grease (Reliance Electric, ~ 1970)

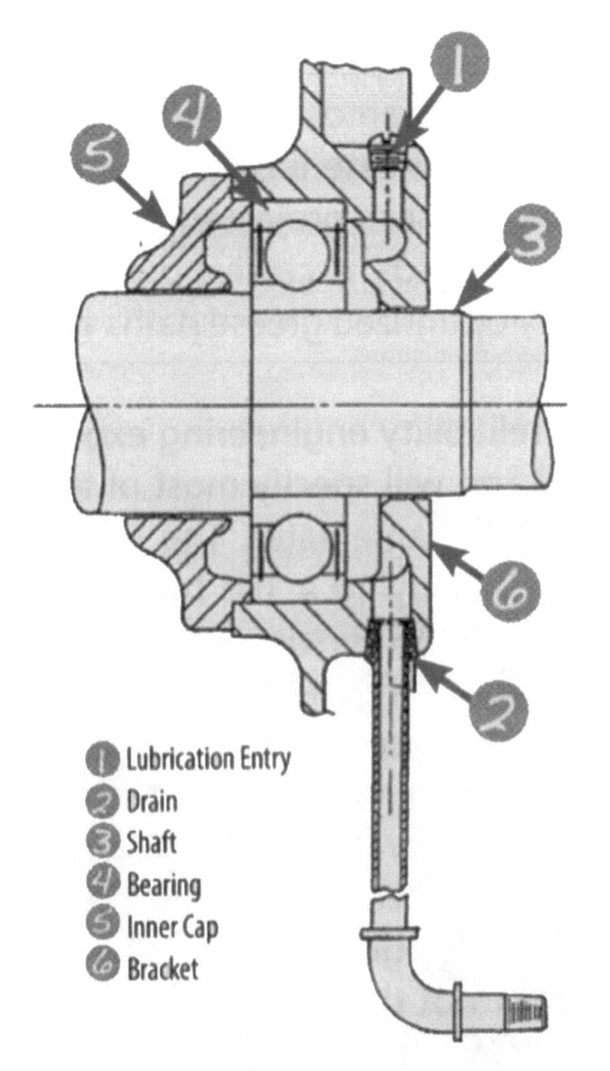

Figure 14-22. Permanent drain opening prevents over-pressuring

United Arab Emirates where a large refinery reported replacing 7 bearings per 1,000 electric motors per year. When asked what magic grease formulation they were using, a senior manager explained that his workers simply followed instructions and that grease-related bearing failures are a rarity. In other words, they know what bearings they have, they remove drain plugs, they re-grease with the prescribed amount of grease, then move on to do the next electric motor. After allowing two to three hours for grease to settle, a worker returns and re-inserts each drain plug.

This simply illustrates that there is no substitute for following a proper work execution procedure. Good supervision and management prevent failures and generate higher profits.

RANKINGS

Some rankings are subjective, and our rankings are among them. One of the authors was asked to provide data wherever possible and that request is commendable. But nobody has data on how many oil rings will malfunction because the maintenance person did not install it right, or the pump design allows the oil ring to slosh around, or because the pump manufacturer supplied oil rings that have not been stress-relief annealed and which, for that reason, lose roundness over time. About the last thing a pump manufacturer would ever want us to know is how many black oil incidents are caused by designs that overlook the need to provide an escape path for small amounts of oil trapped behind bearings (Figure 14-10) or for not seeing to it that there is pressure equalization between the space to the left and right of any particular rolling element bearing. We refer back to Figure 14-9 or, in the case of grease, the vulnerabilities of Figures 14-19 and several others if drain plugs are left in place.

Except for a rather expensive traditional auxiliary lube pump-around system with reservoir, an integrated lube spray system (Figure 14-23, right side) is the best of all worlds. One would no longer worry about oil rings and their many demonstrated flaws, constant level lubricators, installation accuracy, shaft inclination, and so forth. Sooner or later an innovative pump manufacturer will offer a pumping device (Figure 14-23, item "P") and filter (F) which will satisfy many thousands of buyers. The process pump manufacturers unwilling or unable to supply oil spray might console themselves with purchasers who buy only from the lowest initial bidder, or sellers of spare parts and maintenance services. It's a free choice we can all make. Yet, responsible buyers favor sellers who add value, not uncertainty.

Years of obtaining and examining proprietary data from machinery network colleagues (with access to at least 24,000 pump sets) and through empirical observation data repetitive failure data has enabled the production of a ranking table of preferred lubrication methods identified in Table 14-2.

Table 14-2. Preferred Lubrication Method Ranking Table

Method	Rank
OIL LUBRICATION	
• Oil spray, filtered/pressurized. Also oil mist/coalesced oil coats bearing [Highest]	=10
• Circulation, filtered/pressurized	= 9
• Flooded sump, w/debris	= 6
• Sump, with slinger disc [Average Rank]	= 5
• Sump and guided oil rings, well within concentricity specification	= 4
• Sump and non-guided oil rings well within conc. specifications	= 3
• Sump and non-guided oil rings out-of concentricity [Lowest Rank]	= 2
GREASE LUBRICATION	
• Life-time PFPE-PTFE [Highest Rank]	= 7
• Through-flow, low pressure	= 6
• Random-flow, low pressure	= 4
• Life-time EM polyurea	= 3
• Random-flow, random press.	= 2
• Random-flow, over-pressured [Lowest]	= 1

Putting it another way: Decades of field observation must be weighed against manufacturers' test cycles of, typically, a few hours. Factory test durations lasting from 2-8 hours were reported by one major pump manufacturer in 2000. It appears they became the basis of advocating plastic oil rings and marginally thicker oils. However, these changes did not cure the problem of black oil experienced by a disappointed customer, a user company in Canada. Likewise, follow-up talks led nowhere; they left attendees of certain discussion group sessions at a major pump conference with considerable frustration.

Needless to say, the machines on a pump manufacturer's test stand are properly aligned and the lubricant is fresh and clean. In contrast, the degree of inaccuracy encountered in many field environments differs greatly from the accuracy found on test stands. Neither the training nor the abilities of crafts and service personnel will always measure up to expectations. In some installations the piping connected to pumps is pushing and pulling. As a result, bearings are edge-loaded and the oil film can no longer provide adequate separation of parts (Ref. 3).

Shaft alignment is achieved by putting shims under equipment feet which, as a logical consequence, tends to cause shaft systems to be at a slight angle relative to the true horizon. On shipboard, pumps pitch and roll. Equipment surveillance and precautionary oil changes differ on shipboard from what we find at many land-based installations.

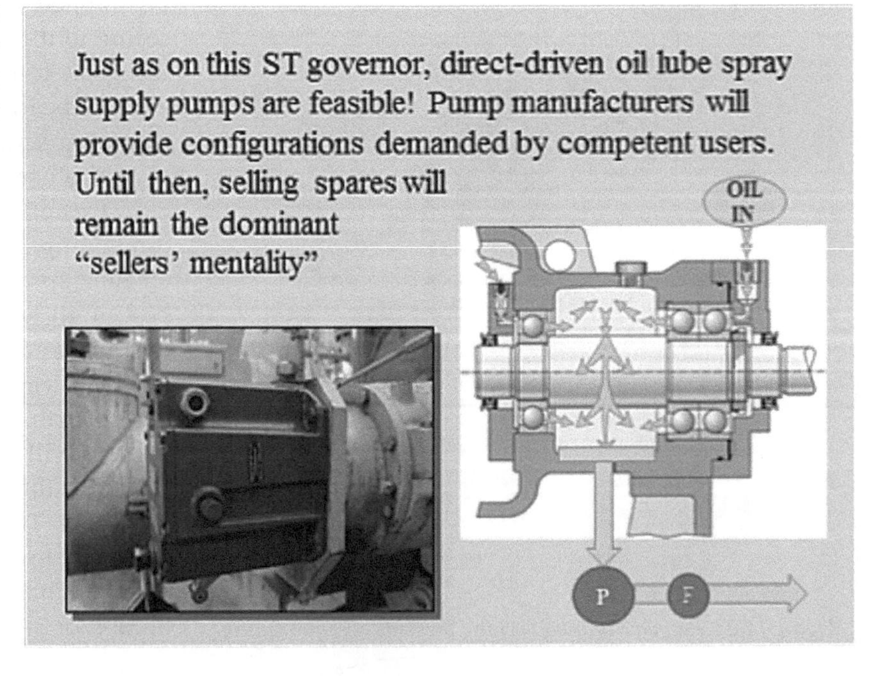

Figure 14-23. Like traditional auxiliary pump-around lube systems, lube spray systems (right side) are ranked best of all worlds. A shaft-drive steam turbine governor (left) demonstrates one of numerous spray pump drive options

These and other experience factors interact; they admittedly shape and skew rankings in the eyes of field-experienced individuals. Their backgrounds differ, their perceptions differ, and we must leave it to our readers and/or our tutorial attendees to judge where to place their trust. We have observed that many process pumps in industry are experiencing repeat failures even as you read this narrative or attend this tutorial session.

The authors do not believe that a reasonable person needs to show data to prove that driving automobiles with worn tires is a greater risk than driving on tires with tread. Also, what looks worn to "A" looks normal to "B." It is no different with oil rings in pumps at location "X" versus location "Y." Gravity being gravity, the sketch in Figure 14-11 is a scientific fact. The ultimate ramifications of a trapezoidal oil ring operating in this manner can be foreseen: A trapezoidal oil ring has two "pointed" or "circumferential" ridges"—one on the left and one on the right side. Pointed ridges have a very small total surface area. As the oil ring slews from side-to-side in its carrier sleeve, it will touch the side of the carrier sleeve. That means the force-per-unit-area, the "pounds-per-square-inch" (commonly known as "pressure" will be rather high. When that happens, the "pointed ridge" will break through the oil film and abrasion will occur.

Whenever a pointed ridge breaks through the oil film, there will be increased friction and the oil ring will slow down. Long-term satisfactory operation will be at risk, unless the pump owners invest heavily in preventive maintenance action. But preventive maintenance costs money, and that is simply an additional reason why oil rings rank very low on our scale.

Reviewers have asked for more detail on oil-air, also called "jet oil lubrication," Figures 14-23 and 14-24 (Ref. 9). Jet oil lube represents the highest-rated application method. Envision, therefore, how jet oil—widely used in aerospace over the past 70 years—will open a window of opportunity for reliability-focused users and/or innovative pump manufacturers. Think of a small oil pump ("P," Figure 14-23), either internal to the process pump bearing housing or incorporated in a small assembly screwed into the bottom drain of your present process pump. This upgrade can provide filtration, metered flow, and proper pressure downstream of the oil sump and upstream of the spray nozzles. Starting in about 2005, motivated and truly reliability-focused users have written this preferred lube application approach into their process pump specifications and are actively pursuing this pump reliability enhancement.

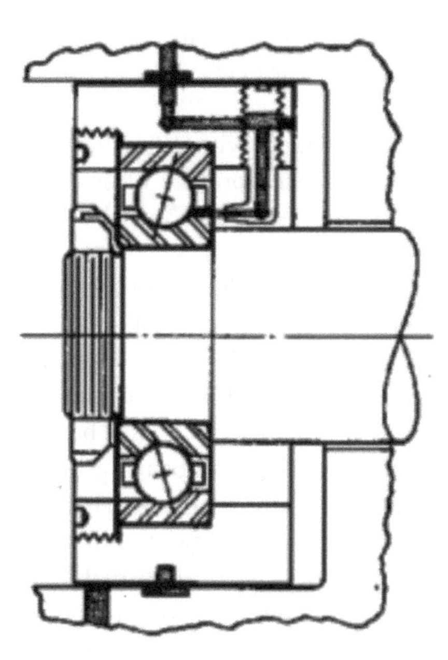

Figure 14-24. Jet-oil lubrication (Ref. 9), whereby liquid oil is sprayed directly into the rolling elements of a bearing

OIL MIST LUBRICATION FOR PUMPS AND MOTORS

Pure ("dry sump") oil mist systems contain no moving parts (Ref. 5). With pure oil mist, service-intensive oil rings, constant level lubricators and the traditional oil refill labor requirements are eliminated. The labor requirement for pumps with oil mist has been estimated as one-tenth that of traditional lubrication. In the past decade the earlier practice of allowing excess oil mist to escape into the atmosphere has been superseded by widespread use of closed systems. Closed systems avoid polluting the environment. Some closed systems have been in highly successful service since the mid-1980s and represent best available technology in all respects.

Primary Advantages of Oil Mist:

- Plant-wide systems are almost completely maintenance-free and fully self-checking. Users do not have to rely on operators or maintenance workers to check and fill housings with oil
- Better lubrication conditions exist because the oil coating on the bearings is always new
- Lower bearing operating temperatures are routinely obtained. Reductions typically range from 10 to 20 degrees F (6-13 degrees C)
- Power requirements are reduced by typically at least 1% and sometimes even 3% since bearings

operate on a thin oil film instead of plowing through a drag-inducing pool of oil

- Oil mist is applied without using oil rings. (Oil rings are subject to abrasive wear and/or slowing down if the shaft system is not absolutely parallel)

The extended mean-time-between-failure (MTBF) benefits of oil-mist over traditional liquid-oil-in-sump lubrication have been well-documented and oil mist was included in the venerable API-610 pump standard, 7th Edition (1989). In describing the basic oil misting process we first note that the bearing housing, Figure 14-25 contains no liquid oil [Ref. 11]. Instead, an oil mist generator (OMG) with no moving parts creates the mist in a central console (Figure 14-26). A typical console and its OMG serve all of the facility's process pumps within about a 600 foot (160 meter) radius. The mist is a mixture of microscopic (<3 micron) oil droplets combined with clean air at a volume ratio of about 1:200,000. The mist is conveyed in 2" pipe headers to virtually any equipment incorporating rolling element-style bearings headers at a low pressure and velocity (< 7 ft/sec to reduce the effects of globules becoming too large for suspension in the carrier air). Near each process pump or electric motor, the oil mist passes through a nozzle or re-classifier—essentially a metering orifice. The mist velocity is thereby greatly increased; also, a bearing in motion further promotes atomized droplets to collide and coalesce into larger liquid drops of oil. Oil mist is supplied to standby equipment as well. In standby or shut-down equipment oil mist serves as a protective blanket. Because its pressure is marginally higher than the surrounding atmosphere, oil mist prevents the entry of airborne dirt and moisture.

An entire system is depicted in Figure 14-26. The oil mist generator (OMG, where oil meets air) is located inside an oil mist cabinet. Piping consists of a delivery header in the foreground and a return oil collection header in the background of Figure 14-26. Oil mist take-offs to and from process pump and motor bearing housings are connected to the top of their respective headers [Ref. 11].

"Old-style Open" and "New-style Closed" Oil Mist Application

In old-style "open" oil mist systems (Figure 14-5) the air/oil mixture fills the bearing housing, but not all of the mist passes through the bearings. A portion of the coalesced droplets takes a straight top-to-bottom path through the bearing housing. Only this portion of the coalesced oil and stray mist then exit near the bottom of the housing and can be collected for disposal at the drain location. For bearings to be properly lubricated the oil mist will have to pass through the bearings and then escape at the two unsealed regions where shafts protrude through the bearing housing.

Efforts to simply provide effective bearing housing seals at the ends labeled "oil mist out" in Figure 14-25 had unexpected consequences for inexperienced users. Oil mist works by causing small globules of lubricating oil to provide an oily coating on the bearing components. However, so as to provide continuous oil replenishment on bearing surfaces, the oil mist must flow and cannot be stagnant, or dead-ended. In "old style" configurations tight-sealing bearing protector seals placed at the "oil mist out" locations in Figure 14-25 very often caused dead-ending. Without oil mist flow there was then neither cooling nor lubricant replenishing on bearing surfaces.

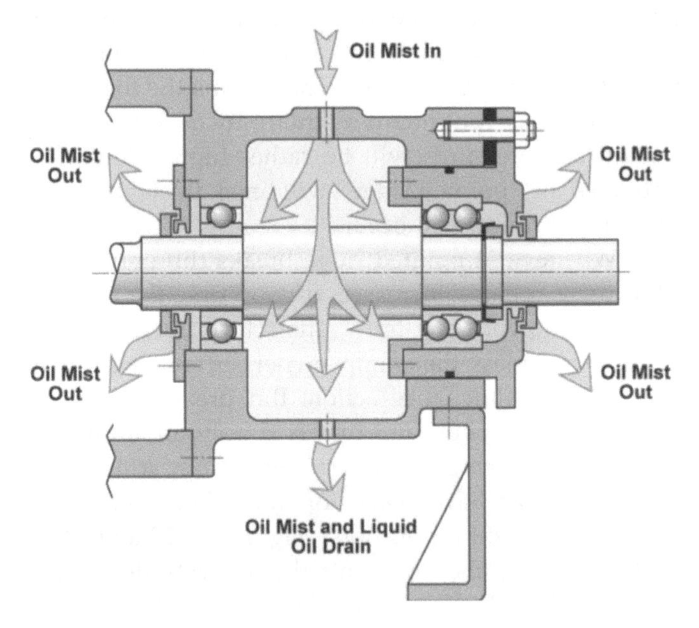

Figure 14-25. Old-style (non-API type) oil mist introduction at midpoint of bearing housing

Reliability leaders at best-of class companies soon implemented the most ideal routing of oil mist through bearings, Figure 14-27 [Refs. 2 and 3]. This routing became an industry standard almost 20 years ago. Here, oil mist is introduced into the space between a modern bearing housing protector seal and the bearing. Any oil reaching the bottom-center of the bearing housing will have first cooled and lubricated the bearing. Also, having a single centrally located exit hole makes it easy to collect the coalesced oil or residual oil mist—the system is now "closed" and excess oil is ready for filtration and re-use.

The workers' health guidelines of many industrialized nations permit the relatively small amount of oil mist released from open systems. However, irrespective of prevailing or mandated clean-air requirements an environmentally conscious user would not allow continual releases of oil mist into the atmosphere. Moreover, from a housekeeping viewpoint, it is clearly advantageous not to have smudges of oil on the ground near pumps and other equipment. Oil smudges or rain run-off will reach a waste oil pit and it takes money to extract that oil before the water can be discharged. An open oil mist system thus does not represent best available technology, which is why proven closed oil mist systems technology, Figure 14-26, is greatly favored today [Ref. 3].

Closed oil mist system technology may differ, but Figure 14-26 incorporates a collecting tank (shown at the far left) to which a return header system is connected. A small blower is provided at the top of the collecting tank and the suction effect of this small blower causes excess or "stray" oil mist to be pulled into the tank. Inserted in the blower is a coalescer maze. Coalesced oil droplets fall out and the oil can be re-used. Only virtually oil-free air is vented to the atmosphere.

Favorable Experience Ascertained

The oil mist routing per Figure 14-27 together with using either advanced dual face-type "fully sealing" (Figure 14-28) or axial O-ring valve-equipped bearing housing seals which allow minor "weepage"

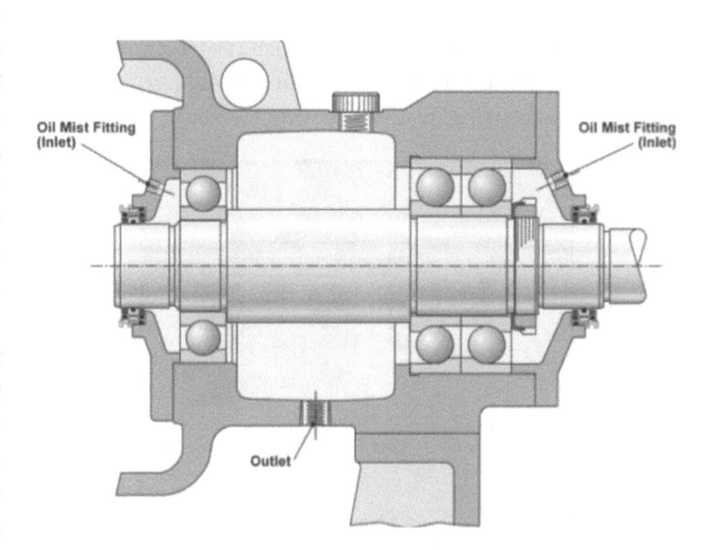

Figure 14-27. API 610-compliant oil mist application at locations between the rolling element bearings and the magnetic dual-face bearing isolators (Source: AESSEAL plc, Rotherham, UK and Rockford, Tennessee, USA)

Figure 14-26. Closed oil mist system supplying continuous lubrication to pumps and drivers (Source: Colfax Lubrication Management, Lube Systems Division, Houston, TX)

through the micro lift gap (Figure 14-29) make closed systems possible. Either way: Forward-looking oil mist users have discontinued "old" oil mist application per Figure 14-25 since the mid-1970s and have enjoyed decades of superior experience with the routing shown in Figure 14-27. The API standards (starting with API 610/8th Edition, released in 2000) have recommended oil mist introduction into the region between the bearing and the bearing housing protector seal.

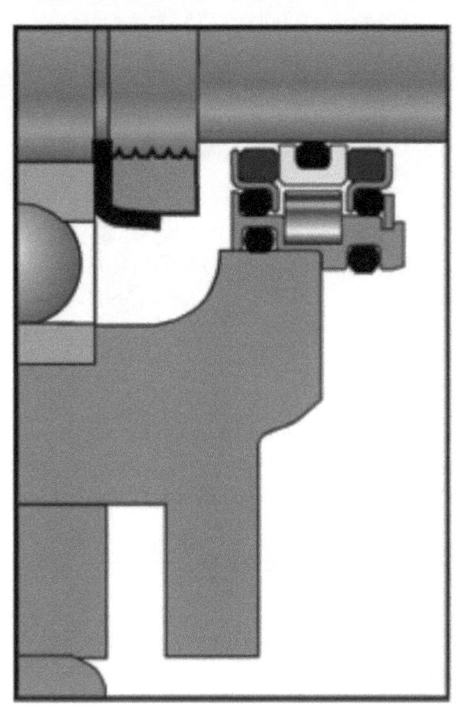

Figure 14-28. Advanced magnetically-closed dual-face bearing housing protector seal. The face geometry is developed for virtually leak-free use on equipment lubricated by closed oil mist systems (Source: AESSEAL plc, Rotherham, UK and Rockford, Tennessee, USA)

Books and articles also describe the unqualified success of oil mist lubrication for electric motors [Refs. 2 through 5]. Contrary to unfounded (and fully refuted) opinions about oil mist attacking electric motor windings or oil mist being a fire hazard, this lubrication method is fully proven to be superior to alternative lube application methods. Closed-loop oil mist best protects industry's physical plant as well as our environmental assets, whether in spare equipment (standstill) or full operating mode. In best technology closed oil-mist systems, from 97 to 99% of the lube oil is recovered and re-used. These systems emit no oil mist into the surrounding atmosphere and, for years, have been praised by refineries and petrochemical plants concerned about the environment.

Absolute Reliability Reaffirmed

In meetings with the primary providers of plant-wide oil mist systems (in 2012 and 2013), the authors ascertained that a mere three shutdown incidents were known to have occurred on oil mist systems in two decades of highly successful operation. An estimated average number of 1,800 plant-wide systems have been in service during that 20-year period.

The first systems interruption involving a modern plant-wide oil mist system occurred at a U.S. Gulf Coast facility in about 1982. At that time, a thorough analysis traced the failure to pipe shavings in the 5-gallon capacity misting chamber reservoir. This is where ferrous debris became attached to a magnetic level switch; it prevented the switch from activating a solenoid, which would have allowed oil from a bulk oil holding tank to replenish the much smaller chamber reservoir. When the small reservoir was depleted, none of the connected equipment received oil mist. The bearings ran dry, but did so without incident or bearing failure. It was known that bearings coated with oil in horizontally installed shaft systems can operate for a few hours after discontinuing oil mist had been reported and explained by Allen Clapp, of Dow Chemical Company, Freeport, TX, in 1973]. These findings were again acknowledged and corroborated in academic research by Shamim and others [Ref. 5]. Allen Clapp made the additional point that a small pool of oil will exist in the five-to-seven o'clock segment of the contoured raceway in a bearing's outer ring. Oil mist lubricated pumps can safely stay in service for approximately eight hours before this small pool of oil is depleted. Given that there is ample supervisory instrumentation to annunciate deviations, no modern oil mist system has ever encountered unavailability in excess of eight hours.

The second unavailability event developed at an oil refinery in Enid, Oklahoma, where a single oil mist console was serving two adjacent process units. When the process unit where the oil mist generator (OMG) was located had to be shut down in preparation for scheduled maintenance and repair downtime, the OMG was inadvertently valved off. A day or so later, the adjoining process unit experienced a pump failure. It was then realized that there had been no oil mist supplied for at least 24 hours. The oil mist supply was restored and there were no other bearing failures on any of the connected pumps. The cause of failure was clearly human error; it could have been averted with a simple advisory note posted at the appropriate switch or valve.

A third incident report relates to a Texas Gulf Coast oil refinery where the OMG did not have an automatic fill option. A reservoir refill line connected a bulk storage tank to the small oil reservoir located inside the main oil mist console. An operator (when in doubt, always blame the operator....) decided to crack open the needle valve in the refill line and let it slowly maintain the oil level in the small reservoir. After a period of time, the entire piping distribution system was full of liquid oil; oil had, in fact, displaced the oil mist. About 10 or 12 pump bearing housings and their respective motor driver bearings were filled with oil. A lot of oil was wasted, but there were again no equipment failures in this incident.

Because modern oil mist systems are provided with suitable supervisory instrumentation, no system has ever been disrupted for more than two hours from 2000 until 2015. Overall reliability and availability was calculated as 99.99962 (in 1998) and is now thought to have reached 99.999997%—a number that has never been approached by any other lubrication method.

What—If Anything—Can Shut Down An Oil Mist System?

The most profitable and reliable plants have found it neither cost nor risk-justified to install oil mist systems with fully connected, ready-to-go spare backup. Nevertheless, some plants ask oil mist suppliers to propose and provide 100% redundancy. In those facilities, a backup or auxiliary oil mist system can be placed in operation on a moment's notice. Switching from the main unit to a full backup takes 30-60 seconds to complete. The switchover procedure calls for one-quarter turn of the handle of a ball valve. A two or three-sentence procedure sheet is posted on the inside of the cabinet door. The sheet explains what to do in the highly unlikely event of such switchovers ever becoming necessary.

In 2002, one of the authors visited eight petroleum refineries located in the U.S. Gulf Coast region. It was quickly discovered during these visits that oil mist lubrication was the predominant method of lubricating pumps throughout the refining industry in the USA. Also in 2002, an equipment sales specialist with over 20 years of experience as a refinery reliability engineer estimated that oil mist was being used by 24 of the 30 refineries in the Beaumont-Port Arthur region of East Texas. He believed that about 80% of the pumps in each facility were lubricated by oil mist systems. One U.S. West Coast consulting engineer with considerable background as a refinery engineer estimated that about 50% of all USA refineries were using oil mist in 2001.

As of 2015, many of these refineries can point to applying closed systems oil mist technology for over four decades and are calling this application method an unqualified success. The refineries consider closed oil mist systems a competitive advantage and have fully endorsed the application routines illustrated in this article. Moreover, these users are doing their part towards achieving a cleaner environment while imparting reliability to their rotating equipment assets.

What else could cause an oil mist unit or system to shut down? Well, one manager added to the three events spelled out earlier. He said running a fork lift into the 2-inch oil mist header would shut the system down. That may be true, but it certainly has never happened on any of the estimated 3,000 plant-wide oil mist systems now in service all over the world. And if it did happen, it would take considerably less than eight hours to repair the low-pressure non-flammable oil mist header.

CHOOSING THE CORRECT OIL VISCOSITY FOR PUMP LUBRICATION

From early experience with automobiles, most of us recall that we selected thicker oils (such as SAE 30) in the warm summer months and switched to thinner oils, perhaps SAE 10, in preparation for winter driving. Table 14-3 illustrates where these motor oils fit in as we compare them with the industrial oil designations in use today.

Thick oils are more viscous and may not readily flow into the bearings. We can heat up the oil or avoid oil rings and other risk-inducing lube application methods by using smarter means. We could use a jet of oil (oil spray) or could convey the oil, mixed with compressed air, in the form of an oil fog—also called oil mist. Whatever we do: We must guard against using the thinnest oil found on the market because we are concerned about potentially inadequate oil film strength and oil film thickness. But we will also not benefit from excessively thick oils, as we will see.

Lube Oils for Process Pumps

The MRC "Engineer's Handbook" (Ref. 10) gives general guidance as it states: "In general, the oil viscosity should be about 100 SUS (Saybolt Universal Seconds) at the operating temperature." If for some reason we had a bearing operating at 210 degrees F, Table 14-1 would apparently call for a lubricant with an ISO Viscosity Grade somewhere between 220 and 320,

Table 14-3. Oil Viscosity Comparison Chart

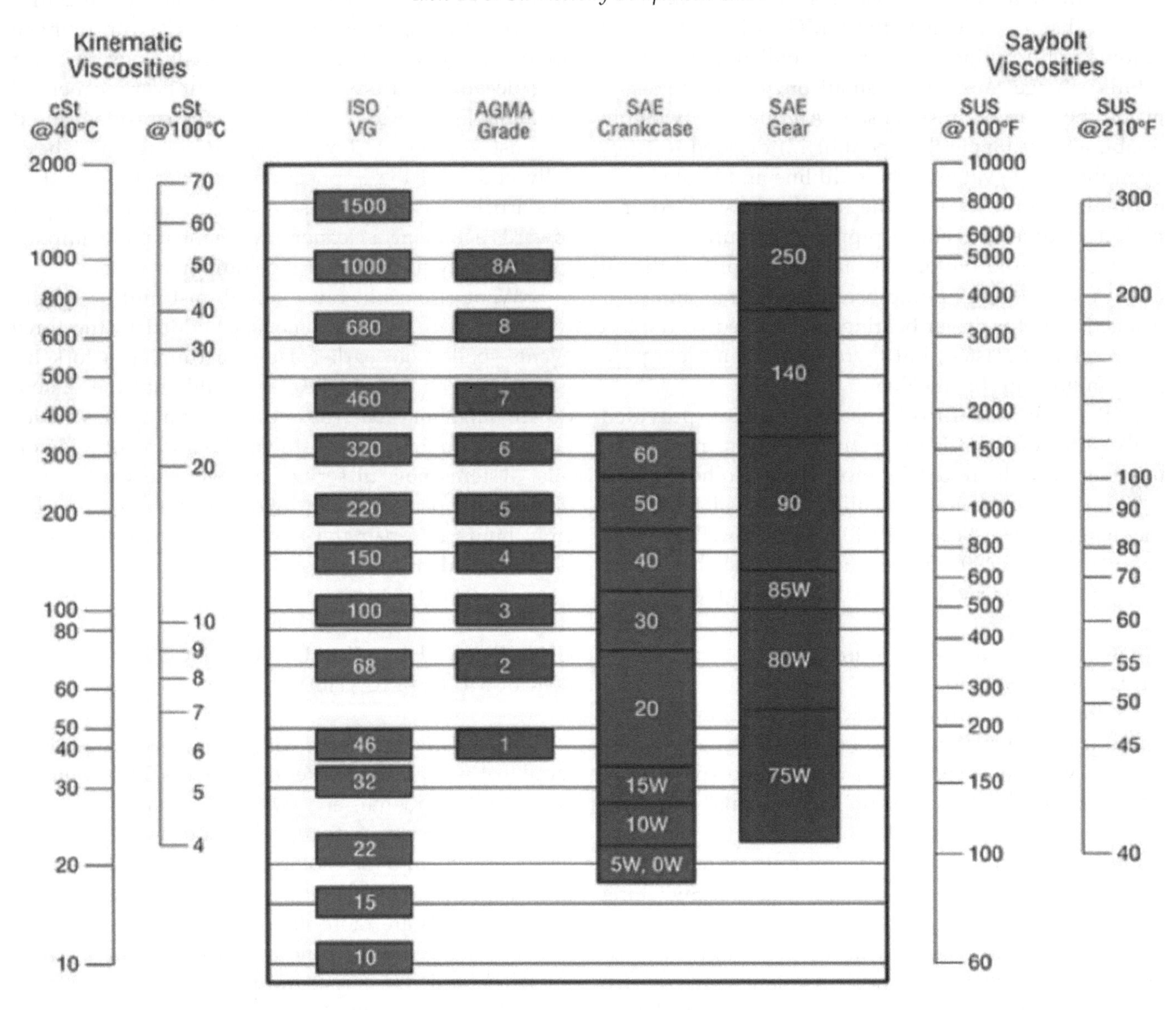

but that would be unrealistically thick for most process pump bearings. Oil rings—if used—would probably slow down and malfunction in such viscous oils. Oil overheating may be an additional concern.

With that in mind, we might wish to consult literature issued by bearing manufacturers with years of process pump experience. Among these is SKF; their graph (Figure 14-29) is time-tested and widely applicable (Ref. 8). Figure 14-29 depicts the required minimum (rated) viscosity _1 as a function of bearing dimension and shaft speed (Ref. 9). A bearing with a mean diameter of 390 mm at a shaft speed of 500 r/min would require__1= 13.2 cSt.

Here's one more example: If we had a bearing mounted on a 70 mm shaft rotating at 3,600 rpm, we might start by assuming that the bearing's outside diameter (OD) is twice its ID, or 140 mm. The bearing's mean diameter would then be 105 mm. To simplify, we might call it 100 mm and (in Figure 14-29) travel up from 100 to a location midway between the 3,000 and 5,000 rpm lines. In this instance, one could operate with a lubricant which, at the bearing operating temperature is somewhere between 8 and 9 cSt.

Note, however, that we would have to know the operating temperature of a bearing to determine what ISO viscosity grade (ISO VG) we needed to use here. The operating temperature derives its combined thermal input from bearing load and from lube oil frictional drag. Unnecessarily viscous oils will become hot. Figure 14-30 is helpful in this regard; but note that Figures 14-29 and 14-30 were drawn years ago and apply to mineral oils. If we today use premium grade synthetic oils, we will

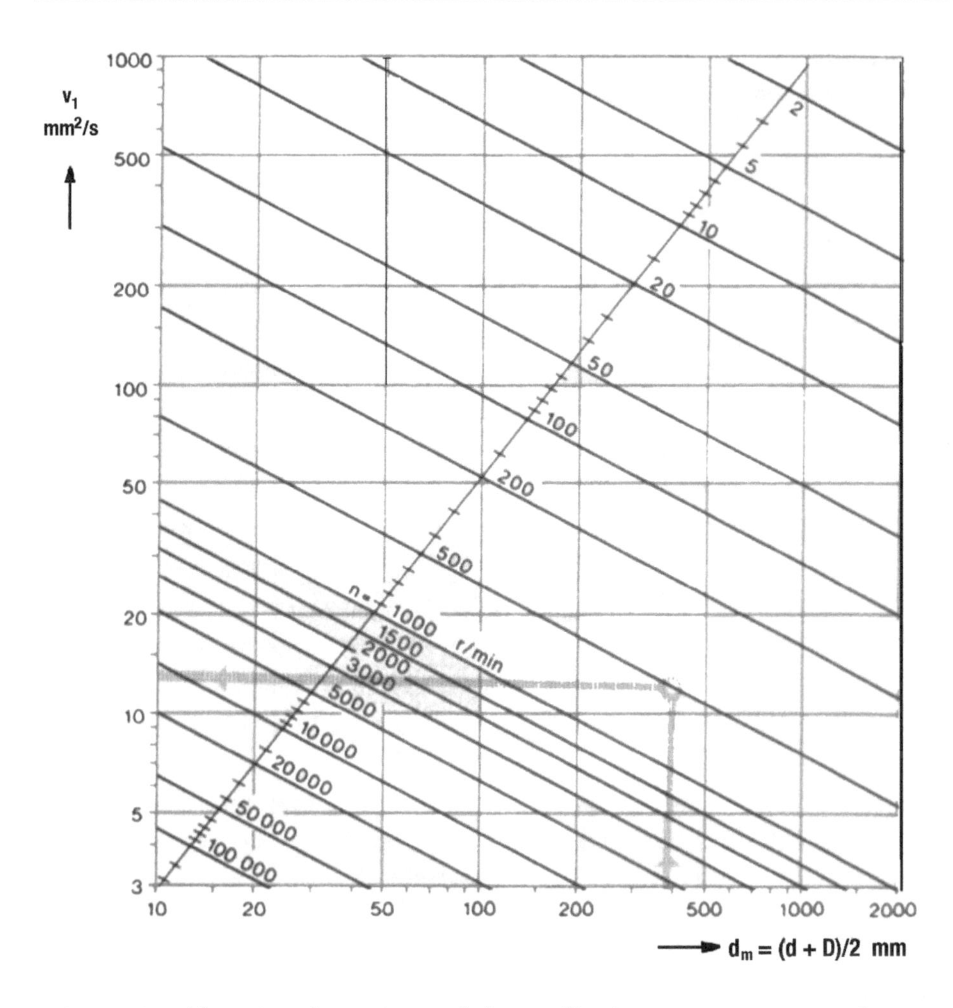

Figure 14-29. Required minimum (rated) viscosity _1 as a function of bearing dimension and shaft speed (Refs. 3 and 9). A bearing with a mean diameter of 390 mm at a shaft speed of 500 r/min will require_1= 13.2 cSt **[Editor: For ??, see symbol on y-scale of graph]**

enjoy a sizeable safety factor in our lube applications.

We would enter the vertical scale in Figure 14-30 near 9 cSt and move towards the right, where we now intersect with oils ranging from ISO VG 22 through ISO VG 320. Suppose we chose ISO VG 32, we might start the pump and verify that its oil temperature had leveled off at no greater than 75 degrees C. Alternatively, we might choose ISO VG 68 and verify that its operating temperature does not exceed 100 degrees C (212 degrees F).

Using lubricants with viscosities in excess of those actually needed may generate excess heat and actually work against us. However, thicker oils quite obviously have their place and we must remember that MRC with their 100 SUS rule-of-thumb had to cover all the bases. That said, a large bearing (200 dm) in a slow speed gearbox (200 rpm) requires an operating_1 of 40 cSt. From Figure 14-30 we could easily see that ISO VG 100 (or higher) oils would be needed here.

Learning from a Recent Case History

In a recent case history, ISO VG 100 was applied to a large pump where ISO VG 68 mineral oil or its equivalent ISO VG 32 would have sufficed. (An ISO VG 32 synthetic is the "bearing life equivalent" of an ISO VG 68 mineral oil. The synthetic ISO VG 32 runs considerably cooler than the mineral oil equivalent).

With ISO VG 100 mineral oil, the oil misted radial bearing ran a few degrees lower in temperature than it had with conventional sump and oil ring lube. The user was pleased but expressed disappointment at a triple-row thrust bearing running as hot as before, 190 degrees F.

To make a long story short: A premium formulation synthetic ISO VG 32 would have been sufficient and would have given the user everything a solid reliability professional could have asked for. Reliability pros would like to see pump bearing housings with no oil rings, no need for constant level lubricators and few if any repeat failures. They start with the right lubricant.

Why, with all that, is the radial bearing nice and cool? After all, it too is presently surrounded by the thick ISO VG 100; it's cool because it has no load. The load is in the triple-row thrust bearing and that creates temperature in addition to the frictional temperature mentioned above

MORE ON PUMP BEARINGS AND ALLOWABLE TEMPERATURES

As we of course know, hundreds of millions of horizontally-oriented process pumps are moving fluids in modern industry. Only the smallest pumps use grease lubrication; the larger ones use oil lubrication. Regardless of bearing type and lubrication options, reliability professionals pay attention to allowable bearing temperatures. With that in mind, Table 14-4

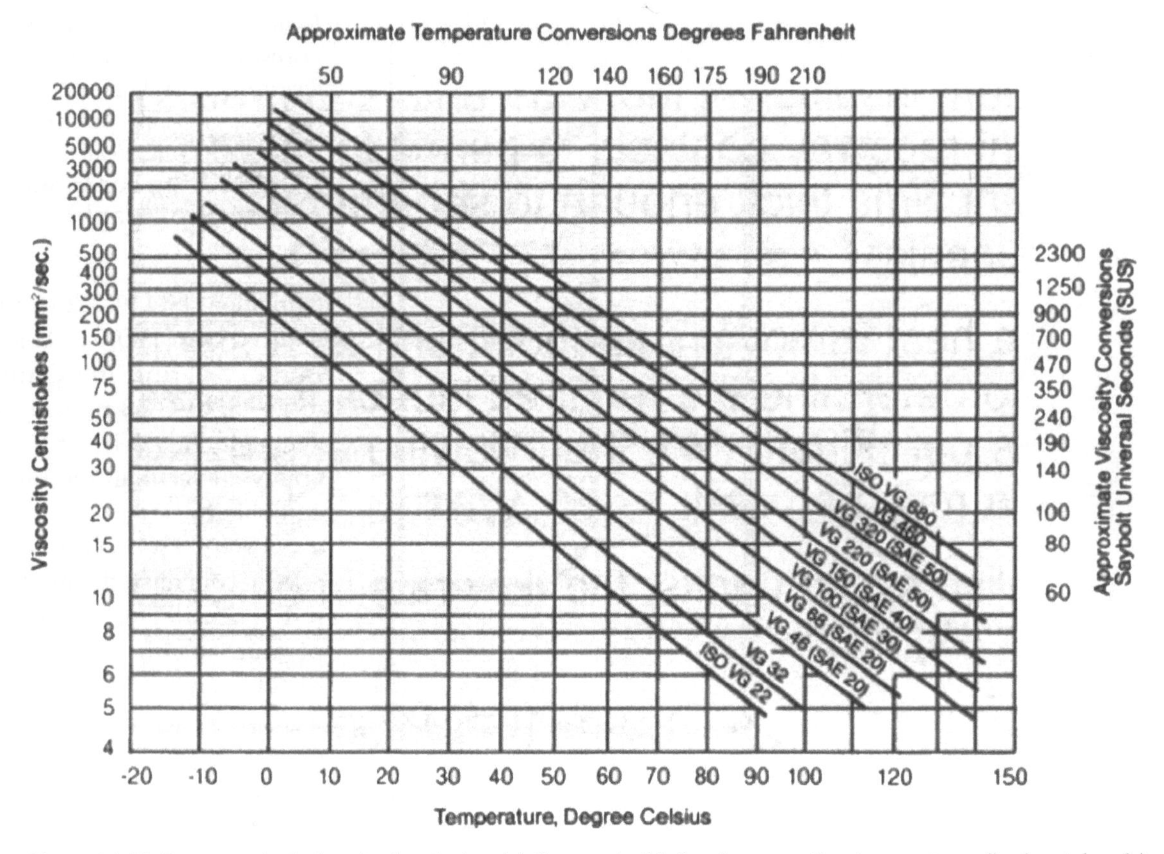

Figure 14-30. For a required viscosity (vertical scale) the permissible bearing operating temperatures (horizontal scale) increase as thicker oils are chosen (diagonal lines)

is of interest. The table illustrates the extended temperature ranges available with modern synthetic lubricants. Properly formulated synthetic lubricants allow continuous operation at 400 degrees F (~205 degrees C). Although no rolling element bearings in process pumps are ever exposed to this temperature, we would be correct in pointing to the information conveyed by Table 14-3. It supports the findings that premium synthetic formulations can be used with every conceivable bearing style or configuration found in process pumps.

Table 14-4. Temperature Ranges of Modern Synthetics

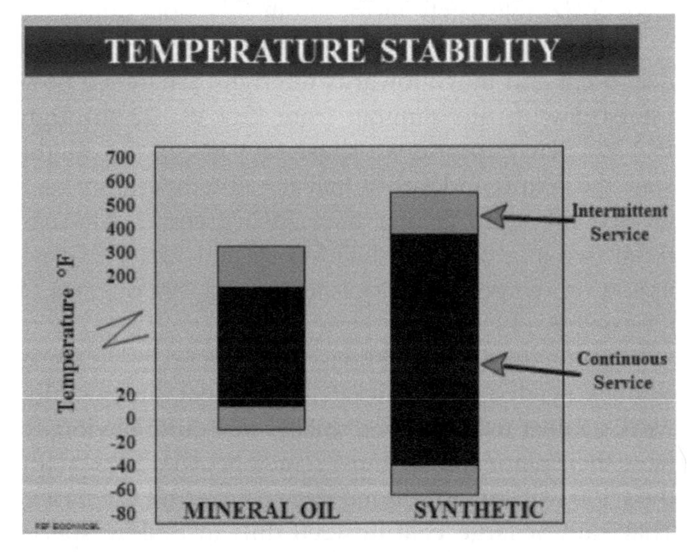

Bearing Styles and Configurations

Process pumps incorporate a radial bearing and a thrust bearing. A radial bearing is usually located near the mechanical seal whereas the thrust bearing is fitted close to the drive end. Fluid pressures acting on impellers cause a net axial force on the shaft system; we call it axial thrust. In operation, thrust acts in one direction, which would make the two "tandem-mounted" bearings in Figure 14-31(a) suitable for thrust acting from right to left. If they are precision-ground, the bearings will share the load and each will carry 50% of the total. If they're not precision ground, there will be issues (Ref. 9).

Expect the thrust action to briefly shift from the normally active to the opposite, or inactive, direction when process pumps are started and accelerating to operating speed. This concern favors using two angular contact thrust bearings mounted in back-to-back fashion, as depicted in Figure 14-31(b). The "back" of an angular contact bearing is the wider outer ring land; the narrower outer ring land is the "face." For the sake of completeness, Figure 14-31(c) is a "face-to-face" mounted set. If, in Figure 14-31(c), the inner ring became hotter than the outer ring, the resulting thermal growth would load up the bearing. This is but one of a number of reasons why face-to-face bearings are rarely used in pumps.

At all times care must be taken not to allow interference at the bearing and shoulder radii ra and rb. Also, precision-ground bearings are, understandably, more expensive than bearings with more liberal manufacturing tolerances. However, as we think through how the thrust bearing set in Figure 14-31(a) works, we realize that with one bearing heavily loaded, the other one might be unloaded. When that happens, the unloaded bearing may skid and wipe the lubricant off the raceways. Metal-to-metal contact on the skidding bearing creates much heat and accelerated bearing failure.

API-610 asks for the contact angles in each bearing making up a set to be equal, which is why we showed them equal in Figure `14-31(b). The API recommendation is influenced by the desire for standardization and by the economics of initial cost and the desire to simplify training of technicians. Economics aside, two angular contact thrust bearings with equal load carrying capacities are not necessarily best for pumps which briefly experience thrust reversal at startup. An unloaded bearing may skid (unintended) while the rolling elements in the loaded bearing are rolling, just as intended. Skidding bearing elements tend to wipe off the oil film, in which case there will be metal-to-metal contact. This contact can initiate failures and often manifests itself as high bearing temperature.

Special sets of back-to-back mounted thrust bearings may be a better choice in certain applications. These sets would use dissimilar angles; as an example, instead of the customary 40/40 degrees, a 40/15 or 29/15 degree design might be used to avoid skidding (Refs. 9 and 10). Even an occasional set of triple-row bearings (Fig. 14-32) can be found in multistage pumps. Like Figure 14-31(b) and related sets, they must be carefully matched and precision-ground at all contacting faces and surfaces.

Finally, there are thrust bearing sets with matched/mated split inner ring (separable ring) and angular contact bearing geometry (Figure 14-33). Because they accommodate the maximum number of rolling elements split inner ring bearings can impart favorable load capabilities but are more expensive to produce and acquire.

Bearing Temperature Limits

Among the many references linking bearing health and life expectancy, both in the plant and in the field, Figure 14-34 stands out because of its general applicability. The illustration pertains to the thermal range of a typical rolling element bearing. Note that

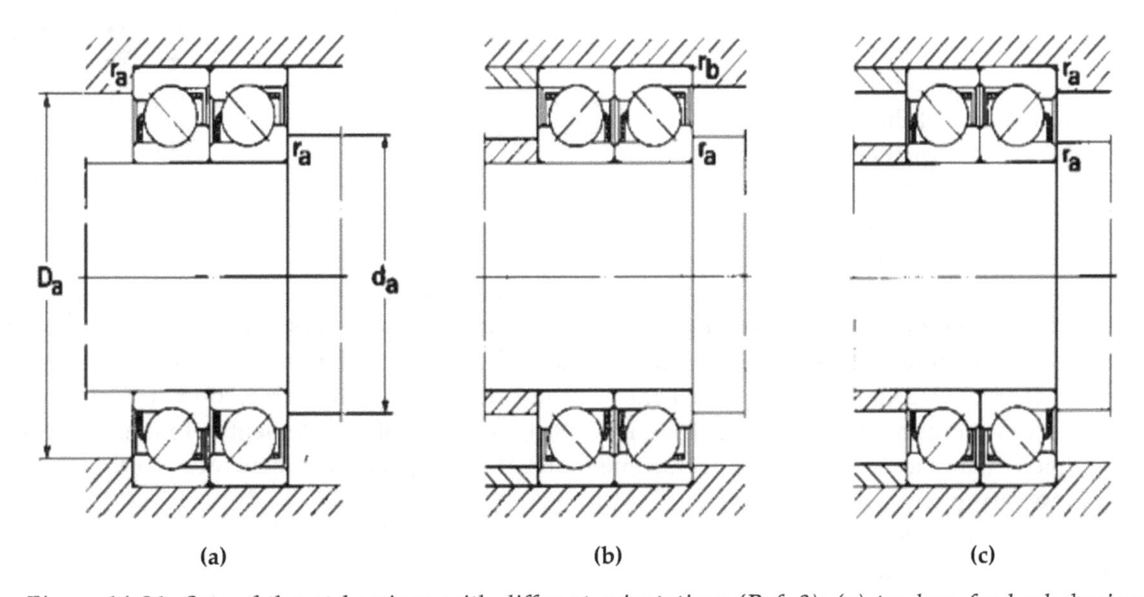

Figure 14-31. Sets of thrust bearings with different orientations (Ref. 9): (a) tandem, for load sharing of a pump shaft thrusting from right-to-left; (b) back-to-back, the customary API-610 recommended orientation with shafts possibly exerting axial load in each direction; (c) face-to-face, rarely desirable in centrifugal process pumps.

bearing metal temperature is often higher (10 to 25 degrees Celsius) than the oil temperature in the bearing within an oil circulation system. In non-circulating systems, bearing metal temperature is more likely 5 to 10 degrees Celsius higher than the oil temperature.

The green zone in Figure 14-34 represents the optimum zone for bearing and lubrication temperature; operating in the yellow zone will reduce lubricant and bearing life. If bearing temperatures are found in the red zone, expect both the bearing and a mineral oil lubricant to be severely compromised. It is of particular interest to note that pump bearing housing temperatures around 160 degrees F are quite normal. This temperature is near the center of the green zone. Yet, bearing housing temperatures of 190 degrees F are allowable and will likely occur in process pumps if an unnecessarily high oil viscosity is selected. Bearing manufacturer MRC states that its brand of rolling element bearings will run in continuous duty at a temperature of 121 degrees C (Ref. 10). Therefore, operating at 190 degrees C, while perhaps wasteful if the oil is unnecessarily viscous, is permissible. Be sure to add a personnel protection guard, though.

Of course, operating with a lower viscosity synthetic oil would be no problem. That said, operation with temperatures in the yellow zone of Figure 14-34 can usually be avoided by switching to a lighter oil, preferably a premium grade synthetic lubricant. As was brought out in Part 1 of this 2-part article, premium grade synthetics tend to remove more heat than equivalent grades of mineral oil. Synthetics run cooler but they, too, will become hot if unnecessarily high viscosities are chosen.

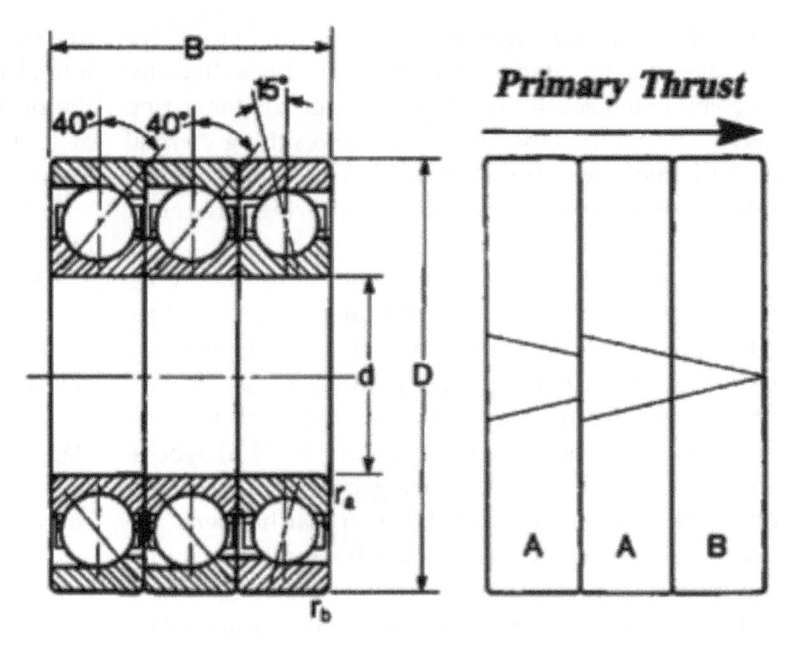

Figure 14-32. Triple-row thrust bearings require special attention

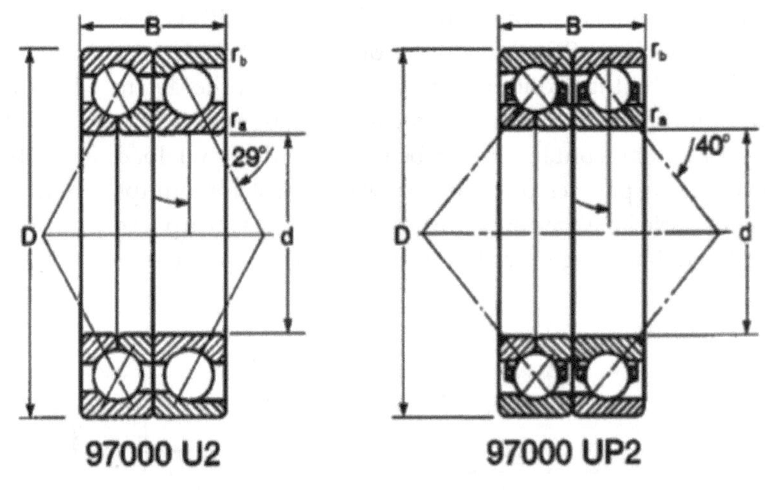

Figure 14-33. Thrust bearing sets with dual inner rings (Ref. 9)

Rolling element bearings will start degrading once a heat-treat-annealing temperature, probably slightly over 121 degrees C (250) degrees F, is exceeded. However, no process pump bearing housing designed or built by reputable manufacturers has ever even come close to this temperature while properly lubricated with the as-purposed quantity and quality of lubricating oil. In a recent case, copious amounts of an excessively viscous lubricant applied as an oil mist leveled off and stabilized at 190 degrees F.

An interesting lesson is again centered on the misunderstanding that the higher the lubricant viscosity, the better the bearing protection. That is far from correct, because rolling elements ploughing through thick oils at high speeds will generate considerable frictional

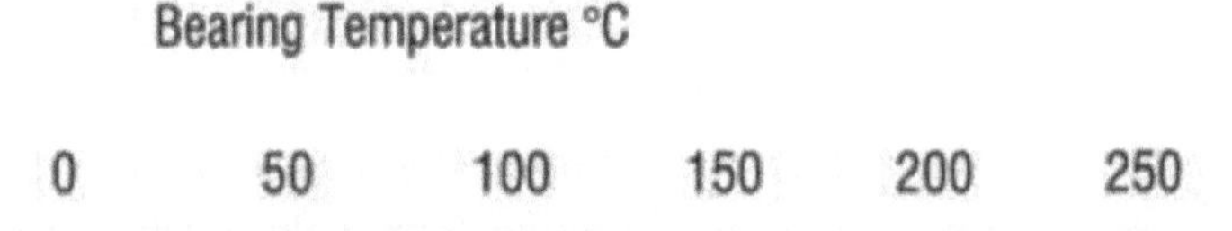

Figure 14-34. Heat Ranges of Bearings (Ref. 9)

heat. Using the correct viscosity lube is always important; the recommendations of major pump manufacturers and/or manufacturers of rolling element bearings are backed by decades of experience and we should not disregard them.

Optimized Lubricant Application

Lube application methods include grease, oil flooding, and oil rings or slinger discs used to bring lubricant from an oil sump into the bearing elements. There are presently thousands of plant-wide oil mist systems and hundreds of these have been in highly successful use since the early 1960s. As of 2015, an estimated 150,000 process pumps and 27,000 electric motors are lubricated by oil mist.

Only jet oil application (Figure 14-35) is ranked higher than oil mist. Jet oil lubrication applies the filtered lubricant precisely where needed and dispenses with constant level lubricators. The oil jet is directed at the space between the outside diameter of the bearing inner ring and the bore of the cage. Metallic cages are used because they have the highest allowable temperature capability. Means for scavenging the oil should be provided on both sides of the bearing; allowing oil to collect and overheat in those spaces must be avoided. Note how the oil system may be used to allow free axial floating of the bearing cartridge in the housing on a thin pressurized oil film (Ref. 3). A clearance of 0.0005 to 0015 inches is typically recommended between the cartridge and the housing. References 12 and 13 are of great value if additional information is of interest to the reader.

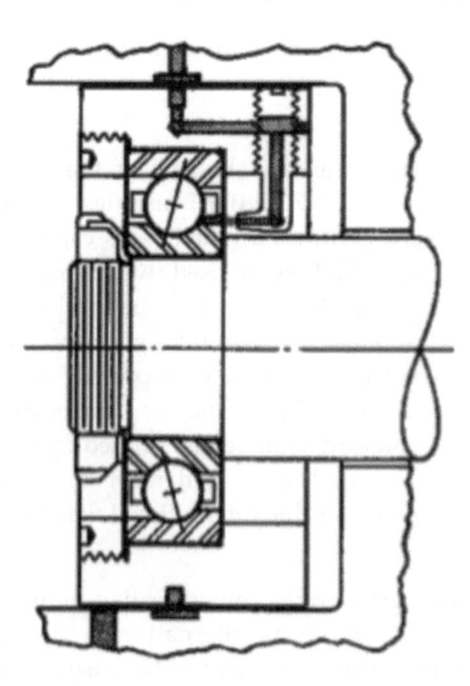

Figure 14-35. Jet oil lubrication is recommended for high speeds and heavy loads (Ref. 10)

Reliability-focused users are generally disinclined to view process pumps with old style, failure-prone oil rings as reliable as pumps with circulating lube systems. Over 100 technical publications agree that oil rings are sensitive to shaft horizontality, depth of ring immersion in the oil, material selection, and machining and roundness parameters.

Envision, therefore, how the lube application of Figure 14-35 will open a window of opportunity for reliability-focused users and/or innovative pump manufacturers. Think of a small oil pump, either internal to the process pump bearing housing or incorporated in a small assembly screwed into the bottom drain of your present process pump. This upgrade can provide filtration, metered flow, and proper pressure downstream of the oil sump and upstream of the spray nozzles. Motivated users have written this preferred lube application approach into their pump specifications and are actively pursuing this pump reliability enhancement.

LUBRICATION OPTIONS FOR MODERN API-610 COMPLIANT VERTICAL PUMPS

Although vertical pumps are ideally suited for multistage (high differential pressure) services, many vertical pumps are of single stage design. They incorporate an end suction "back pull out" type casing with the hydraulic end being mounted below the liquid level. The impeller is connected to the motor by means of an extended shaft; the shaft is housed and supported in a rigid tubular intermediate pipe.

As a rule, vertical pumps in important hydrocarbon processing services are designed to the requirements found in an API (American Petroleum Institute) Standard—API-610. User-purchasers view adherence to the standard's stipulations as a powerful risk reduction step. API-compliant equipment incorporates considerable experience-based content derived from reliability-focused users. While equipment cost is always a concern, the best companies aim for proper balance between initial monetary outlay and long-term reliability. As a rule, life-cycle cost studies show significant advantages for API-compliant pumps over their lower-priced non-API competitors. An API pump's generally higher reliability will translate to reduced catastrophic failure risk and such reductions, of course, are of interest to a multitude of industries.

API 610 HIGHLIGHTS

Special attention is given to bearings in the drive systems of vertical pumps. Regardless of feasible bearing options, these bearings must always be designed for prevailing radial and/or axial loads. These loads, of course, are transmitted from the hydraulic end of the pump and API-610 makes clear that bearings must meet a number of requirements:

- Rolling element bearings must be selected to give a basic rating life, in accordance with ISO 281, equivalent to at least 25,000 hours with continuous operation at pump rated conditions.
- Rolling element bearings shall be selected to give a basic rating life equivalent to at least 16,000 hours when carrying the maximum loads (radial or axial or both) imposed with internal pump clearances at twice the design values and when operating at any point between minimum continuous stable flow and rated flow.
- Concessions are made for vertical motors 750 kW (1,000 HP) and larger that are equipped with spherical or tapered roller bearings which may have less than 16,000 hour life at worst conditions to avoid skidding in normal operation. In such cases, the vendor must state the shorter design life in the proposal.
- For vertical motors and right-angle gears, the thrust bearing must be in the non-drive end and shall limit axial float to 125 μm (0.005 in).
- Thrust bearings shall be designed to carry the maximum thrust which the pump can develop while starting, stopping or operating at any flow rate.
- Hydrodynamic thrust bearings must be selected at no more than 50% of the bearing manufacturer's rating at twice the pump internal clearances specified elsewhere in the pump standard.

More detailed guidelines on bearings are found in the same standard, API-610. As general examples, a number of additional bearing-related items are worthy of note:

- Each shaft shall be supported by two radial bearings and one double-acting axial (thrust) bearing which might or might not be combined with one of the radial bearings. Bearings shall be one of the following arrangements: rolling-element radial and thrust; hydrodynamic radial and rolling-element thrust; hydrodynamic radial and thrust.
- Thrust bearings shall be sized for continuous operation under all specified conditions, including maximum differential pressure and comply with the following.
  - — All loads shall be determined at design internal clearances and also at twice design internal clearances
  - — Thrust forces for flexible metal-element couplings shall be calculated on the basis of the maximum allowable deflection permitted by the coupling manufacturer
  - — A sleeve-bearing motor (without a thrust bearing) is directly connected to the pump shaft with a coupling, the coupling-transmitted thrust shall be assumed to be the maximum motor thrust.
- Single-row, deep-groove ball bearings shall have radial internal clearance in accordance with ISO 5753, i.e., larger than "Normal" internal clearance.
- Single- or double-row bearings shall not have filling slots.
- In addition to thrust from the rotor and any internal gear reactions due to the most extreme allowable conditions, the axial force transmitted through flexible couplings shall be considered a part of the duty of any thrust bearing.
- Thrust bearings shall provide full-load capabilities if the pump's normal direction of rotation is reversed.
- Ball thrust bearings shall be of the paired, single-row, 40° (0.7 radian) angular contact type (7000-series) with machined brass cages. Non-metallic cages shall not be used (we disagree—see below). Pressed steel cages may be used if approved by the purchaser. Unless otherwise specified, bearings shall be mounted in a paired arrangement installed back-to-back. The need for bearing clearance or preload shall be determined by the vendor to suit the application and meet the bearing life requirements of this International Standard.

Experienced professionals realize, however, that in certain applications an alternative bearing arrangement will prove superior. This is because massive machined bronze cages tend to promote smearing. Smearing (or skidding) is often noted where bearings operate continuously with minimal axial loads. If loads exceed the capability of paired, angular-contact bearings, alternative rolling-element arrangements may be proposed.

Pump specialists note that sub-clauses apply to all rolling-element bearings, including both ball and roller types. However, for certain roller bearings (such as cylindrical roller types with separable races) bearing-housing diametric clearance might not be appropriate.

- — Rolling-element bearings shall be located, retained and mounted in accordance with the following:
- — Bearings shall be retained on the shaft with an interference fit (usually in the vicinity of 0.0003 to 0.0007 inches) and fitted into the housing with a diametric clearance, both in accordance with ANSI/ABMA 7 (usually ranging from 0.0005 to 0.0012 inches).
- — Bearings shall be mounted directly on the shaft. Bearing carriers are acceptable only with purchaser approval.
- — Bearings shall be located on the shaft using shoulders, collars or other positive locating devices. Snap rings and spring-type washers are not acceptable.
- — The device used to lock thrust bearings to shafts shall be restricted to a nut with a tongue-type lock washer.

Vertically Suspended Pump-Driver Combinations

Thrust bearings that are integral with the driver are addressed in the same API Standard, API-610. The manufacturer must pay attention to the need of adjusting pump rotors in the axial direction. To allow axial rotor adjustment and oil lubrication, the thrust bearing are mounted with an interference fit on a slide-fit, key-driven sleeve.

Designs which tolerate a certain amount of solids typically incorporate fully recessed, vortex flow impellers with rear mounting shroud incorporating integrally cast back balancing vanes. This design is suitable for handling free-flowing slurries and sludge. In fact, the impeller design is capable of handling solids up to the diameter of the discharge port. In a leading design offered by a Swiss pump manufacturer, a suction strainer fitted to protect the spark arrestor used in explosion-proof designs limits the practical solid size.

Special design combinations. Zone Zero (Lowest possible explosion risk) pumps are required in continuously flammable atmospheres. The casing of one such pump is unique in that it employs an internal helical contour which contributes to improved hydraulic efficiency without impeding free passage. Superior products often facilitate maintenance by using step-machined mating face designs, also known as "rabbeted fits." Thus, when the pump is being re-assembled, it is totally self-aligned.

Recessed impellers are available for many kinds of slurry pumps. Impellers are usually located on the shaft by a parallel key and fixed into position with an impeller screw and washer. Many excellent vertical pumps incorporate a large sole plate (pit cover) allowing the unit to be mounted on top of a tank. The discharge pipe passes through the sole plate and is held in position by means of a weld neck flange; this flange is bolted to the sole plate. A loose discharge flange is often provided which facilitates and simplifies the matching of the pump discharge pipe to the customer's pipe work.

The safety regulations for Zone Zero pumps require provision of a "minimum flow" bypass. In some well-proven designs the bypass pipe is connected from the discharge pipe and directed through the sole plate back into the tank. Two minimum liquid level probes are often mounted into the sole plate; one is for monitoring the liquid level in the tank and the other for monitoring the liquid level in the column pipe. Each represents an explosion-proof enclosure.

On Zero-Zero pumps and as an additional safety feature the discharge pipe from the cover plate is fitted with a spark arrestor. Ball valves are fitted on either side of the arrestor. ISO-compliant pump are usually provided with a single mechanical shaft seal (conforming to DIN 24960 dimensions) behind the impeller. The seal is mounted on a replaceable shaft sleeve or in a readily replaceable cartridge. A bottom journal bearing is fitted; it, too is mounted on a separate replaceable shaft sleeve or in a unitized assembly called a cartridge.

When the pump length dictates, intermediate bearings are employed and located between the flange joints of the intermediate pipes. As with the bottom journal bearing, intermediate bearings are mounted on separate replaceable shaft sleeves. The intermediate column pipe is filled with oil; this liquid column provides lubrication to the journal bearings and also encases the drive shaft of superior pump designs.

More Vertical Pump Design Characteristics

When the pump length dictates, more than one drive shaft may have to be threaded into each other to achieve the needed length between shaft coupling and impeller. Also, two or more sections are then connected by intermediate flanges. To prevent fluid traveling up the shaft, many vertical pump models incorporate a simple disc which acts as a "liquid thrower." At the drive shaft end, the shaft is located by two angular contact ball bearings (Figure 14-36).

The motor is mounted above the sole plate via a motor stand. This stand also has locating registers to guarantee correct alignment of both motor and pump shaft ends. Access to the spark-proof flexible coupling is from opposite sides. These access points are provided with guards so that when the machine is in operation, no rotating part is exposed. Motors can be either a metric frame or NEMA frame, flange mounted.

LUBRICATION OPTIONS FOR "ZONE ZERO" VERTICAL PUMPS

Many vertical pump designs are of single stage design. They incorporate an end suction "back pull out" type casing with the hydraulic end being mounted below the liquid level. The impeller is connected to the motor by means of an extended shaft; the shaft is housed and supported in a rigid tubular intermediate pipe.

Designs which tolerate a certain amount of solids typically incorporate fully recessed, vortex flow impellers with rear mounting shroud incorporating integrally cast back balancing vanes. This design is suitable for handling free-flowing slurries and sludge. In fact, the impeller design is capable of handling solids up to the diameter of the discharge port. However, a suction strainer fitted to protect the spark arrestor limits the practical solid size.

The unique feature of this pump is in the design of the case; it employs an internal helical contour which contributes to improved hydraulic efficiency without impeding free passage. For ease of maintenance, all machined mating faces incorporate locating registers. Thus, when the pump is re–assembled, it is totally self–aligned.

In most process pumps the impeller is located on the shaft by a parallel key and is locked into position with an impeller screw and washer. Vertical pump often incorporate a large sole plate (pit cover) enabling the unit to be mounted on top of a tank. The discharge pipe usually passes through the sole plate and is held in position by means of a weld neck flange; this flange is bolted to the sole plate. A loose discharge flange is provided which facilitates and simplifies the matching of the pump discharge pipe to the customers pipe work.

Some regulations require the pump to be fitted with a "minimum flow" bypass. The bypass pipe is connected from the discharge pipe and directed through the sole plate back into the tank. Two minimum liquid level probes are mounted into the sole plate, one is for monitoring the liquid level in the tank and the other for monitoring the liquid level in the column pipe. Each

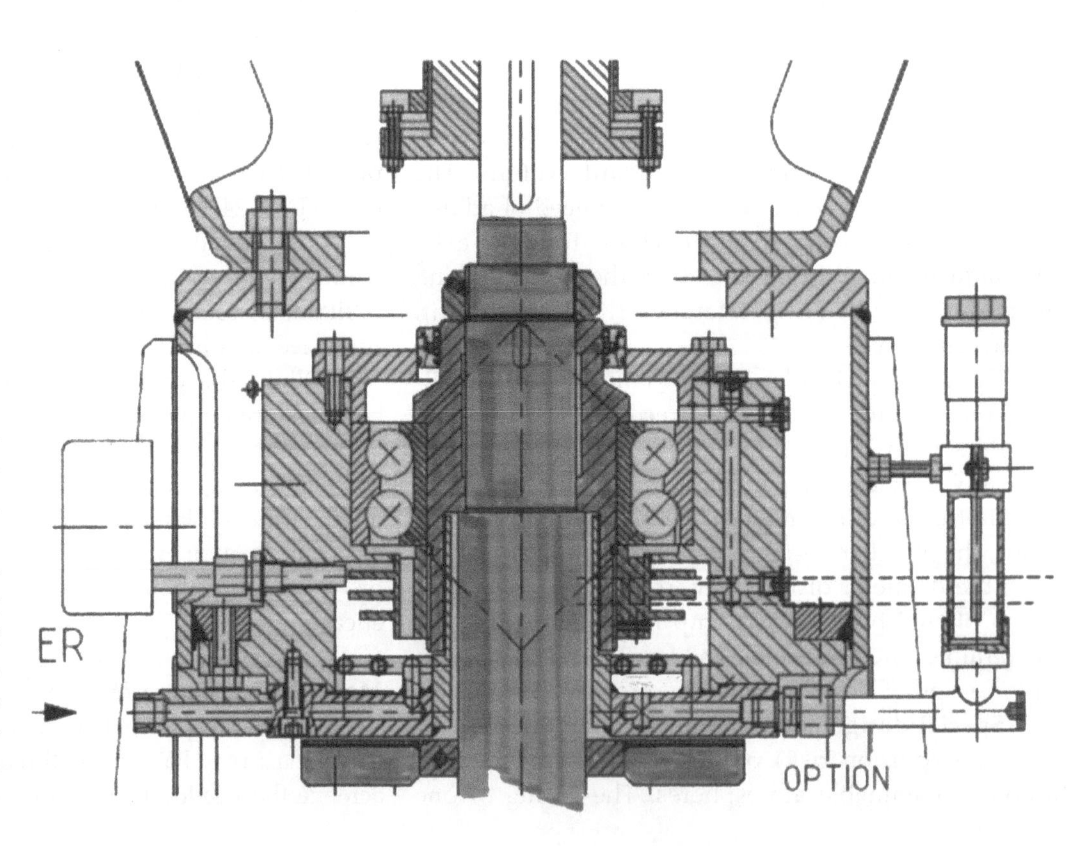

Figure 14-36. Oil-lubricated thrust bearing at drive end of a modern vertical pump with rotating components highlighted in green; cooling water entry is designated as "ER." Liquid oil is fed from the sump (bottom) to the top of the double-row thrust bearing. (Source: Egger Pumps, Cressier, NE, Switzerland)

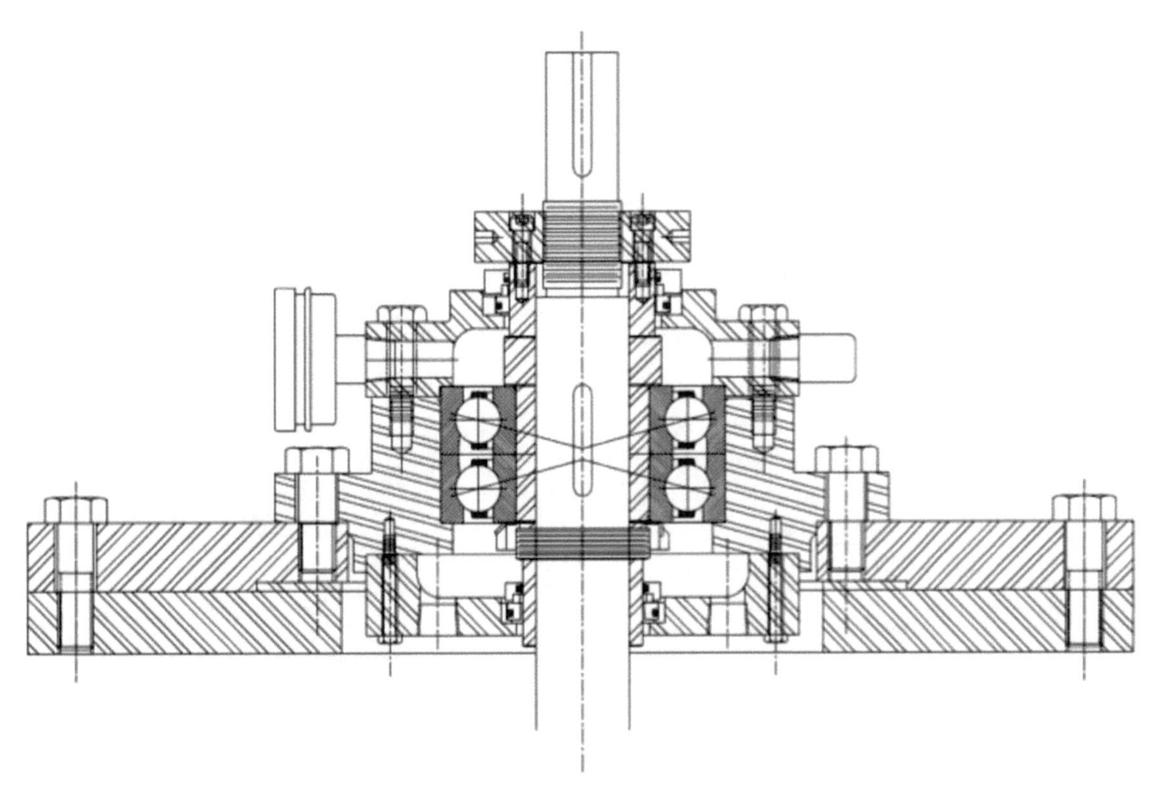

Figure 14-37. Oil mist-lubricated vertical pump thrust bearing The oil mist enters at the upper right. After passing through this back-to-back set of bearings, coalesced oil droplets or "excess oil mist" exit from one of the two drain ports (Source: Afton Pumps, Houston, TX)

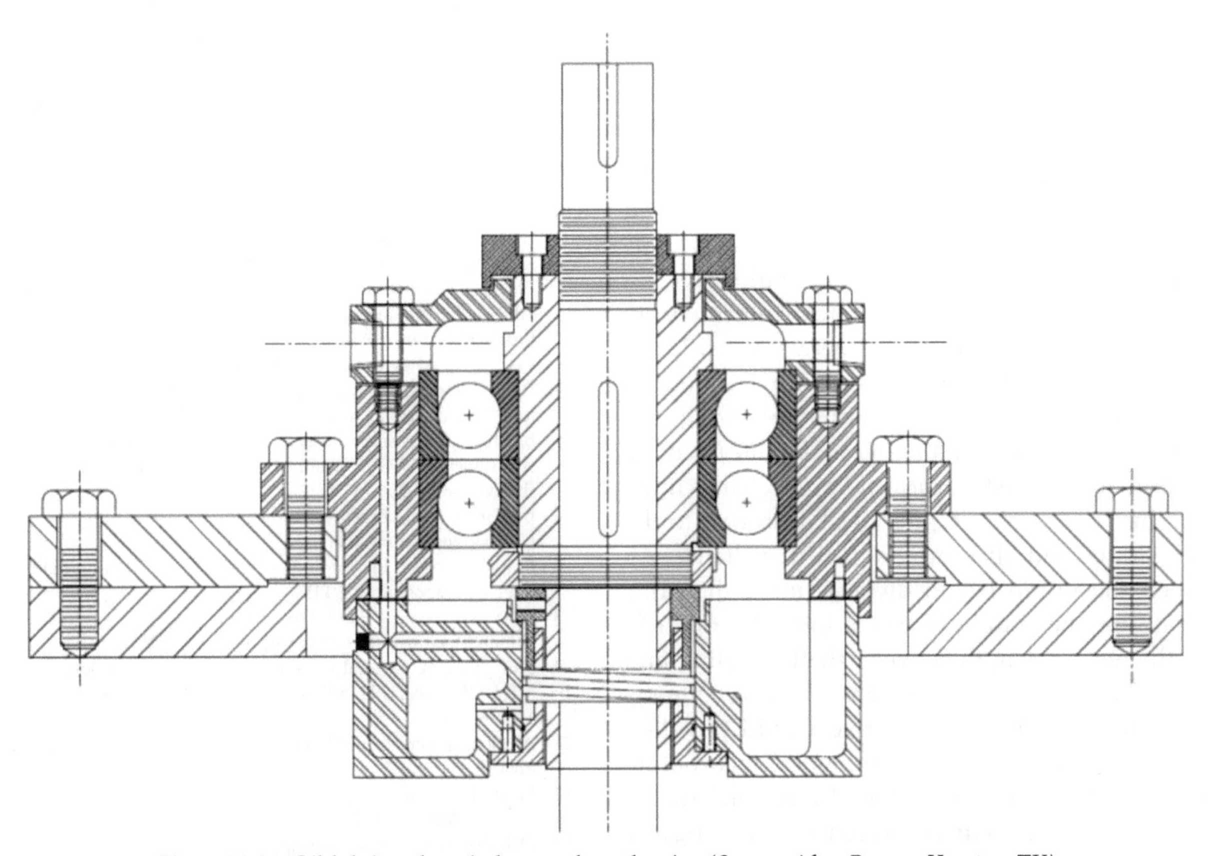

Figure 14-38. Oil-lubricated vertical pump thrust bearing (Source: Afton Pumps, Houston, TX)

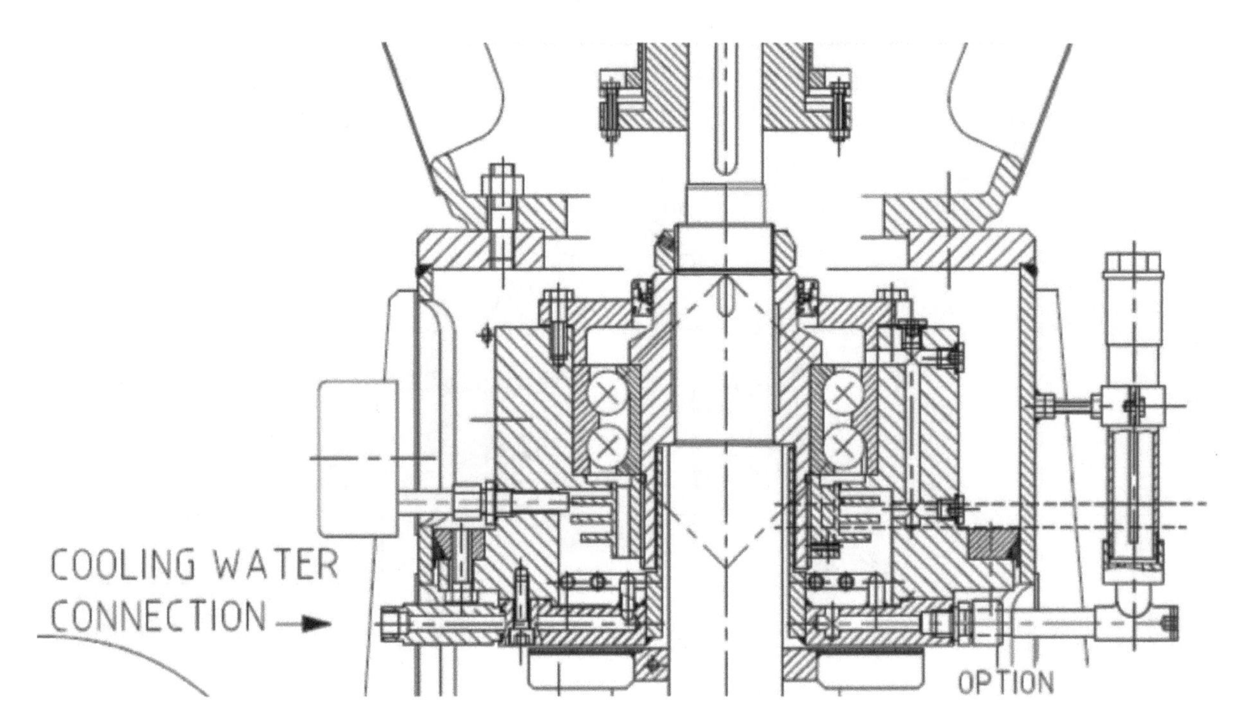

Figure 14-39. Oil-lubricated thrust bearing at drive end of a modern vertical pump. (Source: Egger Pumps, Cressier, NE, Switzerland)

represents an explosion-proof enclosure.

To eliminate all conceivable explosion risks the discharge pipe from the cover plate is fitted with a spark Arrestor, ball valves are fitted either side of the Arrestor. Modern zero-risk vertical pumps are fitted with a single mechanical shaft seal (conforming to DIN 24960 dimensions) behind the impeller; the seal would be a cartridge-type, although some are mounted on a replaceable shaft sleeve. Designs from experienced manufacturers generally include a bottom journal bearing mounted on a separate replaceable shaft sleeve.

When the pump length dictates, intermediate bearings are employed and located between the flange joints of the intermediate pipes. As with the bottom journal bearing, intermediate bearings are mounted on separate replaceable shaft sleeves. The intermediate column pipe in zero-risk vertical pumps is likely filled with oil; this liquid column provides lubrication to the journal bearings and also encases the drive shaft.

When the pump length dictates, more than one drive shaft may have to be threaded into each other to achieve the needed length between shaft coupling and impeller. Also, two or more sections are then connected by intermediate flanges. To prevent fluid traveling up the shaft, many vertical pump models incorporate a simple disc which acts as a "liquid thrower." At the drive shaft end, the shaft is located by two angular contact ball bearings (Figure 14-39).

Acknowledgements

Oil mist systems illustrations and updated statistics were provided by Don Ehlert, Total Lubrication Management, a Colfax Fluid Handling Company, Houston, TX. Additional illustrations and narrative explaining bearing housing protector seals were contributed by Chris Rehmann, AESSEAL, Inc., Rockford, TN.

References

1. Eschmann, Hasbargen, Weigand, *Ball and Roller Bearings,* (1985) John Wiley & Sons, Hoboken, NJ
2. Bloch, H.P., and Budris, A.R., *Pump User's Handbook: Life Extension,* 4th Ed., (2014) Fairmont Press, Lilburn, GA, ISBN 0-88173-720-8
3. Bloch, H.P., *Pump Wisdom: Problem Solving for Operators and Specialists,* (2011) Wiley & Sons, Hoboken, NJ, ISBN 978-1-118-04123-9
4. Bloch, Heinz P.; "Dry Sump Oil Mist Lubrication for Electric Motors," *Hydrocarbon Processing,* March 1977
5. Bloch, Heinz P. and Abdus Shamim; *Oil Mist Lubrication: Practical Applications,* (1998) Fairmont Publishing Company, Lilburn, GA, (ISBN 0-88173-256-7)
6. Bradshaw et al., Proceedings of the 17th TAMU International Pump User's Symposium, (2000)
7. Bradshaw et al., Proceedings of the 30th TAMU International Pump User's Symposium, (2014)
8. Wilcock, Donald F., and Booser, E.R., *Bearing Design and Application,* (1957), McGraw-Hill Company, New York, NY
9. SKF America, General Catalog, Kulpsville, PA (2000)
10. MRC "Engineer's Handbook," General Catalog 60, Copyright TRW, 1982
11. Aronen, Robert (see publications by Boulden Company, Conshohocken, PA and Ellange, Luxembourg)
12. Bloch, H.P.; *Improving Machinery Reliability,* Gulf Publishing Company, Houston, TX, 1993
13. SKF America "Pump Bearings,"—Publication, (2000), Kulpsville, PA"

Chapter 15

Lubricating Paper Machine and Forestry Equipment*

The of the most critical aspects in maintaining the reliability of today's modern paper machine, Figures 15-1 and 15-2, is selecting a high-quality oil to lubricate key machine elements. Such a lubricant yields high productivity by lengthening the life of machine elements, typically bearings, gears, and couplings, while coping with the specific challenges of the machine's environment. Even though it may be operating in extreme temperatures and contamination sources, the paper machine oil is expected to have extended life.

The effect of improper lubricant selection can be significant. In many cases, the loss of a critical bearing can cause unscheduled downtime (always when it's most inconvenient), resulting in a loss of production and hiking equipment operating costs.

Even if a failure occurs when sufficient personnel are available to make a rapid repair, the cost of the premature machine element failure can be high in terms of the cost of replacement parts and added inventory costs incurred to maintain an excessive number of spares.

The current generation of paper machine oils is specially formulated for the unique requirements of the mill environment and ideally maximize machine life and reliability while minimizing maintenance costs.

Brand names are not static; they may be used until a product consolidation or acquisition event takes place. Also, some manufacturers market certain lubricants in different regions of the world with brand designations that differ from each other. However, their standard product formulations have generally remained consistent; moreover, the lubricants are often known under the older, original brand name. This facilitated our decision to retain the 1995-vintage brand names in this text.

In the rare event of a named lubricant no longer being available under its previous, more widely used designation, the reader is encouraged to: (a) ask the supplier for the current name of a particular lubricant, and (b) compare the current "inspections"--the collective name for properties and ingredients--to those found widely accepted and described in this text. The significance of these deviations is highlighted in Chapter 3, Lubricant Testing. If the deviations are judged significant, consider requesting answers or explanations from the lubricant supplier.

Dryer Section Lube Critical

The most critical lubrication function is in the dryer section, where as many as several hundred roller bearings are present. These units are exposed to high temperatures due to their proximity to the superheated steam used in the drying process.

Compounding the effect of these high temperatures is the great likelihood that the oil system can become contaminated. Water is perhaps the most prevalent external contaminant, and the high probability of its presence in the oil system raises many concerns. These include accelerated oil breakdown, rust or corrosion problems, the poor lubrication properties of a water/oil mixture and potential chemical reaction with the additives used in formulating the paper machine oil. To complete this picture, the limited opportunity to change contaminated oil is notable, as the cost of the required downtime would be prohibitive in terms of production. Hence, the oil must maintain its lubricating properties over an extended period (as much as 30 years in some cases) without draining the system.

MODERN PAPER MACHINE (PM) OILS HAVE GROWN SOPHISTICATED

Thus, the modern PM oil has, as a matter of necessity, evolved into a sophisticated product meeting many diverse performance requirements. As a starting point, most commercial oils are blended from high-quality mineral oil basestocks with viscosities in the range of 160-320 cSt/40°C (760-1,760 SUS/100°F). Selected performance

*Source: Exxon Company, U.S.A., Houston, Texas.

additives further enhance the oils in other non-viscosity related properties. A premium PM oil is formulated with most, if not all, of the following properties:

- Appropriate viscosity
- Oxidation stability
- Rust protection
- Good water shedding traits
- Foam resistance
- Filterability through fine filters
- Dependable technical service
- Detergency
- Anti-wear protection
- Reliable local supply

A closer look at these properties is necessary to appreciate their value in a reliable PM operation.

Typically, most mills employ ISO 150 or 220 grade oils, although as machine oil temperatures have increased, there has been a trend toward higher viscosity oils (e.g., ISO 320). In older installations, operators

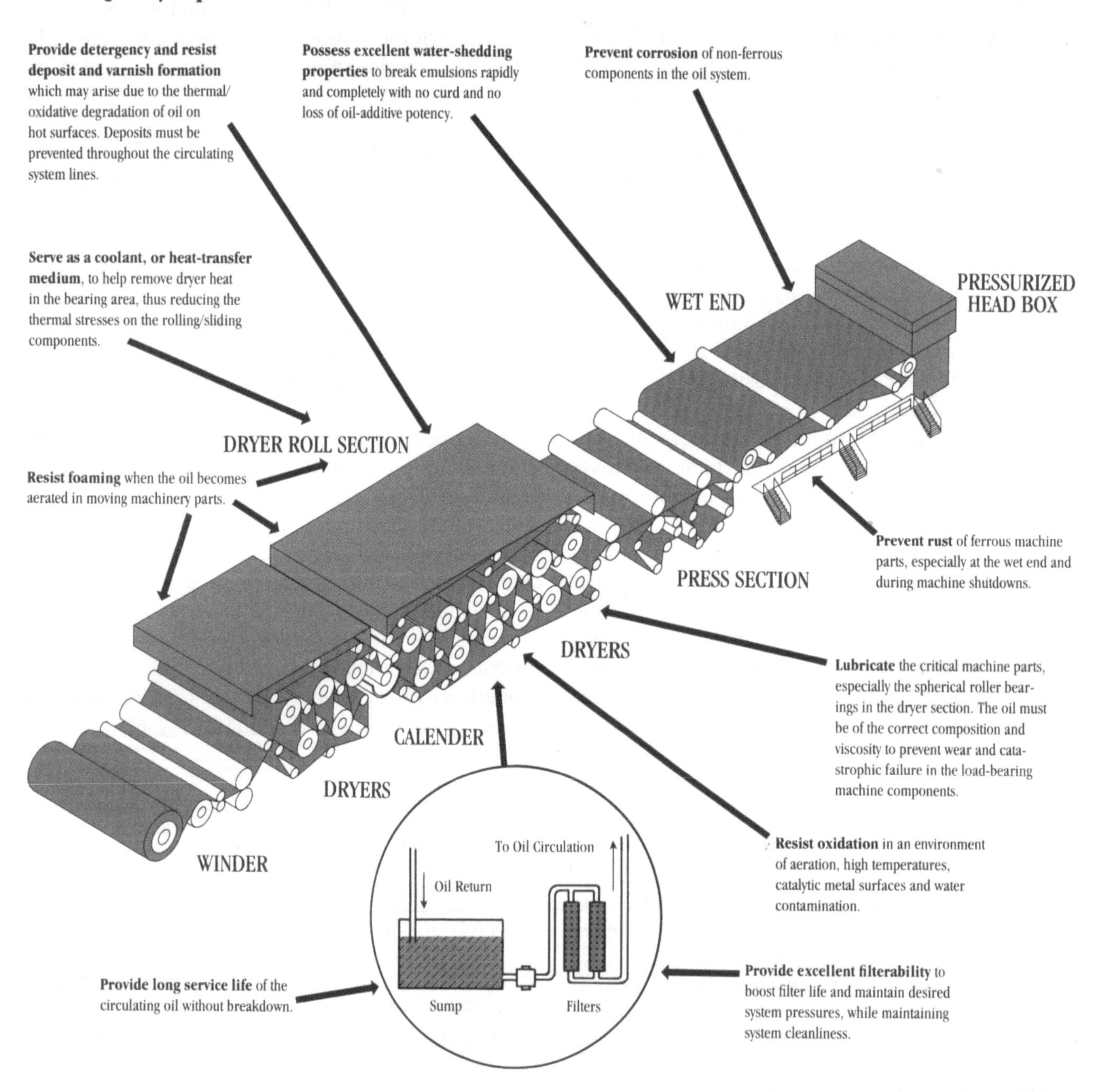

Figure 15-1. Typical paper machine.

Figure 15-2. Paper machine.

should consider the size of the bearing oil exit piping before increasing oil viscosity.

Appropriate Viscosity

The selection of PM oil is crucial because its major function is to obtain the rated life of the bearings or gears. This is accomplished by maintaining an adequate oil film to prevent metal-to-metal contact in PM elements that otherwise would result in wear and metal fatigue.

The viscosity grade of oil should be selected on the basis of such factors as machine speed, oil temperature and bearing size. Bearing and oil suppliers often can supply details about specific methods that have been developed to estimate optimum oil viscosity based on the mill's specific conditions. Higher oil viscosity can provide longer bearing life, but that must be weighed against other considerations such as energy consumption, ease of handling and existing system constraints.

This is because higher viscosity oils present greater resistance to flow, which can be compounded by system piping that is partially obstructed by deposits that have built up over the years. Nonetheless, we find that mills often unknowingly use oils with a viscosity that is too low. This can occur when upgrading of machinery results in changed machine operating conditions whose corresponding impact on system lubrication requirements has not been considered. The optimization of oil viscosity can many times provide a significant reduction in future maintenance cost by lengthening bearing life.

Oxidation Stability

The need for long oil life requires that PM oil formulations not only be based on a high quality base oil, but also be inhibited with oxidation inhibitors to prevent rapid oil breakdown.

Oil oxidation results in the formation of system deposits, varnishes and corrosive by-products that can markedly diminish system performance. Because in-machine testing is both risky and impractical, premium PM oils are generally formulated using long-term laboratory oxidation tests designed to simulate the oxidative conditions in a PM. The most common of these is the ASTM D-943 Oxidation Test which exposes the oil to high temperature, oxygen, catalytic metals and water.

In this severe test, PM oil life typically ranges between 1,000 and 1,600 hours. While there is no direct correlation between this test and PM oil performance, oils with lives in this range have generally provided good field service over the years. As the trend to hotter running dryer action continues, the need for oils with superior oxidation resistance will continue to grow.

Rust Protection

This is an essential property of any circulating oil, but it is especially important in the PM application where local humidity is high and where water is likely to enter the lubrication system. Ideally, rust protection is needed both when the oil is in contact with the metal surfaces and during short periods of PM shutdown.

Suitable rust inhibition capabilities are almost universally provided by most leading PM oils, while the longer-term "humidity environment" often protection often requires the careful selection of an advanced rust inhibitor technology.

Demulsibility

The ability to shed water is a key property of PM oils and is also related to the very wet conditions in which these oils are forced to perform. If significant amounts of water are introduced into the lubrication system, it is essential that the oil release the moisture quickly. This is especially important in wet-end oil circulation systems, where copious quantities of water are present. A good PM oil will provide this water-shedding capability and give a clean break between water and oil in the system sump, where the water can then be drained from the system.

Detergency

Protection against harmful deposit formation cannot solely be provided by oxidation inhibitors in a PM oil. This is because high local temperatures can exist in the steam-heated dryer roll bearings and result in thermal decomposition of the oil.

To combat this decomposition process, quality PM oils are formulated with detergent additives that act at the metallic bearing surface to prevent deposit build-up or coking.

While non-detergent oils have been used in the past, their use may cause long-term bearing life to be sacrificed as a result of carbon build-up in bearing housings. Nowadays, it is often this detergency property which distinguishes a PM oil from similar industrial circulating oil types, such as turbine or hydraulic oils.

Anti-wear Protection

While the lubrication requirements of rolling and sliding contact bearings are primarily dependent on maintaining an adequate oil viscosity, PM oils are frequently used in other applications in and around the PM.

In light- to moderate-duty gear applications, it is sometimes desirable to provide some supplemental anti-wear protection, and PM oils thus are commonly formulated with this need in mind. Selection of the type of anti-wear additive is extremely important because the additive must tolerate the presence of water, i.e., minimize any possible reaction with water.

Foam Resistance

Often, air is carried in the oil system, and if the oil is not appropriately inhibited, foam and possibly an oil overflow situation can develop. For this reason, quality PM oils generally incorporate additives that can minimize foam formation and facilitate the release of captured air.

Filterability

Many mills, in an effort to reduce the level of abrasive contaminants circulating with the oil into bearing housings, routinely filter these oils through full-flow filters. The types and cleaning efficiencies of these filters vary widely among mills.

The smallest pore-size filtration systems, however, can range down to six microns. While most oils will filter without significant additive depletion at this level of filtration, in some instances the life of the filter elements has varied dramatically, depending on oil composition.

Careful selection of PM oil additives can alleviate this concern, and excellent filtration characteristics are now achievable. Oil filtration often depends on oil temperature, the presence of water or other contaminants, and flow rates. When considering oil filtration systems, it is best to consult both the oil supplier and filter manufacturer regarding the most cost-effective route to required system cleanliness.

A second aspect of filterability is the ability of a cold oil to flow through these filter elements upon startup. In many mills, plugged filters are often encountered upon startup due to materials collected on filters from cold oil. This problem can also be addressed by appropriate oil formulation technology.

Technical Service

Even with the highest quality PM oil formulations, the harsh conditions that a PM oil must endure will often lead to unique site-dependent concerns. It is then essential that your oil supplier has experts that can assist in developing solutions. This calls for a dedicated supplier whose people are familiar with the operation and needs of the paper industry customer.

The PM oil supplier also should be willing to provide support even when there are no immediate problems. Your supplier should be able to provide some or all of the following, depending on your mill's specific needs: oil analysis programs, lubricant recommendations, surveys, training programs and a complete line of products for other mill lubrication needs.

Technical support capability of the oil supplier is of extreme importance to the paper industry. The oil is only as good as the organization that stands behind it.

Local Supply

The final factor in the PM oil quality equation must include the capability for effective local supply. Without rapid access to product supply in the event of an emergency, the best PM lubricant is worthless. It cannot protect bearings if it is not in or near your mill. A reputable oil supplier should be willing to work with your mill to establish the most appropriate supply network to meet your needs.

Clearly, dependable production in the modern paper mill requires more than excellent equipment. The machinery, operating under complex conditions of temperature and the threat of contamination, needs oils that are often diverse and complex.

The papermaker is therefore well advised to seek out an oil supplier who not only formulates a sophisticated, premium PM oil with all the above traits in mind, but also places skilled technical staff at the service of the papermaker. The effort pays off in longer machine life and more consistent and economical operations—which ultimately translates into a more profitable end product.

TERESSTIC N PAPER MACHINE OILS

TERESSTIC N 150 and 220 have proven themselves many times over in the paper industry. Advanced oxidation inhibitors help prevent premature bearing failure and increase the life of the oils. Outstanding detergency helps keep deposits from building up inside machinery, and excellent filterability makes TERESSTIC N 150 and 220 oils ideal for today's circulating systems and smaller filters.

Excellent water separability (demulsibility) at typical sump operating temperatures and dependable corrosion protection round out the balanced performance of TERESSTIC N paper machine oils.

TERESSTIC N 320 is *specifically formulated* to maximize the reliability of dryer roll bearings and other equipment where oil temperatures can exceed 93°C(199°F).

TERESSTIC N 460, an ISO 460 viscosity grade lube, equals or surpasses synthetic-base oils in the critical areas of oxidation, rust and corrosion inhibition, demulsibility and detergency—at a fraction of the cost.

See Table 15-1 for typical inspections and test results pertaining to these lubricants. Due to their versatility, TERESSTIC N oils are often recommended for other applications around the mill—gear boxes and pump bearings, among others. This facilitates lube oil consolidation.

TERESSTIC N EP PAPER MACHINE OIL

Although most paper machine wet and dry ends can be satisfactorily lubricated with a single premium paper machine oil, the press section may require an EP gear oil designed to protect heavily loaded components, such as the extended nip presses, crown control roll hydraulic systems, and integral gear systems. The paper industry has expressed growing interest in the convenience of a single product that combines the detergency and high temperature performance of a paper machine oil with the EP performance of a premium

Table 15-1. TERESSTIC N Paper Machine Oil

TERESSTIC N Typical Inspections

Grade	150	220	320	460
ISO viscosity grade	150	220	320	460
Color, ASTM D 1500	3.0	3.0	3.5	3.5
Viscosity:				
cSt @ 40°C, ASTM D 445	147.0	230.0	320.0	427.0
cSt @ 100°C	14.5	19.0	24.3	29.2
SUS @ 100°F, ASTM D 2161	764.0	1163.0	1703.0	2510.0
SUS @ 210°F	77.4	96.8	121.0	146.0
Viscosity index	97	97	97	97
Gravity, ASTM D 1250				
Specific @ 15.6°C(60°F)	0.8905	0.8950	0.899	0.9024
Flash point, ASTM D 92, COC, °C(°F)	254(489)	264(507)	279(534)	288(550)
Pour point, ASTM D 97, °C(°F)	-9(16)	-9(16)	-9(16)	-9(16)
Four-ball wear test, ASTM D 2266 40kg, 600 rpm, 150°C(302°F), 1 hr, scar diam, mm	0.5	0.5	0.5	0.5
Water separability ASTM D 1401	pass	pass	pass	pass
Rust test, ASTM D 665 A/B				
Distilled water	pass	pass	pass	pass
Synthetic sea water	pass	pass	pass	pass
Copper strip corrosion, ASTM D 130, 3 hrs @ 100°C(212°F), rating	1	1	1	1
Detergency, panel coker test, 277°C(530°F)	trace deposit only			

gear oil. This would permit cost-effective consolidation of lubricant inventories and reduce the possibility of lubricant misapplication.

TERESSTIC N EP (Table 15-2) provides heavily loaded systems components with a ***double*** benefit—60-lb Timken EP performance ***plus*** detergent and thermal properties significantly superior to those of conventional gear oils. TERESSTIC N EP meets or exceeds the requirements of the major paper machine manufacturers in anti-wear and EP performance, oxidation and thermal stability, detergency, rust protection, demulsibility and filterability. It also meets AGMA and USS 224 Gear Oil requirements.

CYLESSTIC STEAM CYLINDER AND WORM GEAR OIL

CYLESSTIC is the trademark for a line of steam cylinder oils formulated to meet exacting lubrication requirements in the forest products and paper industries. Of course, they find numerous applications in other industry sectors as well. And, although officially classified as steam cylinder lubricants, the compounded grades also provide excellent protection against wear in worm gear drives and are recommended for engines operating on saturated or slightly superheated steam at either high or low pressures. CYLESSTIC lubricants are also used where cylinder wall condensation occurs. The non-compounded grade is the recommended oil for use with high-pressure superheated steam systems. Because of their high viscosity indexes, the CYLESSTIC oils are well adapted to wide variations in temperature.

Grades

CYLESSTIC steam cylinder oils are available in four viscosity grades. These grades conform to the International Standards Organization (ISO) viscosity classification system. Three grades—CYLESSTIC TK 460, TK 680, and TK 1000—are compounded with acidless tallow and a tackiness agent to provide lubrication under the wet conditions encountered with saturated steam. The fourth grade—CYLESSTIC 1500—is not compounded, but is formulated specifically for the dry, high-temperature operating conditions associated with

Table 15-2. TERESSTIC N EP typical inspections.

TERESSTIC N EP Typical Inspections **Grade**	**150**	**220**	**320**
ISO viscosity grade	150	220	320
Color, ASTM D 1500	3.5	5.0	5.0
Viscosity:			
cSt @ 40°C, ASTM D 445	147	230	320
cSt @ 100°C	14.5	19.0	24.3
SUS @ 100°F, ASTM D 2161	764	1163	1703
SUS @ 210°F	77.4	96.8	121
Viscosity index	97	97	97
Gravity, ASTM D 1250			
Specific @ 15.6°C(60°F)	0.8905	0.8950	0.8990
Flash point, ASTM D 92, COC, °C(°F)	254(490)	264(507)	279(535)
Pour point, ASTM D 97, °C(°F)	-9(15)	-9(15)	-9(15)
Timken OK load, lbs	60	60	60
4-Ball EP, ASTM D 2783			
Load wear index, kg	48	48	48
Weld point, kg	250	250	250
FZD D 5182, pass stage	12	12	12
Water separability, ASTM D 1401	pass	pass	pass
Rust test, ASTM D 665 A/B			
Distilled water	pass	pass	pass
Synthetic sea water	pass	pass	pass
Copper strip corrosion, ASTM D 130, 3 hrs @ 100°C(212°F), rating	lb	lb	lb
Detergency, panel coker test, 277°C(530°F)	———— trace deposit only ————		

super-heated steam. It also meets the requirements for a straight mineral SAE 250 gear lubricant.

Steam Cylinder Lubrication

All four grades are suitable for use where separation of the lubricant from condensate is desirable. CYLESSTIC TK 460 is recommended for low-pressure saturated steam systems. CYLESSTIC TK 680 and CYLESSTIC TK 1000 are recommended for high-pressure saturated steam systems. The tackiness agent incorporated in the compounded grades functions to reduce consumption, to provide better adhesion to the cylinder walls, and to provide better separation from exhaust steam.

Atomization

Unlike most moving parts, which are lubricated by the direct application of grease or oil, steam cylinders are generally lubricated by a mist of oil carried by the steam. Oil is injected into the steam by means of an atomizer inserted into the steam line ahead of the steam chest. As the steam flows past the open end of this atomizer at relatively high velocity, it picks up droplets of oil discharged from the atomizer tube. Under the proper conditions, the oil mist produced in this manner is diffused throughout the incoming steam. All moving parts in contact with the steam receive a share of lubricant.

To be effective, the oil mist must be diffused in minute particles. Oversize droplets settle out of the steam and may not reach the more distant areas to be lubricated. In other locations, they may accumulate in excessive quantities, leaving residues on the wearing surfaces. Thorough atomization is essential, therefore, to complete lubrication of the cylinders. Proper atomization is partly dependent upon characteristics of the oil, such as viscosity. An oil that is too heavy does not break up into droplets that are sufficiently small. On the other hand, an oil that is too light will not carry the required loads.

The CYLESSTIC oils, which have inherently good atomization characteristics and are available in four viscosity grades, Table 15-3, can be applied in the correct viscosity for complete atomization and for effective protection to the lubricated surfaces.

Table 15-3. CYLESSTIC steam cylinder and worm gear oils.

Typical Inspections

The values shown here are representative of current production. Some are controlled by manufacturing specifications, while others are not. All may vary within modest ranges.

CYLESSTIC Grade	TK 460	TK 680	TK 1000	1500
ISO viscosity grade	460	680	1000	1500
AGMA lubricant number	7 comp	8 comp	8A comp	—
Gravity, ° API	21.8	21.7	21.5	20.5
specific at 15.6°C (60°F)	0.9230	0.9236	0.9248	0.9309
Viscosity				
cSt at 40°C	429	627	925	1600
cSt at 100°C	30.1	37.8	46.7	64.2
SSU at 100°F	2289	3384	5017	8802
SSU at 210°F	147	184	227	312
Viscosity index	99	97	93	92
Pour point				
°C	-7	-7	-1	-1
°F	20	20	30	30
Flash point				
°C	271	279	288	313
°F	520	535	550	595
Fire point				
°C	316	321	338	352
°F	600	610	640	665
% compounding	5	5	4	—

Worm Gear Lubrication

In addition to meeting difficult steam engine lubricating requirements, CYLESSTIC, in the compounded grades, is an excellent lubricant for many worm gears. CYLESSTIC TK 460, CYLESSTIC TK 680, and CYLESSTIC TK 1000, respectively, meet the viscosity requirements of the American Gear Manufacturers Association (AGMA) specifications for (7) Compounded, (8) Compounded, and (8A) Compounded gear lubricants.

Worm gears, threaded shafting, and other such lubricant applications are characterized by a high degree of sliding motion under heavy pressure. The compounded CYLESSTIC grades have extra oiliness that provides good lubrication, which minimizes wear in machine elements of this type.

EXXON SAWGUIDE BIO SHP SYNTHETIC LUBRICANT

Here, interestingly, is a lubricant that fits into the categories biodegradable, synthetic, and forest product-oriented. Because of its rather unique application, we have elected to discuss it within the forest product chapter. Figures 15-3 and 15-4 shed light on this fact.

Figure 15-3. Exxon Sawguide Bio SHP keeps equipment running cleaner, longer.

Exxon Sawguide Bio SHP is a synthetic biodegradable lubricant specifically developed to provide superb, trouble-free lubricating performance in demanding sawmill operating environments. Formulated with a synthetic basestock and a proprietary additive package, Exxon Sawguide Bio SHP is readily miscible in water, has exceptional oxidation stability and anti-wear performance, and protects against rust in wet environments. The outstanding performance properties of Exxon Sawguide Bio SHP can improve operational efficiencies and extend equipment life compared to mineral-base lubricants.

Figure 15-4. Exxon Sawguide Bio SHP is specifically formulated for dependable lubrication in demanding sawmill operations.

Reduced Deposits and Wear

Conventional petroleum-base lubricants tend to mix with water and sawdust to produce a gummy residue that clogs the sawguide and overheats the equipment. This often necessitates time-consuming equipment shutdowns. In contrast, the excellent oxidation stability and high solvency of Exxon Sawguide Bio SHP synthetic lubricant help minimize sawguide clogging, thereby reducing unscheduled downtime.

Dependable Performance at Temperature Extremes

Exxon Sawguide Bio SHP provides reliable lubrication protection over a wide range of seasonal and

Table 15-4. Exxon Sawguide Bio SHP synthetic lubricant.

Typical Inspections
The values shown here are representative of current production. Some are controlled by manufacturing specifications, while others are not. All of them may very within modest ranges.

Kinematic viscosity @ 40°C, ASTM D 445	110
Timken OK load, ASTM D 2282, lbs	85
Rust, ASTM D 665	Pass
Emulsion, 15 min, 54°C, ASTM D 1401, ml water	0
Tack	Pass
Biodegradable, calculated	75
Compatibility with Marinus oil	Pass
Storage stability	Pass
Density @ 15°C, ASTM D 4052-95, g/ml	0.9239
Pour Point, °C	N/A
Appearance	Hazy

operational temperatures. It has good flow characteristics in cold weather, excellent oxidation and thermal stability at high operating temperatures, and exceptional storage stability, even at high ambient temperatures.

Reduced Oil Consumption

The synthetic basestock of Exxon Sawguide Bio SHP has a strong natural affinity for metal surfaces. The product's tenacious adherence to lubricated parts, combined with its relatively high viscosity, can significantly reduce lubricant consumption.

Biodegradability

In addition to superb lubricating performance, Exxon Sawguide Bio SHP also has a distinct environmental advantage. It is classified as "readily biodegradable," as defined by the OECD 301B CO_2 Evolution Test (Modified Sturm Test).

TERESSTIC SHP SYNTHETIC PAPER MACHINE OILS

To help meet increasingly more demanding operating conditions it is essential to use circulating oils capable of withstanding increasingly higher operating temperatures.

Exxon's TERESSTIC SHP circulating oil, Table 15-5, are specially formulated to meet the demands of modern calendar rolls and other dry-end paper machine equipment applications where bearing temperatures in excess of 100°C(210°F) are often experienced.

TERESSTIC SHP is a super-premium, synthetic-base ashless circulating oil. Compared with mineral-base oils, TERESSTIC SHP provides superior high-temperature performance and service life. The polyalphaolefin basestock also has an inherently high viscosity index and low pour point, which permit excellent retention of oil film thickness across a wider temperature range and better energy efficiency at cold startup.

TERESSTIC SHP provides excellent anti-wear performance and good extreme-pressure performance under moderate-load conditions. It is fortified to protect against rust, provides excellent demulsibility and resists foaming and air entrainment.

OTHER LUBRICANTS FOR THE PAPER AND FOREST PRODUCTS INDUSTRY

For hydraulic and gear oils used in this industry, see chapters dealing with these lubricants. These oils include SPARTAN Synthetic EP, NUTO H. Greases include RONEX, Polyrex and Lidok; refer to the chapter on grease (Chapter 9).

Table 15-5. Typical inspection and test data for Teresstic *SHP lubricants.*

Teresstic SHP* Typical Inspections Grade	150	220	320	460	680
ISO viscosity grade	150	220	320	460	680
Color, ASTM D 1500	<1	<1	<1	<1	1
Viscosity:					
cSt @ 40°C, ASTM D 445	159	211	333	466	680
cSt@ 100°C	20.0	24.7	34.7	45.0	59.9
SUS @ 100°F, ASTM D 2161	845	1165	1715	2477	3200
Viscosity index, ASTM D 2270	145	147	148	151	150
Gravity, specific, ASTM D 1250					
@ 15.6°C(60°F)	0.863	0.845	0.859	0.861	0.860
Flash point, ASTM D 92,					
°C	276	272	280	286	282
°F	529	522	536	547	540
Pour point, ASTM D 97,					
°C	-42	-42	-36	-30	-30
°F	-44	-44	-33	-22	-22
FZG gear test,					
minimum stages passed	12	12	12	12	12
Water separability,					
ASTM D 1401	pass	pass	pass	pass	pass
Rust test, ASTM D 665 A/B	pass	pass	pass	pass	pass
Copper strip corrosion,					
ASTM D 130,					
3 hrs @ 100°C(212°F), rating	1A	1A	1A	1A	1A

*The grades listed are the most commonly used; Teresstic SHP is also available in ISO viscosity grades 32 through 1500.

HOW HIGH PERFORMANCE OILS AND GREASES EXTEND THE APPLICATION RANGE FOR PAPER MACHINE SEALED BEARING APLICATIONS

The development of high performance perfluoropolyether (PFPE) lubricants dates back a few decades. These developments were both necessitated and accelerated by aerospace and aviation markets in which lubrication low and high temperature extemes is far more important than it would be in the average industrial environment. Even beyond aviation and aerospace, PFPEs have served admirably whenever higher initial cost was easily overcome by the far more important need to consistently meet, and even exceed, performance expectations.

To what extent the traditionally lower initial cost of mineral oil-based lubricants has influenced procurement decisions in process industries is of peripheral interest at best. However, solid cost justification for PFPEs has recently become available. Such cost justifications were derived from a large Canadian Paper Mill (Ref. 1), which struggled with grease-lubricated electric motor bearings. When the mill opted to dispense with re-lubrication of electric motor bearings by purchasing and converting to PFPE grease-filled (sealed, life-time lubricated) bearings, their electric motor bearing life improved drastically. The following cost justification study examines PFPE greases and highlights their applicability and use for many process lubrication services.

Composition of Standard Perfluoropolyether (PFPE) Lubricants

Standard PFE oils and PTFE (poytetrafluoroethylene, "Teflon®") thickeners contain only three elements: carbon, oxygen, and fluorine. The molecular structure provides thermal and chemical stability to lubricants produced in ISO viscosity grades ranging from 2 to 1000. One prominent manufacturer of high-performance chemicals engineered a PFPE molecule with its otherwise degradation-susceptible oxygen atoms fully "encased" by fluorine. The manufacturer's PFPE product bulletins show the degradation temperature or onset of decomposition in air for this grease to be above that of competing products.

A straightforward comparison of PFPE oils to alternatives is available from Reference 2; it is reproduced in Table 15-6.

From a practical point of view, PFPE lubricants excel and surpass in their capability to form an elasto-hydrodynamic- EHD film--an important oil strength-in-service property that explains effectiveness at all temperatures of interest. The film stays in place under the many operating conditions imposed on, for example, the rolling element bearings in electric motors. "Staying in place" is a desirable property; it implies both resistance to water washout and the necessity to use special procedures to remove PFPE lubricant from bearings—if that should ever become necessary. Compatibility concerns require that PFPE lubricants be applied to clean bearings only. In this regard, one may take cues from the Canadian Paper Mill (Ref. 1, which purchased its electric motor bearings from a competent manufacturer. This manufacturer then pre-filled the bearings with the specified grease and applied the bearing seals.

Table 15-6: PFPE Oil Comparison to Alternatives (Ref. 2)

Property	Mineral	PAO	Diester	Silicones	DuPont™ Krytox®
Thermal Stability	Moderate	Moderate	Good	Very Good	Excellent
Oxidation Stability	Moderate	Very Good	Very Good	Very Good	Excellent
Hydrolytic Stability	Excellent	Excellent	Moderate	Good	Excellent
Volatility	Moderate	Very Good	Good	Very Good	Excellent
Viscosity Index (VI)	Moderate	Very Good	Good	Excellent	Good to Very Good
Fire Resistance	Poor	Poor	Moderate	Good	Excellent
Seal Material Compatibility	Good	Very Good	Poor	Good	Excellent
Lubricating Ability	Good	Good	Good	Poor	Excellent
Toxicity	Good	Excellent	Good	Excellent	Excellent
Cost Compared To Mineral Oil	1	3-5	3-7	30-100	60-120

Examining Cost Versus Benefit

It is generally agreed that, polyalphaolefin (PAO) premium grade grease is a baseline competitor of the PFPEs; PAOs are among the leading products presently used in electric motor bearings. The question is: What would be the cost justification for the more expensive PFPEs?

If we assume the PAO grease to cost $1.00 and a certain size bearing sells for $200, the cost of grease equals 0.5% of the total. Based on cost ratio information derived from typical commercial suppliers the PFPE grease would cost $24 per bearing, although we might assume the bearing manufacturer will charge $250. Purchasing the bearing with PFPE sounds reasonable at this relatively small incremental cost. But we need to make a more detailed comparison. Our projected incremental

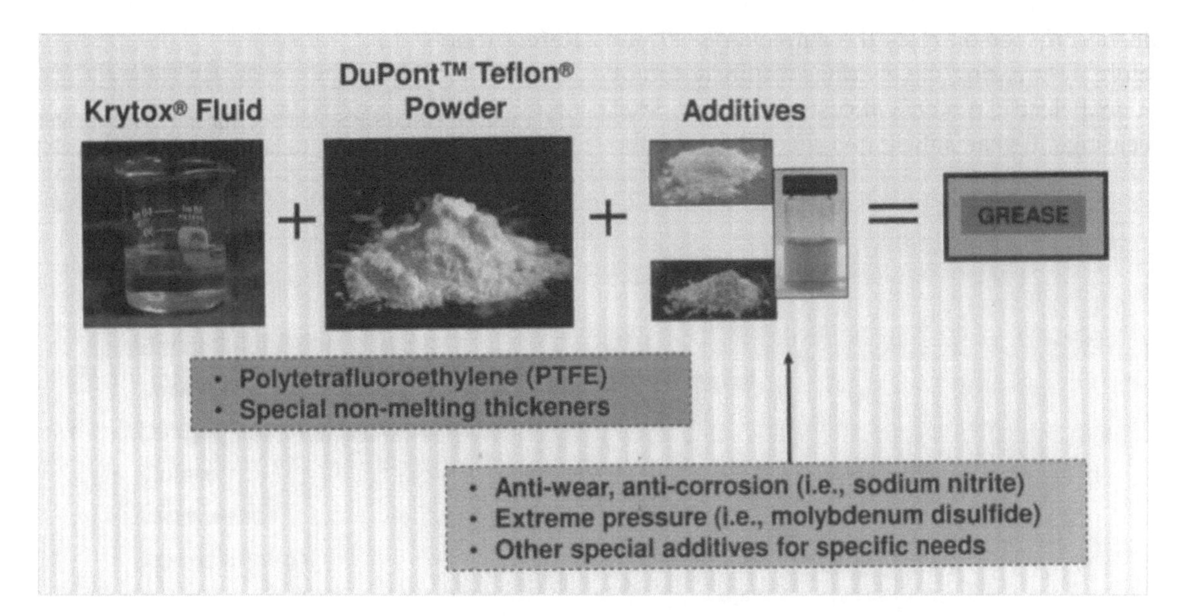

Figure 15-5: A prominent PFPE Grease Recipe (Ref.1)

cost (perhaps $50 per bearing) should convince us to dig a bit further. In a more careful examination we may wish to know what it really costs to periodically re-apply traditional PAO-based greases to electric motor bearings.

The rotational speed, bearing diameter, and the environment in which the bearing operates determine the frequency of grease replenishment. We have to look at a number of plausible scenarios and compare these with simply purchasing and installing life-time, PFPE-prefilled (sealed), motor bearings. Three different scenarios are envisioned below. The purpose is to show the ease with which such cost justifications can be explored and how the results are easily expressed as payback, or benefit-to-cost ratio.

Scenario 1: using bearings with PFPE sealed-in (no re-greasing possible). This is the base case scenario. All comparisons will take into account that a set of sealed-in (no re-greasing possible) electric motor bearings will cost 100 dollars more than customarily supplied (re-greasable) bearings.

Scenario 2: periodic re-greasing. A reasonable assumption would assume that the average bearing is being re-greased 16 times during its (assumed average) 8-year life. A rather optimistic expectation further assumes that the person doing this type of work will do everything just right. They will ascertain that the grease fitting is clean, will not over-grease, will diligently remove the drain plug while adding grease, and will carefully re-insert the drain plug after greasing is done. That person can do 16 electric motors per day. Counting straight salary, overhead, vacations, training time, administrative costs, etc., a trained craftsperson costs the employer $800 per day. Therefore, re-greasing the bearings routinely found in conventional electric motor will costs its owners $800 over the bearings' 8-year anticipated life. However, the incremental cost of two sealed bearings per motor would be only $100.

Subtracting an incremental $100 from $800 = $700; the motor with sealed bearings leads with a payback of 7:1. That simply means that every year, the owner of the electric motor saves $700/8, or about $ 87. An installation with 1200 motors would save approximately $100,000 in labor costs per year. Assume further that 10 motors will require bearing replacement each year. Therefore, bearings would be replaced after 8 years of operation, regardless of bearing style (re-greasable and being re-greased, versus life-time sealed with no need to re-grease).

Scenario 3: standard grease with no periodic re-greasing. A facility with 1200 electric motors and not doing any re-greasing might expect (on average) 200 motors requiring bearing replacement each year. This is to be contrasted against life-time (PFPE sealed-in) bearings. No labor cost is incurred if standard motor bearings are never getting re-greased. However, an incremental number of 190 sets of motor bearings would have to be replaced each year. Replacement bearings and associated labor would cost $2000; 190 x 2000 = 380,000 per year. It might be prudent to assume there would be a process unit outage event---the cost is anybody's guess. In that case, however, the entire Scenario 3 makes even less economic sense than Scenario 2.

It simply pays to reconsider "old" re-greasing strategies in light of recent experience at a Canadian paper mill. High performance oils and greases extend the application range for sealed bearings and call for a re-thinking of the way things were done previously.

References

1. Aronen, Robert; "Krytox® Blog," Boulden Company, Coshohocken, PA, and Ellange, Luxembourg, (2014)
2. Rudnick, Leslie R.; "Synthetics, Mineral Oils, and Bio-Based Lubricants"—Chemistry and Technology, Taylor & Francis (2013)

Chapter 16

Lubricating Steam and Gas Turbines

MECHANICAL DRIVE STEAM TURBINES

Designed for variable speed, steam turbines are used in industry in a multitude of ways to drive compressors, blowers and pumps. As they are both turbomachines, turbines and compressors have similar output-to-speed ratios. This is why steam turbines are particularly suitable as direct drives for variable-speed compressors. Steam turbines, Figures 16-1 and 16-2, are able to make full use of the economic advantages of cogeneration by converting the heat produced by a process into drive power and returning heat to the process in the form of steam at the right pressure and temperature. They must therefore be designed to operate at steam temperatures and pressures which are ideally suited to the process in question. This enables them to be used in a wide range of applications in the chemical and petrochemical industries.

UTILITY STEAM TURBINES

Utilities use constant-speed, high-pressure steam turbines for power generation. As is the case with mechanical drive machines, steam turbines can be designed for condensing, back-pressure, or combination service (Figure 16-3). Also, a variety of exhaust casing designs, Figure 16-4, are available.

PRESSURE LUBRICATION OF MULTISTAGE STEAM TURBINES*

All multistage turbines require cool, clean oil supplied to the journal bearings while the turbine is operating. This oil is supplied by a system provided by one of three parties: the turbine manufacturer, the

*Source: Murray Turbomachinery Corporation, Burlington, Iowa.

Figure 16-1. 3,6 MW packaged backpressure turbine generator. (Source: Siemens Power Generation, Erlangen, Germany)

driven equipment vendor, or the customer/user. If the lubrication system is to be provided by a customer, the turbine manufacturer will specify the applicable flow, pressure, temperatures, and cooler heat load to allow others to design the system. Normally, the turbine manufacturer will provide either "stub" piping at the journal areas, or "manifolded" piping for ultimate connection to the customer provided lube system. (Manifolded piping will minimize the field work required.)

A thrust bearing and an auxiliary drive gear are often provided with steam turbines, and these must also be lubricated. Driven equipment, whether provided by the turbine manufacturer or others, must also be evaluated in the design of an appropriate lubrication system.

A lubrication system may also be required to provide oil for a trip-and-throttle valve, or a governor valve power cylinder. When providing oil to a power cylinder, it is normal practice to provide a combined pressure lubrication and control oil system. This segment of our text will address lubrication systems and combined lubrication/control oil systems both as "lube systems."

Lube systems can be divided into three classifications, each a variant of the others. These classifications, ranging from the simplest (and therefore the least costly) to the more complex are the following:

1.) The basic duty lube system (Figure 16-5) is designed for turbines which can be shut down for maintenance, typically for turbines operating seasonally, such as sugar mill or air conditioning drives. This system includes a single oil filter, and a single oil cooler. While the turbine is operating, the filter element cannot be changed, nor can the cooler be cleaned.

2.) The continuous duty lube system (Figure 16-6) is designed for units

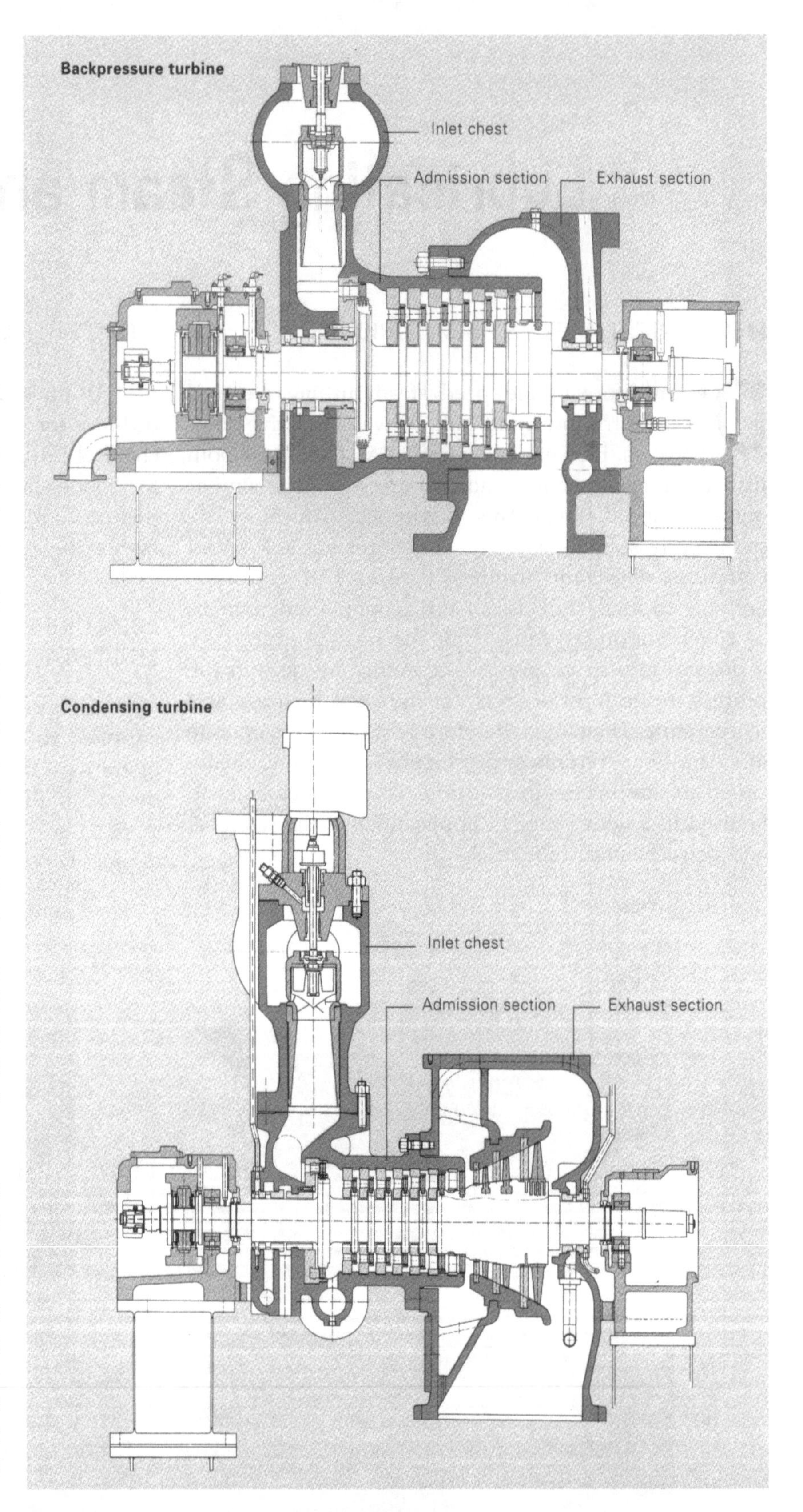

Figure 16-2. Reaction steam turbines. (Source: Siemens Power Generation, Erlangen, Germany)

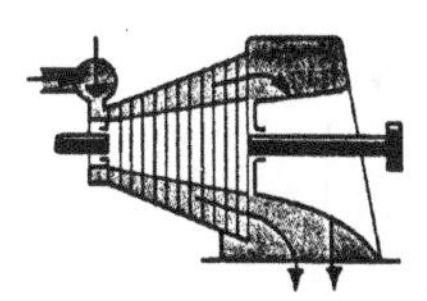

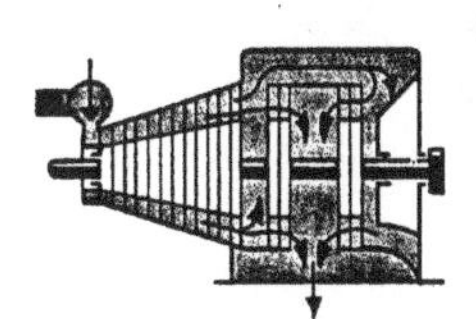

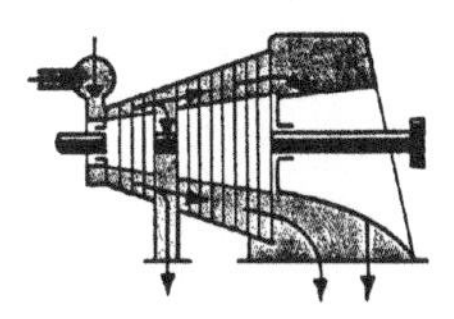

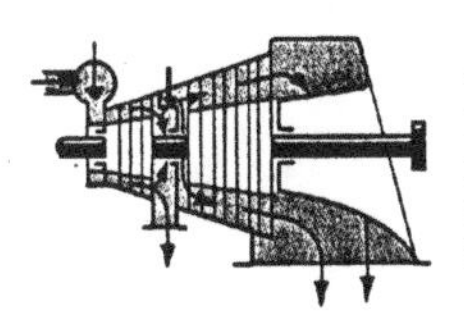

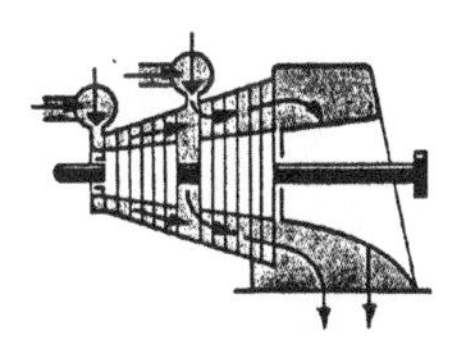

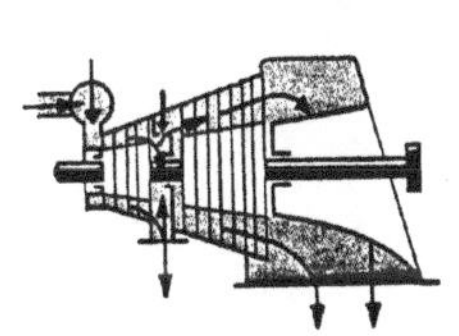

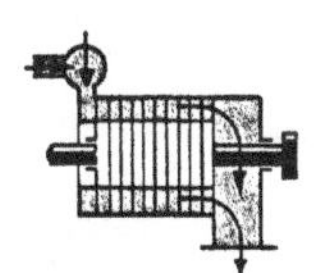

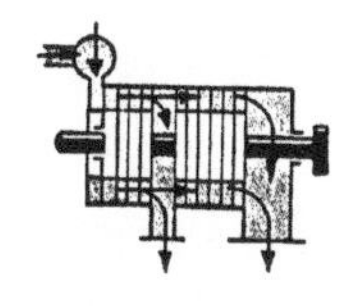

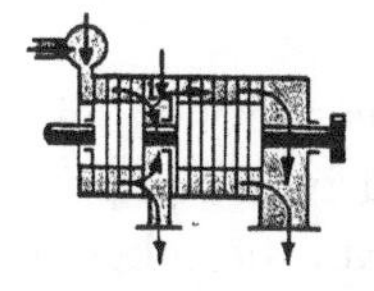

Figure 16-3. Steam turbine design options. (Source: General Electric Company, Fitchburg, Massachusetts)

which run 8000 hours or more per year, with few scheduled shutdowns. Examples of applications for continuous duty lube systems are boiler feedwater pump drives, or process compressors.

3.) The turbine-generator lube system (Figure 16-7) is a variation of the continuous duty lube system, and is also designed for units which run continuously. The addition of a second cooler will allow the lube system to operate for up to three years without shutdown. Since either cooler, or either filter, can be serviced without shutdown, the limiting factor on operating time is the turbine, which must be internally inspected at least once every three years. An optional oil purifier, which will remove water and light hydrocarbons, will often be provided to allow years of operation without an oil change.

Major Components of a Steam Turbine Lube System

The standard main oil pump on direct connected turbines is shaft driven from the turbine by an auxiliary gear on the governor end of the turbine. Due to physical size restrictions, a typical flow is 75 GPM at 100 PSIG. This is more than adequate for the turbine requirements, but may not be sufficient for the requirements of the driven equipment. If flow greater than 75 GPM is required, or if a preference is specified by the purchaser, a separate motor driven or auxiliary steam turbine driven *main oil pump* is generally provided.

On a turbine-speed reducer (gear) application, the *main oil pump* is typically driven by the blind end of the slow speed shaft of the speed reducer. Larger pumps can be mounted on the speed reducer than on the turbine, but a practical limit on speed reducer driven pumps, again due to size restrictions, is 100 GPM. Above 100 GPM, a spacer type coupling becomes unwieldy, and increases the overall length of the train. Therefore, motor driven, or small steam turbine driven *main oil pumps* are provided for applications requiring more than 100 GPM.

Low oil pressure alarm and trip switches are recommended and will be found in modern systems.

Auxiliary oil pumps for any of the three systems options can be AC motor driven, or DC motor driven, or separate steam turbine driven. (The AC motor driven auxiliary oil pump is used as standard). *Auxiliary oil pumps* are required on all pressure lubricated steam turbines for startup and coast down to ensure lubrication of the bearings. All auxiliary oil pumps are provided with a pressure switch to start automatically upon reduction of oil pressure from the main oil pump. An *auxiliary oil pump running pressure switch* is typically wired to the annunciator panel.

A minimum of one oil cooler is required with each pressure lubrication system. Cooling temperatures and heat loads will vary from system to system, but typical

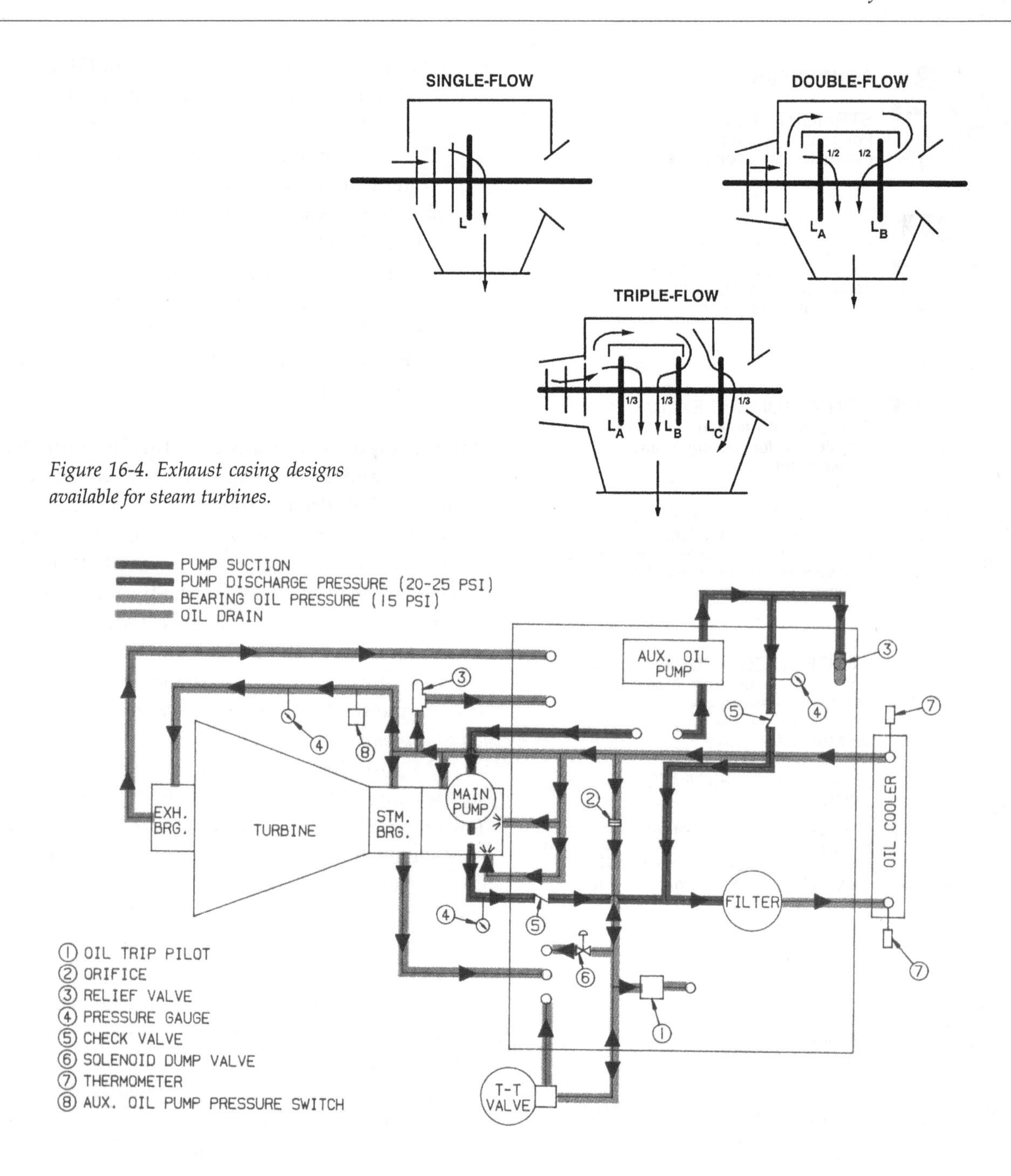

Figure 16-4. Exhaust casing designs available for steam turbines.

Figure 16-5. Basic duty lube system for steam turbines.

temperatures are 140°F oil into the cooler, and 120°F oil from the cooler.

Standard oil coolers are shell-and-tube heat exchangers, with cooling water fed through the tubes, and oil flow cascaded over the tubes. Care must be used in specifying a proper fouling factor for the site specific cooling water. A fouling factor degrades an overall heat transfer coefficient, thereby increasing the size of the heat exchanger, allowing the heat exchanger to operate for longer periods of time, as the tubes become "fouled," without losing capacity. Lube systems supplied with one cooler will require the turbine and driven equipment to be taken off-line to mechanically clean the tubes. When two coolers are provided, the cooler not in actual operation may be cleaned at any time, with the turbine and driven equipment in operation.

Stem or dial type *thermometers* are provided before and after the oil coolers. Operators should periodically read and record these temperatures to ensure that the coolers are operating satisfactorily, and to establish a

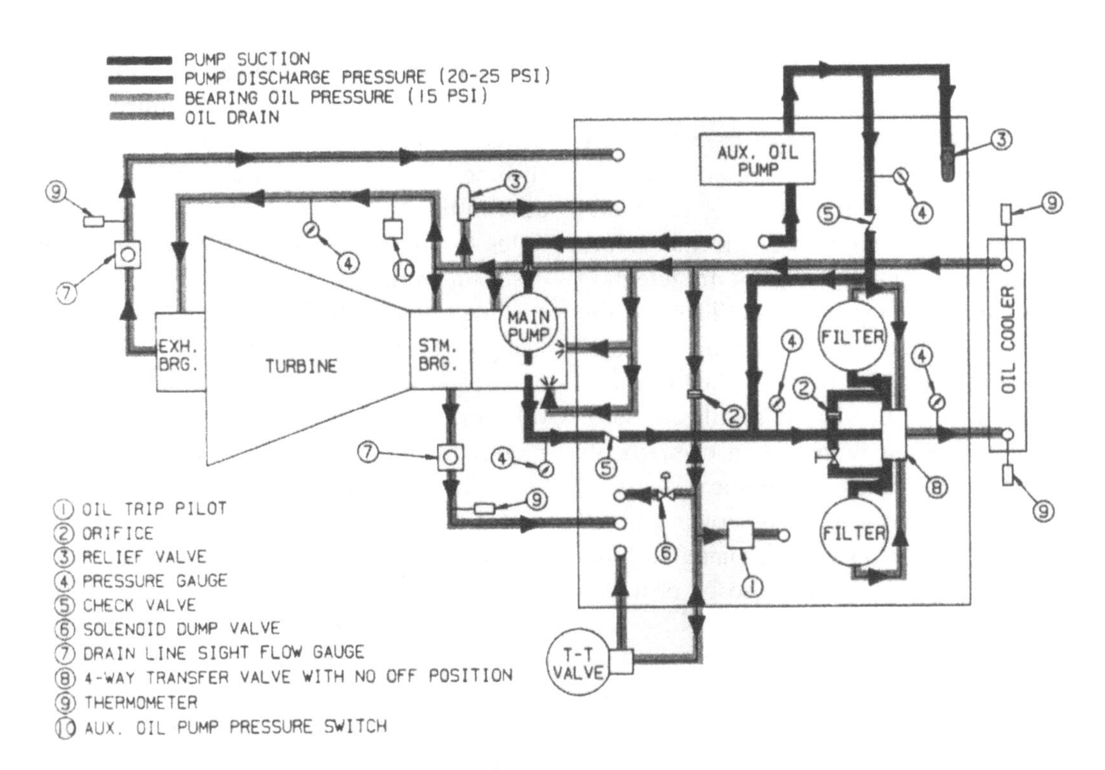

Figure 16-6. Continuous duty lube system for steam turbines.

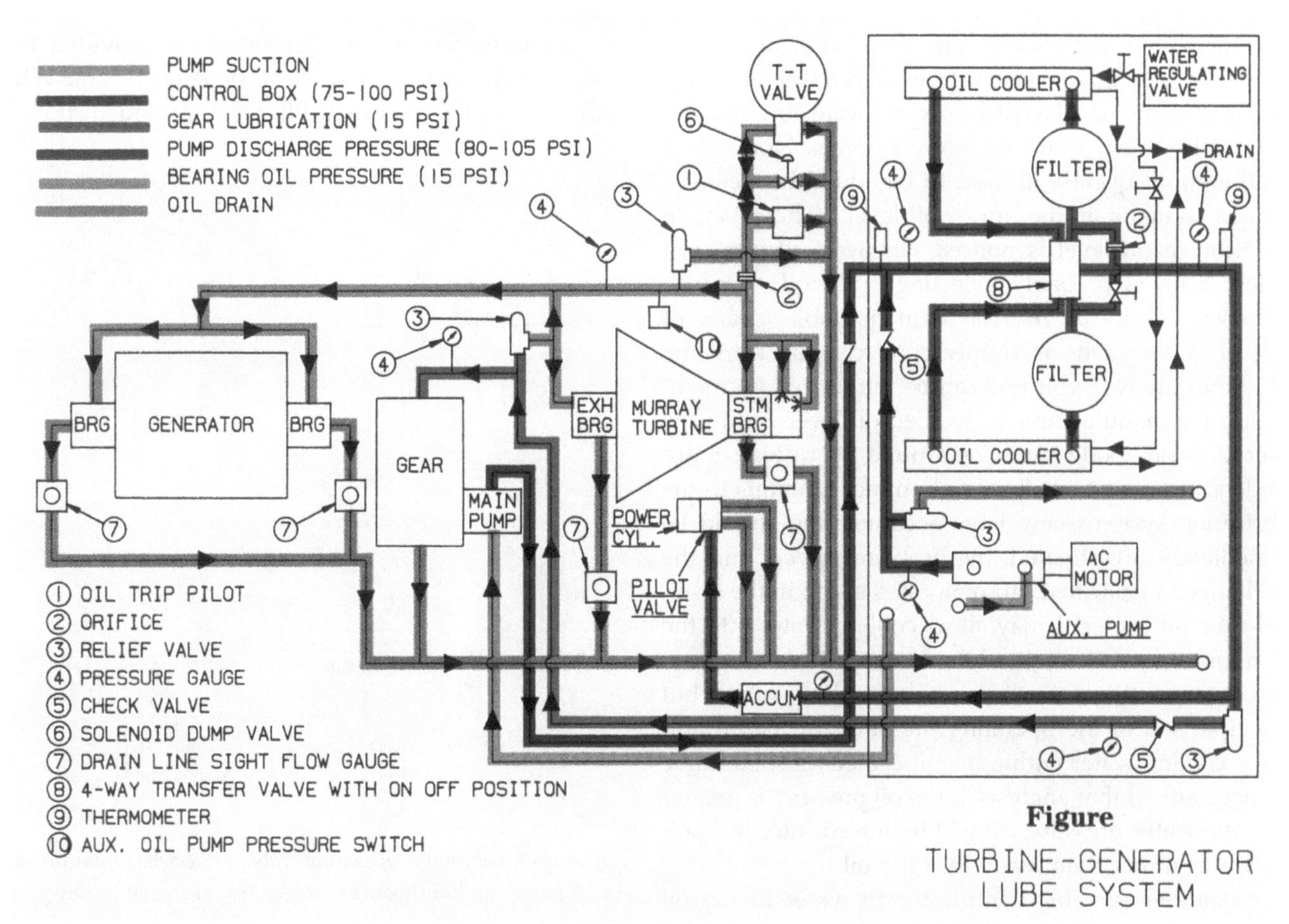

Figure 16-7. Turbine-generator lube system (typical).

basis for cleaning schedules. High oil cooler outlet temperature *alarm and trip switches* are recommended and will be found in most systems.

Prior to about 1980, carbon steel *oil reservoirs* were provided as standard, with stainless steel reservoirs available as an option. Since then, stainless steel reservoirs and piping have become the standard. Intermittent duty lube systems incorporate a 1-minute (minimum) retention time, while continuous duty lube systems utilize a 3-minute (minimum) retention time. Smaller systems (10 to 20 GPM) often utilize a rectangular reservoir, with system components mounted on it. Larger units utilize a separate console type oil reservoir with system components mounted on it. Turbine-gear units, or turbine generators, may incorporate the oil reservoir in the baseplate with all system components mounted on the baseplate. Turbine-gear units, or turbine generators, may also be provided with the reservoir in the baseplate, and filters, coolers, and auxiliary pump(s) on a separate baseplate.

The standard *oil reservoir* design includes a sloped bottom, an oil level sight glass, clean-out openings, and fill, drain and vent openings. Oil heaters mounted in the oil reservoir are supplied if the oil temperature can go below 60°F, or when specified.

Reservoirs are usually provided with an *oil level sight glass*, and this level must be checked periodically. Optional liquid level switches to indicate low and/or high oil level are found on many systems. Normally a small amount of oil will need to be added between oil changes to maintain the proper oil level. If, however, an elevation in oil level is noticed when no oil has been added, water is probably collecting in the oil.

Water in the oil reservoir is attributable to any of several factors. One is simply condensation from the air within the reservoir and can be minimized by maintaining the manufacturer's specified oil level within the reservoir, and good ventilation around the turbine. Since condensation will contribute only minor amounts to the lubrication systems, any large accumulation should be immediately investigated, the problem solved, and the oil changed or purified. In some cases a leak in the shell-and-tube oil cooler(s) may allow cooling water into the oil reservoir. An analysis of the water will determine if the contaminating water is from the cooling source, but a comparison of the operating pressures of the oil and of the cooling water within the oil cooler will determine the necessity of that analysis. If the oil pressure is greater than the water pressure, oil will be forced into the cooling water rather than water into the oil.

Another possible contributor to water in the oil is steam bypassing the steam seals, Figure 16-8. This is especially prevalent in turbines with high back pressures and/or high first-stage pressure, after the seals are worn. It is good practice to minimize this occurrence by providing air purge connections on the bearing seals of turbines. Dry instrument air will provide positive pressure in the oil seal area. This buffers the seal and eliminates the possibility of outside air entering the bearing case. Air purge connections can be retrofitted to existing turbines, as can the bearing protector seals described in Chapter 12. (See Figure 12-66.)

Sight flow indicators provide a visual indication of oil flow through the bearings, and are generally recommended.

A minimum of one oil filter is also required with each pressure lubrication system. The standard filter cartridge will remove particles 25 micron and larger. When specified, filters that will remove particles as small as 5 microns and larger are generally provided, but these cartridges will need to be replaced more frequently. *Oil pressure gauges* before and after the filters are normally supplied, and up-to-date installations incorporate remote sensing and automated data logging as well. A *high oil filter differential pressure alarm switch* is an added safeguard.

Continuous flow *transfer valves* are provided with systems utilizing dual coolers, or dual filters. When both dual coolers and dual filters are provided, transfer

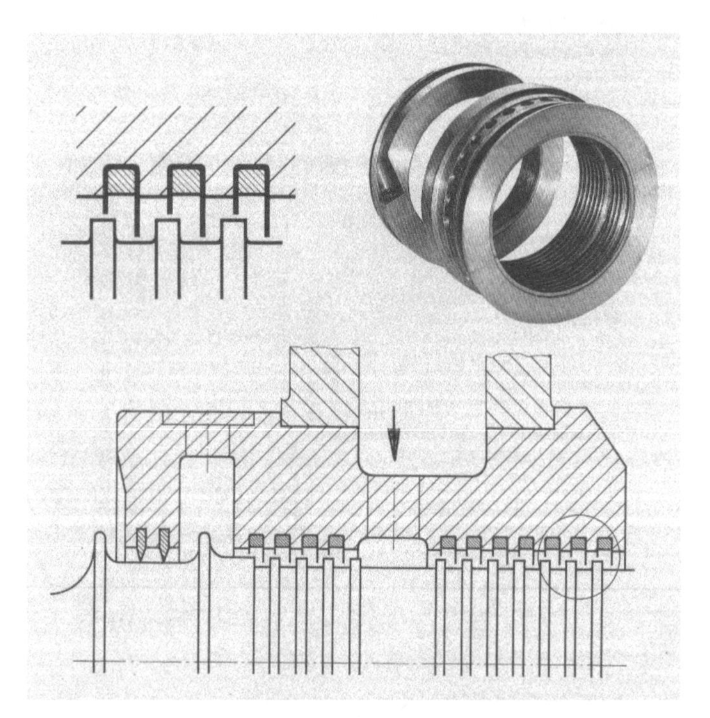

Figure 16-8. Labyrinth-type steam seals for Siemens steam turbines. An air purge can be introduced into the space between sealing groups (arrow).

valves may be provided as common between coolers and filters, or as an individual valve between coolers and between filters. These transfer valves have no off position.

Pressure relief or pressure control valves are used to maintain the proper pressure levels throughout the lube system, and to provide ultimate protection against overpressure within the system.

GAS TURBINES*

Within the context of industrial machinery, the reader is likely to encounter gas turbines as drivers for electric power generators and as mechanical drive turbines for large compressor trains.

Although gas turbines have been on the industrial scene since the late 1920s, large-scale applications had to wait until the 1950s when rapid advancements in aircraft jet engines brought significant improvement and vastly enhanced acceptance of industrial gas turbines. These improvements touched virtually every requirement cited for modern process plants: low initial cost, good efficiency, maintainability, reliability, operational ease, process flexibility, and environmental acceptability.

From a thermodynamic point of view, a gas turbine—or gas turbine engine—is a machine that accepts and rejects heat at different energy levels and, in the process, produces work. While this work is converted to pressure and velocity energy in the aircraft jet engine, the commercial or industrial gas turbine is arranged to convert this work into shaft rotation or, more correctly, torque.

The gas turbine (Figure 16-9) consists of an air compressor and gas combustion, gas expansion, and exhaust sections. The gas turbines cycle is composed of four energy exchange processes: an adiabatic compressor, a constant-pressure heat addition, an adiabatic expansion, and a constant-pressure heat rejection. The four thermodynamic processes can be accomplished either in an open-cycle or a closed-cycle system. The open-cycle gas turbine takes ambient air into the compressor as the working substance that, after compression, is passed through a combustion chamber where the temperature is raised to a suitable level by the combustion of fuel. It is then expanded inside the turbine and exhausted back to the atmosphere. Most industrial-type gas turbines work on this principle, and Figure 16-10 explains their principal components. Enhancements, such as regeneration (Figure 16-11) are employed to increase cycle efficiencies. The use of two or more hot gas expansion stages makes it possible to produce two-shaft turbines, Figure 16-10. This configuration has greater speed flexibility than single-shaft machines.

The closed-cycle gas turbine uses any gas as the working substance. The gas passes through the compressor, then through a heat exchanger where energy is added from a source, then expanded through the turbine and finally back to the compressor through a precooler where some energy may be rejected from the cycle.

Perhaps the most important reasons why process plants use gas turbines are summarized as high system reliability and high combined energy system and process efficiency. Where the forced outages of a single driver can shut down an entire complex, highest reliability is a must. For projects involving process system modifications of a new process design, choosing the most reliable turbine or energy system rather than maintaining an already existing process design can result in significantly higher reliability and reduced financial loss due to excessive process shutdowns.

Lube Systems for Gas Turbines

The gas turbine baseplate houses the oil tank and practically all piping connections ensuring the most at-

*Source: Bloch, H.P., "Process Plant Machinery," Butterworth-Heinemann, Woburn, Massachusetts, 1988/1998.

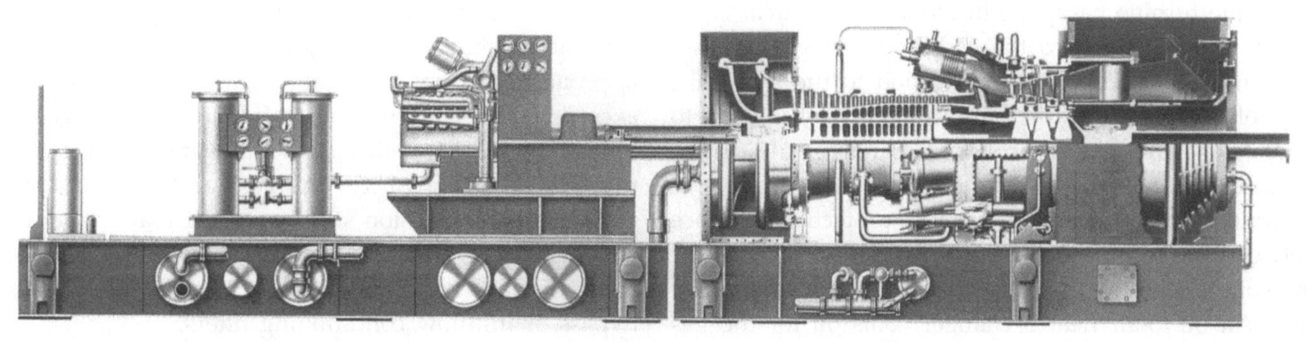

Figure 16-9. General Electric Company Frame 6 gas turbine (35-50 MW range).

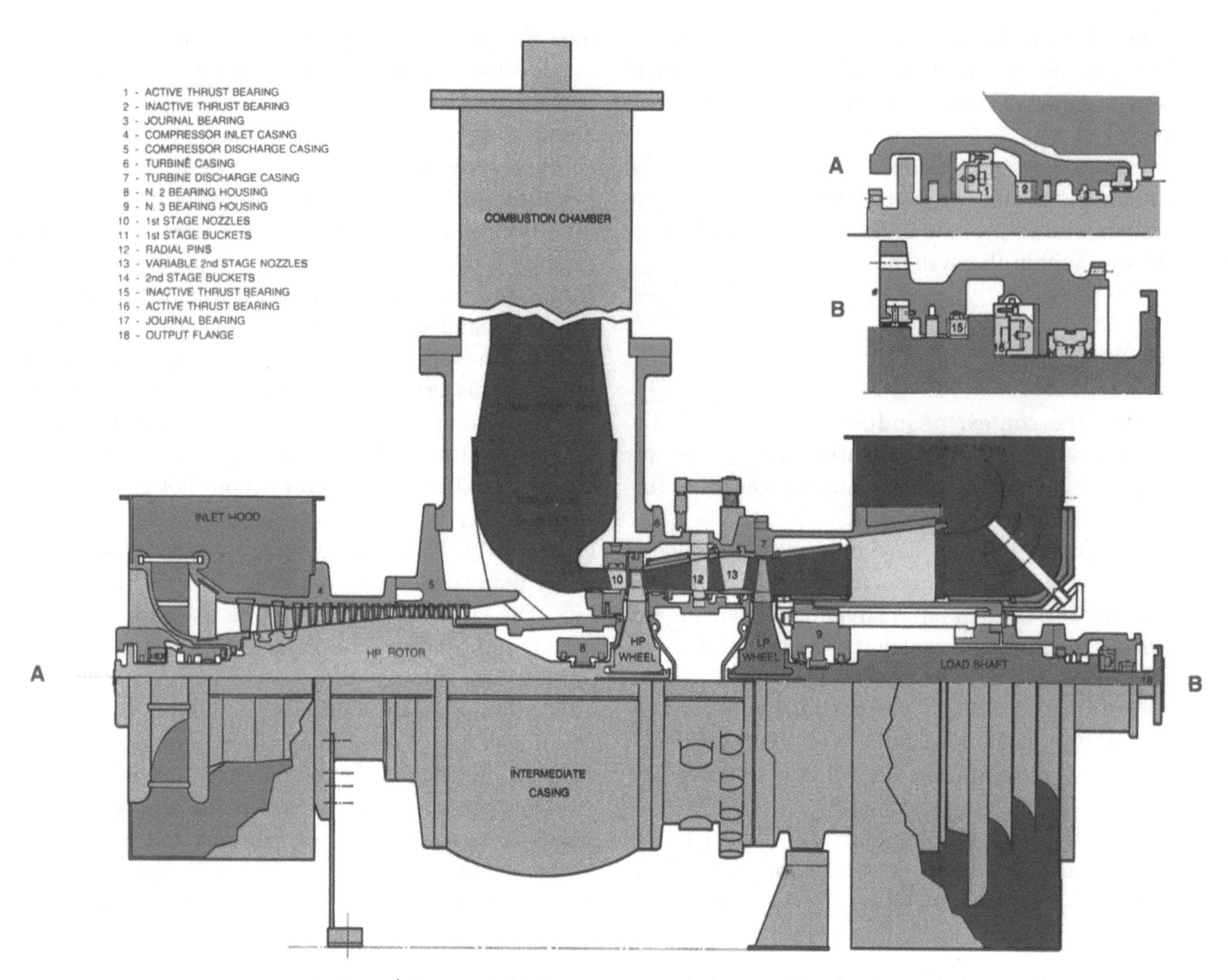

Figure 16-10. General Electric/Pignone Model Series 1002 two-shaft gas turbine showing typical nomenclature.

tractive arrangement and easy access to the machine. Flexible connections are used between the baseplate and the flange-to-flange assembly to allow quick removal of the gas generator or power spool for scheduled maintenance operations. Lube oil pumps, hydraulic oil pumps, filters, pressure control valves and various control devices are mounted on the lube oil console located near the gas turbine in the best position for site requirements, or on the turbine baseplate in the accessory area.

Lube oil is fed to the turbine bearings, accessories and load equipment in addition to the hydraulic control devices. High pressure hydraulic oil is used to operate the fuel gas control valves and for the driven compressor seal oil system in turbocompressor units. A hydraulic trip system is the primary protection interface between the turbine control and protection system and the components on the turbine which admit or shut off fuel. An oil-to-air heat exchanger cools oil for the gas turbine lubricating and hydraulic systems. The cooler is sized to meet oil cooling requirements when operating at the maximum rated temperature.

Lubricants for Steam and Gas Turbines

Whatever their different operating environments, all power generators have a critical requirement in common: dependable-quality turbine lubricants that will provide long, cost-effective, trouble-free service.

Such lubricants must have excellent thermal and oxidation stability at bearing oil temperatures that may approach 200°F in a typical steam or gas turbine and exceed 400°F in modified aircraft (aero-derived) gas turbines. They must readily shed the water that infiltrates turbine systems; control the rust and corrosion that could destroy precision surfaces; resist foaming and air entrainment, which could impair lubrication and lead to equipment breakdown; and filter quickly through bypass or full-flow conditioning filters.

Turbine lubricants should also be versatile, able

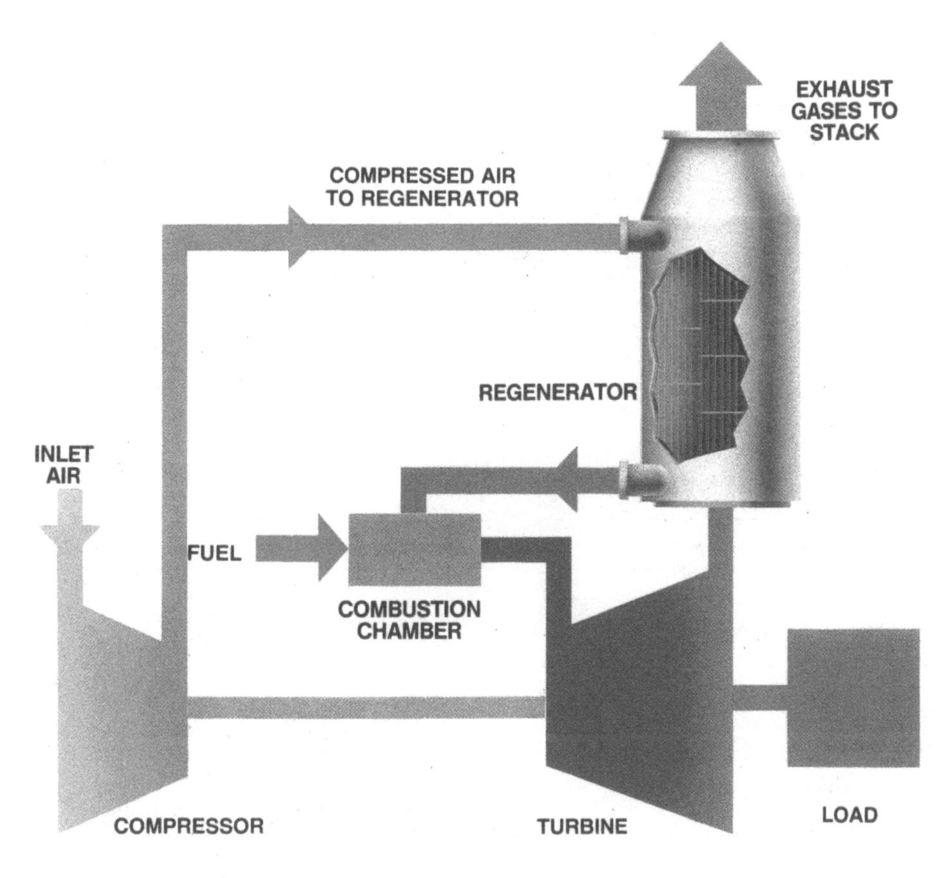

Figure 16-11. Gas turbine regenerative cycle diagram. (Source: Nuovo Pignone, Florence, Italy)

to serve as both hydraulic fluid and lubricating oil for pumps, compressors, and other auxiliary components.

A handful of premium petroleum-base and synthetic turbine oils easily meet these demanding requirements. For example, TERESSTIC GT 32, Exxon's superpremium mineral-oil turbine oil, has served in numerous turbine applications for over 30 years without a changeout. For the higher temperature operation of industrial aero-derived turbines, Exxon offers ETO 2380 synthetic turbine oil, one of the most widely trusted aircraft turbine engine oils in the world. Table 16-1 summarizes some of these lubricants.

Keep in mind that superior products are generally highly versatile, providing satisfactory service in more than one type of plant application. This allows simplifying lubricant inventories to a relatively few multipurpose products, thus minimizing the chances of potentially costly lubricant misapplication.

Recall also, from Chapter 4, that TERESSTIC is Exxon's line of premium circulating oils. These versatile, multipurpose oils are formulated to provide long service life in steam turbines, land-based gas turbines, hydraulic systems, heat transfer systems, gear cases, friction clutches, and other industrial units for which long, trouble-free service is required.

All the TERESSTIC grades have superb thermal and oxidation stability and excellent rust-preventive, antifoam, and water-shedding properties. Their high viscosity indexes allow more uniform lubricating performance over a wide range of ambient and operating temperatures. They are also easily filterable without additive depletion. TERESSTIC GT 32, Table 16-2, is Exxon's first recommendation for steam and industrial gas turbine applications. It has a potent antioxidant incorporated in a carefully selected and refined base oil. This assures exceptionally long life in demanding high-temperature turbine operations. In many cases TERESSTIC GT 32 has lasted more than 30 years in such applications without a changeout.

The extraordinary thermal and oxidation stability of TERESSTIC GT 32 was confirmed in severe laboratory tests in which it was compared with seven premium ISO 32 competitive turbine oils.

The results are shown in Figure 16-13 and Table 16-3.

While most of the oils performed well in at least

Table 16-1. Lubricants for turbines, generators, and industry-associated equipment. (NOTE: The product recommendations given in this table may not apply in all specific instances. Equipment manufacturer recommendations must always take precedence. When in doubt about a particular application, consult the lube manufacturer's representative. The product names listed here are trademarks of Exxon Corporation.)

TURBINES/GENERATORS	
Steam Turbines	TERESSTIC GT 32, 46 TERESSTIC GT EP 32, 46 TERESSTIC SHP 32
Gas Turbines	
General	TERESSTIC GT 32
G.E. Frame 7,9	TERESSTIC GT 32
Geared	TERESSTIC GT EP 32, 46 TERESSTIC SHP 32, 46
Aero-derived	ETO 2380
Hydroturbines	TERESSTIC 6S, 77 TERESSTIC SHP 68
Wind Turbines	SPARTAN Synthetic EP TERESSTIC SHP

MOBILE EQUIPMENT	
Automotive Engines	XD3, XD3 Extra SUPERFLO SUPERFLO Synthetic SUPERFLO Synthetic Blend
Gears	GEAR OIL GX, GEAR OIL ST, SUPERFLO GEAR OIL, SGO
Hydraulic Lifts	UNIVIS J, UNIVIS N
Wheelbearings, Joints, Chassis	RONEX MP, RONEX EP 1 UNIREX SHP 220
Transmissions	SUPERFLO ATF, TORQUE FLUID

AUXILIARY EQUIPMENT	
Air Compressors	TERESSTIC, SYNESSTIC, TERESSTIC SHP
Couplings	RONEX EXTRA DUTY 1
Cranes (open gears)	DYNAGEAR, DYNAGEAR Extra
Electric Motor/Fans	UNIREX N, UNIREX S 2 POLYREX, RONEX MP
Furnace Fans	POLYREX, UNIREX S 2
GEARS	
GEAR TRAINS	SPARTAN EP SPARTAN Synthetic EP
Open Gears (ball mills, cranes, crushers)	DYNAGEAR, DYNAGEAR Extra, SURETT
Worm Gears (Pulverizers)	SPARTAN Synthetic EP TERESSTIC SHP
Hydraulics	NUTO H, SYNESSTIC, TERESSTIC TERESSTIC SHP
Pumps	NUTO H, SYNESSTIC, TERESSTIC TERESSTIC SHP
Transformers	UNIVOLT 60, UNIVOLT N 61
Valve Actuators (Limitorque)	NEBULA EP
Wicket Gates	RONEX EXTRA DUTY 1, CAZAR,

one test, TERESSTIC GT 32 was the only one that achieved excellent performance across the board.

These laboratory results, combined with many years of proven dependability in the field, provide strong assurance of reliable, long-life performance under a wide range of operating conditions.

Where the turbine manufacturer specifies a higher viscosity oil, TERESSTIC GT 46 and TERESSTIC 68 and 77 also provide excellent service in turbine operations.

Lubricants for Geared Turbines

Geared steam and gas turbines are subject to shock loads and occasional overloading. The resulting extreme pressure can force the lubricating film out from between meshing gear teeth, causing metal-to-metal contact and excessive wear. To meet these extreme conditions, Exxon offers TERESSTIC GT EP and TERESSTIC SHP antiwear turbine oils.

TERESSTIC GT EP, a premium antiwear turbine oil, is formulated to meet the special requirements of geared turbines, while offering the same high-quality performance as the other TERESSTIC products. Under shock conditions, the non-zinc antiwear additive in TERESSTIC GT EP reacts with the metal surfaces to form a protective boundary layer, thus minimizing wear. For typical inspections see Table 16-4.

TERESSTIC SHP, a synthetic-base oil incorporates polyalphaolefin (PAO) basestocks and carefully selected zinc-free, ashless additives. It offers outstanding antiwear performance and mild EP characteristics. Com-

Table 16-2. Typical inspections for super-premium mineral-oil-base turbine oils.

Grade	TERESSTIC GT 32	TERESSTIC GT 46
ISO viscosity grade	32	46
Flash point, ASTM D 92, °C(°F)	210(410)	213(415)
Pour point, °C(°F)	–36(–33)	–30(–22)
Viscosity:		
cSt @ 40°C, ASTM D 445	30.7	46.2
cSt @ 100°C	5.2	6.7
SUS@ 100°F, ASTMD2161	158.5	238.7
SUS @ 210°F	43.6	48.7
Viscosity index	96	97
Neutralization number, ASTM D 974, mg KOH/g	.06	.07
Gravity,		
°API	29.9	29.3
Specific @ 15.6°C(60°F)	0.877	0.880
Water separability, ASTM D 1401 54.4°C(130°F), 15 minutes, ml		
Oil	40	40
Water	40	40
Emulsion	0	0
Foam test, ASTM D 892, ml		
Sequence 1	20/0	20/0
Sequence 2	20/0	20/0
Sequence 3	20/0	20/0
Rust test, ASTM D 665 A/B		
Distilled water	pass	pass
Synthetic sea water	pass	pass
Copper strip corrosion, ASTM D 130, rating	1A	1A
CEICO rust test	pass	pass

Table 16-3. Competitive survey of seven turbine oils.

	TERESSTIC GT			Competitor				
	32	A	B	C	D	E	F	G
Turbine oil stability test (TOST), ASTM D 943	E	E	E	P	F	E	F	E
Rotary Bomb Oxidation Test (RBOT), ASTM D 2272	E	P	F	P	F	P	P	F
Staeger oxidation test	E	P	E	P	P	F	F	F
Universal oxidation test	E	P	E	P	F	—	—	—

E = Excellent, F = Fair, P = Poor, — = Data not available

pared with conventional petroleum-base oils, TERESSTIC SHP has superior oxidation control and thermal stability, better lubricity, and lower carbon-forming tendency. These qualities can reduce unscheduled downtime, extend drain intervals, and maximize the life of bearings and other critical components.

The superior lubricity of TERESSTIC SHP versus comparable petroleum-base oils can reduce energy consumption. Laboratory and field tests on synthetic-base lubricants have shown that energy savings of 3.5-8.5%

Table 16-4. Typical inspections, EP-grade super premium turbine oils.
The values shown here are representative of current production. Some are controlled by manufacturing specifications, while others are not. All may vary within modest ranges.

TERESSTIC GT EP Grade	32	46
ISO viscosity grade	32	46
Gravity, =API	29.3	28.9
Specific	0.880	0.882
Viscosity, cSt @ 40°C	30.4	45.2
cSt @ 100°C	5.1	6.6
SUS @ 100°F	157	233
SUS @ 210°F	43.5	48.4
Viscosity index	96	97
Flash point, °C(°F)	210(410)	213(415)
Pour point, °C(°F)	–36(–33)	–30(–22)
Neutralization number, ASTM D 974	0.09	0.08
Color, ASTM	L1.5	L2.0
Copper strip corrosion,		
3 hr at 100°C (212°F)	1b	1b
Rust protection, ASTM D 665		
Distilled water	Pass	Pass
Synthetic water	Pass	Pass
FZG gear test, DIN 51354, A/8.3/90		
Failure load stage (FLS)	10	10
Oxidation life, ASTM D 943, hr	5500+	5500+

are achievable, compared with a petroleum-base oil. In some cases, because of the higher viscosity index of synthetic-base oils, these energy savings were achieved using a *lower* ISO viscosity grade synthetic lubricant. It should be noted, however, that switching to a lower viscosity grade should be done only with the concurrence of the equipment manufacturer.

Lubricants for Aero-derived Gas Turbine Engines

In general, there are two classes of gas turbines used in industrial applications:

- Heavy-duty gas turbines based on steam turbine technology (Figures 16-9 and 16-10).
- Lightweight gas turbines derived from aircraft gas turbine engines (Figure 16-14).

Heavy-duty gas turbine designs are not restricted by size and weight. Standard components are fairly massive and bearings are located at some distance from heat sources. Petroleum-base lubricants like TERESSTIC oils perform satisfactorily under these operating conditions.

By contrast, size and weight are extremely important design considerations in aero-derived gas turbine engines. Equipment is quite compact, with bearings located relatively close to sources of heat. Aero-derived gas engines require that the oil not only lubricate under more severe thermal and oxidative conditions, but that it serve as a heat transfer fluid as well, carrying heat away from the bearings and shafts. Additionally, aero-derived gas turbine engines subjected to repeated, rapid start-ups during peak power demand typically carry higher loads than conventional heavy-duty turbines.

ETO 2380—Unsurpassed Performance in Aero-derived Gas Turbine Engines

These extreme operating conditions usually require a high-quality synthetic-base oil—an oil like Exxon's ETO 2380. This ester-base synthetic oil supplies approximately 50% of the free world's commercial airline requirements for 5-cSt turbo oil. Over 360 airlines entrust the safety of their passengers to ETO 2380. Evidently, the "synthetics option" deserves closer examination. Accordingly, the reader is referred to Chapter 7 of this text.

To summarize, in most applications petroleum-base lubricants provide excellent lubrication. However, modern industrial machinery design is placing unprecedented, severe demands on lubricants. Newer machines are designed for faster speeds and higher unit loads, resulting in higher operating temperatures. Older equipment is being run harder to maximize output. These

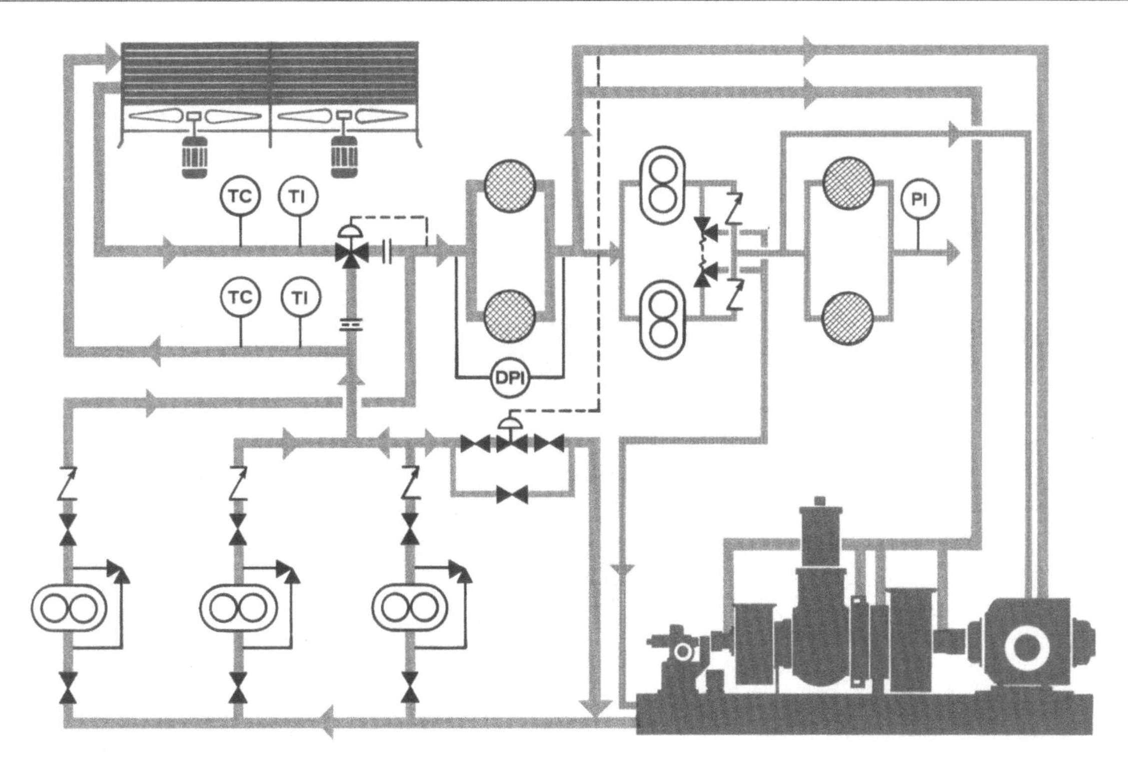

Figure 16-12. Simplified lube and hydraulic oil systems diagrams for industrial gas turbines. (Source: Nuovo Pignone, Florence, Italy)

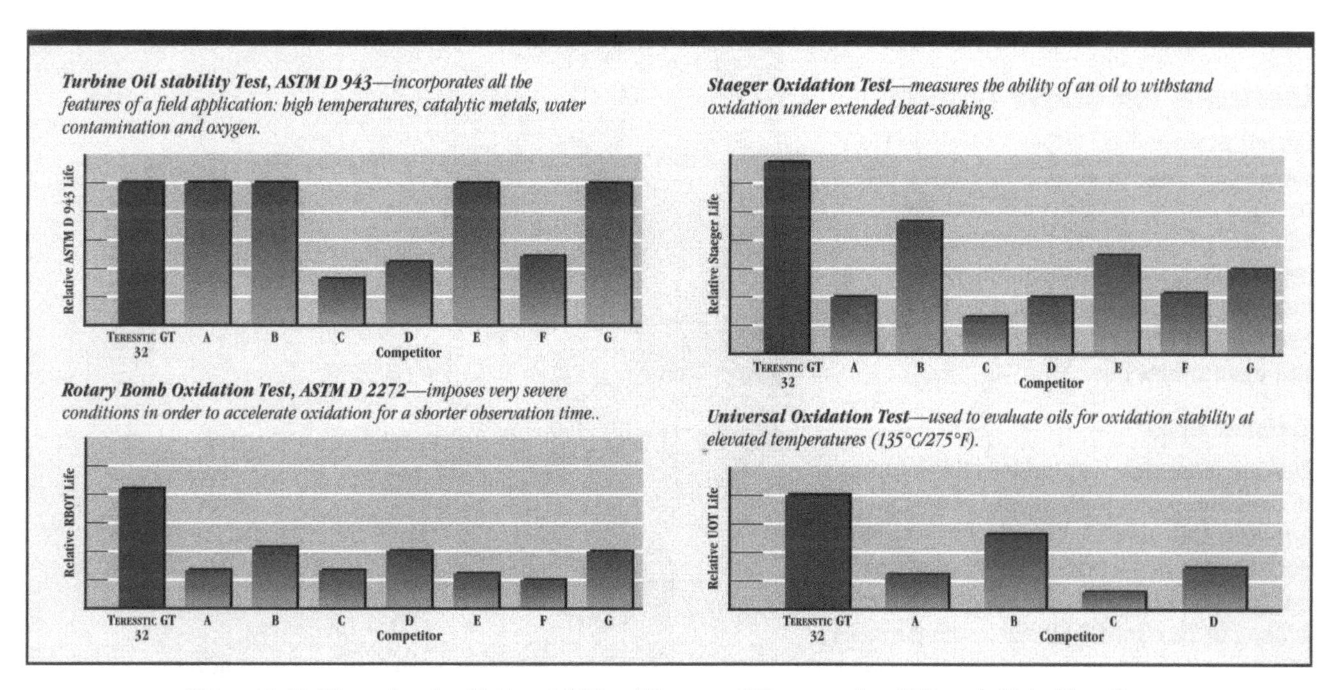

Figure 16-13. Thermal and oxidation stability of Teresstic GT vs. premium ISO grade 32 turbine oils.

punishing operating environments have placed a premium on lubricants that can ensure machine reliability and efficiency in severe operations. Additionally, safety and environmental concerns increasingly are dictating the use of long-life, low-volatility lubricants.

In many cases, these demands have pushed petroleum-base lubricants to the limits of their capabilities, necessitating the development of a new generation of lubricants: synthetics.

Although their initial cost may be higher, syn-

Gas generator data

Applications

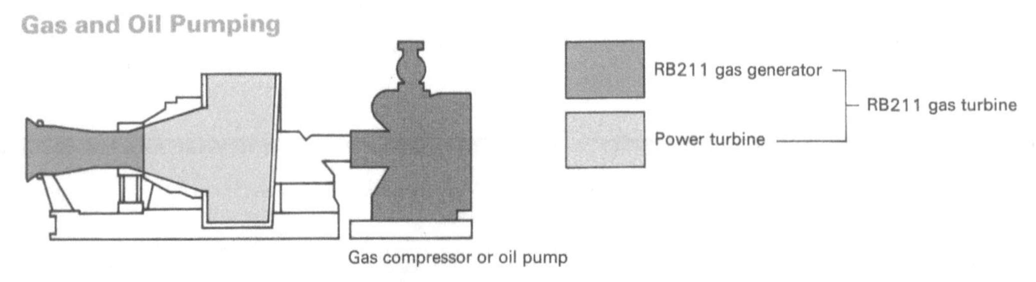

Electrical Generation

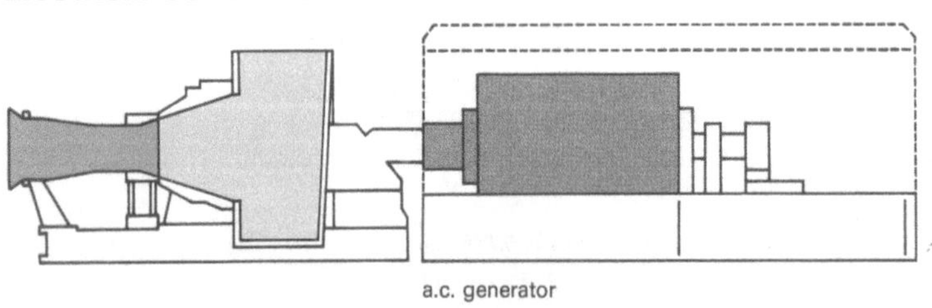

Marine Drive (Electrical)

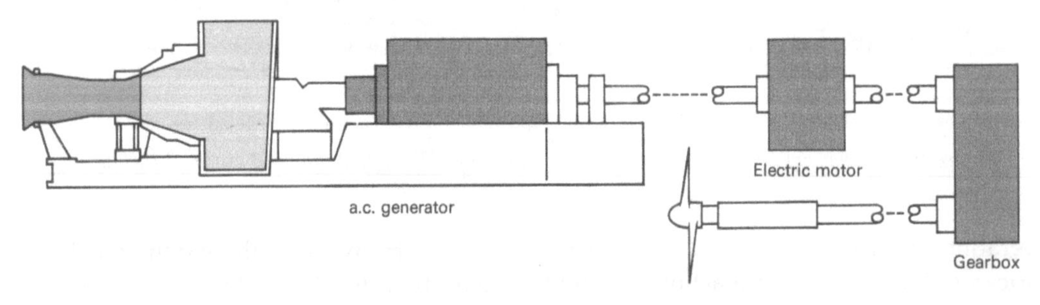

Figure 16-14. Rolls Royce aeroderivative gas turbine—generator and applications.

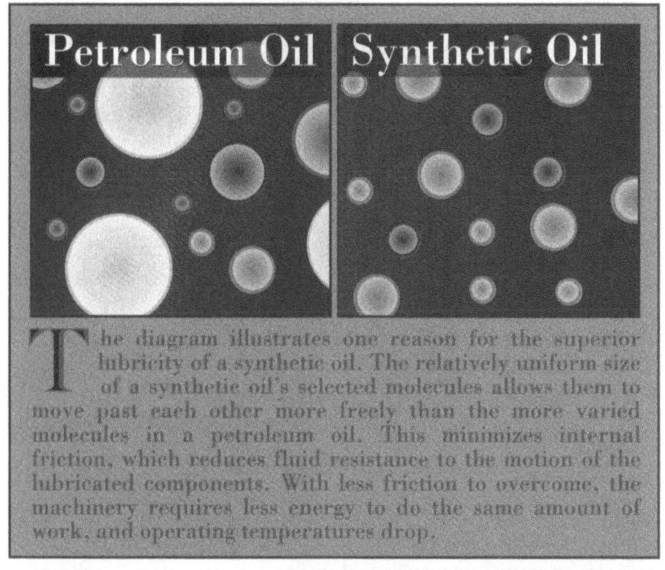

FEATURES AND BENEFITS OF SYNTHETIC LUBRICANTS

FEATURES	BENEFITS VS PETROLEUM-BASE LUBRICANTS
High V.I. **Low pour point** **High flash point**	Superior high- and low-temperature performance over wide operating range
Superior wear protection/ reduced deposit formation	Reduced downtime/lower maintenance costs
Lower volatility	Improved environmental control/reduced lubricant consumption/greater safety
Greater lubricity/ reduced internal friction	Lower energy consumption/lower lubricant system temperature/lower manufacturing costs/ increased productivity
Extended lubricant life	Lower lubricant & labor costs/ improved environmental control

Figure 16-15. Features and benefits of synthetic lubricants summarized.

thetic lubricants can offer numerous advantages over conventional petroleum-base lubricants in severe-service applications—advantages such as longer lubricant life, superior wear protection, greater thermal stability, and lower carbon-forming tendencies (Figure 16-15). These qualities can significantly reduce long-term costs by extending equipment life and minimizing downtime.

Here are the lubricants that merit consideration:

SPARTAN Synthetic EP (polyalphaolefin base)—a line of seven long-life, extreme-pressure industrial gear and bearing lubricants, particularly recommended for gear trains and worm gears.

SYNESSTIC (diester base)—a versatile line of five industrial lubricants for compressors, hydraulic systems, mist lubrication systems, air-cooled heat exchanger drives, and bearings in pumps and electric motors.

SGO (polyalphaolefin base)—a line of three automotive gear oils specially formulated for longer gear life and improved operating economies.

TERESSTIC SHP (polyalphaolefin base)—superpremium circulating, gear, and hydraulic oil, with applications in gear reducers, pumps, marine centrifuge gear boxes, and work gears containing copper alloys where mild EP performance is required.

POLYREX (polyurea soap)—a high-temperature, long-life, multipurpose grease for all types of bearings.

UNIREX SHP (polyalphaolefin base)—a line of five lithium-complex synthetic base oil greases for automotive and industrial applications.

UNIREX S 2 (polyolester base)—high-viscosity, low-volatility lithium complex grease that provides excellent high-temperature performance where frequent relubrication is impractical.

SUPERFLO Synthetic, SUPERFLO Synthetic Blend, XD-3 Elite (polyalphaolefin base)—passenger car and heavy-duty automotive oils.

In-service Monitoring of Turbine Oil Quality

A well-ordered method of surveillance of the lubrication system is essential for trouble-free operation. Several resources are available to the lubrication engineer.

ASTM Recommendations

Steam and gas turbine oils are expected to provide years of trouble-free service. In-service monitoring of turbine oils is a valuable means of assuring optimum oil performance and extended equipment life. ASTM D 4378, "Standard Practice for In-service Monitoring of Mineral Turbine Oils for Steam and Gas Turbines," can be used by the turbine oil user as a basis for developing a monitoring program and interpreting the test results. The essential tests and recommendations of ASTM D 4378 are summarized in Table 16-5. This summary is intended only as a general guide; consult an application specialist for assistance in implementing a monitoring program and in interpreting test results.

Table 16-4. Typical inspections, EP-grade super premium turbine oils.

The values shown here are representative of current production. Some are controlled by manufacturing specifications, while others are not. All may vary within modest ranges.

Teresstic GT EP Grade	32	46
ISO viscosity grade	32	46
Gravity, =API	29.3	28.9
Specific	0.880	0.882
Viscosity, cSt @ 40°C	30.4	45.2
cSt @ 100°C	5.1	6.6
SUS @ 100°F	157	233
SUS @ 210°F	43.5	48.4
Viscosity index	96	97
Flash point, °C(°F)	210(410)	213(415)
Pour point, °C(°F)	–36(–33)	–30(–22)
Neutralization number, ASTM D 974	0.09	0.08
Color, ASTM	L1.5	L2.0
Copper strip corrosion,		
3 hr at 100°C (212°F)	1b	1b
Rust protection, ASTM D 665		
Distilled water	Pass	Pass
Synthetic water	Pass	Pass
FZG gear test, DIN 51354, A/8.3/90		
Failure load stage (FLS)	10	10
Oxidation life, ASTM D 943, hr	5500+	5500+

Table 16-5. In-service monitoring of turbine oil performance.

Test	Warning Limit	Interpretation	Action
Total Acid Number Increase Over New Oil	0. 1-0.2 mg KOH/g	Above-normal degradation for steam turbines up to 20,000 hr oil life and gas turbines up to 3,000-hr oil life. Possible causes: a) system very severe, b) antioxidant depleted, c) wrong oil used, d) oil contaminated.	Increased testing frequency and compared with RBOT data.
	0.30.4 mg KOH/g	Oil at or approaching end of service life; also, c) or d) above may apply.	Look for signs of increased sediment on filters and centrifuge. Check RBOT—if RBOT less than 25% of original oil, change the oil. If oil left in system, increase test frequency.
RBOT	Less than half value of original oil	Above-normal degradation for steam turbines up to 20,000-hr oil life and gas turbines up to 3,000-hr oil life.	Investigate cause. Increased frequency of testing.
	Less than 25% of original value	Together with high TAN, indicates oil at or approaching end of service life.	Resample and retest. If same results, consider oil change.
Water content	Exceeds 0.2%	Oil contaminated, possible water leak.	Investigate and remedy cause. Clean system by purifying/centrifuging. If still unsatisfactory, consider oil change or consult Exxon rep.
Cleanliness	Particulates exceed accepted limits	Source of particulates may be: a) make up oil, b) dust or ash entering system, c) wear conditions.	Locate and eliminate source of particulates. Clean system oil by filtration or centrifuging or both.

Chapter 17

Compressors and Gas Engines

As is implied by the term "compressor," we are dealing here with machinery that elevates the pressure of a compressible process fluid, typically air, or a host of other gases. Dynamic compressors are based on the principle of imparting velocity to a gas stream and then converting this velocity energy into pressure energy. In contrast, positive displacement compressors confine a certain inlet volume of gas in a given space and subsequently elevate this trapped amount of gas to some higher pressure level.

The overwhelming majority of compressors in either category incorporate moving and/or sliding components. Only "static" jet compressors (ejectors) and late 20th- and early 21st-century oil-free machines whose rotors are suspended in magnetic or air bearings are exempt from the need for bearing lubrication.

Dynamic turbomachinery, such as the equipment depicted in Figures 17-1 through 17-3, requires lubrication of bearings and seals. To date, the majority of dynamic compressors continue to utilize liquid-lubricated seals, items 1 through 3 of Figure 17-4. Only labyrinth seals or gas-lubricated seals (item 4, Figure 17-4) operate without a liquid film separating the faces. On the more conventional liquid-lubricated seals, the bearing and sealing lubricant are often the same, i.e., an R&O or hydraulic oil. However, this "seal oil" generally enters the supply port (Figure 17-5) at a higher pressure than would be needed to lubricate the bearings.

Positive displacement compressors are primarily represented by reciprocating piston machines, Figures 17-6 and 17-7. In addition, there are rotary piston blowers (Figure 17-8), sliding vane compressors (Figure 17-9), liquid ring compressors (Figure 17-10), helical screw (Figure 17-11) and perhaps a dozen hybrid machine

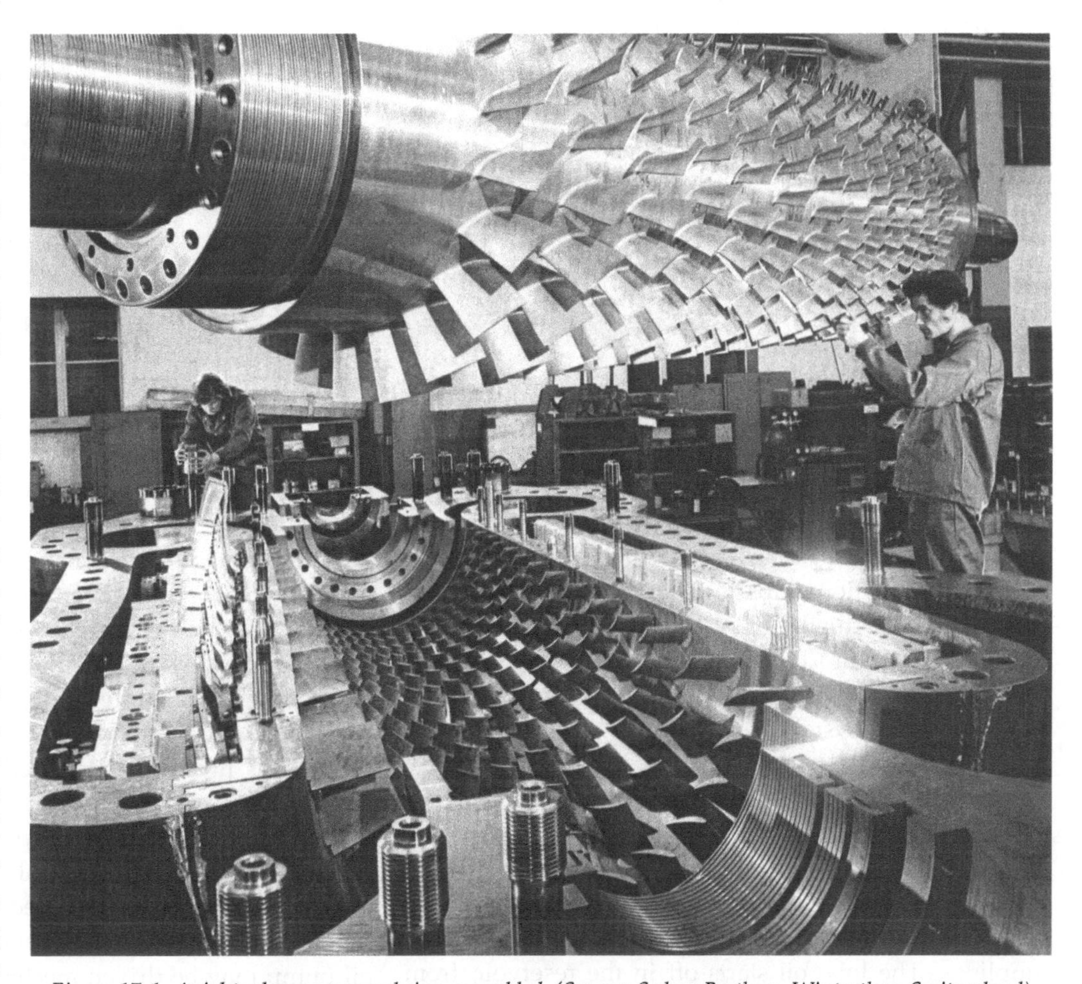

Figure 17-1. Axial turbocompressor being assembled. (Source: Sulzer Brothers, Winterthur, Switzerland)

types deserving the classification "positive displacement." Some of these frequently operate with discharge temperatures exceeding 163°C (325°F); therefore, the lubricating oil must have good oxidation stability. Also, in cases where the lubricant removes moisture that has condensed from the gas, the lubricant must have good demulsibility.

Whenever practical, the lube oil supplied to the bearing system ("running gear") is the same as the lube oil supplied to the compressor working space, i.e., cylinder, piston rod packing, vanes, lobes, etc. Nevertheless, at times it will be more appropriate to provide a different lubricant to this working space. In that case, separate lube supply systems are used for working space and running components.

Figure 17-2. Centrifugal compressor with thrust bearing (inset, lower left) and radial bearing (inset, upper right). Source: GHH-Borsig, Berlin, Germany.

Elsewhere in this text, the reader will find an overview of different means of applying the lubricant. These different means range from simple static sumps to elaborate circulating systems. Still, in virtually every case bearings and sliding components require lubrication, and on major machinery this lubrication is generally applied with the help of a circulating oil system.

LUBRICATION SYSTEM

The lube oil system (Figure 17-12) supplies oil to the compressor and driver bearings and to the gears and couplings. The lube oil starts off in the reservoir, from where it is drawn by the pumps and fed under pressure through coolers and filters to the bearings. On leaving the bearings the oil drains back to the reservoir.

The reservoir is designed to permit circulation of its entire contents between eight to 12 times per hour. Oil level and temperature are constantly monitored. The oil can be preheated electrically or indirectly by steam for starting up at low temperatures. A thermostat with surface temperature limiter prevents overheating of the oil. The reservoir is vented.

Oil is normally circulated by the main oil pump. An auxiliary pump serves as standby, as illustrated in Figure 17-13. These two pumps generally have different types of drive, or power supplies. When both are driven electrically, they are connected to separate supply networks. On compressors with step-up gearboxes the main oil pump may be driven mechanically from the gearbox and the auxiliary pump then operates during the start-up and run-down phases of the compressor train. Relief

Figure 17-3. Turbotrain assembly in progress. (Source: Sulzer Brothers, Winterthur, Switzerland)

valves protect both pumps from the effects of excessively high pressures. Check-valves prevent reverse flow of oil through the stationary pump.

Heat generated by friction in the bearings is transferred to the cooling medium in the oil coolers. The return temperature is monitored by a temperature switch. Air-cooled oil coolers may be employed as an alternative to water as a coolant. The former have long been used in regions where water is in short supply. Twin coolers with provision for changeover have filling and venting connections so that the standby cooler can be filled with oil prior to changing over. This eliminates the possibility of disturbances and damage due to air bubbles in the piping system. Twin oil filters with provision for changeover have the same facilities.

A pressure-regulating valve is controlled via the pressure downstream of the filters and maintains constant oil pressure by regulating the quantity of bypassed oil. The auxiliary oil pump is switched on by a pressure switch if the oil pressure falls. A second pressure switch shuts down the compressor train if the pressure still continues to fall.

The filters clean the lube oil before it reaches the lubrication points. A differential pressure gauge monitors the degree of fouling of the filters.

An overhead oil tank can be provided to ensure a supply of lubricant to the bearings in the event of faults while the compressor is being run down. A continuous flow of oil through an orifice maintains the header oil constantly at operating temperature. Should the pressure in the lube oil system fall, the non-return valve beneath the tank opens to provide a flow of oil.

The flow of oil to each bearing is regulated individually by orifices, particularly important for lubrication points requiring different pressures. Lube oil for the driver and other users is taken from branch lines.

When a hydraulic shaft position indicator is used, this is supplied with oil from the lube oil system.

Temperatures and pressures are measured at all important locations in the system; the readings can be taken locally or transmitted to a monitoring station.

Except for a few components, the lube oil system is conveniently installed in a packaged unit supplied complete and ready for installation. Oil pumps, coolers and filters are grouped around the oil reservoir on a common baseplate. Design and construction of the lube oil system takes into account relevant regulations and any special requirements.

Lube oil systems for reciprocating compressors are generally simpler than the system for dynamic compressors described above. Figure 17-14 represents the oil supply schematic for a conventional reciprocating machine.

SEAL OIL SYSTEM

The seal liquid system (Figure 17-15) supplies mechanical contact and floating ring seals with an adequate flow of seal liquid at all times, thus ensuring that they function correctly. An effective seal is provided at the settle-out pressure when the compressor is not running. The seal oil system may be combined with the lube oil

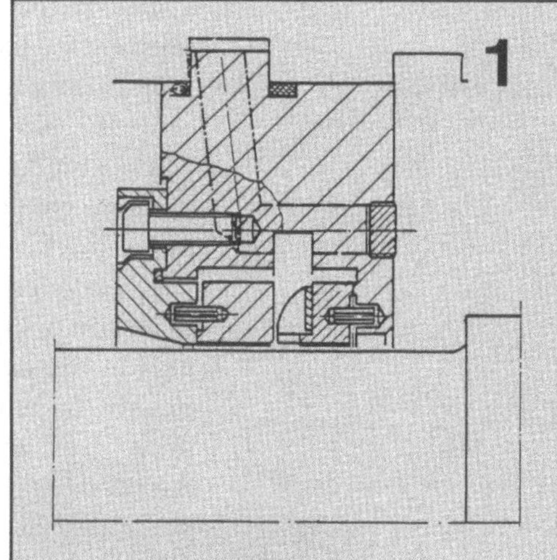

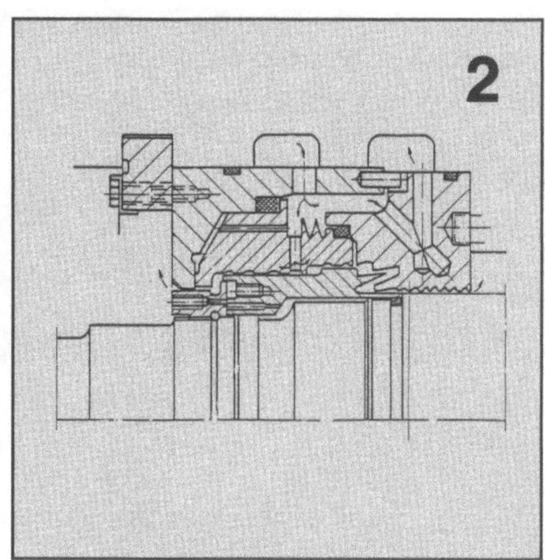

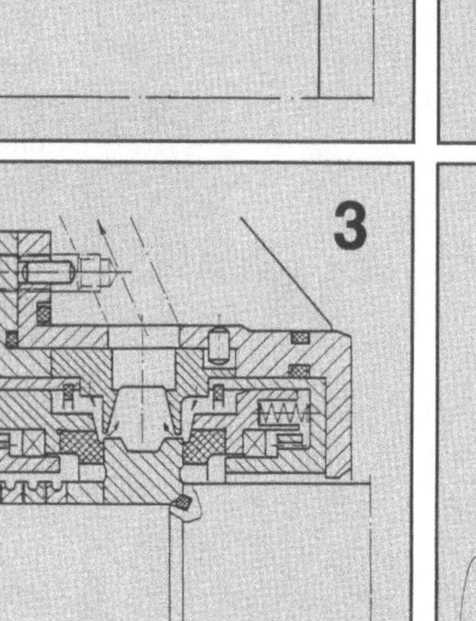

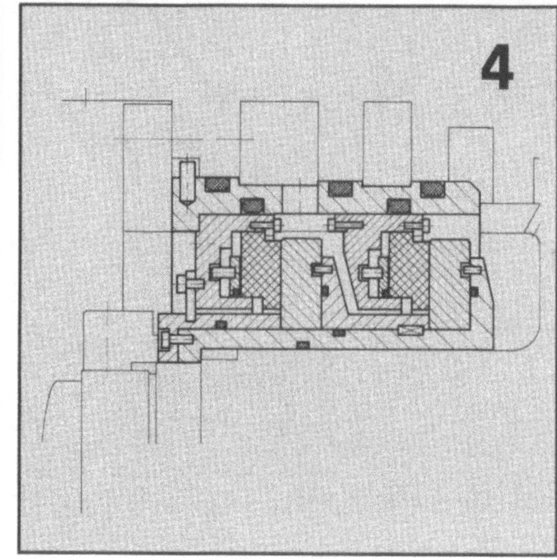

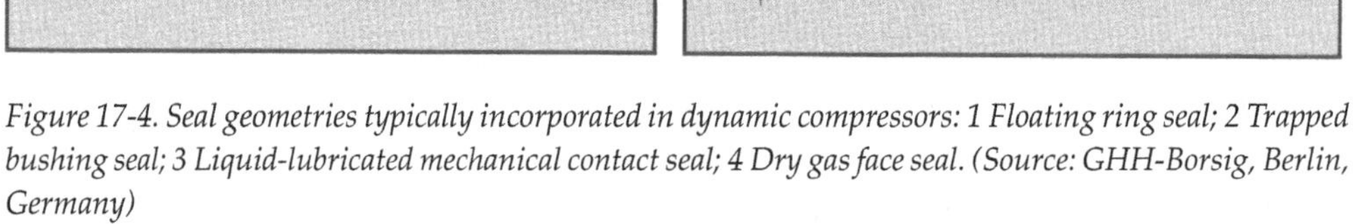

Figure 17-4. Seal geometries typically incorporated in dynamic compressors: 1 Floating ring seal; 2 Trapped bushing seal; 3 Liquid-lubricated mechanical contact seal; 4 Dry gas face seal. (Source: GHH-Borsig, Berlin, Germany)

system if the gas does not adversely affect the lubricating qualities of the oil or provided the oil made unserviceable by the gas does not return into the oil system.

There are two methods of combining lube oil and seal oil systems. In the first of these (Figure 17-16), the oil can be raised to the pressure required for lubrication purposes and part of it then raised further to the pressure needed for sealing (booster system.) Alternatively, all the oil is initially raised to the seal oil pressure and the flow of oil required for lubrication then reduced in pressure (combined system).

Starting in the main oil reservoir, the medium passes to the seals via the pumps, the twin oil coolers and the twin filters.

Instruments for monitoring the oil level and temperature are mounted on the reservoir. If necessary, the seal oil is heated; a thermostat with surface temperature limiter protects against excessively high temperatures.

Every system has a main oil pump and an auxiliary oil pump with independent drives. They are designed for a higher delivery rate than is actually needed by the seals. Safety valves protect the pumps and equipment downstream. Non-return valves after each pump prevent seal oil from flowing back to the reservoir through the non-running pumps.

The coolers dissipate the heat transferred to the seal oil. A temperature switch monitors the permissible temperature range.

The filters retain all impurities, the pressure drop across them being checked by a differential pressure indicator. Mechanical face seals and floating ring seals are supplied with seal oil at a defined differential pressure above the reference gas pressure (pressure within the inner seal drain). The flow of seal oil is regulated by a differential pressure regulating valve which, if there are changes in the reference gas pressure, regulates the pressure of the seal oil or, as shown in Figure 17-16, by a level-control valve that maintains a constant level in the overhead tank. The oil in the overhead tank is in contact with the reference gas pressure via a separate line. The static head provides the required pressure differential. In addition, the oil in the overhead tank compensates for pressure fluctuations and serves as a rundown supply if pressure is lost. If the level in the tank falls excessively, a level switch shuts down the compressor plant. There is a constant flow of oil through the overhead tank, and this heats the oil at all times.

For the mechanical contact seal the seal oil is kept at a constant differential pressure with respect to the

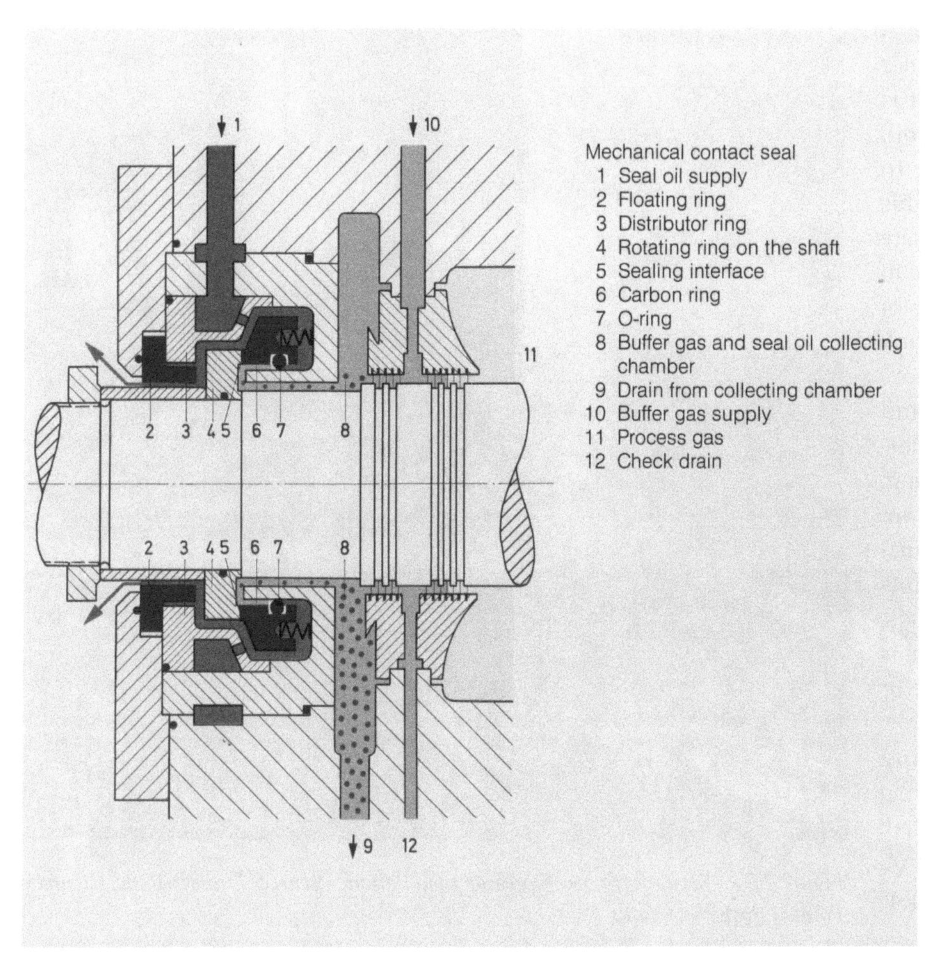

Mechanical contact seal

The mechanical contact seal employs a stationary carbon ring in sliding contact with a rotating ring manufactured from high-quality material with a special finish. A seal liquid is employed. This type of seal is also effective when the compressor is at standstill and the oil pumps have been shut down.

The main components are the carbon ring and the rotating ring for inward sealing. In the outboard direction a floating ring controls the flow of the seal liquid which cools the seal.

The seal liquid enters the seal via the supply pipe 1 and flushes the seal ring components via the holes in the distributor ring. The pressure of the liquid is higher than that of the gas, so that the carbon ring, under constant spring pressure, is always kept in sliding contact with the rotating ring. Some of the liquid wets the sliding surface 5 and reduces wear. Only a very small proportion of this liquid passes to the gas side. A controlled flow of buffer gas flowing from the supply pipe 10 through a labyrinth to port 8 entrains this leakage liquid and leads it via outlet 9 to the separator.

O-rings fitted externally and within the seal reliably separate the buffer gas and seal liquid spaces.

Figure 17-5. Mechanical contact seal showing seal oil inlet at point 1. (Source: Mannesmann-Demag, Duisburg, Germany)

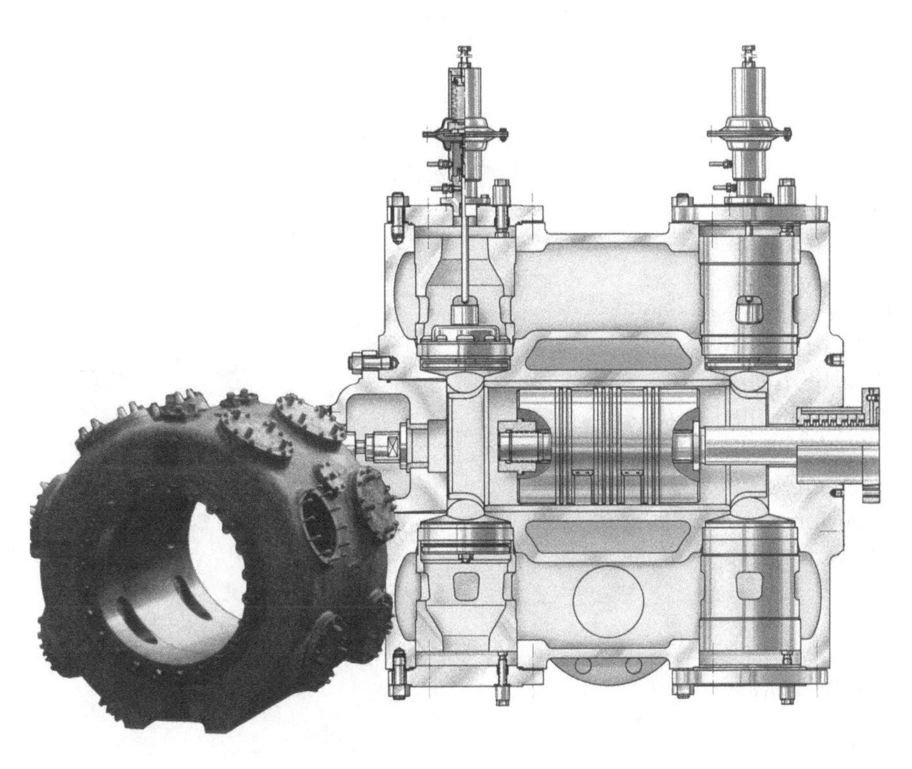

reference gas by a regulating valve. As the name indicates, the mechanical contact seal provides a mechanical standstill seal when the compressor plant is shut down.

To prevent oil from gaining ingress to the compressor, the space between the oil drain and compression space is sealed by a flow of gas. The pressure of this sealing gas is above the pressure of the reference gas. A differential pressure indicator monitors the pressure differential.

The flow of seal oil divides in the compressor seals. Most of the flow

Figure 17-6. Reciprocating compressor cylinders require lubrication in the piston rod packing and piston ring areas. Source: GHH-Borsig, Berlin, Germany)

returns under gravity to the reservoir. A small quantity passes through the inner seal ring to the inner drain, where it is exposed to the gas pressure. This oil, mixed with the buffer gas, is led to the separator system. On each side this consists of a separator and a condensate trap. The separated gas is led either to the flare stack or to the suction side of the compressor. The oil flows into a tank for degassing.

If oil is used as sealing liquid and can be used again, degassing is accelerated by heating or air or N_2 sparging. The oil is then returned to the reservoir. Sparging units perform on-stream purification of oil which can keep lubricants serviceable for very long time periods. Only if the oil becomes unusable is it led away for separate treatment or disposal.

The quantities of oil passing through the inner drain in modern centrifugal compressors are very small. Recall, however, that there is no seal oil system on compressors using dry gas seals (item 4, Figure 17-4).

Temperature and pressure measuring points with local or remote reading are provided at all major points of the seal liquid system.

Figure 17-7. Reciprocating compressor installation. (Source: Dresser-Rand Company, Painted Post, New York)

COMPRESSOR LUBRICANTS

The overwhelming majority of compressors are best served by premium grade turbine oils with ISO viscosities 32 or 46. Many of these lubricants were discussed in Chapters 4 and 5 of this text. However, there are many different types of compressors and each manufacturer is likely to recommend only those lubricants that have been used on his test stand and at controlled user facilities.

Occasionally, compressor lubricants have to be formulated for exceptional severe-service performance. Several of these lubricants are described here for comparison purposes.

Figure 17-8. Large lobe-type rotary piston blower. (Source: Aerzen USA, Coatesville, Pennsylvania, USA)

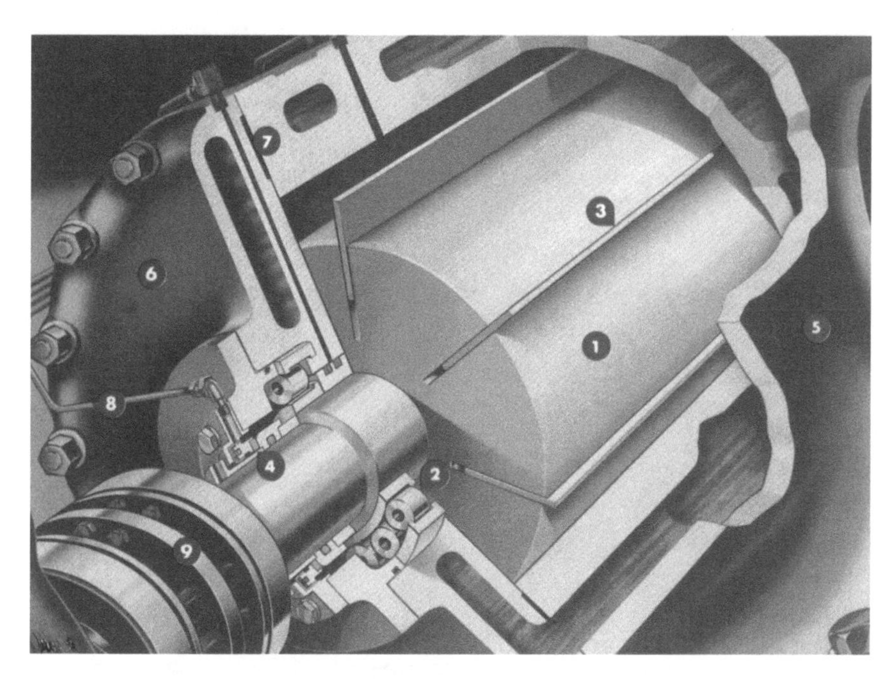

Figure 17-9. Sliding vane compressor and principal components: rotor and shaft (1), bearings (2), blades (3), mechanical seals (4), cylinder/housing (5), heads/covers (6) gaskets (7), lube supply line (8), coupling (9). (Source: A-C Compressor Corporation, Appleton, Wisconsin, USA)

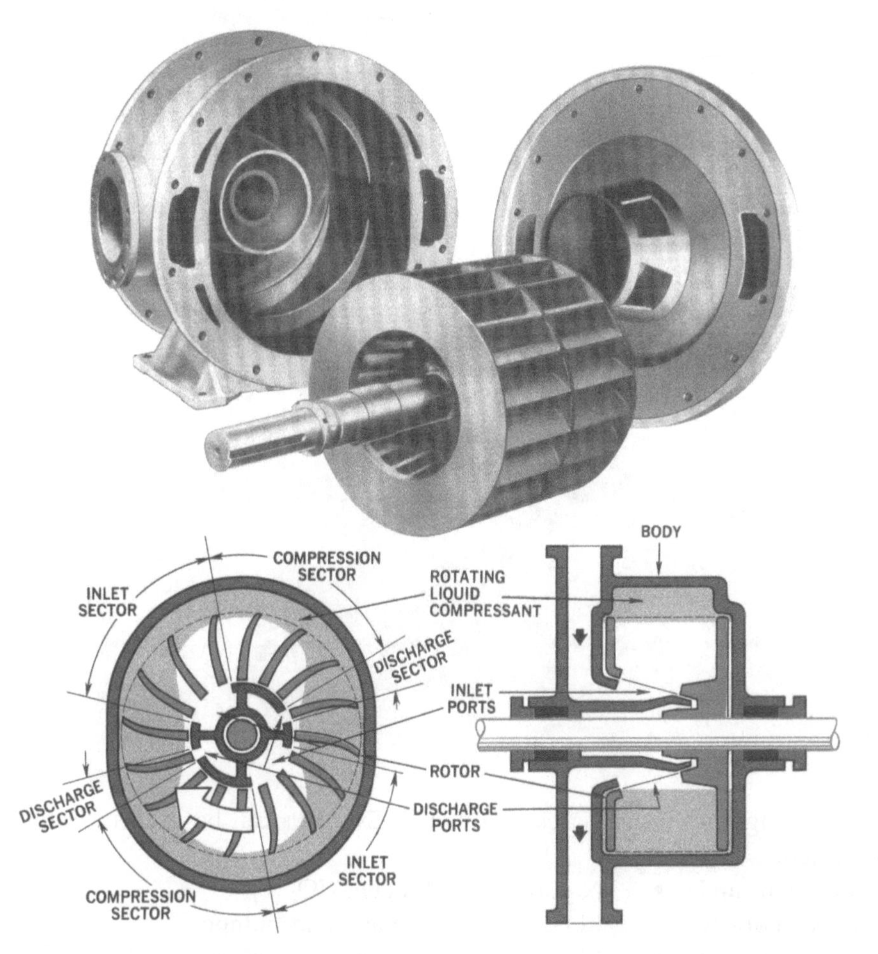

Figure 17-10. Liquid ring compressor with elongated casing (A), and schematic section at inlet and discharge sectors (B). (Source: Nash Engineering, Norwalk, Connecticut, USA)

Exxon Exxcolub SLG is designed for cylinder and packing lubrication of reciprocating compressors, and Exxcolub SRS for rotary screw compressors. Their performance advantages include:

- Low hydrocarbon solubility
- Minimal lubricant viscosity loss
- Exceptional control of sludge, varnish, or lacquer formation
- High viscosity indexes
- Outstanding oxidative and thermal stability
- Excellent wear protection
- Low vapor pressure
- Excellent lubricity
- Good water solubility
- Non-poisoning to most catalysts

Formulated with a polyalkylene glycol (PAG) polymer basestock, Exxcolub exceeds the capabilities of petroleum-base and many synthetic-base lubricants in this severe-service application. The excellent oxidative and thermal stability of Exxcolub lubricants assure long service life in high-temperature operations. Their inherently high viscosity indexes, from 200 to 226, facilitate low-temperature startup and help maintain acceptable viscosity over a wide temperature range. Exxcolub lubricants have outstanding lubricity. Their proven additive technology provides enhanced protection against wear, oxidation, corrosion and foam.

Polyalkylene glycols are highly stable even at sustained high temperatures and thus have very low deposit-forming tendency. Any decomposition products that may form are soluble in

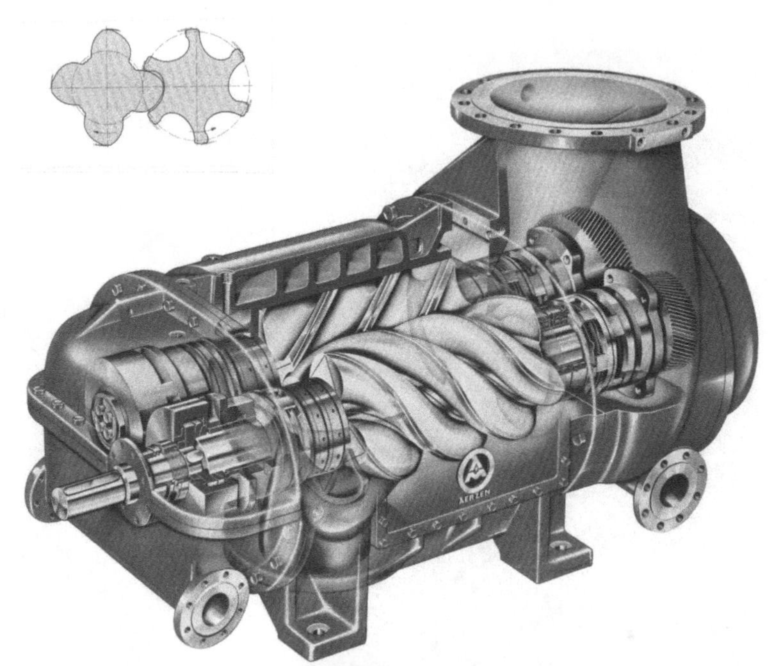

Figure 17-11. Helical screw compressor. (Source: Aerzen USA, Coatesville, Pennsylvania).

Figure 17-12. Lube/seal oil skid for a centrifugal compressor. (Source: Elliott Company, Jeannette, Pennsylvania).

the lubricant and do not tend to separate as sludge or contribute to the formation of varnish or lacquer.

The low solubility of hydrocarbon, nitrogen, and CO_2 gases in EXXCOLUB helps maintain proper viscosity and, along with low vapor pressure, minimizes lubricant consumption. These lubricants also can tolerate significant amounts of moisture with little effect on lubrication efficacy.

Gases with which EXXCOLUB can be used include:

- Natural gas, nitrogen, CO_2
- LPG, such as propane and butane
- LNG, such as methane and ethane
- Hydrocarbon chemical gases, such as ethylene, propylene and butylene
- Landfill gas

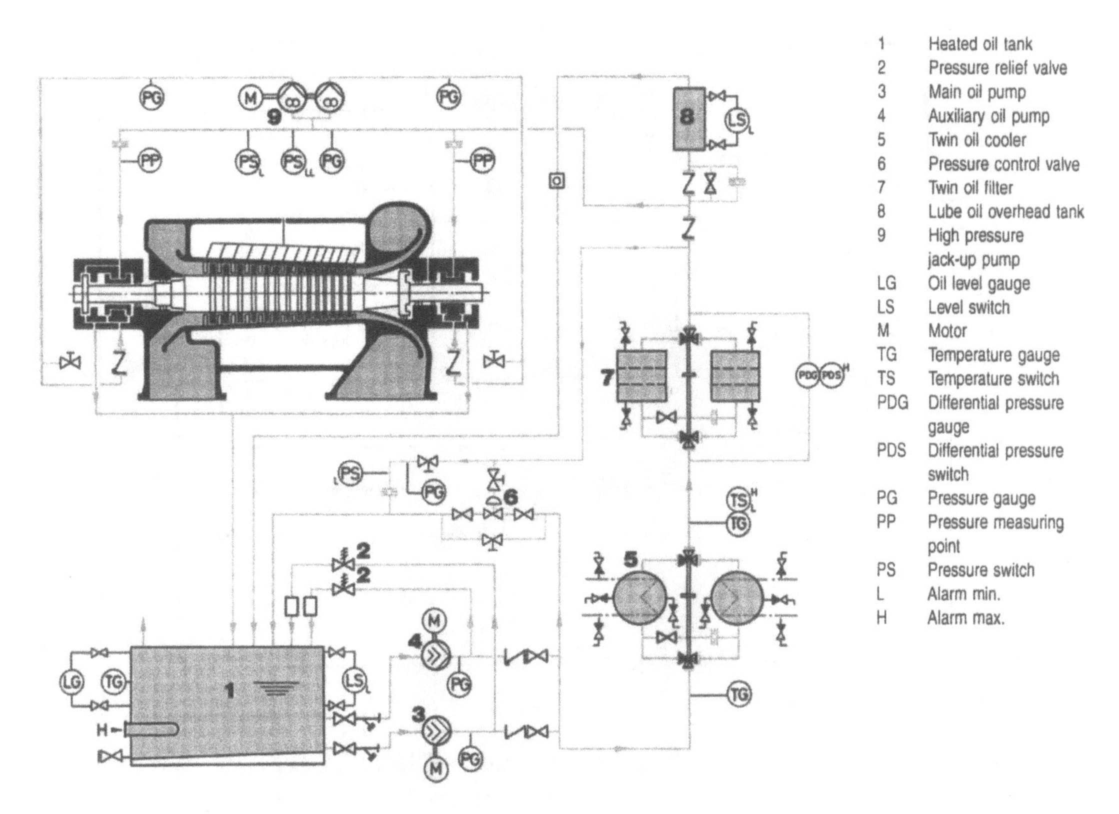

Figure 17-13. Typical lube oil schematic for turbomachinery. (Source: Mannesmann-Demag, Duisburg, Germany)

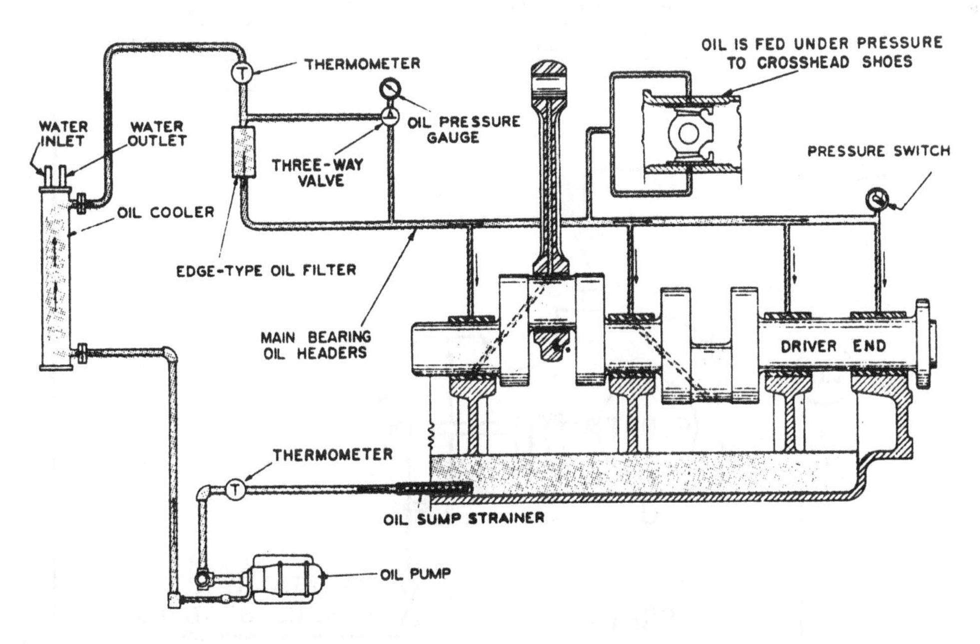

Figure 17-14. Reciprocating compressor lube oil system. (Source: Dresser-Rand, Painted Post, New York)

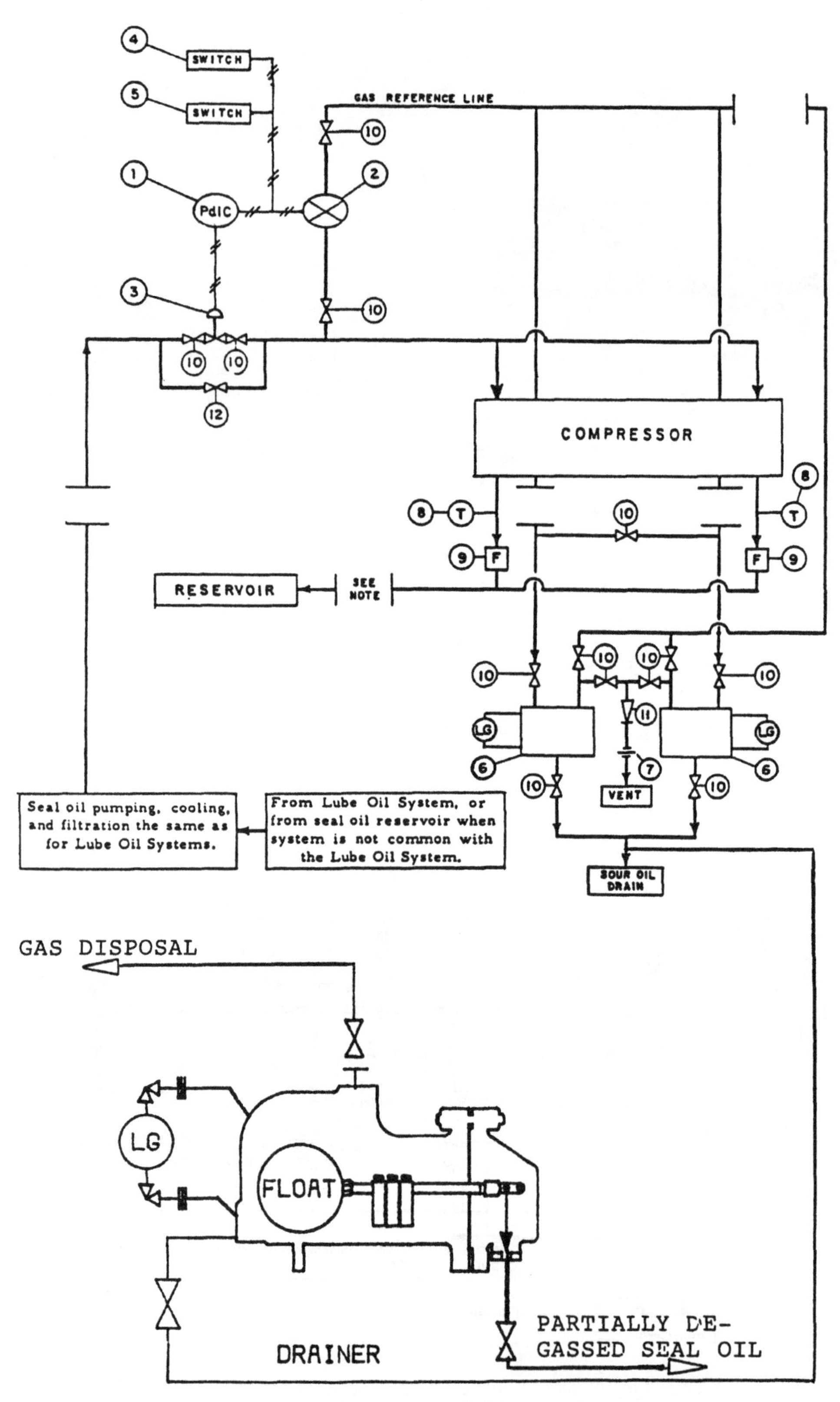

Figure 17-15. Compressor seal oil system schematic.

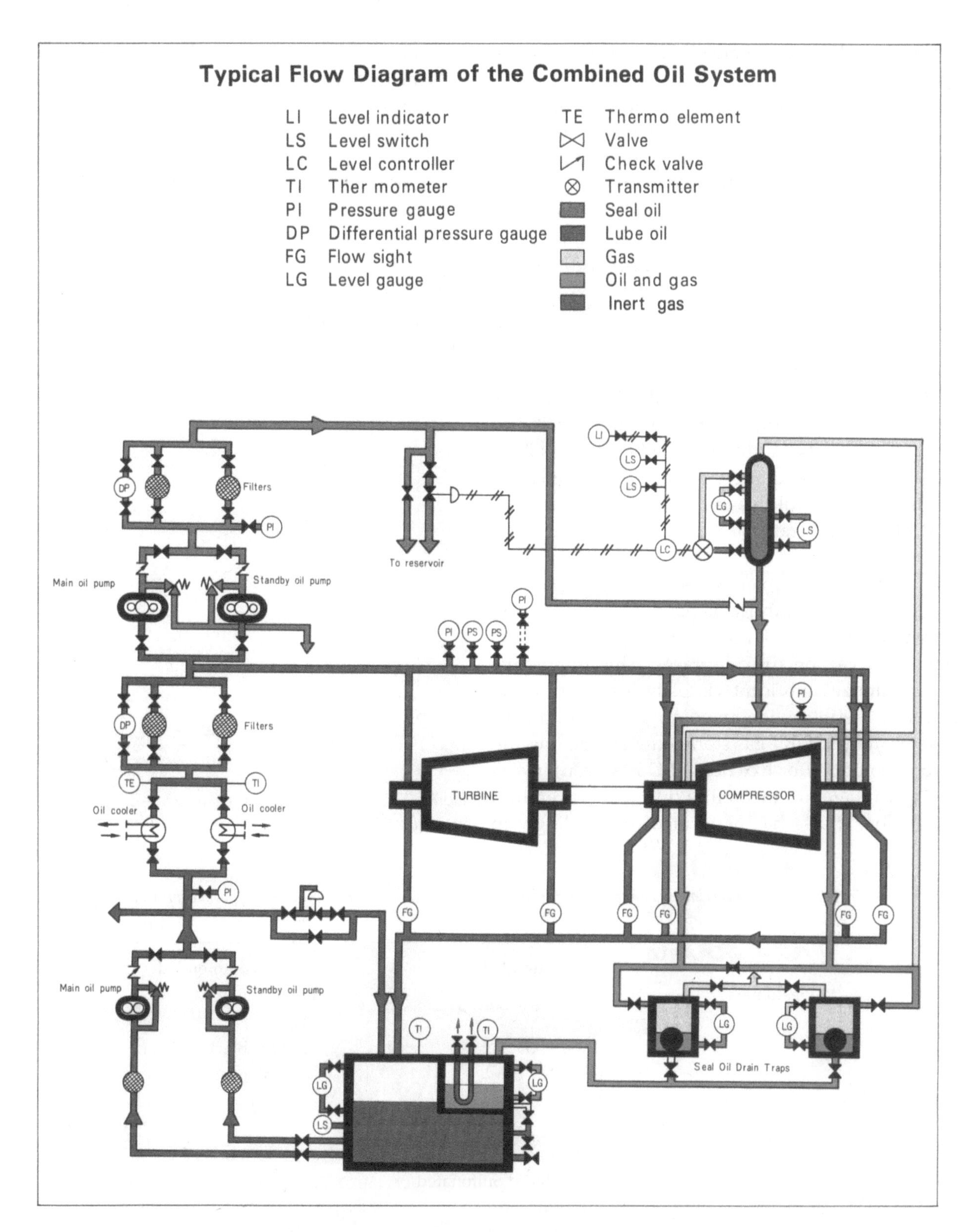

Figure 17-16. Combined lube and seal oil system for turbo compressors. (Source: Hitachi, Tokyo, Japan)

EXXCOLUB lubricants are compatible with the elastomers and coatings listed in Table 17-1. If you are uncertain about lubricant compatibility, consult with the equipment manufacturer.

EXXCOLUB SLG 100 and EXXCOLUB SLG 190 are specifically designed for cylinder and packing lubrication of reciprocating compressors in hydrocarbon and chemical gas service. The viscosity of the 100 grade is equivalent to ISO 100; the viscosity of the 190 grade is intermediate between ISO 150 and ISO 220. Both grades offer the same high-performance advantages. Additionally, EXXCOLUB SLG 100 is specially formulated to be non-poisoning to most catalysts.

Solubility Considerations

With a conventional mineral oil, compressed gas becomes dissolved in the oil, rapidly diluting it and lowering its viscosity. Additionally, lubricant dissolved in the gas can be carried away, depleting the lubricating film in the cylinder. All this can result in cylinder scoring and higher wear rates in the packing, cylinder, and rings and rider bands.

In contrast to petroleum-base oils, EXXCOLUB SLG synthetic compressor lubricants have very low gas solubility and are thus much less affected by hydrocarbon, nitrogen and carbon dioxide gases. This distinctive feature minimizes lubricant viscosity loss, permitting long service life in a wide range of gas environments. In underground natural gas storage fields, the low hydrocarbon solubility of EXXCOLUB SLG has been proven highly advantageous in minimizing carry-over into the formation. It also has performed outstandingly in high-pressure (> 6000 psi) reciprocating compressors in nitrogen and hydrocarbon service.

Figure 17-17 compares the solubility of EXXCOLUB SLG in methane gas, compared with that of a polyalphaolefin (PAO) lubricant and a petroleum-base lubricant; EXXCOLUB SLG exhibits significantly lower solubility. Figure 17-18 compares the effect of methane gas pressure on the viscosity of EXXCOLUB SLG with that of a petroleum-base oil; EXXCOLUB is clearly superior in maintaining viscosity.

EXXCOLUB SLG compressor lubricants further differ from a petroleum-base lubricant in being water soluble. This characteristic is highly desirable in oil field formation gas injection applications, since any lubricant car-

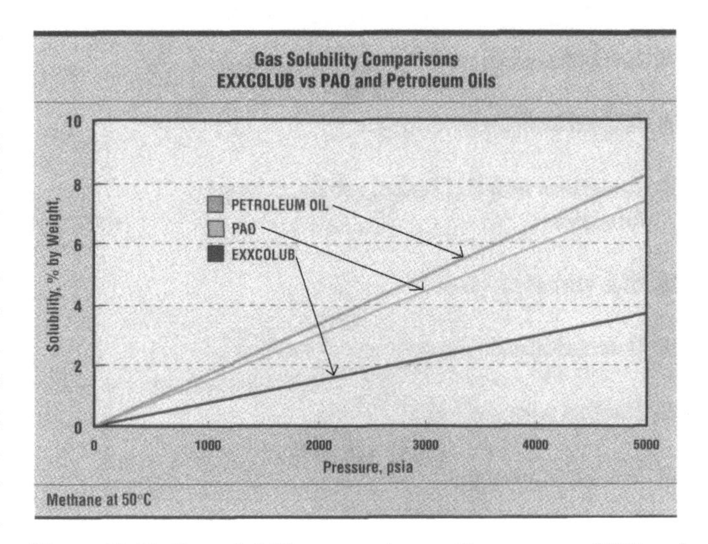

Figure 17-17. Gas solubility comparisons—EXXCOLUB vs. PAO and petroleum oils.

Table 17-1. EXXCOLUB elastomer and coatings compatibility. EXXCOLUB lubricants are compatible with a variety of elastomers and coatings.

	Recommended	Not Recommended
Elastomers	"Viton" A	Buna S
	"Kalrez"	"Hycar"
	Butyl K 53	Natural Black Rubber
	Buna N	"Hypalon"
	EPDM	
	Natural Gum Rubber	
	Neoprene	
	EPR Chloro Sulfonated	
	Polyethylene	
	"Thiokol" 3060	
	Polysulfide	
Coatings	Catalyzed Epoxy-Phenolic	Alkyd
	Modified Phenolic	Vinyl

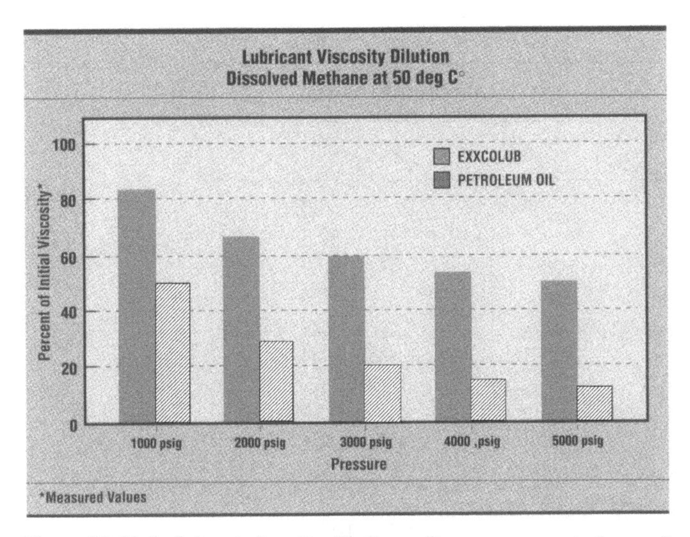

Figure 17-18. Lubricant viscosity dilution—EXXCOLUB vs. petroleum oil.

ried over into the formation will stay dissolved and not plug the oil field formation. In addition to water solubility and low gas solubility, EXXCOLUB SLG also has high VI, good film strength, excellent oxidative and thermal stability and resistance to sludge formation.

Extra Oxidation Resistance Available

EXXCOLUB SRS is a premium lubricant specifically designed for rotary screw compressor applications. It shares the performance features of EXXCOLUB SLG, plus extra oxidation resistance for long-life service in the closed circulatory systems of rotary screw compressors. EXXCOLUB SRS also is an excellent lower-viscosity alternative to EXXCOLUB SLG in reciprocating compressors.

The performance of EXXCOLUB SRS in severe operating environments surpasses that of conventional petroleum-base oils and many comparable synthetic lubricants. The lubricant exhibits exceptional oxidation and thermal stability, excellent film strength and lubricity at high temperatures, and very good viscosity characteristics. Its low hydrocarbon solubility permits it to maintain viscosity even during the violent mixing of lubricant and gas that is characteristic of rotary screw operations. This permits EXXCOLUB SRS to maintain a dependably strong lubricant film for assured wear protection.

Typical Inspections

The values shown in Table 17-2 are representative of 1998 production of the various grades of EXXCOLUB. Some are controlled by manufacturing specifications, while others are not. All may vary within modest ranges.

GLYCOLUBE for Severe-Service Performance

Exxon GLYCOLUBE is a line of premium extreme-pressure, multi-purpose synthetic lubricants designed for dependable performance over a wide range of temperatures and operating conditions. GLYCOLUBE synthetic lubricants are well suited for many types of industrial applications, including gears, pumps, compressors, drivers, mobile equipment and anti-wear hydraulics. Formulated from polyalkylene glycol (PAG) basestock, GLYCOLUBE reliably lubricates and protects against wear in severe-service conditions that may exceed the capabilities of conventional petroleum-base lubricants, as well as many other synthetic-base lubricants.

The exceptionally long and dependable service life of GLYCOLUBE is particularly advantageous in out-of-the-way locations where frequent lubrication is difficult or impractical.

The GLYCOLUBE formulation combines the inherently superior qualities of PAG basestock with proven additive technology to provide the following outstanding performance features:

- Superb oxidative and thermal stability
- High viscosity indexes
- Low pour points for easier cold-temperature startup
- Excellent lubricity for enhanced resistance to friction and wear
- Extreme-pressure lubrication without chorine- or sulfur-containing additives
- Resistance to mechanical breakdown at high shear rates
- High resistance to sludge and varnish formation
- Non-corrosivity to metal surfaces/stain resistant to non-ferrous metals

Polyalkylene glycols are very stable even at sustained high temperatures and thus have a low deposit-forming tendency. Decomposition products that may form are soluble in the lubricant and do not tend to separate as sludge or contribute to the formation of varnish or lacquer.

The excellent oxidative and thermal stability of GLYCOLUBE lubricants assures long lubricant service life, even under heavy load, high-temperature conditions. These performance features are highly cost-effective in their ability to significantly reduce lubricant consumption and maintenance shutdowns.

The inherently high viscosity index of the PAG basestock both facilitates low-temperature startup and helps maintain acceptable viscosity over a wide temperature

Table 17-2. Typical inspections for EXXCOLUB severe service compressor lubricants.

	EXXCOLUB SLG 100
Viscosity	
SUS @ 100°F (38°C)	400
210°F (99°C)	84.5
cSt @ 100°F (38°)	80.0
210°F (99°C)	16.3
Viscosity Index, ASTM D 2270	220
Pour Point, ASTM D 97, °F (°C)	-55 (-48)
Flash Point ASTM D 92*, °F (°C)	450 (232)
Water Content, maximum, %	0.15
Ash Content, maximum, %	0.01
Specific Gravity, 20/20°C	1.049
Weight per Gallon, lb (Liter, g)	
@ 60°F (16°C)	8.76 (1053)
@ 68°F (20°C)	8.73 (1050)

*Cleveland Open Cup

Grade	EXXCOLUB SLG 190
Viscosity, cSt	
@ 40°C	195
@ 100°C	35.1
@ 100°F	985
Viscosity Index	226
Pour point, °C(°F)	-43 (-45)
Flash point,	
Cleveland Open Cup, ASTM D 92, °C(°F)	304(580)
Pensky-Martens Closed Cup, ASTM D 93, °C(°F)	240(465)
Turbine Oil Rust Test, ASTM D 665	Pass
Copper Strip Corrosion Test	1a
FZG Extreme Pressure Test Stages Passed	12
Coefficient of Expansion, at 55°C	0.00079 per °C
Density, lb/gal, 20°C	8.75
Specific Gravity, 20/20°C (68/68°F)	1.05150

Grade	EXXCOLUB SRS 100	EXXCOLUB SRS 150
Viscosity, cSt		
@ 100°F(40°C)	91(84)	152(140)
@ 210°F(100°C)	16(15.6)	26(25.4)
SUS @ 100°F	425	700
Viscosity Index	200	215
Pour Point, °F(°C)	–45(–43)	-40(-40)
Specific Gravity, 20/20°C	1.045	1.047
Flash point, ASTM D 93, °F(°C)	390(199)	345(174)
Fire Point, ASTM D 92, °F(°C)	545(285)	575(302)

range. This eliminates the need for seasonal lubricant changeovers and simplifies lubricant inventories.

In compressor applications, the low solubility of hydrocarbon, nitrogen and CO_2 gases in GLYCOLUBE minimizes lubricant consumption and thereby maintains viscosity at effective levels. While suitable for all types of compressors, GLYCOLUBE is particularly recommended for centrifugal types. For reciprocating and rotary screw compressors, Exxon's EXXCOLUB line of lubricants is the first choice.

GLYCOLUBE lubricants are compatible with the elastomers and coatings listed in Table 17-3. If you are uncertain about lubricant compatibility, consult with the equipment manufacturer.

The exceptional versatility of GLYCOLUBE products may enable a plant to consolidate lubricants, thereby simplifying inventory and reducing the chances of lubricant misapplication. Some of the more common applications for GLYCOLUBE are listed in Table 17-4, while Table 17-5 gives typical specifications. For possible applications not listed here, the lubricant manufacturer's specialist work force can analyze specific lubrication needs and make an appropriate recommendation.

Table 17-3. Compatibility chart for GLYCOLUBE lubricants.

	Recommended	Not Recommended
Elastomers	"Viton" A	Buna S
	"Kalrez"	"Hycar"
	Butyl K 53	Natural Black Rubber
	Buna N	"Hypalon"
	EPDM	
	Natural Gum Rubber	
	Neoprene	
	EPR Chloro Sulfonated	
	Polyethylene	
	"Thiokol" 3060	
	Polysulfide	
Coatings	Catalyzed Epoxy-Phenolic	Alkyd
	Modified Phenolic	Vinyl

Table 17-4. Typical inspections for GLYCOLUBE lubricants. The values shown here are representative of current production. Some are controlled by manufacturing specifications, while others are not. All may vary within modest ranges.

	GLYCOLUBE Lubricants						
Property	**32**	**46**	**68**	**150**	**220**	**460**	**680**
Viscosity, ISO Grade, cSt	32	46	68	150	220	460	680
@ 100°F	39	52	66	143	263	389	615
@210°F	6.8	9.0	11.0	21.9	39.0	55.8	85.9
Viscosity Index	160	170	174	193	212	220	235
Pour point, °C(°F)	-43(-45)	-40(-40)	-40(-40)	-32)-25)	-29(-20)	-29(-20)	-27(-17)
Flash point, °C(°F)							
Pensky-Martens Closed Cup[1]	191(375)	138(280)	218(425)	166(330)	221(430)	183(362)	193(380)
Cleveland Open Cup[2]	246(475)	254(490)	277(530)	279(535)	291(555)	293(560)	282(540)
Water Content, % by wt	<0.25	<0.25	<0.25	<0.25	<0.25	<0.25	<0.25
Weight per Gallon at 60°F, lb	8.24	8.28	8.31	8.36	8.37	8.38	8.41
"Falex" Wear Test[3], Failure Load, psi	2900	2900	2900	2900	2900	2900	2900
Steam Turbine Rust Test[4],	Pass	Pass	Pass	Pass	Pass	Pass	Pass
Copper Strip Corrosion Test[5]	1a	1a	1a	1a	1a	1a	1a
Coefficient of Friction[6]	..........0.0184..........						
Foam Tendency[7]							
@ 75°F	0	0	0	0	0	0	0
@ 200°F	0	0	0	0	0	0	0
@ 75°F	0	0	0	0	0	0	0

[1]ASTM D 92
[2]ASTM D 93
[3]ASTM D 3704
[4]ASTM D 665
[5]ASTM D 130
[6]Phosphorus bronze on steel, determined by David Brown Gear Industries Limited.
[7]ASTM D 892

Table 17-5. Applications for Glycolube lubricants.

		Glycolube Lubricants						
EQUIPMENT	**APPLICATIONS**	**32**	**46**	**68**	**150**	**220**	**460**	**680**
Compressors	Axial	√	√	√				
	Radial	√	√	√				
	Radial Integral Gear	√	√	√				
	Reciprocating Crosshead Cylinder, Packing & Crankcase)				√	√	√	√
	Reciprocating Diaphragm (Crankcase)				√	√	√	√
	Reciprocating Trunk (Crankcase)				√	√	√	√
	Rotary Liquid Ring Rotor	√	√	√				
	Rotary Lobe (Gears)	√	√	√	√	√	√	√
	Rotary Screw (Dry & Oil-Flooded)	√	√	√	√	√	√	√
	Rotary Vane (Rotor)	√	√	√	√	√	√	√
Drivers	Chain Drives	√	√	√	√	√	√	√
	Electric Motors (Fixed & Variable Speed)	√	√	√				
	Expansion Turbine	√	√	√				
	Fluid Drive Couplings	√	√	√				
	Gas Turbine	√	√	√				
	Generators	√	√	√				
	Hydraulic-Driven Turbine	√	√	√				
	Steam-Driven Turbine	√	√	√				
Gears	Gears of All Types	√	√	√	√	√	√	√
Hydraulics	Fluid Drive Couplings	√	√	√				
	Hydraulic-Driven Turbine	√	√	√				
	Hydrodynamic Hydraulic Systems	√	√	√				
	Hydrostatic Drives	√	√	√				
Mobile	Cherry Pickers	√	√	√	√	√	√	√
	End Loaders	√	√	√	√	√	√	√
	Tractors	√	√	√	√	√	√	√
Pumps	Axial Piston	√	√	√	√	√	√	√
	Bent Axial Piston	√	√	√	√	√	√	√
	Centrifugal	√	√	√	√	√	√	√
	Diaphragm	√	√	√	√	√	√	√
	Double Diaphragm Fluid	√	√	√	√	√	√	√
	External Gear	√	√	√	√	√	√	√
	Internal Gear	√	√	√	√	√	√	√
	Liquid Ring	√	√	√	√	√	√	√
	Radial Piston	√	√	√	√	√	√	√
	Reciprocating	√	√	√	√	√	√	√
	Reciprocating Diaphragm	√	√	√	√	√	√	√
	Rotary Lobe	√	√	√	√	√	√	√
	Rotary Screw	√	√	√	√	√	√	√
	Rotary Vane	√	√	√	√	√	√	√
	Triplex	√	√	√	√	√	√	√
Others	Calendars	√	√	√	√	√	√	√
	Conveyors	√	√	√	√	√	√	√
	Crushers	√	√	√	√	√	√	√
	Grinders	√	√	√	√	√	√	√

Procedures for Changing over to Exxcolub or Glycolube

When preparing to change over to Exxcolub or Glycolube, it is important to determine its compatibility with the former lubricant. Exxon can assist you in making this determination. If the two lubricants are shown to be incompatible, employ the following procedures *before* installing either lubricant. At a minimum, drain the old lubricant, clean the system to remove possible sludge and varnish, inspect seals and elastomers and replace the filters or clean the screens. If residual contamination is suspected, wipe or flush with a small amount of solvent or EXXCOLUB/GLYCOLUBE ; in new units, follow the same procedure to remove preservative or coating fluids.

After installing the lubricant, adjust the lubricators to deliver the manufacturer's recommended rate of lubricant. Check the filters or screens frequently during the early stages of operation, since EXXCOLUB and/or GLYCOLUBE will likely loosen residual sludge, varnish and paint. If the manufacturer's recommended grade of lubricant is not available, consult Table 17-6, later.

NOTE: In gearbox applications, after 24 hours of operation, the lubricant should be drained and the gearbox refilled.

It is also important to determine the compatibility of these lubricants with the elastomers and coatings in the system. For assistance in this determination, refer to the earlier Tables 17-1 and 17-3.

For Exxon's 1998-vintage recommendations on compressor lubricants, consult the following pages. Recall, however, that these Exxon lubricant recommendations are based on manufacturer recommendations or specifications contained in Exxon's MAC (Manufacturers' Acceptance and Classification) Sheets. **IMPORTANT: These recommendations are intended as a guide only. If there is a discrepancy between the recommendations given here and those provided in the equipment manufacturer's manual, the latter must take precedence.**

Before using a synthetic lubricant such as SYNESSTIC in any application, consult with the manufacturer to ensure compatibility with paints, elastomers, plastics, filter plugs, O-rings, etc. The Exxon representative can answer any questions regarding the use of SYNESSTIC synthetic industrial lubricants in any compressor not listed below.

Exxon Lubricant Recommendations for Compressors

Ro-Flo* Rotary Compressors
A-C COMPRESSOR CORPORATION
Appleton, Wisconsin

The **type** and **grade** of oil recommended by A-C Compressor for use with Ro-Flo units is usually indicated on the name plate by a letter and SAE number. The SAE number will normally correspond to that in the table below for the extended discharge gas temperature and for inlet gas temperatures above 32°F.

Discharge Gas Temp. °F	SAE Number
Below 200	20
200 to 250	30
250 to 300	40
Below 300	50

If gas handled is expected to condense and dilute the oil, use the next higher viscosity grade. Multi-viscosity grades of oil are recommended for air inlet temperatures below 32°F, as are lubricator reservoir heaters and thermostats. On multi-stage units, use the highest discharge temperature to select the oil viscosity.

A-C Compressor Oil Type	Exxon Lubricant
A	ZERICE N 68 or ZERICE 68
B	XD-3
C	TERESSTIC 100
D	No product
E	XD-3
R	No product
S	No product

Dry Screw Compressors

Rotary, axial flow (Models 67S, 85S, 85L, 105S, 105L, 132S, 132L, 165S, 165L, 200L and 250L)	TERESSTIC 32

** RO-Flo is an A-C Compressor trademark*

ATLAS COPCO COMPRESSORS, INC. Holyoke, Massachusetts	Exxon Lubricant
Air Compressor Type	
BE – BT – IE	
Above -10°C	TERESSTIC 68 or TERESSTIC SHP 68 XD-3 10W or 20W-20
BP 3	TERESSTIC SHP 68
DR – DT – DA – Inclusive Pack Above -10°C	TERESSTIC 68 or TERESSTIC SHP 68 XD-3 10W or 20W-20
GA Pack – GA – GAR	
Above 25°C	NUTO H 68 or UNIVIS N 68
Between -10°C and 25°C	NUTO H 32 or UNIVIS N 32
Below 0°C	UNIVIS N 15
Above -15°C	TERESSTIC SHP 46
GA Pack – GR Above -15°C	TERESSTIC SHP 46
ZR-ZA Pack, ZR-ZA Above -10°C	NUTO H 68 or UNIVIS N 68
Below -5°C	NUTO H 32 or UNIVIS N 32
ST	XD-3 10W
XA – XAH	
Above 25°C	NUTO H 68 or UNIVIS N 68
Between -10°C and 25°C	NUTO H 32 or UNIVIS N 32
Below 0°C	UNIVIS N 15
XR – XRH – XRV Above -15°C	TERESSTIC SHP 46
MA(S) – PN(S) – PT(H)(S) – PTMS – XT Above -10°C	XD-3 20W-20 or 15W-40
Below -5°C	XD-3 10W
EA – ER – ET Inclusive Pack/Package Above -10°C	TERESSTIC 68 or TERESSTIC SHP 68 XD-3 10W or 20W-20

continued next column

Air Compressor Type	Exxon Lubricant
LE – LT (AE, AT, ME, MT)	
-25°C to 40°C	XD-3 10W
-35°C to 50°C	SYNESSTIC 32
-25°C to 40°C	TERESSTIC SHP 32
ZT	
Above 0°C	NUTO H 68 or UNIVIS N 68
Below 5°C	NUTO H 32 or UNIVIS N 32
Above -10°C	XD-3 20W-20
Below -5°C	XD-3 10W
All temperatures	TERESSTIC SHP 68
TE – Line Compressor	
Above 10°C	XD-3 20W-20 or 15W-40
Below 20°C	XD-3 10W-30 or 10W
C-N	XD-3 20W-20 or 30, TERESSTIC 68 or 100

Portable Rotary "Mono-Rotor" Air Compressors
(Equipment no longer manufactured)

Compressor oil reservoir	XD-3 20W-20
Compressed air cleaner	
Above 0°F	XD-3
0 to -40°F	(1)
Miscellaneous oil applications	XD-3
Wheel bearings Suction control shaft All other grease-lubricated points	RONEX MP, LIDOK EP 2, UNIREX N 2 or POLYREX
Engine crankcase Gasoline or Diesel	XD-3 (See engine manual for ambient temperature oil grade recommendation)

(1) *Consult Atlas COPCO*

CHAMPION PNEUMATIC MACHINERY COMPANY, INC.
Princeton, Illinois

Single- and Two-Stage Compressors

Crankcase and cylinders	
32°F to 100°F	TERESSTIC 100

NOTE: Consult manufacturer for compressor operations below 32°F.

CHICAGO PNEUMATIC EQUIPMENT DIVISION
Franklin, Pennsylvania
(Chicago Pneumatic no longer manufactures this equipment, but many units remain in field use.)

Stationary Compressors and Vacuum Pumps

Crankcase lubrication (pressure and flood system)	
Above 90°F	TERESSTIC 150
30 °F to 90°F	TERESSTIC 100
-10°F to 30°F	ESSTIC 68
Crankcase lubrication (splash system)	
Above 90°F	TERESSTIC 100
30°F to 90°F	TERESSTIC 68
-10°F to 30°F	ESSTIC 32

continued next column

Stationary Compressors and Vacuum Pumps	Exxon Lubricant
Cylinder lubrication (force-feed lubricator)	
Pressures 0 to 1000 psi	
Type 1* gas service	TERESSTIC 68
Type 2* gas service	TERESSTIC 100
Type 3* gas service	AROX EP 150
Type 4* gas service	No product
Pressures 1000 to 2500 psi	
Type 1* gas service	TERESSTIC 150
Type 2* gas service	TERESSTIC 220
Type 3* gas service	AROX EP 150
Type 4* gas service	TERESSTIC 220
Type 5 – Chlorine, hydrogen chloride, oxygen, non-contaminated gases	Use no petroleum products for cylinder lubrication
Steam cylinder lubrication	
Steam temperature up to 610°F, pressure below 250 psia	CYLESSTIC TK 460
Steam temperature above 610°F, pressure above 250 psia	
Saturated steam	CYLESSTIC TK 1000
Superheated steam	CANTHUS 1000
Portable Compressors	
Reciprocating types	
Above 32°F	TERESSTIC 100
Below 32°F	TERESSTIC 68
Rotary "Power Vane"	SUPERFLO ATF(1) or TERESSTIC 32
Rotary "Singl-Screw" (Cylindrical)	TERESSTIC 32, 33 or 46(2)
Rotary "Singl-Screw" (Conical)	
Pressure up to 150 psi	SUPERFLO ATF or TERESSTIC 32, 33 or 46
Pressure above 150 psi	ETO 2380

* Type of gas service:
Type 1 – Dry gases, not carrying suspended liquids but may carry liquids that remain in the superheated state during the compression cycle, CO_2, N_2 helium, neon, and other inert gases, air, ammonia, hydrogen, methane
Type 2 – Hydrocarbon gases, such as butane, propane, butadiene, ethylene and nitrous oxide
Type 3 – Where water carry-over is a problem, in instances where wet gases are compressed, and when hydrogen sulfide is compressed
Type 4 – Methyl chloride, ethyl chloride, Freon, sulfur dioxide

(1) SUPERFLO ATF recommended for 2000-hour service.
(2) 1000-hour service maximum.

COMPAIR KELLOGG
Division of Robert Shaw Controls Company
Independence, Virginia

KRS Industrial Rotary Screw and Vane Compressors(1)	
Compressor oil	TERESSTIC 46, XD-3 20W-20, SUPERFLO ATF or SYNESSTIC 68
Reciprocating Compressor	
Compressor oil	
Ambient or room temperature	
55°F to 120°F	TERESSTIC 100
32°F to 55°F	TERESSTIC 68
0°F to 32°F	TERESSTIC 32 or 33
Hydrovane(1)	
Compressor oil	XD-3 40 or SYNESSTIC 100

(1) Equipment manufacturer recommends their private label diester-based lubricants.

DAVEY COMPRESSOR COMPANY Division of KECO Industries, Inc. Florence, Kentucky	Exxon Lubricant
PERMAVANE ROTARY COMPRESSORS **Air-Auto Models 85RA, 85PRAB, 125 PRAP, 160RPAB, 125RA and 160RA.** **Industrial Stationary Models 15BA, 20BA, 25BA, 25BW, 30BA,30BW, 40BA, 40BW, 50BA, 50BW, 75DA, 75DW, 100DA, 100DW, 150DA, 150DW.** Compressor reservoir Over 40°F	XD-3 EXTRA 30 or XD-3 30
Below 40°F	XD-3 EXTRA 30 or XD-3 10W-30
Oil bath air cleaner (if supplied) Speed control	Use same type and grade as compressor
Portable Models 85RPV, 125PVC, 150RPV, 160RPV, 190RPG, 190RPD, 200PVC, 250RPV, 365RPV, 600RPV, and 750RPV. Compressor reservoir Over 40°F	XD-3 EXTRA 30 or XD-3 30
Below 40°F	XD-3 EXTRA 30 or XD-3 10W-30
Engine crankcase	Follow engine manufacturer's recommendations
Speed control Oil filter Air cleaner Other hand oiling applications	Use same type and grade as compressor
Wheel bearings Spring shackles Other grease applications	RONEX MP or LIDOK EP 2
PORTABLE RECIPROCATING COMPRESSORS **Models 66RMP, 64DDC, 84DDC, 33RME** Compressor crankcase Above 90°F	XD-3 30
32°F to 90°F	XD-3 20W-20
15°F to 32°F	XD-3 10W
Engine crankcase	Use same oil as compressor or follow engine manufacturer's recommendations
Compressor and engine air cleaners Starter and generator Other hand oiling applications	Use same type and grade as compressor
Unloader cylinders, water pump, spring shackles, wheel bearings All other grease applications	RONEX MP or LIDOK EP 2
Magneto Oil-lubricated	ESSTIC 32
Grease-lubricated	RONEX

NOTE: If the intake of any of the compressors shown above is in a location of high humidity or if any of the compressors is not equipped with a thermal bypass valve, use TERESSTIC 68 or ESSTIC 68 in the compressor reservoir.

DELAVAL TURBINE DIVISION IMO Industries, Inc. (Formerly TransAmerica Delaval, Inc.) Trenton, New Jersey	
Centrifugal Compressors Direct drive and normally loaded high-speed geared units With and without oil seals and not in NH3 refrigeration service	TERESSTIC 32 or 33
With oil seals and in NH3 refrigeration service	No product
All units Grease-lubricated parts	RONEX MP, LIDOK EP 2, UNIREX N 2 or POLYREX

DEVILBISS AIR COMPRESSOR PRODUCTS Division of ITW Canada Barrie, Ontario, Canada	Exxon Lubricant
Compressor Models No. 120, 121, 123, 130, 220, 225, 230, 235, 330, 342, 432, 445, 1202, 44633, 44642 and 44643 Compressor crankcase Below 40°F	TERESSTIC 32 or 33
Above 40°F	TERESSTIC 46
Above 90°F	TERESSTIC 68

ELLIOTT COMPANY United Technologies Corporation Jeannette, Pennsylvania	
Centrifugal Compressors Bearings Seals Couplings	TERESSTIC 32
Axial Compressors Bearings	TERESSTIC 32

FULLER-KOVAKO CORPORATION Bethlehem, Pennsylvania	
Air* Service **(Normal Room Temperature, 60°F to 90°F)** Discharge temperature, 250°F to 325°F (Usually atmospheric to 30 psig or higher)	XD-3 40
Abnormal discharge air temperature above 325°F	SYNESSTIC 100**

* For gases other than air, refer to Fuller-Kovako.
** Check with Fuller-Kovako for compatibility with various models.

GARDNER DENVER COMPRESSORS MACHINERY, INC. Quincy, Illinois	
Horizontal Air Compressors, Vacuum Pumps, Tandems, Duplex and Balance-Opposed Models RL, RX, RT, HA, HL, ML,MD,MBV,MCY, MDY Cylinders(1)	TERESSTIC 100
Power-end crankcase, main bearings, crankpin, crosshead pin, crosshead guide and shoes Below 32°F	TERESSTIC 32 or XD-3 10W
32°F to 90°F	TERESSTIC 68 or XD-3 20W-20
Above 90°F	TERESSTIC 100 or XD-3 30
Vertical and W Type Air Compressors, Vacuum Pumps, Water-Cooled and Air-Cooled, Models AA, AB, AC, AD, AS, AT, AV, LA, WA, WB, WC, WE, WF, WH, WR, WS, WT, WV, WX Cylinders, pistons, pins, rings, main bearings, and crankpin bearings Above 0°F	TERESSTIC 32
Below 90°F	TERESSTIC 68
Above 90°F	TERESSTIC 100
Rotary Portable Air Compressors, Vane Type, All Models RP Pump reservoir	SUPERFLO ATF
Rota-Screw, Flex-Air, Portable Air Compressors, All Models Pump reservoir	SUPERFLO ATF

continued next column

Electra-Screw, Electra-Saver Compressors, All Models	**Exxon Lubricant**
Pump reservoir	SUPERFLO ATF
Where periods of operation exceed 4 hours at discharge temperatures between 200°F and 210°F	No product

(1) If carbon or Teflon rings are used, no lubrication is required.

GAST MANUFACTURING GROUP
Benton Harbor, Michigan

Lubricated Compressors	
Ambient temperatures below 100°F	XD-3 10W
Ambient temperatures above 100°F	XD-3 20W-20

IMO INDUSTRIES, INC.
Deltex Division
(Formerly TransAmerica Delaval, Inc.)
Houston, Texas

Turbopac Units	
Gas generator, centrifugal compressor Common lube system	TURBO OIL 2380

INGERSOLL-RAND
Air Compressor Group
Davidson, North Carolina

PORTABLE COMPRESSORS – GYRO-FLO, SPIRO-FLO AND SUPER SPIRO-FLO	
XHP, HP, XP, P -10°F to 125°F	SUPERFLO ATF or XD-3 10W
-40°F to 125°F	(1)
XHP (300 psi) -10°F to 125°F	SUPERFLO ATF or TERESSTIC SHP 68
XHP (350 psi) -10°F to 100°F	TERESSTIC SHP 68
70°F to 125°F	TERESSTIC SHP 150
Engine Crankcases: Gasoline, Diesel	See engine manufacturer's recommendations
Air cleaners, Oil-lubricated bearings and parts	Same oil as used in engine
Grease-lubricated bearings and bearings and chassis points (excluding wheel bearings)	RONEX MP, LIDOK EP 2, UNIREX N 2 or POLYREX
CENTAC CENTRIFUGAL COMPRESSOR	
Lubricating Oil[2]	TERESSTIC 32
ROTARY SCREW TYPE COMPRESSORS	
AXI compressors Oil reservoir	TERESSTIC 32, 46 or 68
RECIPROCATING AIR-COOLED COMPRESSORS	
Type 30 and Type 40 Compressors Crankcase lubrication Below 40°F[3]	TERESSTIC 32 or 33, or ESSTIC 32
40°F to 80°F	TERESSTIC 68 or ESSTIC 68
80°F to 125°F	TERESSTIC 150
Grease-Packed Bearings (All Models)	RONEX MP, LIDOK EP 2, UNIREX N 2 or POLYREX

(1) Consult Ingersol-Rand
(2) Centac Model CV-O requires Turbo Oil 2389
(3) Pour point of lubricant should be at least 10°F below ambient temperature at starting time.

THE JAEGER MACHINE COMPANY Columbus, Ohio	**Exxon Lubricant**

(The Jaeger Machine Company no longer manufactures this equipment.)

Portable Rotary Compressors – "Roto-Air Plus" Models 75, 85, 125, 125/150, 250, 365, 600 and 900	
Engine crankcase Gasoline or diesel	XD-3*
Engine air cleaner (wet type)	Same oil as used in crankcase
Wheel bearings and other grease applications	RONEX MP or LIDOK EP 2
Air compressor oil -20°F and above – ambient temperature	NUTO H 32
Air compressor air cleaner	Same as compressor oil
Stationary Rotary Compressors – "Roto-Air Plus" Models 130, 260, 380 and 625	
Compressor motor bearings, Blower motor bearings	ANDOK B, RONEX MP or LIDOK EP 2
Air compressor oil, Air compressor air cleaner	Same as above for portable models
Stationary and Portable Reciprocating Compressors Models 75, 125, 185, 250, 365 and 600	
Engine crankcase Gasoline or diesel	XD-3*
Engine air cleaner	Same oil as used in crankcase
Air compressor oil Below 10°F	XD-3 10W
10°F to 32°F	XD-3 20W-20
32°F to 100°F	XD-3 30
Above 100°F	XD-3 40
Compressor air cleaner	XD-3 20W-20
All grease applications other than electric motor bearings	RONEX MP or LIDOK EP 2

* See engine builder's manual for ambient temperature oil grade recommendation.

JOY MANUFACTURING COMPANY
Quincy, Illinois
(Joy Manufacturing Company is part of Gardner Denver Machinery, Inc.)

WL-80, WL-80H, and WL-81 Small Stationary Compressors	
Models 15, 20, 25, 30, 40, 50, 60, 75, 90, 100, 125 Crankcase and air cleaners -10°F to 32°F ambient	XD-3 10W, 20W-20, TERESSTIC 46 or ESSTIC 68
32°F to 90°F ambient	XD-3 30 or TERESSTIC 100
Above 90°F ambient	XD-3 40 or TERESSTIC 150
Large Stationary Compressors	
Models WG, WGO, WGAP, WGAPO, WN, WNO, WNAP, WNAPO, WM, WMO, WR Crankcase and cylinders -10°F to 32°F ambient	ESSTIC 68 or TERESSTIC 46
32°F to 90°F ambient	TERESSTIC 77
Above 90°F ambient	TERESSTIC 100
Air filters – oil bath, stationary, traveling screen -10°F to 32°F ambient	ESSTIC 68
32°F to 90°F ambient	TERESSTIC 77
Above 90°F ambient	TERESSTIC 100

continued next page

Gas Compressors	**Exxon Lubricant**
Models WB, WBJ, WBF	
Crankcase and cylinders	
(with dry, clean, and sweet gas)	
-10°F to 32°F ambient	ESSTIC 68 or TERESSTIC 46
32°F to 90°F ambient	TERESSTIC 77
Above 90°F ambient	TERESSTIC 100
Rotary Sliding Vane Portable Air Compressors	
Models RP, RPQ, and RPV	
-20°F and above ambient	TERESSTIC 32
Twistair Compressors	
Stationary Helical Screw,	
Models TA0180E thru TA070E	
+50°F to 100°F ambient	XD-3 10W, SUPERFLO ATF, TERESSTIC 32 or SYNESSTIC 32
Portable Helical Screw, Models RPS and RPQ	
Rotary Construction Screw,	
Models RCS0180 thru RCS1535	
-20°F to 120°F ambient	SUPERFLO ATF or SYNESSTIC 32
+30°F to 100°F ambient	TERESSTIC 32 or 33
-10°F to 100°F ambient	XD-3 10W

LeROI INTERNATIONAL, INC.
Sidney, Ohio

Portable and Stationary Screw Compressors	
Compressor Oil	
Single-stage screw – air ends Two-stage screw – air ends	TERESSTIC 32 or 33, NUTO H 32, UNIVIS N 32 or SYNESSTIC 32*
All screw compressors	
Engine oil – portable units Electric motors – portable and stationary units	See manufacturer's recommendations
Portable and Stationary Rotary Sliding Vane Compressors	
Compressor Oil	
With aluminum sliding vanes	SUPERFLO 10W-40
Without aluminum sliding vanes	TERESSTIC 32 or 33, NUTO H 32, UNIVIS N 32 or SYNESSTIC 32*
Engine Oil	
LeROI engines	
Above 60°F	SUPERFLO 10W-40 or XD-3 30
32°F to 60°F	XD-3 20W-20
Below 32°F	XD-3 10W or 10W-30
Other engine makes	See manufacturer's recommendations
Electric motors	See manufacturer's recommendations
Oil bath air cleaners (engines and compressors)	Refer to air cleaner decal or use engine oil
Drive coupling	
Portable compressors with spline drive	RONEX MP or LIDOK EP 2
All other sliding vane compressors	See manufacturer's recommendations
Portable Reciprocating Compressors	
Compressor Oil	
Single-stage units, including Tractair	
Above 60°F	XD-3 30
32°F to 60°F	XD-3 20W-20
Below 32°F	XD-3 10W or 10W-30
Two-stage units	
Above 60°F	TERESSTIC 100
32°F to 60°F	TERESSTIC 68 or ESSTIC 68
Below 32°F	TERESSTIC 32 or 33 or ESSTIC 32

continued next column

Engine Oil	**Exxon Lubricant**
Two-stage units	
LeROI engines	
Above 60°F	XD-3 30
32°F to 60°F	XD-3 20W-20
Below 32°F	XD-3 10W or 10W-30
Other engines	See manufacturer's recommendations
Oil bath air cleaners (engines & compressors)	See air cleaner decal
Large Stationary and Unit Type Reciprocating Compressors	
Compressor Oil	
S1, S2 and SDS units	
Above 60°F	TERESSTIC 100
32°F to 60°F	TERESSTIC 68 or ESSTIC 68
Below 32°F	TERESSTIC 32 or 33 or ESSTIC 32
VC, WC, VS, WS, Y AND G, and all "Series" units	
Above 32°F	TERESSTIC 68 or ESSTIC 68
0°F to 32°F	ESSTIC 32
Below 0°F	Use auxiliary heaters

NOTE: For cold weather operation, the lubricating oil should have a pour point 20°F below the lowest expected operating temperature.

* If SYNESSTIC is to be used, check with manufacturer for seal compatibility.

NORWALK COMPANY, INC.
South Norwalk, Connecticut

Cylinder Lubrication	
Air Service	
Dry suction conditions	TERESSTIC 150
Wet suction conditions	AROX EP 150
Gas Service	
Non-oxidizing and non-reactive gases in general service	
Dry suction conditions	TERESSTIC 150
Wet suction conditions	AROX EP 150
Non-oxidizing and non-reactive gases in refrigeration service	ZERICE 68
Chemically active gases – sulfur dioxide, chlorine, hydrogen chloride, etc.	No product
Oxygen, nitrous oxide	No product
Running Gear	
Lubrication systems	TERESSTIC 150

QUINCY COMPRESSOR DIVISION
Coltec Industries
Bay Minette, Alabama
Quincy, Illinois

Rotary (Helical Screw) Compressors – QS, QSI and QST, QA and QSA Series	
All ambient temperatures	SYNESSTIC 68[1]
Reciprocating Compressors – QR and QT Series	
Ambient temperatures	
Below 32°F	TERESSTIC 32 or 33[2]
32°F to 80°F	TERESSTIC 68, NUTO H 68[2] or SYNESSTIC 68[3]
80°F to 100°F	TERESSTIC 100[2] or SYNESSTIC 68[3]

(1) Mineral oils are no longer recommended by the equipment manufacturer. Equipment manufacturer recommends their own private-label PAO-based lubricant. If SYNESSTIC is to be used, compressor must be ordered from the factory specifying use of this lubricant in new unit. Seals compatible with diester lubricants must be installed.
(2) Equipment manufacturer recommends its own private-label mineral oil lubricants.
(3) If SYNESSTIC is to be installed in the field, compressor should be run-in for 200-300 hours on TERESSTIC 68 to seat rings. Drain completely and replace filters before installing SYNESSTIC. Seal conversion kits are available to ensure that seals are compatible with diester lubricants.

ROOTS BLOWER DIVISION Dresser Industries, Inc. Connersville, Indiana	Exxon Lubricant
SPIRAXIAL COMPRESSORS	
Rolling-contact bearing type	
Ambient temperature	
32°F to 90°F	TERESSTIC 77
Below 32°F	SPINESSTIC 22
Sleeve bearing type	
Ambient temperature	
32°F to 90°F	TERESSTIC 32 or 33
Below 32°F	SPINESSTIC 22

SCHRAMM, INCORPORATED
West Chester, Pennsylvania
(Compressors are no longer in production)

Gasoline and Diesel Engine Crankcase	XD-3 (See chart below for proper SAE grades)

	SAE Grades for Gasoline Engines		
	Above 90°F	32°F to 90°F	Below 32°F
Compressor Models			
20, 35, 50	30	20W-20	10W
60, 85, 100	30	20W-20	10W
125, 160, 250	40	30	20W-20
300, 375	40	30	20W-20

	SAE Grades for Buda Diesel Engines		
	Above 90°F	32°F to 90°F	Below 32°F
Compressor Models			
125, 420	40	30	20W-20

	Detroit Diesel and GM V-71 Engines		
	Above 32°F	0°F to 30°F	-20°F to 0°F
Compressor Models			
250, 300, 350 425/350, 600	30	20W-20	10W

THE SPENCER TURBINE COMPANY
Windsor, Connecticut

Standard Overhung Turbo-Blowers	
Motor bearings	
Rolling-contact bearings	POLYREX(1)
Four-Bearing Overhung Blowers	
Rolling-contact bearings	POLYREX(1)
Four-Bearing Outboard Blowers	
Rolling-contact bearings	POLYREX(1)

(1) Equipment manufacturer recommends a competitive product exclusively for machinery under warranty.

SULLAIR CORPORATION
Michigan City, Indiana

Rotary Screw Air Compressors	
Ambient temperature	
Below 90°F	XD-3 10W, SUPERFLO ATF or SPINESSTIC 32*
Above 90°F	XD-3 20W-20 or SPINESSTIC 32*
Exception: Model 24KT (Unit is filled with silicone fluid which is incompatible with petroleum or SYNESSTIC type fluids)	No product

continued next column

Refrigeration Compressors	ZERICE N 68 or ZERICE 68

* To avoid possible elastomer (seal) compatibility problems when using SYNESSTIC, contact Exxon Marketing Technical Services for a suitable flushing procedure.

THE TRANE COMPANY La Crosse, Wisconsin	Exxon Lubricant
Centrifugal Compressors	TERESSTIC 68, ZERICE N 68 or ZERICE 68
Reciprocating Compressors	
Models G, J and M (for refrigeration)	No product
Models E and F	ZERICE S 68
Models G (for air conditioning)	No product

WORTHINGTON COMPRESSORS
Dresser-Rand
Houston, Texas

Stationary Water-Cooled Compressors	
Compressor Crankcase (frame) Lubrication	
For all frames, horizontal and vertical, BDC, CUB,DC, DYC, HB, ODC, ODP, OXP, VB, YC, and Blocair	TERESSTIC 100 or 150
Compressor Cylinder Lubrication	
For all compressor cylinders, separately lubricated, regardless of the type of frame upon which the cylinder is mounted:	
Discharge pressures 0 to 200 psig	
Cylinder diameters 0 to 20"	TERESSTIC 68
Cylinder diameters above 20"	TERESSTIC 150
Discharge pressures 200 to 1000 psig	
Cylinder diameters 0 to 10"	TERESSTIC 77
Cylinder diameters above 10"	TERESSTIC 150
Discharge pressures 1000 to 2500 psig	
Cylinder diameters 0 to 15"	TERESSTIC 150
Cylinder diameters 15 to 20"	TERESSTIC 320
Discharge pressures 2500 to 4000 psig	
Cylinder diameters 0 to 15"	TERESSTIC 320
Discharge pressures 4000 psig and above	
Cylinder diameters 0 to 15"	TERESSTIC 460

NOTE: Consult Dresser-Rand for the use of synthetic oil such as SYNESSTIC 68 or 100.

Refrigeration Type Compressors	
Cylinder lubrication	ZERICE N 68 or ZERICE 68
Steam-Driven Compressors	
Steam Power Cylinder Lubrication (Separately lubricated)	
Steam inlet temperature 600°F	CYLESSTIC 1500
Steam inlet temperature 500°F	CANTHUS 1000
Steam inlet temperature 400°F	CYLESSTIC TK 1000
Wet Steam	CYLESSTIC TK 1000
Centrifugal Compressors	
Lubrication systems for CAP 14-31, CAP 21-70, and CAP 85-120 packaged plant air centrifugal compressors and for EA air and nitrogen centrifugal compressors.	TERESSTIC 32 or 33

YORK INTERNATIONAL CORPORATION
York, Pennsylvania

Refrigerant compressor oil	
Ammonia compressors	ZERICE N 68 or ZERICE 68*

continued next page

	Exxon Lubricant
Freon 12 compressors; evaporator temperature above -20°F	ZERICE N 68 or ZERICE 68*
All refrigerants; evaporator temperature above -20°F	No product
Hermetic compressors; evaporator temperature below -20°F	No product
Freon 12 automotive compressors	No product
All refrigerants; evaporator temperature below -20°F	No product

* Equipment manufacturer sells its own private-label oil and does not approve other oils. ZERICE N 68 or ZERICE 68 may be used as a substitute for either York "A" or "B" oils.

Cylinder-Oil Feed

In the lubrication of double-acting compressor cylinders, one of the most important factors is the rate of oil feed. The likelihood of over-lubrication is greater than that of supplying too little oil. Many problems associated with compressor operation can be overcome by preventing excessive lubrication. Proper control of the supply of oil to the cylinders is the most effective means of preventing the formation of objectionable deposits around valve ports, in ring grooves, and on cooler surfaces.

Under average conditions, one quart of oil will properly lubricate an operation equivalent to the sweep of a piston over 10,000,000 square feet of cylinder surface. In a 24-hour day, for example, the piston ofa compressor with 5 square feet of cylinder area and operated at 500 rpm would sweep

$$5\text{ft}^2/\text{stroke} \times 2\text{ strokes/rev} \times 500\text{ rev/min} \times 1440\text{ min/24-hr day}$$

or 7,200,000 square feet per day. Such a compressor would normally require, therefore,

$$7{,}200{,}000/10{,}000{,}000$$

or 0.72 quarts of oil per 24-hour day. The oil feed rate for the average copressor, in quarts per 24-hour day, can thus be determined by the following formula:

$$\frac{\text{Bore (inches)} \times \text{Stroke (inches)} \times \text{rpm} \times 62.8}{10{,}000{,}000}$$

The same result can be obtained graphically from the chart, Figure 17-19.

Here, the value, 10,000,000 square feet, is a nominal one representative of average conditions. Under other conditions, it may be necessary to substitute a different value, one that can be expected to lie, however, between 6,000,000 and 15,000,000. Substitution may be made either in the formula or in the graph.

To determine whether oil is being fed to the cylinder at the computed rate, the compressor oil reservoir must first be filled at the beginning of a specified run. After a certain period of operation, the reservoir is refilled from a graduated container, so that the amount of oil consumed during the run can be noted. The feed rate can then be increased or decreased to conform to the predetermined value.

Many lubrication systems are equipped with sight-feed oilers by which the flow of oil to the cylinders, in drop form, can be observed. These devices show whether the lubrication system is operating properly and, in drops per minute, give some indication of the feed rate. While the sight-feed oiler can be very helpful in the making of a trial feed adjustment, it should not be relied upon as the sole determining factor in feed regulation.

Size of the oil drop is subject to considerable variation. The number of drops per quart depends on the oil's viscosity and temperature, the diameter and cleanliness of the oil discharge orifice, and the properties of the sight-feed medium. Since differences in oil-drop size have a pronounced effect on the feed rate, the number of drops that pass per minute may not indicate the feed rate accurately. When the feed rate is checked, therefore, it should be done by actual measurement of the added oil in the manner described above.

Assume, for example, that a new compressor is computed to require 0.72 quarts of oil per 24-hour day. Oil-drop diameter in the sight-feed is estimated at 3/16." Table 17-6 shows that 16,700 of these drops are required for one quart. In 24 hours, therefore, 0.72 × 16,700, or about 12,000, drops should be fed. This is approximately 8 drops per minute. Obviously, if the cylinder were supplied by two oilers, each should be adjusted to feed 4 drops per minute.

How closely the applied feed rate meets actual cylinder lubrication requirements should also be checked. This can be done by examination of internal surfaces, such as cylinder walls or exhaust and intake valve parts. Properly lubricated, these surfaces should be covered with a thin, even film of oil. There should be no evidence of oil accumulation.

Probably the most obvious symptom of over-lubrication is the appearance of little oil puddles in low spots in the valve boxes. It may also be advisable to examine the cylinders.

If the cylinder surfaces are wiped with a piece of cigarette paper, oil should stain the paper evenly, but should not soak it. If the paper is dry or unevenly spotted, the feed rate is too low; if the paper is saturated,

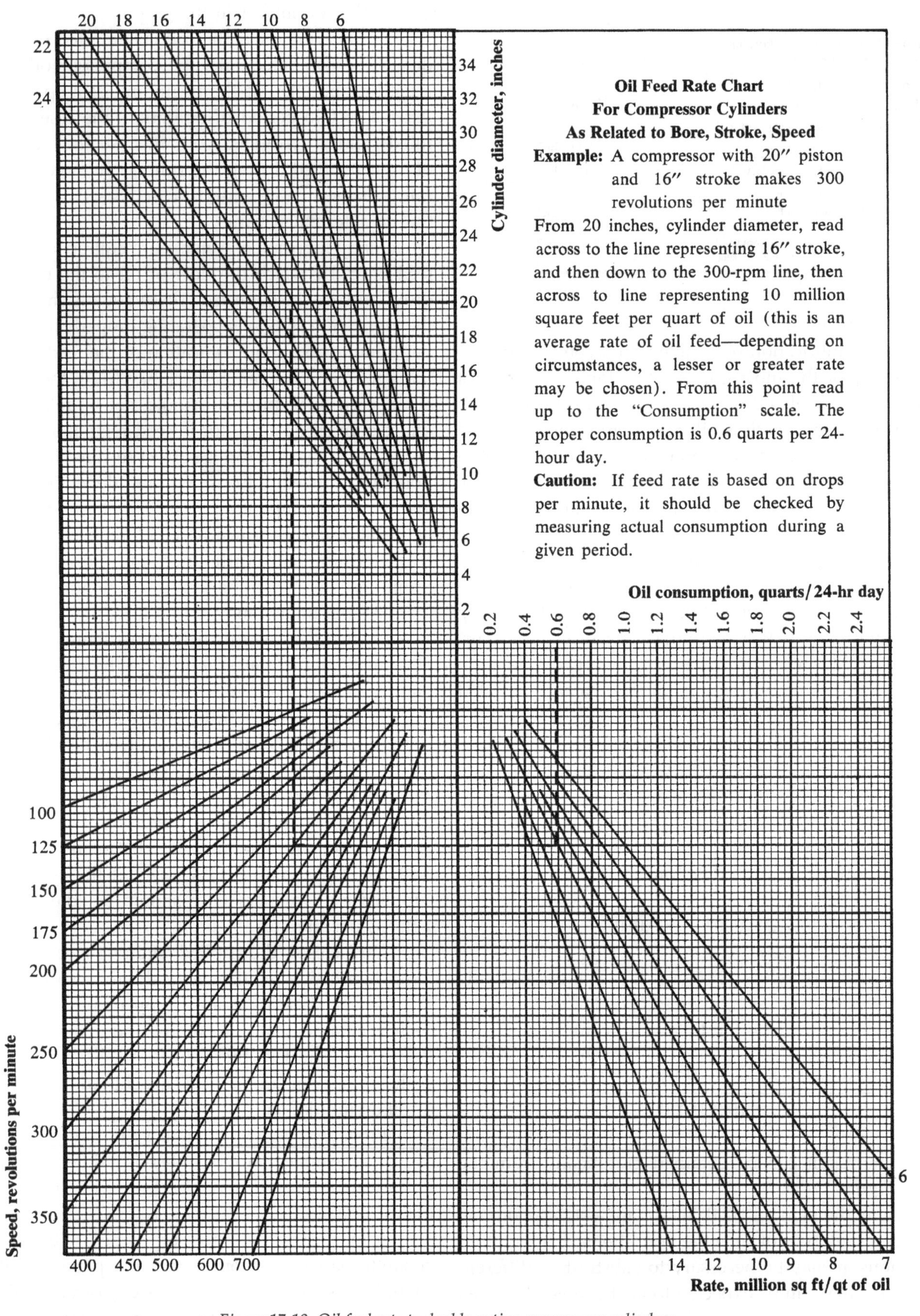

Figure 17-19. Oil feed rate to double-acting compressor cylinders.

Table 17-6. Oil feed rate to double-acting compressor cylinders (approximation only).

GUIDE TO OIL FEED RATE

Compressor Cylinders

1 quart of oil willlubricate the sweep of piston over approximately 10,000,000 square feet of cylinder surface.

$$\frac{B \times S \times N \times 62.8}{10{,}000{,}000} = Q$$

B = bore (inches)
S = stroke (inches)
N= rpm
Q = quarts of oil per 24-hour day

Effect of Oil Drop Size on Feed Rate

Diameter of drop, inches	Volume of drop, cu in	Drops per quart	Drops/minute for rate of 1 qt/24 hr
1/16	0.00013	454,000	315
1/8	0.00102	56,600	39
3/16	0.00345	16,700	11.5
1/4	0.00818	7,070	4.9
5/16	0.01598	3,620	2.5
3/8	0.02761	2,090	1.5

This table should be used only as a guide. It illustrates how greatly drop size influences actual oil feed rate. Since it is almost impossible to measure drop size exactly, feed rate should be checked carefully against actual consumption during a given period.

the feed rate is too high. Where necessary, the feed rate should be adjusted to provide not more than the minimum of lubricant.

LUBRICATION OF GAS ENGINES*

Gas engines with integrally mounted process gas compressors are shown in Figures 17-20 and 17-21. The general maintenance guidelines issues by major manufacturers of gas engines show both similarities and differences. Nevertheless, the equipment engineer or gas engine operator should be guided by customary practices and experience. To that end, let's investigate such issues as oil drain intervals, filter life, port deposits, and component lives.

Oil Drain Intervals

The majority of manufacturers have no specific recommendations regarding lube oil drain intervals because requirements vary considerably with the type of installation. Therefore, good operating procedures dictate that oil samples be taken from the crankcase periodically to determine acidity, type of contaminants, and general condition of the oil. This information should be recorded and used to determine proper oil drain intervals.

Experienced major suppliers and manufacturers are providing a lube oil analysis service that can be shown to represent an excellent, cost-effective means of monitoring the condition of the oil and

Filter Life

As in the oil drain intervals, filter life will vary with operating conditions and the type of filter used. As a guide, the interval between filter changes should not exceed one year. Oil filters should be changed in any case whenever there is an increased pressure differential of 15 to 20 lbs/in^2 across the filter or when necessary to conform to the recommendations of the filter manufacturer.

Ring Life

Ideally, the engine should not have to be overhauled except at five-year intervals. In most low-BMEP engines, it is possible to achieve a five-year ring life. In high-BMEP engines, five years between overhauls is

Source: Exxon Company, U.S.A.; Houston, Texas: Publication DG-6D, 1995.

Figure 17-20. This Cooper-Bessemer model V-250 with 16 power cylinders is rated at 5500 hp. Unit is equipped with eight GT pipeline cylinders and is on transmission duty in New York State.

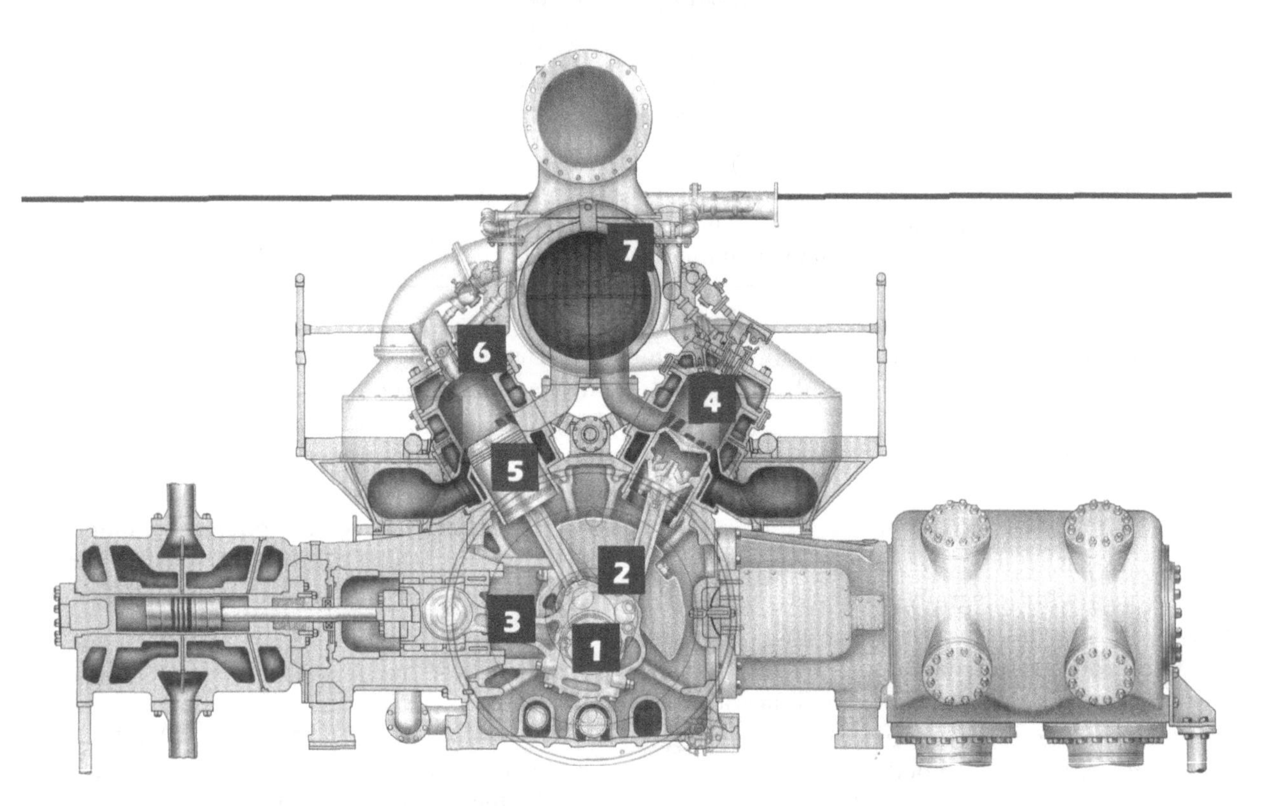

1 Crankshaft The crankshaft is forged from alloy steel and is bedded from the side, allowing easy access without removing the top half of the engine.

2 Main Engine Bearings Bearings are designed to permit maximum bearing area. Materials provide maximum load-carrying capability to give long life with excellent fatigue resistance.

3 Connecting Rods Connecting rods are an articulated design with all forces acting in the same plane. The rod is a one-piece forging with H-column construction for strength.

4 Power Cylinders Power cylinders on smaller-bore engines are one-piece cast iron. For the extra strength required on large-bore engines, the cylinders consist of a one-piece block for each bank of the engine, with individual cylinder liners. The designs incorporate porous chrome bores with intake and exhaust porting located to achieve maximum fuel mixing and high-efficiency scavenging. Two-point oil injection gives an even flow on the bore surface.

5 Power Pistons Each power piston features a one-piece, "cocktail shaker", oil-cooled design with a hemispherical cap and six compression rings.

6 Power Cylinder Heads The one-piece heads are water-cooled, with a hemispherical design to match the piston crown.

7 Turbocharger There are several models of turbochargers, designed and built by Cooper-Bessemer, to provide an exact match with the engine for optimum operating performance.

Figure 17-21. Principal components of Cooper-Bessemer integral gas engine/reciprocating compressor unit.

possible under favorable conditions, but three to four years is more common. Of course, unfavorable conditions—e.g., operating outside engine design parameters, high ambient temperatures, poor maintenance practices - may make more frequent overhauls necessary.

Valve Life

The valves in four-cycle engines, like the piston rings, should not have to be changed within a five-year overhaul period. This, of course, applies under very favorable conditions. Some engines are subject to valve wear that reduces this life, and some manufacturers recommend replacement at earlier intervals. See Figure 17-22.

Port Deposits

It should not be necessary to punch carbon from ports more frequently than 10,000-12,000 hours (17-18 months). In many installations using superior oils, there is no need to punch carbon between major overhauls.

LUBRICANT RECOMMENDATIONS FOR NATURAL GAS ENGINES

The recommendations in Table 17-7 are made by Exxon Company U.S.A., based upon this company's experience with the engines and lubricants involved. The recommendations are those that would normally apply; if experience indicates abnormally severe conditions, it may be necessary to reduce the oil drain interval or recommend an oil of higher detergency. For each type of equipment, the Exxon lubricant listed first is the primary recommendation. Where an OEM (original equipment manufacturer) lubricant recommendation differs from Exxon's, the OEM recommendation must take precedence.

Figure 17-22. Gas engine valve with indications of wear. Recess on the valve face indicates valve beat-in.

Tables 17-8 and 17-9 contain data on the physical characteristics of suitable, premium-grade lubricants. In the case of Exxon, ESTOR is the trademark for a line of natural gas engine lubricants that covers the entire range of gas engine lubrication requirements. ESTOR gas engine oils are suitable for both crankcase and cylinder lubrication. They are highly effective in suppressing ring-zone and port deposits and crankcase sludge.

These products, are formulated from specially selected solvent-extracted naphthenic basestocks, which have inherent resistance to carbon formation.

The ESTOR line of gas engine oils offers a wide selection of viscosities to meet a broad range of OEM and operational requirements. For better operation at low temperatures, ESTOR GA, ESTOR GLX-C, and ESTOR Super are available in an SAE 15W-40 multigrade, as well as an SAE 40 grade, whereas the other ESTOR products are single grades typically at the low end of the SAE 40 viscosity range, which is acceptable to the major engine manufacturers.

The distinct features of each grade can be summarized as follows:

ESTOR Elite—A full-synthetic gas engine oil designed for superb performance under the most severe operational conditions; approved against Waukesha cogeneration requirement (Figures 17-23 and 17-24).

ESTOR Super—API CD universal lubricant recommended for all natural gas engines approved against Waukesha cogeneration requirement (Figures 17-25 and 17-26).

ESTOR Select—premium medium-ash lubricant for lean-burn and cogeneration applications; approved against Waukesha cogeneration requirement (Figures 17-27 and 17-28).

ESTOR AGX—premium ashless lubricant for two-cycle gas engines

ESTOR GLX-C—detergent dispersant lubricant with 0.4% ash; highly recommended for most four-cycle gas engines

ESTOR GA-40—ashless formulation with a highly effective anti-wear additive (Figures 17-29 and 17-30).

ESTOR GMA—Super-premium low ash lubricant formulated with a proprietary balanced additive package for excellent control of oxidation, nitration, deposits and wear

Table 17-7. Lube recommendations for gas engines.

Engine Make and Model	2-Cycle or 4-Cycle	Typical Service	Lube Recommendation
AJAX EA DP C-42 DPC	2	Mild	ESTOR AGX ESTOR GA
CATERPILLAR G349, G348 G346, G343 G333, G399 G398, G379 G353, G342	4	Moderately Severe	ESTOR Super, ESTOR GLX-C ESTOR AGX
G3300, G3400 G3500, G3600	4	Severe Service	ESTOR ELITE ESTOR Super
CLARK			
RA RAS, RAT[1]	2	Mild	ESTOR AGX, ESTOR GA
HRA, HRAT, HSRA, TRA	2	Mild (moderately severe with respect to port plugging)	ESTOR AGX, ESTOR GA
BA, HBA, HBAT[1]	2	Mild (moderately severe with respect to port plugging)	ESTOR AGX, ESTOR GA
HLA, HLAT, TLA, TLAB[1] TLAC[1] TLAD	2	Moderate	ESTOR AGX, ESTOR GA
TVC, TPV[1], TCVA, TCVB[1] TCVC, TCVD	2	Moderate	ESTOR AGX, ESTOR GA
CLIMAX K-67[1] K-75 (F 1850G)[2] V-80[1] V-85 (H 2475G)[2] V-122[1] V-125 (L 371 IG)[2]	4	High-speed engines generally severe on the lubricant	ESTOR Super ESTOR GLX-C

NOTES: (1) Models no longer in production.
(2) Model number change only.

Table 17-7. (Continued)

Engine Make and Model	2-Cycle or 4-Cycle	Typical Service	Lube Recommendation
COOPER BESSEMER			
GMVA, GMVC[1], GMVD[1]			
GMVE, GMVG[1], GMVH			
GMW[1], GMWA[1], GMWC[1]			
GMWD[1], GMWF[1], GMWH[1]	2	Mild (moderately severe with	ESTOR AGX[2]
GMX[1], GMXA[1]		respect to port plugging)	ESTOR GA
GMXC[1]			ESTOR Super
GMXD[1], GMXE[1]			
GMXF[1]			
GMXH[1]			
QUAD 145	2	Mild (moderately severe with	ESTOR AGX
QUAD 155		respect to port plugging)	ESTOR GA
V-250[1] V-275			ESTOR Super
W-330	2	Mild (moderately severe with	ESTOR GLX-C
Z-330		respect to port plugging)	ESTOR GA
FVB[1]			
FVG[1]			
JS[1]	4	Moderately severe	ESTOR Super
JSA[1]			ESTOR GLX-C
JST[1]			
KSV			
LS[1]		Moderately severe	ESTOR Super, ESTOR GLX-C
LST[1]	4		
LSV, LSVB		Duel Fueled	XD-3 Extra 40
LSVT[1]			
INGERSOLL RAND			
PVG[1]			
XVG[1]			
JVG[1]		Moderately severe	ESTOR Super[3]
KVG[1]			ESTOR GA
SVG[1]	4		
KVGR[1]		Moderately severe to severe	ESTOR Super
TVS[1]			
TVR[1]			
KVS[1]			
KVT, KVH[1]			
KVR, SYS			
KVSR, KVSE	4	Very severe	ESTOR Super
KVSF			

NOTES: (1) Models no longer in production
(2) For BMEP >85psi, ESTOR Ashless preferred
(3) ESTOR Super 40 CAT I, II & III approved

Table 17-7. (Continued)

Engine Make and Model	2-Cycle or 4-Cycle	Typical Service	Lube Recommendation
ROILINE			
H 540[1]			
H 8441[1]			
F 1500[1]'			
F 1850			
H 2000[1]	4	High-speed, very severe	ESTOR Super
H 2470[1]			
L 3000[1]			
L 3230[1]			
L 3460[1]			
L 4000[1]			
SUPERIOR (WHITE)			
6G - 825 12G - 825			
6GTL (B) 12GTL (B)			ESTOR Super
8G - 825 12SGT (B)	4	Moderate to high speed, very severe	ESTOR Select[2]
8GTL (B) 16G - 825			ESTOR GLX-C
8SGT (B) 16GTL (B)			
16SGT (B)			
1700, 2400 Series	4	Moderate to high speed, very severe	Refer to Superior's Engineering Std-1005
WAUKESHA			
VSG, F11G, GSI, GSID			GLX-C
VGF F18, H24, L36, P48			
G. GL. GLD			ESTOR Select
Intermediate F817, F1187G			ESTOR Super, ESTOR Elite
VHP 2895, 3521, 5108,			
5790, 7042, 9390 G, GSI			
(Rich Burn or Lambda 1)			High-speed, very severe
Estor GLX-C			
VHP 2895, 5108, 5790, GL			
(Lambda 1.5-2)			
VHP 3521, 5115, 7042, 9390 GL Turbo &			
Lean Burn (Lambda 1.5-2)			Estor Select
AT25/27 8L 12VGL0.35% Min			ESTOR GLX-C
WORTHINGTON			
CGG	4	Generator engine, severe service	ESTOR Super, ESTOR GLX-C
SLHC	4	Moderately severe	ESTOR Super, ESTOR GLX-C
LTC			
ML			
SUTC	2	Mild service	ESTOR GA
UTC			ESTOR AGX
VEE			

NOTES: (1) Models no longer in production.
(2) Recommended for lean/clean burn applications.

Table 17-8. Typical inspections for four gas engine lubricants.

Typical Inspections	ELITE	SUPER		SELECT	GLX-C	
ESTOR Grade	20W40	40	15W-40	40	40	15W-40
Viscosity,						
cSt @ 100°C	13.2	13.2	14.4	13.5	13.2	14.3
cSt @ 40°C	99	126	108	133	128	108
cP @ -15°C, CCS	3500	—	3250	—	—	3300
Viscosity index	132	98	136	95	96	135
Flash point, COC,						
°C	330	246	205	240	246	205
°F	446	475	401	464	475	401
Pour point,						
°C	<-50	-15	-24	-15	-21	-24
°F	<-58	5	-11	+5	-5	-11
Sulfated ash, mass %	0.45	0.45	0.45	0.95	0.40	0.40
Total Base Number,						
ASTM D 2896	5.2	5.2	5.4	9.8	3.2	3.4
Phosphorus, ppm	280	280	280	290	700	700
Density, lb/gal	6.99	7.37	7.33	7.43	7.38	7.34
Gravity, °API	36.9	28.4	29.1	27.0	28.2	29.1

Table 17-9.

ESTOR Grade	GLX-C	GA-40	AGX-40	GMA-40
Viscosity,				
cSt@ 100°C	13.2	13.5	13.5	13.5
cSt @ 40°C	124	133	135	133
cP @ -15°C, CCS	—	—	—	—
Viscosity index	96	96	95	95
Flash point, COC,				
°C	246	246	250	240
°F	475	475	482	464
Pour point,				
°C	-21	-18	-21	-21
°F	-5	0	-5	-5
Sulfated ash, mass %	0.40	0.14	<0.1	0.45
Total Base Number,				
ASTM D 2896	3.2	1.0	2.5	5.0
Phosphorus, ppm	270	690	940	300
Density, lb/gal	7.38	7.39	7.41	7.40
Gravity, °API	28.2	27.9	27.5	27.6

FIGURE A. ESTOR Elite Provided Outstanding Wear Protection in a Waukesha Engine Field Test

Engine Model: Waukesha 3521 GSIU / Rated Power: 1067 BHP @ 1200 rpm / Operating Conditions: 900 BHP @ 1200 rpm

Test Conditions	
3965 Hours	
Coolant Temperature	250°F
Oil Temperature	160°F

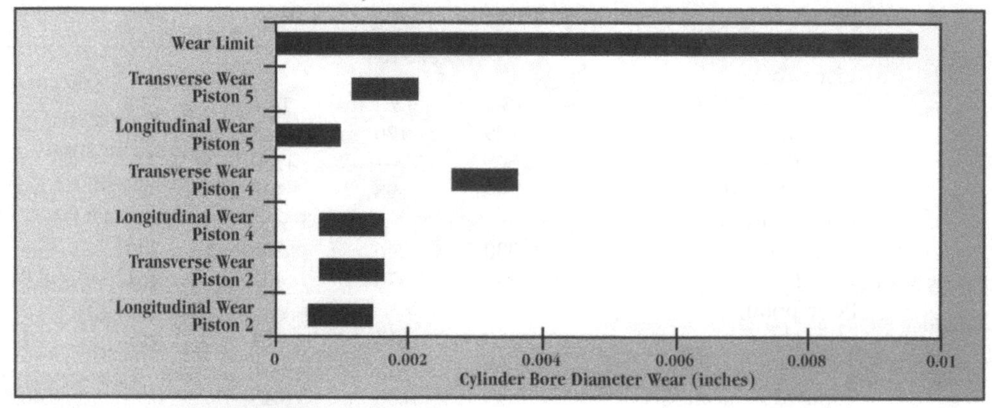

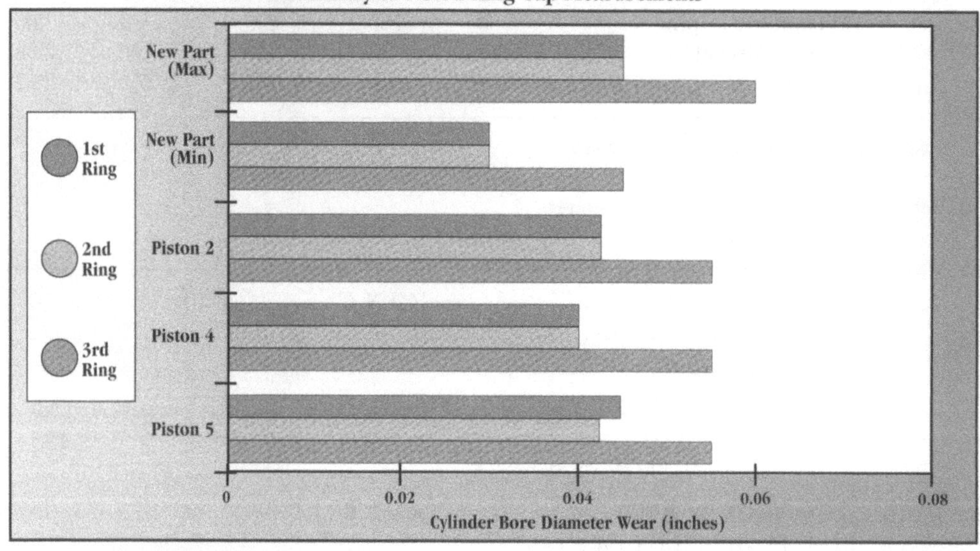

FIGURE B. ESTOR Elite Provided Superior Deposit Control in a Caterpillar 3304 Engine Test, Compared with Two Competitive Gas Engine Oils

Test Conditions
250 Hours
Full head
Reduced oil sump capacity
Air/Fuel ratio to maximize Nitration

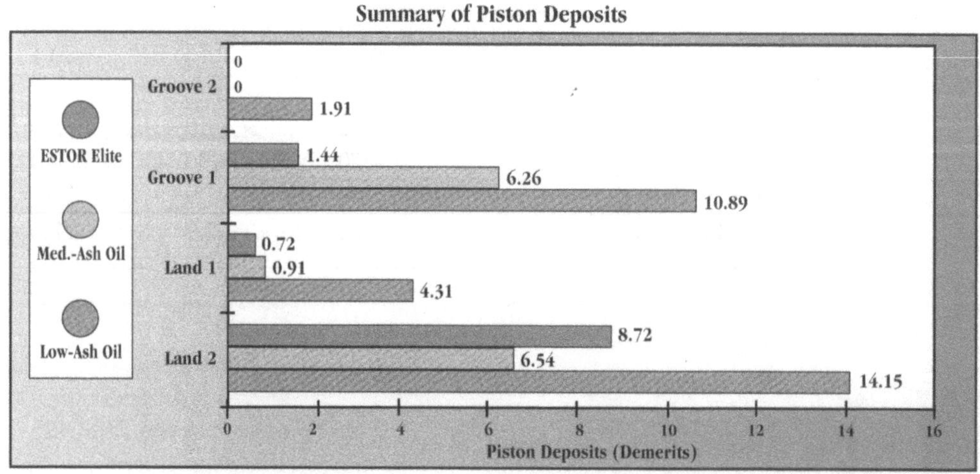

Figure 17-23. Performance comparison, ESTOR Elite gas engine oil.

FIGURE C. ESTOR Elite Waukesha F 3521 GSIU Engine Field Test Inspection Photos

Cylinder #2
Piston Undercrown

The piston undercrown is deposit-free. Excellent deposit control is particularly critical at the high operating temperatures typical of GSIU engines in cogeneration service.

Piston #4
Rings Removed

There were minimal-to-no deposits in the ring grooves.

Piston #5

ESTOR Elite eliminated piston skirt deposits and minimized land deposits.

Figure 17-24. Performance photos, Estor Elite gas engine oil.

ESTOR Super

Commercial Gas Engine Oil "A"

Figure 17-25. ESTOR Super demonstrates outstanding piston deposit control. Pistons removed from a single-cylinder CLR engine after a 140-hour Nitration/Oxidation Inhibition Test clearly show the superiority of ESTOR Super versus a competitive oil in maintaining piston cleanliness.

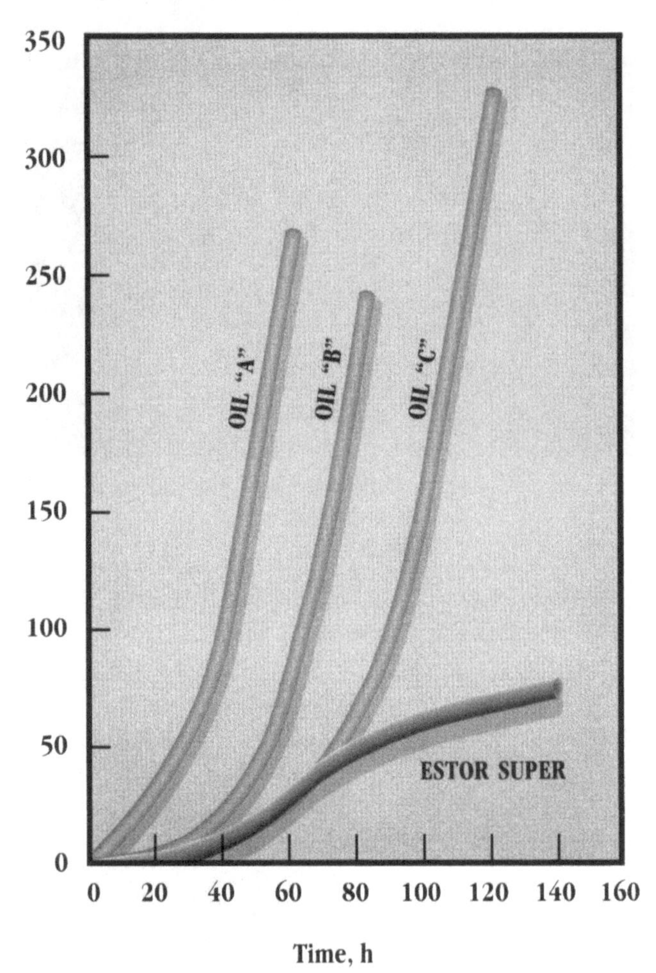

Figure 17-26. Nitration and oxidation inhibition viscosity increase laboratory engine test. The graph illustrates the excellent viscosity control demonstrated by ESTOR Super compared with three competitive commercial gas engine oils in the severe Nitration/Oxidation Engine Test.

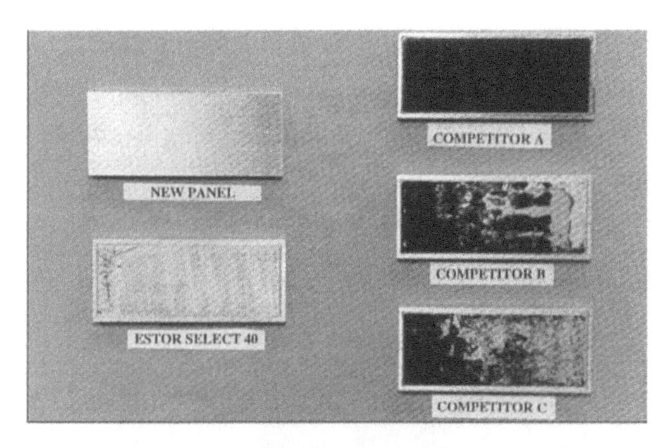

Deposit Control

Testing was completed by heating metal panels to 315°C in order to demonstrate how oxidation of the oil might lead to piston deposits. The panel looks almost like new when tested with ESTOR Select 40, a significant improvement over the competition.

Waukesha L7042 GSIU Piston

As verified by extensive field testing, ESTOR Select 40 produced only minimal carbon deposits, despite being run in the high-temperature environment of cogeneration service.

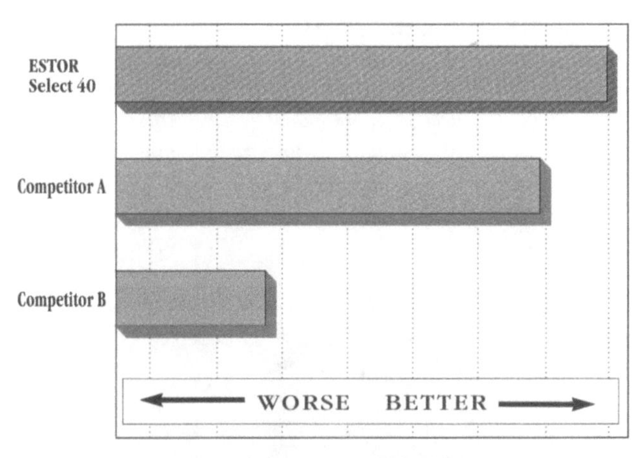

Oxidation Control

ESTOR Select 40 has a proprietary formulation which is designed to resist oxidation and nitration. Testing has demonstrated that this formulation offers excellent oxidation control, which leads to extended oil life.

Waukesha L7042 GSIU Control Panel

ESTOR Select 40 was proven to maintain outstanding oxidation control in a cogeneration application, where jacket water temperatures were pushed to 265°F.

Figure 17-27. Performance comparison for ESTOR Select gas engine oil.

Waukesha L7042 GSIU Exhaust Valve

Valve shows a controlled level of ash deposit with minimal valve recession.

Waukesha L7042 GSIU Cylinder Liner

Good anti-wear performance allows ESTOR Select 40 to protect metal surfaces, as shown by this cylinder liner, which has maintained good crosshatching.

Waukesha L7042 GSIU Piston Under crown

Piston undercrowns exhibit low varnish/lacquer to ensure good heat dissipation.

Waukesha L7042 GSIU Valve Deck

Valve deck is sludge-free.

Figure 17-28. Performance photos, ESTOR Select gas engine oil.

Clark TLA 10
9156 Hours on ESTOR GA 40
Ports, Cylinder #3

Intake and exhaust ports show minimal deposits, thereby eliminating any need for cleaning.

Clark TLA 10
9156 Hours on ESTOR GA 40
#7 Piston, RIng Zone

Crownland carbon is low. Rings move freely and show minimal wear, thus preventing the need for premature overhaul.

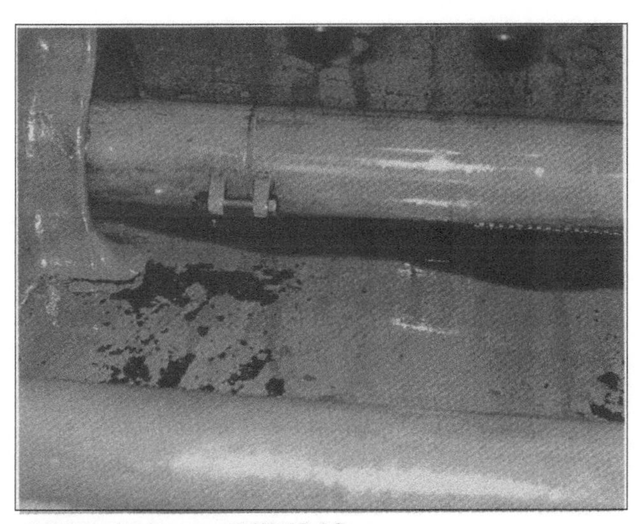

Cooper-Bessemer GMWC 10
8000 Hours on ESTOR GA 40
Crankcase

Crankcase is sludge-free.

Cooper-Bessemer GMWC 10
8000 Hours on ESTOR GA 40
Exhaust Ports, #2L Cylinder

Port deposits are minimal – no cleaning needed.

Figure 17-29. Performance photos, ESTOR GA 40 gas engine oil.

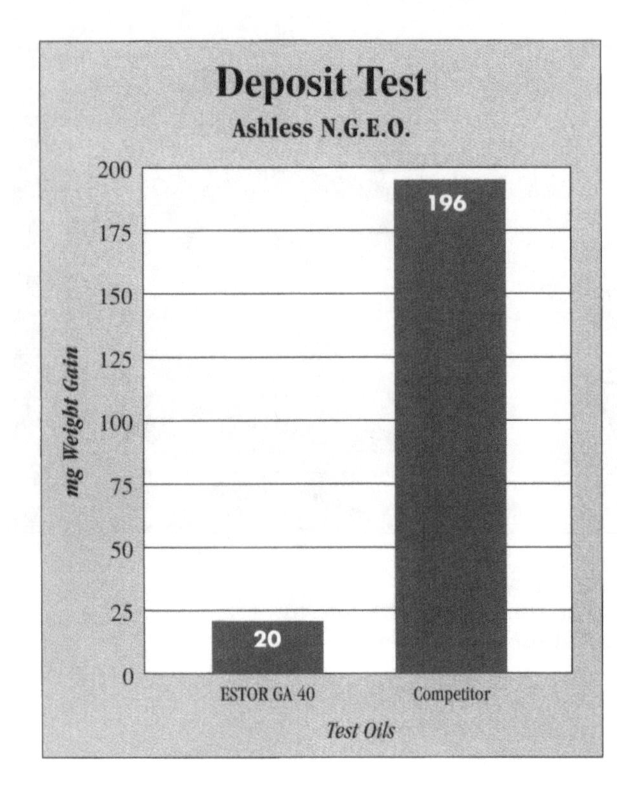

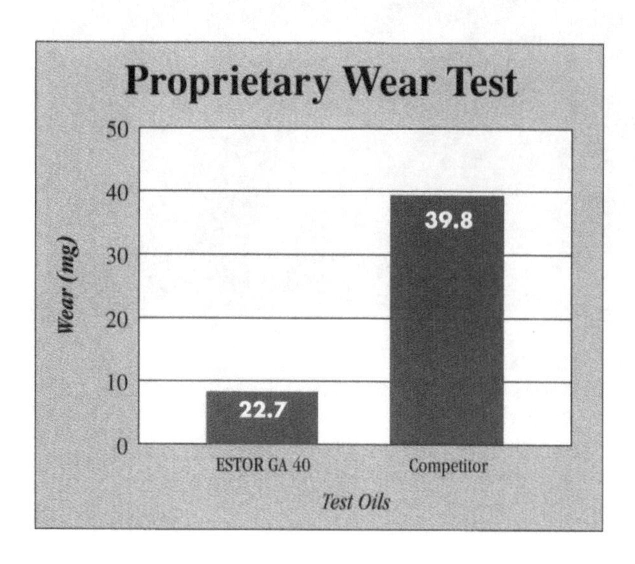

Figure 17-30. Deposit and wear test comparisons, ESTOR GA 40 vs. competitor.

Chapter 18

Lubricant Purchase, Handling and Storage Methods

Lubricants are often thought of as a machine's lifeblood and if they are to protect your machine investment and perform as designed their needs and requirements must be understood and met with diligence. Getting the best out of your lubricants requires a cradle-to-cradle (C2C) lifecycle management strategy!

IMPLEMENTING A CRADLE-TO-CRADLE LUBRICANT MANAGEMENT PROGRAM

In a cradle-to-cradle lubricant management program a lubricant commences life as base oil refined from mineral crude or synthesized stocks, in which is then blended a variety of additives to make up a proprietary lubricant product ready for delivery to market.

From the refiner/manufacturer the finished lubricant is transferred into bulk containers to continue its journey to the market supplier who offloads the bulk oil into their own storage tanks. The oil is then purchased and delivered to the end user in pre-packaged pails and drums, or transferred once again into a bulk container for bulk delivery and transfer into the end user's bulk storage totes. At the end user's site, the lubricant is then stored and transferred into smaller containers used to fill up machine reservoirs.

Once the lubricant is no longer deemed fit for use, ideally it is collected in select used oil containers, classified and stored ready to sent to a re-refiner to be recycled once again into marketable oil stocks.

Providing we purchase the correct lubricants from the onset, each time a lubricant is stored or transferred it is at risk for solids and water contamination, both highly detrimental to bearings. As an end user, we gain control of lubricant cleanliness upon receipt from the supplier through an effective in-house lubricant purchase, storage and handling program. Prior to receipt, we can exercise a degree of cleanliness control through an audited cleanliness control lubricant purchase program set up in conjunction with the lubricant supplier(s).

Designing and building an end user world class lubricant management storage and handling facility requires the end user to view and treat lubricants with reverence using simple but effective management processes from the time lubricants are purchased, received, transferred, stored, dispensed, drained and collected for recycling. Adopting the following approach the end user will have the tools needed to put in place the best program for their facility and budget.

Introduce a Lubricant Purchase Policy/Program

Maintenance departments may perform their own purchasing/expediting of MRO materials, or rely on a formal relationship with the corporate purchasing department to perform that task for them. In either case, it behooves the maintenance department to partner with purchasing to put in place clear and concise purchasing process to ensure the timely purchase of the exact lubricants required is successfully ordered and received every single time.

If any type of management software is used to track purchases within the corporation there will likely be a formal purchase request/purchase order system in place. As purchasing is a partner provider to the maintenance department, if purchases are to be received in a timely manner maintenance must meet with purchasing to review or establish and understand the workflow procedure for both the request and work order process.

It is the purchasing department's job to attain requested items at the best price possible in the shortest time. If the request is non-specific, vague or unclear, for example: please purchase 2 totes of ISO 22 Air tool oil, you will likely receive the least expensive product they can find that may not entirely suit your needs. The following automotive assembly plant case study illustrates the importance of specifying your purchasing needs exactly, especially when purchasing lubricants!

Automotive Assembly Plant Purchasing Case Study

This case study involves the maintenance and purchasing departments of an automotive assembly plant located in North America. At the time of the lubrication failure occurrence, the plant was operating a 7 day - 24 hr. three continual shift operation to accommodate a running four week backlog of vehicle orders. With an average unit selling price of $35,000 and a line speed averaging 60 units per hour, downtime losses were unrecoverable at the rate of over $2 million sales losses per hour!

Despite the successful order book, the corporation was in the middle of a corporate purchasing efficiency program that required the purchasing staff to individually review every purchase request and seek out less expensive alternatives wherever possible, in exchange for a percentage personal bonus on all accumulated savings. A motivated purchasing department would now slow down every purchase as they eagerly took on the role of a cut price negotiator, which only served to reduce the level of service to the maintenance department and to upset regular suppliers. If the original supplier didn't lower their price, the purchasing agent would then venture to seek out an alternative source based on the order specification. If the order specification wasn't exacting they could seek out a "similar" product, and so began the start of a very expensive lubrication failure!

Brewing a Perfect Storm

The maintenance purchase requisition simply stated "two totes of ISO 22 air tool oil." Because the lubrication program had recently been through a lubricant consolidation program they had consolidated their different lubricant types/SKU's by half to single digits producing six figure savings. Lubricating oils were now purchased in bulk totes from a reputable manufacturer through a reputable local supplier.

The new lubrication program had ensured that lubricants were stored separately in a lubrication crib and were adequately spill protected with a concrete berm and drain system. In addition to the lubricating oils and greases a number of release chemical agents were also stored in the lubricant crib.

To accommodate lubricant transfer the maintenance department requested multiple "pogo" style pneumatic drum pumps to transfer lubricants to smaller containers. At the time of the failure, the lubrication crib had only one operating transfer pump in use for all lubricants and chemical fluids. Upon later questioning, purchasing would admit they had made a conscious decision to only purchase one pump to save money.

With no purchasing confines in place, only the lubricant viscosity specification to go on and a motivated purchasing department with little understanding or experience in lubrication, lubricants or lubrication equipment, the perfect storm was in place.

The Consequence of Saving Money in the Wrong Places

On the day of the lubrication failure it was a productive morning for the lubrication staff within the maintenance department who had just received a new shipment of ISO 22 grade viscosity air tool oil that appeared to have new SKU numbers and markings on the tote. Thinking nothing of it they went about their business of dispensing a number of release chemicals and filling their corresponding reservoirs. Next, they dispensed ISO 220 gearbox oil and filled the corresponding gearbox reservoirs around the plant while the second lubrication crew now used *the same* pogo transfer pump to dispense the newly received ISO 22 air tool oil and went about their business diligently filling up air tool FRL units (Filter, Regulator, Lubricator) around the plant.

An assembly plant utilizes a lot of air tools and clean lubricated air is essential for the air tools to work properly. Within hours, one by one, assembly stations started to shut down as the air assembly tools started to fail. Maintenance was called out as the assembly line started to shut down and was shocked to find all air tools, lines and fittings were coated in a gummy plaque that couldn't be flushed or cleaned without mechanical assistance. The assembly line came to a complete halt and the second shift was sent home when the enormity of the situation was evaluated.

The third shift was told to stay home as engineers and maintenance staff frantically worked together to put in place a series of rented air compressors, new tools and flexible hose to get the assembly stations back up and working and get the line started, which they did in time for the third shift.

After suffering the loss of approximately one and a half unrecoverable production shifts additional cost was absorbed in the dismantling and cleaning of the original air tool oil system and interim rental costs.

A RCFA (Root Cause Failure Analysis) investigation found that the combination of the new air tool oil additive chemistry, purchased from a recycled oil blender was different to the original air tool oil and when combined with the gearbox oil and release chemical it acted as a catalyst that caused the resultant fluid to become highly viscous and gummy, which in turn caused the air tool bearings to overheat, degrade and rapidly fail.

Purchasing had managed to save thousands of dollars in lubricant and lube pump costs but had inadvertently cost the corporation tens of millions of dollars in lost sales!

Lessons Learned

1. Many lubricants are not compatible with one another or other chemical fluids—Always perform a compatibility test prior to changing out lubricants for a different manufacturers product.
2. With engineered products and processes in place, always insist on "like for like" replacement/replenishment.
3. When developing lubricant purchase specifications, do not rely on viscosity alone, spell out the manufacturer and full lubricant specification on the purchase request and accept no alternatives without a lubricant trial.
4. When a lubrication management program is in place, accept no new lubricants in the plant without them first going through a lubricant trial exercise to see how it/they perform(s). This will require you to develop a lubricant trial process designed to find out where, and why the lubricant is needed, what it will replace and how it will benefit the department.
5. Always use dedicated transfer equipment for each and every lubricant and chemical
6. Never store chemicals and lubricants in the same immediate area or allow their effluent to mix as they could create harmful gases

Introduce a Service Level Agreement (SLA)

IN the end, your lubricant purchases are only as good as your purchase specification and your business partnership service level agreement (SLA). An SLA is merely a signed agreement between two partners stating what services each will provide within the partnership, and what process each will follow to best facilitate each others jobs.

A typical purchasing SLA would first include all relevant workflow processes, purchase request form and its minimum information requirements highlighted, vendor requirements statements, etc.

In addition, to further facilitate the SLA process, the maintenance department must first understand the exact lubricants it needs for its equipment by performing a lubricant consolidation process that will determine the exact specification listing for the entire plant. Maintenance must then implement a lubricant trial process for any new chemical or lubricant introduction to the plant.

These processes, along with the consolidated lubricant specification, and reasons shown in the above case study "lessons learned" can be used to formulate an effective SLA.

Lubricant Consolidation

Foundational to any Lubrication Management program are four cornerstones commonly known as the four "R's" of lubrication; these dictate that the Right Lubricant be placed in the Right place, in the Right amount, at the Right time.

A basic tenet of lubrication-management is to ensure the right lubricants are employed for specific equipment and, more important, specific operating conditions—"right" being an engineered choice based on performance and economy. Lubricant choice should never be based on price alone, but rather be the result of an engineered process that takes into account the bearing surface types, the ambient conditions, the working conditions, performance requirements, life cycle expectations, and economy.

Because every manufactured lubricant is a unique blend of base oil and additive packages, choosing the right lubricant is a specialized task. A task best left to the lubricant engineering specialists employed by the lubricant manufacturing companies to provide a lubrication consolidation service or program in which the least number of lubricant products are determined to fulfill the end user's needs and requirements.

Why Consolidate?

According to the Canadian Oxford dictionary, to consolidate means, "to make strong or combine." In the case of a lubricant consolidation program the chosen lubricant manufacturer will perform an end user plant wide audit to determine the least amount of lubrication products required to effectively provide the highest degree of lubrication protection for all moving elements within the plant or facility.

In a facility with no lubrication management program in place it is usual to find dozens of different lubricants in use—and no longer in use but still stocked, purchased from multiple lubricant vendors often resulting in many duplicate purchases of similar classification products.

The consolidation audit team groups and classifies the current lubricants in use and examines their suitability for use in the current plant-working environment. With redundancy and duplication taken out of the equation, the remaining list is used as the starting point for cross matching to the new vendor's suite of products. Further consolidation is often possible through understanding the composition and design attributes of the new lubricants that allows the lubrication engineer to better group and match the products to the client's needs and requirements. This process usually drives the number of required products into the single digits resulting in considerable benefits that include:

- **Reduced Onsite Inventories**: with less product SKUs, valuable inventory space is freed up and less operating capital is tied up in inventory,
- **Reduced Purchasing Demand/Cost**: consolidation results in a single source supplier operating on a single blanket purchase order that in turn results in reduced purchase cost, expediting costs, and administration costs, thereby speeding up the purchasing process and virtually eliminating stock out delays,
- **Improved Handling and Storage**: less products require less dedicated transfer equipment and result in less real estate to manage, thereby reducing the chances of product cross contamination,
- **Improved Health and Safety**: less products result in less MSDS (Material Safety Data Sheet)/WHMIS (Workplace Hazardous Material Information System) requirements and simplify/standardize safety procedures,
- **Improved Maintenance Control**: with less lubricants to manage, preventive and predictive programs and job tasks are simplified,
- **Improved Troubleshooting and Problem solving**: dedicated supplier relationships are more likely to produce invested working relationships in which both parties work together to solve lubrication related problems.

The cost of providing this type of engineered consolidation service is most often waived in return for an exclusive purchase contract for a mutually agreed time period (usually one to two years) so the question should not be why consolidate, but rather, why not consolidate?

Choosing a Consolidation Partner

With equipment availability, reliability, and uptime at stake, choosing a reputable lubrication vendor partner is paramount. Choosing a National or Multi-national lubricant manufacturer is most likely to assure a major commitment to product support and research and development of the widest variety of lubricant products. However, this does not preclude that small specialized lubricant manufacturers cannot provide an equal or better commitment to product development and service. Always interview a number of lubricant suppliers to see how they have handled previous consolidation programs for other and similar clients. Reputation is made one client at a time, regardless of the size of the company; lubricants are major purchases, therefore ALWAYS follow through with reference checks. When making your choice take the following into consideration:

- **Value Added Services**: many services can be provided for free and must be taken into consideration when making a partner choice. Services can include:
  - **Engineering**—this would include the consolidation audit process, 24 hr general technical assistance, product manuals and technical datasheets/bulletins, personalized technical problem solving assistance,
  - **Training**—In house training programs on general lubrication practices, web based training programs, You tube®/Vimeo® training videos, etc.
  - **Product Identification Labeling**—implementing lubrication management programs always require sumps, applicators and storage reservoirs to be labeled with the product identification. Are these provided? If so, are they provided free of charge as part of the program?
- **Service Turnaround**: with reduced inventories on hand, turnaround on product delivery is critical and should be measured in hours. Plants that run 24/7 require 24/7 assistance,
- **Pricing**: with longer-term contracts you may be asked to negotiate a price escalation clause. Pricing should be balanced with your service needs when making your final decision,
- **Warm and Fuzzy Factor**: when entering into partnerships the warm and fuzzy factor cannot be underestimated. Test the manufacturers claims before you buy—are they nice to you after hours? Does the representative get back to you to answer silly questions in a timely manner?

Preparing for the Consolidation Process

With a partner successfully chosen, the next step is to facilitate the consolidation process by preparing a list of current lubricants; all lubricants used and stored inside and outside the plant are to be listed on the chart as shown in Table 18-1.

Table 18-1. Lubricant Listing Chart

Lubricant Manufacturer	Lubricant Type	Location	Container	Amount on hand
Petro-Canada	32 Hydraulic Oil	Main Lube Crib	55 gal Drum	2.5
Esso	Unirex 2 grease	Cabinet 2	Cartridge	15
Shell	Omala 220 Gear oil	Main Lube Crib	20 Litre Pail	4

Identifying the how and where lubricants are used further speeds up the consolidation process as depicted in Table 18-2. Each equipment piece is identified on the form and each of its lubrication application requirements are listed complete by current lubricant type, specification, and manufacturer. The delivery application method is also identified on the form. Note; if oil analysis is performed, it too should be noted on the form.

Ideally the plant lubricant technician/specialist should be the person charged with the responsibility of making the list and reviewing it with the consolidation auditor. This process will make the data gatherer much more familiar with the equipment and its lubrication needs. Once the information is gathered, forms are handed over to the consolidation lubrication engineer who then performs the audit requirements analysis to match the correct lubricant for each application. A report is then delivered to the end user, in which all applications are now matched to the new lubricant choice. If any previously used lubricants are identified as incompatible with the new lubricant, a change out procedure must be provided. A material safety data sheet (MSDS) for every lubricant recommended must be included in the report.

Implementing Change

Throughout the consolidation process all unwanted and unused open and closed lubricant containers can be collected for recycling as only the consolidated lubricants are to be allowed on site once the program is activated.

With the report complete and new lubricants on order, the maintenance department must prepare for their delivery and perform a series of updates to manage the new lubricants, these will include:

- Arranging for the collection and return of old lubricants if containers are unopened and undamaged. Note: All lubricants have a shelf life that rarely surpasses 12 months—be prepared to either pay a restocking fee if accepted for return,

Table 18-2. Consolidation Data Gathering Form example

Equipment #: ABC123			Description: 1000 Ton Straight Side Stamping Press		
Lubrication Application		Lubricant Type	Delivery Method	Lubricant Specification	Lubricant Manufacturer
Bearings	Motor	Grease	Manual	Lithium EP2	Exxon
	Mains	Oil	Automatic - Total Loss	ISO 220	Petro-Canada
Slides, Ways		Oil	Automatic - Total Loss	ISO 220	Shell
Compressed Air	Line Lubricator	Oil		ISO 32	ABC - Complube
Hydraulic System	Cushions	Oil	Reservoir	ISO 68	Petro-Canada

- Arranging for the disposal of old lubricants unacceptable for return. Employees could be offered the opportunity to take products (grease tubes) home for personal use with left over lubricant to be picked up by a reputable lubricant disposal company – be prepared to pay for this disposal service,
- Update the Asset Management PM system to reflect the new lubricant choices on all lubrication PM work orders,
- Update all Asset Bill of Materials in the Asset Management System
- Update the Inventory portion of the Asset management system to reflect all new lubricant products. Purge out old lubricant information
- Rearrange Lubricant storage space to accommodate any new container or dispensing equipment. Clean and free up space for other use
- Update the inventory storage labeling
- Update new lubricant labeling on equipment reservoirs and dispensing equipment
- Update MSDS manuals
- Update Equipment manuals
- Update Purchasing records

DESIGNING AND PREPARING A LUBRICANT STORAGE AREA

With a new consolidated lubricant list in place, work must begin on proper storage facilities designed to protect lubricants from the elements and allow for easy stock control and product filtration and dispensing. Utilizing the consolidation report as a starting point, review the following:

- Number of different lubricant products to be carried, stored and dispensed,
- Anticipated usage amounts for each lubricant—used to recommend the most economical purchase and storage container size. This data is also used to determine the real estate requirement for *three to six months* supply of each lubricant at the most (to ensure lubricant is always fresh), and the type of filtering, dispensing and transfer equipment needed to ensure only contamination free lubricant ends up in the bearing surface area,
- In-plant geography of where each lubricant is used, to help determine economical lube routes and the logistical requirements (number of lubrication technicians required, lube truck, fork lift, lubricant delivery equipment, etc. For getting the right lubricant(s) to the right machine(s),
- MSDS sheets and staff training requirements.

With this information, the end user can now decide how best to design their lubricant storage and handling facility. Table 18-3 depicts typical attributes of storage and handling facility that must be factored into the design, a best practice and a reasonable alternative design are shown.

Location/Size

As with any real estate, a good location is paramount; logistically important for receiving lubricants and dispatching them throughout the plant, larger facilities may also require controlled satellite locations. As equally important is protecting the virgin stock lubricants from the outside elements of extreme hot and cold temperatures (Large temperature swings promote moisture condensation in the containers), wind and rain (bringing in lots of solid and water contamination possibility if the containers are not stored or handled with extreme care prior to opening for lubricant transfer).

All locations must be rated for and large enough to allow fork lift truck traffic. Indoor locations are to be temperature controlled, large enough to house 3-6 months of inventory and contain a plant exterior wall for a shipping and receiving dock/door

Ventilation

Because lubricants discharge vapors that might be harmful if allowed to accumulate, a good cross flow ventilation consisting of fresh air units and exhaust fans units is very important. Outdoor facilities may require their own cross flow ventilation if due to its location the building cannot take advantage of prevailing wind vents open to air (vents should be filtered using a furnace style filter) complimented with exhaust fan units.

Fixtures

Based on the lubricant turn over rate and storage container economics, storage and handling facilities may contain 300 gallon poly totes for bulk oil distribution, custom color coded steel tank bulk oil storage and dispensing units systems, drum racks designed to take palletized drums in the upright position or drum dispensing racks set on spill control platforms with the drum positioned on its side in the rack complete with a dispensing/metering valve system. Pails can also be used with the lesser used lubricants and stacked on pallets similar to the drums

Table 18-3. Attributes of Lubricant Storage and Handling Facility

Attribute	Best Practice Design	Alternate Design
Location	Indoors in a contained temperature controlled environment	Outdoors in a contained environment protected from the elements
Size	Big enough to house 3-6 months supply of lubricants based on expected number and turnover rate of all lubricant products. Must be rated for fork lift truck use and big enough to store waste lubricants	Big enough to house 2-3 months supply of lubricants based on expected number and turnover rate of all lubricant products Must be rated for fork lift truck use and big enough to store waste lubricants
Ventilation	Motorized cross flow ventilation system c/w fresh air and exhaust ventilation/fan units	Prevailing wind vents and exhaust fan units
Fixtures	Storage containers and racking custom designed to suit the consolidation inventory flow, dispensing requirements and space limitations	Shelving designed to take stacked drum and pails containers
Transfer equipment/filtration	Dedicated filter transfer equipment for each lubricant type	Dedicated filter transfer equipment for each lubricant type
Spill Control	In floor central collection tank backed up with dry spill collection	Bermed or dyked areas with dry spill collection
Safety	Permanent plumbed eye wash station and portable eye wash station Post MSDS Binders at entrance	Portable eye wash station Post MSDS Binders at entrance
Stock control	FIFO* Stock control	FIFO stock control
Waste oil control	Dedicated waste tanks for each lubricant product in the plant	Dedicated waste oil lay down area to store clearly labeled waste oil containers ready for recycling
Identification	Dedicate specific areas for each lubricant and waste oils and clearly label	Dedicate specific areas for each lubricant and waste oils and clearly label
Processes and procedures	Develop processes and procedure for sustainable operation of the facility and train all relevant staff	Develop processes and procedure for sustainable operation of the facility and train all relevant staff

*First In-First Out

Transfer/Filtration Equipment

To move the lubricant to the machine maintenance must transfer the lubricant from one container to another in the most non-contaminable way possible. For bulk containers, use of dedicated transfer/filter cart style dispensing units will ensure lubricant is moved from the bulk and pre packaged supplier containers to the machine reservoir and/or dedicated closed pour containers similar to that shown in Figure 18-2 for transfer to the lube system reservoirs.

Spill Control

Best practice is to slope storage room floors to a low point drain where spilled product can be collected into an easy accessible common tank for recycling. Local spills can be managed with dry spill absorbent products.

Safety

A permanently plumbed eye wash station is a must when dealing with petroleum based products. Up to date MSDS binders and Standard Operating Procedures are to be posted at the entrance to the facility

Stock Control

As most lubricants are only rated for a shelf life between 6-12 months, stock must be rotated on a regular basis following a FIFO (First In First Out) approach to stock control. Pass due date lubricants should be returned to the supplier or recycled and the stock purchasing/usage history be evaluated and adjusted accordingly.

Identification

Having gone to the trouble of consolidating your lubricants clearly identify a dedicated area in the facility for each and every lubricant clearly label marking in large letters (2-3 inches high) the lubricant ID. Develop a plan drawing of the facility identifying locations of each lubricant and waste tanks and post at the facility entrance.

Processes and Procedures

Best practices are not only rooted in the design but in a sustainable operation of the facility. Be sure to develop, map put, and train all staff on all processes and procedures related to use of the storage and handling facility

Program Monitoring

Not everyone gets it right the first time, a lubricant consolidation process is an exercise in compromise to a degree, and there may be occasions in which lubricant performance is less than desirable. Work with the lubricant vendor to detail a performance-monitoring program suitable for your work environment, and meet on a regular basis in the first three to six months to ensure all lubricants are performing at their intended level. Any exceptions can be caught early and a remedy put in place before any equipment damage is experienced.

Performing a lubricant consolidation program is a valuable exercise that results in optimized performance at the lowest cost and should be reassessed every three to five years. A consolidation exercise can considerably reduce costs by reducing the amount of different products carried in stock, which saves on operating capital, storage and handling, and administration costs. A typical company can carry over 20 lubricant types in stock that can be reduced by up to 60% depending on the type of operation, which should make the program mandatory for every plant!

Outdoor Storage

Outdoor storage should be avoided if possible. Weathering can obliterate the labels on containers, leading to possible mistakes in selecting lubricants for specific applications. Furthermore, widely varying outdoor temperatures, with consequent expansion and contraction of seams, may lead to leakage and wastage. The likelihood of contamination is also increased. Water can leak into even tightly closed drums by being sucked in past the bung as the drum and its contents expand and contract. See Figure 18-2.

Extremely cold or hot weather can also change the nature of some compounded oils and emulsions, making them useless.

When containers must be stored outside, the following precautions are advised:

- Keep bungs tight
- Lay drums on their sides (Figure 18-2). Position the drums so that the bungs are at 9 and 3 o'clock, to ensure that they are covered by the drum contents, thus minimizing moisture migration and drying out of the seals.
- If drums must be placed upright without weather protection, tilt them slightly to prevent water from collecting around the bungs, or use drum covers, or spread a tarpaulin over the drums.
- Before removing the bungs, dry the drum heads and wipe them clean of any contaminant which might get into the lubricant later. The importance of keeping grit and sand out of oil used in expensive bearings must be kept in mind.

Figure 18-1. Outside Storage Facility with unprotected and incorrectly positioned drums allowing water to cover bungs—Source: ENGTECH Industries Inc.

Figure 18-2. Twelve-drum pallet rack showing bungs at 6 and 12 o'clock. The preferred bung location is 3 and 9 o'clock. (Source: MECO, Omaha, NE)

Indoor Storage

Storage temperatures should remain moderate at all times. The oilhouse should be located away from such possible sources of industrial contamination as coke dust, cement dust, textile mill fly, and similar forms of grit or soot. It should be kept clean at all times, with regular cleaning schedules being maintained. This applies above all to the dispensing equipment, which must never be allowed to become fouled, since this results in contamination and poor functioning.

Contamination and confusion of brands are the two main things to be avoided in the handling of partially emptied containers and dispensing equipment. Thus orderliness is essential. Dispensing equipment should bear a label that matches the container from which it was filled. Labels on all equipment and containers should be kept legible at all times, as seen in Figures 18-3 and 18-4. Drying oils, such as linseed oil, should not be stored in the oilhouse. If they get into a lubrication system, the result, of course, is faulty lubrication and stoppage.

One should never use the same dispensing equipment for both detergent engine oils and R&O turbine and hydraulic oils. Contamination of the rust- and oxidation-inhibited industrial oils with detergent engine oil substantially impairs the quality of the industrial oils. Trace amounts of the detergent and other alkaline contaminants can react with the acidic rust inhibitor and cause operational problems like foaming, filter plugging, and emulsion formation.

Also, never use the same dispensing equipment for oils containing zinc additives and those that are zinc-free. For instance, contamination of Exxon's premium railroad diesel engine crankcase lubricant ("Diol RDX") with zinc additives could lead to catastrophic engine failure of EMD diesel engines, in which the zinc additives attack the silver wrist-pin bushings. Similar concerns exist in certain centrifugal compressors where zinc additives may adversely affect the reliability of sealing components.

Galvanized containers (Figure 18-5) should never be used for transporting oil. Many of the industrial oils used today contain additives that would react with the zinc of the galvanizing to form metal soaps, which would then clog small oil passages, wicks, etc. Moreover, contamination of zinc-free diesel-engine oils could be disastrous, as mentioned in the preceding paragraph.

Practical Dispensing Equipment

A number of procedures, practices, and equipment are already well documented which serve to mitigate the ingress and effects of contaminants in process machinery lubrication systems. These include:

Figure 18-3. Orderly lubricant storage with adequate spill containment—Source: ENGTECH Industries Inc.

Figure 18-4. Fully labeled and color-coded lubricant storage/dispensing management system—Source: Fluid Defense systems

Figure 18-5. Rusty, or galvanized containers, are unsuitable for reliability-conscious lube transfer.

- hermetic sealing of bearing housings
- oil mist systems
- optimum bearing selection
- optimum lubricant/additive selection
- documented maintenance procedures
- personnel training & development
- condition monitoring —(vibration, oil sampling & analysis, thermography, etc.)

The extent to which organizations are adopting any or all of the above approaches is as always a function of available resources, inclusive of capital, labor and time.

Nevertheless, not all solutions to the lubricant contamination issue need be expensive or time-consuming. All too often, substantial improvements can be obtained by simply paying attention to such basics as cleanliness of transfer (dispensing) containers.

Walking through a process plant, an alert observer will often see oil cans or substitute containers with open spouts, equipment with open fill-ports ready to accept rainwater and airborne dirt, and rusty vessels which we would not dare to use as feeding bowls for farm animals. (See Figure 18-6.) Even at the better facilities, the observer may find that transfer containers are not always marked with the proper oil grade, or that responsibilities and accountabilities for transfer equipment are largely undefined.

Since even the best available lubricant or hermetically sealed bearing housing will not perform under these conditions, the replacement of any questionable transfer containers with rust-proof, well-designed transfer tools should be a priority issue for modern industrial plants. With payback periods often measured in *days*, the cost-effectiveness of these tools is utterly self-evident.

Figure 18-6. A variety of open style transfer containers full of contamination— Source: ENGTECH Industries Inc

The Chicago, USA, based company Fluid Defense Systems manufactures and distributes the Oil Safe brand of lubrication dispensing (Figure 18-7, 18-8) and oil transfer containers shown in 18-11 and 18-12. These are available in a variety of styles and colors. Their unique lid and spout design keeps oil in and contaminants out. A quick-action push-pull valve incorporated in the spout allows for the adjustment of oil flow to task demands (Figure 18-8). Responsible reliability professionals consider these devices essential lubrication management tools and, surely, our readers will appreciate their significance. It makes much economic sense to go back to basic steps of ensuring lube oil cleanliness before contemplating any of the other, more glamorous high-tech approaches to optimized lubrication.

Lubricant Transfer Policy

Similar to the storage policy, a policy/protocol is required for transferring fluids

- All Industrial equipment Lubricants to be transferred and dispensed by a trained Lubrication Specialist
- All transportation vehicle lubricants to be transferred and dispensed by trained lift truck mechanics

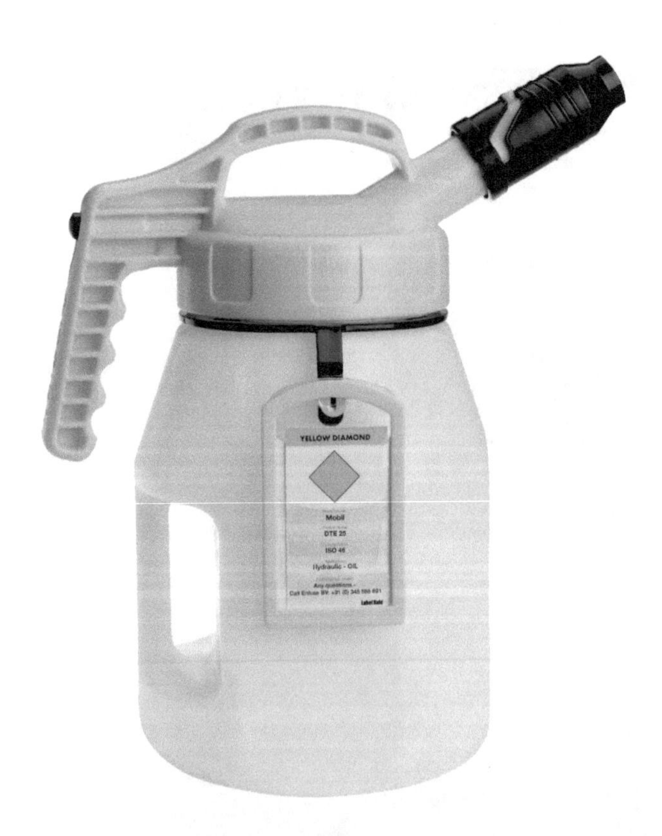

Figure 18-7. State-of-the-art utility oil transfer containers can offer payback in mere days! (Source: Fluid Defense Systems)

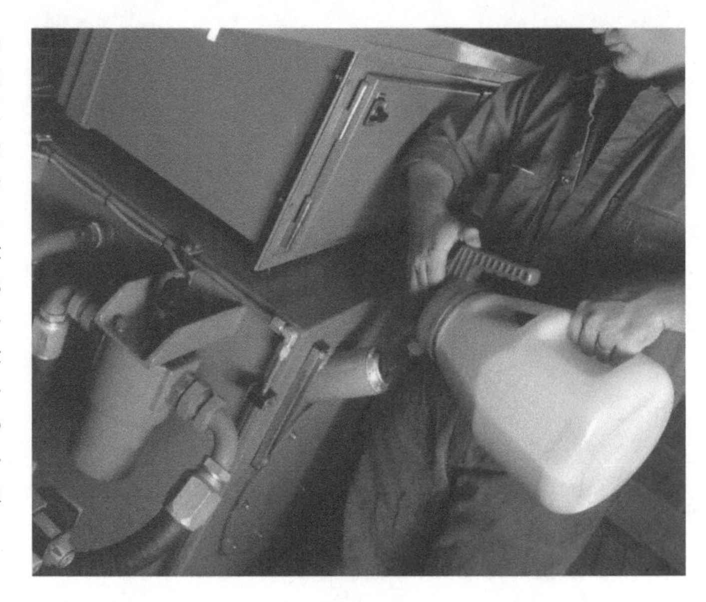

Figure 18-8. Transferring lubricant with the "new style" oil transfer container

- Transfer all lubricants using the recommended safety equipment designated on relevant MSDS sheets, and in accordance with current plant health and safety regulations; to include gloves, protective eyewear, footwear, coveralls, etc.
- Transfer all tote managed hydraulic fluids using a dedicated filter cart pumping system (one for each hydraulic lubricant type), designed to be close coupled to both the tote and the receiving hydraulic system reservoir—allowing no contamination to enter during the transfer process
- All other fluid lubricants to be transferred using color-coded, dedicated, Oil-safe storage/transfer devices. These devices are to be labeled with the lubricant type
- Transfer bulk grease lubricants using dedicated air operated grease/transfer pump units
- If any pumping/transfer device is to be used to transfer lubricant from a bulk container to a dedicated Oil safe type container, it must also be a dedicated pumping/transfer device, and labeled as such
- Clean all transfer equipment on a regular basis
- Use only XYZ Corporation approved pumping/transfer devices

On larger reservoir fills a lubricant transfer/filter cart (shown in Figure 18-13) can be employed to not only perform the lubricant transfer in a timely manner, but also to "polish" the lubricant clean through a filter as the reservoir is filled. Always ensure that if a transfer cart is used, it has been dedicated for use with only that lubricant you are transferring!

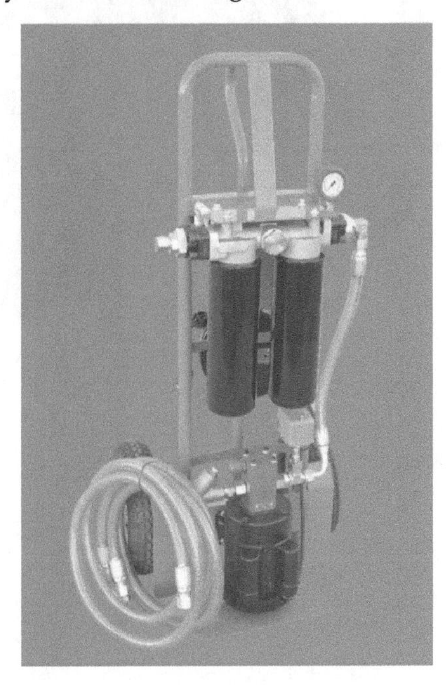

Figure 18-9. Lubricant transfer/filter cart

HANDLING LUBRICANTS

Most lubricating oils and greases are a relatively harmless class of material. No unusual hazard is involved in their use, provided care is taken to avoid ingestion, keep them off the skin, and avoid inhalation of their vapors and mists.

Company policy must ensure that all its products in their prescribed use and subsequent disposal shall not create a significant hazard to the public health or environment.

The following preventive measures are recommended for personnel who regularly handle petroleum products:

- Avoid all unnecessary contacts, and use protective equipment to prevent contact.
- Remove promptly any petroleum product that gets on the skin.
- Do not use gasoline, naphtha, turpentine, or similar solvents to remove oil and grease from the skin.
- Use waterless hand cleaner or mild soap with warm water and a soft brush. Use only clean towels, not dirty rags.
- Remove all contaminated clothing immediately. Launder or dry-clean it thoroughly before reuse.
- Use protective hand cream on the job, and reapply it each time hands are washed. After work hours, use simple cream to replace fats and oils removed from the skin by washing.
- Wash hands and arms at the end of the work day and before eating.
- Get first aid for every cut and scratch.
- Avoid breathing oil mist or solvent vapors.
- Keep work area clean.
- Clean up spilled petroleum products immediately. Keep them out of sewers, streams, and waterways.
- Contact the medical staff on all potential health-hazard problems.

LABELING LUBRICANT STORAGE DEVICES

To prevent cross contamination of lubricants during the transfer process all lubricants in non-OEM marked storage container are required to be labeled for environmental, safety, and good management reasons. This includes machine reservoirs and all storage, transfer and handling containers.

There are numerous color code methods available for use that adopt a visual management approach with use of shapes and colors, as well as letters and numbers to easily identify and match lubricants correctly. Figure 18-10 and 18-11 show numerous labels that utilize the color/shape/language method of display that are easily attached to storage/transfer devices.

Receiving Lubricants

In smaller facilities, most purchase lubricants will be received in smaller containers and drums as shown in figures 18-3 and 18-4. These are relatively easy to receive and with proper handling are put into stock with relative ease.

Larger facilities, or facilities that consume a lot of one particular lubricant due to their process, will consider purchasing their lubricants in bulk totes similar to

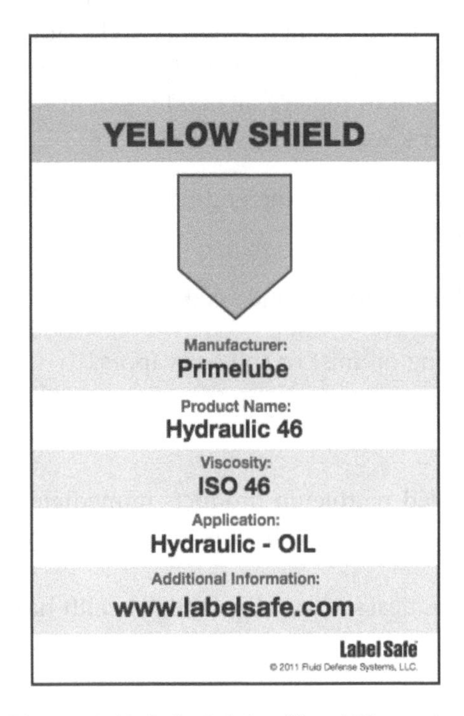

Figure 18-10. Lube Label utilizes Yellow color, Shield shape and written description of lubricant

Figure 18-11. Lube Labels hang off a dispensing tank utilizing a blue circle and red hour glass with their written descriptions to identify two different lubricants metered side by side Source: Fluid Defense Systems

that shown in Figure 18-12. Smaller totes can be filled by transfer from a drum, but larger totes like the 400 gallon (1600 litre) style, and usually filled on sit by a tanker truck or replaced full on an exchange program.

Prior to initial shipment, lubricant is tested for base oil viscosity and flash point, additive package composition and concentration known as the "treat rate," and its level of contamination or cleanliness; if the batch meets the lubricant design requirements it is given a Certificate of Analysis (COA), which is copied to the lubricant purchaser. This COA document is important as it acts as a baseline measurement for all corresponding quality checks prior to machine "point of use."

To ensure your supplier provides you with the cleanest bulk lubricant product when filling tote container follow these simple rules before implementing bulk lubricant delivery in your new storage and handling facility.

- Insist on receiving a lubricant Certificate of Analysis (COA) for each lubricant delivered and keep this document on file until the batch of lubricant has been used,

Figure 18-12. 400 gallon refillable tote sitting on a dedicated transfer pump station

- Never assume all lubricants are delivered as per their COS document specification,
- Set up a delivery acceptance agreement with the supplier to deliver lubricant based on the COA and/or a set of internal minimum cleanliness usually a minimum ISO cleanliness level of 18/15/12 and viscosity specifications (within +/- 10% of COA specification),
- Establish an oil quality analysis test acceptable to end user and supplier, and a develop a service level agreement that outlines the lubricant condemning levels and remedial action requirements should the lubricant fail the quality test on delivery,
- Perform quality testing regularly taking a bulk sample after the tanker truck lines have been flushed prior to transfer, and from the center of any supplier pre-filled containers.

Now that you have set up a bulk delivery program, the totes and bulk lubricants must be monitored on a regular basis following a policy/procedure similar to that shown below.

- Upon purchase order issue, the first lubricant delivery is to be delivered in clean virgin sealed totes. The filled lubricant will have a lubricant cleanliness rating of ISO 18/15/11 or cleaner,
- Totes are to be dedicated specifically to your company only, and marked clearly with:
  — Product name
  — Viscosity
  — Approval or Tox # details

  — Date the tote was put in service
- The tote cap is to be sealed directly after filling and the seal number recorded on the accompanying invoice,
- The driver to inspect the cap seals on all full totes and note any damage. If a compromised seal, the driver will notify the user Lubricant Specialist and together they will inspect the tote for contamination. If the tote is believed to be contaminated the tote is to be removed from service and a credit issued for the unused lubricant. The tote is to be taken off site, drained, and thoroughly cleaned, inspected, refilled, sealed and returned to service. If the cap is determined to be acceptable, the driver will reseal the cap and date accordingly,
- Upon subsequent bulk filling of dedicated totes situated at the user's plant, the supplier's driver will assess each tote for physical damage and both internal/external cleanliness prior to filling. IF the driver believes the current condition of the tote will compromise the 18/15/14 ISO lubricant cleanliness during the filling process, the tote will be removed from service. The driver will notify the user lubrication specialist and inspect the tote together and a credit will be issued for any substantial residual fluid found in the tote. The tote is to be taken off site, drained, and thoroughly cleaned, inspected, refilled, sealed and returned to service,
- On occasion, the user lubrication Specialist will perform lubricant analysis for quality control purposes, checking for fluid cleanliness level and water presence in the lubricant totes. If significant water or cleanliness levels higher than 20/18/14 are found, the Lubricant Specialist will contact the vendor to show the results and the tote will be removed from service. The tote is to be taken off site, drained, and thoroughly cleaned, inspected, refilled, sealed and returned to service,
- Once the totes are determined fit for refilling, they will receive product from a clean bulk fill nozzle. All product delivered is to be dispensed through a 5 micron absolute filter and metered through a temperature compensated pumping system,
- Once filled, the totes are to be immediately capped and sealed, unless a fluid sample is to be taken by the Lubrication Specialist at this time. The seal #'s are to be recorded on the invoice and for FIFO stock control purposes. Driver may be asked to assist the Lubrication Specialist in the arranging and maintaining of the tote FIFO inventory control,
- All poly totes installed for bulk fluid containment will not remain in service longer than 6 months, and are to be replaced when emptied after 6 months with clean fresh totes,
- All metal totes are to have a protected exterior tank level sight gauge/tube installed,
- The driver is to maintain a record of dates, volumes, deliveries, and removals of totes, product and seals. The record is to include any problems, requests, and proposed solutions incorporated, and to whom they were addressed. The driver is expected to share copies of this record with the user lubrication specialist.

Figure 18-13. this tote should not be accepted for filling before it is cleaned

In addition, the lubrication specialist should perform a weekly stock rotation check based on the following procedure

Stock Rotation Procedure

- Check lubricant tote manifest for all new product arrivals during the previous week,
- If no new product arrival, check to ensure totes are still lined up chronologically by delivery date, with the last tote to be delivered (newest delivery date) staged furthest from the door, and the "First In" (Oldest) delivery date closest to the door ready for immediate delivery in to the plant
- If new product has been shipped, or existing totes have been rearranged, check all tote delivery dates and use a fork lift truck to move and position the totes into delivery date chronological order by staging the newest product arrival furthest from the door, and the "First In" (Oldest) delivery dated tote closest to the door, ready for immediate delivery in to the plant
- Complete FIFO check work order and return to Planner
- If any product is over six months old it should be taken out of the stock rotation

WASTE OIL MANAGEMENT

There will be waste oil generated through oil changes that in some jurisdictions will require a waste oil proof of disposal certificate to prove waste oil was managed and disposed of in a responsible and eco friendly manner according to municipal and state requirements.

Numerous companies specialize in recycled oil and waste oil management that will assist you in understanding the legislation in your area and how best to collect the oil to keep the costs as low as possible. Mixing oils in one big tank can inflate the disposal cost more than ten times as this oil cannot be recycled easily.

Now that all lubricants on site have been consolidated it is much easier to determine what is in your plant and have dedicated collection tanks (these can be barrel or tote style tanks) in the oil crib and clearly marked for each waste lubricant type, the only exception being spilled oils.

Oils spill materials and Oily rags must also be managed and disposed of in the correct manner according to your local environmental legislation.

With waste oil taken care of in a responsible manner, the Cradle to cradle oil life cycle is complete.

Chapter 19

Lube Oil Contamination, Filtration and On-Stream Purification Control*

Lubricants rarely get the respect they deserve until they are recognized as an integral part of a machine's design.

Original equipment manufacturer's (OEM) machine designers have long recognized and respected the value of choosing the correct lubricant for the job and spend considerable time in conjunction with numerous lubricant company engineers to choose a lubricant selection that will adequately perform in a given set of ambient and machine conditions and allow the machine to function as per its design specification.

The machine designer also recognizes that a lubricant is a consumable weak link in the design and as such is subjected to many stresses, such as temperature and contamination that can significantly degrade its ability to protect the machine. To combat temperature excesses, a lubricant with a suitable viscosity and viscosity index is chosen. To ensure the lubricant has a chance of an extended life cycle, a good designer will incorporate a variety of contamination control devices in the lubrication system design. Equally important in the lubricant life cycle equation is the role of the equipment owner, who must also recognize the lubricant as an integral part of the equipment design and be diligent in controlling contamination elements such as dirt and water around the machine, and care for the contamination control devices built into the machine.

THE CONTAMINATION EFFECT

In 1986, the National Research Council of Canada (NRC) published a landmark study titled "A Strategy for Tribology in Canada." While it primarily focused on natural resource industries and agriculture its findings were remarkably similar to the Jost report, published some twenty years earlier in the UK. One of the standout statements made by the report was the recognition of failure due to ineffective lubrication practices regarding lubricant cleanliness in its statement *"In primary industries 82% of wear induced failure was particle induced failure from use of dirty lubricants, with the greatest wear caused by particles whose size equaled the oil film thickness."* (See Chapter 1 – Wear – Abrasive wear – 3-body abrasion).

The study went on to illuminate the apathy surrounding lubrication failure with the statement *"77% of respondents believed their current level of friction and wear could be reduced but no action was being taken as current levels were erroneously accepted as normal!"*

Three Types of Lube Oil Contamination Identified

There are three primary contamination threats to lubricants, and by extension, the bearings they are designed to protect. Of these, the first one, dirt, is usually filterable; hence, it can be readily controlled. However, dirt is often catalyzed into sludge if water is present. Experience shows that if water is kept out of lube oil, sludge can be virtually eliminated.

The second contaminant, process-dependent dilution, is seen in internal combustion engines and gas compressors where hydrocarbons and other contaminants blow past piston rings or seals and are captured within the lube and/or seal oils. Dilution results in reduced viscosity, lower flash points, and noticeable reduction of lubrication efficiency.

The last one, water, is perhaps both the most elusive and vicious of rotating machinery enemies. In lube oil, water acts not only as a viscosity modifier but also actively erodes and corrodes bearings through its own corrosive properties and the fact that it dissolves acid gases such as ones present in internal combustion engines. Moreover, water causes corrosion of pumps, and rusts cold steel surfaces where it condenses.

Also, in some systems water promotes biological growth which, in itself, fouls oil passages and produces corrosive chemicals. Ironically, the maintenance department directly and indirectly causes much of the contaminant ingression into its lubricants.

*Based on technical papers co-authored by Judith L. Allen, Heinz P. Bloch, and Tom Russo. Adapted by permission. Based on technical articles authored by Kenneth E. Bannister for Maintenance Technology magazine's Lubrication Strategies series.

CONTAMINATION SOURCES

The maintenance department as described below, can manage contamination, in all forms, once it recognizes the contamination source. The three main contamination sources are built in, ingested, and generated

Built-In

Not all original equipment manufacturers (OEM) are diligent in ensuring all the manufacturing process debris and assembly fluids have been cleaned prior to first time equipment start up. The same is true for any service / upgrade work performed on the equipment throughout its life cycle. It is the responsibility of the maintenance department to ensure a thorough clean up has taken place prior to lubricant fill on any equipment when any work affecting lubricated areas has been performed.

Ingested

Ingestion is the primary source for any second type process-dependent dilution contamination, often the result of the ambient conditions the host machine works in. In specific cases machine design may dictate that ingested contamination occurrences will take place requiring the lubricant cleanliness strategy to take this into account if the design cannot be change (more regular oil changes). This is also the case when excessive wear (wear induced combustion chamber blow-by) allows contamination to accelerate.

Ingested contamination is also the result of carelessness through forgetting to reinstall breathers, or tank opening caps.

Generated

The third source is a result of how the lubricant performs to protect the bearing surfaces that comes down to correct lubricant choice, preventive and predictive maintenance strategies and effectiveness of the maintenance efforts.

We can see from the type of contaminants that affect lubricant and equipment performance listed in Figure 19-1, most can be prevented from entering in the system and those that can't be prevented can be managed with the right strategy.

SOLIDS CONTAMINATION

Solids contaminants usually present in the form of dust, silica, silt, sediment, lint, fibers, etc., more commonly referred to as "dirt". The larger the particulate size, the greater damage they will cause if allowed to accumulate in the lubrication system and lubricated bearing areas.

The Macpherson contamination curve Figure 19-2

Built In
OEM & Service Debris

- Burrs
- Swarf
- Filings
- Assembly lube
- Protective lubricants
- Dust
- Service fluids
- Lint/fibers

Ingested
Process, Atmosphere & Combustion

- Process Chemicals
- Process Debris
- Raw Material
- Radiation
- Tank Opening contaminated air
- Airborne Dust
- Air
- Blow by
- Soot
- Fly Ash
- Fuel
- Glycol
- Water

Generated
Surfaces & Oil

- Mechanical Wear
- Corrosion
- Fibers
- Break In Debris
- Elastomers
- Paint Chips
- Silt
- Sediment
- Additive depletion
- Oxidized base oil

Figure 19-1. Contamination ingression causes

clearly shows the relationship between particulate size and the expected life cycle of a bearing. We can see that filtering contaminants to 4 micron in size will triple the bearing life cycle has we only filtered to 14 microns.

One of the primary tests performed with an oil sample is the ISO 4406: 1999 test to count and measure the size of contaminant particulate into three groups of > 4 microns, > 6 microns, and >14 microns. These counts are then used to determine the lubricant cleanliness by measuring the number and size groupings of particles found in a 1ml representative sample of oil, which are then compared to Figure 19-3 to determine the representative sample range and come up with the lubricant sample's ISO cleanliness number.

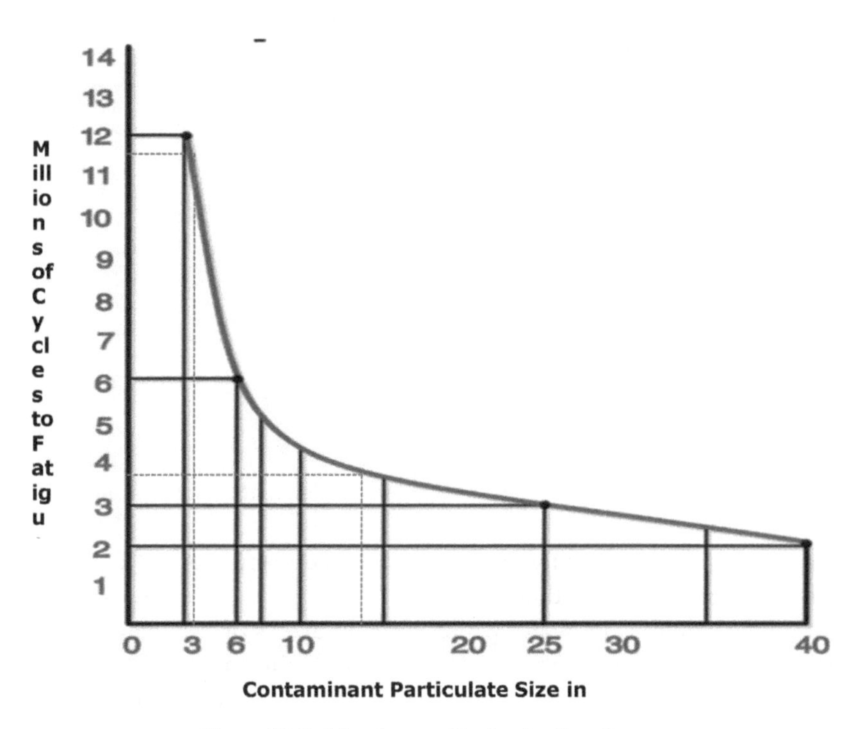

Figure 19-2. Macpherson Contamination Curve

The example count findings shown in Figure 19-4 are 2130 of the > 4 micron, 620 of the > 6 micron and 72 of the >14 micron. When applied to the range chart we end up with an 18/16/13 ISO fluid, making it reasonably clean oil close to what one would expect to find when new. Note: one micron = 1/1000th mm or 1/100,000th inch.

Particles per mL		
>	< including	Range #
80,000	160,000	24
40,000	80,000	23
20,000	40,000	22
10,000	20,000	21
5,000	10,000	20
2,500	5,000	19
1,300	2,500	18
640	1,300	17
320	640	16
160	320	15
80	160	14
40	80	13
20	40	12
10	20	11
5	10	10
2.5	5	9
1.3	2.5	8
0.64	1.3	7
0.32	0.64	6
0.16	0.32	5
0.08	0.16	4
0.04	0.08	3
0.02	0.04	2
0.01	0.02	1

Figure 19-3. ISO particle concentration range table

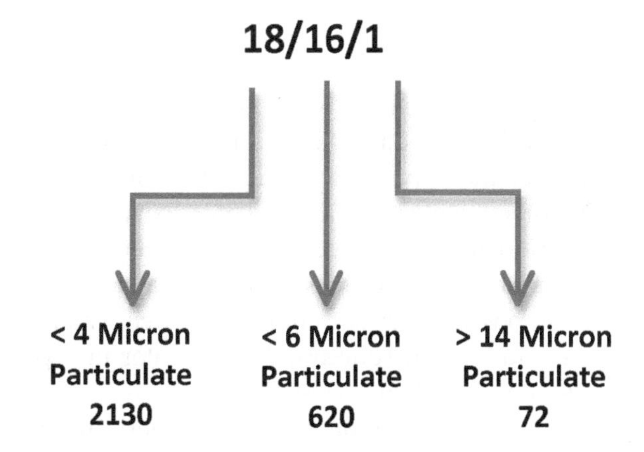

Figure 19-4. Particulate count sample

Note: Not all lubricants are as clean when delivered and regular cleanliness checks of virgin stock oils should be part of any lubrication management program—especially when bulk oils are used. Figure 19-5, shows the physical difference under a microscope between a very dirty oil (26/25/23), a dirty oil (21/10/17), and a clean oil (16/14/10).

Water Contamination

In oil systems associated with process machinery, water can, and will, often exist in three distinct forms: free, emulsified, and dissolved. But, before examining

Images courtesy of Fluid Defense

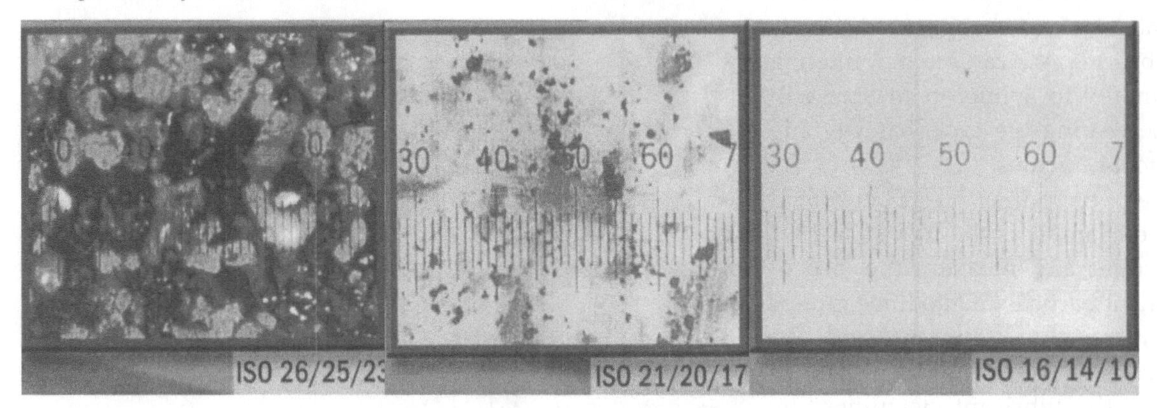

Figure 19-5. Fluids cleanliness under a microscope

the effects of water contamination, it may be useful to more accurately define these terms.

Free water is any water which exists in excess of its equilibrium concentration in solution. This is the most damaging water phase. Free water is generally separable from the oil by gravity settling.

Emulsified water is a form of free water which exists as a colloidal suspension in the oil. Due to electro-chemical reactions and properties of the oil/water mixture in a particular system, some or all of the water that is in excess of the solubility limit forms a stable emulsion and will not separate by gravity even at elevated temperatures. In this respect, emulsified water behaves as dissolved water, but it has the damaging properties of free water and modifies the apparent viscosity of the lubricant.

Dissolved water is simply water in solution. Its concentration in oil is dependent upon temperature, humidity and the properties of the oil. Water in excess of limits imposed by these conditions is free water.

The equilibrium concentration of water in typical lube oils is give in Figure 19-6. Dissolved water is not detrimental either to the oil or the machinery in which it is used.

For corrosion to occur, water must be present. Free water, in particular, will settle on machinery surfaces and will displace any protective surface oil film, finally corroding the surface. Emulsified water and dissolved water may vaporize due to frictional heat generated as the lube oil passes through bearings. Very often, though, the water vapors recondense in colder pockets of the lube oil system. Once recondensed, the free water continues to work away at rusting or corroding the system.

Larger particles generated by corrosion slough off the base metal surface and tend to grind down in the various components making up the lube system, i.e. pumps, bearings, control valves, and piping. The mixing of corrosion products with free and emulsified water in the system results in sludge formation which, in turn, can cause catastrophic machinery failures. Suffice it to relate just one of many examples of water-related damage to major machinery.

When a steam turbine at a medium-sized U.S. refinery failed catastrophically, the initial problem was attributed to coupling distress and severe unbalance vibration. When the coupling bolts sheared, the steam turbine was instantly unloaded and the resulting over-speed condition activated a solenoid dump valve. Although the oil-pressurized side of the trip piston was thus rapidly depressurized, the piston stem refused to move and the turbine rotor sped up and disintegrated. The root cause of the failure to trip was found to be water contamination of the turbine control oil. Corrosion products had lodged in the trip cylinder and, although enveloped in control oil, the compression spring pushing on the trip piston had been weakened by the presence of water.

As mentioned earlier, water is an essential ingredient for biological growth to occur in oil systems. Biological growth can result in the production of acidic ionic species and these enhance the corrosion effects of water. By producing ionic species to enhance electro-chemical attack of metal surfaces, biological activity extends the range of corrodible materials beyond that of the usual corrodible material of construction, i.e., carbon steel.

While corrosion is bad enough in a lube oil system, erosion is worse as it usually occurs at bearing surfaces. This occurs through the action of minute free water droplets explosively flashing within bearings due to the heat of friction inevitably generated in highly loaded bearings.

Additive loss from the lube oil system is another issue to contend with. Water leaches additives such as anti-rust and anti-oxidant inhibitors from the oil. This occurs through the action of partitioning. The additives partition themselves between the oil and water phase in proportions dependent upon their relative solubilities. When free water is removed from the oil by gravity, coalescing or centrifuging, the additives are lost from the oil system, depleting the oil of the protection they are designed to impart.

The severity of the effects of water on bearing life due to a combination of the above is best illustrated by Figure 19-7 (Ref. 2) where we see that bearing life may be extended by 240% if water content is reduced from 100 wppm to 25 wppm. However, if water content is permitted to exist at 400 wppm, most of which will be free water, then bearing life will be reduced to only 44% of what it could be at 100 wppm.

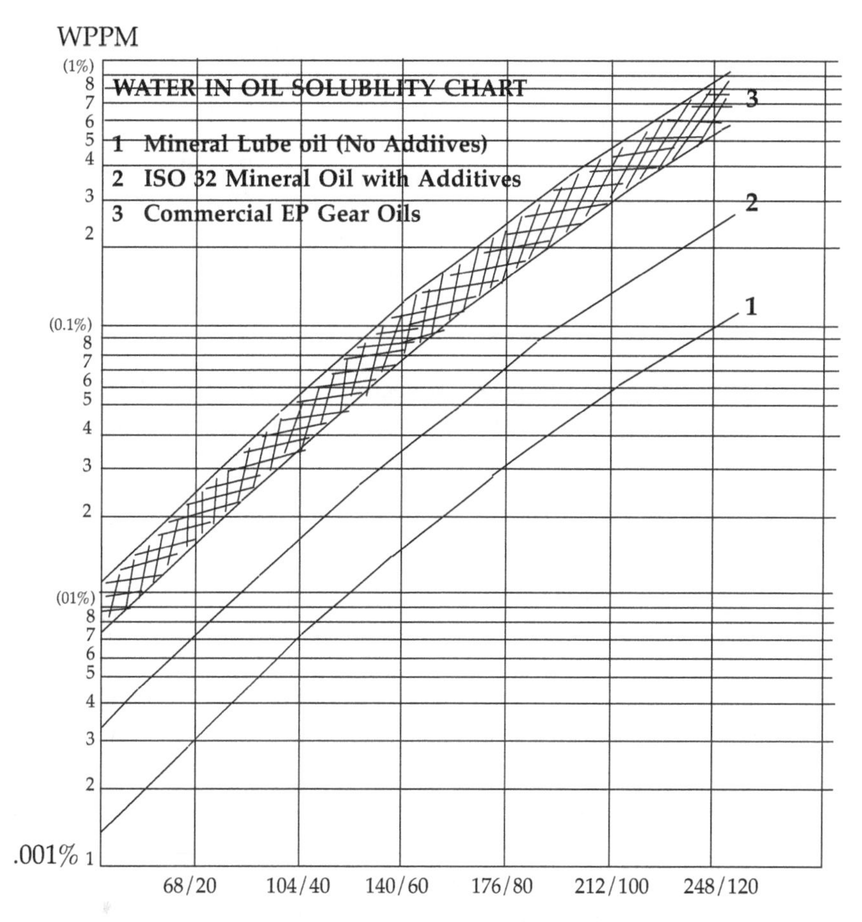

Figure 19-6. Solubility chart for water in oil.

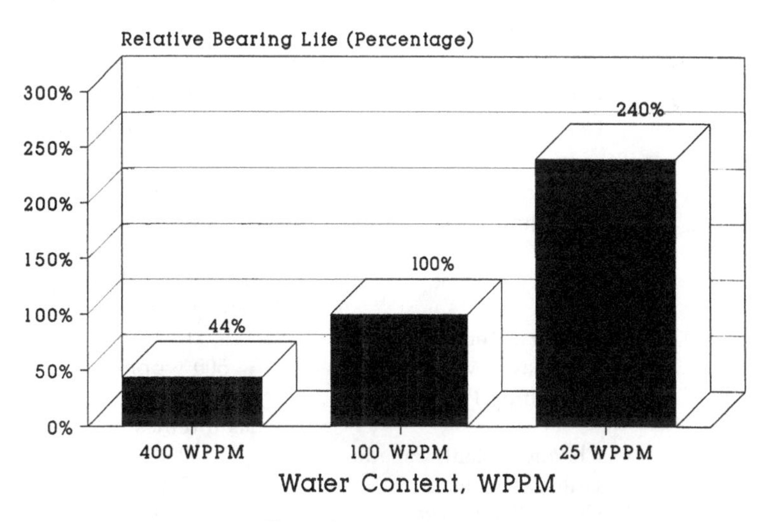

Figure 19-7. Effects of water in lube oil on bearing life.

Water Ingression

Water can affect the base oil and the additive package in different ways as depicted in the Tables 19-1, and 19-2.

Ideal Water Levels Difficult to Quantify

Manufacturers, operating plants and the U.S. Navy all have their own water condemnation limits as illustrated in Table 19-3. Extensive experience gained by a multi-national petrochemical company with a good basis for comparison of competing water removal methods points the way here. This particular company established that ideal water levels are simply the lowest practically obtainable and should always be kept below the saturation limit. In other words, the oil should always have the ability to take up water rather than a propensity to release it.

FILTRATION AND CONTAMINATION CONTROL DEVICES

Methods Employed to Remove Solids

The primary contamination control device used in an oil lubrication system is the fluid filter, which can be separated into two categories, 1) Surface type filter and, 2) Depth type filter. For semi-fluid systems using grease as a lubricant only surface type filters, known as strainers and mechanical filters are used.

Surface Fluid Filter—Oil

Surface filters are the most common style of oil fluid filter in use and can be found in many design configurations. They are primarily designed to work in the direct flow path of the lubricant and capture any dirt particles (contaminants) held in colloidal suspen-

Table 19-1. Water Ingression effects on base oils

Result	Symptoms
Increased Aeration	Foaming, Air entrapment, Cavitation
Increased Hydrolysis / Oxidation	Acid formation, Viscosity thickening, Varnish, Sludge, Rust, Poor filterability, Flash vaporization, EHD film strength loss
Reduced Dielectric Strength	Reduced insulating property

Table 19-2. Water Ingression effects on lubricant additives

Additive	Result
Antioxidant, Rust Inhibitor	Acid, Sediment (floc), premature depletion
Rust Inhibitor, Detergents, De-Emulsifying agent	Sludge, Sediment, Bacteria, Poor oil/water separation, Premature depletion
Borate EP	Loss of EP capability
ZDDP Anti Wear	Hydrogen sulfide, Sulfuric acid

Table 19-3. How guidelines on allowable water contamination vary. The table applies to turbomachinery lubricants.

Guidelines Issued by User or Manufacturer	Max. Allowable Water Content Quoted	Sampling Frequency	Purifier Operation
Chemical Plant, Texas	4000 wppm	Monthly	Intermittent
Consulting Firm, Europe	2000 wppm (steam turbines only)	—	—
Refinery, Europe	1000 wppm (BFW and cooling water pumps)	Monthly	Intermittent
Refinery, Kentucky	1000 wppm	—	As Required
U.S. Navy (MIL-P-20632A)	500 wppm	—	Continuous
Consulting Firm, USA	500 wppm gas turbines only	—	—
Chemical Plant, Canada	200	Visual	As Required
Utility, Michigan	100	Monthly	Continuous
Chem. Plant, Texas	100	6 Months	Intermittent
Chem. Plant, Louisiana	100	Monthly	Intermittent
Refinery, Texas	100	—	Intermittent
Consulting Firm, USA	100	Weekly	As Required
Utility, Europe	100	—	Continuous
Refinery, Texas	100	—	Continuous
Utility, Michigan	100	—	—
Turbine Mfr., Japan	100	—	—
Compressor Mfr., Pennsylvania	40	—	—
Refinery, Louisiana	—	—	Intermittent

sion as the lubricant flows through or across the filter media. Typically, this filter element is very porous so as minimize any differential fluid pressure loss across the filter media; because of this, the filter exhibits low capture efficiency, dirt holding capacity, and a moderate life expectancy requiring that it be inspected, cleaned or replaced on a regular basis.

The actual design and capture capability of the filter will depend on its location within the lubrication system and the desired functional effect of the filter. For example, in a typical re-circulative hydraulic or lubrication system a pump suctions oil from a lubricant reservoir and pumps it under pressure to a moving device such as a valve, cylinder or bearing. Once the lubricant has performed its job it is allowed to return, under gravity, to the reservoir where it can cool and be cycled again. In this typical type of system we can expect to find six to seven types of surface filter designs and one depth filter (see Figure 19-8).

Filter #1—Pump protection: is a suction filter positioned low in the reservoir and connected to the pump suction inlet via a suction tube. Typically the filter media is wire mesh gauze. Paper, or felt designed to capture and stop any large debris or metallic wear particles from entering the pump.

Filter #2—Primary system protection: is a pressure filter positioned directly in the pump output delivery between the pump and the first moving device in the lubrication system. This filter is the primary system protection filter and uses pleated paper, cellulose or fine porous metal media designed to withstand the pump system pressure and capture small micron particulate that have managed to move through the suction filter and pump gear set.

Filter #3—Wear metal and Colloidal Debris collection: is positioned on the gravity return piping side of the lubrication system just before the reservoir lubricant return inlet. Known as a gravity return filter this filter generally employs a low-pressure paper medium deigned to capture wear metal and debris washed from the moving parts of the machine by the lubricant.

Filter #4—Ferritic Metals and Debris: is a more passive filter designed to attract and hold any ferritic (magnetic) debris and wear metals that might of by-

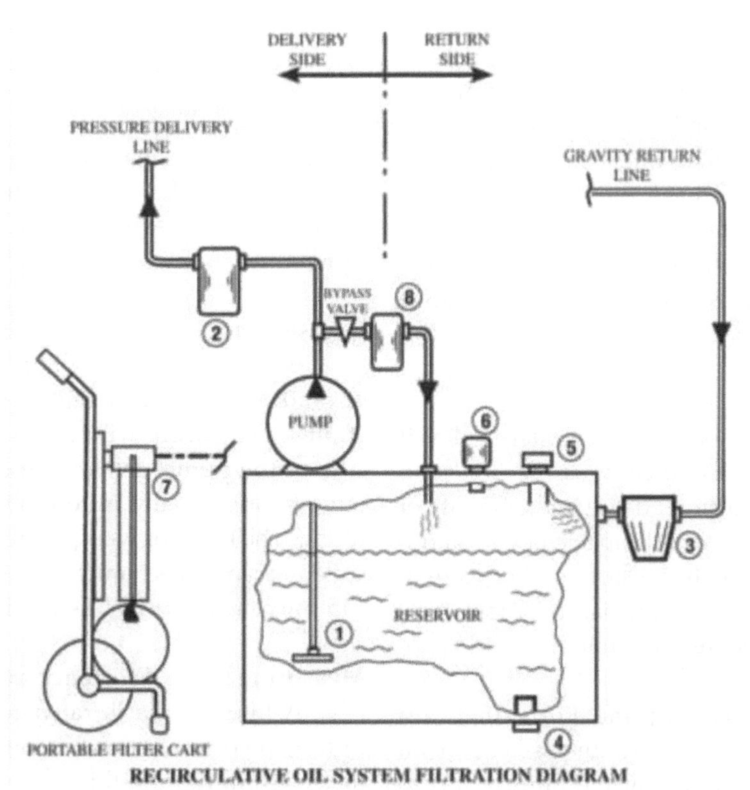

Source: Industrial Lubrication Fundamentals – Lubricant Lifecycle Management. *Lubrication Management & Technology* magazine - Ken Bannister

Figure 19-8. Different filter types found in a lubrication system

passed the return filter or have been left in the reservoir from start up. The filter is magnetic and often serves second duty as a reservoir drain plug.

Filter #5—Solids Ingress: is a large pore metal mesh strainer sock positioned in the inlet mouth of the reservoir fill port designed to capture any errant large particulate from making its way into the reservoir when filling is taking place, or if the fill cap has been left off the reservoir.

Filter #6—Airborne Contaminants and Moisture: is a breather filter designed to equalize pressure in the reservoir and in its simplest form utilizes a wire wool media to keep out any particulate of 40 micron and above in the air from entering the reservoir. In a more sophisticated design the breather employs a desiccant like silica gel hydrophilic material again allows the reservoir to breathe and prevent outside airborne particulate 3 micron and above from entering the reservoir, but with the added advantage of being able to wick and capture moisture from inside the reservoir and prevent outside moisture from entering the reservoir. Once saturated, the gel turns from blue to pink indicating visually its need to be changed, see Figure 19-9.

Filter #7—All Contaminents: is an optional filter utilized in larger systems employing a filter cart that is hooked up to the reservoir and used to extract the oil from the reservoir, filter it using filters similar to the #2 pressure filter, or through a "bag" filter that employs a thick felt type sock in the form of a bag to provide fine filtration of the lubricant, which is then returned to the reservoir clean for further use extending the oil change frequency.

Depth Fluid Filter—Oil

A depth filter differs from a surface filter in that it takes the lubricant through an indirect maze like flow path designed to capture and accommodate large amounts of dirt. The filter is highly efficient and capable of withstanding high differential pressure. If the surface filter were to use a sheet of toilet paper as its media, the depth filter equivalent would be the entire toilet roll of paper!

Filter #8—Depth Cleaning and Polishing; is a depth filter typically placed within a bypass circuit on the output pressure delivery side of the pump prior to the pressure filter. The filter medium can be made from cellulose, fiberglass, felt, or diatomaceous earth, designed to deep clean and polish the lubricant.

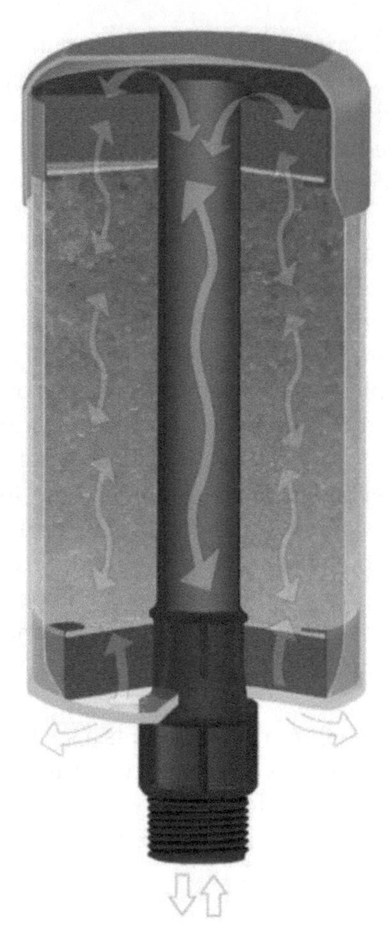

Figure 19-9. Des Case Silica gel hydrophilic reservoir breather unit (Courtesy Des Case)

When choosing an oil filter medium and size, always consider the low temperature conditions as fluid viscosity increase may cause an increase in pressure differential through the filter

Grease Filtration

Because grease does not typically flow the same as oil and is pumped at higher pressures than oil, metal strainers or wedge wire filters are fitted to the pressure side of the pump delivery system and are used to trap large debris usually introduced into the system during filling. The mesh or wedge wire (looks similar to a tightly coiled spring) mechanically trap contaminants down to 145 microns and must be cleaned regularly.

Measuring Filter Efficiency and Beta Ratio

A filter media is rated to capture particles down to a defined minimum micron size and its efficiency is measured in how well it captures those particles. Under lab conditions the volume of upstream particles (of a defined micron size) are counted before entering the filter and counted again after exiting the filter on the downstream side, the difference is calculated as a percentage to arrive at the filter efficiency (see Table 19-4).

A filter's Beta ratio or Filtration ratio is calculated using the ratio of upstream versus downstream particulates. For example, if 100,000 particles of the same micron size were flowed through a filter that only stopped half or 50,000 particles the efficiency would calculate to 50% and the beta ratio would be 100,000/50,000 = **2**. Similarly, if the filter was successful in stopping 99,000 of the 100,000 it would be 99% efficient and have a beta ratio of 100,000/1000 = 100. Therefore, a higher Beta ratio number equates to a more efficient and higher quality filter medium.

Table 19-4. Beta Ratio / Efficiency Table

Beta Ratio	Efficiency
$\frac{100{,}000}{50{,}000} = \mathbf{2}$	**50.0%**
$\frac{100{,}000}{5{,}000} = \mathbf{20}$	**95.0%**
$\frac{100{,}000}{1{,}333} = \mathbf{75}$	**98.7%**
$\frac{100{,}000}{1{,}000} = \mathbf{100}$	**99.0%**
$\frac{100{,}000}{500} = \mathbf{200}$	**99.5%**
$\frac{100{,}000}{100} = \mathbf{1000}$	**99.9%**

Methods Employed to Remove Water

Centrifuges have been used for decades. They operate on the principle that substances of different specific gravities (or densities) such as oil and water can be separated by centrifugal force. Centrifuges achieve a form of accelerated gravity settling, or physical separation. At a given setting, centrifuges are suitable for a narrow range of specific gravities and viscosities. If they are not used within a defined range, they may require frequent, difficult readjustment. They will not remove entrained gases such as hydrogen sulfide, ethane, propane, ethylene, etc., or air. Although centrifuges provide a quick means to separate high percentages of free water, they are maintenance intensive because they are high-speed machines operating at up to 30,000 RPM. More importantly, they can only remove free water to 20 wppm above the saturation point in the very best case, and none of the dissolved or emulsified water. In fact, centrifuges often have a tendency to emulsify some of the water they are intended to remove.

Coalescers are available for lube oil service and have found extensive use for the dewatering of aircraft fuels. Unfortunately, coalescers remove only free water and tend to be maintenance-intensive. More specifically, a coalescer is a type of cartridge filter which operates on the principle of physical separation. As the oil/water mixture passes through the coalescer cartridge fibers, small dispersed water droplets are attracted to each other and combine to form larger droplets. The larger water droplets fall by gravity to the bottom of the filter housing for drain-off by manual or automatic means. Since coalescers, like centrifuges, remove only free water, they must be operated continuously to avoid long-term machinery distress. The moment they are disconnected, free water will form and begin to cause component damage. Again, because they are based on a physical separation principle, they are only efficient for a narrow range of specific gravities and viscosities.

As mentioned earlier, coalescers are used in thousands of airports throughout the world to remove water from jet fuel. Water, of course, freezes at high altitudes. Refining of jet fuels is closely controlled to rigid specifications which allows successful water removal by this method. There are several disadvantages to coalescers. They are only efficient over a narrow range of specific gravities and viscosities. They do not remove dissolved water which means they must be operated continuously, and it is expensive to change elements.

Filter/Dryers are also cartridge type units which incorporate super-absorbent materials to soak up the water as the wet oil passes through the cartridges. They remove free and emulsified water, require only a small capital expenditure, and are based on a very simple technology. However, they do not remove dissolved water, and their operation might be quite costly because the anticipated usage rate of cartridges is highly variable due to the changing water concentrations. The amount of water contamination at any given time would be difficult, if not impossible, to predict. Additionally, high cartridge use creates a solid waste disposal problem.

Next, we examine *vacuum oil purifiers* which have been used since the late 1940s. Low cost versions generally tend to suffer reliability problems. Therefore, the user should go through a well-planned selection process and should use a good specification. For a properly engineered vacuum oil purifier schematic, see Figure 19-3. Good products will give long, trouble-free service.

A vacuum oil purifier operates on the principle of simultaneous exposure of the oil to heat and vacuum while the surface of the oil is extended over a large area. This differs from the other methods we have discussed in that it is a chemical separation rather than a physical one. Under vacuum, the boiling point of water and other contaminants is lowered so the lower boiling point constituents can be flashed off. Typical operating conditions are 170°F (77°C) and 29.6" Hg (10 Torr). Because water is removed as a vapor in a vacuum oil purifier, there is no loss of additives from the oil system. The distilled vapors are recondensed into water to facilitate rejection from the system. Non-condensables such as air and gases are ejected through the vacuum pump.

Vacuum Oil Purifiers More Closely Examined

As illustrated in Figure 19-10, the typical components of a vacuum oil purifier are an inlet pump; a filter typically rated at 5 microns; some method of oil heating such as electric heaters, steam, hot water, or heat transfer fluid; a vacuum vessel; and a vacuum source such as a mechanical piston vacuum pump or water eductor. Vacuum oil purifiers may, or may not, incorporate a condenser depending upon the application. A discharge pump is employed to return the oil to the tank or reservoir, and an oil-to-oil heat exchanger may be employed for energy conservation.

Vacuum oil purification is applied across a broad spectrum of industries: power generation and transmission, automotive, aluminum, refining and petrochemicals, steel, mining, construction, plastic injection molding, metalworking and food processing.

Vacuum oil purification is the only extended range method capable of removing free, emulsified, and dissolved water. Since vacuum oil purifiers can remove dissolved water, they can be operated intermittently without the danger of free water forming in the oil. Furthermore, they are the only method of oil purification which will simultaneously remove solvents, air, gases, and free acids.

In virtually all instances, a cost justification study by medium and large users of industrial oils will favor well-engineered vacuum oil purifiers over centrifuges, coalescers, and filter/dryers. Cost justification is further

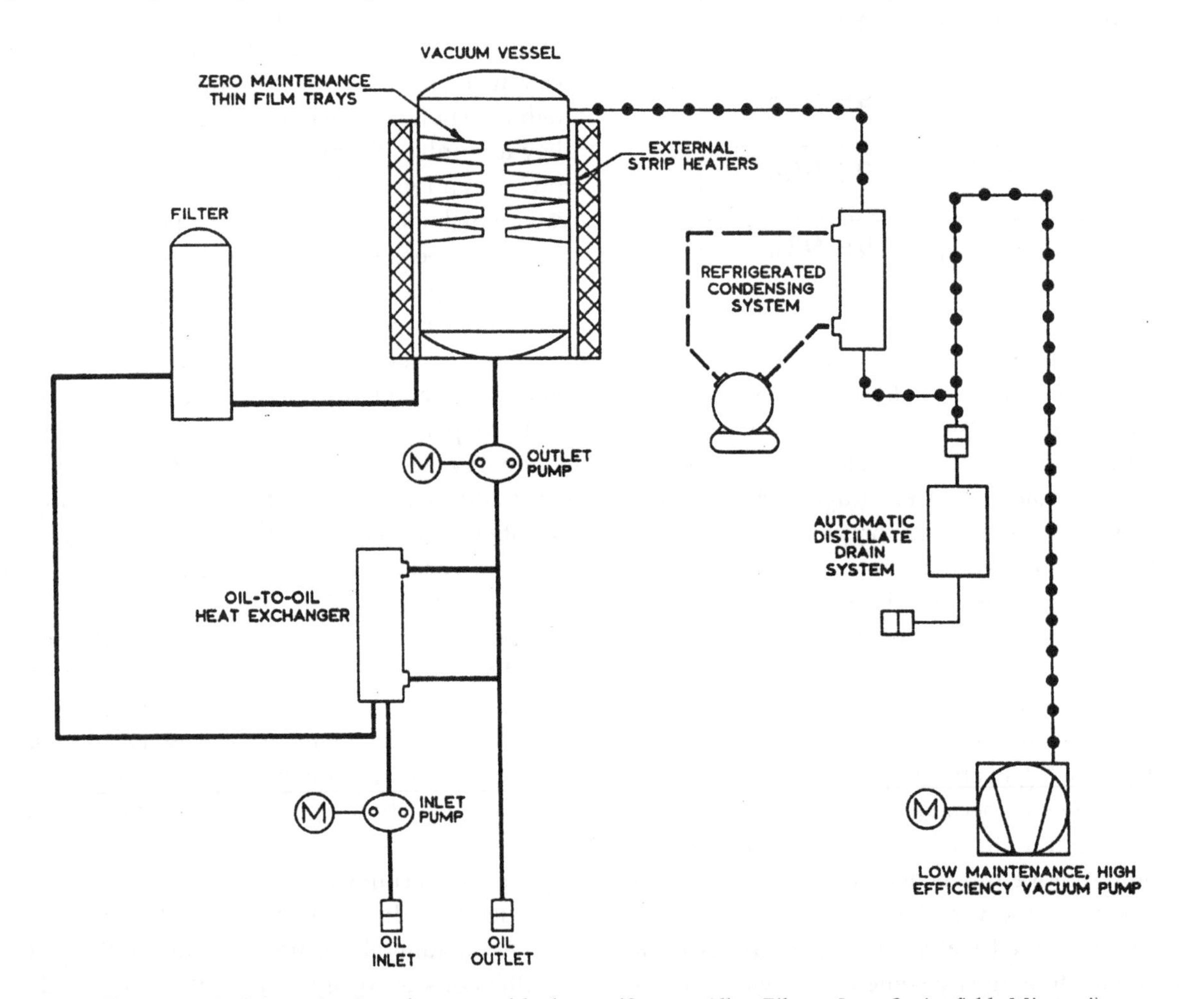

Figure 19-10. Schematic view of vacuum dehydrator. (Source: Allen Filters, Inc., Springfield, Missouri)

influenced in favor of vacuum oil purifiers by bottom-line analyses which look at the cost of maintenance labor and parts consumption.

Sound Alternatives Available for On-stream Purification

Not every application needs or can justify a vacuum oil purifier. Other technologies are available and deserve to be considered. Also, some users due to monetary constraints are willing to give up flexibility, but not effectiveness. Other users may have troublesome machines which are continuously subject to free water contamination and require an inexpensive, dedicated dehydration device operated continuously to purify the lube oil system. These users may be well served by the very latest method of oil purification which operates on the chemical separation principle of air stripping. It is important to recognize that this chemical separation principle, too, removes free, dissolved, and emulsified water. It, therefore, ranks as a viable alternative and close second to vacuum oil purification as the preferred dehydration method, particularly for smaller systems.

Ambient air stripping units are intended for light duty application on small, or medium-sized reservoirs. They are specifically designed for dedicated use on individual machines for water removal only. No operator attention is required, and such units are simple to install. Since they are compact and lightweight, they may be set on top of a lube oil reservoir. They are available at low to moderate initial cost and are extremely easy to maintain. These modern self-contained stripping units remove water at or above atmospheric pressure in the vapor phase and, therefore, conserve oil additives. Units similar to the one shown in Figure 19-4 are capable of removing free, emulsified and dissolved water to well below the saturation levels, and yield a product which is, in most cases, as good as fresh lubricant. Such units can reduce water concentration to below 15 wppm, and as can be seen, for the figures presented, this is a good working level from all considerations.

Operating Parameters for State-of-the-art Stripping Units

As depicted in Figure 19-11, air stripping units draw oil from the bottom of an oil reservoir by a motor-driven gear pump, representing the only moving components of the unit. The oil is then forced through a filter which removes particulates and corrosion products; and then to a steam or electric heater for temperature elevation. From there the oil goes to a very efficient mixer/contactor (jet pump) where ambient air, or low pressure nitrogen, is aspirated into the wet oil mixture. The air is humidified by the water in the oil during its period of intimate contact, and this is the method by which the oil is dehydrated. Since even relatively humid air can absorb even more moisture when heated, ambient air is usually a suitable carrier gas for this water stripping process. The wet air is then vented to atmosphere while the oil collects in the bottom of a knock-out vessel. A gravity or pump-equipped return loop allows the now-dehydrated oil to flow back to the reservoir.

The choice of air versus nitrogen depends on the oxidation stability of the oil at typical operating temperatures between 140°F (60°C) and 200°F (93°C). The choice is also influenced by flammability and cost considerations.

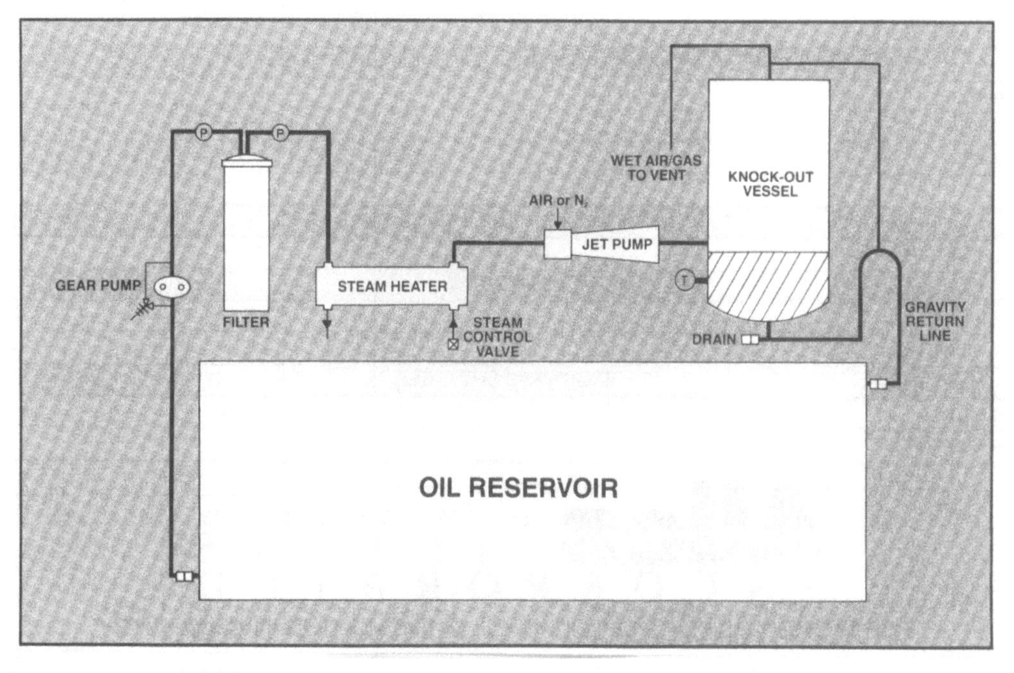

Figure 19-11. Simplified diagram of air stripping unit. (Source: Ausdel Pty, Cheltenham, Victoria, Australia)

Normally, low pressure or waste steam is used to heat the oil as it passes through the unit. Several model sizes are available from experienced suppliers in the U.S. and Australia. Also, a number of different design options are available. For example, if electricity or steam costs are expensive, or, if hot oil returning to the reservoir is a problem, then an oil feed/effluent exchanger can be added to the basic model to significantly reduce heating costs.

Narrowing the Choice to State-of-the-art Units

Vacuum oil purifiers and modern self-contained stripping units produce comparable oil quality, and are generally preferred over the less effective and maintenance intensive physical separation methods. In general, the cost of a self-contained stripping unit is only a fraction of the cost of a vacuum oil purifier, and operating and maintenance costs are markedly lower due to its much simpler concept and construction.

The comparable performance of vacuum oil purification versus air stripping is illustrated in Figures 19-12 and 19-13. After a performance test conducted by a major multi-national oil/chemical company, the results were plotted as shown in Figure 19-12. The test conditions were vacuum oil purifier circulating flow rate of 375 U.S. gallons per hour on a 75 U.S. gallon oil sample which had been contaminated with 2% water (20,000 wppm), 162°F (72°C) average operating temperature, and 29.96" Hg (1 Torr) vacuum level. The exact test conditions were then computer-simulated for an air stripping device, and the results plotted as shown in Figure 19-13.

Looking at the Figure 19-12 and 19-13 and curves, the most striking point is the speed with which a vacuum oil purifier will dehydrate gross amounts of free water versus an air stripper. After 40 minutes of circulation time, water was reduced from 20,000 wppm to 734 wppm with the vacuum unit. This compares to the air stripper simulated performance after 39 minutes in reducing 20,000 wppm water down to 9,413 wppm. Ultimately, both units showed their capability in reducing water to less than 40 wppm with the oil from the vacuum oil purifier performance test measured by the Karl Fischer titration method.

While machinery engineers generally advocate dedicated air/gas stripping purifiers, mobile vacuum oil purifiers offer substantial performance flexibility, especially where rapid de-watering is of importance. Vacuum units would be required in critical situations

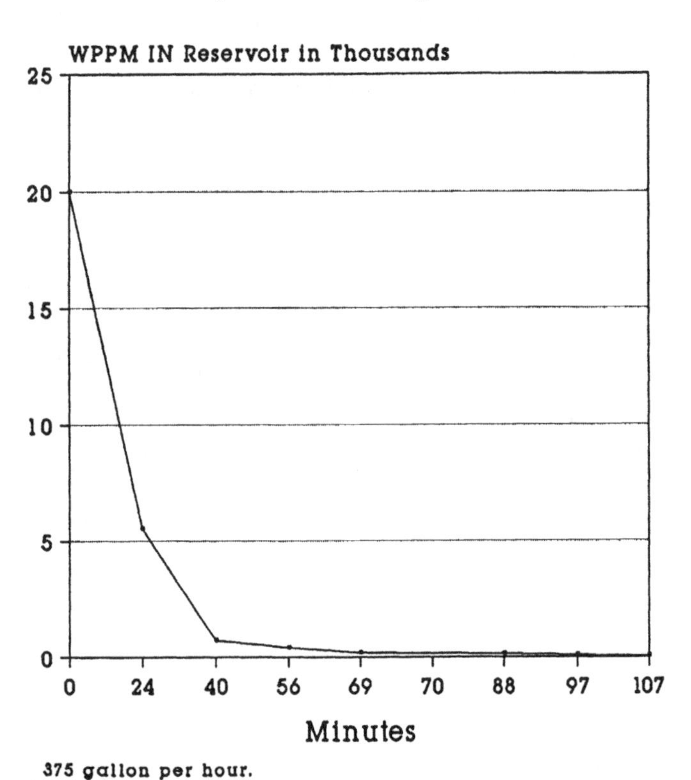

Figure 19-12. Typical residual water vs. time performance of vacuum dehydrators. (Source: Allen Filters, Inc., Springfield, Missouri)

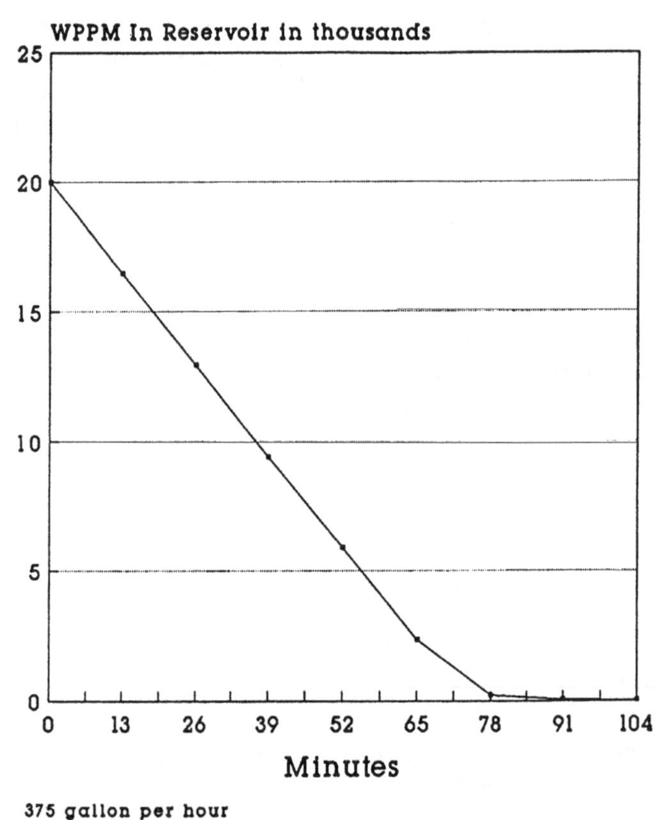

Figure 19-13. Performance simulation for air stripping unit. (Source: Ausdel Pty, Cheltenham, Victoria, Australia)

when an unexpected rate of contaminant ingression into an oil system is encountered beyond the capabilities of a dedicated air/gas stripper. This is because the fixed geometry of the air stripper mixer/contactor does not provide the same flexible contaminant handling capabilities inherent in a vacuum oil purifier. The vacuum units excel also in cases where combustible gases are present such as compressor seal oil applications. In these cases the availability and cost of nitrogen must be considered for the air strippers; but with a vacuum oil purifier, nitrogen use is a non-issue. Plants which are intent on being in the forefront of technology would optimize machinery reliability by dedicating an air stripper (see Figure 19-14 for a modern unit) to each reservoir, and utilizing one or more mobile vacuum units as required to facilitate uptime on critical equipment until the next turnaround/planned maintenance period. All the foregoing considerations permit the user to make a clear separation on when he might favor one type over the other.

Cost Justifications Will Prove Merits of On-stream Purification

The cost justification table which follows will enable potential users to determine for themselves the value of dedicated on-stream oil purification units. On many occasions, plants were astonished at the credits shown by filling in their numbers on the following tabular justification matrix.

BEARING LONGEVITY AND CONTAMINATION*

As if we didn't know, lubricant contamination will reduce bearing life. However, a thorough understanding of the contamination/bearing life relationship is essential if improvements are to be achieved in bearing life and equipment reliability.

*Source: Dr. Richard Brodzinski, BP Oil, Kwinana, Western Australia, and Michael T. Kilian, Oil Safe Systems Pty Ltd, Mount Lawley, Western Australia.

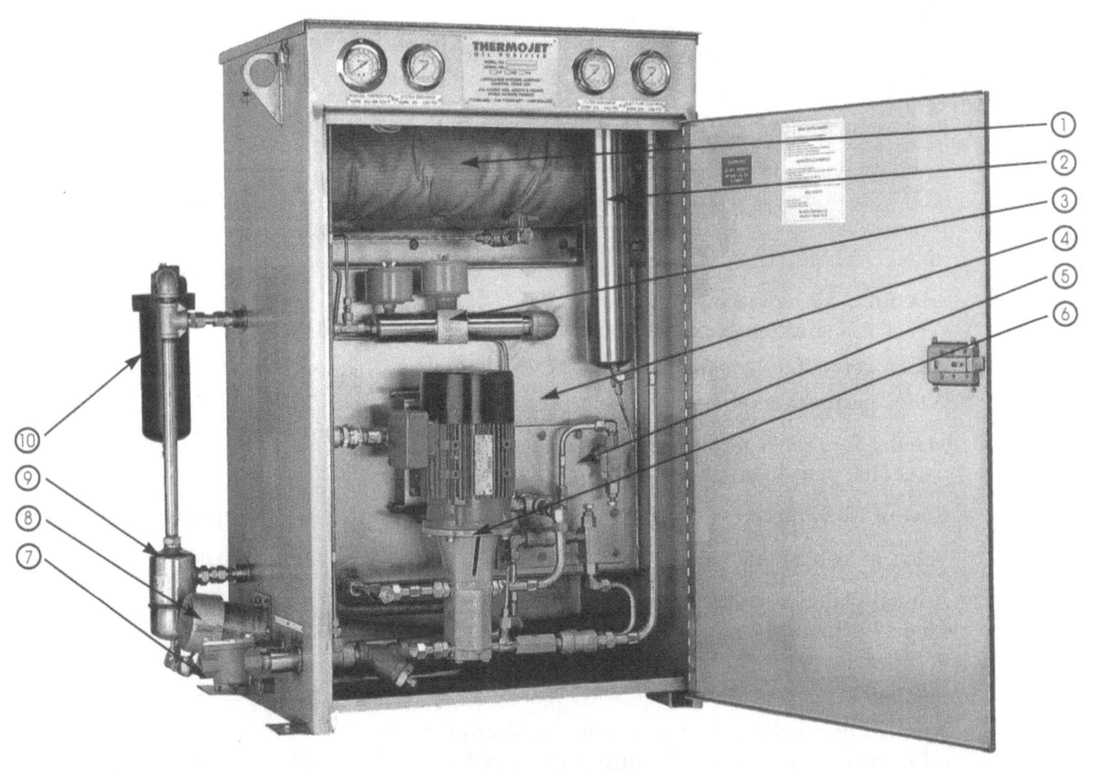

1. 12 kW Oil Heater
2. Filter
3. Dual-Stage Jet Mixer
4. Separation Tank
5. Level Controller
6. Oil Pumps & Motor
7. Oil In/Out & Sight Glass
8. Condensate Purifier Assembly
9. Condensate Trap
10. Oil Mist Eliminator/Vapor Exhaust

Figure 19-14. Modern air stripping unit and principal components. (Source: Lubrication Systems Company, Houston, Texas)

AIR STRIPPER RETURN ON INVESTMENT

	ANNUAL COST SAVINGS		EXAMPLE PLANT	YOUR PLANT
A.	OIL			
	* Oil inventory in reservoir (liters)	1.	2,000	________
	* Annual number of oil changes WITHOUT purifier (once per 2 years)	2.	0.5	________
	* Annual number of oil changes WITH purifier (once per 10 years)	3.	0.1	________
	* Annual liters of oil saved—Line 1 x (Line 2 - Line 3)	4.	800	________
	* Cost/liter of oil	5.	$3.00	________
	* Annual value of oil saved	6.	$2,400	________
	* Labor charge to change oil (30 man-hours @ $40/hr including overhead)	7.	$1,200	________
	* Annual value of labor saved —Line 7 x (Line 2 - Line 3)	8.	$480	________
	* Sub-total (Line 6 + Line 8)	9.	$2,880	________
B.	FILTERS			
	* Annual number of filter changes WITHOUT purifier (3 per year).	10.	3	________
	* Annual number of filter changes WITH purifier (once per 2 years)	11.	0.5	________
	* Replacement filter cost (parts and labor)	12.	$1,200	________
	* Annual savings in replacement filter cost—Line 12 x (Line 10 -Line 11)	13.	$3,000	________
C.	PLANNED MAINTENANCE			
	* Scheduled major machine shutdown WITHOUT purifier (once per 2 years)	14.	0.5	________
	* Scheduled major machine shutdown WITH purifier (once per 10 years)	15.	0.1	________
	* Maintenance labor to inspect and overhaul worn parts (100 man hours @ $40/hr) per shutdown	16.	$4,000	________
	* Cost of worn parts	17.	$20,000	________
	* Value of lost production per shutdown	18.	$20,000	________
	* Annual planned maintenance credit (Line 16 + Line 17 + Line 18) x (Line 14 – Line 15)	19.	$17,600	________
D.	UNPLANNED MAINTENANCE			
	* Number of unscheduled major machine shutdowns for oil-related wear WITHOUT purifier (once per 5 years)	20.	0.2	________
	* Number of unscheduled major machine shutdowns for oil-related wear WITH purifier	21.	0	________
	* Value of lost production, parts and labor per event	22.	$200,000	________
	* Annual unplanned maintenance credit—Line 22 x (Line 20 – Line 21)	23.	$40,000	________

TOTAL ANNUAL SAVINGS			
* Line 9 + Line 13 + Line 19 + Line 23	24.	$63,480	________
E. OPERATING COSTS			
* Annual on-stream hours	25.	8,000	________
* Power consumption (kW)	26.	1.1	________
* Cost of power ($/kWh)	27.	0.08	________
* Annual power cost—Line 25 x Line 26 x Line 27	28.	$704	________
* Steam consumption (kg/hr)	29.	75	________
* Cost of steam ($/ton)	30.	8.0	________
* Annual steam cost—Line 25 x Line 29 x Line 30/1000	31.	$4,800	________
* Nitrogen consumption (m^3/hr)	32.	$0.08	________
* Cost of Nitrogen ($/$m^3$/Zero if air is used)	33.	$2.5	________
* Annual nitrogen cost—Line 25 x Line 32 x Line 33	34.	$1,600	________
* Sub-total operating costs—Line 28 + Line 31 + Line 34	35.	$7,104	________
F. NET ANNUAL BENEFIT			
* Total benefits (Line 24)	36.	$63,480	________
* Total operating costs (Line 35)	37.	$7,104	________
* Net Annual Benefit (Line 36 – Line 37)	38.	$56,376	________
G. PROFITABILITY CALCULATION			
* Cost of Air Stripper Purifier	39.	$35,000	________
* Payback period (Months) (Line 39/Line 38) x 12	40.	7.5	________
* Annual Return on Air Stripper Purifier	41.	>100%	________

Once we quantify and understand how even a small amount of lubricant contamination can adversely affect bearing life, we will better appreciate why significant thought and investment must be made in machinery and support equipment design and maintenance to ensure the utmost cleanliness in machinery lubrication systems.

Quantifying the Effects of Lubricant Contamination on Bearing Life

In order to fully appreciate the lubricant contamination/bearing life relationship it must be quantified.

In recent years bearing materials, design and calculation methods of fatigue life have been substantially improved [5,6]. The major improvements include:

- a more accurate method of fatigue life prediction; the new approach takes into account operating temperature, lubricant viscosity and lubricant contamination.

- it is now possible to obtain almost infinite bearing life; provided that loads are lower than fatigue limit (these are given in bearing catalogues) and utmost cleanliness of the lubricant is being assured.

- new spherical roller bearing designs offer substantial increases in dynamic load capacity and consequently much longer service life when compared with ball bearings.

Bearing Life Equations

Over the passage of time bearing life equations have been developed in order to quantify the actual effects various factors have upon bearing life. Contamination is only one factor. The following will outline the evolution and development of bearing life equations.

The simplest equation for bearing life is based on ISO (International Standards Organization) guidelines:

$$L_{10} = (C/P)^P$$

where:

L_{10} = basic rating life
C = basic dynamic load rating

P = equivalent dynamic bearing load
p = exponent (p=3 for ball bearing, p=10/3 for roller bearing)

The above equation was primarily developed by Lundberg and Palmgren in 1947-1952 [7,8]. It has been used extensively for prediction of rolling-element bearing life. Over the years improvements in manufacturing methods, design and steel quality resulted in considerable extension of bearing life when compared with the calculated life. In addition, the above equation does not take into account the lubrication conditions. In 1977 ISO introduced a revised life equation:

$$L_{an} = a_1 \, a_{23} \, L_{10}$$

where:

a_1 = life adjustment factor for reliability
a_{23} = life adjustment factor for material and lubrication.

The values of a_{23} are calculated as a function of the viscosity ratio. This ratio is defined as the actual viscosity divided by the viscosity required for adequate lubrication at the operating temperature. This parameter represents the size of the oil film thickness relative to the surface irregularities of the bearing stationary and rotating elements.

The recently introduced SKF life equation is a major improvement over previous mathematical expressions. It takes into account contamination, lubrication and also introduces a concept of a fatigue load limit. The adjusted rating life according to the new theory is calculated by the following equation:

$$L_{10aa} = a_{SKF} \, (C/P)^p$$

where the factor a_{SKF} can be written as:

$$a_{SKF} = f\,(\kappa,\ \eta_c,\ P_u/P)$$

Here,

κ represents the lubricant film thickness
η_c takes into account solid contaminants
P_u the fatigue load limit.

Influence of Contamination on Bearing Life

The understanding of lubricant contamination has been developed to such an extent that it is now possible to quantify its effect on bearing life, provided that the operating conditions and type of contamination are known.

The damage mechanism can be rather complicated but in the case of relatively large hard particles it usually occurs in two steps. First, the hard particles induce permanent indentations. As a result the smooth surfaces of the bearing components are destroyed. Secondly, the rough surfaces will produce higher contact stresses resulting in shorter bearing life. Abrasive wear caused by contamination can also change the load zones in the bearing.

It is also known that hard particles larger than the oil film thickness decrease the bearing life. Typical oil film thickness is on the order of 0.1 to 3 μm. Dalal *et al.* [9] conducted tests under ultra clean conditions where the oil was filtered through a 3 μm filter. The bearing life was found to increase by a factor of 40 when compared to calculated values. Under standard test conditions the bearings were known to have 4 to 5 times their theoretical lives.

As the wear rates are usually high when lubricating oil is contaminated, it is not only the filter rating that is important but also the flow rate the filter can accept. It was also observed that the damage to the bearing by particles during the first half-hour of operation was enough to cause early failures. Also, even if the contaminated oil was replaced the bearing did not "recover" and its life was significantly reduced.

It should be noted that only a very small number of hard particles is needed to reduce the bearing life to a fraction of its undamaged life. The tests conducted by FAG [10] on 7205B angular contact bearings showed a reduction of bearing life by a factor of 10 resulting from plastic indentations of 0.1 mm diameter.

A contamination factor η_c, is used in the new life equation. This somewhat complicated parameter depends on size, hardness, shape and quantity of solid particles, bearing size, lubricant film thickness, loads etc. The factor can be expressed as follows:

$$\eta_c = f\,(\kappa_1\ d_m,\ P_u,\ P,\ R_t,\ D_p,\ HV,\ S)$$

where:

dm = mean diameter of bearing
Rt = a contamination balance factor, it takes into account the amount of contaminants entering and removed from the system
Dp = particle size
HV = particle hardness
S = safety factor.

The effect of water on bearing life is well documented but not well understood. Tests conducted by various researchers [11,12,13] showed that a concentration of water as small as 0.01% can decrease the bearing life to half of its original value. Interestingly, a change in failure mode from ball failures to raceway failures occurred when water content was increased.

For these reasons the New SKF Life Theory does not at this stage take into account the influence of water on bearing life. It is assumed that the water content does not exceed 0.05%. Rough guidelines, applicable to fully oxidation-inhibited lubricants, suggest that the calculated life may be halved by water content of 0.1 % and further 50% reduction may be assumed if the water content increases to 0.01%. The curtailment of bearing fatigue life is considerably more severe for pure, uninhibited mineral oils. Here, 0.002 percent water (20 ppm, or roughly one drop of water per quart of oil) has been found to reduce bearing life by 48 percent (Table 19-5).

Figure 19-15 shows an example of the influence of contamination on bearing life in more general terms. A "real life example" best illustrates the various aspects of bearing life calculations. It shows the significance of operating loads, bearing selection and contamination on bearing life.

The pump considered here is a single stage, centrifugal type. Figure 19-16 shows the performance curves and Figure 19-17 radial load on the inboard bearing as a function of flow.

Since the pump is of single volute design the load is highly dependent on flow. The radial force is induced by an unbalanced pressure distribution in the pump volute. Knowing the weight of the impeller, shaft dimensions, operating temperature, bearing type and lubricant properties the expected fatigue life of the bearing can be calculated. The results are shown in Figure 19-18. Both ball and spherical roller bearings are represented.

It can be seen that the bearing life could be increased significantly with spherical roller bearings. Figure 19-19 shows bearing lives as calculated by simple, adjusted, and the new life equations. Also, the effect of contamination is shown. The benefits of clean lubricants are clearly evident.

Conclusions Drawn From Bearing Life Equations

A significant improvement in bearing life can be achieved by paying more attention to bearing selection and ensuring the cleanliness of lubrication systems.

In some cases almost infinite life can be achieved provided the appropriate bearings are selected, operating conditions are known and the utmost cleanliness of the lubricant is assured.

The influence of contamination on bearing life is a function of many parameters. Nevertheless the following major factors have been identified and are listed below:

- wear is proportional to the amount of contaminants.
- the particles larger than the oil film thickness are the most significant.
- particles with hardness greater or equal to the hardness of the bearing material will result in significant wear of the bearing.
- extreme cleanliness pays off, giving 15 to 35 times longer life when compared to the expected or calculated lives.
- water can decrease bearing life and a concentration of water of 0.01% is enough to decrease bearing life to half its original value, however the effect of water on bearing life is not well understood.

References:

1. Allen, J.L.; "On-Stream Purification of Lube Oil Lowers Plant Operating Expenses," *Turbomachinery International*, July/August 1989, pp. 34, 35, 46.
2. Bloch, H.P., *Improving Machinery Reliability*, Volume 1, Gulf Publishing, Houston, 1998.

Table 19-5. Fatigue life reduction of rolling element bearings due to water contamination of lubricant (Mineral Oil). [Source: Ref. 11]

Base Oil Description	Water Content of Wet Oil, %	Fatigue-Life Reduction, %	Test Equipment and Hertzian Stress
Mineral	0.002	48	Rolling 4-Ball
Oil Dried	0.014	54	Bearing Tester
Over Sodium	3.0	78	8.6 GPa
	6.0	83	GPa (1.25×10^4 PSI)

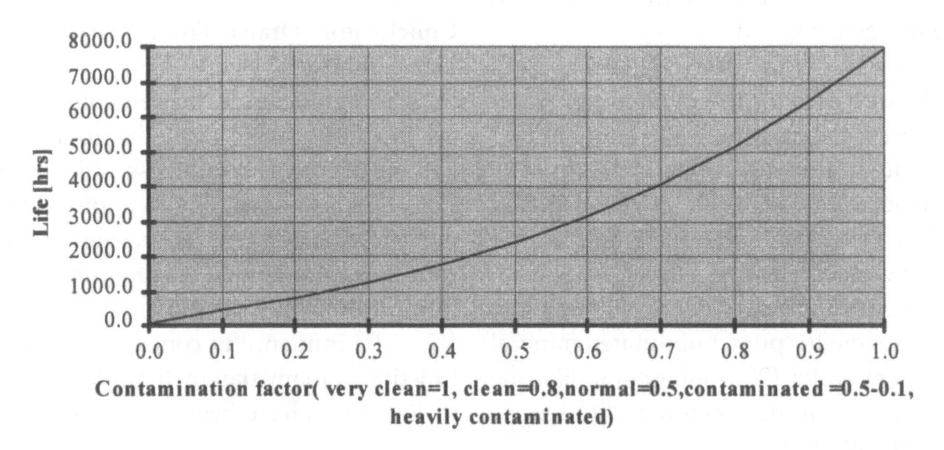

Figure 19-15. Influence of contamination on bearing life.

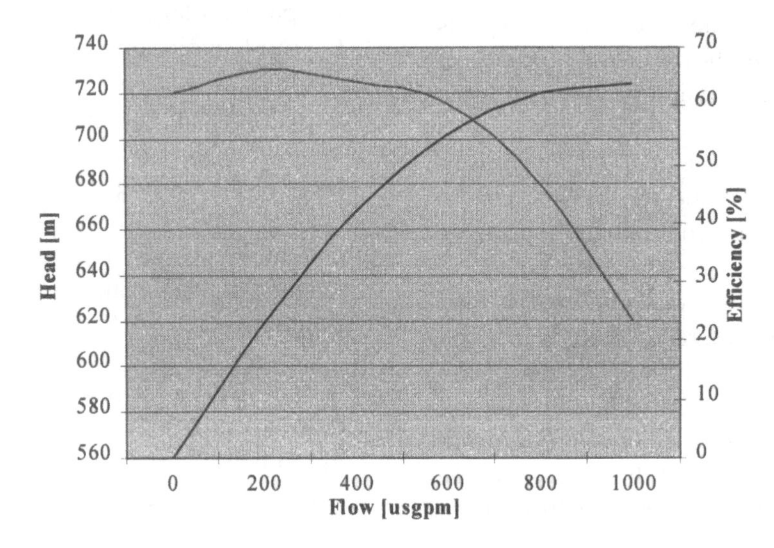

Figure 19-16. Performance curves.

3. Bloch, H.P., Geitner F.R.; *Machinery Failure Analysis and Troubleshooting*, Volume 2, Gulf Publishing Houston, 1998.
4. Symposium on Steam Turbine Oils, ASTM Special Publication No. 211, September 17, 1956.
5. E. Ioannides, T.A. Harris, "A new fatigue life model for rolling bearings," *SKF Ball Bearing Journal* 224,1985.
6. CADalog C - SKF Bearing Calculation and Selection Program-version 4.
7. Lundberg, G. and Palmgren, A., "Dynamic capacity of rolling bearings," *Acta Polytechnica*, Vol. 1, No. 3, 1947.
8. Lundberg, G. and Palmgren, A., "Dynamic capacity of rolling bearings" *Acta Polytechnica*, Vol. 2, No. 4, 1952.
9. Dalal, H.M. et al., "Progression of surface damage in rolling contact fatigue," US Navy Office of Naval Research, Report No. N00014-73-C-0464.
10. Rolling Bearings in Power Transmission Engineering- FAG Publ. No. WL 04202EA
11. Schatzberg, P. and Felsen, I.M., "Effects of water and oxygen during rolling contact lubrication," *Wear*, 12, 1986
12. Schatzberg, P. and Felsen, I.M., "Influence of water on fatigue failure location and surface alteration during rolling contact lubrication," *Journal of Lubrication Technology*, ASME Trans., F91, 2, 1969.
13. Felsen, I.M., Mcquaid, R.W. and Marzani, J.A., "Effect of seawater on the fatigue life and failure distribution of flood-lubricated angular contact bearings," ASLE Trans., Vol. 15, 1, 1972.

Bibliography

Bannister, Kenneth E., Industrial Lubrication Fundamentals Article series, Lubrication Management & Technology magazine, 2014

Bannister, Kenneth E., Industrial Lubrication Fundamentals-Certification Preparatory Training-Level 1 Workshop Manual, ENGTECH Industries Inc., 2012-2016 .

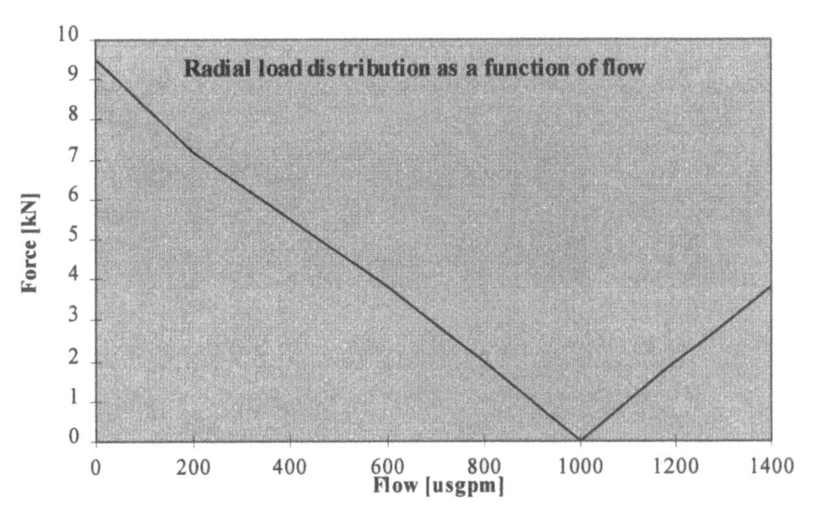

Figure 19-117. Radial load.

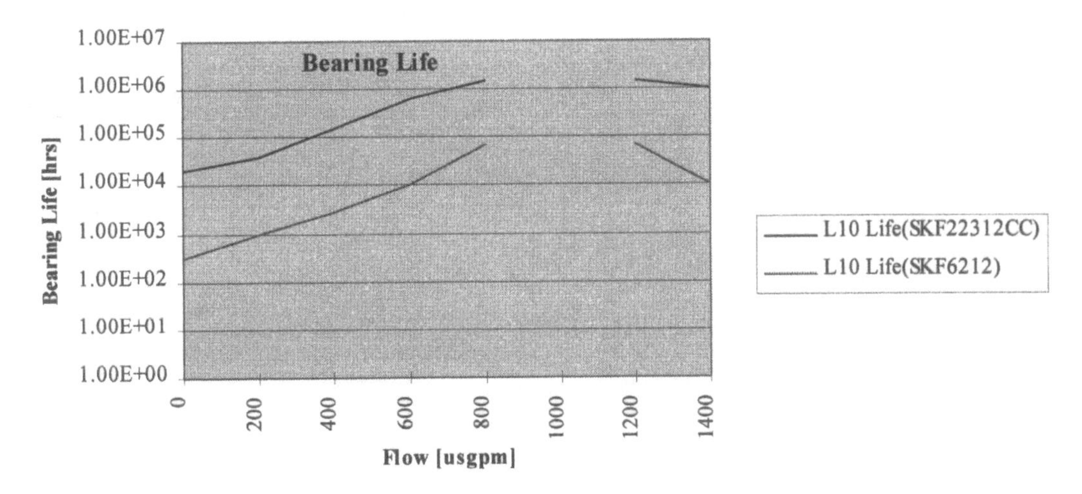

Figure 19-18. Comparison of bearing lives.

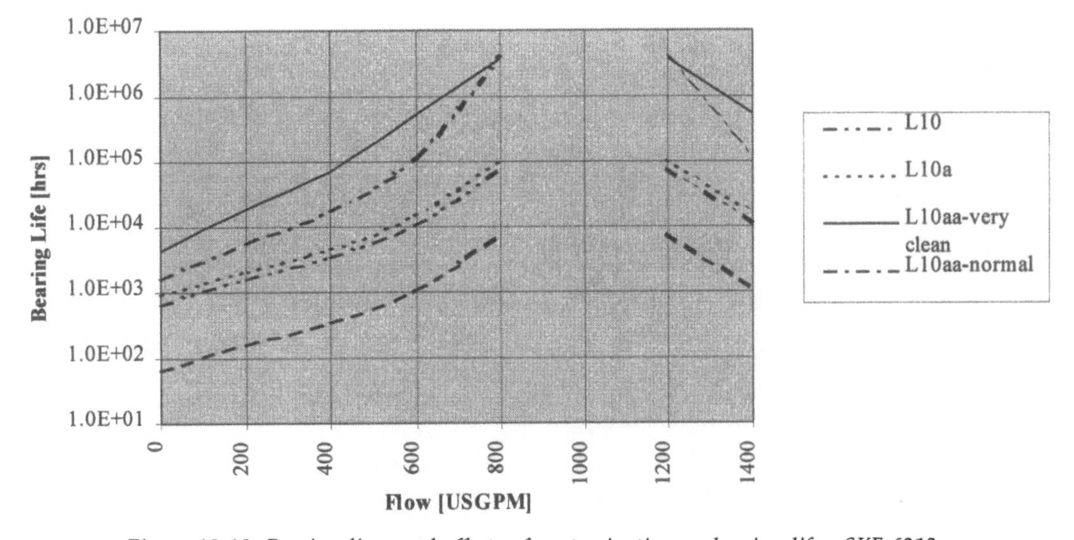

Figure 19-19. Bearing lives and effects of contamination on bearing life—SKF 6212.

Chapter 20

Equipment Storage Lubrication

STORAGE PROTECTION AND LUBRICATION MANAGEMENT*

Storage Protection

Preservation or corrosion inhibiting of inactive process machinery depends on the type of equipment, expected length of inactivity, and the amount of time required to restore the equipment to service.

Petrochemical companies will usually develop their standards to take these criteria into account. One recent typical mothballing program for indefinite storage in a northern temperature climate zone was planned and executed as follows and forms the basis for our recommendations.

Centrifugal and Rotary Pumps

1. Flush pumps and drain casing.
2. Neutralizing step required on acid or caustic pumps.
3. Fresh water flush and air dry all cooling jackets.
4. Fill pump casing with mineral oil containing 5 percent rust preventive concentrate.
5. Plug cooling water jackets—bearing and stuffing box—but keep low point drain valve cracked open slightly.
6. Coat space where shaft protrudes through bearing or stuffing box housings with Product 1 (see Table 20-1) and cover with tape.
7. Coat all coupling parts except elastomers with Product 1.
8. Coat all exposed machined surfaces with Product 1.
9. Fill bearing housing completely with mineral oil containing 5 percent rust preventive concentrate.
10. Pumps do not require rotation.
11. Close pump suction and discharge block valves.

*Excerpted, by permission, from Bloch/Geitner text "Major Process Equipment Maintenance and Repair," 2nd Edition, ISBN 0-88415-663-X, Gulf Publishing Company, Houston, Texas, 1997. See also "Lubrication Program," Appendix A.

Reciprocating Pumps

1. Flush and drain pump casing.
2. Neutralizing step required—if caustic or acid.
3. Blind suction and discharge nozzles of pump.
4. Fill liquid end with mineral oil containing 5 percent rust preventive concentrate. Bar piston to coat all surfaces. Allow some space for thermal expansion.
5. Fill steam end with mineral oil containing 5 percent rust preventive concentrate. Bar piston to coat all surfaces.
6. Close inlet and outlet valves.
7. Coat all joints where shaft protrudes from casings with Product 1. Cover with tape.
8. Coat exposed piston rod, shafts, and machined parts with Product 1.
9. Fill bearing housing and gearbox with mineral oil containing 5 percent rust preventive concentrate.
10. Fill packing lubricator with mineral oil containing 5 percent rust preventive concentrate.

Turbines

1. Isolate from steam system.
2. Seal shaft openings with silicone rubber caulking* and tape.
3. Dry out with air.
4. Fill turbine casing with oil containing 5 percent rust preventive concentrate including steam chest. Hold governor valve open as necessary to ensure chest is completely full. Vent casing, as required, to remove trapped air. Fill trip and throttle valve completely with oil.
5. Install a valved pipe on casing which can serve as filler pipe for adding oil to fill casing. Allow space for thermal expansion of oil in pipe.
6. Coat all external machined surfaces, cams, shafts, levers, and valve stems with Product 1.

* Sealastic® or equal—black, to discourage pilfering.

Table 20-1. Corrosion inhibiting material for machinery protection.

PRODUCT	TYPE	APPLICATION	TRADE NAME
1	Solid Film Corrosion Inhibitor	Hot Dip Hot Brush	RUST-BAN 326 * or equal
2	Solvent Cutback Rust Preventive	Spray After Thinning	RUST-BAN 392 * or equal
3	Solvent Cutback Rust Preventive	Spray Brush	RUST-BAN 343 * or equal
4	Rust Preventive Concentrate	Mix or Full Strength	
5	Barrier Material-Grade C Waxed Paper	Wrap	US Govt. Spec. MIL-B121-D or equal
6	Oil and Moisture Resistant Coating (Aluminum Paint)	Spray	Aluminum Phenolic
7	Petrolatum—(Neutral Unctuous)	Hand Apply	Vaseline or equal

* EXXON / IMPERIAL OIL Products

7. Coat space between case and protrusion of shaft with Product 1. Cover space with tape.
8. Fill bearing housing completely with oil.
9. Coat casing bolts with Product 1.

Large Fans

1. Coat coupling and all external machined surfaces with Product 1.
2. Spray Product 2 on fan wheel.
3. Crack open casing low point drain valve.

Gearboxes

1. Fill gearbox and piping completely with oil containing 5 percent Product 1.
2. Plug all vents. Allow space for thermal expansion.
3. Install a valved pipe on casing which can serve as filler pipe for adding oil to fill casing.

Large motors

1. Blank oil return line.
2. Seal shaft openings with silicone rubber caulking and tape.
3. Fill bearing housing with oil containing 5 percent rust preventive concentrate.
4. Install a valved standpipe such that the inlet is higher than the bearing housing.
5. Coat all exposed machined parts with Product 1.
6. Do not rotate motor.

Centrifugal Process Compressors

1. Purge compressor casing of hydrocarbons.
2. Flush internals with solvent to remove heavy polymers.
3. Pressurize casing with nitrogen.
4. Mix 5 percent rust preventive concentrate to existing lube and seal oil. Circulate oil through the entire system for one hour.
5. Blank oil return header.
6. Seal shaft openings with silicone rubber caulking and tape.
7. Fill bearing housing with oil containing 5 percent rust preventive concentrate by running turbine-driven pump at reduced speed.
8. Fill oil console with mineral oil containing 5 percent rust preventive concentrate.
9. Filling should be done when compressor is at ambient temperature. Turn off all heat tracers.
10. Coat all exposed machined parts, including couplings, with Product 1.

Lube and Seal Oil System

1. Add 5 percent rust preventive concentrate to lube and seal oil.
2. Circulate oil throughout piping system. Open and close control and bypass valves so that all piping and components will receive oil circulation and become coated. Circulate for one hour. Vent trapped air from all components and high points.
3. Block in filters and coolers. Fill completely with oil containing 5 percent rust preventive concentrate but allow small space for thermal expansion. Water side of coolers should be drained and air dry. Plug all vents. Lock drain connections in slightly open position.
4. Fill reservoir with oil containing 5 percent rust preventive concentrate. Blind or plug all connections to tank including vent stack.
5. Coat exposed shaft surfaces and couplings of oil pumps with Product 1.

Reciprocating Compressors

1. Purge compressor casing of hydrocarbons.
2. Blank compressor suction and discharge.
3. Fill crankcase, connecting rod and valves with oil containing 5 percent rust preventive concentrate. Install a valved standpipe. Allow space for thermal expansion.
4. Coat all exposed machined parts with Product 1.
5. Top up oil level in the cooling jacket.

Another occasion would be the three to 12 months' storage of machinery at a construction site. Usually termed a preventive maintenance program, storage protection plans would look like this, again in a northern, dry climate:

Rotation

Rotate all motors, turbines, compressors, pumps, excluding deep well pumps with rubber bushings, fin fans, blowers, aerators, mixers and feeders every two weeks.

Visual Inspection

Check when rotating exposed machined surfaces, shafts and couplings to see that protective coating has been applied and has not been removed. Reapply if needed.

Check all lubricating lines to see if any tubing, piping, tank, or sump covers have been removed. Retape ends and cover. Do this when discovered. If flanges are open on machinery, notify pipefitter general foreman or other designated responsible person in unit.

Inspect the interior of lube oil consoles on a six week schedule. Check to see if clean, and rust and condensate-free. Clean and dry out if needed, then fog with rust preventive concentrate.

Draining of Condensate

Drain condensation from all bearing housings, sumps, and oil reservoirs on a once a month schedule. If an excessive amount of condensation is found, recheck once a week, or at two week intervals depending on amount of condensate present.

Bearings

Fill all bearing housings that are oil lubricated but not force-fed with rust preventive concentrate, bringing the oil level up to the bottom of the shaft. For bearings that are force-fed the upper bearing cap and bearing will be removed. A heavy coat of STP® can be applied to the journal and bearing surfaces. This should be reapplied as needed.

Turbines

Turbines should be spot checked by removing the upper half of the turbine case and visually inspected. Plan to open a sampling of these turbines, selecting from the first preserved and in the worst condition. This should be done on a three month schedule. Other turbines may be inspected by the manufacturer's field service engineer on his periodic (monthly) visits. Small, general purpose turbines should be fogged with rust preventive concentrate through the opening in the top case as the rotor is being rotated. This should be done on a three month schedule.

Compressors

Manufacturers' representatives should inspect the compressors on a monthly visit basis. Preservative needed can be applied under their supervision. Centrifugal process compressors should be fogged and consideration be given to placing dessiccant bags in these machines. They should be inspected on a two month schedule. High speed air compressors should be inspected on a three month schedule. Axial compressors should be inspected and fogged on a three month schedule.

Pumps

Reciprocating pumps should be opened and inspected on a two month schedule. Centrifugal and in-line pumps should be fogged with rust preventive concentrate. Volute cases need not be filled unless it is anticipated that they will remain out of service for over one year.

Electric Motors

Electric motors having grease type bearings need not be lubricated. If received with a grease fitting it should be removed and plugged or capped. For other lubrication type bearings, see "bearings."

Reducing or Speed Increasing Gears

The interior of the housing should be fogged with rust preventive concentrate. Tooth contact points should be coated with STP®. Gears and interiors should be visually inspected on a three month schedule by removing inspection plates.

Blowers

Blowers should be inspected on a three month schedule for rust.

Mixers

Mixers should be filled with rust preventive concentrate.

Fin Fans

Drive belts should stay on. Run several minutes at least every two weeks or whenever snow load dictates.

Miscellaneous Equipment

Miscellaneous equipment should be lubed as applicable and should be rotated on a two week schedule.

Other Considerations

In a warm, high precipitation climate it would be wise to look for alternate solutions to the problem of field storage during construction and prior to start-up. If oil mist lubrication is not already part of the original design, it should be installed with top priority and activated as soon as possible. Figure 20-1 shows temporary field tubing to supply oil mist to the bearing points of a turbine drive pump row. Figure 20-2 shows a similar installation, feeding oil mist to pump and motor bearings, and Figure 20-3 illustrates construction site storage oil mist supply lines to a vertical mechanical drive turbine as well as to a large feed pump motor.

The third and last case of machinery storage protection arises when stand-by capability of laid-up equipment is desired. Reference 1 describes such a case. It appears as though there are no limits to the ingenuity displayed by operators—as long as a "do nothing and

Figure 20-1. Temporary oil mist lines protect equipment prior to installation.

Figure 20-2. Open storage yard with oil mist tubing protecting pump and motor bearings.

Figure 20-3. Construction site storage oil mist supply lines to a vertical mechanical drive turbine and large electric motor.

take your chances" stance is not taken. By the way, all preventive maintenance applied should be logged by item number in the maintenance log.

While the case of extended stand-by protection does not seem to present a problem for process pumps and other general purpose equipment—especially where oil mist lubrication is installed and operating—it might well be a challenge to operators of steam and gas turbines as well as reciprocating engines and compressors. One company[1] has had excellent success with their in-house developed program for stand-by storage of critical machinery, particularly gas turbines. One manufacturer[2] recommends the following procedure for the stand-by protection of gas engines or gas engine driven compressors:

Drain the water jackets and then circulate the proper compound* through the jackets making sure that all surfaces in the jacket are reached. Drain the system and plug all openings.

Lubrication System

On engine lubrication systems, proceed as follows:

1. Drain the lubricating oil system, including filters, coolers, governors, and mechanical lubricators. Flush the complete system with standard petroleum solvent that will take the oil off the surfaces. Use an external pump to force the solvent through the system. Spray the interior of the crankcase thoroughly. Then drain.

2. Refill to the minimum level—just enough to ensure pump suction at all times—in each case with the proper compound.* Crank the mechanical lubricator by hand until all lines are purged. Where compressors are used, be sure to flood the compressor rod packing.

3. Using air pressure or any other convenient means turn the engine at sufficient speed and for a sufficient length of time to thoroughly circulate the compound through the engine.

4. Stop and drain the engine sump, filters, coolers, governor, lubricators, etc. Plug all openings.

5. Remove the spark plugs or gas injection valves and spray with Product 2 inside the cylinders, covering all surfaces. While doing this rotate the engine by hand so that each piston is on bottom dead center when that particular cylinder is being sprayed.

6. After this operation the engine should not be turned or barred over until it is ready to be placed in service. Tag the engine in several prominent places with warning tags.

7. Where compressors are involved—including scavenging air compressors—remove the valves and spray inside the cylinder so as to cover all surfaces. Dip the compressor valves in Product 1 and drain off the excess. Reassemble valves in place.

*Product 4, Table 12-1

ANTICORROSION AGENTS

Hundreds of millions of dollars are spent every year as a result of damage caused by rust and corrosion, especially on raw materials and energy. The corrosion system is shown in Figure 20-4.

Special anticorrosion agents, Table 20-2, consist of fluid or coherent hydrocarbons and special additives. The combination of raw materials is decisive for the product's performance, particularly its anticorrosion properties. Anticorrosion agents also contain a solvent for dilution purposes and an emulsifier to make it water-miscible. The dilution or mixing ratio determines the thickness of the anticorrosion film. All hydrocarbon-based anticorrosion agents have the task of providing temporary anticorrosion protection.

The protective film may vary depending on the product and the requirements. It may be

- colorless
- invisible
- colored
- oily
- ready-to handle
- non-tacky
- dry like wax
- similar to vaseline

The *advantage* of hydrocarbon based temporary anticorrosion agents over metallic, inorganic or resin-type coatings is their easy and cost-efficient application.

The *disadvantage* is that they only function up to one year (sometimes a bit longer) and that removing the protective film, if necessary, requires a solvent.

Application Overview

Anticorrosion agents are suitable for all metallic materials, especially for components made of ferrous metals:

- bolts • springs • chains • bearings
- rings • screws • ropes • pins • tools

Good lubricity is required in addition to anticorrosion protection, for example

- start-up and long-term lubrication of chains
- start-up lubrication under mixed friction conditions
- lubrication of screws ensuring a uniform friction coefficient

Product Data and Selection Guidelines

RUST-BAN products are widely used for equipment protection. Three of these products are compared in Table 20-3.

Rust-BAN 326 is a heavy, amber-colored, petroleum-base rust preventive of a type classified as *hot-dip* because it is generally heated before application. With the temperature raised to the recommended level, these products become sufficiently fluid to be applied easily by dip or, under controlled conditions, by brush.

After application, RUST-BAN 326 leaves a semi-soft film that gives long-lasting protection under various conditions of severity. The film resists abrasion and has a degree of self-healing ability. In the course of time, the exterior of the film forms a crust, while the layer next to the protective film remains soft and plastic. If the film is ruptured, the soft inner layer tends to re-form over the break to restore protection.

This product is recommended for machine parts and other steel products exposed to conditions ranging from moderate to severe. The film has lubricating properties and is compatible with conventional lubricants, and it does not necessarily have to be removed when the part is used. In some cases, however, the film might be thick enough to interfere with the assembly of mating parts. If a different lubricant is required, or the surface is to be painted and removal of the film is necessary, it can be accomplished simply by wiping the surfaces or other areas with a petroleum solvent, such as VARSOL®. Being readily removable represents a major advantage of RUST-BAN 326 over the varnishes sometimes used for protective purposes.

A desirable feature of hot dipping is assurance of adequate film thickness on all areas. The film firms up rapidly upon cooling, and the protected part can be stored or packed with minimum delay. However, if cold application is preferred for other reasons, similar protection against rusting can be obtained from the *solvent-cutback* RUST-BAN grades described later.

RUST-BAN 326 is safe when applied according to recommended procedures. It has a high flash point which is considerably above the recommended application temperatures. It forms a soft film suitable for many severe types of service. Withstanding appreciable exposure to salt-water spray, chemical fumes and normal weathering, RUST-BAN 326 generally gives six months or more of outdoor protection, and it gives indoor protection for prolonged periods.

It is recommended for machined steel surfaces, threaded pipe and parts, castings, forgings, dies, steel bar stock, machinery and equipment. Figures 20-5 and 20-6 depict turbomachinery rotors in humidity-controlled indoor storage. These rotors would be protected with RUST-BAN 326. The

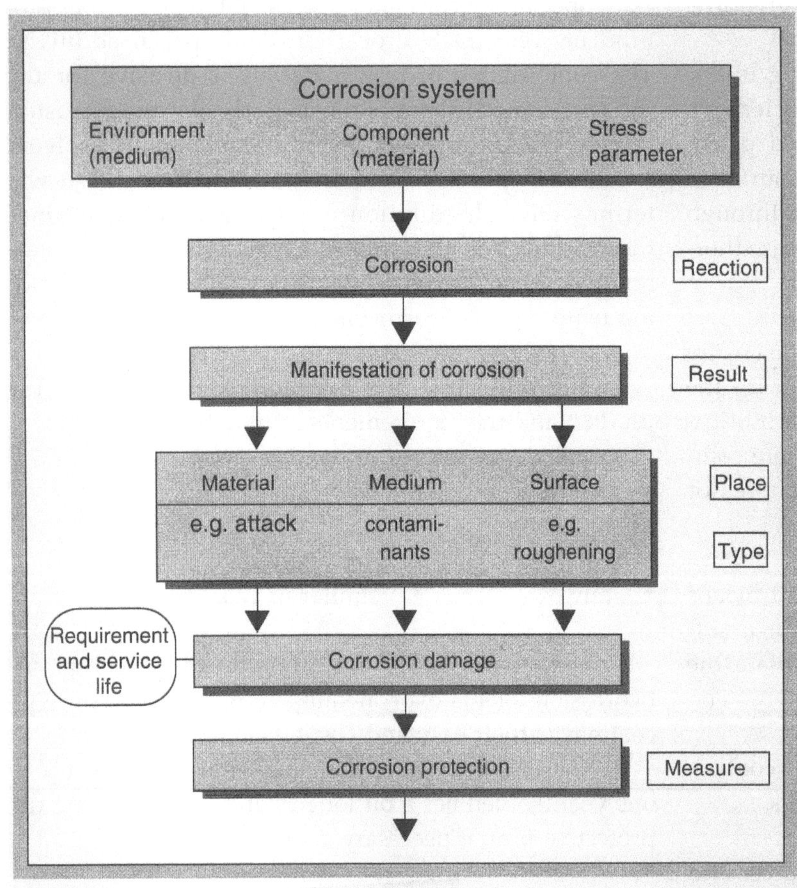

Figure 20-4. Corrosion system. (Source: Klüber Lubrication North America, Londonderry, New Hampshire)

Table 20-2. Product selection chart, anticorrosion agents.

Type	Product name	Viscosity approx.	Density at 20°C (g/ml) DIN 51757 ≈	Flash point DIN ISO 2592 ≈	Corrosion protection	Notes
Anticorrosion oils	CONTRAKOR A 40 K	35mm²/s/ 40°C DIN 51561	0.94	160	No corrosion after 40 cycles (≘ 960 h)*1,3 — Corrosion rating = 0*2,3	Brown, clear anticorrosion oil. A colorless, non-tacky, almost invisible film forms after the solvent naphtha has evaporated. Film thickness: Approx. 5 µm at a ratio of 5% of CONTRAKOR A 40 K, approx. 10 µm at a ratio of 20%, and approx. 20 µm at a ratio of 50%.
	CONTRAKOR	480 mm²/s/ 50°C DIN 51561	0.93	180	No corrosion after 40 cycles (≘ 960 h)*1,2 — Corrosion rating ≘ 0*2,3	Brown, transparent anticorrosion oil. Under boundary conditions more efficient than CONTRAKOR A 40 K. A very thin, non-tacky film forms after the solvent naphtha has evaporated Film thickness: Approx. 5 µm at a ratio of 5 % of CONTRAKOR A 100, approx.15 µm at a ratio of 20 %, and approx. 45 µm at a ratio of 50 %.
Watermiscible emulsions of anti-corrosion wax and vaseline	CONTRAKOR WM	100 mm²/s/ 170°C DIN 51551	1.00	190	No corrosion after 20 cycles (≘ 480 h)*1,4 — Corrosion rating = 0*2,4	Light brown, turbid, homogeneous product. Dilute with 70% water. A soft, non-tacky film will form after the water has evaporated. Film thickness: Approx. 40 µm at a ratio of 30% of CONTRAKOR WM1, approx. 20 µm at a ratio of 20%, and approx. 80 µm at a ratio of 50%.
Watermiscible anticorrosion wax emulsion	Klüberplus SK 01-205	20-30 s DIN 53211 with 4 mm nozzle/20°C	0.98	none	No corrosion after 20 cycles (≘ 480 h)*1,5 —	Fluid, ivory-colored product. The ready-to-use water-containing product forms a film that is fast to handle and has good lubricating properties. At room temperature the film surface is dry after 15 min., and the entire layer is fully dry after 8 h. Film thickness: 10 µm.
	Klüberfluid 4 C 90	90 mm²/s/ 70°C DIN 51561 (of the water-free product)	1.00	190	No corrosion after 20 cycles (≘ 480 h)*1,5 — rating = 0*1,5	Light yellow, milky product. The ready-to-use product contains approx. 70% of water and forms a soft, non-tacky film with good lubricity. Film thickness: approx. 40 µm.
lacquers		20°C			(≘ 168 h)	approx. 20 µm.
One component anticorrosion	Klübertop KA 103	4.5 - 5.5 mm²/s/	1.00	none	No corrosion after 20 cycles	The ready-to-use product is blue and thin-bodied and forms a dry protective layer. Film thickness:

*1 Tested in a condensation water alternating atmosphere, DIN 50017 - KFW at 40 °C and material St 14
*2 Salt spray test, DIN 50021 SS, 5% NaCl solution at 35°C and material QQS-698 grade 1009, test duration > 50 h
*3 Mixed with 70 % solvent naphtha
*4 Mixed with 70 % water and after 24 h drying time at room temperature
*5 After 24 h drying time at room temperature

		DIN 51561			—	
	Klübertop KB 118	2300 mPa s 25°C	1.19	35	— — Corrosion rating = 0*2,5	Brown, high-viscosity product for the preservation of toothed units. The protective film is fully dry after 24 h and is soluble in oil and fat. Recommended film thickness: 60 µm.

Table 20-3. Typical inspections for popular rust preventatives.

	Rust-Ban 326	Rust-Ban 343	Rust-Ban 392
Density, lb/gal at 60°F	7.7	7.3	6.5
Viscosity, cSt @ 40°C	—	20.5	1.14
cSt @ 100°C	33.1	4.0	—
SSU @ 100°F	—	1-7	—
SSU @ 210°F	151	39.8	—
Flash point, °C	260	190	71
°F	500	374	160
Melting or pour point, °C	62	-4	-36
°F	144	+25	-32
Unworked penetration at 25°C (77°F)	237	—	—
Film thickness, mil	0.05	0.9	0.01
Approximate coverage, ft^2/gal	900	1700	9000
Non-volatiles, %	98	—	7.4
Methods of application/ temperature, °C	dip/77[1], swab/18-27[2]'	Roller coating, brush, or misting	spray, dip or brush/ambient
°F	dip/170[1], swab/65-80[2]		
Characteristics	Added chemical inhibitor, self-healing effect	Very thin oily film	Displaces water, neutralizes fingerprints, transparent film

[1]preferred;
[2] suitable
NOTE: The values shown here are representative of current production. Some are controlled by manufacturing specifications, while others are not. All may vary within modest ranges.

same product is also recommended for the protection of anti-friction bearings and precision parts of this sort during shipment and storage. Though hot dipping gives the most effective protection, the product is soft enough to be applied by brush at moderate-to-high temperatures. This latter procedure can be facilitated by preheating.

Rust-Ban 326 has also been applied successfully with suitable heated spray equipment. Where necessary, it can be thinned with a small amount of petroleum solvent, such as Varsol. When sufficiently thinned, it can be sprayed at room temperature. CAUTION: Addition of a volatile solvent to Rust-Ban significantly increases the fire hazard, due to the lower flash point of the solvent. Do not prepare or apply a mixture of Rust-Ban and solvent in the presence of flames or sparks. Note also that rust preventatives are not rust removers, and they should be applied only to surfaces that are dean, dry and free of rust. Application should be made as promptly as possible after cleaning. For tighter bonding and more thorough coverage when applying heated rust preventive, the temperature of the treated part should be near that of the applied coating. Hot-dipped parts should remain in the dip tank until they reach the temperature of the heated rust preventive.

Unless testing proves them to be compatible with nonmetallic materials, petroleum-base rust preventatives should not be applied to parts such as those fabricated of rubber, cork, paper, leather, fabric or plastics. Some of these materials may swell, soften or be otherwise affected.

Slushing Oil Application

Rust-Ban 343 is an oil-type rust preventive that can serve in a variety of applications. It is designed to provide protection under conditions of indoor exposure where the oily film gives adequate protection, or in situations where the oily film is preferred to one of a more durable nature. Rust-Ban 343 meets the exacting require-

Figure 20-5. Turbomachinery rotors in humidity-controlled indoor storage at Sulzer-Hickham Industries, Inc., LaPorte, Texas.

ments for use as a steel mill slushing oil and is widely accepted by major steel producers and users.

Slushing oils are petroleum-base coatings that are applied to many steel mill products. They are usually applied to cold-rolled sheets and coils to protect them from rusting during storage and shipment. Application of slushing oil is usually by spraying, hand painting, or misting. Slushing oils must be formulated to provide extended corrosion protection to cold-rolled sheet supplied to the automotive and appliance industries. They must have inherent anti-staining characteristics and be easily removed.

RUST-BAN 343 is formulated to provide outstanding service as a slushing oil. It contains a proven rust-inhibiting package and is unsurpassed in protection against staining. In addition, it is readily removed by conventional cleaning processes and, consequently, does not interfere with subsequent application of coatings or adhesives.

Protecting Oil Circulation Systems

Oil-type RUST-BAN 343 is particularly adapted to the protection of equipment designed to contain or circulate lubricating oil or hydraulic oil. This includes such applications as gear cases, chain drives, turbines, pumps, instruments, and many others. Turbine manufacturers recommend that turbine lubricating oil systems be flushed to remove all contaminants before being put in operation. Although additives in oil-type rust preventatives may impair demulsibility characteristics of premium turbine oils such as TERESSTIC®, the use of RUST-BAN 343 should cause no problem because it is effectively removed by the flushing operation.

Figure 20-6. Turbomachinery rotor protected by RUST-BAN 326. (Source: Sulzer-Hickham Industries, Inc., LaPorte, Texas)

It can be used for protection of interior surfaces of equipment in storage or in transit and, where heavier-bodied films are not required, it can be used to protect exterior surfaces of small parts. This product is also widely used for protection of thin-gauge steel. Its high fluidity makes it suitable for low-temperature applications. It also is recommended for protection of internal combustion engines laid up for storage. The rust preventive should be thoroughly flushed from the crankcase with engine oil prior to start-up.

As mentioned above, rust preventatives are not rust removers; they should be applied only to surfaces that are clean, dry, and free of rust. If the removal of all water is impractical, Rust-Ban 392, which has special water-displacing properties, may be applied as a base coat. This should be done as soon as possible after cleaning, and in such a way as to completely coat the surfaces to be protected. Once the Rust-Ban 392 is dry, oil-type Rust-Ban 343 should be applied.

What about removal? From a practical point of view, removal of slushing oils from steel sheets or coils is usually necessary to permit subsequent painting, plating, or other treatment. Rust-Ban 343 can be removed easily by conventional cleaning methods. For other applications, there is seldom need to remove the film. Removal may be required only in the rare instances in which the slight emulsifying tendencies of the preservative might be undesirable. In these cases, it is necessary only to flush the system with the regular lubricating oil, drain, and refill with a fresh batch of the new oil.

Solvent Cut-back Products

Rust-Ban 392 is composed of rust-preventative, film, forming organic materials in Exxsol D 80 solvent. It is suitable for indoor protection only. It can be applied without heating. After application, the solvent evaporates, leaving a thin, translucent film that remains flexible, thus providing protection to both rigid and non-rigid surfaces. Turbomachinery rotors placed in a shipping container (Figure 20-7) benefit from Rust-Ban 392 as well as a pressurized nitrogen environment.

Rust-Ban 392 meets the performance requirements of MIL-C-16173C, Grade 3; however, it does not comply with the discernibility requirements of this specification. The specification requires that the color of the finished compound be black or brown and that the protective film be discernible. Rust-Ban 392 is clear and essentially not discernible and, for many types of finished parts, does not require removal.

Distinctive characteristics of Rust-Ban 392 include its fingerprint-neutralizing and water-displacing properties. When applied to surfaces freshly marked with fingerprints, it prevents the staining that otherwise would occur. When it is applied to surfaces wet with rinse water, soluble cutting oil emulsion, or water from other sources, its metal wetting properties displace the water from the surfaces and allow it to drain off. The thin film formed by Rust-Ban 392 minimizes buildup; consequently, Rust-Ban 392 can be applied periodically to protect the finished-steel surfaces of machinery in service with minimal change in appearance. Rust-Ban 392 is especially recommended for use on parts in process. During machining, subassembly, storage, inspection, etc., a machined part typically is exposed to moisture, workmen's fingerprints, and corrosive atmospheres such as SO_2, SO_3, or H_2S, or acid or caustic fumes from pickling baths and degreasing or other operations. A freshly machined steel part exposed to any of these elements will begin to rust immediately. Even nonferrous metals—copper and aluminum—will tarnish and corrode. Rust-Ban 392, applied at one or more stages of manufacture, provides effective protection for parts awaiting assembly, wrapped for indoor storage, or packed for shipment.

Because this product ensures a moisture-free surface, it is sometimes used as a prime coating before the application of a hot-dip or oil-type rust preventive. Its metal-wetting properties also are the basis of its excel-

Figure 20-7. Shipping container for turbomachinery rotor. Although immersed in pressurized nitrogen, the rotor is protected by Rust-Ban 392. (Source: Sulzer-Hickham Industries, Inc., LaPorte, Texas)

lent performance as a penetrating fluid.

RUST-BAN 392 gives maximum protection only when applied to clean surfaces. It should be applied as promptly as possible after cleaning. Complete coverage is essential. Application can be by brush, spray, dip, or squirt can. Dipping is the most effective method, in which case excess rust preventive should be allowed to drain off—never wiped off. In all dip applications, evaporation of solvent from the dip tank and the progressive thickening of the rust-preventive material can be compensated for by adding EXXSOL D 80 solvent to the tank as required.

LUBRICANT CONSOLIDATION

A well-implemented program of lubricant consolidation is best described by its written scope, as was done for "TXYZ," a Texas utility in the late 1980s:

- To recommend optimum lubricants for all equipment at TXYZ. An optimum lubricant would satisfy the reliability requirements implied for a modern utility while at the same time recognizing existing manpower and budgetary constraints.
- To achieve a reasonable consolidation of lube oil types or grades, i.e., minimizing the number of different lubes kept in stores without sacrificing equipment reliability.

This scope was later extended to include the development of purchase specifications for the various lubricants. These specifications will allow TXYZ to obtain competitive bids from a number of lubricant suppliers without, however, sacrificing on the quality which we must consider of foremost importance.

The consolidation results consist primarily of two sets of survey tabulations. The first one gives the lubricants which the equipment manufacturers originally suggested for TXYZ and which were presumably stocked at this generating station. The second tabulation presents new recommendations and reflects the deletion of a number of lubricant grades. This consolidation and review procedure resulted in the following:

- Industrial R&O 32: Retained on critical machinery, occasionally changed to AW 32 or upgraded to AW 68 in general purpose equipment.
- AW Hydraulic Oil 32: Retained on critical turbomachinery and hydraulic units, occasionally upgraded to AW 68 in backstops, pump bearings, etc.
- Industrial R&O 46: *Deleted*. Replaced by AW 46 in critical machinery, upgraded to AW 68 in general purpose machinery.
- AW Hydraulic Oil 46: Retained on critical machinery, occasionally upgraded to AW 68
- AW Hydraulic Oil 68: Retained throughout.
- AW Hydraulic Oil 100: Retained, but note that this lubricant is used only in the secondary crushers. (The review identified a high probability that this product could be replaced by an ISO-100 PAO or diester-base synthetic lubricant and advocated further communication with the equipment manufacturer and two or three competent lubricant formulators. These contacts proved affirmative, allowing the utility to capture the life, extension credits mentioned in the chapter on Synthetic Lubricants.)
- Industrial R&O 68: *Deleted*. Replaced by AW 68
- Industrial R&O 100: Replaced by Syn 100 in soot blowers. It was noted that this difficult chain lubrication duty is a good candidate for oil mist application.
- Industrial R&O 115: *Deleted*. Replaced by Syn 100.
- Industrial R&O 150: *Deleted*. Replaced by Syn 220.
- Industrial R&O 220D: *Deleted*. Replaced by Syn 220.
- NL Gear Compound 150: *Deleted*. Replaced by Syn 220.
- NL Gear Compound 220: Replaced by Syn 220.
- NL Gear Compound 320: Replaced by Syn 320.
- NL Gear Compound 460: Replaced by Syn 460.
- NL Gear Compound 680: Replaced by Syn 680.
- NL Gear Compound 1500: Retained. Used primarily in Air Heater Rotor Drives
- RPM Delo 30: Retained
- Tractor Hydraulic Fluid: Retained
- Universal Gear Lube 80/90: *Deleted*. Replaced by Syn 220
- Torque Fluid 5: Retained
- Automatic Transmission Fluid Dexron II; Retained
- Multi-Motive Grease #1: Retained
- Premium Lubcote EP#2: Retained
- Ultra-Duty Grease #2: Retained
- Moly Grease: Retained

In summary, the utility was advised that the deletion of 7 lubricant grades will result in warehouse stocking and field labor economies. The wider use of superior synthetic lubricants for gear units would save money because considerably longer drain intervals will be achieved and also because these lubricants exhibit greater film strength, cooler operation, better film adhesion, and thus greater load carrying capability than the mineral lubricants they would be replacing.

As a final note, our review stated that frequent foaming incidents in only one of several identical bearing housing oil sumps on large vertical motors shows mechanical agitation as the most likely root cause. It was recommended that bearing housing internal components be carefully compared with those where far less foaming was observed. An appropriate component adjustment or modification should then be made.

Basis for Consolidation of Lubricants

The basis for lube consolidation closely parallels that of lubricant selection. In principle, the general lubrication guidelines given in Table 20-4 are used in this task, and the equipment vendors asked to comment. If there are no valid, experience-based objections from both equipment manufacturers and lube oil formulators, consolidation should proceed.

Table 20-5 illustrates the original vendor recommendations for a small portion of the hundreds of machines at TXYZ. The consolidated selections are listed in Table 20-6.

Table 20-4. General lubrication guidelines.
(See also notes and clarifications, next page)

Lubrication Service	Lubricant Properties Should Match
Compressors, Reciprocating	
Portable Air Compressors	Essolube HDX 30
Gas Engine Driven	
Cylinders and Crankcase	Estor AGX
Mixing Valves and Governors	Teresstic 68
"One-shot" Lubricators	Teresstic 68
Electric Motor Driven	
Cylinders, Ordinary Gas Service	Synesstic 68
Cylinders, Extreme Gas Purity	Consult Vendor
Crankcase	Teresstic 68
Compression Cylinder, Extremely Wet Gas	Cylesstic TK 460
Compression Cylinder, Extremely Wet/Dirty	Synesstic 100
Compressors, Centrifugal	
Suction Temperature Over 10°F	Teresstic 32
Suction Temperature Under 10°F	Esstic 32
Compressors, Air Conditioning & Refrigeration	Zerice S
Electric Motors	
Oil Mist Lubricated	Special
Oil Lubricated	Teresstic 100
Grease Lubricated	Ronex MP
Pumps	
Centrifugal	
Oil Mist Lubricated	Synesstic 68
Oil Lubricated	Synesstic 68
Grease Lubricated	Ronex MP
Rotary	
Bearings, Oil Lubricated	Teresstic 100
Bearings, Grease Lubricated	Ronex MP
Drive Gears	Teresstic 100
Reciprocating	
Steam Valves & Cylinders	Cylesstic TK 460
Turbines, Gas	Teresstic 32
Turbines, Expansion	Teresstic 32

(*Continued*)

Table 20-4. General lubrication guidelines. (Concluded)

Lubrication Service	Lubricant Properties Should Match
Turbines, Steam	
With Pressure Lubricating System	
Compressor Drives	Teresstic 32
G.P. Equipment Drives	Teresstic 100
Without Pressure Lube System	
Oil Temperature to 140F	Teresstic 68
Oil Temperature 140F to 200F	Teresstic 150
Oil Temperature Above 200F	Teresstic 150
Grease Lubricated	Ronex MP
Woodward Governors	Teresstic 68
Steam Engines	
Bearings,	
Oil Lubricated	Teresstic 68
Grease Lubricated	Ronex MP
Cylinders	
Wet Steam, 175 psi or less	Cylesstic TK 460
Superheated, or over 175 psi	Cylesstic 680
Diesel and Gasoline Engines	Essolube HDX 30
Seal Oil Systems	
Specific requirements, consult Machinery Specialist or Compressor Contact Engineer	
Grease Lubricated Couplings	Koppers KHP
Gear Boxes	
Parallel Shaft Units	
General Purpose	Synesstic 100
Special Purpose High Speed	Synesstic 32
Sundstrand	Synesstic 32
Centac	Synesstic 32
Bevel Gears, Spiral And Straight	Teresstic 100 Teresstic 150 Teresstic 220
Cylindrical Worms	
Use Cylesstic product—see Manufacturers recommendations	
Grease Fittings	Ronex MP
Mixers	Teresstic 100

Clarification:

This chart delineates very general guidelines appropriate for use with typical plant equipment in the United States. However, users are cautioned that in some instances, special lubrication requirements may be defined by the manufacturer or users' experience.

Table 20-5. Vendor recommendation—lubrication survey for a utility (partial only).

I.D. Number		Equipment	Industrial Oil R&O 32	AW Hydraulic Oil 32	Industrial Oil R&O 46	AW Hydraulic Oil 46	AW Hydraulic Oil 68	AW Hydraulic Oil 100	Industrial Oil R&O 115	Industrial Oil R&O 150	Industrial Oil R&O 220D	NL Gear Compound 150	NL Gear Compound 220	NL Gear Compound 320	NL Gear Compound 460	NL Gear Compound 680	NL Gear Compound 1500	NL Gear Compound 460	RPM DELO 30		Tractor Hydraulic Fluid	Universal Gear Lub 80/90	Torque Fluid 5	Automatic Trans DEXRON II	Ultra-Duty Grease #2	Multi-Motive Grease #1	Premium Lubcote EP #2	Ultra-Duty Grease #2	Moly Grease #2	
1-FM-001-ABA	~ 50 HP	Auxiliary Boiler Feed Fan																										x		1
1-PM-001A-ABF	~ 10 HP	Pumps Auxiliary Boiler Dearator	x																											2
1-PM-001B-ABF	~ 10 HP	Pumps Auxiliary Boiler Dearator	x																											2
1-PM-002A-ABF	~ 75 HP	Auxiliary Boiler Feed Pump					x																							3
1-PM-002B-ABF	~ 75 HP	Auxiliary Boiler Feed Pump					x																							3
3-PM-001-ABM	Milroyal	MilroyalChemical Feed Pump																x												4
1-PM-002-ABM	Milroyal	MilroyalChemical Feed Pump - Hydrozene																x												5
1-PM-003-ABM	Milroyal	Chemical Feed Pump - Phosphate																x												6
1-SBM-001-ABS	Box 1: NL 150, Box 2: NL 460	Auxiliary Boiler Soot Blower							x			x			x											x				7
1-SBM-002-ABS	Carriage Box MM #1, Chain R&O 100	Auxiliary Boiler Soot Blower							x			x			x											x				7
1-PM-001A-ACW	Vert. Turb R&O 3Z; Motor TB	Auxiliary Cooling Water Pump	x				o		x																					8
1-PM-001B-ACW	R&O 115, Guide Brg. R&O 68	Auxiliary Cooling Water Pump	x				o		x																					9
1-DM-001-AFS	Gear Reducer	Feeder Motor											x																	10
1-PM-002-AFS	Milroyal	Metering Pump		x																										11
1-GM-001A-AHB		Clinker Grinders	x			x							x																	12
1-GM-001B-AHB	Falk Gear: NL 220	Clinker Grinders	x			x							x																	12
1-GM-001C-AHB	Fluid Drive: R&O 32	Clinker Grinders	x			x							x																	12
1-GM-001D-AHB	Chain Guard: AW 46	Clinker Grinders	x			x							x																	12
1-GM-001E-AHB		Clinker Grinders	x			x							x																	12
1-GM-001F-AHB		Clinker Grinders	x			x							x																	12
1-PM-001A-AHB	200 HP, Warman	Bottom Ash Slurry Pump																							x					13
1-PM-001B-AHB	Slurry Pump	Bottom Ash Slurry Pump																							x					13
1-PM-002A-AHB	10 HP	Pump - Hydraulic Oil		x																										14
1-PM-002B-AHB		Pump - Hydraulic Oil		x																										14
1-CM-001A-AHF		Flyash Blower Compressors	x			x																								15
1-CM-001B-AHF	Centac, 1000 HP	Flyash Blower Compressors	x			x																								15
1-CM-001C-AHF	Comp: R&O 32	Flyash Blower Compressors	x			x																								15
1-CM-001D-AHF	Motor: AW 46	Flyash Blower Compressors	x			x																								15
1-DM-001A-AHF		Hydromix Unload Drive Motor								x																				16
1-DM-001B-AHF	Open gear & chain	Hydromix Unload Drive Motor								x																				16
1-DM-002A-AHF		Hydromix Unload Drive Motor								x																				16
1-DM-002B-AHF		Hydromix Unload Drive Motor								x																				16

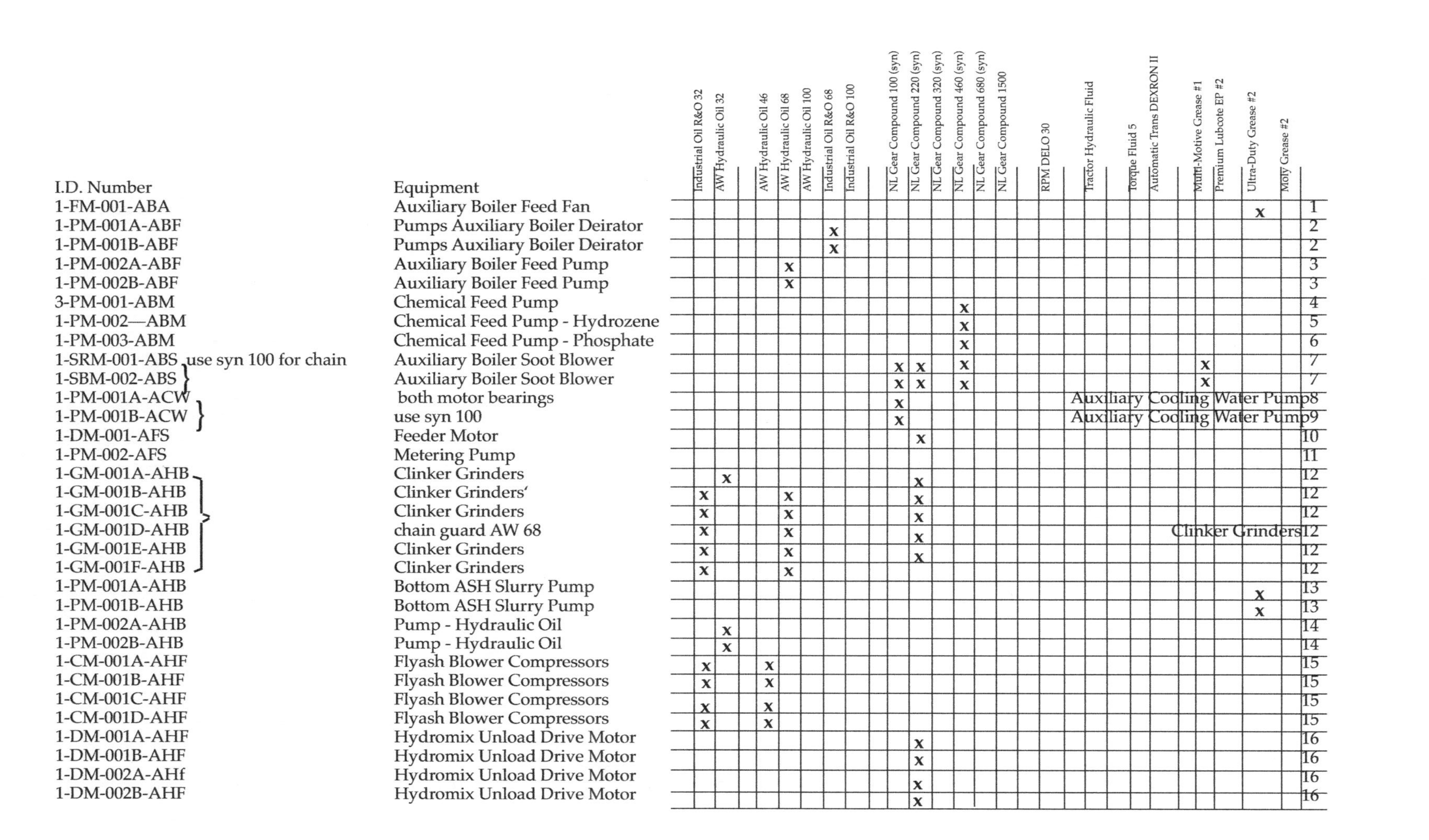

Table 20-6. Consolidation/lubrication survey for a utility (partial only).

I.D. Number	Equipment	Industrial Oil R&O 32	AW Hydraulic Oil 32	AW Hydraulic Oil 46	AW Hydraulic Oil 68	AW Hydraulic Oil 100	Industrial Oil R&O 68	Industrial Oil R&O 100	NL Gear Compound 100 (syn)	NL Gear Compound 220 (syn)	NL Gear Compound 320 (syn)	NL Gear Compound 460 (syn)	NL Gear Compound 680 (syn)	NL Gear Compound 1500	RPM DELO 30	Tractor Hydraulic Fluid	Torque Fluid 5	Automatic Trans DEXRON II	Multi-Motive Grease #1	Premium Lubcote EP #2	Ultra-Duty Grease #2	Moly Grease #2	
1-FM-001-ABA	Auxiliary Boiler Feed Fan																				x		1
1-PM-001A-ABF	Pumps Auxiliary Boiler Deirator						x																2
1-PM-001B-ABF	Pumps Auxiliary Boiler Deirator						x																2
1-PM-002A-ABF	Auxiliary Boiler Feed Pump				x																		3
1-PM-002B-ABF	Auxiliary Boiler Feed Pump				x																		3
3-PM-001-ABM	Chemical Feed Pump											x											4
1-PM-002—ABM	Chemical Feed Pump - Hydrozene											x											5
1-PM-003-ABM	Chemical Feed Pump - Phosphate											x											6
1-SRM-001-ABS use syn 100 for chain	Auxiliary Boiler Soot Blower								x	x		x							x				7
1-SBM-002-ABS	Auxiliary Boiler Soot Blower								x	x		x							x				7
1-PM-001A-ACW	both motor bearings								x							Auxiliary Cooling Water Pump							8
1-PM-001B-ACW	use syn 100								x							Auxiliary Cooling Water Pump							9
1-DM-001-AFS	Feeder Motor									x													10
1-PM-002-AFS	Metering Pump																						11
1-GM-001A-AHB	Clinker Grinders		x							x													12
1-GM-001B-AHB	Clinker Grinders'	x			x					x													12
1-GM-001C-AHB	Clinker Grinders	x			x					x													12
1-GM-001D-AHB	chain guard AW 68	x			x					x									Clinker Grinders				12
1-GM-001E-AHB	Clinker Grinders	x			x					x													12
1-GM-001F-AHB	Clinker Grinders	x			x																		12
1-PM-001A-AHB	Bottom ASH Slurry Pump																				x		13
1-PM-001B-AHB	Bottom ASH Slurry Pump																				x		13
1-PM-002A-AHB	Pump - Hydraulic Oil		x																				14
1-PM-002B-AHB	Pump - Hydraulic Oil		x																				14
1-CM-001A-AHF	Flyash Blower Compressors	x		x																			15
1-CM-001B-AHF	Flyash Blower Compressors	x		x																			15
1-CM-001C-AHF	Flyash Blower Compressors	x		x																			15
1-CM-001D-AHF	Flyash Blower Compressors	x		x																			15
1-DM-001A-AHF	Hydromix Unload Drive Motor									x													16
1-DM-001B-AHF	Hydromix Unload Drive Motor									x													16
1-DM-002A-AHf	Hydromix Unload Drive Motor									x													16
1-DM-002B-AHF	Hydromix Unload Drive Motor									x													16

Chapter 21
Lubricant Condition Testing—Oil Analysis*

Every industrial organization has experienced the consequences of shoddy maintenance: contract penalties, junked parts, injuries, catastrophic damage, ballooning costs, missed shipping dates, irate customers, and sickly quarterly financial reports. Gone are the days when a machine had a predictable service life, after which it was replaced, continuing the cycle. Today, machinery and equipment can be maintained to achieve useful operating lives many times those attainable just a few years ago. For oil lubricated machinery there are many opportunities in what is commonly referred to as proactive maintenance.

By carefully monitoring and controlling the conditions of the oil (nurturing), many of the root causes of failure are systematically eliminated. Case studies of highly successful organizations show that oil analysis plays an important, central role in this nurturing activity. But first, in order for oil analysis to succeed the user organization must define what the goals will be.

Some people see oil analysis as a tool to help them time oil changes. Others view it in terms of its fault detection ability. Still, others apply it to a strategy relating to contamination control and filter performance monitoring. In fact, when a program is well designed and implemented, oil analysis can do all of these things and more. The key is defining what the goals will be and designing a program that will effectively meet them. One might refer to it as a ready-aim-fire strategy. The *ready* has to do with education on the subject of oil analysis and the development of the program goals. The *aim* uses the knowledge from the education to design a program that effectively meets the goals. The *fire* executes the plan and fine-tunes through continuous improvement.

DETECTING MACHINE FAULTS AND ABNORMAL WEAR CONDITIONS

In the past, success in fault detection using oil analysis has been primary limited to reciprocating engines, power train components, and aviation turbine applications. The generally small sumps associated with this machinery concentrated wear metals and the rapid circulation of the lubricating oils kept the debris in uniform suspension making trending more dependable.

In recent years, there has been widespread reported success with wear debris analysis for detecting machine anomalies in stationary industrial lubrication oils and hydraulic fluids as well. There are many explanations for this but much of it has to do with a rapidly growing base of knowledge coming from the burgeoning oil analysis and tribology community. Table 21-1 provides a simplistic overview of the application of oil analysis, specifically wear debris analysis, in machine health monitoring. The various specific methods are discussed in later sections of this chapter.

PERFORMING CONDITION-BASED OIL CHANGES

Each year huge amounts of oil are disposed of prematurly, all at a great cost to the world's economy and ecology. This waste has given rise to a growing number of companies to discontinue the practice of scheduled oil changes by implementing comprehensive condition-based programs in their place. This, of course is one of the principal roles of oil analysis. One might say your oil is talking, but are you listening?

By monitoring the symptoms of oil when it tires and needs to be retired we are able to respond to the true and changing conditions of the oil (see Table 21-2). And, in some cases it might be practical to consider reconditioning the oil, including the reconstructing depleted additives. Some oil analysis tests even provide a forward-looking prediction of residual life of the oil and additives. Distressed oils, in cases, can be conveniently fortified or changed without disruption of production. And, those fluids that degrade prematurely can be reviewed for performance capability in relation to the machine stressing conditions.

*Chapter contributions by James C. Fitch, Noria Corporation, Tulsa, Oklahoma

Table 21-1. Application of lube oil analysis.

	Root Cause Detection	Incipient Failure Detection	Problem Diagnosis	Failure Prognosis	Post Mortem
What Oil Analysis Is Telling You	When something is occurring that can lead to failure - root cause conditions	When an early-stage fault exists that is otherwise going unnoticed - e.g., abnormal wear	What the nature of a problem is that has been observed. - Where is it coming from? - How severe is it? - Can it be fixed?	That a machine is basically worn out and needs to be fixed or replaced	What caused the machine to fail? Could it have been avoided?
What You Monitor	Particles, moisture viscosity, tempera- additives, oxidation, TAN/TBN, soot, glycol, FTIR, RBOT	Wear debris density temperature, particle count, moisture, elemental analysis, viscosity, analytical ferrography	Wear debris, elemental analysis, moisture, particle count, temperature viscosity, analytical ferrography, vibration analysis	Elemental analysis, analytical ferrography, vibration analysis, temperature	Analytical ferrography, ferrous density, elemental analysis
Mainten-ance Mode	Proactive	Predictive	Predictive	Breakdown	Breakdown

Courtesy Noria Corporation

Table 21-2. Changing param-eters and their effect on remaining life of lube oil.

↓ Decreasing Value ↑ Increasing Value	Percent Oil Life Remaining 100	50	25	10	5	0
Viscosity*	—	—	—	↑	↑	↑
TBN (crankcase oil)	—	↓	↓	↓	↓	↓
TAN	—	↓	↓	—	↑	↑
FTIR Oxidation	—	↑	↑	↑	↑	↑
Darkening	—	↑	↑	↑	↑	↑
Pungent Odor	—	—	↑	↑	↑	↑
Surface Tension	—	—	—	↓	↓	↓
Oxide Insolubles	—	—	↑	↑	↑	↑
RBOT Life	—	↓	↓	↓	↓	↓
Filterability	—	↓	↓	↓	↓	↓
Specific Gravity	—	—	—	↑	↑	↑
Sludge	—	—	—	↑	↑	↑

Courtesy Noria Corporation

MONITORING AND PROACTIVELY RESPONDING TO OIL CONTAMINATION

While the benefits of detecting abnormal machine wear or an aging lubricant condition are important and frequently achieved, they should be regarded as low on the scale of importance compared to the more rewarding objective of failure avoidance (see Figure 21-1).

Whenever a proactive maintenance strategy is applied, three steps are necessary to ensure that its benefits are achieved. Since proactive maintenance, by definition, involves continuous monitoring and controlling of machine failure root causes, the first step is simply to set a target, or standard, associated with each root cause. In oil analysis, root causes of greatest importance relate to fluid contamination (particles, moisture, heat, coolant, etc.) and additive degradation.

However, the process of defining precise and challenging targets (e.g., high cleanliness) is only the first step. Control of the fluid's conditions within these targets must then be achieved and sustained. This is the second step and often includes an audit of how fluids become contaminated and then systematically eliminating these entry points. Often better filtration and the use of separators will be required.

The third step is the vital action element of providing the feedback loop of an oil analysis program. When exceptions occur (e.g., over target results) remedial actions can be immediately commissioned. Using the proactive maintenance strategy, contamination control becomes a disciplined activity of monitoring and controlling high fluid cleanliness, not a crude activity of trending dirt levels.

Finally, when the life extension benefits of proactive maintenance are combined with by the early warning benefits of predictive maintenance, a comprehensive condition-based maintenance program results. While proactive maintenance stresses root-cause control, predictive maintenance targets the detection of incipient failure of both the fluid's properties and machine components like bearings and gears. It is this unique, early detection of machine faults and abnormal wear that is frequently referred to as the exclusive domain of oil analysis in the maintenance field.

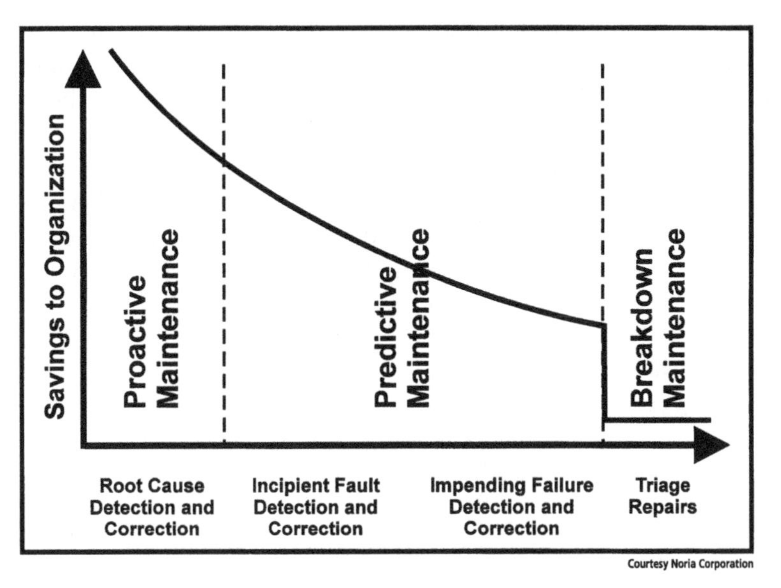

Figure 21-1. Some maintenance strategies are more costly than others.

OIL SAMPLING METHODS EXAMINED

As in all aspects of life, the end result of any endeavor is only as good as the effort put in to the exercise and the quality of elements used to create the result. Such is the case in lubricant and wear particle analysis in which results accuracy is highly dependent upon the care and method used to procure and deliver a quality used oil sample into the hands of a laboratory to commence analysis.

Procuring and delivering an analysis ready, superior quality used oil sample requires discipline and consistency as depicted in Table 21-3: 7-Best Practice Principles of Oil Sampling, and is the result of choosing the best procedure, method, hardware, and sample location. The samples choices for each piece of sampled equipment will likely differ based on whether the sample is taken from a pressurized or non-pressurized system, whether the machine or gearbox is designed or set up for best practice sampling techniques, the consistency of the sampling methods and techniques, the training of the person taking the sample, and the sampling cleanliness protocol used.

The best sample choices for the equipment being sampled are driven by three main objectives to 1) maximize sample data density, 2) minimize sample data disturbance, and 3) maximize sampling consistency.

1) Maximize Data Density through Sampling Point Selection

Each and every oil sample carries a unique time stamped composition signature of base oil chemistry, additive package level and chemistry and wear particle type, size and count used to compare against a virgin oil sample to determine the oil's chemical condition and the machine's moving parts condition at that moment in time—the more representative the sample, the more accurate the diagnosis. Therefore, every sample must contain the maximum amount of data density (representative data) it can, best achieved by extracting the sample in the most appropriate place.

For pressurized systems, e.g. hydraulic, and recirculating lubricating oil systems, oil is pumped from a reservoir under pressure, through a series of filters in a piping distribution system to the bearing surface areas form where it is returned back to the reservoir to be once again filtered and cooled ready for recirculation. Maximum data density is always found downstream of the lubricated bearings and upstream of the return filter where it is laden with any contaminants just washed from the bearing surfaces. To assure the most representative sample, take it:

- When the machine is running at temperature and under regular working condition load,
- From a live fluid zone meaning no dead pipe legs (static areas) or line ends,
- From a sample port connected to an elbow used to create a turbulent zone and ensure a colloidal (well mixed) sample

Table 21-3. 7-Best Practice Principles of Oil Sampling

1	Sample when machine is running under load
2	Sample Upstream of filters, and downstream of lubricated bearings
3	Before taking actual sampling, use the 10x flush rule
4	Always pre flush sampling equipment – pump/hose
5	Only use virgin clean sample bottles
6	Always sample at the same spot, using the same procedure and equipment
7	Forward sample and completed sample details to lab within 24 hours

In circulating oil systems such as the one shown in Figure 21-2, the best (primary) location is a live zone of the system upstream of filters where particles from ingression and wear debris are the most concentrated. Usually, this means sampling on fluid return or drain lines. Figures 21-3 and 21-4 show different options for sampling low pressure return lines. In the case of vented vertical drains from bearing housings there is not a solid flow of oil (air and oil share the line) making sampling more difficult. In such cases, a hardware adapter called a sample trap can be effectively installed to "trap" the oil for easy sampling (see Figure 21-5).

In those applications where oil drains back to sumps without being directed through a line (e.g., a diesel engine and wet-sump bearing and gear casings), the pressure line downstream of the pump (before filter) must be used. Fig-

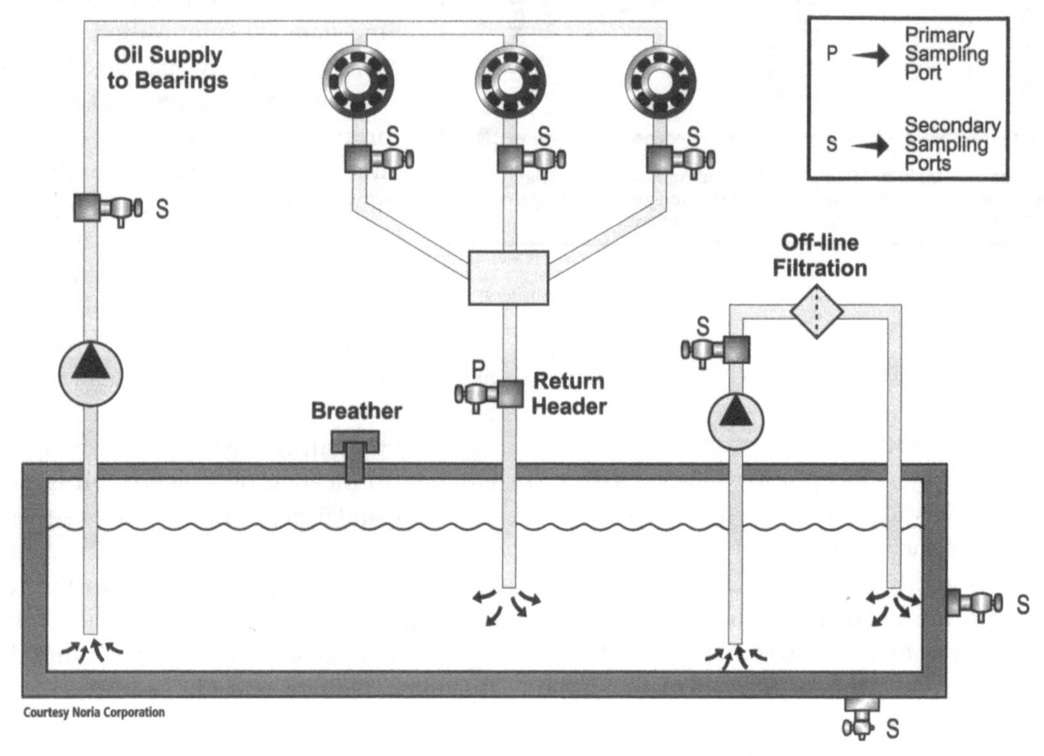

Figure 21-2. Circulating oil system indicating recommended sampling points.

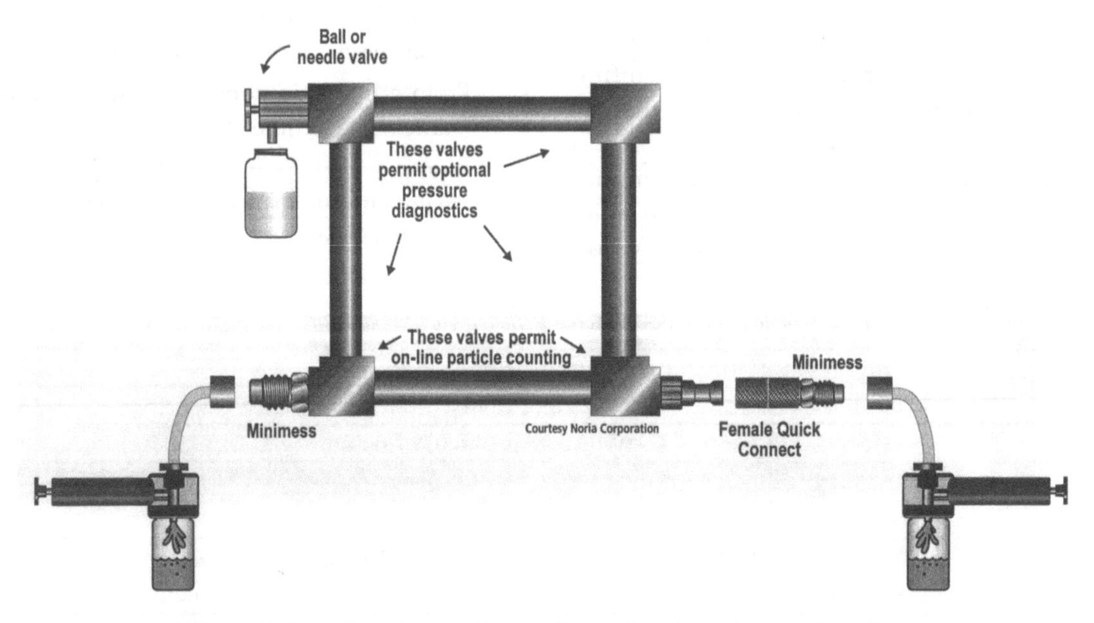

Figure 21-3. Different options for sampling oil from low pressure return lines.

ure 21-4 shows various options for sampling pressurized fluid lines. Where possible, always avoid sampling from dead zones such as static tanks and reservoirs. Splash, slinger ring, and flood-lubricated components are best sampled from the drain or casing side using a short inward-directed tube attached to a sample valve (see Figure 21-6). It may be necessary to use a vacuum pump to assist the oil flow for high viscosity lubricants.

Procedure for Extracting the Sample

Once a sampling point is properly selected and validated, a sample must be extracted without disturbing the integrity of the data. When a sample is pulled from turbulent zones such as at an elbow, particles, moisture, and other contaminants enter the bottle at representative concentrations. In contrast, it is well-known that sampling from ports positioned at right angles to the path of the

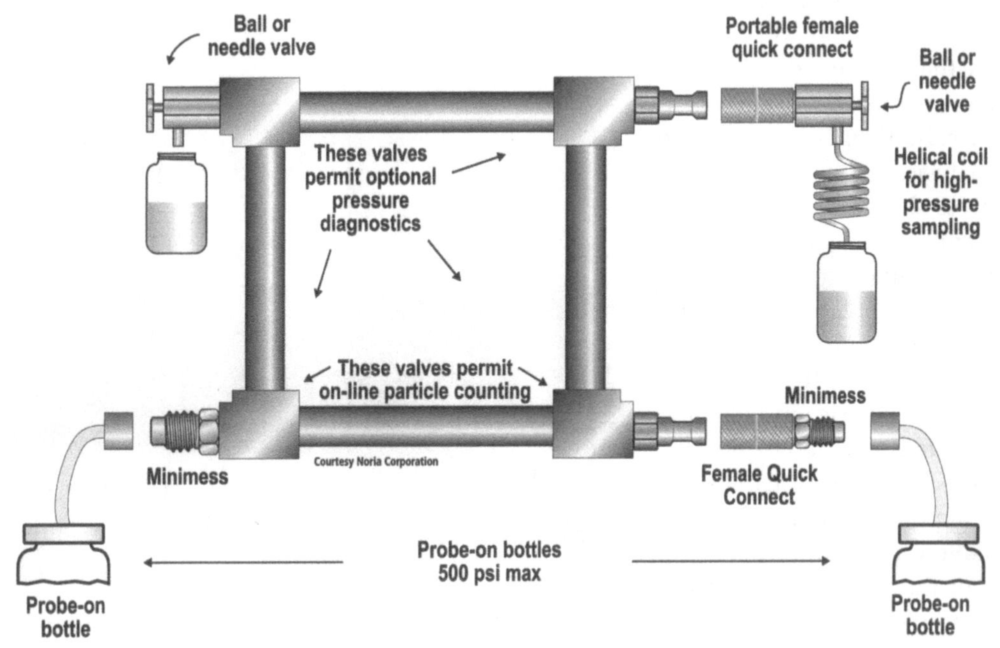

Figure 21-4. More options for taking lube oil samples.

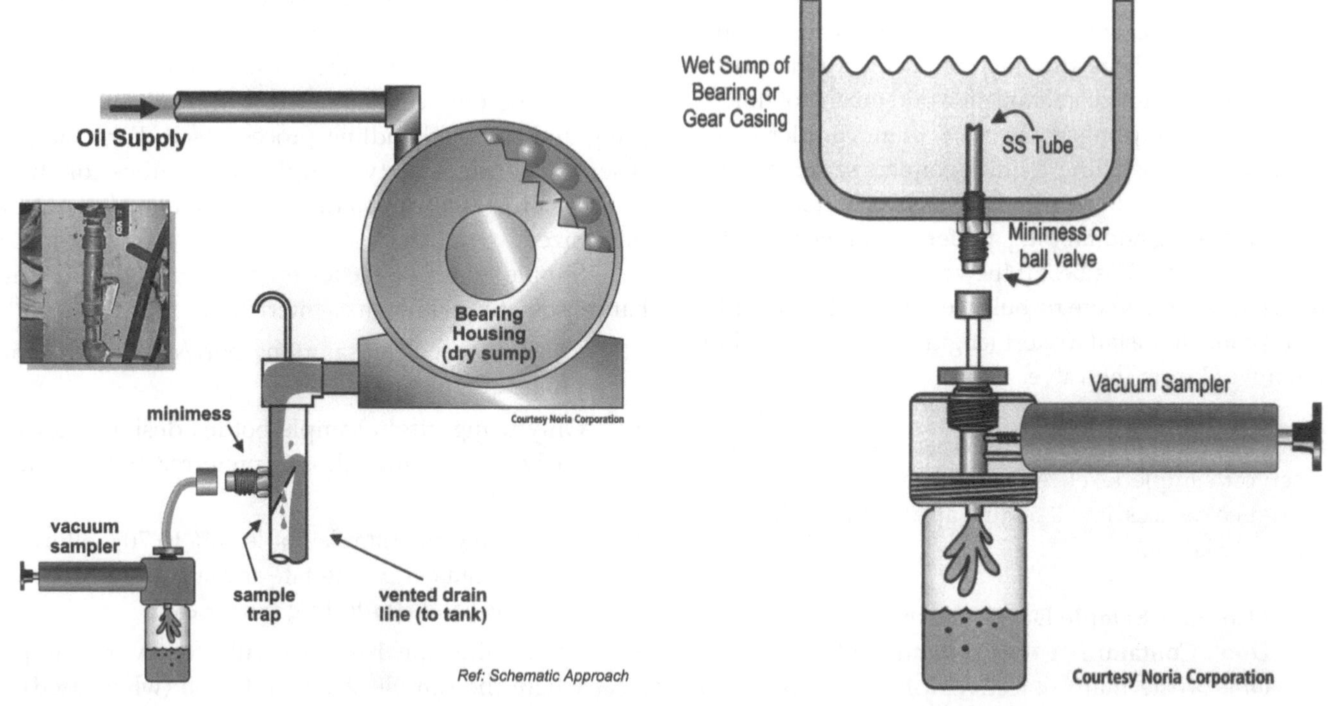

Figure 21-5. Sample trap installed in drain line below bearing housing.

Figure 21-6. Vacuum sampler arrangement.

fluid flow in high velocity, low viscosity fluids causes particle fly-by. In such cases, the higher density particles follow a forward trajectory and fail to enter the sampling pathway.

Machines should always be sampled in their typical work environment, ideally while they are running with the lubricant at normal operating temperature. Likewise, during (or just prior to) sampling, machines should be run at normal loads, speeds, and work cycles. This helps to ensure that the wear debris that is typically generated in the usual work environment and operating conditions is present in the fluid sample for analysis.

Sampling valves should be flushed thoroughly prior to sampling. If other portable sampling hardware is employed, these devices need to be flushed as well. Once the flushing is complete the sample bottle can be filled. However, never fill a sample bottle more than three-fourths full. The headspace in the bottle (ullage) permits adequate agitation by the lab.

With many non-circulating systems, static sampling may be the only option. Often this can be done effectively from drain ports if a sufficient volume of fluid is flushed through prior to the actual sample (see Figure 21-6). Alternatively, drop-tube vacuum samplers could be used (Figure 21-7). Care should be taken to always sample a fixed distance into the sump. Using a rod with a marked standoff from the bottom of the tank is a reliable way to do this. Flushing of the suction tube is also important. Never reuse suction tubes to avoid cross contamination and mixing of fluids.

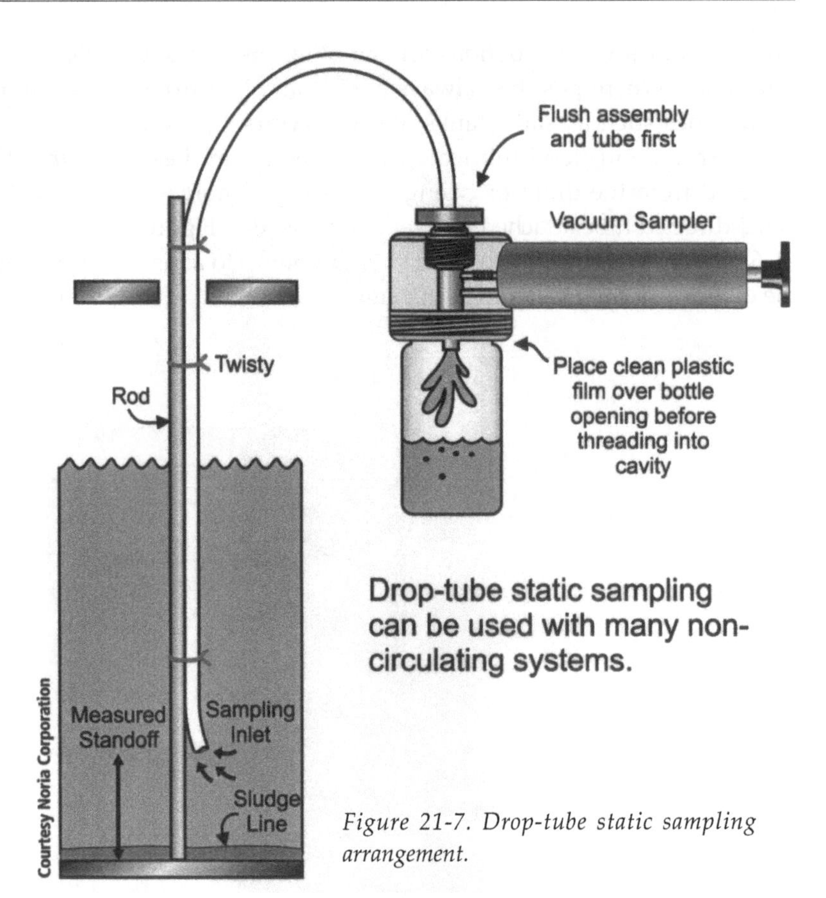

Figure 21-7. Drop-tube static sampling arrangement.

Static sampling using a vacuum sampler can be improved by installing a quick-connect sampling valve to which the vacuum tube is attached. Often this will require drilling and tapping, preferably in the wall of the sump or casing. It is best if the valve can be located near return lines and where turbulence is highest. Generally, it is desirable to install a short length of stainless steel tubing inward from the valve.

Another ideal sample method employs a combination Pilot tube/level gauge device affixed at the correct reservoir sample level. As most reservoirs do not come with such devices, it will require an after market purchase and installation.

2) Minimize Sample Data Disturbance—Don't Contaminate the Contaminant!

One of the main objectives of oil analysis is the routine monitoring of oil contamination. Therefore, in order to do this effectively, considerable care must be taken to avoid "contaminating the contaminant." Once atmospheric contamination is allowed to contact the oil sample, it cannot he distinguished from the original contamination.

It is important that the oil sample data is not disturbed or become contaminated as a result of the sampling and sample handling process itself. For example, reservoir sludge, dirty sample/drop tubes, or dirty sample bottles, etc. can all distort the data readings if not minimized.

Simple, effective tactics for managing data disturbance, sometimes known as interference include:

- Clean hands, clean sampling port/area, clean sampling equipment,
- Only using virgin sample bottles designed for oil analysis sampling, glass is preferred but the most expensive
- Only filling the sample bottle 60% to 70% allowing the lab headspace to agitate and successfully re-suspend the solids for testing purposes
- Performing the 10x flush rule for every sample where the sample valve and tube (when used) is flushed with approximately 10 x the sample volume

space required with the system oil being sampled into a non sample container before the real sample is taken

Avoid sampling methods that involve removing the bottle cap, especially where significant atmospheric contamination is present. One effective method that ensures that particles will not enter the bottle during sampling is a procedure called "clean oil sampling." It involves the use of common zip-lock sandwich bags and sampling hardware such as vacuum pumps and probe devices. Below is an outline description of this procedure:

Step One

Obtaining a good oil sample begins with a bottle of the correct size and cleanliness. It is understandable that the bottle must be at a known level of cleanliness and that this level should be sufficiently high so as not to interfere with expected particle counts. Some people refer to this as signal-to-noise ratio, i.e., the target cleanliness level of the oil (signal) should be several times the expected particle contamination of the bottle (noise). For more information on bottle cleanliness refer to ISO 3722.

Step Two

Before going out into the plant with the sample bottles place the capped bottles into very thin zip-lock sandwich bags, one per bag (Figure 21-8). Zip each of the bags such that air is sealed into the bag along with the bottles. This should be done in a clean-air indoor environment in order to avoid the risk of particles entering the bags along with the bottles. After all of the bottles have been bagged, put these small bags (with the bottles) into a large zip-lock bag for transporting them to the plant or field. Sampling hardware such as vacuum pumps and probe devices should be placed in the large bag as well.

Step Three

After the sampling port or valve has been properly flushed (including the sampling pump or probe if used) remove one of the bags holding a single sample bottle. Without opening the bag, twist the bottle cap off and let the cap fall to the side within the bag. Then move the mouth of the bottle so that it is away from the zip-lock seal. Do not unzip the bag.

Step Four

Thread the bottle into the cavity of the sampling device (vacuum pump or probe). The plastic tube will puncture the bag during this process, however, try to avoid other tears or damage to the bag (turn the bottle, not the probe or pump, while tightening). If a probe device is used, it is advisable to break a small hole in the bag below the vent hole with a pocket knife. This permits air to escape during sampling.

Step Five

The sample is then obtained in the usual fashion until the correct quantity of oil has entered the bottle. Next, by gripping the bottle, unscrew it from the cavity of the pump or probe device. With the bottle free and still in the bag, fish the cap from the bottom of the bag onto the mouth of the bottle and tighten.

Step Six

With the bottle capped it is safe to unzip the bag and remove the bottle. Confirm that the bottle is capped tightly. The bottle label should be attached and the bottle placed in the appropriate container for transport to the lab. Do not reuse the zip-lock bags.

Three levels of bottle cleanliness are identified by bottle suppliers: clean (fewer than 100 particles >10 μm/ml), superclean (fewer than 10), and ultraclean (fewer than 1). Selecting the correct bottle cleanliness to match the type of sampling is important to oil analysis results.

3) Maximize Sample Data Consistency— Oil Sample Frequency

To ensure high quality sample results that can be trusted the sampling protocol must assure sample consistency through:

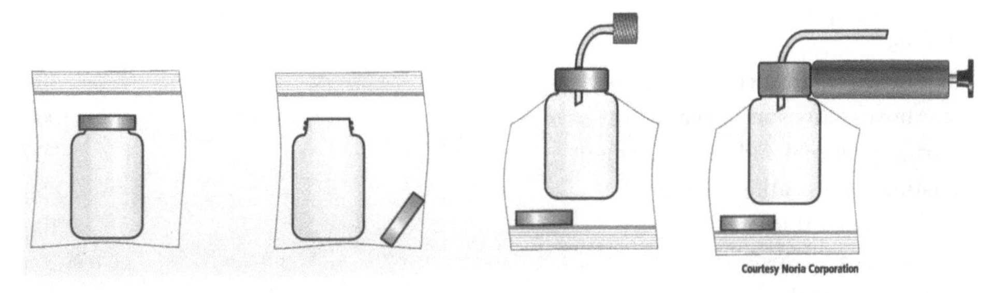

Figure 21-8. Zip-Lock® bags prevent contamination of samples.

- Development of an engineered oil sampling program in which every sampling port and method is documented and regular sampling frequencies are set up in a work order system. Commencing a program, bearings are usually start sampled on a 500 hour frequency, industrial hydraulic systems on a 700 hour frequency, with light duty gearboxes on a 1000 hour frequency and heavy duty gearboxes on a 300 hour frequency,
- Use of the oil sampling program to develop and train oil sampling standard operating procedures,
- Always sampling from the same location,
- Regularly sample virgin oil stocks when new lubricant stock arrives on site,
- Use of the same laboratory for sampling and ask for a dedicated lab technician(s) to perform your sampling
- Always filling in the sample data form, including sampling date and time stamp
- Sending off the sample to the lab within 24 hours of sampling. Over 24 hours, retake the sample.

OIL SAMPLING FREQUENCY

The objective of oil analysis, like condition monitoring in general, is to find bad news. The objective of proactive maintenance is not to have any bad news to find. The machine and oil will generally give off silent alarms when problems first occur. In time, as the severity increases, these alarms are no longer silent and even the most rudimentary condition monitoring methods can reveal the problem. Of course, at this point, a great deal of damage may have already occurred. And, it is likely too late to arrest the problem on the run; the machine may have to be taken apart and repaired.

One of the extraordinary benefits of oil analysis is its incredible sensitivity to these silent alarms and the detection of incipient failures and faults. The methods of doing this—successfully—are still to be discussed. However, it is a very basic principle that you cannot hear an alarm unless you are listening for an alarm—restated, you can't catch a fish unless your hooks in the water. Too often we hear about oil samples being taken every six months or annually; yet, on the same machinery we see vibration readings taken every month.

Scheduled sampling intervals are common in oil analysis. The frequency may be keyed to drain intervals or operating hours. Table 21-4 conservatively recommended intervals based on operating hours for different machine classes. Proper selection of sampling frequencies considers machine and application-specific criteria such as those below:

Penalty of Failure

Safety, downtime costs, repair costs, and general business interruption costs must be considered.

Fluid Environment Severity

Operation and fluid environment conditions influence both frequency and rate of failure progression. Influencing factore include pressures, loads, temperature, speed, contaminant ingression, and system duty.

Table 21-4. Conservatively recommended oil sampling intervals for different equipment categories.

	Hours
Diesel engines - off-highway	150
Transmissions, differentials, final drives	300
Hydraulics - mobile equipment	200
Gas turbines - industrial	500
Steam turbines	500
Air/gas compressors	500
Chillers	500
Gear boxes - high speed, heavy duty	300
Gear boxes - low speed, heavy duty	1000
Bearings - journal and rolling element	500
Aviation reciprocating engines	25-50
Aviation gas turbines	100
Aviation gear boxes	100-200
Aviation hydraulics	100-200

Machine Age

In general, the chances of failure are greatest for machines going through break-in and after major repairs and overhauls. Likewise, the risk increases as a machine approaches the end of its expected life.

Oil Age

Infant oils and old oils are at highest risk. Infant oils are those that have just been changed and are less than 10% into expected life. Old oils are showing trends that suggest additive depletion, the onset of oxidation, or high levels of contamination.

SELECTION OF TYPE OF OIL ANALYSIS

Once proper oil sampling has been mastered it is time to analyze the oil. Because each test that is conducted by an oil lab adds cost to the program, it is important that an optimum selection of tests be defined. There are generally two types of tests; routine and exception. A routine test is a scheduled test that is repeated with each scheduled sample such as tests for viscosity, moisture, and particle count.

An exception test is triggered by a previously non-complying condition or test result. It is conducted to either confirm a conclusion (diagnosis/prognosis) or seek further information that might identify the cause or source of the problem. Exception tests might, for instance, include specialized tests for confirming oil oxidation or abnormal machine wear. Table 21-5 shows how routine tests can be combined with exception tests to provide comprehensive test bundles by machine application.

To be thoroughly effective, a well-designed oil analysis program must encompass three categories of routine tests: (1) fluid properties, (2) fluid contamination, and (3) fluid wear debris.

Fluid Properties Analysis

This essential type of oil analysis helps ensure the fundamental quality of the lubricating fluid. The standard to which a used oil's properties should be routinely compared are the new oil's properties; a listing of each of the new oil properties should be a standard feature of used oil analysis reports. Examples of common tests include viscosity, total acid number, total base number, infrared for oxidation, emission spectroscopy for additive elements, flash point, specific gravity, and rotating bomb oxidation test (RBOT).

Table 21-5. Selecting oil analysis tests by application.

Test or Procedure	Paper Machine Oils	Motor & Pump Bearings	Diesel & Gas Engine	Hydraulics	Air & Gas Compressors	Chillers and Refrigeration	Transmissions, Final Drives, Differentials	Industrial Gear Oils	Steam Turbine Oils	Gas Turbine Oils	EHC Fluids
1. Particle Count	O,L	O,L	L	O,L	O,L	O,L	L	O,L	O,L	L	O,L
2. Viscosity											
a. 40°C	O,L	O,L	-	O,L	O,L	O,L	L	O,L	L	L	O,L
b. 100°C	-	-	L	-	-	-	-	-	-	-	-
3. TAN	L	E(5a)	-	L	L	L	L	L	L	L	O,L
4. TBN	-	-	L	-	-	-	-	-	-	-	-
5. FTIR											
a. Ox/Nit/Sul	L	L	L	L	L	L	L	L	L	L	-
b. Hindered Phen	-	L	-	L	L	-	-	L	L	-	-
c. ZDDP	-	L	-	L	L	-	L	L	-	-	-
d. Fuel Dil/Soot	-	-	L	-	-	-	-	-	-	-	-
6. Flash Point	-	-	E(2b,5d)	-	L*	-	-	-	-	E(2b,5d)	-
7. Glycol-ASTM Test	-	-	E(14b)	-	-	-	-	-	-	-	-
8. Ferrous Density	E(1)	E(1)	L	O,L	O,L	O,L	L	O,L	E(1)	E(1)	O,L
9. Analytical Ferrography	E(8,14a)	E(8,14a)	E(8,14a)	E(8,14a)	E(8,14a)	E(8,14a)	E(8,14a)	E(8,14a)	E(8,14a)	E(8,14a)	E(8,14a)
10. RBOT	-	-	-	-	L	-	-	-	L	L	-
11. Crackle	O,L	O,L	L	O,L	O,L	L	O,L	O,L	O,L	-	O,L
12. Water by KF	E(11)	E(11)	E(11)	E(11)	E(11)	E(11)	E(11)	E(11)	E(11)	-	E(11)
13. Water Separability	L	-	-	-	L**	-	-	-	L	-	-
14. Elemental Analysis											
a. Wear Metals	L,E(1)	L,E(1)	L	L,E(1)	L,E(1)	L,E(1)	L	L,E(1)	L,E(1)	L	L,E(1)
b. K, Na, B, Si	L	L	L	L	L	L	L	L	L	L	L
c. Additives	L	L	L	L	L	L	L	L	L	L	L

*Gas Compressors Only **Air Compressors Only

Courtesy Noria Corporation

O = On-site routine test (small on-site lab or portable instrument)
L = Fully equipped oil analysis laboratory
E = Exception test keyed to a positive result from the test in parenthesis

Fluid Contamination Analysis

Despite the use of filters and separators, contaminants are the most common destroyers of machine surfaces that ultimately lead to failure and downtime. For most machines, solids contamination is the number one cause of wear-related failures. Likewise, particles, moisture, and other contaminants are the principal root cause of additive and base stock failure of lubricants. It is important to perform basic tests such as particle counting, moisture analysis, glycol testing, and fuel dilution as directed by a well-designed proactive maintenance program.

Fluid Wear Debris Analysis

Unlike fluid properties and contamination analysis, wear debris analysis relates specifically to the health of the machine. Owing to the tendency of machine surfaces to shed increasing numbers of progressively larger particles as wear advances, the size, shape, and concentration of these particles tell a revealing story of the internal or state condition of the machine.

Cost-effective oil analysis can generally be done when on-site oil analysis tools are available. For many machines, the particle counter serves as the best first line of defense. Only when particle counts exceed preset limits is exception testing performed. The best exception test is ferrous density analysis, such as a ferrous particle counter. When ferrous levels are high, a failure condition exists, triggering yet further testing and analysis. In addition to on-site particle counting, on-site moisture analyzers and viscometers also assess important root cause contributors.

MONITORING CHANGING OIL PROPERTIES

Today there are a growing number of organizations transforming their lube programs from scheduled to condition-based oil changes. In fact, many companies claim that they easily pay for the cost of oil analysis from savings achieved through reduced lubricant consumption. Such progressive goals as these place a greater burden of precision on the selection of oil analysis tests and alarm limits to reveal non-complying lubricants.

It is not uncommon for plants to interpret oil analysis results independent of the lab. Essentially, the lab is relied on to provide accurate and timely data, leaving both interpretation and response to the plant; i.e., personnel close to the equipment with knowledge of application and operating conditions. The use of modern oil analysis software can greatly assist in this such programs.

In order to reduce oil consumption, two plans must be implemented. The first plan is proactive in nature and relates to the operating conditions the oil lives in. By improving the oil's operating conditions its expected life can increase many fold. For instance, with mineral oils the reduction of operating temperature of just 10 degrees C can double the oil's oxidation stability and double the oil change interval in many instances. An upcoming section discusses how proactive maintenance by controlling oil contamination can lead towards oil life extension.

The second plan to reducing oil consumption is predictive in nature and relates to the timing of oil changes. Basically, through oil analysis, key physical properties can be trended to help forecast the need of a future oil change. Restated, by listening to the oil, it will tell us when it needs to be changed. And, if the need of an oil change occurs prematurely, then an assessment of the oil's operating conditions (cleanliness, dryness, coolness, etc.) and oil formulation should be made. The nature of the degradation will provide basic clues defining the solution.

There are numerous modes of degradation of lubricating oil. These change many of the properties of the fluid. In order to recognize the change it is important that the correct properties be monitored, realizing that overkill is wasteful. What follows is a discussion of common oil degradation modes and the properties that can best reveal them. In all cases, it is important to get a base signature of the properties of the new oil so as to benchmark the trended change. These reference properties should remain as a permanent record on the oil analysis report and include additive elements, neutralization numbers, -infrared units (unless spectral subtraction is used), RBOT minutes, viscosity, flash temperature, VI, and color.

Viscosity Stability

Viscosity is often referred to as the structural strength of a liquid. It is critical to oil film control plus is a key indicator to a host of ailing conditions relating to the oil and machine. As such, it is often considered a critical "catch-all" property in oil analysis. Essentially, when viscosity remains in a controlled narrow band one can assume that a great many things that could be going wrong are, in fact, not going wrong. Conversely, when viscosity falls outside of the band an exception test is usually needed to identify the nature and cause of an abnormality. Monitoring viscosity thus represents a first line of defense against many problems.

Because viscosity is so important it is often monitored on-site by the reliability team. It is used as an acceptance test for new oil deliveries and to verify the correct lubricant is in use. When viscosity changes with in-service lubricants, the cause is either oil degradation or oil contamination. Oil degradation relates to changes to the base oil and additive chemistry (molecular changes). Contamination of an oil can either thicken or thin the oil depending on the viscosity and emulsifying characteristics of the contaminant (see Table 21-6).

In oil labs, viscosity is typically measured using kinematic viscometers. ISO viscosity grades shown on lubricant spec sheets are based on kinematic viscosity in centistokes (cSt) at 40 degrees C. Another way to represent kinematic viscosity is Saybolt Universal Seconds (SUS). Figure 21-9 shows a photo of a common U-tube kinematic viscometer. In this device, the oil is allowed to drain by gravity through a capillary at constant temperature. The drain time (efflux time) is measured and translated into centistokes. Viscosity varies nearly proportionally to drain time. Because gravity is involved, kinematic viscosity characterizes both the oil's resistance to flow (absolute viscosity) and specific gravity.

On-site oil analysis labs frequently use absolute viscometers to obtain a precise indication of base oil condition. Unlike kinematic viscometers, absolute viscosity measures only an oil's resistance to shear or flow (not specific gravity). Figure 21-10 shows an absolute viscometer designed for plant-level use. It employs a capillary in its tip, through which the oil flows under constant pressure and temperature. An inline plunger moves outward with the flow. The speed of this plunger, measured electronically, varies nearly proportionally to absolute viscosity.

Viscosity is typically trended at 40 degrees C although for high temperature applications, such as crankcase lubricants, a 100-degree C trend is sometimes preferred. Both temperatures are needed to determine the oil's Viscosity Index (VI). However, the VI rarely is trended for routine condition monitoring. Monitoring viscosity at 40-degrees C, for most industrial applications, will provide the most reliable early indication of base oil degradation and oxidation.

Oxidation Stability

When an oil oxidizes the base oil thickens and discharges sludge and acidic materials; all detrimental to good lubrication. Oxidation is uncommon in applications when conditions are such that oils are relatively cool, dry,

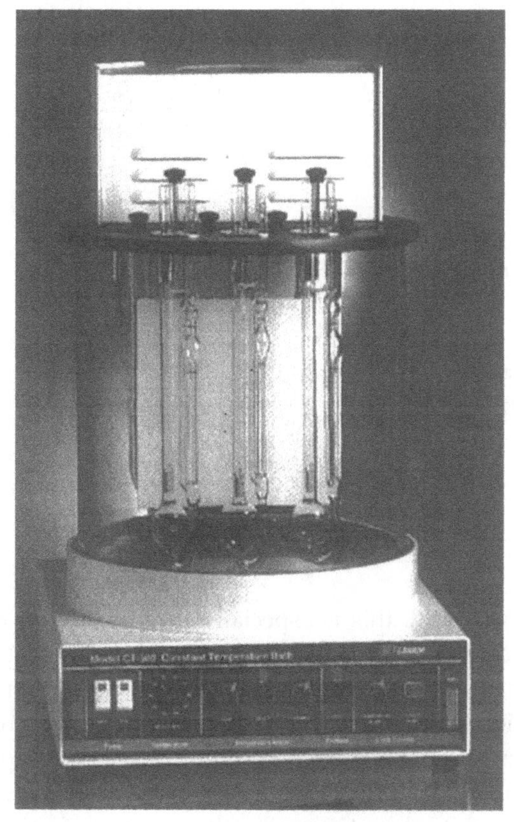

Courtesy Cannon Instruments

Figure 21-9. U-tube kinematic viscometer.

Table 21-6. The origin of viscosity changes.

	Decreases Viscosity	Increases Viscosity
Changes to base oil (molecular changes)	• Thermal cracking of oil molecules • Shear thinning of VI improvers	• Polymerization • Oxidation • Evaporative losses • Formation of carbon & oxide insolubles
Additions to base oil (contamination)	• Fuel • Refrigerant • Solvents • Wrong oil (low viscosity)	• Water (emulsions) • Aeration • Soot • Antifreeze (glycol) • Wrong oil (high viscosity)

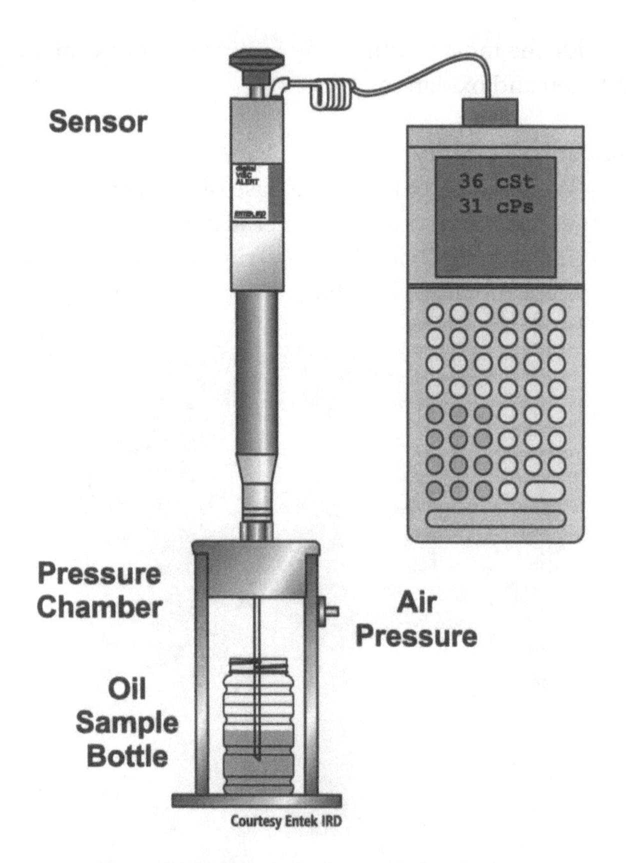

Figure 21-10. Absolute viscometer for plant use.

and clean. And, this is especially true for low viscosity oils such as hydraulic fluids and turbine oils that have higher oxidation stability. However, when operating conditions are severe, oil oxidation can be a recurring problem. Where a proactive solution cannot be applied (controlling oxidation root causes or the use of resistant synthetics) it is best to monitor the progress of oxidation. Monitoring the depletion of oxidation inhibitors provides an early, forecastable trend, but this may not be practical in some applications.

The technologies used to monitor the depletion of oxidation inhibitors include:

1. Infrared spectroscopy (FTIR) can pick up trendable changes in phenolic and ZDDP inhibitors. However, only a few of the laboratories report additive depletion with FTIR because of unreliable reference oils and occasional inferences from contaminants. See Figure 21-11.
2. Total acid number (TAN) is sensitive to both mass-transfer and decomposition depletion of ZDDP inhibitors. Interpretation of the trend takes practice and a good new-oil reference.
3. Elemental spectroscopy can show reliable mass-transfer depletion trends in ZDDP inhibited oils.
4. Rotating Bomb Oxidation Tests (RBOT) provide a highly forecastable trend on additive depletion. Because of the time needed to run this test it is expensive and usually saved for exception testing or special circumstances. (See Table 21-7)
5. Voltametry is a new technology that has shown particular promise in trending the depletion (mass transfer and decomposition) of phenolic and ZDDP inhibitors.

If trending the depletion of oxidation inhibitors is not practical then oxidation itself must be monitored. The problem with this approach relates to the fact that oxidation can progress rapidly in stressful conditions once the antioxidant has depleted. Simply stated, with oxidation, "the worse things get the faster they get worse." If the

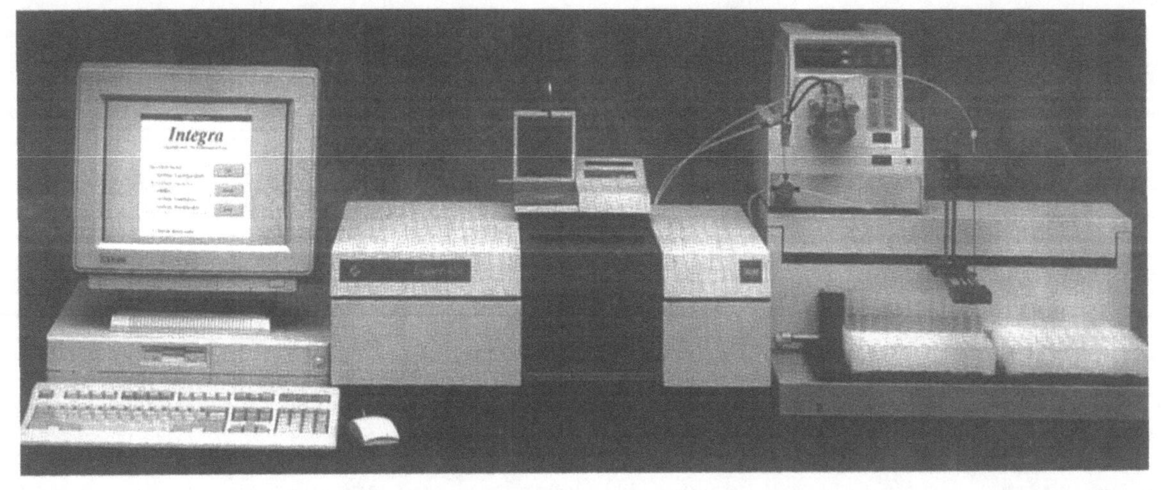

Courtesy Biorad

Figure 21-11. Infrared spectroscopy (FTIR) is used to determine additive depletion.

Table 21-7.

Additive	Monitoring Method	Effectiveness
ZDDP (antioxidant, antiwear, and corrosion inhibitor)	FTIR - (@ ≈950 wavenumbers) FTIR - oxidation (1750 wn) RBOT TAN (downward trend) TAN (upward trend) Voltametry Elemental Spectroscopy	Fair (early detection) Good (later detection) Excellent (early detection) Fair (early detection) Good (later detection) Excellent (early detection) Excellent (early detection)
Rust Inhibitors	Elemental Spectroscopy	Fair (early detection)
Foam Inhibitors	Elemental spectroscopy	Fair (interferences from dirt)
Sulfur Phosphorus - EP	Elemental spectroscopy	Excellent
Molybdenum Disulfide - EP	Elemental spectroscopy	Good
Borate - EP	Elemental spectroscopy	Excellent
VI Improver	Viscosity at 40°C and 100°C	Excellent
Dispersants	Blotter spot test	Good
Detergents	TBN Elemental Spectoscopy	Excellent Excellent
Hindered Phenol (antioxidant)	FTIR	Fair

Courtesy Noria Corporation

goal is a condition-based oil change, this translates to the need to monitor with sufficient frequency to catch the problem in the incipient stages, not after the oil throws sludge and destructive lubrication has occurred.

The most common and reliable methods to detect and trend oil oxidation are the following:

1. If a reliable new oil reference is available to the laboratory, infrared analysis (FTIR) is dependable for mineral oils and many synthetics including organic and phosphate esters. The acids, aldehydes, esters, and ketones formed during oxidation are detected by FTIR in mineral oils and PAO synthetics.

2. Total acid number (TAN) will quantify the growing acid constituents in oxidizing oils.

3. Because oxidation results in polymerization of the base oil and the discharge of oxide insolubles, the viscosity will increase.

4. Color-bodies form in oxidized oils resulting in a marked darkening of the oil's color.

5. Oxidized oils give off a sour or pungent odors, similar to the smell of a rotten egg.

Thermal Stability and Varnish Tendency

The thermal failure of an oil can be localized or uniform. Localized thermal failure occurs when the bulk oil temperature remains generally suitable for the selected lubricant but oil is exposed to hot surfaces such as the discharge valves of reciprocating compressors, or hot surfaces in internal combustion engines and turbomachinery. Another common cause of localized thermal failure is associated with entrained air that is permitted to compress, such as occurs to air bubbles passing through a high-pressure hydraulic pump. The air bubble implosion causes heat to concentrate generating microscopic specks of carbon. These carbon insolubles later condense on machine surfaces, forming what is commonly called varnish.

The varnish tendency of an oil is often difficult to detect due to the fact that the majority of the physical properties of the oil are unaffected. For instance, oxidation may occur without change in viscosity, TAN, or FTIR. However, sophisticated labs having experience with hydraulic fluids will employ specialized tests such as ultracentrifuge, FTIR for nitration, and submicron membrane tests. Other, less reliable, indicators include oil color and paper chromatography (blotter spot test).

The uniform thermal failure of an oil results from excessively high operating temperatures due to any of a number of reasons. However, the most common reasons include overloading, inadequate oil supply, failure of a heat exchanger, and the use of a high watt-density tank heater. When any of these conditions occur, the oil fails by evaporation (thickening), carbonization (coking, carbon stones, etc.), or cracking (thinning) in extreme cases. Regardless of origin, the uniform thermal failure of the oil is serious and threatens the reliable operation of the lubricated machine.

An oil's thermal stability it often measured using the Cincinnati Milacron test (ASTM D 2070-91). However, because this test takes a week to complete, it is generally impractical for routine used-oil analysis. Other ways to evaluate thermal failure include viscosity analysis, ultracentrifuge, total insolubles, and oil color. Less reliable indicators include oil odor (either a burnt, rancid odor or no odor at all) and paper chromatography.

Additive Stability

Additive monitoring is one of the most challenging and evasive areas of used-oil analysis. The reasons for this are many and complex. As a starting point, it is worthwhile to review how additives deplete during normal use and aging.

It is generally accepted that there are two forms of additive depletion, both are common and can occur simultaneously. The first form of depletion is known as decomposition. Here the additive mass stays in the oil but its molecular structure changes resulting in an assortment of transformation products (other molecules). In some instances, the transformation products may possess properties similar to the original additive, but in most cases performance is degraded or is completely lost. This sacrificial form of depletion is common to what happens over time to oxidation inhibitors, as described previously under "oxidation stability."

The second form of additive depletion is called mass transfer. This type of depletion is often the most easy to detect because the entire mass of the additive transfers out of the bulk oil. And, as such, any measurable property of the additive leaves as well. For instance, if the additive is constructed with phosphorous, a downward trend of phosphorous in the used oil is a reliable indication of its mass transfer depletion. Conversely, an unchanging level of phosphorous in used oil in no way confirms that decomposition depletion has not occurred. With decomposition the elements of the additive remain suspended in the oil.

Mass transfer of additives occur in normal operation, usually as a result of the additive doing the job it was designed to do. For instance, when a rust inhibitor attaches itself to internal machine surfaces it depletes by mass transfer. It is common for additives to cling to various polar contaminants in the oil, such as dirt and water. The removal of these contaminants by filters, separators, and settling action causes a removal of the additive as well. And, over time, aging additives can form floc and precipitate out of the oil due to decomposition and long cold-temperature storage. The insolubles formed will migrate out, often ending up on the bottom of the sump or reservoir.

Table 21-6 describes common methods used to monitor additive depletion. It is worth restating that the use of elemental spectroscopy to trend additive depletion is only effective where mass transfer is involved. It is not uncommon, therefore, for oil labs to condemn an oil with only a 25 percent reduction in the concentration of telltale additive elements, e.g., zinc and phosphorous in the case of ZDDP.

MONITORING OIL CONTAMINATION

Contamination can be defined as any unwanted substance or energy that enters or contacts the oil. Contaminants can come in a great many forms, some are highly destructive to the oil, its additives, and machine surfaces. It is often overlooked as a source of failure because its impact is usually slow and imperceptible yet, given time, the damage is analogous to eating the machine up from the inside out. While it is not practical to attempt to totally eradicate contamination from in-service lubricants, control of contaminant levels within acceptable limits is accomplishable and vitally important.

Particles, moisture, soot, heat, air, glycol, fuel, detergents, and process fluids are all contaminants commonly found in industrial lubricants and hydraulic fluids. However, it is particle contamination and moisture that are widely recognized as most destructive to both oil and machine.

Particle Contamination

There is no single property of lubricating oil that challenges the reliability of machinery more than suspended particles. It would not be an exaggeration to refer to them as a microscopic wrecking crew. Small particles can ride in oil almost indefinitely and because they are not as friable (easily crumbled) as their larger brothers, the destruction can be continuous. Many studies have shown convincing evidence of the greater damage associated with small particles compared to larger. Still, most maintenance professionals have misconceptions about the size of particles and the associated harm caused.

These misconceptions relate to the definition people apply to what is clean oil and what is dirty oil. And, it is this definition that becomes the first of the three steps of proactive maintenance; the need to set appropriate target cleanliness levels for lubricating oils and hydraulic fluids. The process is not unlike a black box circuit. If we want a change to the output (longer and more reliable machine life) then there must be a change to the input (a lifestyle change, i.e., improve cleanliness). For instance, it's not the monitoring of cholesterol that saves us from heart decease, instead it's the things we do to lower the cholesterol. Therefore the best target cleanliness level is one that is a marked improvement from historic levels.

This leads us to the second step in proactive maintenance, the lifestyle change. By effectively excluding the entry of contaminants and promptly removing contaminants when they do enter, the new cleanliness targets are frequently easily achieved. Concerns that filtration costs will increase may not materialize due to the greater overall control, especially from the standpoint of particle ingression.

The third step of proactive maintenance is the monitoring step, i.e., particle counting. If this is done frequently enough, not only is proactive maintenance achieved but also a large assortment of common problems can be routinely detected. As such, particle counting is another important "catch all" type test, like viscosity analysis. Because of the obvious value, the particle counter is probably the most widely used on-site oil analysis instrument. It is not uncommon to find organizations testing the cleanliness of their oils as frequently as weekly.

The activity of routine particle counting has a surprising impact on step number two. When the cleanliness levels of lubricants are checked and verified on a frequent basis a phenomenon known as the "invisible filter" occurs, which is analogous to the saying, "what gets measured gets done." Because a great deal of dirt and contamination that enters oils comes from the careless practices of operators and craftsmen, the combined effect of monitoring with a modicum of training can go a long way towards achieving cleanliness goals.

The ISO Solid Contaminant Code (ISO 4406) is probably the most widely used method for representing particle counts in oils. The current standard employs a two-range number system (see Figure 21-12). The first range number corresponds to particles larger than 5 microns and the second range number for particles larger than 15 microns. From the chart, as the range numbers increment up one digit the represented particle count roughly doubles. At this writing, the ISO Code is undergoing revision that will likely add a third range number plus a change to the particle size the three range numbers will relate to.

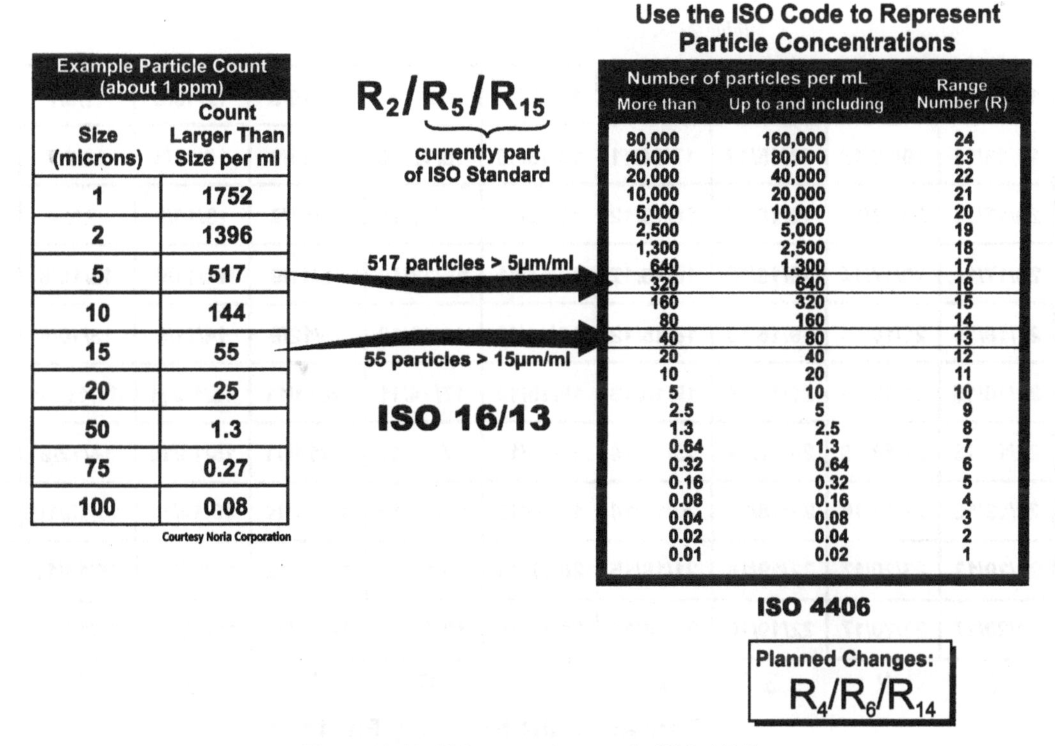

Example Particle Count (about 1 ppm)

Size (microns)	Count Larger Than Size per ml
1	1752
2	1396
5	517
10	144
15	55
20	25
50	1.3
75	0.27
100	0.08

Courtesy Noria Corporation

Use the ISO Code to Represent Particle Concentrations

Number of particles per mL More than	Up to and including	Range Number (R)
80,000	160,000	24
40,000	80,000	23
20,000	40,000	22
10,000	20,000	21
5,000	10,000	20
2,500	5,000	19
1,300	2,500	18
640	1,300	17
320	640	16
160	320	15
80	160	14
40	80	13
20	40	12
10	20	11
5	10	10
2.5	5	9
1.3	2.5	8
0.64	1.3	7
0.32	0.64	6
0.16	0.32	5
0.08	0.16	4
0.04	0.08	3
0.02	0.04	2
0.01	0.02	1

Figure 21-12. ISO contaminant code (ISO 4406).

While there are numerous different methods used to arrive at target cleanliness levels for oils in different applications, most combine the importance of machine reliability with the general contaminant sensitivity of the machine to set the target. This approach is shown in Table 21-8. The Reliability Penalty Factor and the Contaminant Severity Factor are arrived at by a special scoring system that is included with the Target Cleanliness Grid.

There are many different types of automatic particle counters used by oil analysis laboratories. There are also a number of different portable particle counters on the market. The performance of these instruments can vary considerably depending on the design and operating principle. Particle counters employing laser or white light are widely used because of their ability to count particles across a wide range of sizes (see Figure 21-13). Pore blockage-type particles counters have a more narrow size range sensitivity, however, they are also popular because of their ability to discriminate between hard particles of other impurities in the oil such as water, sludge, and air bubbles (see Figure 21-14).

Figure 21-15 shows how particle count trends vary depending on the machine application and the presence of a built-in filter. Because particle counters monitor particles in the general size range controlled by filters, equilibrium is usually achieved, i.e., particles entering the oil from ingression minus particles exiting from filtration will leave behind a steady state concentration. When filters are properly specified and ingression is under control this steady state concentration will be well within the cleanliness target. Systems with no continuous filtration, e.g., a splash-fed gearbox, an equilibrium is not really established (there is no continuous particle removal). This causes the particle concentration to be continuously rising. Still, contamination control can be achieved by periodic use of portable filtration systems, such as a filter cart.

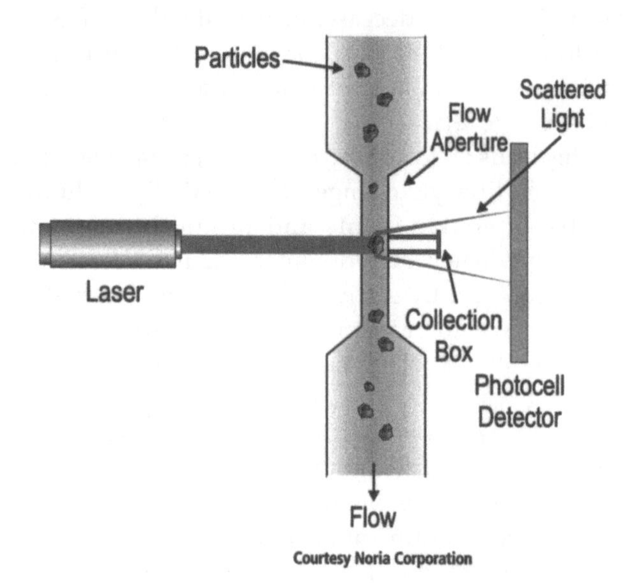

Figure 21-13. Particle counter.

Table 21-8. Contaminant severity factor (CSF).

Reliability Penalty Factor (RPF) — Cost, Safety, and Business Interruption Penalty From Failure	1	2	3	4	5	6	7	8	9	10
10	19/16/13	18/15/12	17/14/12	16/13/11	15/12/10	14/11/9	13/10/8	12/10/8	11/9/7	10/9/7
9	19/16/13	19/16/13	18/15/12	17/14/11	16/13/10	15/12/10	14/11/9	13/10/8	12/9/7	11/9/7
8	20/17/14	20/17/13	19/16/13	18/15/12	16/13/11	15/12/10	14/11/9	13/10/8	12/9/8	12/9/7
7	20/17/14	20/17/14	19/16/13	18/15/12	17/14/11	16/13/10	15/11/9	14/11/9	13/10/8	12/10/8
6	21/18/15	21/18/14	19/16/13	18/15/12	17/14/11	16/13/10	15/12/9	14/11/9	13/10/8	12/10/8
5	21/18/15	21/18/15	20/17/14	19/16/13	18/15/12	17/14/11	16/14/11	15/13/11	14/11/10	13/11/9
4	22/19/16	22/19/16	20/17/14	19/17/14	18/15/13	17/14/11	16/14/11	15/13/10	14/12/9	13/11/9
3	22/19/16	22/19/16	21/18/15	20/17/14	19/16/13	18/15/12	17/14/11	16/14/11	15/13/10	14/12/9
2	23/20/17	23/20/17	22/19/16	21/18/15	20/17/14	19/16/13	18/15/12	17/14/11	16/14/11	15/13/10
1	24/20/17	23/20/17	22/19/16	21/18/15	20/17/14	19/16/13	19/16/12	18/15/11	17/14/11	16/14/11

Contaminant Severity Factor (CSF)
Sensitivity of Machine to Contaminant Failure

Courtesy Noria Corporation

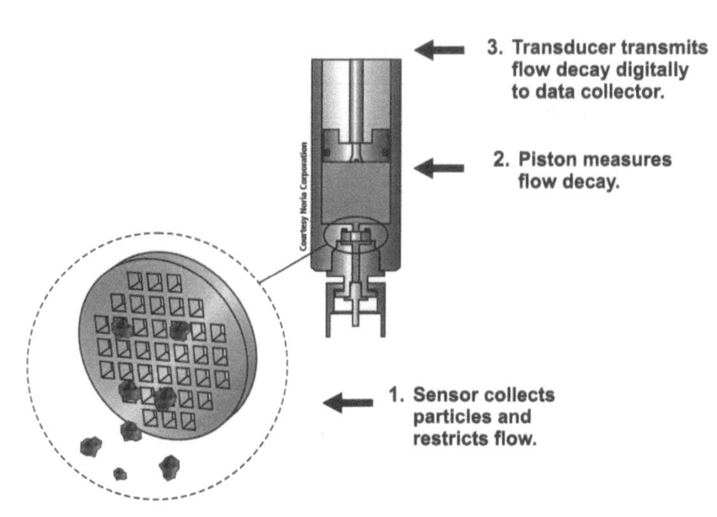

Figure 21-14. Pore blockage-type particle counter.

Moisture Contamination

Moisture is generally referred to as a chemical contaminant when suspended in lubricating oils. Its destructive effects in bearings, gearing, and hydraulic components can reach or exceed that of particle contamination, depending on conditions. Like particles, control must be exercised to minimize water accumulation and resulting destruction to the oil and machine.

Once in the oil, water is in constant search of a stable existence. Unlike oil, the water molecule is polar, which greatly limits its ability to dissolve. Water may cling to hydrophilic metal surfaces or form a thin film around polar solids co-existing in the oil. If a dry air boundary exists, water molecules may simply choose to migrate out of the oil to the far more absorbent air interface. If water molecules are unable to find polar compounds on which to attach, the oil is said to be saturated. Any additional water will create a supersaturated condition causing the far more harmful free and emulsified water. The temperature of the oil as is shown in Table 21-9 also influences the saturation point.

With few exceptions, the chemical and physical stability of lubricants is threatened by small amounts of undissolved suspended water. In combination with oxygen, heat, and metal catalysts, water promotes oxidation and hydrolysis. An overall degradation of the base oil and its additives results. The harmful effects of water on the life of rolling element bearings and other contact zones when boundary lubrication prevails are well documented. According to SKF, "free water in lubricating oil decreases the life of rolling element bearings by ten to more than a hundred times... " And, it is well know that water promotes corrosive attack on sensitive machine surfaces discharging harmful abrasives into the oil.

The omnipresence of water in the environment makes it difficult to completely exclude it from entering and combining with the oil. However, its presence can be greatly minimized and controlled through good maintenance practices. And, just like particle contamination, a proactive maintenance program needs to be established to control water. This should start with the setting of a target dryness level for each different oil application. By

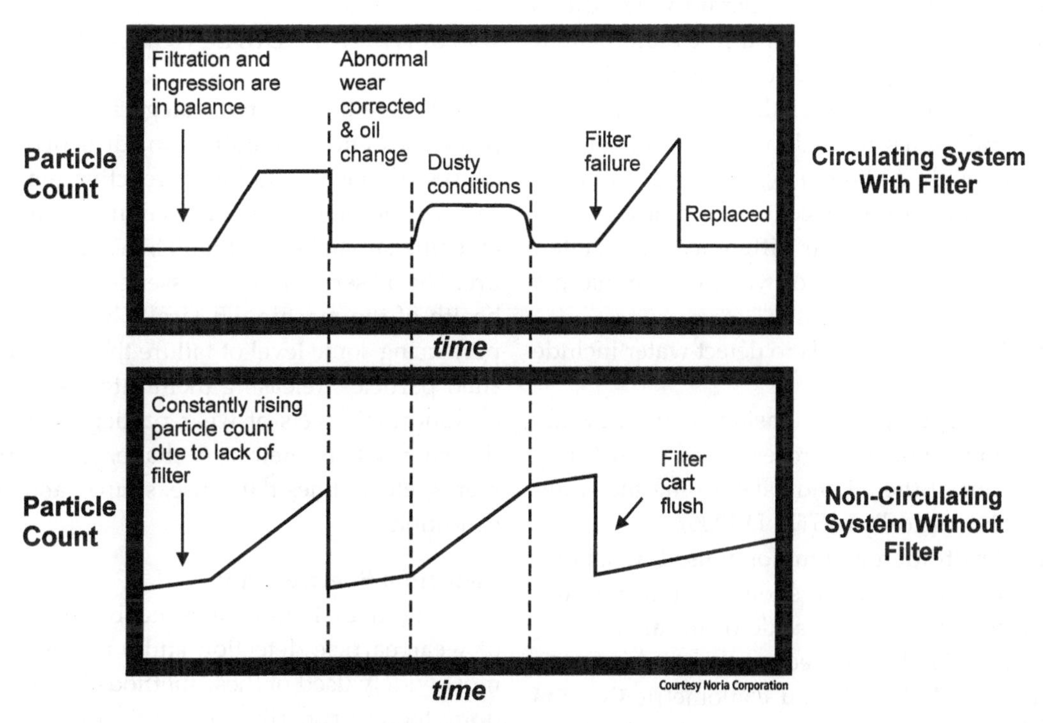

Figure 21-15. Particle count trend graphs.

Table 21-9. Make-up of water concentration in oil at different temperatures. (Courtesy Noria Corporation)

Temperature °F	°C	Oil	100 ppm D	100 ppm F&E	500 ppm D	500 ppm F&E	1000 ppm D	1000 ppm F&E
32	0							
		R&O	28	72	28	472	28	972
		Gear	100	0	100	400	100	900
68	20							
		R&O	72	28	72	428	72	928
		Gear	100	0	200	300	200	800
104	40							
		R&O	100	0	170	330	170	830
		Gear	100	0	500	0	500	500
140	60							
		R&O	100	0	350	150	350	650
		Gear	100	0	500	0	1000	0
158	70							
		R&O	100	0	500	0	520	480
		Gear	100	0	500	0	1000	0

D = Dissolved Water Amount (ppm)
F&E = Free and Emulsified Water (ppm)
R&O = ISO 32 Rust and Oxidation Inhibited oil (e.g., turbine oil).
Gear = Gear oil

investigating the sources of water ingression a plan can be implemented to exclude the water. Occasional removal by water absorbent filters, vacuum dehydrators or air stripping units (see Chapter 19) may also be necessary.

A simple and reliable test for water is the crackle test (a.k.a. the sputter test). In the laboratory two drops of oil are placed on the surface of a hot plate heated to approximately 320 degree F. The presence of free or emulsified water in the oil will result in the formation of vapor bubbles and even scintillation if the water concentration is high enough. Although generally used only as a go/no-go procedure, experienced lab technicians have learned to recognize the visual differences associated with progressive concentrations of water contamination, see Figure 21-16.

Other widely used methods to detect water include the following:

1. Dean & Stark apparatus is occasionally used by laboratories and involves a procedure of co-distilling the water out of the oil and establishing the water content volumetrically (ASTM D 4006).
2. Karl Fischer titration is commonly used by laboratories as an exception test should initial presence of water be detected by crackle or infrared analysis. Two Karl Fischer procedures exist, volumetric titration (ASTM D 1744) and coulometric titration (ASTM D 4928).
3. Infrared spectroscopy can reliably measure water concentrations down to about 0.1 percent. This lower limit may not be adequate for many oil analysis programs.

WEAR PARTICLE DETECTION AND ANALYSIS

Where the first two categories of oil analysis (fluid properties and contamination) deal primarily with the causes of machine failure (proactive maintenance), this category emphasizes the detection and analysis of current machine anomalies and faults, i.e., the symptoms of failure. The oil serves as the messenger of information on the health of the machine. Basically, when a machine is experiencing some level of failure the affected surfaces will shed particles, releasing them into the oil. The presence of abnormal levels of wear particles serves as problem detection where their size, shape, color, orientation, elements, etc. defines the cause, source, and severity of the condition.

Elemental Spectroscopy

Figure 21-17 illustrates the three common categories of wear particle detection and analysis. The oldest and most widely used of these methods is elemental analysis, done today primarily with optical emission spectrome-

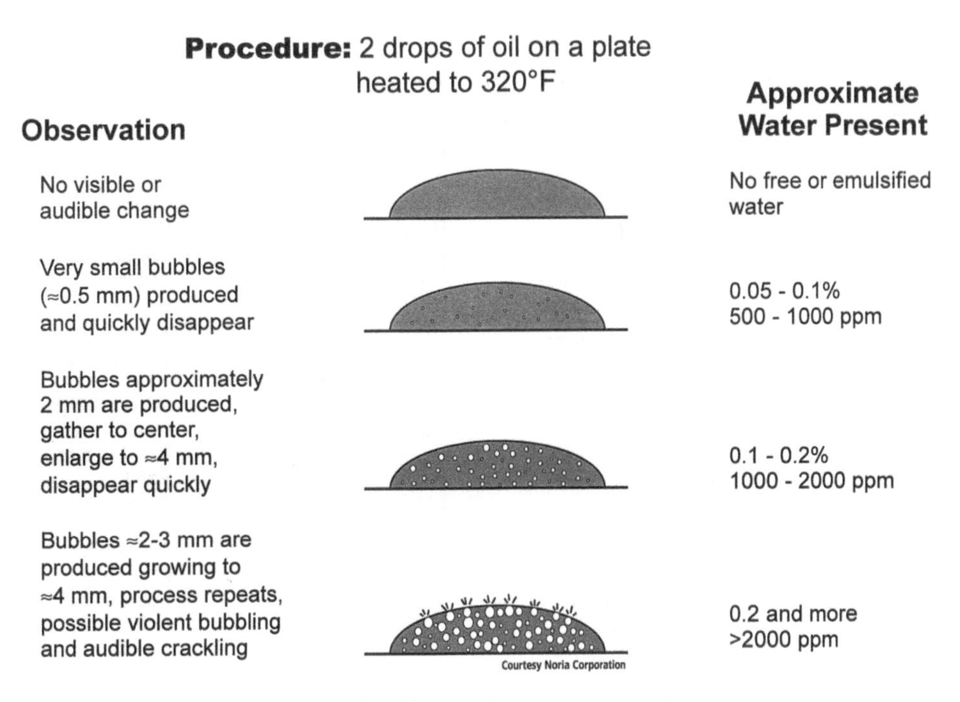

Figure 21-16. Crackle test for water contamination.

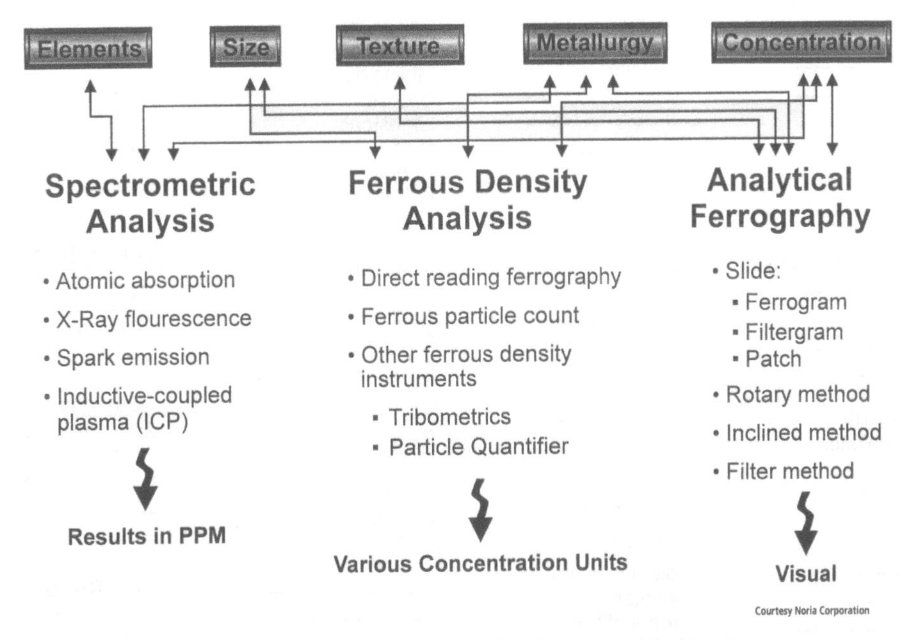

Figure 21-17. Three common categories of wear particle detection.

ters. The procedure involves applying high heat to the oil. Particles in the oil will totally or partially vaporize in the presence of heat, producing incandescent emission of light. The light is diffracted such that spectral intensities at different wavelengths can be measured. Specific wavelengths are associated with certain elements and the spectral intensities define the concentration of the elements.

The typical output from elemental spectroscopy is concentration units (parts per million) across 10 to 25 common elements such as iron, copper, lead, aluminum, etc. By comparing the major, minor, and trace metals to the metallurgical chart of the machine a fingerprint of the probable sources of the wear can be established. Many of the laboratories do wear metal interpretation with the help of sophisticated software programs and extensive metallurgical databases (see Table 21-10).

Most oil analysis laboratories offer elemental spectroscopy as standard with all samples analyzed. Both

Table 21-10. Potential sources of metals in oil.

Iron
Steel
Cast iron
Rust
Wear debris
Mill scale
Ore dust
Fly ash
Paint
Paper mill dust
Asbestos
Talc
Zeolite
Cleaning detergent

Nickel
Alloy of stainless steel
Plating
Stellite (Cobalt-Nickel)
Alloy of hard steels

Silver
Bearing overlay
Solder
Some needle bearings

Potassium
Coolant inhibitor
Fly ash
Paper mill dust
Road dust
Granite

Boron
Coolant inhibitor
EP additive
Oil drum cleaning agent
Boric acid (water treatment)

Chromium
Ring plating
Chrome plating
Paint
Stainless Steel

Silicon
Road dust
Sealant
Antifoam additive
Steel alloy metal
Synthetic lubricant
Wet clutch
Glass mfg
Coolant additive
Foundry dust
Filter fibers (glass)
Fly ash
Slag
Mica
Cement dust
Asbestos
Granite
Limestone
Talc

Lead
Babbitt
Journal bearing overlay
Gasoline additive
Paint
Solder

Sodium
Coolant inhibitor
Saltwater
Some additives
Grease
Base stocks (trace)
Dirt
Road dust
Salt (road salt)
Fly ash
Activated alumina
Paper mill dust

Copper
AW Additive
Bronze
Brass
Bearing cage
Cooler cores
Copper mining
Paint
Babbitt

Molybdenum
EP additive
Alloying metal
Rings

Calcium
Hard water
Salt water
Engine oil additive
Mining dust
Grease
Limestone
Slag
Rubber
Fuller's earth
Lignite
Cement dust
Road dust
Gypsum
Rust Inhibitor
Detergent

Zinc
AW additive
Brass
Plating
Galvanizing
Grease

Phosphorus
AW/EP additive
Surface finish on some gears
Cleaning detergent

Tin
Bearing cage (bronze)
Solder
Babbitt

Aluminum
Road dust
Bearing metal
Paint
Abrasives
Aluminum mill
Coal contaminant
Fly ash
Foundry dust
Activated alumina
Bauxite
Granite
Catalyst

Barium
Engine additive
Grease

Vanandium
Turbine blades
Valves

Titanium
Gas turbine bearings
Paint
Turbine blades

Cadmium
Journal bearings
Plating

Magnesium
Hard water
Engine additive
Turbine metallurgy
Seawater
Fuller's earth
Road dust

Courtesy Noria Corporation

spectrometers and technology vary somewhat, which translates to variations in detection range and sensitivity. The precision of these instruments is also influenced by the size of the wear particles suspended in the oils. During analysis, small particles vaporize more completely while large particles (> 10 microns) are almost not measurable. This particle-size bias leads to occasional errors, some serious (false negatives).

One popular way to reduce the particle-size error is to use electrode filter spectroscopy. This capability is available with spark-emission spectrometers at many of the large commercial laboratories. By pushing the particles into the interstices of the disc electrode a more complete vaporization of larger particles is achieved (possible sensitivity to 20 microns). A special fixture is required to process the sample through the electrode prior to analysis. Because the oil is washed through the electrode during preparation, a separate test is performed on the oil alone to measure dissolved metals and additive elements.

Ferrous Density Analysis

The most serious wear particles of all are generated from iron and steel surfaces. In fact, in most oil-lubricated pairs, at least one of the two surfaces is a ferrous surface. And, it is usually the ferrous surface that is the most important from the standpoint of machine reliability.

This means that the oil analyst requires a dependable reading of the ferrous particle concentration at all sizes, an important issue considering the particle-size bias associated with elemental spectrometers. Therefore, in order to ensure that abnormal wear of iron and steel surfaces doesn't go undetected, ferrous density analyzers are widely employed, both in commercial and on-site laboratories. These instruments provide a first line of defense by reliably detecting free-metal ferrous debris. Example instruments include:

1. Direct Reading Ferrograph: reports results in Wear Particle Concentration units
2. Particle Quantifier: reports an index scale
3. Wear Particle Analyzer: output in micrograms/ml
4. Ferrous Particle Counter: assigns a percent ferrous to particle count ratio

Analytical Ferrography

Elemental spectroscopy and ferrous density analysis are just two of many different ways to detect problems in machinery. Thermography and vibration monitoring are also effective at detecting specific faults and modes of failure. Once there is an initial indication of a fault by any of these methods, the process must continue to:

1. Isolate it to a single component
2. Identify the cause
3. Assess how severe or threatening the condition is, and, finally
4. Determine the appropriate corrective action

When problems are detected and analyzed early they can often be arrested without downtime or expensive repair. In fact, root causes of the most common problems are usually correctable on the run. The key is the timing of the detection. An important part of timing is a regimen of frequent sampling.

Successful analysis of a current wear-related problem requires many pieces of information and a skilled diagnostician. To this end, the practice of analytical ferrography has received recently prominence. Unlike other common instrumentation technologies, analytical ferrography is qualitative and requires visual examination and identification of wear particles. Numerous properties and features of the wear debris are inventoried and categorized. These include size, shape, texture, edge detail, color, light effects, heat treatment effects, apparent density, magnetism, concentration, and surface oxides.

This information is combined with other information obtained by particle counting, ferrous density analysis and elemental spectroscopy in defining a response to items 1-4 above. Figure 21-18 presents a general overview of the combined detection and analysis process. Here are analytical ferrography is represented by microscopic analysis. Two methods are commonly used to prepare the particles for viewing by the microscope.

If a high level of ferromagnetic debris is detected by ferrous density analysis then a ferrogram is typically prepared. The process involves slowly passing solvent-diluted oil down the surface of an inclined glass slide. The instrument involved is called a ferrogram maker. Beneath the slide is a strong magnet. Ferromagnetic particles become quickly pinned down onto the slide and oriented to the vector lines of the magnetic field.

Non-magnetic debris deposit gravimetrically in random fashion, although larger and heavier particles settle first. Approximately 50% of the non-magnetic particles wash down the slide and do not deposit. A ferrogram of wear is shown in Figure 21-19.

In those cases where low levels of ferromagnetic particles are detected but high nonferrous debris is found (by a particle counter or elemental analysis) a filtergram is preferred. Unlike the ferrogram, the filtergram does not use a magnet and therefore all particles are randomly deposited regardless of size, weight, or magnetic attraction. This is accomplished by passing an exact quantity of sol-

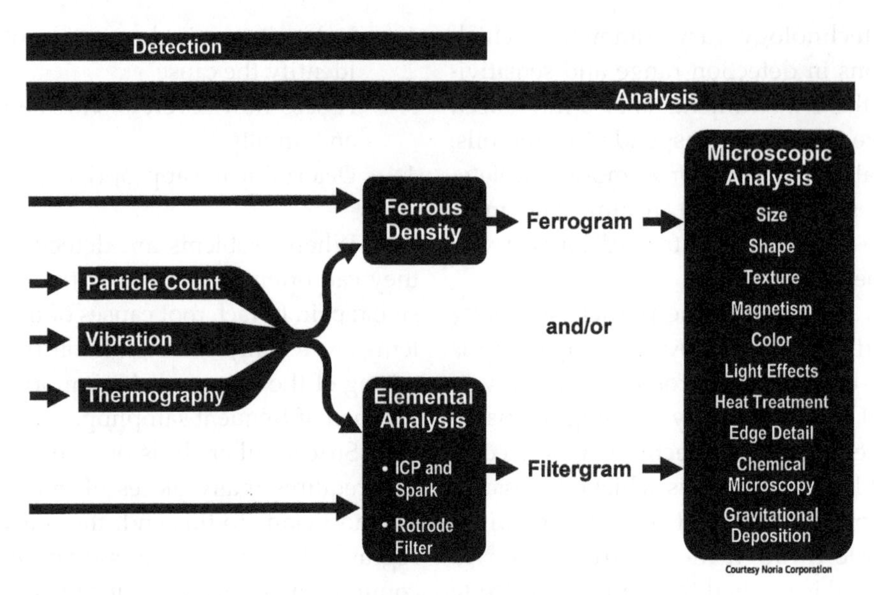

Figure 21-18. Combined detection and analysis process.

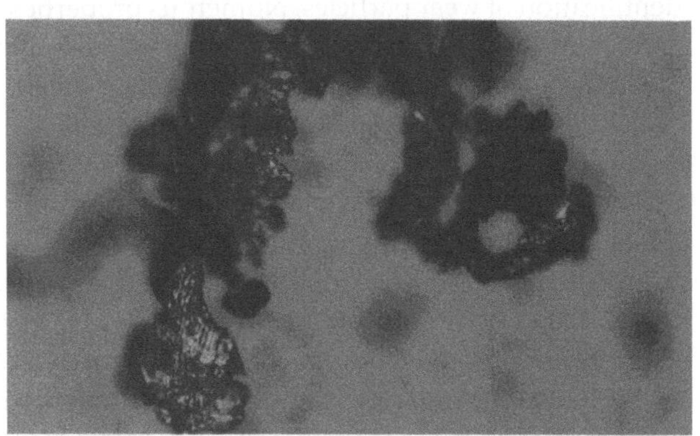

vent-diluted oil through a membrane of about three-micron pore size. No particles are hidden from observation except those too small to be retained by the membrane. The single disadvantage of the filtergram is the difficulty of distinguishing ferrous from non-ferrous debris. The skillful eye of an experienced technician can usually overcome this drawback.

INTERPRETING TEST RESULTS

Most machines are highly complex, consisting of exotic metallurgy and intricate mechanisms. The numerous frictional and sealing surfaces usually employ varying contact dynamics and loads, all sharing a common lubricant. Failure to gain knowledge of the many internal machine details and using them as a reference base for interpreting data may lead to confusion and indecision in response to oil analysis results. A good approach is to build a three-ring binder with index tabs for each machine type.

Include in this binder photocopied pages from the service and operation manuals plus other accumulated information. The following are examples of data and information to include:

1. Identify types of bearings in use and their metallurgy

Figure 21-19. Pictured at right are examples of particles identified by analytical ferrography: cast iron, dense ferrous and products of corrosion magnified to 1,000 times normal size.

2. Identify input and output shaft speeds and torques
3. Identify type of gears in use, speeds, and loads. Determine gear metal hardness, surface treatments, alloying metals
4. Locate and identify all other frictional surfaces, such as cams, pistons, bushings, swash-plates, etc. Determine metallurgy and surface treatments
5. Locate and identify coolers, heat exchangers and type of fluids used
6. Obtain fluid flow circuit diagrams/schematics
7. Locate and determine the types of seals in use, both external and internal
8. Identify possible contacts with process chemicals
9. Record lubricant flow rates, lubricant bulk oil temperatures, bearing drain and inlet temperatures, and oil pressures
10. Record detailed lubricant specification and compartment capacity
11. Record filter performance specification and location

In many cases oil analysis data can be inconclusive when used alone. However, combined with sensory inspection information, a reliable and more exact determination is possible. Likewise, the application of companion maintenance technologies (like vibration and thermography) can help support a conclusion prior to expensive machine tear-down or repair. Table 21-11 represents a two-page summary of combined analytical and inspection/sensory indications of frequently encountered problems.

SIMPLE STEPS TO IMPLEMENTING AN INDUSTRIAL OIL ANALYSIS PROGRAM

When doctors need to know more regarding the condition of your health, they are likely to start with a series of primary diagnostic blood tests. Similarly, oil analysis, sometimes referred to as "wear particle analysis", is a mature condition based / predictive maintenance approach used to determine the lubricated machine and it's lubricating oil current condition by testing a small sample of oil from the machine's operational lubricating system. The test laboratory then compares its working sample to a virgin stock sample through a series of laboratory tests, and secondly to previous (history) sample results from the same sample point (predictive trending) to determine the current "wellness" of both machine and oil.

Oil analysis is a highly effective and inexpensive method used to determine when to change oil based on its condition, to predict incipient bearing failure so that relevant action can be taken in a timely manner to avert failure and to diagnose bearing failure should it occur. Despite being around since the 1940's and its proven track record for effectiveness, oil analysis is still misunderstood and overlooked as a proactive / predictive maintenance strategy in many of today's industrial plants! Setting up and implementing an industrial oil analysis program is relatively easy and should be an included strategy in any industrial plant that purchases, stores, dispenses, changes, uses, or recycles lubricants as part of its manufacturing or maintenance process.

A successful oil analysis program can pay for itself in a matter of weeks as it delivers benefits in multiple ways that include:

- Oil change intervals (often extended) optimized to the machine's ambient conditions and operational use requirements,
- Probable reduction in lubricant inventory purchase costs and spent lubricant disposal costs,
- An understanding of how bearings can, or are failing in their operating environment so that bearing failure can be controlled or eliminated,
- Increased asset reliability, availability, and production throughput.

Note: program success is better assured if a work management approach is already in place that can assure completion of corrective action in a timely manner whenever the oil analysis report recommends such action.

Similar to all successful change management programs that must be rolled out across the organization an oil analysis program will benefit from a piloted phased implementation. Taking a stepped approach allows management and workforce alike to become accustomed to the new sampling and reporting processes and to quickly iron out any problems prior to rolling out the program en masse.

Step One: Appoint a program champion

All programs require a "go to" decision making person who advocates on the program behalf and is committed to making the program implementation a success; the champion should be at a supervisor or manager level.

Step Two: Choosing a suitable pilot area/machine

Oil analysis begins with sampling the oil and can include both lubricating and hydraulic fluids. Choosing a suitable program pilot will depend on the type of industry and business operation. Typical starting points to look at can include:

Table 21-11. Oil analysis data interpretation and problem identification.

Problem Area	Analytical Indications[a]	Inspection/Sensory Indications[b]
Wrong Lubricant	• Change in viscosity, VI, flash point, additive elements, FTIR[b] specta, TAN[d]/TBN[d] • Change in wear patterns	• Change in oil gage or bearing temperature • Bearing distress or noise • Hard turning of shaft
Antioxidant Depletion	• Decreasing TAN[d], RBOT oxidation life, and Zn/P content • Increasing viscosity, TAN[d], particle count • FTIR: decreasing antioxidant, increasing oxidation, sulphation, and/or nitration	• Oil darkening • Pungent odor • Hot running
Dispersancy Failure	• FTIR,[b] low TBN[d] • Increasing particle count, pentane insolubles • Defined inner spot on blotter test	• Filter inspection: sludge on media, filter in bypass • Black exhaust smoke • Deposits on rings and valves
Base Oil Deterioration	• Increasing viscosity, TAN[d], particle count, and/or ferrous particles • Decreasing TBN[d] • Change in VI and lower dielectric strength	• Poor oil/water separability • Air entrainment/foaming • Pungent odor, sludge/varnish formation • Blotter spot yellow/brown, oil darkening
Water Contamination	• Increasing viscosity, TAN[d], Ca, Ma, and/or Na • Rapid additive depletion/failure • Crackle test, VISA[e], KF[f], FTIR[b] • Reduced dielectic strength • Blotter test: sharp or star burst periphery on inner spot	• Oil clouding/opacity, water puddling/ separating, sludging, foaming • Evidence of fretting wear/corrosion • Filter: paper is wavy, high pressure drop, short life. Ferrogram shows rust. • Valve stiction, orifice silting, bearing distress/failure, noisy pump/bearings
Coolant Contamination	• Increasing viscosity, copper, particle count, wear metal, Na, B, and/or K • FTIR[b]: glycol • Crackle test, VISA[e], KF[f]	• Bearings dark charcoal color, distressed • Dispersancy failure, sludging, varnishing • Blotter test: sticky, black center • Filter plugs prematurely, oil has mayonnaise consistency, white exhaust smoke
Fuel Dilution	• Low oil viscosity, flash point • Additive and wear metal dilution (elemental analysis) • FTIR[b]/gas chromatography for fuel • Rising particle count and wear metals	• Rising oil levels and oil gage temperatures • Blotter test: halo around center spot • Blue exhaust smoke (collapsed rings), plugged air filter, defective injectors • Oil has diesel odor, overfueling conditions

(a) Not all of the identified indications would be expected for each problem area. (b) Fourier Transform Infrared Spectroscopy (c) Total Acid Number (d) Total Base Number (e) Vapor-Induced Scintillation Analysis (f) Karl Fischer

Table 21-11. (Continued)

Problem Area	Analytical Indications [a]	Inspection/Sensory Indications[a]
Air Entrainment	• Increased viscosity, TAN[d], water, and/or FTIR[b] for oxidation • Silicon defoamant levels too high/low • Blotter test: coke-like carbon on patch	• Oil clouding/foaming, increase in oil gage temp. • Spongy/slow hydraulics, cavitation of pump/bearing, noisy operation
Abrasive Wear Conditions	• Increased silicon, aluminum, particle count and/or ferrous particles • Water contamination • Ferrogram has cutting wear, silica particles	• Scratch marking or/polishing of frictional surfaces • Cutting wear on blotter/patch/filter media • Filter/breather/seal failure
Corrosive Wear Conditions	• Increased TAN[d], particle count, spectrographic iron & bearing metals, water • Decrease in TBN[d] • Ferrogram shows submicron debris at ferrogram tail, rust particles, metal oxides	• Fretting, pitting and etching on contact surfaces • Transient electric currents, high engine blowby • Rust on patch or filter media
Failed Filter	• Increasing silicon/aluminum, particle count, ferrous particles, and/or elemental iron • Ferrograms show green looking particles, cutting wear, filter fibers	• Valve silt lock, noisy bearings • Unchanging or high delta P of filter • Frequent bearing failures, high levels of bottom sediment
Overheating	• Increasing ferrous particles, particle count, flash point, viscosity, or oil specific gravity • Ferrograms show friction polymers, oxides, bluing/tempering of particles, sliding wear particles, bearing particles, e.g., babbitt	• Bearing distress/failure • Hot spots and high bearing metal temp. • Evidence of coking/sludge • Burnt/rancid odor, high oil gage temp.
Misalignment, Imbalance, Overloading	• Ferrograms densely loaded with black-iron oxides, dark metallo oxides, severe cutting and sliding wear, tempered particles, large, chunky particles, or bearing metals • Increase in viscosity, TAN[d], particle count, and/or ferrous particles • Depletion of Zn and P	• Engine lugging/stalling, black exhaust • Raised oil, bearing metal, or jacket-water temperature • Dark, foul smelling oil, bearing distress/failure, hard turning of shaft • Abnormal vibration, noise • Blotter test: coke, metal chips • Metal chips on filter , highly loaded chip detectors
Impending Failure of Bearing, Gear, Pump, etc.	• Exponential increase in particle count and number of wear particle concentration • Increase in iron or bearing metals • Ferrogram shows rate increase in spheres, dark metallo oxides, particle bluing, spalling/chunks, severe sliding/galling particles, cutting wear	• Shaft wobble, vibration, acoustic changes, blue exhaust smoke, hot spots, hard turning shaft and/or high bearing metal temperatures • Patch/blotter shows coking

- Critical product, process, line, or major piece of equipment. *Criticality determined by constraint and/or lack of back up, downtime costs, product quality, etc.*
- Mechanical equipment with moving components that include lubricant reservoirs for recirculating oil transmission systems, mechanical and / or hydraulic in design.

Step Three: The lubricant audit:

A lubricant audit is required to identify what lubricants are currently employed in service at the plant and will require the following checks:

- Check work order system PM job plans for lubricant specification(s),
- Check on or near the lubricant reservoir for lubricant identification labels or stickers,
- Check for matching MSDS (Material Safety Data Sheets).

Note: If a Lubrication Operation Effectiveness Review (LOER) or Lubricant consolidation program has already taken place (as described in Chapter 18: Lubricant Purchase, Handling & Storage) refer to lubricant listing charts.

If a discrepancy is found at this stage, outside assistance from a lubrication expert or lubricant supplier may be required to identify lubricants in use are the correct lubricants specified for the application.

Step Four: Choose a Laboratory

Not all oil analysis laboratories are created equal, making your choice of laboratory an important step. Most oil analysis reports are divided into three major sections that report on 1), the sampling and virgin oil specification data 2), the spectral analysis testing results for wear elements identified as lubricant additives or contaminants 3), additional physical test results for viscosity, water, glycol, fuel, soot, acidity, etc. and 4), the resulting conclusions and recommendations.

Some laboratories specialize in engine oil testing that focus more heavily on the physical tests for water, glycol, fuel and soot; some laboratories will specialize in industrial sample testing that focus more heavily on the wear particle testing and some physical testing for viscosity, water and acidity and post mortem testing for root failure causes using ferrographic testing. In addition, there are also laboratories that have lab technicians specialized in both areas of testing.

Interview and tour a number of different laboratories to understand their specialty area and how they will look after and process your samples. Key to any testing program is receiving results in a timely and consistent manner, especially when critical equipment is involved. When interviewing laboratories be sure to rate their sample "turnaround" time and how they can assure testing consistency (usually through use of dedicated technicians testing your samples). Working with a laboratory should be viewed as a long-term relationship as the laboratory builds and analyzes your complete data history and makes conclusions and recommendations based not only on your current sample versus its virgin sample counterpart, but also on an understanding of your plant ambient conditions and overall trending history of each sample.

Step Five: Setting up the pilot sampling program

A good laboratory will work with you to set up your sampling program, sell or attain sampling point hardware, extraction pumps, quality sample bottles and train your staff to take consistently "clean" oil samples.

The best oil samples contain maximum data density with minimum data disturbance, meaning the sample should best represent the oil condition and it's particulate levels as it flows through the system, or as it sits in the reservoir. For example, extracting a sample from the bottom of a reservoir in a non-pressurized gearbox lubrication system the particulate fall out will be dense due to large wear particles and / or sludge accumulation and not correctly represent the remaining 80% - 90% of reservoir lubricant that performs the actual lubrication of the gears.

In pressurized recirculating lubrication systems samples are best taken when the machine is running and at operating temperature from a "live" fluid zone where the lubricant is freely flowing. Whenever possible, the sample should be extracted from an elbow taking advantage of the data density caused by the fluid turbulence. Sample points are best located downstream of the lubricated areas to catch any wear elements before they are filtered out by inline pressure or gravity filters. The following sections in this chapter will detail the where and how of sampling techniques.

You will be required at this time to sample and send off virgin oil samples of all lubricants to be checked in the pilot program. These samples are used as a benchmark for the laboratory to measure and understand what additive ingredients and lubricant state represents a normal status so that additive depletion and wear elements can be easily identified in operating samples.

Outside assistance / training from a lubrication expert or oil analysis laboratory is advisable when setting up the pilot sample points.

Step Six: Setting up your work management approach to sampling

Lubricant sampling must be performed in a consistent manner on a frequent basis lending itself well to work management through a maintenance/asset management work order system. Utilizing the written sampling procedure as a job plan, physical sampling can be scheduled effectively through the Preventive Maintenance scheduling software.

Extracting and sending a sample to the laboratory is only the first half of the oil analysis process, someone (usually the Planner, if one exists) has to receive the results electronically by email, read the recommendations and take any necessary corrective action and/or file the laboratory report electronically to history—usually as an attachment to the PM sampling work order. This will require a workflow procedure to be developed and trained to all maintenance staff involved in the program.

Step Seven: Commence sampling and your program roll-out

An oil analysis program will immediately identify major problems with contamination and wear in the first sample set. Sample trending can begin on the third set of samples with which the user can start to identify/predict any negative trend toward potential failure and schedule corrective action before failure occurs. Ideally, a pilot program should be allowed to run for approximately three months or more to show basic results before tweaking the process and rolling out to the next area within the plant.

Once your program is working and providing results, a larger corporation may wish to investigate investing in an in-house staffed laboratory that will deliver faster results turnaround.

IMPORTANCE OF TRAINING

When a well-intentioned oil analysis program fails to produce the expected benefits it is often thought that a main contributing factor is an attitude of indifference among those involved. While this is occasionally true, the problem is generally much more fundamental and deep-rooted. Unless maintenance professionals have an understanding of the purpose and goals of oil analysis and are literate in the language of oil analysis, they cannot be expected to carry out its mission.

This mission is accomplished through a liberal amount of training and education. However, the effort should not simply be concentrated on a single individual but should involve all those that benefit from and contribute to machine reliability. Instead, training and education should be directed at several different functions including craftsmen, operators, engineering, and management. Below are a few subjects for which seminars and training classes are generally available:

1. Lubrication fundamentals and their application
2. Mechanical failure analysis
3. Proactive maintenance and root cause
4. Analysis and toubleshooting of hydraulic systems
5. Lubrication and maintenance of bearings and gear units
6. Oil analysis fundamentals
7. Oil analysis data interpretation
8. Filtration and contamination control
9. Wear particle analysis and machine fault detection

Once these fundamentals are in place, oil analysis can move forward enthusiastically, beginning with the development of its mission and goals. And, instead of indifference to oil analysis exceptions, rapid-fire corrections are carried out and measures are taken to preempt their recurrence. In time, unscheduled maintenance is rare and oil analysis exceptions become fewer as the machine operating environment becomes more controlled.

Finally, as the many elements of oil analysis and proactive maintenance merge together into a cohesive maintenance activity, the benefits should not be allowed to go unnoticed. Unlike many applications of new technology, proactive maintenance seeks non-events as its goal and reward. These non-events include oil that continues to be fit-for-service, machines that don't break down, and inspections that don't have to be performed. This quiet existence is the product of a highly disciplined activity but, let us remember, the activity risks being thought of by the casual observer as unneeded. Therefore, the plant's expenditures for proactive maintenance and the highly attractive benefits of proactive maintenance must be measured, monitored, and the outcome displayed for all to view.

Bibliography

Fitch, J.C., Jaggernauth, Simeon, "Moisture…the Second Most Destructive Lubricant Contaminant, and its Effects on Bearing Life," *P/PM Magazine*, Dec. 1994

Fitch, J.C., "Clean Oil Sampling," *Practicing Oil Analysis Magazine*, July 1998

Fitch, J.C., "The Ten Most Common Reasons Why Oil Analysis Programs Fail and the Strategies That Effectively Overcome Them," *Diagnetics*, 1995

Chapter 22

Implementing a Quality Lubrication Management Program

Since the 1990s there has been much discussion surrounding terms such as "World Class," "Best in Class," "Best of Breed," and more recently "Best Practice," all similarly related to the performance of a corporation or maintenance department. It can be argued that these terms are primarily self-appointed marketing terms that are difficult to substantiate due to the lack of qualifying body, registry, or standard that audits under these terms.

In recent past, lubrication management may have come under partial review through the implementation of a program to the ISO 9000 quality assurance standard or the ISO 14000 environmental management standard—both world standards—or the more recent IAM's (Institute of Asset Management) United Kingdom based PAS 55 (Publically Available Specification #55) maintenance management standard audited by the British Standards Institute.

Fortunately as of 2014, the world of asset management—of which lubrication management is a major part—has been recognized with its very own ISO (International Organization for Standardization) world standard with the introduction of the ISO 55001—Asset Management standard (Interestingly, PAS55 was used as the starting model for the broader ISO 55001 standard). Furthermore, the field of lubrication now has three major certification bodies dedicated specifically to the training and certification of lubrication professionals, these being ICML—International Council for Machinery Lubrication, STLE—Society of Tribologists and Lubrication Engineers, and ISO itself whose certification programs are detailed later in this chapter.

DEVELOPING A LUBRICATION MANAGEMENT PROGRAM TO THE ISO 55001 ASSET MANAGEMENT STANDARD

ISO 55001 views an asset as "an item, thing or entity that has potential or actual value to an organization." It views asset management as something that "involves the balancing of costs, opportunities and risks against the desired performance of assets, to achieve the organizational objectives."

Traditionally, lubrication has always been defined as an act of performing lubrication, or as physical sub-system of a machine consisting of a reservoir, pump, metering valves, lubes lines and control belonging to a parental asset. Under ISO 55001 the definition takes on new meaning when associated with life cycle management of the asset. For example: maintenance and set up of the lubricant delivery system; procurement, storage, and handling of lubricants; analysis and reporting of lubrication related failures; competence and training of persons performing and managing the lubrication program; lubrication job plans, SOPs (standard operating procedures) and work flow will all come under the ISO audit microscope if they affect the operation and life cycle of equipment assets—as they should in any lubrication management program. Typical subject areas/activities pertaining to lubrication that ISO would express interest in are:

- Data management
- Condition monitoring
- Risk management
- Quality management
- Environmental management
- Life cycle costing
- Sustainable development
- Non-destructive testing
- Inspection
- Value management
- Shock and vibration
- Quality and assessment of personnel
- Equipment management
- Energy management
- Financial management

Figure 22-1 depicts how a typical plant can benefit from using the ISO standard framework as a foundation for developing a "best practice" environment.

Best Practice Deliverable	ISO 55000 in Practice
Strategic Asset Management Plan • Accountability • Increased availability • Increased reliability • Increased throughput • Enhanced regulatory compliance • Due diligence	• Corporate objective alignment • Blueprint for the development of tactical and operational processes and procedures • Management action plan for improvement • Focus on critical equipment
Asset Life Cycle Management • Increased availability • Increased reliability • Increased throughput • Enhanced regulatory compliance • Due diligence	• Controlled planning & scheduling • Predictive and Condition based maintenance emphasis • Increased measurement and data trending analytics • Failure analytics
Value Based Asset Management • Increased availability • Increased reliability • Increased throughput • Enhanced regulatory compliance • Due diligence	• Enhanced diagnostics • PM based on failure consequence and asset/component criticality • Planned run to failure • Visual/instrument Go / No Go condition checking
Materials Management • Increased availability • Increased reliability • Increased throughput • Enhanced regulatory compliance	• Demand forecasting • Rationalized reorder levels and lot sizes • Vendor managed inventories • Kitting and staging • Validated inventory
Document and Data Management • Enhanced regulatory compliance • Due diligence	• Consistent and accurate data • Retrievable data • Integrated data systems • Data analytics
Improved Maintenance and Production Staff Skill Sets • Safety • Due diligence	• Technology enhancement o Communication devices o Diagnostic / Measurement instruments • Visual/instrument Go / No Go condition checking
Safety • Enhanced regulatory compliance • Due diligence	• Improved work instruction sets • Workflow, process and procedures • Enhanced diagnostics and measurement
Sustainability • Energy reduction • Carbon footprint reduction • Emission reduction	• Enhanced lubrication practice • Enhanced diagnostics and measurement • Analytics • Asset alignment and set up

Figure 22-1. The ISO 55000 Typical Plant Organization

Interpreting the Standard

There are many practical reasons for adopting the ISO 55001 standard as a framework for continuous improvement, arguably none more important than making sense of maintenance data. Many existing management systems and data repositories are choked full of Meaningless Unrelated Data (MUD) making it difficult to mine the data and turn it into meaningful information that allows management to make confident and effective decisions regarding the management of people and assets. If we lose confidence in our data it is of little value or relevance for historical or due diligence purposes, which means we must address how we collect, compile and use that data. It also means we must assess and address the validity and value of our current asset management and maintenance activities, which include the use of resources, materials, tools, instruments, data, processes, procedures, plans, methods and activities and formulate an asset management approach that allows us to mitigate risk and meet our corporate objectives.

LUBRICATION PROGRAM OBJECTIVES

When an engineer takes on a machine design project in which moving parts are involved, the engineer is expected to choose the appropriate bearings. Optimal bearing choice is dependent on many factors that include:

- Budget
- Application (radial, axial, and planar)
- Load
- Speed
- Space
- Clearance and fit
- Length of machine warranty
- Bearing reliability specification
- Lubrication entry design and method
- Expected operating conditions
- Expected ambient conditions

When a machine goes into service, the reliability baton is passed on to the end user's maintenance and reliability group who must work with the engineer's final design and an all too often-vague machine operations and maintenance (O & M) instruction manual that rarely spells out good lubrication requirements/instructions.

Many maintainers will recognize the catchall work order instruction "lubricate as necessary," placing full responsibility on the end user to singularly develop a lubrication strategy for each bearing that is suitable for the conditions in which the machine is expected to perform. In the unlikely event the maintenance group literally starves the bearing of lubricant in the first year, most bearings will surpass their warranty period and the manufacturer will rarely get to learn if their design was/is truly robust and reliable.

When design conditions are set, machine availability and reliability is dependent on how well the maintenance group understands 1) the current bearing design and its lubrication requirements, 2) how the machine bearings will fail in their working environment, and 3) their own ability to design, implement and execute a relevant asset/bearing lubrication management program.

What's in a Bearing?

When we think of a bearing, many shapes, sizes, and materials come to mind. A bearing can take on many forms to support a sliding or rotating part.

For example, sliding contact bearings, commonly known as plain, sleeve, or journal bearings allow full sliding contact between mating surfaces in three specific ways: *radially*, in which the bearing provides a 360°support for a rotating shaft or journal; *axially*, in which the bearing supports any side thrust load from the end of the rotating shaft; and *planar*, in which a flat bearing surface acts like a slipper to guide moving parts in a straight line. To help carry static and kinetic loads with minimized friction and wear, lubricant is introduced and channeled into the engineered clearance between the mating surfaces to generate a hydrostatic or hydrodynamic fluid full film separation of the surfaces, allowing them to slide freely over one another. Sliding bearings are most commonly manufactured in yellow metal or composites of brass, bronze, and copper, the bearing material designed to be softer (and therefore sacrificial) than the supported component.

Rolling contact bearings, commonly known as ball bearings, roller bearings, and needle bearings, provide a rolling contact that supports both radial and axial thrust—often simultaneously. These bearings were initially marketed as "anti-friction" bearings due to their minimal point contact area, where the lightly lubricated rolling elements (balls rollers and needles) carry the static and kinetic load with the help of an elastohydrodynamic lubricant film.

Rolling-element bearing manufacturers measure the reliability of their bearings using a load-life calculation rating known as its L^{10} rating life, defined in the ISO 281:2007 standard as the expected operating hour life that 90% of that bearing design will achieve in operating conditions. To achieve its reliability/life design rating, the manufacturer assumes the bearing will be operated within its load and speed design limits in a clean operating environment and that an adequate lubricant film of the correct viscosity is applied on a regular basis (adequate described as a film equal to, or greater than, the composite roughness of the two mating surfaces).

Combating Bearing Failure

In a recent 2014 "State of the Nation's Lubrication Practices" study conducted by ATP network's *Maintenance Technology* magazine, found that "60% of respondents did not track or report their lubrication related machine failures or downtime occurrences."

Quite simply, knowledge and awareness of how

bearings fail in an operating environment is the first step to implementing any lubrication management program! Thirty-five years of empirical bearing failure data analysis has shown that failure can occur due to one or a combination of the following top ten conditions/causes (not in any order):

1. Lack of lubrication training,
2. Lack of lubrication application engineering,
3. Lack of cleanliness—use of dirty or contaminated lubricant transfer equipment and/or lubricants
4. Bearing is over lubricated
5. Bearing is under lubricated
6. Bearing is dirt or water contaminated by manufacturing or asset cleaning process
7. Infrequent oil/filter changes
8. Bearing lubricant is cross contaminated with an incompatible lubricant
9. Bearing is lubricated with the incorrect lubricant
10. Bearing is mounted out of square or misaligned when set up

Nine of the top 10 bearing failures are due directly or indirectly to ineffective lubrication practices! Implementing a root cause analysis of failure (RCAF) program to weed out the real reasons your bearings are failing will give immediate direction and focus for your lubrication management program.

Regular PM/Operator Maintenance

Daily Lubrication system checks are essential for ensuring the lubrication system is operating as designed and that there is lubricant in the system. This is often best performed on a daily basis by the equipment operator who visually checks the entire system in a quick system walk-around and only notifies the maintenance department when an exception is found. Check functions can include:

- Check reservoir fill level—is the level between the Lo and HI mark on the reservoirs (some systems use a RAG—Red/Amber/Green indicator system to denote when to fill the reservoir—see figure 22-2: RAG indicator on a grease reservoir The photo shows three colored tapes around the reservoir; the green line denotes the high fill line, amber denotes the reservoir requires filling but has a lubricant reserve good for X (user determined) number of days, and a red line to denote the lubricant reserve is only good for Y hours r days (user determined) before the lubricant runs dry.
- Check for and notify immediately of any system leaks.
- Check for apparent system damage including line crush and any overpressure indicator signal denoting backpressure in the system caused by a damaged or blocked bearing or line.
- Check for controller warning signals/lights and report immediately.
- Check any pressure filter (used in recirculating oil and hydraulic systems) is not showing a red flag signal indicating the filter is full and in bypass mode.

Of course, the type of lubrication system and lubricant will dictate the level of checking required. For example, recirculating oil systems are prime candidates for oil analysis allowing the lubricant to be changed only when its condition requires it.

LUBRICATION TECHNICIAN CERTIFICATION

A major success component of any best practice program is having trained staff who understand the impact of their actions on asset reliability. The past 20 plus years have witnessed a massive shift in the mainte-

Figure 22-2. RAG indicator on a grease reservoir. Courtesy ENGTECH Industries Inc.

nance industry toward a total asset reliability approach. More importantly, this revised approach has embraced and recognized the role and impact Good Lubrication Practices (GLP) now plays in achieving and sustaining asset reliability and lifecycle stewardship. Ironically, lubrication management is a discipline and subject matter known to directly, and indirectly affects up to 70% of all mechanical failures, yet there still is no lubrication specific trade, and likely never will be as vocational schools and apprenticeship programs in their current format continue to shrink and disappear. To meet this challenge many of the world's leading lubrication experts and organizations that include scientists, consultants, practitioners, and suppliers have supported and worked diligently on three fronts to develop industry recognized world wide lubrication certification programs.

These certification programs have been painstakingly designed to certify qualified individuals to meet and surpass the escalating reliability demands of industry in the proactive field of lubrication application, lubricant selection, and lubricant analysis.

Why Certify?

As an individual, lubrication certification is about telling the world that you have worked diligently to study and assimilate lubrication knowledge from an accredited body and domain of knowledge, and have demonstrated that learning by successfully passing a closed book exam that draws questions from that entire knowledge base. In addition, you have also had to meet numerous training and practical requirements to earn the right to take the exam.

ISO 17204, a world standard on the certification process, likens certification to a level of competency that "demonstrates [an individual's] ability to apply knowledge, skills, and attributes" in a chosen field.

Individuals embark on the certification process for numerous reasons:

- Personal qualification to set themselves apart from the many others they may have to compete with for an internal or external job posting in the maintenance reliability and lubrication fields. Many potential employees offer substantial pay grade increases for certified personnel. The US lubrication magazine "Lubes & Greases" recently reported that in a lubricant sales representative salary survey, certified personnel earned on average $30,000 more than their non-certified counterparts in the industry!
- The certification may be a current employment requirement. Many lubrication service providers that include lubricant sales, lubrication systems engineering and sales, and oil analysis laboratories now demand their employees be appropriately certified as a condition of sustained employment, or as a basis for hiring and / or promotion.
- Industry maintenance and reliability departments are rapidly recognizing the value of lubrication certified individuals as pivotal members of their reliability team to facilitate and manage proactive approaches and initiatives designed to manage and eliminate preventable mechanical failures.
- Corporations are now turning to certification programs to meet prescribed workforce competence levels while simultaneously raising their level of due diligence.

Whatever the driving factor, the advent of the lubrication professional here, and is reflected in the ever-accelerating number of individuals certifying in a variety of professional lubrication disciplines.

Certification Choices

Almost every professional group now has a certification process; doctors, lawyers, engineers, and accountants have long had to go through a board certification process to achieve professional status and practice their profession. Similarly, the maintenance industry has in the recent past chosen to recognize its own through a variety of general certification programs. Canada took the lead in the 1990s through the PEMAC (Plant Engineering and Maintenance Association of Canada) organization with its Certified Maintenance Manager—CMM designation designed to be a national maintenance management certification and delivered through the Canadian provincial technical college system. The Society of Maintenance and Reliability Professionals, a US based international organization, soon followed suit in 2001 with its CMRP (Certified Maintenance Reliability Professional) certification designation.

In the specific field of lubrication the US based Society of Tribologists and Lubrication Engineers—STLE, has continued to offer its Certified Lubrication Specialist (CLS) certification—originally designed for lubrication engineers—since 1993.

To meet the practical needs of lubrication practitioners and analysts the International Council of Machinery Lubrication (ICML) was formed in the year 2000 and has aggressively converted lubrication certification into a reliability mindset thanks to a broader range of certification disciplines, and its recognition of "hands-on" lubrication practitioners through its Machinery

Lubrication Technician—MLT certifications.

There are now three active certification bodies in the lubrication field that include the previously described STLE, ICML, and more recently, the International Organization for Standardization (ISO), who has worked in close collaboration with the ICML to develop and model their lubrication analyst LCAT certification programs.

STLE (Society of Tribologists and Lubrication Engineers)

Since 1993 to 2016, the STLE has accrued a roster of over 1900 current lubrication certifications in four disciplines, these being:

- **CLS—Certified Lubrication Specialist**
  A lubrication specialist is typically recognized as an individual trained in the fundamentals of lubrication that may be called upon to perform plant wide lubrication surveys and to evaluate, select, and specify lubricants for use within the plant. The CLS would also be responsible for troubleshooting any lubrication related problems, development of in-house quality assurance and lube analysis programs, and the development, implementation and training of staff for any in house lubrication management program.
- **OMA I—Oil Monitoring Analyst Level I**
  The OMA I analyst is a predictive maintenance professional responsible for sampling oil at the equipment level with the ability to review and interpret the oil sample reports and make decisions about the overall care of equipment being sampled.
- **OMA II—Oil Monitoring Analyst Level II**
  The OMA II individual can be found in the test laboratory responsible for managing the appropriate tests, interpreting data and supervising the oil analysis program.
- **CMFS—Certified Metalworking Fluids Specialist**
  This is a certification for individuals experienced in the specialized field of metalworking and cutting fluids. Similar to the CLS, this individual would be typically responsible for evaluating, specifying, troubleshooting and managing all aspects of metal removal, forming and cutting fluid issues.

The STLE currently offers its certification exams in English only, but is working to make available its program and examinations in other languages in the near future. All certifications have a published recommended reading list that potential certificate designates are encouraged to read and familiarize themselves with well in advance of examination. A tailored list for each certificate designation is published on the STLE website and can be viewed at www.stle.org/certifications

The CLS program is by far the STLE's most popular certification with over two thirds of certifications listed in this discipline. Because many large North American oil companies and oil analysis laboratories—and their employees—are active members in the STLE, it is understandable that the CLS and OMA designations are popular with this industrial demographic, but the STLE is working to expand its certification program to appeal to a wider demographic. The CLS certification requires 3 years job experience in the field of lubrication; the OMA certification requires one year of laboratory experience backed up with 16 hours of formal oil analysis training, whereas the CFMA certification requires between 3-5 years of experience and 20 hours of training.

STLE certification exams are proctored, 150 question, close book, three-hour, multiple-choice exams requiring a 70% or higher mark to be successful. The certification is good for 3 years, and recertification is easily obtained by completing and providing proof of four of twelve options that include professional training program credits, conference attendance, volunteer STLE work, etc. or by writing a recertification exam. STLE certification exams are the highest of all three at $475 for non-STLE members, with a 25% discount offered to STLE members.

In addition to its certification programs, the STLE offers scholarship and fellowship awards to students interested in pursuing a career in the field of tribology and lubrication engineering

ICML (International Council of Machinery Lubrication)

In comparison, to date 2016, the ICML—a non-profit organization dedicated to helping lubrication practitioners succeed in their professional careers—has accrued a roster of close to 7000 professional lubrication certifications in 70+ countries worldwide over a 15-year period, in seven occupation-specific disciplines:

- **Machinery Lubrication Technician Level I—MLT**
  A typical MLT is a plant technician responsible for daily lubrication tasks that can include oil top up and changes, filter changes, manual greasing, lubricant receiving and general care of lubricating dispensing and transfer equipment

- **Machinery Lubrication Technician Level II—MLT II**
  An MLT II designation is designed for the technician or engineer responsible for managing the lubrication program and its staff. The individual would also be responsible for appropriate lubricant selection, lubrication troubleshooting and supporting machine design activities.
- **Machinery Lubricant Analyst Level I—MLA (ISO 18436-4)**
  This designation targets in plant technicians associated with performing basic lubrication tasks and machine condition monitoring/analysis. Activities can include oil changes, greasing bearings, lubricant receiving, storage and transfer; basic oil sampling duties and contamination control.
- **Machinery Lubricant Analyst Level II—MLA II (ISO 18436-4)**
  The MLA II is for the plant technician responsible for the condition monitoring/oil-sampling program that includes such duties as oil sampling, sample management, simple on site testing and managing test results/diagnostics.
- **Machinery Lubricant Analyst Level III—MLA III (ISO 18436-4)**
  A level III MLA targets technicians and engineers responsible for managing the lubricant analysis function. Responsibilities include team management; test slate selection, setting alarms and limits sampling systems design and advanced diagnostics.
- **Laboratory Lubricant Analyst Level I—LLA (ISO 18436-5)**
  The LLA designation targets in house laboratory technicians performing and reporting on basic lubricant sample testing
- **Laboratory Lubricant Analyst Level II—LLA II (ISO 18436-5)**
  A level II laboratory technician is responsible for daily activities supporting the production of lubricant analysis data for machine condition monitoring performing tasks that include testing, diagnosing lubricant failure mechanisms
- **Laboratory Lubricant Analyst Level III—LLA III (ISO 18436-5)**
  A level III laboratory technician is responsible for daily activities supporting the production of lubricant analysis data for machine condition monitoring performing tasks that include testing, diagnosing lubricant failure mechanisms
- **Machinery Lubrication Engineer—MLE**
  The MLE designation is the highest designation offered and specifically targets staff, contract or consulting lubrication engineers responsible for the development and management of all aspects related to the implementation and sustainability of lubrication management programs

The most popular designation with over 55% of certifications is the base MLT certification designation.

The ICML has always prided itself as a lubrication end user/practitioner organization reflecting the needs of the asset reliability industry in management offices, on the shop floor, and in the laboratory. In addition, the ICML now examines and promotes its certifications in nine languages across the world.

Similar to the STLE, the ICML certification examinations are proctored exams. The ICML exam utilizes a 100 close book, multiple-choice set of questions with three-hours to complete, also requiring a 70% pass rate to be successful. Depending on the certification discipline the ICML requires up to 3 years job experience in the appropriate lubrication field and 16-32 hours of formal lubrication training based on the corresponding bodies of knowledge (depending on the discipline being examined for) as a pre-requisite to take the exam. Initial certification is good for 3 years and recertification is based on a points system accrued based on training, experience and service as with the STLE, or by taking a recertification exam; recertification requirements are an excellent way of ensuring each individual stays current. ICML examination costs are considerably lower at $200 US. Individuals are encouraged, but not required to be an ICML member to certify. See Table 1 for STLE/ICML program offering comparison.

All ICML certifications are in compliance with the ISO 18436-4/5 standard for condition monitoring and diagnostics of machines—requirements for qualifications and assessment of personnel—Part 4—field lubricant analysis and Part 5—laboratory analysis.

ISO (International Organization for Standardization)

The ISO organization is a relative newcomer to lubrication certification and offers the following three levels of international standard certification:

- ISO 18436-4 Lubrication Analyst ISO Category I—LCAT I
- ISO 18436-4 Lubrication Analyst ISO Category II—LCAT II

Attribute	STLE	ICML
Web site	www.stle.org	www.lubecouncil.org
First Certification	1993	2001
Active Certifications	Approximately 1900 (2016)	Approximately 6900 + (2016)
Experience requisite	1- 5 years (discipline dependent)	1-3 years (discipline dependent)
Formal Lubrication Training requisite	0-20 hours (discipline dependent)	16-32 hours (discipline dependent
Exam time	3 hours	3 hours
Exam type	Multiple choice	Multiple choice
# Exam questions	150	100
Exam registration	2 weeks advance	1 week advance
Exam Pass Score	70% +	70%+
Renewal cycle	3 years	3 years
Exam failure - wait to rewrite	1 year	1 month
2016 Exam cost	$450 US non STLE member $340 US STLE member	$200 US
Domain of Knowledge*	Yes	Yes
Exam language	English	English, French, Dutch, German, Spanish, Italian, Portuguese, Japanese, Korean, Mandarin, plus requests.

* Recommended textbook library

Figure 22-3. Lubrication Certification Comparison Table

- ISO 18436-4 Lubrication Analyst ISO Category III—LCAT III

ISO collaborated closely with ICML to develop their certification designation requirements and Bodies of Knowledge resulting in the current ICML MLA level I and II being adopted by the ISO as their LCAT level II and III certification designations, with the LCAT Level I being derived from the ICML MLT I curriculum in which the qualification ensures personnel are trained and certified to perform simple tasks related to the proper lubrication of machines according to established and recognized procedures.

Participants who have attended the requisite preparatory formal training for MLT I, MLA I and II are also eligible to write the corresponding ISO LCAT exams with payment of the appropriate examination fee.

Regardless of which certification route you choose, any choice to certify is an excellent choice demanding a lot of preparatory hard work and assimilation of lubrication knowledge prior to writing the examination(s). All certification designations discussed above are industry recognized, and potential designate candidates can be assured their efforts will be rewarded both intrinsically and extrinsically for the rest of their working life.

CERTIFICATION BODY OF KNOWLEDGE

Each certification body publishes both a *domain of knowledge* and a *body of knowledge* for the benefit of potential designees and its certified designates. The domain of knowledge is recognized as the subject matter knowledge base represented in a library of reference books, texts and papers written by lubrication subject matter experts. This domain of knowledge is then used to develop the body of knowledge, or areas of knowledge expertise, in each certification level that designees are expected to study and understand for certification purposes. Exam questions for each of the body of knowledge requirements are based on the domain of knowledge reference material.

Tables 22-1 and 22-2 depict the body of knowledge study requirements for the most popular designation certifications for both STLE and the ICML. The ICML tables also show the % weight of each section by designation certification.

These tables are based on information deemed to be correct at the time of publishing and may have changed. For the most up-to-date information visit the STLE website at www.stle.org, and view the "Career Development" tab. For ICML information, visit www.lubecouncil.org and view the " Certifications and Exams" tab.

Table 22-1. STLE Examination Body of Knowledge (BOK) Requirements Matrix

Certification Designation	Body of Knowledge requirements
CLS	• Lubrication Fundamentals • Fluid Conditioning • Storage, Handling and Application of Lubricants • Monitoring and Reducing Consumption of Lubricants • Gears
	• Bearings • Seals • Fluid Power • Lubricant Manufacturing • Pneumatics • Transportation Lubricants • Metalworking • Solvents and Cleaners • Problem Solving • Lubricant Analysis • Lubrication Programs
OMA I	• Lubrication Fundamentals • Lubricant Sampling • Lubricant Application/Test Method • Oil Analysis Data Interpretation • Troubleshooting
OMA II	• Lubricant Selection • Analysis Program Configuration • Oil Sampling and Analytical Methods • Sampling Intervals • Lubricant Suppliers • Analysis Program Logistics • Training • Sample Baseline Data • Sample Limits and Alarms • Test ValidationTest Data Interpretation • Actions • Maintenance • Analysis • Quality • Safety • Documentation • Failure Analysis • Tribological Factors Affecting Design • Lubricant Analysis Program Management
CMFS	• Metalworking Operations • Metal Removal and Forming Fluid Chemistry • Metal Removal and Forming Fluid Condition Monitoring • Central and Stand-Alone Systems: Function and Design • Plant Conditions/Operations Affecting Metalworking Fluid Performance • Metalworking Fluid Condition Management • Enclosures and Ventilation • Health and Safety • Metalworking Fluid Waste Treatment • Communications: Data Interpretation, Training, Cost/Benefit Analysis

*Table 22-2. ICML MLT and MLT II/MLA and MLA II Examination Body of Knowledge (BOK) Requirements Matrix**

<table>
<tr><th colspan="3">BOK Section 1: Maintenance Strategy</th></tr>
<tr><th>Certification Designation</th><th>Exam Weight</th><th>Body of Knowledge requirements</th></tr>
<tr><td>MLT I</td><td>5%</td><td rowspan="2">• Why Machines fail
• The impact of poor maintenance on corporate profits
• Failure avoidance through effective lubrication
• Lube routes and scheduling
• Oil Analysis and technologies that assure Lubrication effectiveness
• Identifying and tagging equipment for lubrication</td></tr>
<tr><td>MLA I</td><td>10%</td></tr>
<tr><td>MLT II</td><td>5%</td><td>• Lubrication impact on machine reliability
• Lubrication Impact on lubricant life and consumption
• Strategies for achieving lubrication excellence</td></tr>
<tr><td>MLA II</td><td>4%
Oil Analysis Maintenance Strategies</td><td>• RCM (Reliability Centered Maintenance) fundamentals
• CBM (Condition Based Maintenance) fundamentals:
o Predictive Maintenance strategies
o Proactive Maintenance strategies
• PBM (Perimeter Based Maintenance) fundamentals</td></tr>
<tr><th colspan="3">BOK Section 2: Lubrication Theory/Fundamentals/Lubricants</th></tr>
<tr><th>Certification Designation</th><th>Exam Weight</th><th>Body of Knowledge requirements</th></tr>
<tr><td>MLT I</td><td>25%
Lubrication Theory/Lubricants</td><td rowspan="2">• Fundamentals of Tribology
o Types of Friction
o Wear modes and influencing factors
• Functions of a Lubricant
• Mechanisms of Lubrication Regimes:
o Hydrodynamic lubrication (HDL)
o Elasto-Hydrodynamic lubrication (EHD)
o Mixed film lubrication (MF)
o Boundary Layer lubrication (BL)
• Base Oils – Function and Properties
• Additives and their function:
o Surface active type
o Bulk oil active type
• Oil Lubricant physical and chemical performance properties and classifications
• Grease lubrication:
o How grease is made
o Thickener types
o Thickener compatibility
• Grease lubricant physical and chemical performance properties and classifications
•</td></tr>
<tr><td>MLA I</td><td>18%
Lubrication Theory/Fundamentals</td></tr>
<tr><td>MLT II</td><td>5%
Lubrication Theory</td><td>• Lubricant categories:
o Gaseous
o Liquid
o Cohesive
o Solid</td></tr>
<tr><td>MLA II</td><td>0%</td><td></td></tr>
</table>

*Table 22-2 (Cont'd). ICML MLT and MLT II/MLA and MLA II Examination Body of Knowledge (BOK) Requirements Matrix**

BOK Section 3: Lubricant Formulation		
Certification Designation	**Exam Weight**	**Body of Knowledge requirements**
MLT I	**0%**	
MLA I	**0%**	
MLT II	**10%**	• Base-oil refining methods and API categories: o Solvent refined o Hydro treated o Severely hydro treated o Hydrocracked • Mineral based oils: o Naphthenic o Paraffinic o Aromatic • Vegetable base oils and bio-lubes • Synthetic Lubricant characteristics/application/compatibility: o Synthesized Hydrocarbons (Polyalphaolifins) o Dibasic acid esters o Polyol esters o Phosphate esters o Polyalkalene esters o Silicones o Flourocarbons o Polyphenil esters • Food grade lubricant classification • Solids additives and their function • Modes of additive depletion
MLA II	**4%**	• Synthetic Lubricant characteristics/application/compatibility: o Synthesized Hydrocarbons (Polyalphaolifins) o Dibasic acid esters o Polyol esters o Phosphate esters o Polyalkalene esters o Silicones o Flourocarbons o Polyphenil esters
BOK Section 4: Lubricant Selection		
Certification Designation	**Exam Weight**	**Body of Knowledge requirements**
MLT I	**15%**	• Viscosity selection/ adjustments: o Machinery condition factors o Environmental condition factors • Base oil type selection: o When to use synthetic lubricants o When to use biodegradable lubricants • Additive system selection

*Table 22-2 (Cont'd). ICML MLT and MLT II/MLA and MLA II Examination Body of Knowledge (BOK) Requirements Matrix**

		• Machine specific lubricant selection: o Hydraulic systems o Rolling element bearings o Journal bearings o Reciprocating engines o Gearing and gearboxes • Application and environment related adjustments •
MLA I	**10%**	• Viscosity selection/ adjustments: o Machinery condition factors o Environmental condition factors • Base oil type selection: o When to use synthetic lubricants o When to use biodegradable lubricants • Additive system selection • Machine specific lubricant selection: o Hydraulic systems o Rolling element bearings o Journal bearings o Reciprocating engines o Gearing and gearboxes • Application and environment related adjustments
MLT II	**15%**	• Viscosity selection/ adjustments: • Machinery condition factors • Environmental condition factors • Base oil type selection: o When to use synthetic lubricants o When to use biodegradable lubricants • Lubricant Consolidation • Selecting Lubricating oils for: o Fire resistant applications o Hydraulics – mobile/industrial o Turbines o Compressors o Bearings o Chains/conveyors o Mist applications o Gears – industrial/automotive o Engines – diesel/gas/gasoline (petrol) o Pneumatic tools o Spindles o Ways/slides • Selecting greases for: o Chassis o Couplings o Anti-friction bearings o Journal bearings o Automotive bearings o Automatic lubrication systems • Lubricant selection standards development • Procedures for testing and quality assurance of incoming lubricants

*Table 22-2 (Cont'd). ICML MLT and MLT II/MLA and MLA II Examination Body of Knowledge (BOK) Requirements Matrix**

		• Procedures for approval of candidate lubricants
MLA II	**0%**	
BOK Section 5: Lubricant Application		
Certification Designation	**Exam Weight**	**Body of Knowledge requirements**
MLT I	**25%**	• Basic calculations to determine required lubricant volume • Basic calculations to determine lubrication frequency o Oil o Grease • When to select oil vs. grease • Effective use of manual delivery techniques • Automated lubricant delivery systems: o Progressive divider o Single line resistance o Positive displacement injection ▪ Single line ▪ Dual line o Pump to point ▪ Box cam ▪ Annular o Oil mist o Air/Oil o Single point ▪ Gravity (drip/wick) ▪ Chemical ▪ Electro chemical o Electro mechanical • When to employ automated lubrication • Maintaining automated systems
MLA I	**18%**	
MLT II	**15%**	• Procedures for: o Oil drain o Reservoir/system flushing o Disassembling/cleaning reservoirs and sumps o Filling o Top-up o Grease packing o Re-greasing o Grease change out • Basic calculations to determine lubrication frequency o Oil o Grease • Select and manage optimum equipment/systems for lubricant application according to machinery requirements • Safety/health requirements for lubricant application • Manage proper maintenance of lubrication equipment • Manage proper maintenance of automatic lubrication systems

*Table 22-2 (Cont'd). ICML MLT and MLT II/MLA and MLA II Examination Body of Knowledge (BOK) Requirements Matrix**

<table>
<tr><td></td><td></td><td>• Create/update lube survey
• Record execution of lube program
• Proactive management and detection of leaks
• Waste oil/filters management/disposal
• Developing and Writing an objective lubrication PM job plan</td></tr>
<tr><td>**MLA II**</td><td>**0%**</td><td></td></tr>
<tr><td colspan="3">**BOK Section 6: Preventive and Predictive Maintenance**</td></tr>
<tr><td>**Certification Designation**</td><td>**Exam Weight**</td><td>**Body of Knowledge requirements**</td></tr>
<tr><td>**MLT I**</td><td>**10%**</td><td>• Lube routes and scheduling
• Oil analysis and technologies to assure lubrication effectiveness
• Equipment tagging and identification</td></tr>
<tr><td>**MLA I**</td><td>**0%**</td><td></td></tr>
<tr><td>**MLT II**</td><td>**10%**</td><td>• Creating and managing lubrication PM's and routes
• Creating and managing lubrication check lists
• Used oil analysis to determine optimum condition based oil changes
• Used oil analysis to troubleshoot abnormal lubricant degradation conditions
• Used oil analysis to troubleshoot abnormal wear related to lubricant degradation/contamination
• Procedures and methods for identifying root cause of lubricant failure
• Use of technology aids to determine optimum re-grease frequency/quantity:
  o Ultrasonic
  o Temperature monitoring
  o Shock pulse</td></tr>
<tr><td>**MLA II**</td><td>**0%**</td><td></td></tr>
<tr><td colspan="3">**BOK Section 7: Lubricant Condition Control**</td></tr>
<tr><td>**Certification Designation**</td><td>**Exam Weight**</td><td>**Body of Knowledge requirements**</td></tr>
<tr><td>**MLT I**</td><td>**10%**</td><td rowspan="2">• Filtration and separation technologies
• Filter ratings
  o Beta ratio
• Filtration system design and filter selection</td></tr>
<tr><td>**MLA I**</td><td>**10%**</td></tr>
<tr><td>**MLT II**</td><td>**20%**</td><td>• Proper sampling procedures
• Proper sampling locations
• Proper selection of breathers/vents
• Proper selection of filters according to cleanliness objectives
• Sump/Tank Management to reduce:
  o Air entrainment/foam
  o Particles
  o Water</td></tr>
</table>

Table 22-2 (Cont'd). ICML MLT and MLT II/MLA and MLA II Examination Body of Knowledge (BOK) Requirements Matrix*

Certification Designation	Exam Weight	Body of Knowledge requirements
		o Sediments o Heat o Silt/sediments o Unnecessary lubricant volume • Proper selection of lubricant reconditioning systems for: o Water o Air/gas o Particles o Oxidation products o Additive depletion • Lube reclamation: o Requirements o Feasibility o Procedures for reclaiming/reconditioning o Use of oil analysis to approve reclaimed/reconditioned lubricants
MLA II	**0%**	
BOK Section 8: Lubricant Storage and Management		
Certification Designation	**Exam Weight**	**Body of Knowledge requirements**
MLT I	**10%**	• Lubricant receiving procedures • Proper storage and inventory management • Lube storage containers • Proper storage of grease guns and other lube dispensing devices • Health and safety assurance - MSDS
MLA I	**10%**	
MLT II	**5%**	• Optimum lubricant storage room design • Define maximum storage time according to environmental conditions /lubricant life • Proper sampling procedures / locations for sampling stored lubricants • Lubricant transfer • Procedures for reconditioning/filtering stored lubricants • Health and safety assurance - MSDS
MLA II	**0%**	
BOK Section 9: Oil Sampling		
Certification Designation	**Exam Weight**	**Body of Knowledge requirements**
MLT I	**0%**	
MLA I	**10%**	• Objectives for Oil Sampling • Sampling methods: o Non-pressurized systems o Pressurized systems - Low o Pressurized systems - High • Managing sampling interference

*Table 22-2 (Cont'd). ICML MLT and MLT II/MLA and MLA II Examination Body of Knowledge (BOK) Requirements Matrix**

		○ Bottle cleanliness and management ○ Flushing ○ Machine conditions appropriate for sampling
MLT II	**0%**	
MLA II	**29%**	• Equipment specific sampling: ○ Gearboxes with circulating systems ○ Engines ○ Single and multi-component circulating oil systems with separate reservoirs ○ Hydraulic systems ○ Splash, ring and collar lubricated systems • Sampling process management: ○ Sampling frequency ○ Sampling procedures ○ Sample processing
BOK Section 10: Lubricant Health Monitoring		
Certification Designation	**Exam Weight**	**Body of Knowledge requirements**
MLT I	**0%**	
MLA I	**10%**	• Lubricant failure mechanisms: • Oxidative degradation ○ The oxidation process ○ Causes of oxidation ○ Effects of oxidative degradation • Thermal degradation ○ The thermal failure process ○ Causes of thermal failure ○ Effects of thermal degradation • Additive depletion/degradation ○ Additive depletion mechanisms • Additives at risk for depletion/degradation by the various mechanisms • Testing for wrong or mixed lubricants: ○ Base lining physical and chemical properties tests ○ Additive discrepancies • Fluid properties test methods and measurement units: ○ Kinematic Viscosity (ASTM D445) ○ Absolute (Dynamic) Viscosity (ASTM D2893) ○ Viscosity Index (ASTM D2270) ○ Acid Number (ASTM D974 et al) ○ Base Number (ASTM D974 et al) ○ Fourier Transform Infrared (FTIR) analysis ○ Rotating Pressure Vessel Oxidation Test

*Table 22-2 (Cont'd). ICML MLT and MLT II/MLA and MLA II Examination Body of Knowledge (BOK) Requirements Matrix**

		(ASTMD2272) ○ Atomic Emission Spectroscopy
MLT II	**0%**	
MLA II	**21%**	• ***See MLA I requirements***
BOK Section 11: Wear Debris Monitoring and Analysis		
Certification Designation	**Exam Weight**	**Body of Knowledge requirements**
MLT I	**0%**	
MLA I	**4%**	• Common Wear Mechanisms: ○ Abrasive wear ○ Two body ○ Three body ○ Surface fatigue (contact fatigue ○ Adhesive wear ○ Corrosive wear ○ Cavitation wear
MLT II	**0%**	
MLA II	**17%**	• Detecting abnormal wear: ○ Atomic emission spectroscopy methods ▪ Inductive coupled plasma (ICP) ▪ Arc-spark emission • Wear particle density measurement ○ Wear debris analysis: ○ Ferrogram preparation ○ Filtergram preparation ○ Light effects ○ Magnetism effects ○ Heat treatment ○ Basic morphological analysis
BOK Section 12: Lubricant Contamination Measurement and Control;		
Certification Designation	**Exam Weight**	**Body of Knowledge requirements**
MLT I	**0%**	
MLA I	**0%**	
MLT II	**0%**	
MLA II	**25%**	• Particle contamination: ○ Effects on the machine ○ Effects on the lubricant ○ Methods and units for measuring particle contamination ○ Techniques for controlling particle contamination

*Table 22-2 (Cont'd). ICML MLT and MLT II/MLA and MLA II Examination Body of Knowledge (BOK) Requirements Matrix**

		• Moisture contamination: o Effects on the machine o Effects on the lubricant o States of coexistence o Methods and units for measuring moisture contamination o Demulsibility measurement o Techniques for controlling moisture contamination • Glycol coolant contamination o Effects on the machine o Effects on the lubricant o Methods and units for measuring glycol contamination o Techniques for controlling glycol contamination • Soot contamination o Effects on the machine o Effects on the lubricant o Methods and units for measuring soot contamination o Techniques for controlling soot contamination • Fuel contamination (fuel dilution in oil) o Effects on the machine o Effects on the lubricant o Methods and units for measuring fuel contamination o Techniques for controlling fuel contamination • Air contamination (air in oil) o Effects on the machine o Effects on the lubricant o States of coexistence • Methods for assessing air contamination o Air release characteristics (ASTM D3427) o Foam stability characteristics (ASTM D892) • Techniques for controlling air contamination

*Table Courtesy ENG**TECH** Industries Inc.

Appendix A

Lubrication Program*

Work Process Manual

Lubrication Program Work Process Manual

Overview

Introduction

The purpose of this guide is to provide a detailed process for use by plants in developing, implementing, maintaining and improving their industrial lubrication programs. It is a compilation and organization of existing published material, input from maintenance and reliability professionals, and experience gained by the author during his tenure as a Reliability Engineer and consultant.

Audience

This guide is intended for all individuals who are involved with any of the four phases necessary to establish industrial lubrication programs.

Purpose

To significantly reduce the amount of time required by maintenance and reliability professionals to establish effective lubrication programs.

Program Goal

The goal of every lubrication program should be to ensure that all equipment receives and maintains the required levels of lubrication such that no equipment fails due to inadequate or improper lubrication.

Note to the Reader

The step descriptions in this manual should be read in conjunction with the work process diagrams.

In this document

This document contains the following information:

*Source: Richard P. Ellis, Pearland, Texas (richard.ellis@armic.com)

Step 1.1 – Create Equipment List

Purpose

To establish a preliminary list of equipment to be included in the lubrication program.

Description

Before a plant can begin implementing (or thoroughly reviewing) a lubrication program, it is necessary to create or obtain a current list of all equipment that requires lubrication. The list should include all types of equipment requiring lubrication, i.e. mobile equipment, valves, HVAC equipment, etc., and not just the usual pumps, compressors and fans.

Inputs

- process flow diagrams (PFD's)
- piping and instrument diagrams (P&ID's)
- plant maintenance files
- Computerized Maintenance Management System
- physical survey of the equipment

Outputs

A Master Lubrication Schedule with the following information completed:

- item number
- process description

Tools

- Master Lubrication Schedule template: master lubrication schedule.xls

Functions & Responsibilities

This section defines the responsibilities of each function involved in this step of the work process.

Function	Responsibilities
Maintenance	1. Using the inputs defined above, create an equipment list containing, as a minimum, the item number and process description of all equipment to be surveyed in Step 1.2.

Continued on next page

Step 1.2 – Conduct Lubrication Survey

Purpose

To collect and record on the Master Lubrication Schedule, lubrication related equipment information that is required to make a lubricant selection.

Description

The lubrication survey will consist of a detailed lubrication inspection of all plant equipment. Each machine will be studied and its lubrication related characteristics recorded on the Master Lubrication Schedule.

Obtaining this information is time consuming and may take several days to complete a survey for a typical hydrocarbon or chemical processing plant. However, such a survey is the only way of obtaining an accurate picture of current lubrication practices and it is the basis upon which future steps to select lubricants and improve lubrication practices will be made.

Since a general knowledge of the design of a machine is required for making decisions about its lubrication requirements, it may be necessary to make frequent reference to machine drawings and OEM manuals.

Inputs

- process flow diagrams (PFD's)
- plant maintenance files
- physical survey of the equipment
- OEM manuals

Outputs

A current Master Lubrication Schedule with the following information completed:

- manufacturer
- model
- equipment orientation
- bearing type
- lubricant type (oil, grease)
- method of lubrication (bath splash, circulation system, oil mist, etc.)
- normal operating temperature
- reservoir capacity
- Horsepower
- RPM
- copies of lubricant sections out of the OEM manuals for each piece of equipment

Tools

- Master Lubrication Schedule template: master lubrication schedule.xls

Functions & Responsibilities

This section defines the responsibilities of each function involved in this step of the work process.

Function	Responsibilities
Maintenance	1. Using the inputs defined above, fill in the remaining fields on the Master Lubrication Schedule.
Equipment Manufacturer	2. Provide equipment design and configuration information as required.

Continued on next page

Step 1.3 – Select Lubricants

Purpose

To define the recommended lubricant for each piece of equipment on the Master Lubrication Schedule.

Description

Once equipment configuration and operating conditions have been collected and organized into the Master Lubrication Schedule, review the information with your lubricant vendor and a request they provide a recommended lubricant and lubrication frequency.

Inputs

- Master Lubrication Schedule
- OEM manuals

Outputs

An updated Master Lubrication Schedule with the following information completed:

- lubricant name
- lubrication frequency

Functions & Responsibilities

This section defines the responsibilities of each function involved in this step of the work process.

Function	Responsibilities
Maintenance	1. Give a copy of the Master Lubrication Schedule and associated equipment information to your lubricant supplier and request they make lubricant recommended.
Lubricant Supplier	2. Using the information provided by maintenance, make lubricant selections for each piece of equipment on the Master Lubrication Schedule. 3. Make lubrication frequency recommendations based on equipment information and supplier experience.

Continued on next page

Step 1.4 – Consolidate Lubricants

Purpose

To reduce the total number of lubricants used in the Lubrication Program.

Description

Once lubricants have been selected for each piece of equipment on the Master Lubrication Schedule, it is important to review the list and determine if there are opportunities to reduce the total number of lubricants that will be used in the program.

In some instances you may find that there are only a few pieces of equipment that use a particular brand or grade of lubricant, and by allowing for a change in lubricant viscosity, it is possible to eliminate the use of the lubricant entirely.

Reducing the number of lubricants has the following effect on the program:

- Reduce the number of lubricants that have to be purchased
- Reduces the number of lubricants that have to be stored
- Reduces the chance of error that the wrong lubricant will be used in a piece of equipment
- Reduces the number of lubricants that have to be documented and kept track of as part of environmental compliance

Inputs

- Master Lubrication Schedule

Outputs

- A reviewed and/or revised Master Lubrication Schedule with fewer lubricants

Functions & Responsibilities

This section defines the responsibilities of each function involved in this step of the work process.

Functions	Responsibilities
Maintenance	1. Reviews and approves the list of recommendations made by the lubricant supplier on which lubricants to eliminate from the program.
Lubricant Supplier	2. Reviews Master Lubrication Schedule for opportunities to consolidate lubricants without impacting equipment reliability.
Equipment Manufacturer	3. Provides input on changes to lubricant recommendations when requested by Maintenance or Lubricant Supplier.

Continued on next page

Step 1.5 – Create Lubrication Manual

Purpose

To provide a place where all of the information collected for development of the Lubrication Program can be stored for future reference.

Description

The process of developing a lubrication program requires the collection of a significant amount of equipment data, usually found in disperse locations. After all the time and effort expended to locate and collect the data, it is worth while to consolidate that information into a Lubrication Manual that can be referenced over time.

Inputs

- Master Lubrication Schedule
- Copies of lubricant sections from OEM manuals
- Vendor furnished lubricant product data sheets
- Vendor furnished Material Safety Data Sheets

Outputs

An assembled Lubrication Manual with the following contents:

- Master Lubrication Scheduled sorted by tag number
- Lubricant product data sheets
- Material Safety Data Sheets
- Copies of lubricant sections from OEM manuals

Note: The storage location of the Material Safety Data Sheets is most likely dependent on the environmental and industrial hygiene policies for your particular plant. Be sure and discuss this with your Environmental Specialist and/or Industrial Hygienist.

Functions & Responsibilities

This section defines the responsibilities of each function involved in this step of the work process.

Functions	Responsibilities
Maintenance	1. Collect the material and assemble it into the Lubrication Manual. 2. Store the Lubrication Manual in the plant library.
Lubricant Supplier	3. Provide lubricant product data sheets for all lubricants in the Lubrication Program. 4. Provide Material Safety Data Sheets for all lubricants in the Lubrication Program.
EH & S	5. Provide input and direction on plant or company policies regarding storage of Material Safety Data Sheets.

Continued on next page

Step 1.6 – Purchase Lubrication Equipment

Purpose

To generate a Lubrication Program Equipment List defining the equipment that will be required to carry out the work involved with the Lubrication Program.

Description

The processes of lubricating equipment involves the use of equipment to both store and apply lubricants as defined by the scheduled service reports generated by the Computerized Maintenance Management System (CMMS). The equipment includes items such as:

- grease guns
- bulk lubricant storage facilities
- drum handling equipment (dollies, drum tilters, bung removal tools, etc.)
- shop towels
- garden type sprayers for topping-off lube levels or oil changes

Inputs

- The consolidated Master Lubrication Schedule

Outputs

- Lubrication Program Equipment List

Functions & Responsibilities

This section defines the responsibilities of each function involved in this step of the work process.

Functions	Responsibilities
Maintenance	1. Create the Lubrication Program Equipment List to be used by purchasing to buy the necessary equipment.
Purchasing	2. Procure equipment as listed on the Lubrication Program Equipment List.

Continued on next page

Step 1.7 – Set PM Tasks & Frequency

Purpose

To define the lubrication related tasks for each equipment item and the frequency with which the tasks are to be carried out.

Description

Prior to entering the lubrication tasks into the CMMS, it is necessary to define the frequency at which the lubrication tasks will be repeated. This information, along with the data collected in earlier steps, will be input in the CMMS and used to generate the scheduled service reports.

Inputs

- The consolidated Master Lubrication Schedule
- Lubrication Program PM Task Picklist
- OEM Manuals
- Lubricant Supplier Recommendations

Outputs

- A completed Master Lubrication Schedule with all fields necessary to generate a scheduled service report

Functions & Responsibilities

This section defines the responsibilities of each function involved in this step of the work process.

Function	Responsibilities
Maintenance	1. Using the inputs shown above, define the lubrication tasks and their frequency
Equipment Manufacturer	2. Provide input into the selection of PM tasks and frequency as requested by maintenance.
Lubricant Supplier	3. Provide input into the selection of PM tasks and frequency as requested by maintenance.

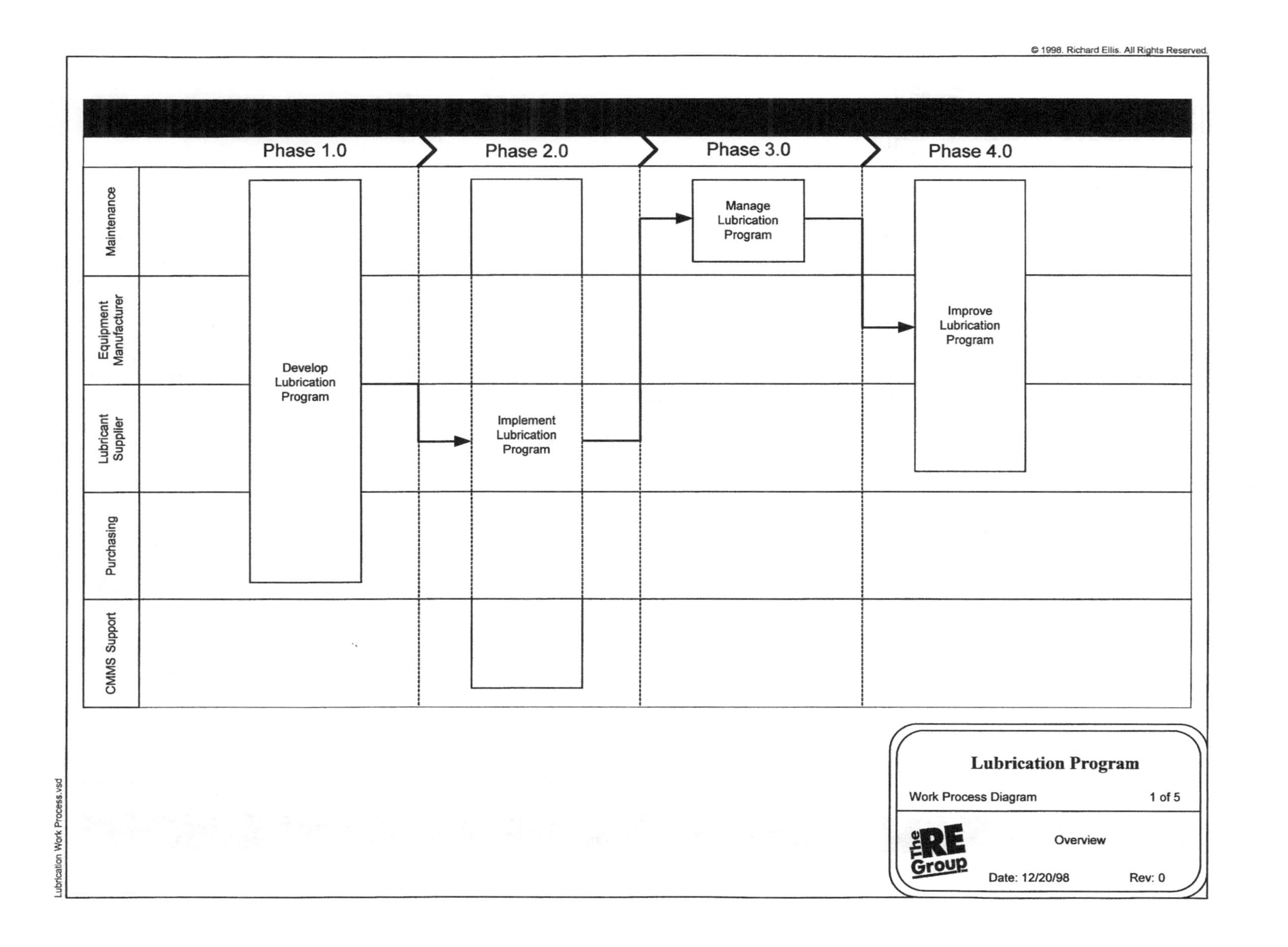
© 1998. Richard Ellis. All Rights Reserved.
Phase 1.0
Phase 2.0
Phase 3.0
Phase 4.0
Maintenance
Equipment Manufacturer
Lubricant Supplier
Purchasing
CMMS Support
Develop Lubrication Program
Implement Lubrication Program
Manage Lubrication Program
Improve Lubrication Program
Lubrication Program
Work Process Diagram
1 of 5
The RE Group
Overview
Date: 12/20/98
Rev: 0
Lubrication Work Process.vsd

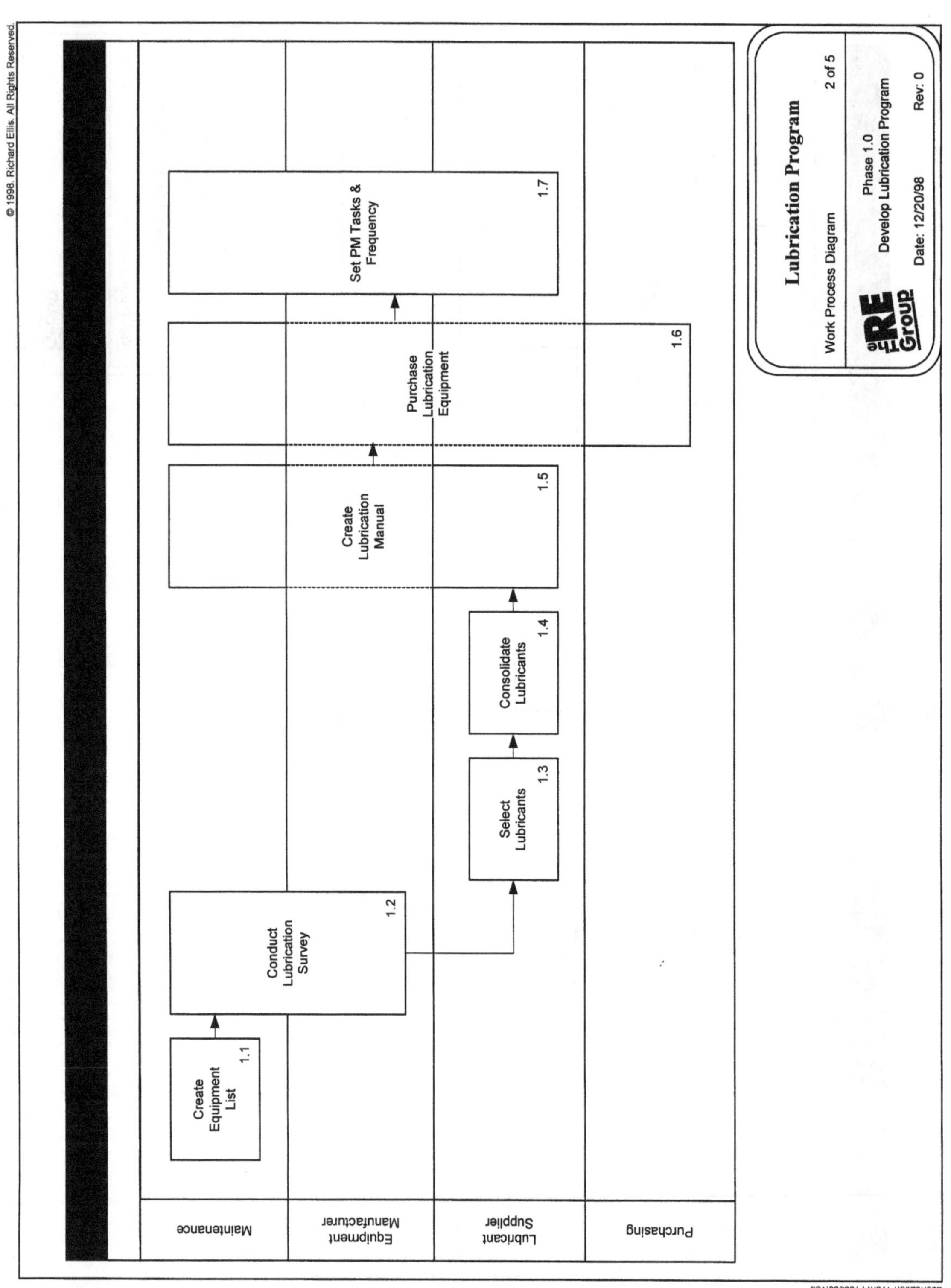
© 1998. Richard Ellis. All Rights Reserved.
Maintenance
Equipment Manufacturer
Lubricant Supplier
Purchasing
Create Equipment List
1.1
Conduct Lubrication Survey
1.2
Select Lubricants
1.3
Consolidate Lubricants
1.4
Create Lubrication Manual
1.5
Purchase Lubrication Equipment
1.6
Set PM Tasks & Frequency
1.7
Lubrication Program
Work Process Diagram
2 of 5
The RE Group
Phase 1.0
Develop Lubrication Program
Date: 12/20/98
Rev: 0
Lubrication Work Process.vsd

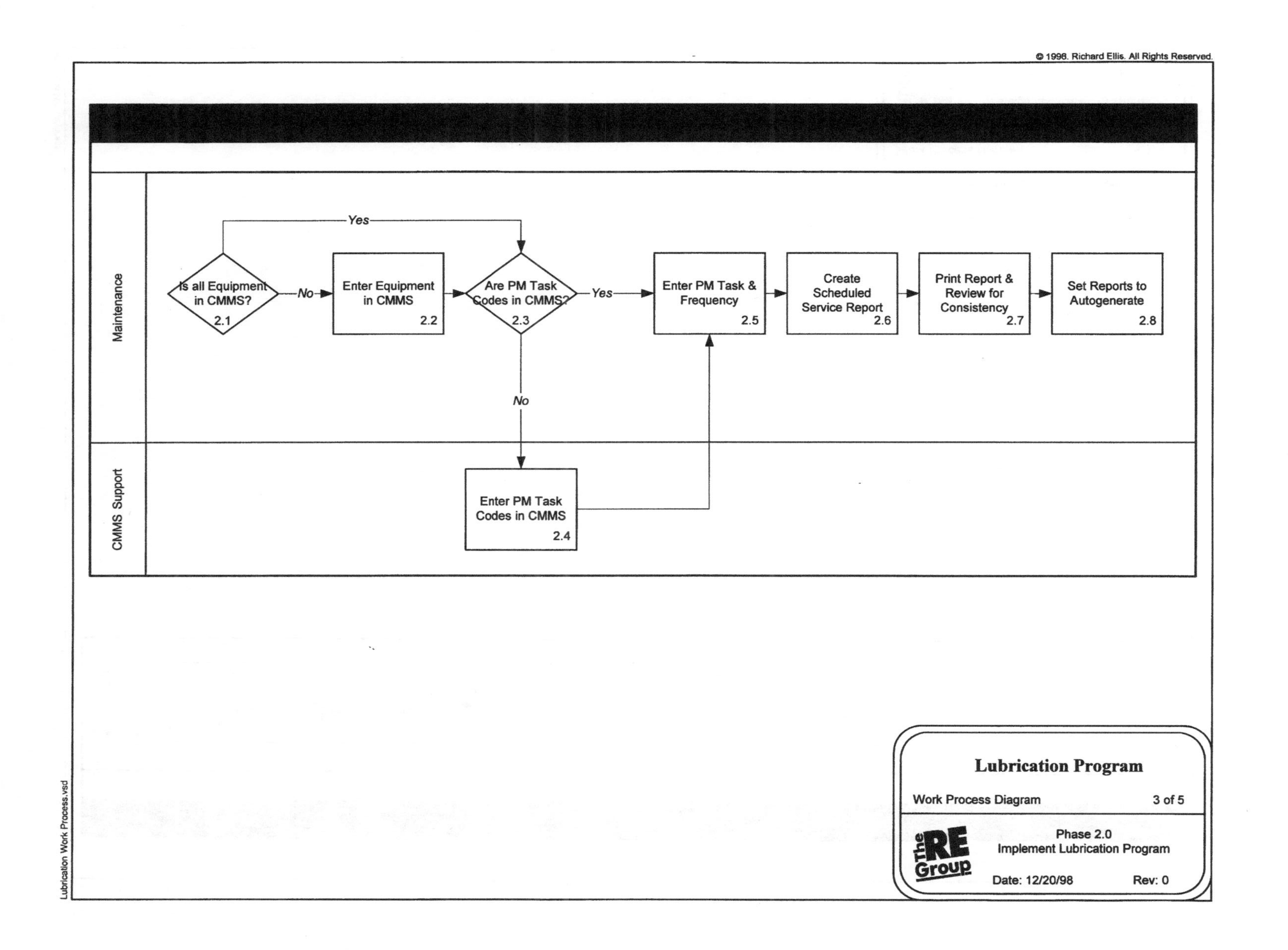
© 1998. Richard Ellis. All Rights Reserved.
Maintenance
CMMS Support
Is all Equipment in CMMS? 2.1
No
Yes
Enter Equipment in CMMS 2.2
Are PM Task Codes in CMMS? 2.3
Yes
No
Enter PM Task Codes in CMMS 2.4
Enter PM Task & Frequency 2.5
Create Scheduled Service Report 2.6
Print Report & Review for Consistency 2.7
Set Reports to Autogenerate 2.8
Lubrication Program
Work Process Diagram
3 of 5
The RE Group
Phase 2.0
Implement Lubrication Program
Date: 12/20/98
Rev: 0
Lubrication Work Process.vsd

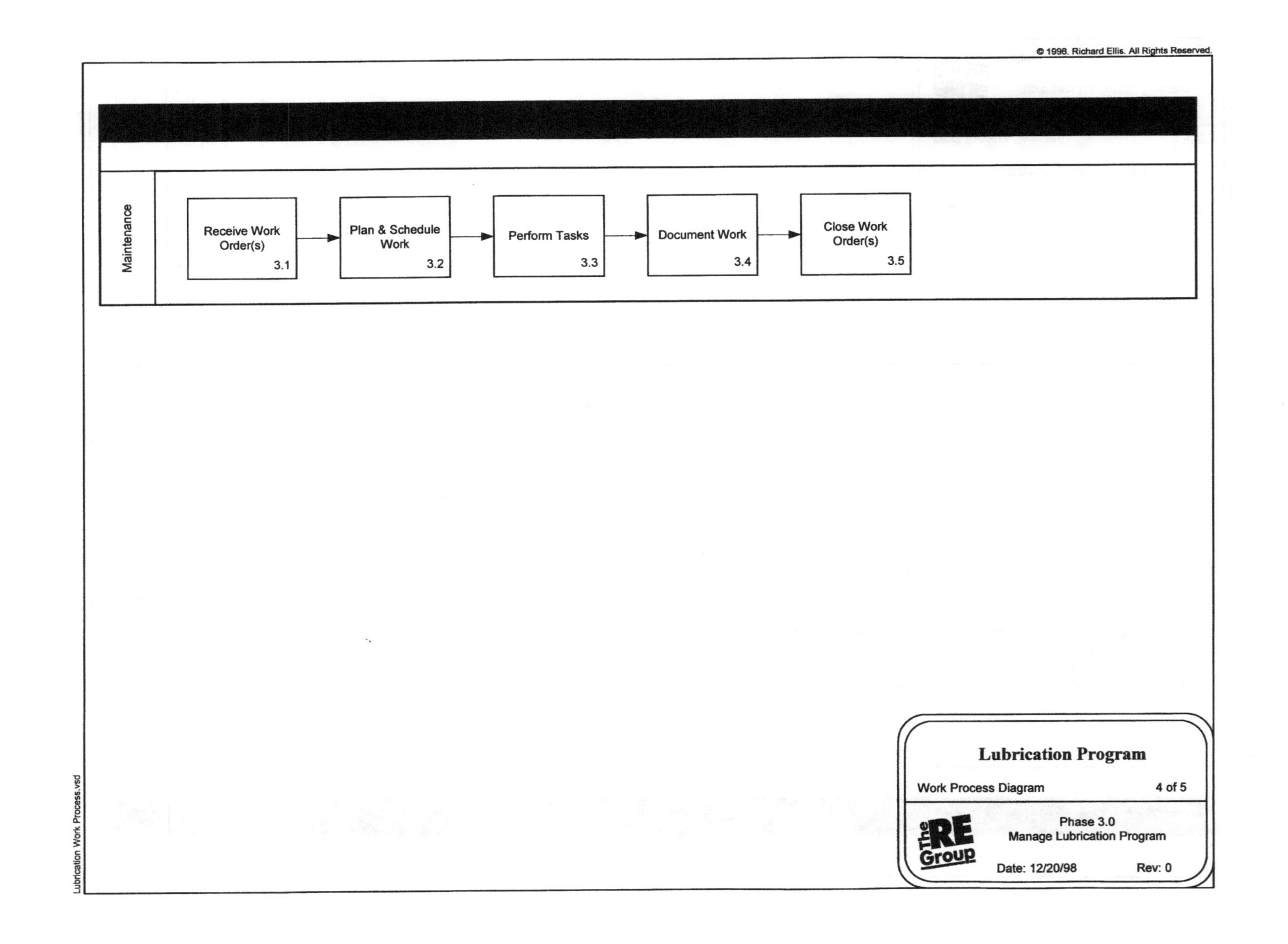
© 1998. Richard Ellis. All Rights Reserved.
Maintenance
Receive Work Order(s) 3.1
Plan & Schedule Work 3.2
Perform Tasks 3.3
Document Work 3.4
Close Work Order(s) 3.5
Lubrication Program
Work Process Diagram
4 of 5
The RE Group
Phase 3.0
Manage Lubrication Program
Date: 12/20/98
Rev: 0
Lubrication Work Process.vsd

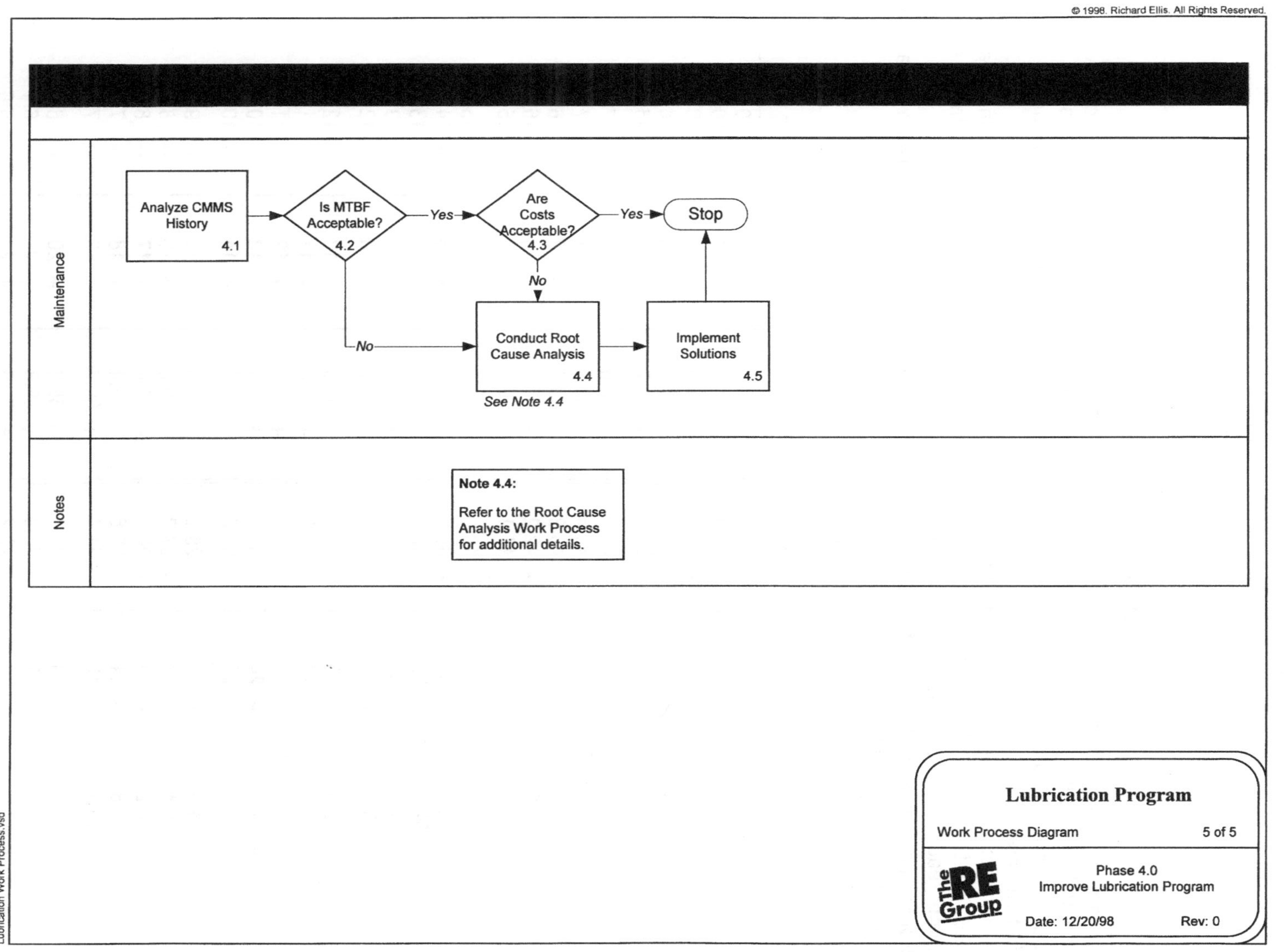
© 1998, Richard Ellis. All Rights Reserved.
Maintenance
Notes
Analyze CMMS History
4.1
Is MTBF Acceptable?
4.2
Yes
No
Are Costs Acceptable?
4.3
Yes
No
Stop
Conduct Root Cause Analysis
4.4
See Note 4.4
Implement Solutions
4.5
Note 4.4:
Refer to the Root Cause Analysis Work Process for additional details.
Lubrication Program
Work Process Diagram
5 of 5
Phase 4.0
Improve Lubrication Program
Date: 12/20/98
Rev: 0
The RE Group
Lubrication Work Process.vsd

Appendix B

- Temperature Conversion Table
- Viscosity Index Charts
- ASTM Viscosity Blending Chart
- Approximate Color Scale Equivalents
- Representative Masses of Petroleum Products
- Abridged Gravity, Volume, and Mass Conversion Table
- Miscellaneous Conversion Factors

Temperature Conversion Tables

Locate in the center column of the three columns the temperature in °C or °F to be converted. For given Celsius temperatures, read Fahrenheit equivalent in the left-hand column. For given Fahrenheit temperatures, read the Celsius equivalent in the right-hand column.

Examples: (a) Given − 30°C; convert to Fahrenheit. Locate − 30 in the center column; read − 22.0°F in the left-hand column. (b) Given − 30°F; convert to Celsius. Locate − 30 in the center column; read − 34.4°C in the right-hand column.

For temperatures not shown, interpolate from the values listed, or use the following equations:

Celsius to Fahrenheit: °C x 9/5 + 32 = °F

Fahrenheit to Celsius: (°F − 32) x 5/9 = °C

°F	°C or °F	°C	°F	°C or °F	°C
−148.0	−100	− 73.3	+ 23.0	− 5	− 20.6
−130.0	− 90	− 67.8	+ 24.8	− 4	− 20.0
−112.0	− 80	− 62.2	+ 26.6	− 3	− 19.4
− 94.0	− 70	− 56.7	+ 28.4	− 2	− 18.9
− 76.0	− 60	− 51.1	+ 30.2	− 1	− 18.3
− 58.0	− 50	− 45.6	+ 32.0	0	− 17.8
− 54.4	− 48	− 44.4	+ 33.8	+ 1	− 17.2
− 50.8	− 46	− 43.3	+ 35.6	+ 2	− 16.7
− 47.2	− 44	− 42.2	+ 37.4	+ 3	− 16.1
− 43.6	− 42	− 41.1	+ 39.2	+ 4	− 15.6
− 40.0	− 40	− 40.0	+ 41.0	+ 5	− 15.0
− 36.4	− 38	− 38.9	+ 42.8	+ 6	− 14.4
− 32.8	− 36	− 37.8	+ 44.6	+ 7	− 13.9
− 29.2	− 34	− 36.7	+ 46.4	+ 8	− 13.3
− 25.6	− 32	− 35.6	+ 48.2	+ 9	− 12.8
− 22.0	− 30	− 34.4	+ 50.0	+ 10	− 12.2
− 18.4	− 28	− 33.3	+ 51.8	+ 11	− 11.7
− 14.8	− 26	− 32.2	+ 53.6	+ 12	− 11.1
− 11.2	− 24	− 31.1	+ 55.4	+ 13	− 10.6
− 7.6	− 22	− 30.0	+ 57.2	+ 14	− 10.0
− 4.0	− 20	− 28.9	+ 59.0	+ 15	− 9.4
− 0.4	− 18	− 27.8	+ 60.8	+ 16	− 8.9
+ 3.2	− 16	− 26.7	+ 62.6	+ 17	− 8.3
+ 6.8	− 14	− 25.6	+ 64.4	+ 18	− 7.8
+ 10.4	− 12	− 24.4	+ 66.2	+ 19	− 7.2
+ 14.0	− 10	− 23.3	+ 68.0	+ 20	− 6.7
+ 15.8	− 9	− 22.8	+ 69.8	+ 21	− 6.1
+ 17.6	− 8	− 22.2	+ 71.6	+ 22	− 5.6
+ 19.4	− 7	− 21.7	+ 73.4	+ 23	− 5.0
+ 21.2	− 6	− 21.1	+ 75.2	+ 24	− 4.4

Temperature Conversion Tables

°F	°C or °F	°C	°F	°C or °F	°C
+ 77.0	+ 25	− 3.9	+158.0	+ 70	+ 21.1
+ 78.8	+ 26	− 3.3	+159.8	+ 71	+ 21.7
+ 80.6	+ 27	− 2.8	+161.6	+ 72	+ 22.2
+ 82.4	+ 28	− 2.2	+163.4	+ 73	+ 22.8
+ 84.2	+ 29	− 1.7	+165.2	+ 74	+ 23.3
+ 86.0	+ 30	− 1.1	+167.0	+ 75	+ 23.9
+ 87.8	+ 31	− 0.6	+168.8	+ 76	+ 24.4
+ 89.6	+ 32	0.0	+170.6	+ 77	+ 25.0
+ 91.4	+ 33	+ 0.6	+172.4	+ 78	+ 25.6
+ 93.2	+ 34	+ 1.1	+174.2	+ 79	+ 26.1
+ 95.0	+ 35	+ 1.7	+176.0	+ 80	+ 26.7
+ 96.8	+ 36	+ 2.2	+177.8	+ 81	+ 27.2
+ 98.6	+ 37	+ 2.8	+179.6	+ 82	+ 27.8
+100.4	+ 38	+ 3.3	+181.4	+ 83	+ 28.3
+102.2	+ 39	+ 3.9	+183.2	+ 84	+ 28.9
+104.0	+ 40	+ 4.4	+185.0	+ 85	+ 29.4
+106.8	+ 41	+ 5.0	+186.8	+ 86	+ 30.0
+107.6	+ 42	+ 5.6	+188.6	+ 87	+ 30.6
+109.4	+ 43	+ 6.1	+190.4	+ 88	+ 31.1
+111.2	+ 44	+ 6.7	+192.2	+ 89	+ 31.7
+113.0	+ 45	+ 7.2	+194.0	+ 90	+ 32.2
+114.8	+ 46	+ 7.8	+195.8	+ 91	+ 32.8
+116.6	+ 47	+ 8.3	+197.6	+ 92	+ 33.3
+118.4	+ 48	+ 8.9	+199.4	+ 93	+ 33.9
+120.2	+ 49	+ 9.4	+201.2	+ 94	+ 34.4
+122.0	+ 50	+ 10.0	+203.0	+ 95	+ 35.0
+123.8	+ 51	+ 10.6	+204.8	+ 96	+ 35.6
+125.6	+ 52	+ 11.1	+206.6	+ 97	+ 36.1
+127.4	+ 53	+ 11.7	+208.4	+ 98	+ 36.7
+129.2	+ 54	+ 12.2	+210.2	+ 99	+ 37.2
+131.0	+ 55	+ 12.8	+212.0	+100	+ 37.8
+132.8	+ 56	+ 13.3	+213.8	+101	+ 38.3
+134.6	+ 57	+ 13.9	+215.6	+102	+ 38.9
+136.4	+ 58	+ 14.4	+217.4	+103	+ 39.4
+138.2	+ 59	+ 15.0	+219.2	+104	+ 40.0
+140.0	+ 60	+ 15.6	+221.0	+105	+ 40.6
+141.8	+ 61	+ 16.1	+222.8	+106	+ 41.1
+143.6	+ 62	+ 16.7	+224.6	+107	+ 41.7
+145.4	+ 63	+ 17.2	+226.4	+108	+ 42.2
+147.2	+ 64	+ 17.8	+228.2	+109	+ 42.8
+149.0	+ 65	+ 18.3	+230.0	+110	+ 43.3
+150.8	+ 66	+ 18.9	+231.8	+111	+ 43.9
+152.6	+ 67	+ 19.4	+233.6	+112	+ 44.4
+154.4	+ 68	+ 20.0	+235.4	+113	+ 45.0
+156.2	+ 69	+ 20.6	+237.2	+114	+ 45.6

Temperature Conversion Tables

°F	°C or °F	°C	°F	°C or °F	°C
+239.0	+115	+ 46.1	+392.0	+200	+ 93.3
+240.8	+116	+ 46.7	+395.6	+202	+ 94.4
+242.6	+117	+ 47.2	+399.2	+204	+ 95.6
+244.4	+118	+ 47.8	+402.8	+206	+ 96.7
+246.2	+119	+ 48.3	+406.4	+208	+ 97.8
+248.0	+120	+ 48.9	+410.0	+210	+ 98.9
+251.6	+122	+ 50.0	+413.6	+212	+100.0
+255.2	+124	+ 51.1	+417.2	+214	+101.1
+258.8	+126	+ 52.2	+420.8	+216	+102.2
+262.4	+128	+ 53.3	+424.4	+218	+103.3
+266.0	+130	+ 54.4	+428.0	+220	+104.4
+269.6	+132	+ 55.6	+437.0	+225	+107.2
+273.2	+134	+ 56.7	+446.0	+230	+110.0
+276.8	+136	+ 57.8	+455.0	+235	+112.8
+280.4	+138	+ 58.9	+464.0	+240	+115.6
+284.0	+140	+ 60.0	+473.0	+245	+118.3
+287.6	+142	+ 61.1	+482.0	+250	+121.1
+291.2	+144	+ 62.2	+491.0	+255	+123.9
+294.8	+146	+ 63.3	+500.0	+260	+126.7
+298.4	+148	+ 64.4	+509.0	+265	+129.4
+302.0	+150	+ 65.6	+518.0	+270	+132.2
+305.6	+152	+ 66.7	+527.0	+275	+135.0
+309.2	+154	+ 67.8	+536.0	+280	+137.8
+312.8	+156	+ 68.9	+545.0	+285	+140.6
+316.4	+158	+ 70.0	+554.0	+290	+143.3
+320.0	+160	+ 71.1	+563.0	+295	+146.1
+323.6	+162	+ 72.2	+572.0	+300	+148.9
+327.2	+164	+ 73.3	+590.0	+310	+154.4
+330.8	+166	+ 74.4	+608.0	+320	+160.0
+334.4	+168	+ 75.6	+626.0	+330	+165.6
+338.0	+170	+ 76.7	+644.0	+340	+171.1
+341.6	+172	+ 77.8	+662.0	+350	+176.7
+345.2	+174	+ 78.9	+680.0	+360	+182.2
+348.8	+176	+ 80.0	+698.0	+370	+187.8
+352.4	+178	+ 81.1	+716.0	+380	+193.3
+356.0	+180	+ 82.2	+734.0	+390	+198.9
+359.6	+182	+ 83.3	+752.0	+400	+204.4
+363.2	+184	+ 84.4	+932.0	+500	+260.0
+366.8	+186	+ 85.6	+1112.0	+600	+315.6
+370.4	+188	+ 86.7	+1292.0	+700	+371.1
+374.0	+190	+ 87.8	+1472.0	+800	+427.0
+377.6	+192	+ 88.9	+1652.0	+900	+482.0
+381.2	+194	+ 90.0	+1832.0	+1000	+538.0
+384.8	+196	+ 91.1	+2012.0	+1100	+593.0
+388.4	+198	+ 92.2	+2192.0	+1200	+649.0

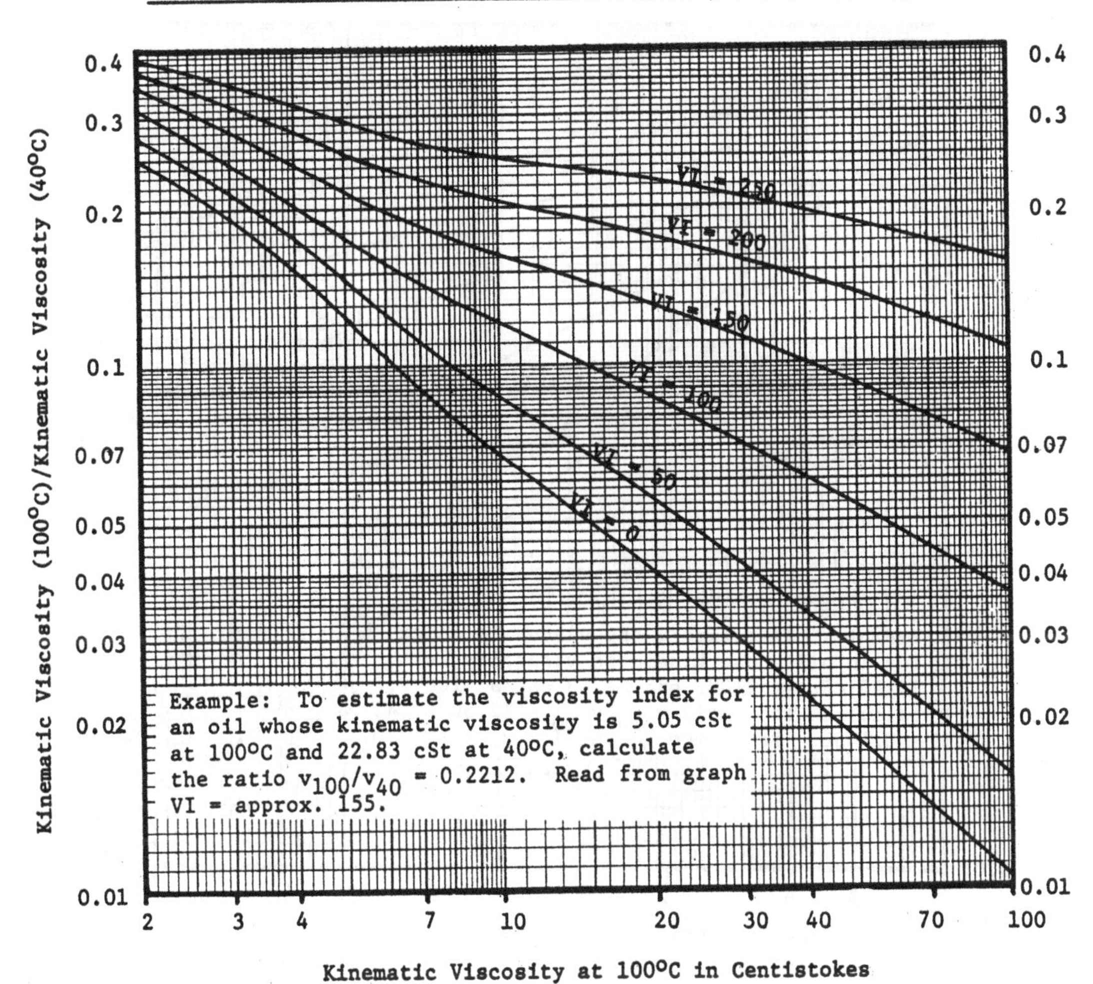
VISCOSITY INDEX FROM KINEMATIC VISCOSITIES AT 40°C AND 100°C
Kinematic Viscosity (100°C)/Kinematic Viscosity (40°C)
0.4
0.3
0.2
0.1
0.07
0.05
0.04
0.03
0.02
0.01
VI = 250
VI = 200
VI = 150
VI = 100
VI = 50
VI = 0
Example: To estimate the viscosity index for an oil whose kinematic viscosity is 5.05 cSt at 100°C and 22.83 cSt at 40°C, calculate the ratio v100/v40 = 0.2212. Read from graph VI = approx. 155.
2
3
4
7
10
20
30
40
70
100
Kinematic Viscosity at 100°C in Centistokes

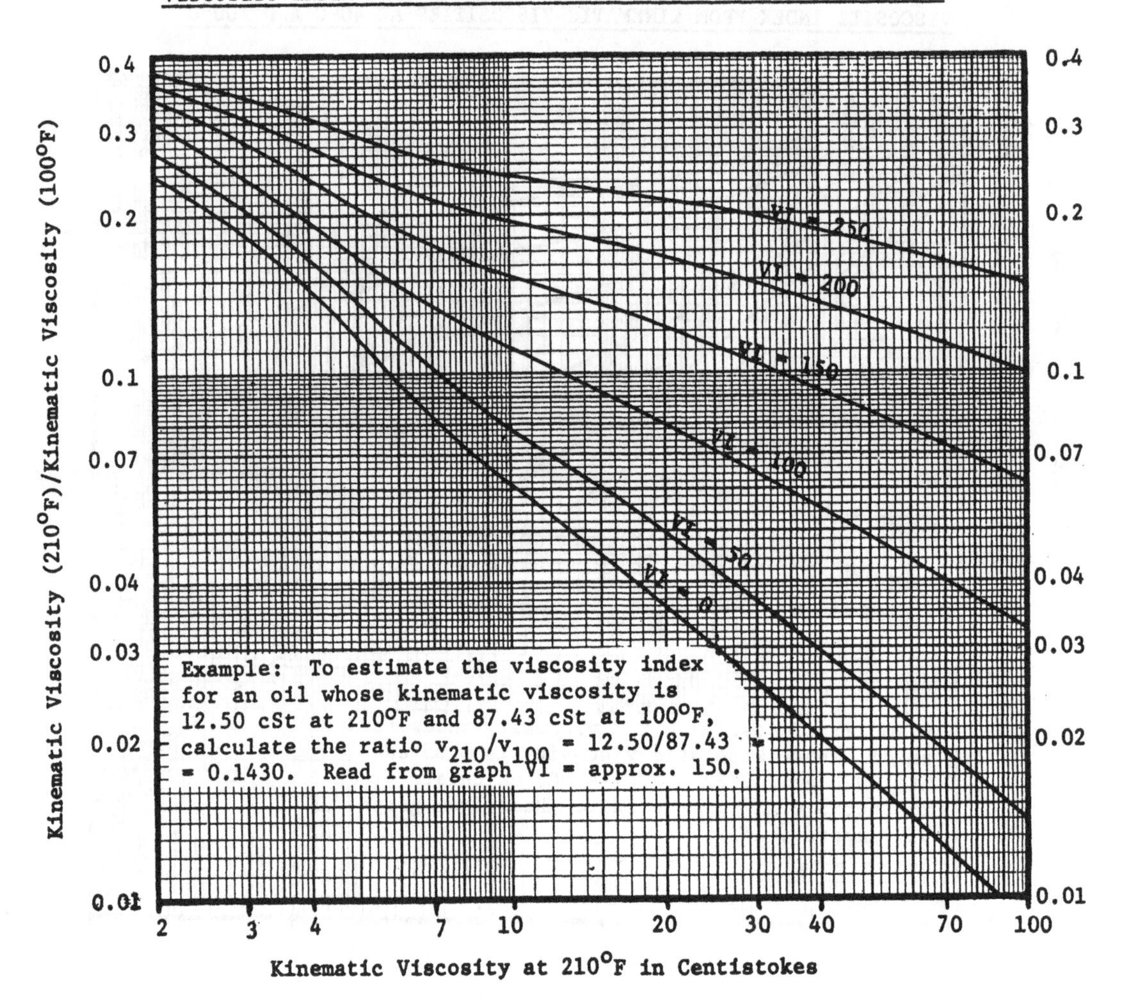
VISCOSITY INDEX FROM KINEMATIC VISCOSITIES AT 100°F AND 210°F
Kinematic Viscosity (210°F)/Kinematic Viscosity (100°F)
0.4
0.3
0.2
0.1
0.07
0.04
0.03
0.02
0.01
VI = 250
VI = 200
VI = 150
VI = 100
VI = 50
VI = 0
Example: To estimate the viscosity index for an oil whose kinematic viscosity is 12.50 cSt at 210°F and 87.43 cSt at 100°F, calculate the ratio v210/v100 = 12.50/87.43 = 0.1430. Read from graph VI = approx. 150.
2
3
4
7
10
20
30
40
70
100
Kinematic Viscosity at 210°F in Centistokes

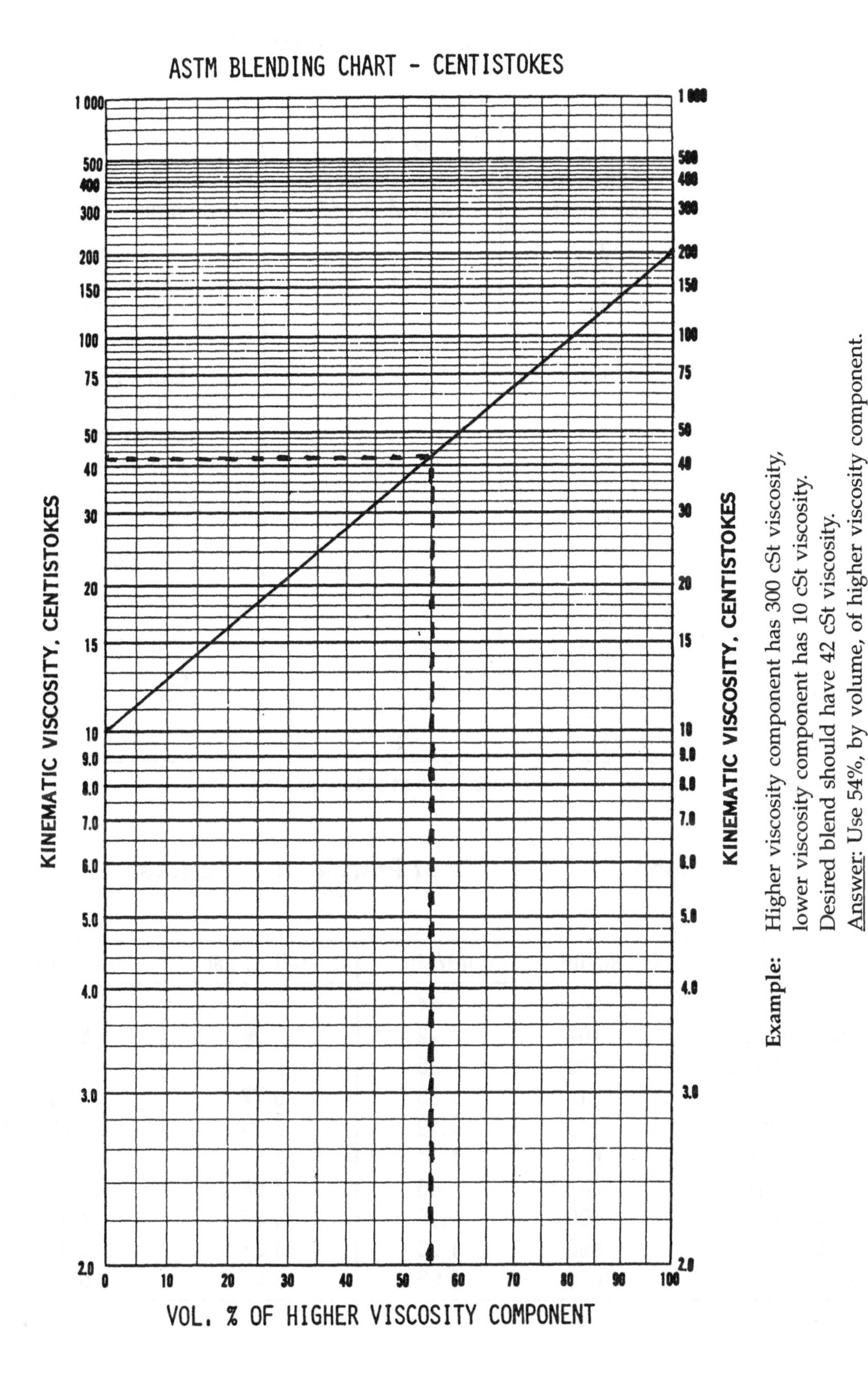

Example: Higher viscosity component has 300 cSt viscosity, lower viscosity component has 10 cSt viscosity.
Desired blend should have 42 cSt viscosity.
Answer: Use 54%, by volume, of higher viscosity component.

Approximate Color Scale Equivalents

DIN 51-578* ASTM D 1500	TAG-Robinson Colorimeter	Union (N.P.A.) Colorimeter (ASTM D 155)	N.P.A. Common Terms
0	22¾	—	—
0.5	21½	1	Lily White
1.0	17½	1½	Cream White
1.5	13½	1¾	—
2.0	11¾	2	Extra Pale
2.5	10½	2½	Extra Lemon Pale
3.0	9¾	3	Lemon Pale
3.5	9½	3½	Extra Orange Pale
4.0	9	3¾	Orange Pale
4.5	7	4¼	—
5.0	5½	4½	Pale
5.5	4½	4¾	Light Red
6.0	3¼	5½	
6.5	2½	6	Dark Red
7.0	2	7	Claret Red
7.5	1½	7½	—
8.0	1	8	—

*Color scales for ASTM D 1500 and the German standard DIN 51-578 are the same.

Representative Masses of Petroleum Products

	kg/m³	m³/Mg	lb/U.S. gal	bbl/tonne
LPG	542	1.84	4.53	11.60
Aviation Gasoline	707	1.42	5.90	8.90
Motor Gasoline	740	1.35	6.18	8.50
Kerosine	812	1.23	6.77	7.75
Distilate Fuel Oil	843	1.19	7.04	7.46
Lubricating Oil	899	1.11	7.50	7.00
Residual Fuel Oil	944	1.06	7.88	6.66
Paraffin Wax	799	1.25	6.67	7.87
Grease	998	1.00	8.33	6.30
Asphalt	1038	0.96	8.66	6.06

ABRIDGED GRAVITY, VOLUME, AND MASS CONVERSION TABLE

Approximate Conversions at 15°C Except Where Noted

Density 15°C kg/m³	Specific Gravity 60/60°F	API Gravity 60°F	dm³/Mg	U.S. gal/Mg	Barrels/Mg	Imp. gal/Mg
1075	1.0757	0.04	930.2	245.77	5.852	204.63
1050	1.0507	3.18	952.4	251.62	5.991	209.50
1025	1.0256	6.46	975.6	257.76	6.137	214.61
1000	1.0006	9.92	1000.0	264.20	6.291	219.97
990	0.9906	11.34	1010.1	266.87	6.354	222.20
980	0.9806	12.80	1020.4	269.59	6.419	224.46
970	0.9706	14.29	1030.9	272.37	6.485	226.78
960	0.9606	15.81	1041.7	275.21	6.553	229.14
950	0.9505	17.36	1052.6	278.11	6.622	231.55
940	0.9405	18.95	1063.8	281.06	6.692	234.01
930	0.9305	20.56	1075.3	284.09	6.764	236.53
920	0.9205	22.22	1087.0	287.17	6.838	239.10
910	0.9105	23.91	1098.9	290.33	6.913	241.73
900	0.9005	25.64	1111.1	293.56	6.989	244.42
890	0.8905	27.40	1123.6	296.85	7.068	247.16
880	0.8805	29.21	1136.4	300.23	7.148	249.97
870	0.8705	31.06	1149.4	303.68	7.231	252.84
860	0.8605	32.95	1162.8	307.21	7.315	255.78
850	0.8504	34.89	1176.5	310.82	7.401	258.79
840	0.8404	36.87	1190.5	314.52	7.489	261.87
830	0.8304	38.90	1204.8	318.31	7.579	265.03
820	0.8204	40.98	1219.5	322.20	7.671	268.26
810	0.8104	43.11	1234.6	326.17	7.766	271.57
800	0.8004	45.29	1250.0	330.25	7.863	274.97
790	0.7903	47.53	1265.8	334.43	7.963	278.45
780	0.7803	49.83	1282.1	338.72	8.065	282.02
770	0.7703	52.19	1298.7	343.12	8.170	285.68
760	0.7603	54.61	1315.8	347.63	8.277	289.44
750	0.7503	57.10	1333.3	352.27	8.387	293.30
725	0.7252	63.61	1379.3	364.41	8.677	303.41
700	0.7002	70.59	1428.6	377.43	8.986	314.25
675	0.6751	78.08	1481.5	391.41	9.319	325.89
650	0.6501	86.16	1538.5	406.46	9.678	338.42
625	0.6251	94.88	1600.0	422.72	10.065	351.96
600	0.6000	—	1666.7	440.33	10.484	366.62

MASS CONVERSIONS

	mg	g	lb	kg	short ton	Mg (tonne)
mg	1	0.001	0.000 002	0.000 001	---	---
g	1 000.0	1	0.002 205	0.001	---	---
lb	453 592.4	453.592 4	1	0.453 592	0.000 5	0.000 454
kg	1 000 000	1 000.0	2.204 623	1	0.001 102	0.001
short ton	---	---	2 000.0	907.184 7	1	0.907 185
Mg (tonne)	---	---	2 204.623	1 000.0	1.102 311	1

VOLUME CONVERSIONS

	U.S. qt	dm^3 (litre)	U. S. gal	Imp. gal	Barrel	m^3 (kL)
U.S. qt	1	0.946 353	0.25	0.208 168	0.005 952	0.000 946
dm^3 (litre)	1.056 688	1	0.264 172	0.219 969	0.006 290	0.001
U.S. gal	4.0	3.785 412	1	0.832 674	0.023 810	0.003 785
Imp. gal	4.803 802	4.546 092	1.200 950	1	0.028 594	0.004 546
Barrel	168.0	158.987 3	42.0	34.972 30	1	0.158 987
m^3 (kL)	1 056.688	1 000.0	264.172 0	219.969 2	6.289 811	1

PRESSURE CONVERSIONS

	mm Hg	in. H_2O	kPa	psi	kg/cm^2	atm.
mm Hg	1	0.535 776	0.133 322	0.019 337	0.001 360	0.001 316
in. H_2O	1.866 453	1	0.248 84	0.036 091	0.002 537	0.002 456
kPa	7.500 615	4.018 647	1	0.145 038	0.010 197	0.009 869
psi	51.714 92	27.707 59	6.894 757	1	0.070 307	0.068 046
kg/cm^2	735.559 1	394.094 6	98.066 50	14.223 34	1	0.967 841
atm.	760.0	407.189 4	101.325 0	14.695 95	1.033 227	1

POWER CONVERSIONS

	W	hp	kW	Btu/s	kcal/s	MW
W	1	0.001 341	0.001	0.000 948	0.000 239	0.000 001
hp	745.699 9	1	0.745 700	0.706 787	0.178 227	0.000 746
kW	1 000.0	1.341 022	1	0.947 817	0.239 006	0.001
Btu/s	1 055.056	1.414 853	1.055 056	1	0.252 164	0.001 055
kcal/s	4 184.0	5.610 836	4.184	3.965 666	1	0.004 184
MW	1 000 000	1 341.022	1 000.0	947.817 0	239.005 7	1

ENERGY, WORK CONVERSIONS

	kJ	Btu①	kcal②	MJ	hp·h	kW·h
kJ	1	0.947 817	0.239 006	0.001	0.000 373	0.000 278
Btu	1.055 056	1	0.252 164	0.001 055	0.000 393	0.000 293
kcal	4.184 0	3.965 666	1	0.004 184	0.001 559	0.001 162
MJ	1 000.0	947.817 0	239.005 7	1	0.372 506	0.277 778
hp·h	2 684.520	2 544.434	641.615 7	2.684 520	1	0.745 700
kW·h	3 600.0	3 412.141	860.420 7	3.60	1.341 022	1

① International Table
② Thermochemical

LENGTH CONVERSIONS

	mm	cm	in.	ft	m	km	mile
mm	1	0.1	0.039 370	0.003 281	0.001	0.000 001	---
cm	10.0	1	0.393 701	0.032 808	0.01	0.000 01	---
in.	25.4	2.54	1	0.083 333	0.0254	0.000 025	---
ft	304.8	30.48	12.0	1	0.304 8	0.000 305	0.000 189
m	1 000.0	100.0	39.370 08	3.280 840	1	0.001	0.000 621
km	1 000 000	100 000.0	39 370.08	3 280.840	1 000.0	1	0.621 371
mile	---	---	63 360.0	5 280.0	1 609.344	1.609 344	1

AREA CONVERSIONS

	mm^2	cm^2	$in.^2$	ft^2	m^2	acre	ha
mm^2	1	0.01	0.001 550	0.000 011	0.000 001	---	---
cm^2	100.0	1	0.155 000	0.001 076	0.000 1	---	---
$in.^2$	645.16	6.451 6	1	0.006 944	0.000 645	---	---
ft^2	92 903.04	929.030 4	144.0	1	0.092 903	0.000 023	0.000 009
m^2	1 000 000	10 000.0	1 550.003	10.763 91	1	0.000 247	0.000 1
acre	---	---	---	43 560.0	4 046.856	1	0.404 686
ha	---	---	---	107 639.1	10 000.0	2.471 054	1

SAE J300

Rev. March 1993

SAE VISCOSITY GRADES FOR ENGINE OILS[1]

SAE Viscosity Grade	Low-Temperature (°C) Cranking Viscosity[2], cP Max	Low-Temperature (°C) Pumping Viscosity,[3] cP Max With No Yield Stress	Kinematic Viscosity[4] (cSt) at 100 °C Min	Kinematic Viscosity[4] (cSt) at 100 °C Max	High-Shear Viscosity[5] (cP) at 150 °C and 10^6 & [1] Min
0W	3250 at -30	30 000 at -35	3.8	—	—
5W	3500 at -25	30 000 at -30	3.8	—	—
10W	3500 at -20	30 000 at -25	4.1	—	—
15W	3500 at -15	30 000 at -20	5.6	—	—
20W	4500 at -10	30 000 at -15	5.6	—	—
25W	6000 at -5	30 000 at -10	9.3	—	—
20	—	—	5.6	<9.3	2.6
30	—	—	9.3	<12.5	2.9
40	—	—	12.5	<16.3	2.9 (0W-40, 5W-40, and 10W-40 grades)
40	—	—	12.5	<16.3	3.7 (15W-40, 20W-40, 25W-40, 40 grades)
50	—	—	16.3	<21.9	3.7
60	—	—	21.9	<26.1	3.7

Note — 1 cP = 1 mPa's; 1 cSt = 1 mm^2/s

1 All values are critical specifications as defined by ASTM D 3244

2 ASTM D 5293

3 ASTM D 4684 Note that the presence of any yield stress detectable by this method constitutes a failure regardless of viscosity.

4 ASTM D 445

5 ASTM D 4683, CEC L-36-A-90 (ASTM D 4741)

SAE J306

Rev. October 1991

SAE Gear Viscosity Grades Axle and Manual Transmission Lubricants

	70W	75W	80W	85W	90	140	250
Viscosity at 100°C							
Min. (cSt)	4.1	4.1	7.0	11.0	13.5	24.0	41.0
Max. (cSt)	No requirements				<24.0	<41.0	No req
Viscosity at 150,000 cP							
Max. Temp. °C	-55	-40	-26	-12	No requirements		

MIL-L-2105D

	75W	80W-90	85W-140
Viscosity at 100°C			
Min. (cSt)	4.1	13.5	24.0
Max. (cSt)	—	<24.0	<41.0
Viscosity at 150,000 cP Max. Temp. °C	-40	-26	-12
Channel Point, Min. °C	-45	-35	-20
Flash Point, Min. °C	150	165	180

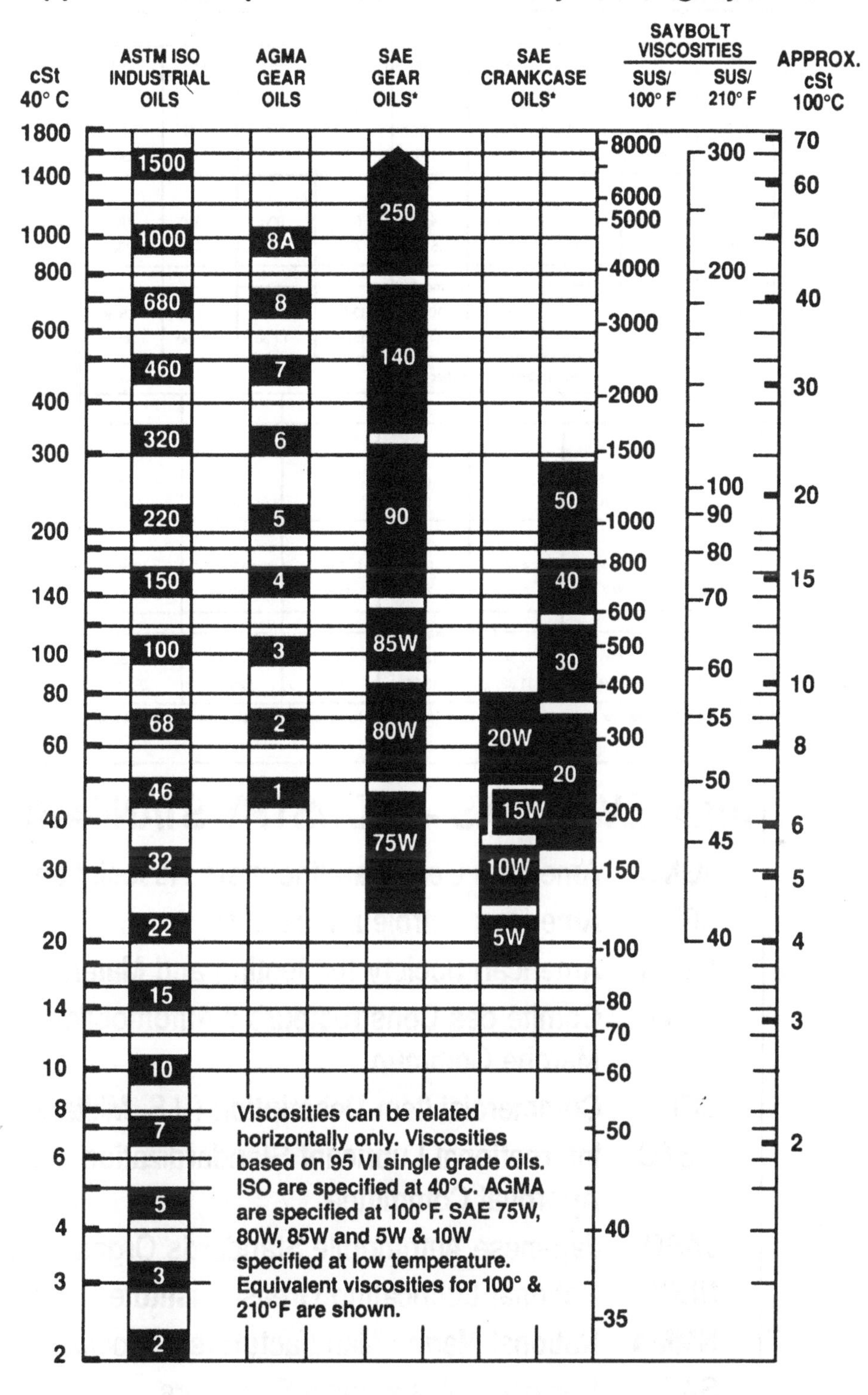
Approximate Equivalence of Viscosity Grading Systems
cSt 40° C
ASTM ISO INDUSTRIAL OILS
AGMA GEAR OILS
SAE GEAR OILS*
SAE CRANKCASE OILS*
SAYBOLT VISCOSITIES
SUS/ 100° F
SUS/ 210° F
APPROX. cSt 100°C
1800
1400
1000
800
600
400
300
200
140
100
80
60
40
30
20
14
10
8
6
4
3
2
1500
1000
680
460
320
220
150
100
68
46
32
22
15
10
7
5
3
2
8A
8
7
6
5
4
3
2
1
250
140
90
85W
80W
75W
50
40
30
20W
20
15W
10W
5W
8000
6000
5000
4000
3000
2000
1500
1000
800
600
500
400
300
200
150
100
80
70
60
50
40
35
300
200
100
90
80
70
60
55
50
45
40
70
60
50
40
30
20
15
10
8
6
5
4
3
2
Viscosities can be related horizontally only. Viscosities based on 95 VI single grade oils. ISO are specified at 40°C. AGMA are specified at 100°F. SAE 75W, 80W, 85W and 5W & 10W specified at low temperature. Equivalent viscosities for 100° & 210°F are shown.

COMPARATIVE VISCOSITY CLASSIFICATIONS

	ISO Viscosity Grade								
	22	32	46	68	100	150	220	320	460
Vis. @ 40° C									
Min. cSt	19.8	28.8	41.4	61.2	90	135	198	288	414
Max. cSt	24.2	35.2	50.6	74.8	110	165	242	352	506
Vis. @ 100° F									
Min. SUS	96	135	191	280	410	615	900	1310	1880
Max. SUS	115	164	234	345	500	750	1110	1600	2300
Vis. @ 100°C	← No Requirement* →								
Min. cSt	4.0	4.97	6.22	7.96	10.30	13.56	17.50	22.40	28.30
Max. cSt	4.5	5.61	7.05	9.09	11.82	15.51	19.96	25.50	32.30

	AGMA Lubricant Number						
	1	2	3	4	5	6	7
Vis. @ 40° C							
Min. cSt	41.4	61.2	90	135	198	288	414
Max. cSt	50.6	74.8	110	165	242	352	506
Vis. @ 100° C							
Min. SUS	193	284	417	626	918	1335	1919
Max. SUS	235	347	510	765	1122	1632	2346
Vis. @ 100° C	← No Requirement* →						
Min. cSt	6.22	7.96	10.30	13.56	17.50	22.40	28.30
Max. cSt	7.05	9.09	11.82	15.51	19.96	25.50	32.30

*Viscosities given are not required; however, for "informational purposes," viscosities listed are calculated assuming an oil with a V.I. of 95.

ABBREVIATIONS – INDUSTRY STANDARDS

AGMA	American Gear Manufacturers Association
API	American Petroleum Institute
ASTM	American Society for Testing and Materials
CCMC	Comité des Constructeurs d'Automobiles du Marché Commun
CID	Commercial Item Description (U.S. Military)
ILSAC	International Lubricant Standardization and Approval Committee
JASO	Japanese Automobile Standards Organization
NLGI	National Lubricating Grease Institute
NMMA	National Marine Manufacturers Association
SAE	Society of Automotive Engineers
STLE	Society of Tribologists and Lubrication Engineers

Viscosity Index, 0-100

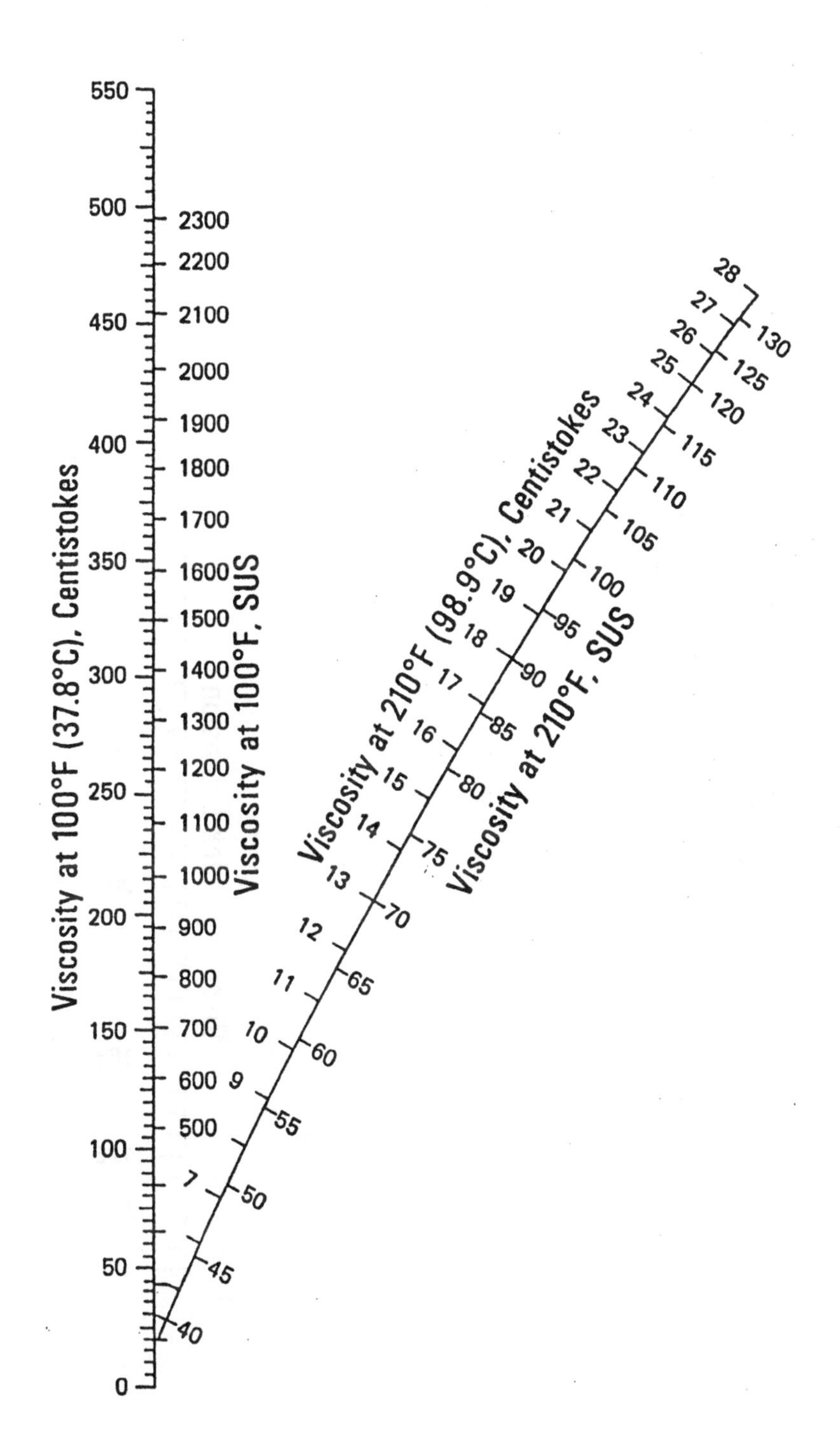

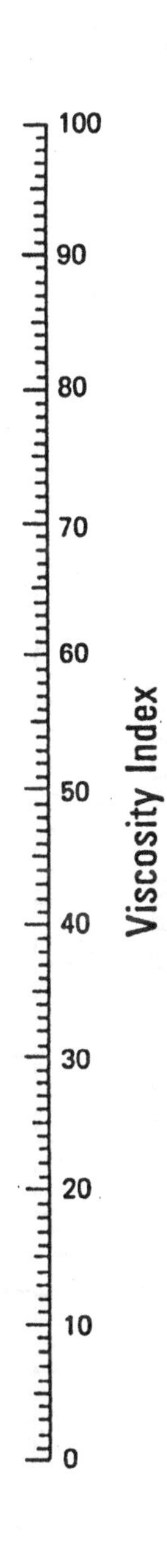

Viscosity Index, 100-400 (Based on ASTM D 2270)

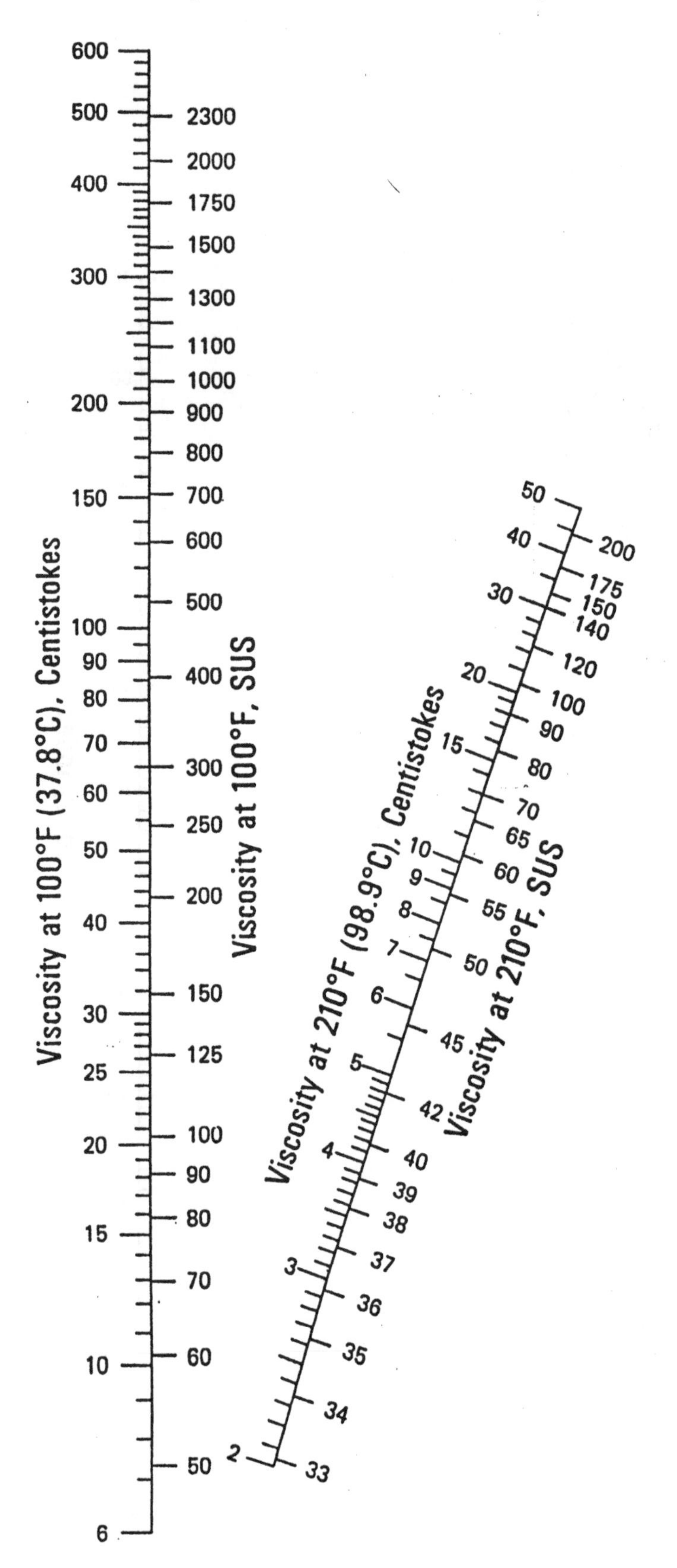

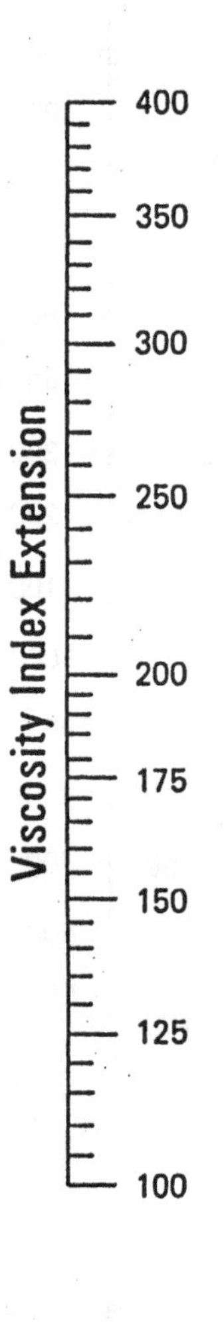

Glossary of Terms

—A—

AAR—Abbreviation for "American Association of Railroads"

Absolute viscosity—A term used interchangeably with viscosity to distinguish it from kinematic viscosity or commercial viscosity. It is occasionally referred to as dynamic viscosity. Absolute viscosity and kinematic viscosity are expressed in fundamental units. Commercial viscosity such as Saybolt viscosity is expressed in arbitrary units.

Acidity—In lubricants, acidity denotes the presence of acid-type constituents whose concentration is usually defined in terms of acid number. The constituents vary in nature and may or may not markedly influence the behavior of the lubricant. (See also Total Acid Number.)

Acid Number—See Strong Acid Number and Total Acid Number.

Additive—A chemical compound or compounds added to a lubricant for the purpose of imparting new properties or to improve those properties that the lubricant already has.

AGMA—Abbreviation for "American Gear Manufacturers Association," an organization serving the gear industry.

Aluminum Soap Base Greases—Greases containing aluminum soap and mineral oils. They are mainly used in gearboxes for gear lubrication.

Aniline Point—The Aniline Point of a petroleum product is the lowest temperature at which it is completely miscible with an equal volume of freshly distilled aniline.

Anti Friction Bearing—A rolling contact type bearing in which the rotating or moving member is supported or guided by means of ball or roller elements. Does not mean without friction.

Antioxidant—A substance which retards the action of oxidation.

Anti-Stick-Slip Additives -They prevent stick-slip operation, e.g. carriage tracks and guideways in machine tools.

Antiwear Additives—Additives to reduce wear in the mixed friction range:

- Mild additives, e.g. fatty acids, fatty oils
- EP additives, e.g. lead, sulphur, chlorine and phosphorus compounds
- Dry lubricants, e.g. graphite and molybdenum disulphide.

API— American Petroleum Institute

API Gravity—A gravity scale established by the API and in general use in the petroleum industry, the unity being called the "API degree." This unit is defined in terms of specific gravity as follows:

$$\text{sp.gr.}(@60°F) = \frac{141.5}{(\text{Degrees API} + 131.5)}$$

or

Degrees API = [141.5 / sp.gr.(@60—°F)]—131.5

Apparent viscosity—A measure of the resistance to flow of a grease whose viscosity varies with shear rate. It is defined as the ratio of the shear stress to the shear rate calculated from Poiseulle's equation at a given rate of shear and is expressed in poises.

Aromatics—Unsaturate hydrocarbons with a molecular ring structure (benzene, toluol, naphthalene). Aromatics have poor viscosity temperature properties and affect the oxidation stability of lubricants.

Ash Content—refers to the incombustible residues of a lubricant. The ash can be of different origin: it can stem from additives dissolved in the oil; graphite and molybdenum disulphide, soaps and other grease thickeners are ash producers. Fresh, straight refined mineral oils must be ash free. Used oils also contain insoluble metal soaps produced during operation, incombustible residues from contaminants, e.g. wear particles from bearing components and seals. Sometimes, incipient bearing damage can be diagnosed from the ash content.

Asphaltic—Essentially composed of or similar to asphalt. Frequently applied to naphthenic base lubricating oils derived from crudes that contain asphalt.

ASTM—Abbreviation for "American Society for Testing Materials," a society for developing standards for materials and test methods.

ATF—Abbreviation for Automatic Transmission Fluid. Special lubricants adapted to the requirements in automatic transmissions.

Axial Load Bearing—A bearing in which the load acts in the direction of the axis of rotation.

—B—

Babbitt—A soft, white, non-ferrous alloy bearing material composed principally of copper, antimony, tin and lead.

Ball Bearing—An antifriction bearing comprising rolling elements in the form of balls.

Barium Complex Soap Base Greases—Greases consisting of barium complex soaps and mineral oils or synthetic oils. They are water repellent, retain their consistency, and form a lubricating film with a high load carrying capacity.

Base Oil—is the oil contained in a grease. The amount of oil varies with the type of thickener and the grease application. The penetration number and the frictional behavior of the grease very with the amount of base oil and its viscosity.

Base Stock—A fully refined lube oil, which is a component of lubricant formulations.

Bearing—A support or guide by means of which a moving part such as a shaft or axle is positioned with respect to the other parts of a mechanism.

Bentonites—Minerals (e.g., aluminum silicates) which are used for the production of thermally stable greases with good low-temperature properties.

Bleeding—The oil contained in the grease separates from the soap. This can be caused e.g. by low resistance to working and/or a low thermal stability of the greases.

Block Grease—Generally, a grease of high soap content, which, under normal temperatures is firm to the touch and can be handled in block or stick form.

Bloom—A sheen or fluorescence evident in some petroleum oils when viewed by reflected light.

BMEP—Brake mean effective pressure (in gas engine power cylilnders).

Boundary Lubrication—A condition of lubrication in which the bulk viscosity characteristics of the lubricant do not apply or in which partial contact takes place between the mating surfaces.

Also refers to a thin film, imperfect, or non-viscous lubrication.

Bright Stock—A term referring to high viscosity lubricating oils which have been refined to make them clear products of good color.

By-Pass Filtration—A system of filtration in which only a portion of the total flow of a circulating fluid system passes through a filter or in which a filter, having its own circulating pump, operates in parallel with the main flow.

—C—

Calcium Soap Base Greases—Calcium soap base greases are water repellent and are therefore excellent sealants against the ingress of water. Since their corrosion protection is limited, they are usually treated with corrosion inhibitors. Additive-treated calcium soap base greases are appropriate for rolling mill bearings, even in chocks which are exposed to roll cooling water. Calcium soap base greases are normally suitable for temperatures from—20 to +50°C.

Carbon Residues—The residue remaining after the evaporation of a sample of mineral oil under specified conditions, i.e., Ramsbottom and Conradson.

Centipoise (cP.)—A unit of absolute viscosity. 1 centipoise = .01 poise.

Former unit for the dynamic viscosity.

1 cP= 1 mPa s

Centistoks (cSt.)—A standard unit of kinematic viscosity = 0.10 stoke. Former unit for the kinematic viscosity.

1 cSt= 1 mm^2/s

Cetane Number—A number that expresses the ignition quality of diesel fuel and equal to the percentage by volume of cetane ($C_{16}H_{34}$) in a blend with methyl naphthalene, which blend has the same ignition performance as the test fuel.

Channeling—The tendency of grease to form an unobstructed path or channel following the movement of the rolling elements in a bearing.

Channel Point—Lowest safe temperature at which a gear lubricant can be used.

Characteristics—The following are the most important characteristics of lubricating oils: flash point, density, viscosity at 40°C, setting point, and additive data. Greases are defined by: saponification basis, drop point, worked penetration and, where present, additives.

Circulating Effect—If grease is carried along by rotating parts the rotation causes lumps of grease to be pulled between rolling elements and raceways with a corresponding increase in friction due to grease working. High-speed applications therefore require greases which are not likely to be carried along. The circulating effect depends on the type of thickener, penetration, temperature and the bearing type. Sodium soap base greases tend to participate in the circulating movement.

Circulating Lubrication—A system of lubrication in which the lubricant, after having passed through a bearing or group of bearings, is re-circulated by means of a pump.

Cleveland Open Cup—See Flash Point, Fire Point.

Coefficient of Friction—The ratio of the friction force between two bodies to the normal, or perpendicular, force between them.

Color of Oils—Spent oils are usually judged by their color. However, caution should be exercised in using this criterion because even fresh oil can be more or less dark. Whether the discoloration is due to oxidation can only be confirmed by comparing it with a fresh sample of the same oil type. Contamination by dust and soot may also be the cause of discoloration.

Complex Greases—Besides metal soaps of high-molecular fatty acids, complex soap base greases contain metal salts of low-molecular organic acids. These salts and the soap form a complex compound which outperforms conventional greases as far as thermal stability, water resistance, anti-corrosive action and load carrying capacity are concerned.

Compounded Oil—A petroleum oil to which have other chemical substances added .

Consistency—A term used synonymously with the term Penetration Number of a grease.

Copper Corrosion Test—Method for determining active sulphur in mineral oils (DIN 51759, ASTM D 130-75) and greases (DIN 51 811, ASTM D 130-68, IP154/69).

Corrosion—The attrition or wearing away of a substance by acid or electrochemical action.

Corrosion Inhibiting Greases, Corrosion Inhibiting Oils—They protect corrodible metal surfaces against moisture and atmospheric oxygen.

Cup Grease—An early term for a calcium or lime base grease, practically obsolete now but meant originally to designate a degree of quality suitable for grease cup application, etc.

Cutting Fluid or Oil—Any fluid applied to a cutting tool to assist in the cutting operation by cooling, lubricating or other means.

CVD—Chemical Vapor Deposition—A method of thin coating (3-5 microns) metal parts with metallic alloys through a gaseous medium. The coating adds to the hardness while reducing wear and increasing lubricity of base metal.

—D—

Demulsibility—The ability of a non-water-miscible fluid to separate from water with which it may be mixed. The higher the demulsibility rating, the more rapidly the fluid separates from water. Demulsibility is sometimes expressed as the rate, in cubic centimeters per hour, or settling out of a fluid from an emulsion under specified conditions See Steam Emulsion Number.

Density—The mass of a unit volume of a substance Its numerical value varies with the units used.

Detergent—In lubrication, either an additive or a compounded lubricant having the property of keeping insoluble matter in suspension, thus preventing its deposition where it would be harmful. A detergent may also re-disperse deposits already formed.

Deterioration—is the chemico-physical alteration of lubricants under the effect of atmospheric oxygen, heat, pressure, humidity, metallic debris etc. Deterioration of mineral oils is indicated by a change in color and viscosity and by sludge formation. Grease deterioration: change in color, consistency and structure. Oxidation life test ASTM D-942.

Dewaxing—Process which removes wax from a lube distillate by solvent means (physical separation) or catalytic means (conversion).

Dielectric Strength—A measure of the ability of an insulating fluid to withstand electric stress (voltage) without failure. Fluids with high dielectric strength (usually expressed in volts or kilovolts) are good electrical insulators.

Dispersing—In lubrication, usually used interchangeably with detergent. An additive which keeps fine particles of insoluble materials in a homogeneous solution. Hence, particles are not permitted to settle out and accumulate.

Dispersion lubrication—Here the grease is dispersed in a suitable solvent, e.g., toluol; it is contained in the liquid in a finely dispersed undissolved state. The cleaned dry bearing is dipped into the compound and dried in a dust free environment, a thin grease film remaining on the bearing. These bearings excel by an extremely low lubricant shearing friction.

Distillate—A term applied to a liquid collected when condensing distilled vapors such as naphtha, kerosene, fuel oil and light lubricating oils.

Dopes—see additives.

Doped Lubricants— see Additive-treated lubricants.

Drop Feed Lubrication—A system of lubrication in which the lubricant is applied to the bearing surfaces in the form of drops at regular intervals.

Drop Point—Temperature at which a grease sample, when heated under standard test conditions, passes into a liquid state, flows through the opening of a grease cup and drops to the bottom of the test tube. Grease: DIN51 801T1, ASTMD-566

Dry Lubricants—Substances, such as graphite, molybdenum disulphide, or PTFE suspended in oils and greases, or applied directly.

Dynamic Viscosity—See Absolute Viscosity.

—E—

Elastic Behavior of Greases—The elastic properties of lubricating greases indicate the suitability of a grease for centralized lubrication systems (DIN 51 816T2).

Emcor Method—Testing of corrosion preventing properties of rolling bearing greases according to DIN 51 802.

Emulsibility—Tendency of an oil to emulsify with water.

Emulsifiers—Additives which help to form an emulsion.

Emulsion—Mixture of insoluble substances, usually mineral oils with water, which is activated by emulsifiers.

EP Lubricants—Lubricants that have been fortified with additives that appreciably increase the load carrying properties of the base lubricant, thus reducing excessive wear.

Esters (Synthetic Lubricating Oils)—Compounds of acids and alcohols with water eliminated. Esters of higher alcohols with divalent fatty acids form the diester oils (synthetic lubricating oils). Esters of polyhydric alcohols and different organic acids are particularly heat stable.

Evaporation Loss—Lubricating oil losses occurring at higher temperatures due to evaporation. It can lead to an increase in oil consumption and also to an alteration of the oil properties.

Extreme-Pressure Lubricants—see EP lubricants.

—F—

Fatty Acid—An organic acid of aliphatic structure originally derived from fats and fatty oils.

Fiber Grease—Grease having a distinctly fibrous structure which is noticeable when a sample of the grease is pulled apart. Greases having this fibrous structure tend to resist being thrown off gears and out of bearings.

Filler—Any solid substance such as talc, mica, or various powders, etc., which is added to a grease to increase its weight or consistency.

Filter—Any device or porous substance used as a strainer for cleaning fluids by removing suspended matter.

Fire Point (Cleveland Open Cup)—The flash point of an oil is the temperature to which it must be heated to give off sufficient vapor to form a sustained flammable mixture with air when a small flame is applied under specified conditions.

Flash Point—Flash point is that temperature to which an oil must be heated for sufficient vapor to be given off to form, briefly, a flammable mixture with air. The flash point is one of the characteristics of oils; it is not a criterion for their quality.

Flinger Disk—Disk dipping in lubricant sump.

Flow Pressure—Pressure required to press grease in a continuous stream from a nozzle. It is a measure of the consistency and fluidity of a grease. It is determined according to DIN 51 805.

Foam—A froth produced by whipping air into a lubricant. Foaming in mineral oils should be avoided. Foaming promotes deterioration of the oil. Excessive foaming can lead to an overflow and, consequently loss of oil.

Force Feed Lubrication—A system of lubrication in which the lubricant is supplied to the bearing surface under pressure.

Form Oil—A compound or an oil used to lubricate wooden or metal concrete forms in order to keep cement from sticking to them.

Four Ball Test Rig—Machine for lubricant testing (DIN 51 350, ASTM D 2266-67, ASTM D 25 9669, ASTM D 2783-71, IP 239/73). Four balls are arranged in a pyramid shape, with the upper ball rotating. The load applied can be increased until welding occurs between the balls (welding load). The load, expressed in N, is the four ball welding load. The diameter of the weld scar on the stationary balls measured after one hour of testing is the four ball wear value which is used for wear evaluation.

Fretting Corrosion—A process of mechanical attrition combined with chemical reaction taking place at the common boundary of loaded contact surfaces having small oscillatory relative motion.

Friction—The resisting force encountered at the common boundary between two bodies when, under the action of an external force, one body moves or tends to move over the surface of the other.

Full Flow Filtration—A system of filtration in which the total flow of a circulating fluid system passes through a filter.

—G—

Gear Greases—Gear greases are usually sodium soap based, stringy, soft to semifluid greases (NLGI 0 and 00) for gears and gear motors. Some greases are treated with EP additives.

Gel Greases—They contain an inorganic-organic thickener made up of finely dispersed solid particles; the porous surface of these particles tends to absorb oil. Gel greases are suitable for a wide temperature range and are water resistant. Caution is recommended at high speeds and loads.

Graphite—A crystalline form of carbon either natural or synthetic in origin, which is used as a lubricant.

Gravity—see Specific Gravity, API Gravity

Grease—A lubricant composed of an oil or oils thickened with a soap, or other thickener to a solid or semi-solid consistency.

Grease Service Life—Life of a grease charge determined in laboratory and field tests. The individual life values scatter by 1: 10 even under comparable test and operating conditions.

Gum—A rubber-like, sticky deposit black or dark brown in color, which results from the oxidation of lubricating oils in service.

—H—

HD Oils—Heavy-duty oils are additive-treated engine oils particularly adapted to the rugged conditions in internal combustion engines.

High-Temperature Greases—Lithium greases can be used at steady-state temperatures up to 130°C and bentonite greases up to 140°C. Special MoS_2, silicone, and synthetic greases can be used up to 260° C.

Homogenizing—Final step in grease production. In order to obtain a uniform structure and fine dispersion of the thickener, the grease is thoroughly worked in a special machine.

Hydraulic Fluids—Fire-resistant pressure fluids for hydraulic load transmission and control.

Hydraulic Oil—An oil specially suited for use as a power transmission medium in hydraulically operated equipment. Non-aging, thin-bodied, non-foaming, highly refined hydraulic fluids produced from mineral oil, with a low setting point, for use in hydraulic systems.

Hydrodynamic Lubrication—A system of lubrication in which the shape and relative motion of the sliding surfaces causes the formation of a fluid film having sufficient pressure to separate the surfaces.

Hydrotreating—A process which converts and removes undesirable components with the use of a catalyst.

Hypoid Gear Lubricant—A gear lubricant having extreme pressure characteristics for use with a hypoid type of gear as in the differential of an automobile.

HVI—High Viscosity Index, typically from 80 to 110 VI units.

—I—

Inhibitor—Any substance which slows or prevents chemical reaction or corrosion.

Interfacial Tension (I.F.T)—The energy per unit area present at the boundary of two immiscible fluids. It is commonly measured as the force per unit length necessary to draw a thin wire or ring through the interface.

Intermediate Base Crude—See Mixed Base Crude.

ISO—International Standards Organization, sets viscosity reference scales.

ISO-equivalent—Consistency of greases at 25°C measured by the penetration depth of a standard cone, after treatment of the grease sample in a grease worker (DIN 51 804).

—J—

Journal Bearing—A sliding type of bearing in conjunction with which a journal operates. In a full or sleeve type journal bearing, the bearing surface is 380. in extent. In a partial bearing, the bearing surface is less than 360° in extent.

—K—

Kinematic Viscosity—The absolute viscosity of a fluid divided by its density. In a c.g.s. system, the standard unit of kinematic viscosity is the stoke and is expressed in sq. cm. per. sec. In the English system, the standard unit of kinematic viscosity is the newt and is expressed in sq. in. per sec.

—L—

Lacquer—A deposit resulting from the oxidation and polymerization of fuels and lubricants when exposed to high temperatures. Similar to but harder than varnish.

Lard Oil—An animal oil prepared from chilled lard or from the fat of swine.

Lime Base Grease—A grease prepared from a lubricating oil and a calcium soap.

Lithium Soap Base Greases—have definite performance merits in terms of water resistance and temperature range. Frequently, they incorporate oxidation inhibitors, corrosion inhibitors and EP additives. Lithium soap base greases are widely used as rolling bearing greases. To a limited extent, lithium greases emulsify with water. They can, to a certain degree, tolerate moisture, but severe moisture or ingress of water should be prevented, because this would cause the grease to become extremely soft and escape from the bearing. Standard lithium soap base greases can be used for temperatures ranging from –35 to +130° C.

Low-Temperature Properties—see Setting Point.

Lubricant—Any substance interposed between two surfaces in relative motion for the purpose of reducing the friction between them. Less exactly, any substance interposed between two surfaces in relative motion to facilitate their action.

Lubricant Analysis Data—The analyzed data are density, flash point, viscosity, setting point, drop point, penetration, neutralization number, saponification number. These physical and chemical properties of lubricants indicate the fields of application of the lubricants. See also Specifications.

Lubricating Greases—Greases are mixtures of thickeners and oils:

- Metal soap base greases consisting of metal soaps as thickeners and oils
- Non-soap greases comprising inorganic gelling agents or organic thickeners and oils
- Synthetic greases consisting of organic or inorganic thickeners and synthetic oils

LVI—Low Viscosity Index, typically below 40 VI units.

—M—

Mineral Oil—Oils derived from a mineral source, such as petroleum, as opposed to oils derived from plants and animals.

MIL Specifications—Specifications of the US Armed Forces indicating the minimum mandatory requirements for the materials to be supplied. Some engine and machine builders apply the same minimum mandatory requirements to the lubricants. The MIL minimum mandatory requirements are taken as a quality standard.

Mineral Oils—Crude oils and/or liquid oil products.

Miscibility of Oils—Oils of different grades or from different manufacturers should not be mixed. The only exception are HD engine oils which can generally be mixed. If fresh oils are mixed with used oils, sludge can deposit. Whenever there is the risk of sludge formation, samples should be mixed in a beaker.

Multigrade Oils—Engine and gear oils with improved viscosity-temperature behavior.

—N—

Naphthenic Base Oils—A characterization of certain petroleum product prepared from naphthenic type crudes (crudes containing a high percentage of ring type hydrocarbon molecules).

Neatsfoot Oil—A pale yellow animal oil made from the feet and shinbones of cattle.

Needle Bearing—A bearing comprising rolling elements in the form of rollers that are relatively long compared to their diameter.

Neutralization Number—A term still used in the petroleum industry, but rapidly becoming obsolete in the lubrication field. See Strong Acid, Strong Base, Total Acid, and Total Base Numbers.

Nitration—chemical attack on the lube oil by nitration oxides that are formed in the process of combustion. The nitrogen-bearing products that are formed degrade the lube oil, hasten additive depletion, and contribute to deposit formation. Nitration results from operating at air-fuel mixtures that give 1-5% excess oxygen.

NLGI—An abbreviation for "National Lubricating Grease Institute," a technical organization serving the grease industry.

NPRA—National Petrochemical & Refiners Association

—O—

Oil—A viscous, unctuous liquid of vegetable, animal, mineral or synthetic origin.

Oil Ring—A loose ring, the inner surface of which rides a shaft or journal and dips into a reservoir of lubricant from which it carries the lubricant to the top of a bearing by its rotation with the shaft.

Oil Separation—Oil can separate from the greases if they are stored for a longer time or temperatures are high. Oil separation is determined according to DIN 51 817, ASTM

D 1742, IP 121/63. For-life lubrication requires a small oil separation rate which must, however, be large enough to lubricate all contact areas.

Operating Viscosity—Kinematic viscosity of an oil at operating temperature. The operating viscosity is termed v. It can be determined by means of the viscosity-temperature diagram if the viscosity values at two temperatures are known. For determining the operating viscosity of oils with average viscosity-temperature behavior, diagrams can be used.

Oxidation—see Deterioration

Oxidation Inhibitors—see Anti-oxidants

Oxidation Stability—Ability of a lubricant to resist natural degradation upon contact with oxygen.

—P—

Pad Lubrication—A system of lubrication in which the lubricant is delivered to a bearing surface by a pad of felt or similar material.

Paraffin Base Oil—A characterization of certain petroleum products prepared from paraffinic type crudes (crudes containing a high percentage of straight chain aliphatic hydrocarbon molecules). Lubricating oils made from these crudes are normally distinguished from similar oils from other crudes (both oils equally well refined) by higher API gravity and higher viscosity index.

Penetration or Penetration Number—The depth, in tenths of a millimeter that a standard cone penetrates a solid or semisolid sample under specified conditions. This test is used for comparative evaluation of grease and grease-like materials. (See Worked Penetration.)

Petrolatum—A jelly-like product obtained from petroleum and having a microcrystalline structure. Often used in rust preventives.

Plain Bearing—Any simple sliding type bearing as distinguished from tapered land, tilting pad, or antifriction bearings, etc.

Poise—The standard unit of absolute viscosity in the c.g.s. system expressed in dyne sec. per sq. cm.

Polyalkylene Glycol—A base fluid prepared by polymerizing one or more alkylene oxides, most usually ethlene oxide and/or propylene oxide.

Polyalphaolefin—A base fluid prepared by polymerizing alpha olefinic hydrocarbons and hydrogenating the polymer.

Polyurea Base Grease—A grease prepared from a lubricating oil and a polyurea thickener.

Pour Point—The pour point of a lubricant is the lowest temperature at which the lubricant will pour or flow when it is chilled without disturbance under specified conditions.

Power Factor—A measure of the dielectric loss, or ability to perform as an electrical insulating oil.

Pour Point Depressant—An additive that retards wax crystallization, and lowers the pour point.

Process Oils—A lube base stock that receives additional processing to impart a very specific hydrocarbon composition in addition to viscometrics. Process oils are not used as lubricants; they are used as chemical components in the manufacture of rubber, plastics, and other polymeric materials.

PVD—Physical Vapor Deposition—A thin metal-plasma coating (2-5 microns) that is applied in a low heat temperature environment (350°F to 600°F) which can be applied to standard metal surfaces to help resist wear while increasing lubricity and hardness.

—R—

Rated Viscosity—This is the kinematic viscosity attributed to a defined lubrication condition. It is a function of speed and can be determined with diagrams by means of the mean bearing diameter and the rotational speed. By comparing the rated viscosity v_1 with the operating viscosity v the lubrication condition can be assessed.

RCFA—Root Cause Failure Analysis

Refined Oils—A positive resistance to aging of lubricating oils is obtained by refining the distillates. Unstable compounds which can incorporate sulphur, nitrogen, oxygen and metallic salts are removed. Several refining processes are used, the most important being the treatment with sulphuric acid and the extraction of oil-insoluble unstable compounds with solvents.

Relubrication Interval—Period after which lubricant is replenished. The relubrication interval should be shorter than the lubricant renewal interval.

Ring Lubrication—A system of lubrication in which the lubricant is supplied to the bearing surfaces by an oil ring.

R&O—An additive inhibitor package which contains Rust and Oxidation Inhibitors.

Roller Bearing—An antifriction bearing comprising rolling elements in the form of rollers.

Rust Prevention Test (Turbine Oils)—A test for determining the ability of an oil to aid in preventing the rusting of ferrous parts in the presence of water.

—S —

SAE—An abbreviation for "Society of Automotive Engineers," an organization serving the automotive industry.

SAE Numbers—Numbers applied to motor, transmission and rear axle lubricants to indicate their viscosity range. Conversion of the SAE values for engine oils are indicated in DIN 51511, and for automotive gear oils in DIN 51512.

Saponification Number—The state of straight oil deterioration can be assessed by means of the saponification number. It is expressed in milligrams of potassium hydroxide which are required to neutralize the free and bonded acids contained in one gram of oil.

Saybolt Furol Viscosity—The time in seconds required for 60 cubic centimeters of a fluid to flow through the orifice of a Saybolt Furol Viscometer at a given temperature under specified conditions The orifice of the furol viscometer is larger than that of the universal viscometer, the former instrument being used for more viscous fluids.

Saybolt Universal Viscosity—The time in seconds required for 60 cubic centimeters of a fluid to flow through the orifice of the Standard Saybolt Universal. Viscometer at a given temperature under specified conditions. (ASTM Designation D 88-56.)

Seals, Seal Compatibility—The reaction of sealing materials with mineral oils and grease differs widely. They can swell, shrink, embrittle or even dissolve, operating temperatures and lubricant composition playing a major role. Seal and lubricant manufacturers should be consulted for seal compatibility.

Sediments—Sediments are usually formed by soot and dirt particles. They are caused by oil deterioration, mechanical wear, excessive heating, too long oil renewal intervals. They settle in the oil sump, in the bearings, in filters, and in lubricant feed lines. Sediments are hazardous to the operational reliability.

Semi-fluid Greases—These are lubricating greases of semi-fluid to pasty consistency, e.g. aluminum, calcium and sodium soap base greases with a mineral base oil of a viscosity > 70 mm^2/s at 40°C. To improve their load carrying capacity, semi-fluid greases which are generally used for gear lubrication, can be doped with EP additives or solid lubricants.

More generally, any substance in which the force required to produce a deformation depends both on the magnitude and on the rate of deformation.

Setting Point—The setting point of a lubricating oil is the temperature at which the oil ceases to flow if cooled under specific conditions. The low-temperature behavior of the oil slightly above the setting point may be unsatisfactory and must therefore be determined by measuring the viscosity.

Shear Stress—The force per unit area acting tangent to the surface of an element of a fluid or a solid.

Silicone Oils—Synthetic oils which are used for special operating conditions. They have better physical data than mineral oils, but have poor lubricating properties and a low load carrying capacity

Sleeve Bearing—A journal bearing, usually a full journal bearing. (See journal bearing.)

Sludge—Insoluble material formed as a result either of deterioration reactions in an oil or by contamination of an oil, or both.

Air and water can effect the formation of oxidation material and polymerizates in mineral oil products. They settle as sludge.

Slushing Oil—An oil or grease-like material used on metals to from a temporary protective coating against rust, corrosion, etc.

Sodium Soap Base Greases—These greases adhere well to the bearing surfaces and form a uniform and smooth lubricating film on the functional surfaces. They are more prone to emulsifying with water than lithium soap base greases, i.e. they are not water resistant.

The grease is able to absorb minor quantities of water; larger amounts of water would liquefy the grease and make it run out of the bearing. Sodium soap base greases have poor low-temperature properties. They can be used for temperatures ranging from –30 to +120°C.

Solid Foreign Particles—All foreign contaminants insoluble in naphtha and benzene. Solid foreign particles in oils are evaluated according to DIN 51 592, in greases to DIN 51 813.

Solid Lubricants—see Dry Lubricants

"Soluble" Cutting Oil—A mineral oil containing an emulsifier which makes it capable of mixing easily with water to form a cutting fluid.

Solvates—Mineral oils refined with solvents.

Solvency—Ability of a fluid to dissolve organic materials and polymers, which is a function of aromaticity.

Specifications—Military and industrial standards for lubricants which stipulate physical and chemical properties as well as test methods.

Specific Gravity—The ratio of the weight in air of a given volume of a material to the weight in air of an equal volume of water at a stated temperature.

Sperm Oil—A fixed nondrying pale yellow oil obtained from the head cavities and blubber of the sperm whale. Formerly used as an oil additive but now prohibited from use by law in the United States.

Spindle Oil—A light-bodied oil used principally for lubricating textile spindles and for light, high speed machinery.

Splash Lubrication—A system of lubrication in which parts of a mechanism dip into and splash lubricant onto themselves and/or other parts of the mechanism.

SSU—An abbreviation for Saybolt Seconds Universal used to indicate viscosity, e.g., SSU @ 100°F. Also SUS

Stability—Ability of a lubricant to resist natural degradation reactions upon exposure to UV radiation, heat, or oxygen.

Static Friction—The friction between two surfaces not in relative motion but tending to slide over one another. The value of the static friction at the instant relative motion begins is termed break-away friction.

Strong Acid Number (S.A.N.)—The quantity of base, expressed in milligrams of potassium hydroxide, required to titrate the strong acid constituents present in 1 gram of sample.

Strong Base Number (S.B.N.)—The quantity of acid, expressed in terms of equivalent number of milligrams of potassium hydroxide, required to titrate the strong base constituents present in 1 gram of sample.

Sulfurized Oil—Oil to which sulfur or sulfur compounds have been added.

Surface Tension—The tension exhibited at the free surface of liquids, measured in force per unit length.

Suspension—Colloidal suspension of solid particles dispersed in liquids, e.g. the oil-insoluble additives in lubricants.

Swelling Properties—The swelling properties of natural rubber and elastomers under the effect of lubricants are tested according to DIN 53 521.

Synthetic Ester—Oil Molecule prepared by reacting an organic acid with an organic alcohol and possessing some lubricant properties

Synthetic Hydrocarbon—Oil Molecule prepared by reacting paraffinic materials.

Synthetic Lubricant—A lubricant produced from materials not naturally occurring in crude oil by either chemical synthesis or refining processes. Lubricants produced by chemical synthesis; their properties can be adapted to meet special requirements: very low setting point, good V-T behavior, small evaporation loss, long life, high oxidation stability.

—T—

Tacky—A descriptive term applied to greases which are particularly sticky or cohesive.

Tallow—Animal fat prepared from beef and mutton.

Thermal Conductivity—Measure of the ability of a solid or liquid to transfer heat.

Thickener—Thickener and base oil are the constituents of lubricating greases. The percentage of the thickener and the base oil viscosity determine the consistency of the grease.

Thixotropy—The property of a grease to become softer when mechanically stressed and to return to its original consistency when left to rest. Preserving oils with special additives are also thixotropic.

TOST- Turbine Oil Oxidation Stability Test, ASTM D-943

Total Acid Number (TAN)—The quantity of base, expressed in milligrams of potassium hydroxide, that is required to titrate all acidic constituents present in 1 gram of sample.

Total Base Number (TBN)—The quantity of acid that is required to titrate all basic constituents present in 1 gram of sample.

Turbine Quality—Lube base stocks suitable for turbine applications, finished with severe hydrotreating. TQ base stocks exhibit improved oxidation stability over other base stocks.

—U—

Unworked Penetration—The penetration at 77°F of a sample of grease that has received only the minimum handling in transfer from a sample can to the test apparatus and which has not been subjected to the action of a grease worker.

—V—

Varnish—When applied to lubrication, a deposit resulting from the oxidation and polymerization of fuels and lubricants. Similar to but softer than lacquer.

Viscometer—Viscosimeter—An apparatus for determining the viscosity of a fluid.

Viscosity—That property of a fluid or semi-solid substance characterized by resistance to flow and defined as the ratio of the shear stress to the rate of shear of a fluid element. The standard unit of viscosity in the c.g.s. system is the poise and is expressed in dyne sec. per square centimeter. The standard unit of viscosity in the English system is the reyn and is expressed in lb. sec. per square in.

Viscosity Classification—The standards ISO 3448 and DIN 51519 specify 18 viscosity classes ranging from 2 to 1,500 mm^2/s at 40°C for industrial liquid lubricants

Viscosity Grade—Any of a number of systems that characterize lubricants according to viscosity for particular applications, such as industrial oils, gear oils, automotive engine oils, automotive gear oils, and aircraft piston engine oils.

Viscosity Index (VI)—A measure of a fluid's change of viscosity with temperature. The higher the viscosity index the smaller the change in viscosity with temperature.

Viscosity Index Improver—Additive that increases lubricant viscosity index, necessary for formulation of multi-grade engine oils.

Viscosity-Pressure Behavior—Viscosity of a lubricating oil as a function of pressure. With a rise in pressure the viscosity increases.

V-T Behavior—The viscosity-temperature behavior refers to the viscosity variations with temperatures. The V-T behavior is good if the viscosity varies little with changing temperatures.

—W—

Water Content—If an oil contains water, the water droplets disrupt the lubricating film and reduce lubricity. Water in oil accelerates deterioration and leads to corrosion. The water content can be determined by distillation or by settling in a test tube; due to its higher specific gravity the water settles at the bottom. Samples of emulsifying oil must be heated. A small amount of water (0.1% or less) is identified by a crackling noise which is produced when the oil is heated in a test tube. A higher water content will cause the oil to boil over.

Water Resistance—The water resistance of greases is tested according to DIN 51807 (static test); it is not indicative of the water resistance of the grease when used in the field. The test merely shows the effect which static, distilled water has on an unworked grease at different temperatures.

Water Separation Ability—Ability of an oil to separate water. The test is carried out according to DIN 51589.

"Wetting Bearings"—The pre-lubrication of bearing surfaces prior to starting a machine that has been idle for an elongated time period. Prevention of possible Brinell damage to bearing components upon sudden dry start of a machine.

Wet Fuel—Gas containing heavier products, such as ethane or propane, and having a heat content greater than normal (above 1000 Btu / cubic foot).

White Oils Light—Colored and usually highly-refined mineral oils, usually employed in medicinal and pharmaceutical preparations and as a base for creams, salves, and ointments, but also used as lubricants.

Worked Penetration—The penetration of a sample of lubricating grease immediately after it has been brought to 77°F ±1°F and then subjected to 60 strokes in the ASTM standard grease worker.

Index

T